OXFORD GRADUATE TEXTS IN MATHEMATICS

Series Editors

M. BRIDSON
G.G CHEN
S.K. DONALDSON
T.J. LYONS
M.J. TAYLOR

OXFORD GRADUATE TEXTS IN MATHEMATICS

Books in the series

1. Keith Hannabuss: *An Introduction to Quantum Theory*
2. Reinhold Meise and Dietmar Vogt: *Introduction to Functional Analysis*
3. James G. Oxley: *Matroid Theory*
4. N.J. Hitchin, G.B. Segal, and R.S. Ward: *Integrable Systems: Twistors, Loop Groups, and Riemann Surfaces*
5. Wulf Rossmann: *Lie Groups: An Introduction through Linear Groups*
6. Qing Liu: *Algebraic Geometry and Arithmetic Curves*
7. Martin R. Bridson and Simon M. Salamon (eds): *Invitations to Geometry and Topology*
8. Shmuel Kantorovitz: *Introduction to Modern Analysis*
9. Terry Lawson: *Topology: A Geometric Approach*
10. Meinolf Geck: *An Introduction to Algebraic Geometry and Algebraic Groups*
11. Alastair Fletcher and Vladimir Markovic: *Quasiconformal Maps and Teichmüller Theory*
12. Dominic Joyce: *Riemannian Holonomy Groups and Calibrated Geometry*
13. Fernando Villegas: *Experimental Number Theory*
14. Péter Medvegyev: *Stochastic Integration Theory*
15. Martin A. Guest: *From Quantum Cohomology to Integrable Systems*
16. Alan D. Rendall: *Partial Differential Equations in General Relativity*
17. Yves Félix, John Oprea, and Daniel Tanré: *Algebraic Models in Geometry*
18. Jie Xiong: *Introduction to Stochastic Filtering Theory*
19. Maciej Dunajski: *Solitons, Instantons, and Twistors*
20. Graham R. Allan: *Introduction to Banach Spaces and Algebras*
21. James Oxley: *Matroid Theory, Second Edition*
22. Simon Donaldson: *Riemann Surfaces*
23. Clifford Henry Taubes: *Differential Geometry: Bundles, Connections, Metrics and Curvature*
24. Gopinath Kallianpur and P. Sundar: *Stochastic Analysis and Diffusion Processes*
25. Selman Akbulut: *4-Manifolds*
26. Fon-Che Liu: *Real Analysis*
27. Dusa Mcduff and Dietmar Salamon: *Introduction to Symplectic Topology, Third Edition*
28. Chris Heunen and Jamie Vicary: *Categories for Quantum Theory: An Introduction*
29. Shmuel Kantorovitz and Ami Viselter: *Introduction to Modern Analysis, Second Edition*
30. Stephan Ramon Garcia, Javad Mashreghi, and William T. Ross: *Operator Theory by Example*
31. Maciej Dunajski: *Solitons, Instantons, and Twistors, Second Edition*
32. Ibrahim Assem and Flávio U. Coelho: *An Introduction to Module Theory*
33. Stanisław Goldstein and Louis Labuschagne: *Noncommutative Measures and Lp and Orlicz spaces, with Applications to Quantum Physics*

Noncommutative measures and L^p and Orlicz Spaces, with Applications to Quantum Physics

STANISŁAW GOLDSTEIN

Professor, Faculty of Mathematics and Computer Science,
University of Lodz

LOUIS LABUSCHAGNE

Professor of Mathematics and Director of the Focus Area for Pure
and Applied Analytics, Faculty of Natural and Agricultural Sciences,
North-West University

OXFORD
UNIVERSITY PRESS

OXFORD
UNIVERSITY PRESS

Great Clarendon Street, Oxford, OX2 6DP,
United Kingdom

Oxford University Press is a department of the University of Oxford.
It furthers the University's objective of excellence in research, scholarship,
and education by publishing worldwide. Oxford is a registered trade mark of
Oxford University Press in the UK and in certain other countries

Published in the United States of America by Oxford University Press
198 Madison Avenue, New York, NY 10016, United States of America

British Library Cataloguing in Publication Data

Data available

Library of Congress Control Number: 2024949885

ISBN 9780198950202
ISBN 9780198950219 (pbk.)

DOI: 10.1093/oso/9780198950202.001.0001

Preface

This monograph has been written so that it can serve the needs of both graduate students and young researchers seeking entry to the field of noncommutative L^p and Orlicz spaces, and advanced researchers who require a fairly comprehensive reference book. To facilitate such flexibility, several exit points have been built into the book, which we shall elucidate shortly. For the sake of emphasising the utility of the theory, Chapter 14 (detailing the applications) has been written in such a way that readers with a basic grounding in operator algebras may read it with benefit before tackling any of the other chapters. (To fully comprehend all the subtleties in that chapter, readers will of course first need to wade through much of the rest of the book.) We assume the reader to have a basic knowledge of functional analysis, in particular that he or she is acquainted with the spectral theory and functional calculus of both bounded and unbounded self-adjoint operators. Knowledge of the theory of operator algebras is not strictly indispensable but would be very helpful. In the *Preliminaries* chapter, we have gathered results from the basic theory of operator algebras needed for the rest of the book. All this material is standard, and we highly recommend the prospective reader to have on her or his shelves at least one of the following excellent sources: *Operator Algebras and Quantum Statistical Mechanics 1* by Bratteli and Robinson, the two volumes of *Fundamentals of the Theory of Operator Algebras* by Kadison and Ringrose [**KR83**, **KR86**], *Lectures on von Neumann Algebras* by Strătilă and Zsidó [**SZ79**] or *Theory of Operator Algebras I* by Takesaki [**Tak02**]. For more advanced material, Takesaki's *Theory of Operator Algebras II* [**Tak03a**] and Strătilă's *Modular Theory in Operator Algebras* [**Str81**] are among the best. Blackadar's encyclopedic *Operator Algebras* [**Bla06**] is excellent for those who would like to find a piece of information quickly.

In chronological order, the precursors to this book were, of course, Marianne Terp's notes [**Ter81**], which have long served the mathematical community, and a brief introduction to tracial L^p-spaces in

Section IX.2 of [**Tak03a**]. There is also an excellent survey paper by Pisier and Xu [**PX03**] aimed at specialists interested in the interface of the noncommutative theory with the deep theory of geometry of Banach spaces. The 2020 notes of the authors [**GL20**] provides a more up to date account of the theory but, as with Terp's notes, focusses more on introducing readers to the rudiments of the theory of Haagerup L^p-spaces as quickly as possible. So, there is no advanced theory in these notes, and the notes end as soon as the basic theory of Haagerup L^p-spaces has been laid down. Real interpolation, though more accessible than the advanced theory, is not discussed at all. More recently, 2021 and 2023, respectively, have seen the appearance of the monographs of Hiai [**Hia21**] and Dodds, de Pagter and Sukochev [**DdPS23**], both of which also deal with noncommutative integration. These two monographs and the present book actually complement each other very well. To see this, note that the present book is intended to be a comprehensive monograph on noncommutative L^p and Orlicz spaces (presented from the Haagerup prespective) including much of the advanced theory as well as nontrivial applications to Quantum Theory. By contrast, the monograph of Hiai is intended to be a fast-track introduction to von Neumann algebras and noncommutative integration and will be especially interesting for those wishing to compare Haagerup's and Connes–Hilsum's L^p-spaces. By contrast, it lacks much of the advanced theory and does not address Orlicz spaces at all. The monograph of Dodds, de Pagter and Sukochev [**DdPS23**], on the other hand, focusses specifically on tracial semifinite von Neumann algebras and in particular on the theory of noncommutative symmetric and rearrangement invariant Banach function spaces, which that category of algebras allows. In Chapter 7 of the present monograph, we also very briefly touch on the topic of noncommutative Banach function spaces. However, we arrive at the definition of such spaces by means of a very different route from the one followed in [**DdPS23**].

As was the case with the notes by the authors, one distinctive feature of the present work is the extent to which we have incorporated, on the one hand, the technology of noncommutative decreasing rearrangements, as developed by Fack and Kosaki [**FK86**], and, on the other, the fairly recent theory of Orlicz spaces for general von Neumann algebras [**Lab13**]. The theory of Orlicz spaces we present here stems from the research interests of the second-named author. The importance of Orlicz spaces is explained in the Introduction and is further justified by the applications in Chapter 14. In addition to the issues mentioned in the Introduction, the refinement brought about by the development of this theory has in fact enabled us to come up with a much smoother, more streamlined path through the theory of Haagerup L^p-spaces. As a case in point, we mention the complex interpolation theory presented in Chapter 11. It is precisely the use of Orlicz spaces that here enables us to construct a proof *using only Haagerup L^p-structures* that non-σ-finite Haagerup L^p-spaces form a complex interpolation scale. (All other known proofs use Connes–Hilsum spaces.) Other unique features of the book are:

- a full presentation of Maharam's theorem in Chapter 1 from a von Neumann algebra perspective;
- a full description of the interrelation of various notions of measurability for operators in Section 3.3 (p. 162);
- the novel presentation of the real interpolation theory of the pair $(L^1(\mathcal{M}, \tau), L^\infty(M, \tau))$ as a consequence of the theory of monotone interpolation spaces (see Chapter 7);
- a proof of the existence of the Plancherel weight for general group algebras using only group constructs rather than left Hilbert algebras (see Subsection 8.6.3 (p. 355));
- a proof that the dual weight construction is valid for merely normal weights (Theorem 9.24 (p. 387));
- an existence proof of conditional expectations using only Haagerup L^p-structures and basic modular theory (Section 10.5 (p. 458));
- a very general theory of Markov operators in Section 12.3 (p. 543);
- a proof of the Haagerup reduction theorem valid for possibly non-σ-finite von Neumann algebras.

The *Preliminaries* chapter revises the essential background. As scene-setters, Chapters 1 and 2 play a very important role in that Chapter 1 clearly identifies von Neumann algebras as the appropriate framework for noncommutative L^p-spaces, with Chapter 2 then giving full details of the theory of Schatten–von Neumann classes, which may be regarded as the genesis of noncommutative L^p-spaces. This then represents the first exit point. As regards L^p and Orlicz spaces for semifinite algebras with a trace, Chapters 3–5 then go on to present what may be regarded as the noncommutative theory of measures and measurable functions, with Chapters 6 and 7, respectively, introducing the reader to L^p and

Orlicz spaces and 'Monotone Riesz–Fischer' spaces (alias rearrangement invariant Banach function spaces) for semifinite algebras. Readers wishing to do no more than get to the tracial theory of noncommutative L^p-spaces as quickly as possible should at least master the material on traces and τ-measurable operators in Chapters 3 and 4, and then read Chapter 5 and Sections 6.1 (p. 245) and 6.2 (p. 266) of Chapter 6. Chapters 8 and 9, respectively, present the aspects of modular theory and the theory of crossed products that form the foundation on which the theory of Haagerup L^p-spaces is built. The actual theory of Haagerup L^p-spaces is ultimately presented in 10. For this theory to be comprehensible, all of Chapters 5 and 6 need to be covered (including Section 6.3 (p. 276)) and a high degree of familiarity achieved with at least Theorems 9.17 (p. 379), 9.28 (p. 399), 9.31 (p. 401), 9.38 (p. 408), 9.40 (p. 412), and Propositions 9.27 (p. 399), 9.33 (p. 404), and 9.36 (p. 406). Comprehending Haagerup L^p-spaces is then in principle a third exit point. The theory of Haagerup L^p-spaces is deeply intertwined with the theory of crossed products, and so readers wishing to ultimately master the deeper subtleties of Haagerup L^p-spaces and move on to the advanced theory are therefore advised to at some stage take the time to master Chapter 9 in its entirety.

Manuscripts which greatly assisted in galvanising our thoughts regarding this monograph include the iconic notes of Terp [**Ter81**], the extremely useful paper of Fack and Kosaki [**FK86**], the more recent very elegant set of notes by Xu [**Xu07**], the paper [**BGL22**] of the authors with David Blecher and the monograph of Dodds, de Pagter and Sukochev [**DdPS23**]. (We are deeply grateful to the authors for sharing an early draft copy with us.) There are of course many people who have in different ways directly or indirectly contributed to getting the notes to the point where they are now. These include people like Jurie Conradie, Pierre de Jager, and Claud Steyn, who read large tracts of the preliminary draft of these notes. Graduate students who served as test subjects for parts of this work include Christiaan Pretorius. However, some individuals deserve special mention.

We are both deeply grateful to Marek Bożejko, Gilles Pisier and Quanhua Xu for the notable part they played in consistently including us in the ongoing debate around noncommutative analysis and its applications by way of invitations to conferences. On a more personal level, we would like to extend our gratitude in the manner described below.

SG: First and foremost, I would like to thank Martin Lindsay and Jill Anderson for their long-standing friendship and absolutely outstanding hospitality. The time I spent with Martin was also pivotal to my mathematical life. It is an honour to recognise the inspiring passion for mathematics of the late Robin Hudson and the enduring friendship with both him and his wife Olga Hudson. Warm thanks to those who have offered me their exceptional hospitality: Urszula Ledzewicz and Heinz Schättler, Carlo Cecchini, Sergio Albeverio, David Appelbaum, Graham Vincent-Smith and Gerhard Racher. Warm thanks to my teacher and master, Ryszard Jajte, and my colleagues, friends and collaborators, Andrzej Łuczak and Adam Paszkiewicz.

LL: I would like to acknowledge my supervisor Ron Cross who guided me through the ups and downs of graduate life. Also various people like Heinrich Raubenheimer, Hans Jarchow and Joe Diestel whose advice helped me to cut the graduate umbilical cord thereby setting me on a road of ongoing mathematical exploration. Mathematically, it was Jurie Conradie who helped me get started with Operator Algebras and Anton Ströh who first introduced me to noncommutative L^p-spaces. People in the Operator Algebra community who deserve special mention are Ola Bratteli and David Blecher. Their engagement and support were invaluable. David Blecher's many valuable and deep contributions to the paper that form the foundation of Chapter 1 also deserve mention. I am deeply grateful to Adam Majewski who, with deep wisdom and warm hospitality, helped me to find my feet in the algebraic approach to Quantum Theory. The many other people who helped me in my journey are all in my thoughts.

On a more personal level, I am deeply indebted to my wife Sheila who helped carry my burdens and gave me the room to pursue mathematics. Through the entire journey, it was the gracious hand of God who not only blessed me with moments of great success, but also carried me through dark valleys of disappointment and physical challenges.

We are both grateful to the anonymous reviewers whose insightful comments influenced the final shape of the book.

Contents

PART 1 FOUNDATIONAL EXAMPLES

PART 2 TRACIAL CASE

PART 3 GENERAL CASE

PART 4 ADVANCED THEORY AND APPLICATIONS

Introduction

The theory of operator algebras originated from a paper of John von Neumann [**vN30**] from 1930, followed by a series of papers of Francis Murray and himself ([**MvN36**], [**MvN37**], [**vN40**], [**MvN43**], [**vN49**]) from 1936 to 1949 on 'rings of operators'. The measure-theoretic or probabilistic aspects of such rings equipped with trace-like functionals were clear to von Neumann from the very beginning. In [**MvN36**], the authors write about the trace value $T(A)$ of an operator A as an '*a priori* expectation value of the observable A'. This is even more strongly pronounced in Section 8 of the same paper, where the dimension function is used to measure projections, just as we measure sets in classical measure theory. The next major steps forward were made by Irving Segal [**Seg53**] and Jacques Dixmier [**Dix53**] in 1953. For a semifinite von Neumann algebra with a faithful normal semifinite trace, Segal introduced the algebra of measurable operators and introduced L^1, L^2 and L^∞, while Dixmier defined all the L^p-spaces. The term 'von Neumann algebra' was coined by Dixmier in [**Dix57**], a first book on the subject of 'rings of operators'. (Now, one can use either the second French edition [**Dix96a**] or the English one [**Dix81**].) The 1975 paper of F.J. Yeadon then provided a very complete discussion of L^p-spaces in the tracial case [**Yea75**]. The important notion of a τ-measurable operator was introduced by Edward Nelson in [**Nel74**]. The significance of this concept is that it enables one to realise the tracial L^p-spaces as concrete spaces of operators. Further progress would have been impossible without the modular theory of Minoru Tomita and Masamichi Takesaki [**Tak70**]. It was this theory that formed the bedrock for the first constructions of noncommutative L^p-spaces for general von Neumann algebras due to Uffe Haagerup [**Haa79a**] and Alain Connes [**Con80**]. For a fuller history of general L^p-spaces including an account of more recent developments, we refer the reader to Section 10.7 (p. 468).

Noncommutative L^p-spaces feature in a variety of applications, of which we only mention one of the first ones, by Ray Kunze [**Kun58**], to L^p-Fourier transforms on locally compact unimodular groups. This very early paper is interesting, since it also combines the results of Segal and Dixmier on L^p-spaces to, for the first time, realise these

Noncommutative measures and L^p and Orlicz Spaces, with Applications to Quantum Physics. Stanisław Goldstein and Louis Labuschagne, Oxford University Press. © Stanisław Goldstein and Louis Labuschagne (2025). DOI: 10.1093/oso/9780198950202.003.0001

spaces as spaces of measurable operators. Among other results, Kunze proved a Hausdorff–Young inequality in this setting. One would expect a generalisation of his results to general L^p-spaces and non-unimodular groups, and in fact, such results were obtained by Terp [**Ter17**]. The application of noncommutative harmonic analysis of Hausdorff locally compact groups also clearly shows that these spaces occur naturally, and not as some very exotic pathological phenomenon. Specifically, given a Hausdorff locally compact group G, one may form the group von Neumann algebra, which is the von Neumann algebra generated by the left-shift operators on $L^2(G)$. If one is interested in quantum harmonic phenomena, it then makes sense to do Fourier analysis on the noncommutative L^p-spaces associated with this algebra as Kunze did. The nature of the algebra one has to deal with depends on the nature of the group one starts with, with 'wilder' groups leading to wilder group von Neumann algebras. There is in fact now renewed interest in noncommutative harmonic analysis with a lot of attention being given to Quantum Groups and Fourier multipliers. There are a large number of researchers currently working on this topic—too many to list them all here. We, therefore, content ourselves with mentioning a mere sampling of papers ([**Cas13**, **CdlS15**, **CFK14**, **DKSS12**, **DFSW16**, **FS09**, **JMP14**, **JNR09**, **CPPR15**, **NR11**]) involving Martijn Caspers, Matt Daws, Mikael de la Salle, Pierre Fima, Uwe Franz, Marius Junge, Pawel Kasprzak, Tao Mei, Mathhias Neufang, Stefan Neuwirth, Javier Parcet, Mathilde Perrin, Éric Ricard, Zhong-Jin Ruan, Adam Skalski, Piotr Soltan and Stuart White. At the same time, we offer our sincere apologies to authors who may feel slighted by their exclusion from this incomplete list.

Noncommutative Orlicz spaces were first introduced by Muratov in 1978 [**Mur78**, **Mur79**] in the context of von Neumann algebras equipped with a faithful normal tracial state. However, it was the later papers of Wolfgang Kunze (see [**Kun90**], [**ARZ07**]) and Peter Dodds, Theresa Dodds and Ben de Pagter [**DDdP89**] that sparked a wider interest in this topic. Although these two approaches can be shown to be equivalent (see [**LM11**]), they did spark traditions which initially developed independently of each other. The definition of Orlicz spaces for type III algebras is a much more recent phenomenon [**Lab13**]. We refer the reader to Section 6.4 (p. 303) and the final part of 10.7 (p. 468) for a fuller account of the history and development of noncommutative Orlicz spaces.

At this point, we should note that the paper of Dodds, Dodds and de Pagter [**DDdP89**] in no small way helped synthesise ideas of several authors that had been brewing behind the scenes for some time and as such helped to kick-start a very successful and burgeoning theory of noncommutative rearrangement invariant Banach function spaces, which has attracted a very large number of adherents. Readers wishing to know more should consult either the monograph of Dodds, de Pagter and Sukochev [**DdPS23**] or the survey paper of de Pagter [**dP07**]. Yet, despite the great success of this theory, it is at present only known to hold in the semifinite setting. The only known extension of this theory to the type III setting is the theory of Orlicz spaces presented in this monograph. It is our hope that a deeper understanding of this theory by the mathematical community will help pave the way for the eventual extension of the theory of rearrangement invariant Banach function spaces to the type III setting.

Much of the current motivation for studying these spaces comes from Physics. Although we shall not cover any of these applications in these notes, it is nevertheless instructive to review them. The issue of return to equilibrium is, for example, still not fully settled in Quantum Statistical Mechanics (QSM). For this issue to be settled, a better understanding of entropy for QSM is required. At a naive level, one may consider for semifinite algebras the formal quantity $\tau(f \log(f))$ as starting point for a quantum theory of entropy. The problem with the current quantum mechanical formalism where the pair (L^1, L^∞) is used as 'home' for states and observables is that the L^1 topology is notoriously bad at distinguishing states with 'good' entropy. In this topology, one may have a sequence (f_n) of positive elements of L^1 converging to some f for which $\tau(f \log(f))$ is a well-defined finite quantity, but with $\tau(f_n \log(f_n))$ infinite for each n. So, a better topology is required to study entropy. In addition to the above, log-Sobolev inequalities also play an important role in studying the 'return to equilibrium' issue (see the concluding remarks in, for example, [**Zeg90**]). So, such a theory should also be well suited to such inequalities. These two factors already strongly suggest the use of noncommutative Orlicz spaces as the appropriate technology. However, the classical theory of entropy itself also suggests Orlicz spaces as the appropriate tool.

Let us quote from [**LM20**]: *The origins of a quantity representing something like entropy may be found in the work of Ludwig Boltzmann. In his study of the dynamics of rarefied gases, Boltzmann formulated* *the so-called spatially homogeneous Boltzmann equation as far back as 1872, namely*

$$\frac{\partial f_1}{\partial t} = \int d\Omega \int d^3 v_2 I(g,\theta) |\mathbf{v}_2 - \mathbf{v}_1| (f_1' f_2' - f_1 f_2)$$

where $f_1 \equiv f(\mathbf{v}_1, t)$, $f_2' \equiv f(\mathbf{v}_2', t)$, ..., are velocity distribution functions, $I(g, \theta)$ denotes the differential scattering cross section, $d\Omega$ the solid angle element, and $g = |\mathbf{v}|$. *The natural Lyapunov-type functional for this equation is the so-called Boltzmann H-function, which is*

$$H_+(f) = \int f(x) \log f(x) dx$$

where f is a postulated solution of the Boltzmann equation. The connection to entropy may be seen in the fact that the classical description of continuous entropy S differs from the functional H only by sign. Hence, Boltzmann's H-functional may be viewed as the first formalisation of the concept of entropy. Lions and DiPerna were the first to rigorously demonstrate the existence of solutions to Boltzmann's equation. (Lions later received the Fields medal for his work on nonlinear partial differential equations.) Their solution was for the density of colliding hard spheres, given general initial data (see, for example, [DL88, DL89] for a sampling of this work). Villani subsequently announced, see [Vil02, Chapter 2, Theorem 9], that for particular cross sections (collision kernels in Villani's terminology), weak solutions of the Boltzmann equation are actually in $L \log(L+1)$. So one consequence of the work by these authors was to give a strong indication that the Orlicz space $L \log(L+1)$ is the appropriate framework for studying entropy-like quantities like the Boltzmann H-functional.

Physicists who on the basis of these facts strongly advocated the use of noncommutative Orlicz spaces for studying QSM include Ray Streater [Str04], Boguslaw Zegarlinski [ARZ07] and Adam Majewski [Maj17]. The 1995 paper of Giovanni Pistone and Carlo Sempi [PS95] added another strand of thought to this mix of ideas, namely the concept of regular observables. In [PS95], the authors introduce a moment generating class of random variables which they call *regular random variables*, and then go on to show that the weighted Orlicz space $L^{\cosh - 1}(X, \Sigma, f \, d\nu)$ forms the natural home for these regular random variables. The significant fact regarding this concept is that the space $L^{\cosh-1}$ is (up to isomorphism) the Köthe dual of $L\log(L+1)$. One may therefore expect that, at the quantum level, noncommutative versions of $L^{\cosh-1}$ would similarly be home to *regular observables*. This was formalised in [**LM11**]. So, the picture that begins to emerge is that the (dual) pairing $(L\log(L+1), L^{\cosh-1})$ may be better suited to studying and refining QSM (and ultimately clarifying the issue of return to equilibrium) than the more classical pairing of (L^1, L^∞). Readers should note that such a paradigm shift will in no way impact the well-established paradigm for elementary quantum mechanics pioneered by Paul Dirac et al., since in the case of $B(H)$ the two approaches agree (as was shown in [**LM11**]). The utility of this pairing for the noncommutative context was strongly demonstrated in [**ML14**].

Thus far, all the theory we have presented has been developed in the context of semifinite von Neumann algebras. That in itself is a problem since it is known that many of the most important von Neumann algebras in Quantum Physics are necessarily type III algebras (see [**Yng05**]). One may also note the work of Robert Powers. In [**Pow67**], Powers studied representations of uniformly hyperfinite algebras. The types of algebras Powers studied may in physical terms be regarded as thermodynamic limits of an infinite number of sites, with the algebra $M_2(\mathbb{C})$ associated with each site. (See [**Maj17**] for details of the physical interpretation of Powers' result.) However, despite the simplicity of these 'local' algebras, the algebra obtained in the limit turned out to be of type III. To appreciate the significance of this fact, readers should take note of the fact that type III algebras exhibit markedly different behaviour than their semifinite cousins. Thus, for a theory of noncommutative Orlicz spaces to fully address the challenge emanating from Physics, one dare not ignore the type III setting. A theory of Orlicz spaces for type III algebras therefore had to be developed. This was finally done in [**Lab13**] and then slightly refined in [**LM20**]. However, type III algebras do not admit f.n.s. traces. Hence, in passing to type III algebras, an alternative prescription for entropy to the naive one of $\tau(f\log(f))$ needed to be found. This was ultimately done in [**ML20**]. The contribution of the paper [**LM20**] was to show that complete Markov dynamics canonically extends to even the most general noncommutative $L^{\cosh-1}$ spaces. The theory of noncommutative Orlicz spaces is therefore now well set for an onslaught on the challenge of further refining and developing QSM.

It is of interest to note that in a recent preprint [**LM22**], noncommutative Orlicz spaces were also shown to naturally occur in Algebraic Quantum Field Theory. The significance of these spaces for Physics therefore reaches beyond just QSM. Readers wishing to know more about these applications to physics and also about what still needs to be done should consult not just the references mentioned above but pay careful heed to the paper [**Maj17**] and the references mentioned therein. This paper clearly outlines some of the remaining challenges and the development they require.

Although the theory of Haagerup L^p-spaces goes through for general von Neumann algebras, there is a marked increase in the degree of complexity of the proofs when one ventures beyond σ-finite algebras. Alongside this fact, there is also the pervading conviction in Physics that separable Hilbert spaces (in which setting all von Neumann algebras are σ-finite) are sufficient for modelling quantum phenomena. This then raises the question of why, in the type III setting, we need to bother venturing beyond σ-finite algebras. In response to this concern, we first note that the incorporation of Orlicz spaces goes some way to assuaging the challenges inherent in the type III case. For example, access to this technology enables us to construct a very regular superspace into which all these Orlicz spaces naturally embed (Proposition 10.22 (p. 429)). Inside this superspace, we can give naive meaning to, for example, the notion of intersection and sum of two such spaces in a manner which mimics the classical situation. Second, the nature of the canonical Hilbert space associated with some particular quantum mechanical system can be shown to (in a mathematical sense) be a property of the system rather than some a priori restriction we are compelled to impose. We pause to justify this fact. The objective of canonical quantisation, as described in Chapter 14, is to identify the von Neumann algebra and Hamiltonian of a given quantum mechanical system. However, von Neumann algebras can be expressed abstractly without any dependence on some *a priori* given Hilbert space (Theorem 0.79 (p. 31)). Moreover, the canonical Hilbert space on which concrete representations of such algebras live is determined by the algebra itself (see Proposition 10.14 (p. 423) and Theorem 10.59 (p. 467)). Similarly, the σ-weak and σ-strong topologies (which describe the convergence of the time evolution determined by the Hamiltonian) are algebraic invariants (Proposition 0.89 (p. 34)). So, *a priori*, there is no reason to make any assumptions about the nature of the Hilbert space on which this algebra lives, since ultimately the structure of that Hilbert space is determined by the system itself. Taking these factors into account, responsible analysis therefore demands that we develop a theory applicable to all von Neumann algebras.

Preliminaries

In this chapter, we gathered various facts from functional analysis and the theory of operator algebras that we will freely use in the sequel. There are many excellent books on functional analysis, so the reader will find the material we use, for example, on spectral theory, without any problems. One book that stands out for future operator-algebraists is Gert Pedersen's *Analysis Now* [**Ped89**].

Section 0.3 sets the stage for future material on operator algebras. In particular, it identifies various classes of elements of a C^*-algebra and introduces functional calculi that will be used in the sequel. In Section 0.1, we deal mainly with various topologies in $B(H)$ and with the Borel functional calculus. Section 0.4 gathers the most important notions and results on von Neumann algebras, together with a Structure Theorem 0.127. Sections 0.2 and 0.5 deal with general unbounded operators and with those unbounded operators that 'almost belong' to a von Neumann algebra. Finally, Section 0.6 introduces a useful notion of generalised positive operators, corresponding to not-necessarily densely defined unbounded positive self-adjoint operators.

0.1 Bounded operators

In this section, we gather important information on bounded operators on a Hilbert space H. For functional calculi for self-adjoint (or normal) operators, we recommend the first volume of Kadison and Ringrose [**KR83**], Strătilă and Zsidó [**SZ79**] and Arveson [**Arv02**].

Definition 0.1 Let H be a complex Hilbert space with the inner product $\langle \cdot, \cdot \rangle$, linear in the first argument and antilinear in the second. The norm on H denotes the norm given by the inner product. A linear map $a\colon H \to H$ is bounded if

$$\|a\| := \sup\{\|a\xi\|\colon \|\xi\| \le 1\} < \infty.$$

Then, $B(H)$ consists of bounded (or continuous) linear operators on H.

Noncommutative measures and L^p and Orlicz Spaces, with Applications to Quantum Physics. Stanisław Goldstein and Louis Labuschagne, Oxford University Press. © Stanisław Goldstein and Louis Labuschagne (2025). DOI: 10.1093/oso/9780198950202.003.0002

It is clear that $B(H)$, with the usual operations of addition and multiplication by a scalar, and with the composition of operators as multiplication, forms a (complex) algebra.

Lemma 0.2 *Let a be a bounded operator on H. There exists a unique operator $a^* \in B(H)$ such that $\langle \xi, a^*\eta \rangle = \langle a\xi, \eta \rangle$ for all $\xi, \eta \in H$. Moreover, $\|a^*\| = \|a\|$.*

Proof. Take $\eta \in H$. Note that $f_\eta : H \ni \xi \mapsto \langle a\xi, \eta \rangle \in \mathbb{C}$ is an element of the Banach space dual H^* of the Hilbert space H. By the Riesz representation theorem, there exists a unique $\zeta \in H$ such that $\langle \xi, \zeta \rangle = \langle a\xi, \eta \rangle$. Put $a^*\eta = \zeta$. The linearity of a^* follows easily from the uniqueness of $a^*\eta$ for all $\eta \in H$. We also have $\|a^*\eta\| \leq \|f_\eta\| \leq \|\eta\|\|a\|$, so that $\|a^*\| \leq \|a\|$ for each $\eta \in H$ and, for all $\xi \in H$,

$$\|a\xi\|^2 = \langle a\xi, a\xi \rangle = \langle \xi, a^*a\xi \rangle \leq \|\xi\|\|a^*\|\|a\xi\|$$

which implies $\|a\xi\| \leq \|\xi\|\|a^*\|$, so that $\|a\| \leq \|a^*\|$. $\qquad\square$

Definition 0.3 For each $a \in B(H)$, the operator a^* from Lemma 0.2 is called the *adjoint* of a.

We shall need the concept of a *-algebra.

Definition 0.4 An *algebra with involution* or a *-algebra* $\mathcal{A}$ is a (complex) algebra with a map $a \mapsto a^*$ from $\mathcal{A}$ into itself satisfying $(\lambda a)^* = \bar{\lambda}a^*$, $(a + b)^* = a^* + b^*$, $(ab)^* = b^*a^*$ and $a^{**} = a$. The natural morphisms between two *-algebras $\mathcal{A}$ and $\mathcal{B}$ are algebraic homomorphisms $\mathscr{I}$, which preserve the involutive structure in that $\mathscr{I}(a^*) = \mathscr{I}(a)^*$ for all $a \in \mathcal{A}$. Such homomorphisms are called *-homomorphisms (∗-isomorphisms if they are injective).

It is clear that $B(H)$, endowed with adjoint operation, is a *-algebra. We have a special notation for the most important classes of bounded operators on H.

Definition 0.5 The operators satisfying $a = a^*$ are called *self-adjoint* or *Hermitian*, and the real subspace of self-adjoint operators is denoted by $B(H)_h$. A bounded operator a is *positive* if $\langle a\xi, \xi \rangle \geq 0$ for all $\xi \in H$ (or equivalently, if a is positive as an element of the

C^*-algebra $B(H)$), and the pointed cone of positive operators is denoted by $B(H)_+$. For $a, b \in \mathcal{M}_h$, we say that $a \leq b$ if $\langle a\xi, \xi \rangle \leq \langle b\xi, \xi \rangle$ for any $\xi \in H$.

Definition 0.6 An (*orthogonal*) *projection* p is a bounded operator on H satisfying $p = p^* = p^2$, and the complete lattice of projections (cf. Definition 1.3) is denoted by $\mathbb{P}(B(H))$. A *subprojection* of p is a projection q satisfying $q \leq p$. We write $p^\perp$ for an (*orthogonal*) *complement* of p. Projections p and q are *orthogonal*, which is written as $p \perp q$, if $pq = 0$. An *orthogonal family* is a family of non-zero mutually orthogonal projections. It can be empty.

When there is no danger of confusion, we will, when referring to orthogonal projections hereafter, simply speak of projections.

Definition 0.7 An operator $u \in B(H)$ is *unitary* if $u^*u = uu^* = \mathbb{1}$, an *isometry* if $u^*u = \mathbb{1}$ and a *partial isometry* if $p := u^*u$ is a projection. Then, $q := uu^*$ is also a projection, and p and q are called, respectively, the *initial projection* and *final projection* of u.

Besides the *norm* (or *uniform*) topology on $B(H)$, there are several other topologies that are constantly used in the theory of operator algebras. Here are a few of the most important ones.

Definition 0.8

(1) *Weak (operator)* topology is given by the family of seminorms $a \mapsto |\langle a\eta, \xi \rangle|$ for $\xi, \eta \in H$.

(2) *Strong (operator)* topology is given by the family of seminorms $a \mapsto \|a\xi\|$ for $\xi \in H$.

(3) *Strong* (operator)* topology is given by the family of seminorms $a \mapsto (\|a\xi\|^2 + \|a^*\xi\|^2)^{1/2}$ for $\xi \in H$.

(4) σ-*weak* (or *ultraweak*) topology is given by the family of seminorms $a \mapsto |\sum_{n=1}^\infty \langle a\eta_n, \xi_n \rangle|$ indexed by all sequences $(\xi_n), (\eta_n)$ of vectors from H with $\sum_{n=1}^\infty \|\xi_n\|^2 < \infty$ and $\sum_{n=1}^\infty \|\eta_n\|^2 < \infty$.

(5) σ-*strong* (or *ultrastrong*) topology is given by the family of seminorms $a \mapsto (\sum_{n=1}^\infty \|a\xi_n\|^2)^{1/2}$ indexed by all sequences (ξ_n) of vectors from H with $\sum_{n=1}^\infty \|\xi_n\|^2 < \infty$.

(6) σ-*strong** (or *ultrastrong**) topology is given by the family of seminorms $a \mapsto (\sum_{n=1}^{\infty}(\|a\xi_n\|^2 + \|a^*\xi_n\|^2))^{1/2}$ indexed by all sequences (ξ_n) of vectors from H with $\sum_{n=1}^{\infty} \|\xi_n\|^2 < \infty$.

We will put all the above topologies under one collective name of *non-uniform topologies*.

Proposition 0.9 *The following diagram shows how the topologies relate to each other:*

$$
\begin{array}{ccccc}
weak & \subseteq & strong & \subseteq & strong^* \\
\cap & & \cap & & \cap \\
\sigma\text{-}weak & \subseteq & \sigma\text{-}strong & \subseteq & \sigma\text{-}strong^* & \subseteq & uniform.
\end{array}
\tag{0.1}
$$

On bounded subsets of $B(H)$, the weak topology coincides with the σ-weak (respectively, strong with σ-strong, strong with σ-strong*) topology. For a convex subset of $B(H)$, each of the σ-weak, σ-strong and σ-strong* topologies yield the same closure.*

Proposition 0.10 *A net (a_i) converges to a weakly (respectively, strongly, strongly*) if for each $\xi, \eta \in H$ (respectively, for each $\xi \in H$), we have $\langle a_i\eta, \xi \rangle \to \langle a\eta, \xi \rangle$ (respectively, $a_i\xi \to a\xi$ in norm, $a_i\xi \to a\xi$ and $a_i^*\xi \to a^*\xi$ in norm).*

Proposition 0.11

(1) *The adjoint operation * is continuous in the weak, σ-weak, strong* and σ-strong* topologies, but in general not in the strong or σ-strong topology.*

(2) *With* ball$(B(H))$ *denoting the unit ball of $B(H)$, multiplication $(a,b) \mapsto ab$ is:*

    (1) *continuous from* ball$(B(H)) \times B(H)$ *to $B(H)$ for each of the σ-strong and strong topologies;*

    (2) *continuous from* ball$(B(H)) \times$ ball$(B(H))$ *to $B(H)$ for each of the σ-strong* and strong* topologies;*

    (3) *separately but not jointly continuous from $B(H) \times B(H)$ to $B(H)$ for each of the σ-weak and weak operator topology.*

Proposition 0.12 *If $(a_i)_{i \in I}$ is an increasing net from $B(H)_+$ bounded above by an operator $b \in B(H)_+$, then $a_i \nearrow a$ for some $a \in B(H)$*

(i.e., $\langle a_i\xi,\xi\rangle \nearrow \langle a\xi,\xi\rangle$ for all $\xi \in H$), a is the supremum of a_is and $(a_i)_{i\in I}$ converges to a strongly (and σ-strongly).

Corollary 0.13 *Any family of projections$\{p_i\}_{i\in I}$ possesses both a supremum $\bigvee_{i\in I} p_i$ and an infimum $\bigwedge_{i\in I} p_i$. Moreover, any increasing net of projections $(p_i)_{i\in I}$ is strongly convergent to $\bigvee_{i\in I} p_i$, and any decreasing net of projections $(p_i)_{i\in I}$ is strongly convergent to $\bigwedge_{i\in I} p_i$. Finally, the sum $\sum_{i\in I} p_i$ of a family of (non-zero) projections $\{p_i\}_{i\in I}$ exists in the strong topology and is a projection if and only if the elements of the family are mutually orthogonal.*

Definition 0.14 We write $\mathrm{n}(a)$ for the *null projection* of a, that is the projection onto the *null space* or *kernel* $\{\xi\colon a\xi = 0\}$ of a. The *right support* of a is $\mathbf{s}_r(a) := \mathbb{1} - \mathrm{n}(a)$, and the *left support* or the *range projection* $\mathbf{s}_l(x)$ is the projection onto the closure (in H) of $a(H)$. If $a \in B(H)_h$, then $\mathbf{s}(a) := \mathbf{s}_l(a) = \mathbf{s}_r(a)$ is called the *support* of a.

Proposition 0.15 *The right support is the smallest projection p satisfying $ap = a$, and the left support is the smallest projection p satisfying $pa = a$.*

Definition 0.16 A family of projections $(e_\lambda)_{\lambda\in\mathbb{R}}$ that is increasing: $e_\lambda \leq e'_\lambda$ for $\lambda \leq \lambda'$, continuous from the right in the sense of strong convergence: $e_\lambda = \bigwedge_{\lambda'>\lambda} e_{\lambda'}$ for each $\lambda \in \mathbb{R}$ and satisfies both $\bigwedge_{\lambda\in\mathbb{R}} e_\lambda = 0$ and $\bigvee_{\lambda\in\mathbb{R}} e_\lambda = \mathbb{1}$ is called a *resolution of the identity*. A resolution of the identity is *bounded* if there is a $\lambda_0 > 0$ such that $e_\lambda = 0$ for $\lambda < -\lambda_0$ and $e_\lambda = \mathbb{1}$ for $\lambda > \lambda_0$; otherwise, it is called *unbounded*.

Theorem 0.17 (Spectral decomposition) *Each $a \in B(H)_h$ has a unique spectral decomposition*

$$a = \int_{-\|a\|}^{\|a\|} \lambda de_\lambda \tag{0.2}$$

where $\{e_\lambda\}_{\lambda\in\mathbb{R}}$ is a bounded resolution of the identity satisfying $e_\lambda = 0$ for $\lambda < -\|a\|$ and $e_\lambda = \mathbb{1}$ for $\lambda \geq \|a\|$ and

$$ae_\lambda \leq \lambda e_\lambda, \; ae_\lambda^\perp \geq \lambda e_\lambda^\perp \; \text{for all } \lambda \in \mathbb{R}$$

and the integral is understood as a norm limit of approximating Riemann sums. These sums can be chosen as finite linear combinations of projections $e_{\lambda'} - e_\lambda$ with coefficients in $\mathrm{sp}(a)$. We call $(e_\lambda)_{\lambda \in \mathbb{R}}$ the spectral resolution of the operator a *and the formula (0.2)* the spectral decomposition of a.

We use the Borel functional calculus for bounded operators in the following form:

Theorem 0.18 (Borel functional calculus for bounded operators) *Let $a \in B(H)_h$. There exists a unique injective *-homomorphism $f \mapsto f(a)$ from the *-algebra $\mathcal{B}_b(\mathrm{sp}(a))$ of bounded Borel functions on the spectrum of a into the *-algebra $B(H)$, mapping the identity function $\lambda \mapsto \lambda$ to a and satisfying the following continuity condition:*

if $f, f_n \in \mathcal{B}_b(\mathrm{sp}(a))$, $\sup \|f_n\| < \infty$, and $f_n \to f$ pointwise,

then $f_n(a) \to f(a)$ strongly.

One can write a spectral decomposition of $f(a)$ as

$$f(a) = \int_{-\infty}^{\infty} f(\lambda) de_\lambda,$$

to be understood in a weak sense: for any $\xi, \eta \in H$,

$$\langle f(x)\xi, \eta \rangle = \int_{-\infty}^{\infty} f(\lambda) d\langle e_\lambda \xi, \eta \rangle.$$

We have

$$\|f(x)\xi\|^2 = \int_{-\infty}^{\infty} |f(\lambda)|^2 d\langle e_\lambda \xi, \xi \rangle.$$

It should be noted that for $a \in B(H)_+$ and the function $\lambda \mapsto \lambda^{1/2}$, the operator $f(a)$ is exactly the element $a^{1/2}$, as defined in Definition 0.58.

Proposition 0.19 *For any $a \in B(H)$, we have $\mathsf{s}_l(a) = \mathsf{s}_r(a^*)$ and $\mathsf{s}_r(a) = \mathsf{s}_l(a^*)$. Moreover, $\mathsf{s}_l(a) = \mathsf{s}(aa^*)$ and $\mathsf{s}_r(a) = \mathsf{s}(a^*a)$. For positive a, $\mathsf{s}(a) = \mathsf{s}(a^{1/2})$, so that $\mathsf{s}(|a|) = \mathsf{s}_r(a)$ and $\mathsf{s}(|a^*|) = \mathsf{s}_l(a)$.*

Theorem 0.20 (Polar decomposition) *Let $a \in B(H)$. There exists a partial isometry with initial projection $\mathsf{s}_r(a)$ and final projection*

$\mathbf{s}_l(a)$ *such that* $a = u|a| = |a^*|u$. *If* $a = vb$ *with* $b \in B(H)_+$ *and* v *a partial isometry with initial projection* $\mathbf{s}(b)$, *then* $v = u$ *and* $b = |a|$. *If both* a *and* a^* *are injective, then* $u \in \mathbb{U}(B(H))$, *the unitary group of* $B(H)$.

Definition 0.21 The unique representation of a in the form $a = u|a|$ is called *the polar decomposition of* a.

0.2 Unbounded operators

A good acquaintance with unbounded operators on a Hilbert space is indispensable for dealing with noncommutative L^p and Orlicz spaces. This material is not known as well as that on bounded operators, and we try to prove whatever is possible. Much of the material in this and the following section has been adapted from [**SZ79**].

Let H be a (complex) Hilbert space with an inner product $\langle\,\cdot\,,\cdot\,\rangle$, linear in the first argument and antilinear in the second one.

Definition 0.22 By an (*unbounded*) *operator* on H, we understand a linear map x from a linear subspace $\mathrm{dom}(x) \subseteq H$ into H. We call $\mathrm{dom}(x)$ the *domain* of x and denote by $\mathcal{G}(x)$ the set $\{(\xi, x\xi): \xi \in \mathrm{dom}(x)\} \subseteq H \oplus H$, the *graph* of x. 0 denotes an operator x with $\mathrm{dom}(x) := H$ and $x\xi = 0$ for all $\xi \in H$.

Definition 0.23 (Operations on unbounded operators) We say that:

(1) x and y are *equal* and write $x = y$ if $\mathcal{G}(x) = \mathcal{G}(y)$. Then obviously $\mathrm{dom}(x) = \mathrm{dom}(y)$ and $x\xi = y\xi$ for all $\xi \in \mathrm{dom}(x)$.

(2) y is an *extension* of x and write $x \subseteq y$ or $y \supseteq x$ if $\mathcal{G}(x) \subseteq \mathcal{G}(y)$. Then obviously $\mathrm{dom}(x) \subseteq \mathrm{dom}(y)$ and $x\xi = y\xi$ for all $\xi \in \mathrm{dom}(x)$.

(3) x is *positive* if $\langle x\xi, \xi \rangle \geq 0$ for all $\xi \in \mathrm{dom}(x)$.

(4) For any $\lambda \in \mathbb{C}$ and any operator x, we define operator λx with $\mathrm{dom}(\lambda x) := \mathrm{dom}(x)$ by $(\lambda x)\xi := \lambda(x\xi)$ for all $\xi \in \mathrm{dom}(x)$.

(5) For any operators x and y, we define the *sum* of x and y as the operator $x + y$ with $\mathrm{dom}(x + y) := \mathrm{dom}(x) \cap \mathrm{dom}(y)$ by $(x + y)\xi := x\xi + y\xi$ for all $\xi \in \mathrm{dom}(x + y)$. We define the *difference* of x and y as the operator $x - y := x + (-1)y$.

(6) For any operators x and y, we define the *product* or *composition* of x and y as the operator xy with $\mathrm{dom}(xy) := \{\xi \in \mathrm{dom}(y)\colon y\xi \in \mathrm{dom}(x)\}$ by $(xy)\xi := x(y\xi)$ for all $\xi \in \mathrm{dom}(xy)$.

(7) For any *injective* operator x, we define the *inverse* operator x^{-1} with $\mathrm{dom}(x^{-1}) := x\,\mathrm{dom}(x)$ by $x^{-1}(\eta) := \xi$ whenever $\eta = x\xi$ for all $\eta \in \mathrm{dom}(x^{-1})$.

Proposition 0.24 *The addition of operators is commutative and associative, and the multiplication of operators is associative. We also have, for any operators x_1, x_2, y,*

$$(x_1 + x_2)y = x_1 y + x_2 y$$

$$y(x_1 + x_2) \supseteq yx_1 + yx_2.$$

If x is injective, then $(x^{-1})^{-1} = x$. If additionally y is injective and $x \subseteq y$, then $x^{-1} \subseteq y^{-1}$.

Proof.   Obvious from the definitions. $\square$

Definition 0.25 Operator x is:

(1) *densely defined* if its domain is dense in H;

(2) *closed* if its graph is closed in $H \oplus H$;

(3) *closable* of *preclosed* if the closure $\overline{\mathcal{G}(x)}$ of the graph of x is itself a graph of some operator y. We write then $[x] := y$ and call $[x]$ the *closure* of x. It is the smallest closed extension of x, in the sense that $x \subseteq [x]$ and if, for some closed z, we have $x \subseteq z$, then $[x] \subseteq z$. x is preclosed if for any sequence (ξ_n) from $\mathrm{dom}(x)$, whenever $\xi_n \to 0$ and $x\xi_n$ converges, then $x\xi_n \to 0$.

(4) *bounded* if it is everywhere defined and $\|x\| := \sup\{\|x\xi\|\colon \xi \in H, \|\xi\| \leq 1\} < \infty$. In this case, $\|x\|$ is the *norm* of x. The set of all bounded operators on H is denoted by $B(H)$.

Proposition 0.26

(1) *x is closed if whenever (ξ_n) is a sequence from $\mathrm{dom}(x)$ such that $\xi_n \to \xi \in H$ and $x\xi_n \to \eta \in H$, then $\xi \in \mathrm{dom}(x)$ and $\eta = x\xi$.*

(2) *x is preclosed if for any sequence (ξ_n) from $\mathrm{dom}(x)$, whenever $\xi_n \to 0$ and $x\xi_n$ converges, then $x\xi_n \to 0$.*

(3) *if x is densely defined and $\sup\{\|x\xi\|\colon \xi \in \mathrm{dom}(x),\ \|\xi\| \le 1\} < \infty$, then x is closable and $[x]$ is bounded.*
(4) *If x is closed and $\mathrm{dom}(x) = H$, then x is bounded.*
(5) *If x is closed and injective, then x^{-1} is closed.*
(6) *If x is closed, then its kernel is closed.*

Proof.

(1) The condition guarantees that the graph $\mathcal{G}(x)$ of x is closed.
(2) The condition guarantees that the closure of $\mathcal{G}(x)$ is the graph of a function. We can then define $\mathrm{dom}([x]) := \{\xi \in H\colon$ there is an $\eta \in H$ such that $(\xi, \eta) \in \overline{\mathcal{G}(x)}\}$ and $[x]\xi$, for $\xi \in \mathrm{dom}([x])$, as a unique element $\eta \in H$ such that $(\xi, \eta) \in \overline{\mathcal{G}(x)}$.
(3) If $\xi \in H$, there is a sequence (ξ_n) in $\mathrm{dom}(x)$ such that $\xi_n \to \xi$. The boundedness condition guarantees that the image $(x\xi_n)$ of the Cauchy sequence (ξ_n) is itself a Cauchy sequence and $x\xi_n \to \xi$. Consequently, the condition of closability is satisfied and the closure of x is everywhere defined and satisfies the boundedness condition; hence, it is bounded.
(4) This is the famous *closed graph theorem*.
(5) Immediate from $\mathcal{G}(x^{-1}) = \{(\eta, \xi)\colon (\xi, \eta) \in \mathcal{G}(x)\}$.
(6) Immediate from (1). $\qquad\qquad\qquad\qquad\qquad\qquad\qquad\square$

Definition 0.27 For a closed densely defined operator x on H, we define:

(1) the *null projection* $\mathrm{n}(x)$ as the projection onto the null space $\{\xi\colon x\xi = 0\}$, i.e. the kernel of x;
(2) the *right support* $\mathrm{s}_r(x) := \mathbb{1} - \mathrm{n}(x)$;
(3) the *left support* $\mathrm{s}_l(x)$ as the projection onto the closure (in H) of $x(\mathrm{dom}(x))$.

Lemma 0.28 *Let x be a densely defined operator on H. Let $f_\eta :\mathrm{dom}(x) \ni \xi \mapsto \langle x\xi, \eta\rangle \in \mathbb{C}$. Put $D := \{\eta\colon f_\eta \text{ is bounded }\}$. Then, D is a linear subspace of H. If $\eta \in D$, then there exists a unique $\zeta \in H$ such that $\langle \xi, \zeta\rangle = \langle x\xi, \eta\rangle$.*

Proof. The density of $\mathrm{dom}(x)$ implies that f_η extends to a bounded linear form on the whole of H. By the Riesz theorem, there exists a unique $\zeta \in H$ such that $\langle \xi, \zeta\rangle = \langle x\xi, \eta\rangle$. $\qquad\qquad\square$

Definition 0.29 We define the *adjoint* of x as an operator x^* with domain $\mathrm{dom}(x^*) = D$ from the previous lemma such that $x^*\eta = \zeta$. In other words, we have $\langle x\xi, \eta \rangle = \langle \xi, x^*\eta \rangle$ for $\xi \in \mathrm{dom}(x)$ and $\eta \in \mathrm{dom}(x^*)$. We say that a densely defined operator x is *self-adjoint* if $x = x^*$.

Proposition 0.30 *For any densely defined operator x on H:*

- (1) *x^* is closed;*
- (2) *x is preclosed (closable) if and only if x^* is densely defined, in which case $[x] = x^{**}$;*
- (3) *$\mathsf{s}_r(x^*) = \mathsf{s}_l(x)$;*

Proof.

(1) Let $\eta_n \in \mathrm{dom}(x^*)$, $\eta_n \to \eta$ and $x^*\eta_n \to \zeta$. Then, for $\xi \in \mathrm{dom}(x)$,

$$\langle x\xi, \eta \rangle = \lim_{n \to \infty} \langle x\xi, \eta_n \rangle = \lim_{n \to \infty} \langle \xi, x^*\eta_n \rangle = \langle \xi, \zeta \rangle.$$

Therefore, $\eta \in \mathrm{dom}(x^*)$ and $x^*\eta = \zeta$, which shows that x^* is closed.

(2) Let u be an operator on $H \oplus H$ given by $(\xi, \eta) \mapsto (\eta, -\xi)$. It is easy to check that u is a unitary, and $u^* = u^{-1} = -u$. If $(\eta, x^*\eta) \in \mathcal{G}(x^*)$ and $(\xi, x\xi) \in \mathcal{G}(x)$, then

$$\langle u(\xi, x\xi), (\eta, x^*\eta) \rangle = \langle (x\xi, -\xi), (\eta, x^*\eta) \rangle = \langle x\xi, \eta \rangle + \langle -\xi, x^*\eta \rangle = 0.$$

Hence, $\mathcal{G}(x^*) \subseteq (u\mathcal{G}(x))^\perp$. To see that equality holds, observe that if $(\eta, \zeta) \in (u\mathcal{G}(x))^\perp$, then

$$0 = \langle u(\xi, x\xi), (\eta, \zeta) \rangle = \langle (x\xi, -\xi), (\eta, \zeta) \rangle = \langle x\xi, \eta \rangle + \langle -\xi, \zeta \rangle.$$

Thus, $\xi \mapsto \langle x\xi, \eta \rangle$ then corresponds to the continuous mapping $\xi \mapsto \langle \xi, \zeta \rangle$, and hence, by definition, $\eta \in \mathrm{dom}(x^*)$ with $\zeta = x^*\eta$. Consequently, $\mathcal{G}(x^*) = (u\mathcal{G}(x))^\perp$, whence

$$\mathcal{G}(x^*)^\perp = (u\mathcal{G}(x))^{\perp\perp} = \overline{u\mathcal{G}(x)} = u\overline{\mathcal{G}(x)}.$$

It can now easily be verified that $\eta \perp \mathrm{dom}(x^*)$ if and only if $(\eta, 0) \in (\mathcal{G}(x^*))^\perp$ if and only if $(0, \eta) = u^*(\eta, 0) \in \overline{\mathcal{G}(x)}$. The only way that $(0, \eta)$ can belong to $\overline{\mathcal{G}(x)}$ is if there existed a sequence (ξ_n) in $\mathrm{dom}(x)$ such that $\xi_n \to 0$ while $x(\xi_n) \to \eta$. This clearly shows that x fulfils the criteria for closability if and only if $\mathrm{dom}(x^*)$ is dense in H. Finally,

$$\mathcal{G}(x^{**}) = (u^*\mathcal{G}(x^*))^\perp = (u^*(u\mathcal{G}(x))^\perp)^\perp = \mathcal{G}(x)^{\perp\perp} = \mathcal{G}([x])$$

whence $x^{**} = [x]$.

(3) Since x^* is closed, its null space is closed, and $\mathbf{s}_r(x^*)^\perp(H) = \mathfrak{n}(x^*)(H) \subseteq \mathrm{dom}(x^*)$. If $\eta \in \mathfrak{n}(x^*)(H)$, then $x^*\eta = 0$ and $\langle x\xi, \eta\rangle = \langle \xi, x^*\eta\rangle = 0$ for all $\xi \in \mathrm{dom}(x)$. Hence, $\eta \perp \mathbf{s}_l(x)(H)$ and $\mathbf{s}_l(x)(H) \subseteq \mathbf{s}_r(x^*)(H)$. If, on the other hand, $\eta \perp \mathbf{s}_l(x)(H)$, then $\langle x\xi, \eta\rangle = 0$ for all $\xi \in \mathrm{dom}(x)$ implies that $\eta \in \mathrm{dom}(x^*)$ and $x^*\eta = 0$, which means that $\eta \perp \mathbf{s}_r(x^*)$. Consequently, $\mathbf{s}_r(x^*)(H) \subseteq \mathbf{s}_l(x)(H)$, which ends the proof. $\qquad\square$

Proposition 0.31   *Let $\lambda \in \mathbb{C}$, and assume that $x, y, x+y$ and xy are densely defined operators on H, and that $a \in B(H)$. Then:*

(1) $(\lambda x)^* = \bar{\lambda} x^*$;

(2) *if $x \subseteq y$, then $x^* \supseteq y^*$;*

(3) $(x+y)^* \supseteq x^* + y^*$;

(4) $(xy)^* \supseteq y^* x^*$;

(5) *if x is injective and $x(\mathrm{dom}(x))$ dense in H, then $(x^{-1})^* = (x^*)^{-1}$;*

(6) $(x+a)^* = x^* + a^*$;

(7) $(ax)^* = x^* a^*$.

Proof.   (1)–(4) follow easily from the definitions. We will show (4) to indicate the way. Assume $\eta \in \mathrm{dom}(y^*x^*)$ and $\xi \in \mathrm{dom}(xy)$. Then

$$\langle xy\xi, \eta\rangle = \langle y\xi, x^*\eta\rangle = \langle \xi, y^*x^*\eta\rangle$$

From the density of $\mathrm{dom}(xy)$ and continuity of the map $\xi \mapsto \langle xy\xi, \eta\rangle$, we infer $\eta \in \mathrm{dom}(xy)^*$, hence $\mathrm{dom}(y^*x^*) \subseteq \mathrm{dom}((xy)^*)$ and $\langle \xi, y^*x^*\eta\rangle = \langle \xi, (xy)^*\eta\rangle$, so that $(xy)^* \supseteq y^*x^*$.

(5) If $\eta \in \mathrm{dom}((x^*)^{-1})$, then $\eta = x^*\zeta$ for some $\zeta \in \mathrm{dom}(x^*)$, so that, for $\xi \in \mathrm{dom}(x)$,

$$\langle x\xi, (x^{-1})^*\eta\rangle = \langle x\xi, (x^{-1})^*x^*\zeta\rangle = \langle \xi, x^*\zeta\rangle = \langle x\xi, \zeta\rangle$$
$$= \langle x\xi, (x^*)^{-1}x^*\zeta\rangle = \langle x\xi, (x^*)^{-1}\eta\rangle.$$

Hence, $\eta \in \mathrm{dom}((x^{-1})^*)$, so that $(x^*)^{-1} \subseteq (x^{-1})^*$. Put now $y := x^{-1}$. Then, by what we have just proved, $(y^*)^{-1} \subseteq (y^{-1})^* = x^*$. By Proposition 0.24, $(x^{-1})^* \subseteq (x^*)^{-1}$.

(6) We have $\mathrm{dom}(x + a) = \mathrm{dom}(x)$. Since $\langle (x + a)\xi, \eta \rangle = \langle x\xi, \eta \rangle + \langle a\xi, \eta \rangle$, the domains of x^* and $(x+a)^*$ coincide. Now, if $\xi \in \mathrm{dom}(x+a) = \mathrm{dom}(x)$ and $\eta \in \mathrm{dom}((x + a)^*) = \mathrm{dom}(x^*)$, then

$$\langle \xi, (x + a)^* \eta \rangle = \langle (x + a)\xi, \eta \rangle = \langle x\xi, \eta \rangle + \langle a\xi, \eta \rangle$$
$$= \langle \xi, x^* \eta \rangle + \langle \xi, a^* \eta \rangle = \langle \xi, (x^* + a^*)\eta \rangle.$$

Hence, $(x + a)^* = x^* + a^*$.

(7) Note that $\mathrm{dom}(ax) = \mathrm{dom}(x)$ and by (4), $(ax)^* \supseteq x^* a^*$. Take $\eta \in \mathrm{dom}((ax)^*)$ and $\xi \in \mathrm{dom}(x)$. We have

$$\langle \xi, (ax)^* \eta \rangle = \langle ax\xi, \eta \rangle = \langle x\xi, a^* \eta \rangle.$$

Hence, $a^* \eta \in \mathrm{dom}(x^*)$ and $\eta \in \mathrm{dom}(x^* a^*)$. Consequently, $(ax)^* \subseteq x^* a^*$ and $(ax)^* = x^* a^*$. $\square$

The following useful result can be found, among others, in [**KR83**, Theorem 2.7.8(v)].

Proposition 0.32   *If an unbounded operator x is closed and densely defined, then $x^* x$ is self-adjoint.*

Theorem 0.33 (Spectral decomposition for unbounded operators)   *Every self-adjoint x acting on a Hilbert space H has a unique spectral decomposition*

$$x = \int_{-\infty}^{\infty} \lambda de_\lambda \tag{*}$$

where $\{e_\lambda\}$ is a resolution of identity, that is a family of projections satisfying $e_\lambda \leq e'_\lambda$ for $\lambda \leq \lambda'$ with strong convergence of $e_\lambda \to 0$ as $\lambda \to -\infty$, $e_\lambda \to \mathbb{1}$ as $\lambda \to \infty$ and with $e_{\lambda+} = e_\lambda$ for each $\lambda \in \mathbb{R}$ (continuity from the right in the sense of strong convergence). We call $e_\lambda = e_\lambda(x)$ the spectral resolution of x.

The integral $(*)$ can be understood in a weak sense:

$$\langle x\xi, \xi \rangle = \int_{-\infty}^{\infty} \lambda d\langle e_\lambda \xi, \xi \rangle.$$

The domain of x is given by

$$\mathrm{dom}(x) = \{\xi \in H \colon \int_{-\infty}^{\infty} \lambda^2 d\langle e_\lambda \xi, \xi \rangle < \infty\}$$

and

$$\|x\xi\|^2 = \int_{-\infty}^{\infty} \lambda^2 d\langle e_\lambda \xi, \xi \rangle.$$

We shall use the functional calculus of unbounded operators only for positive self-adjoint ones.

Theorem 0.34 (Borel functional calculus for unbounded operators)   *Let x be a positive self-adjoint operator on H and $f \in \mathcal{B}([0,\infty))$, the set of complex Borel measurable functions on $[0,\infty)$ that are bounded on compact sets. The following equation defines a unique operator $f(x)$:*

$$\langle f(x)\xi, \xi \rangle = \int_0^{\infty} f(\lambda) d\langle e_\lambda \xi, \xi \rangle$$

which is written further as

$$f(x) = \int_0^{\infty} f(\lambda) de_\lambda,$$

with domain

$$\mathrm{dom}(f(x)) = \{\xi \in H \colon \int_0^{\infty} |f(\lambda)|^2 d\langle e_\lambda \xi, \xi \rangle < \infty\}.$$

Moreover, there is a dense subspace D of H contained in $\mathrm{dom}(f(x))$ for any $f \in \mathcal{B}([0,\infty))$, and $\overline{f(x){\restriction}D} = f(x)$, i.e., D is a core for all $f(x)$. The subspace D can be obtained as a union of a countable number of ranges of spectral projections of x. We have

$$\|f(x)\xi\|^2 = \int_0^{\infty} |f(\lambda)|^2 d\langle e_\lambda \xi, \xi \rangle \text{ for } \xi \in D.$$

The above theorem immediately yields a square root of an unbounded positive self-adjoint operator, thus extending the notion of a square root of a positive bounded operator (cf. Definition 0.58 and the comment after Theorem 0.18). Similarly, we can define the absolute value of an unbounded closed densely defined operator.

Definition 0.35 For any closed densely defined x on H, we define $|x| := (x^*x)^{1/2}$ and call it the *absolute value* (or *modulus*) of x.

The following proposition extends the results of Proposition 0.19 to unbounded operators (cf. Proposition 0.30(3) for a part that is true with weaker assumptions).

Proposition 0.36 *For any closed densely defined operator x on H, we have $\mathbf{s}_l(x) = \mathbf{s}_r(x^*)$ and $\mathbf{s}_r(x) = \mathbf{s}_l(x^*)$. Moreover, $\mathbf{s}_l(x) = \mathbf{s}(xx^*)$ and $\mathbf{s}_r(x) = \mathbf{s}(x^*x)$. For positive x, $\mathbf{s}(x) = \mathbf{s}(x^{1/2})$, in particular $\mathbf{s}(|x|) = \mathbf{s}_r(x)$ and $\mathbf{s}(|x^*|) = \mathbf{s}_l(x)$.*

One also has polar decomposition of unbounded operators (cf. Theorem 0.20).

Theorem 0.37 (Polar decomposition) *Let x be a closed densely defined operator on H. There exists a partial isometry $u \in B(H)$ with initial projection $\mathbf{s}_r(x)$ and final projection $\mathbf{s}_l(x)$ such that $x = u|x| = |x^*|u$. If $x = vy$ with y positive and $v \in B(H)$ a partial isometry with initial projection $\mathbf{s}(b)$, then $v = u$ and $y = |x|$. Furthermore, $u(x^*x)u^* = xx^*$.*

Definition 0.38 The unique representation of a closed operator x in the form $x = u|x|$ is called *the polar decomposition of x*.

The following easy technical result will be used in the sequel:

Lemma 0.39 *Let $\{p_n\}_{n \in \mathbb{N}}$ be an orthogonal family of projections on H and let $\{\lambda_n\}_{n \in \mathbb{N}}$ be a family of positive numbers. Then, the operator x defined on $D := \bigcup_{n \in \mathbb{N}} p_n H$ by $p_n \xi = \lambda_n \xi$ is closable and its closure is a positive self-adjoint operator on H.*

Notation 0.40 We will denote by $\sum_{n \in \mathbb{N}} \lambda_n p_n$ the positive self-adjoint operator from the previous proposition.

0.3 C^*-algebras

In this section, we will give the definition and the most basic properties of C^*-algebras. In addition to the monographs already mentioned in the Introduction, the reader interested in the theory of C^*-algebras could learn a lot from the classical book of Naimark [Naĭ172] and the books of Sakai [**Sak71**], Dixmier [**Dix96b**], Pedersen [**Ped18**], Murphy [**Mur90**], Arveson [**Arv76**], Fillmore [**Fil96**] and Davidson [**Dav96**].

Definition 0.41 A *Banach algebra* is a (complex) algebra that is also a Banach space, with a submultiplicative norm: $\|ab\| \leq \|a\|\|b\|$. A *Banach $*$-algebra* is a Banach algebra with involution additionally satisfying $\|a^*\| = \|a\|$. A Banach $*$-algebra is said to be an (abstract) C^*-algebra if the norm satisfies the C^*-condition: $\|a^*a\| = \|a\|^2$. If the C^*-algebra is *unital*, we denote its unit by $\mathbb{1}$; then $\|\mathbb{1}\| = 1$.

Remark 0.42 It is well known that any $*$-homomorphism from one C^*-algebra into another is automatically contractive (see, for example, [**Tak02**, Proposition I.5.2]). Hence, the $*$-homomorphisms are also the natural morphisms for the category of C^*-algebras.

As seen above, C^*-algebras constitute a subclass of the class of Banach algebras. The importance of this particular subclass stems from the following:

Example 0.43 There are two prototypical examples of C^*-algebras.

(1) $\mathcal{C}_0(X)$, where X is a locally compact Hausdorff space, and $\mathcal{C}_0(X)$ denote continuous functions on X vanishing at infinity. With the supremum norm and the natural involution $f \mapsto (\bar{f}\colon x \mapsto \overline{f(x)})$, it becomes a commutative C^*-algebra. For compact X, we get a unital commutative C^*-algebra $\mathcal{C}(X)$, with unit $\mathbb{1} := \mathbb{1}_X$.

(2) Endowed with the operator norm, $B(H)$ becomes a Banach space. With the usual algebraic operations and the adjoint map $a \mapsto a^*$, $B(H)$ becomes a $*$-algebra. The operator norm satisfies all the conditions of a C^*-norm, turning $B(H)$ into a C^*-algebra with unit $\mathbb{1} := \mathbb{1}_H$. It is obvious that $a \in B(H)$ is self-adjoint (respectively, positive) as a bounded operator

if and only if it is self-adjoint (respectively, positive) as an element of the C^*-algebra $B(H)$. Similarly, it is clear that $a \leq b$ means the same for a and b treated as bounded operators on H and elements of the C^*-algebra $B(H)$.

To show the importance of the first example, we will introduce the notion of a *spectrum* of a commutative C^*-algebra $\mathcal{A}$.

Definition 0.44 The *spectrum* of $\mathcal{A}$, denoted by $\mathrm{Sp}(\mathcal{A})$, is the set of *characters* of $\mathcal{A}$, i.e., non-zero homomorphisms of $\mathcal{A}$ into $\mathbb{C}$.

Since $\mathrm{Sp}(\mathcal{A})$ is contained in the Banach space dual $\mathcal{A}^*$ of $\mathcal{A}$, we can endow it with the (restriction of) the weak*-topology.

Proposition 0.45 *The space* $\mathrm{Sp}(\mathcal{A})$ *endowed with the restriction of the* $\sigma(\mathcal{A}^*, \mathcal{A})$*-topology is locally compact Hausdorff. It is compact if* $\mathcal{A}$ *is unital.*

Definition 0.46 For any $a \in \mathcal{A}$, let $\widehat{a}$ denote a map from $\mathcal{C}_0(\mathrm{Sp}(\mathcal{A}))$ into $\mathbb{C}$ given by $\widehat{a}(\chi) := \chi(a)$. The map $a \mapsto \widehat{a}$ from $\mathcal{A}$ into $\mathcal{C}_0(\mathrm{Sp}(\mathcal{A}))$ is called the *Gelfand transform*.

We pause to remind the reader that the iconical morphisms which identify two Banach spaces as being 'the same' in all respects are the surjective linear isometries which we shall refer to as *isometric isomorphisms*. When Banach dual spaces are in view, the appropriate class of morphisms that identify two such spaces as being the same are the isometric isomorphisms which are also weak* homeomorphisms. When dealing with Banach algebras, we additionally demand that these morphisms which identify two objects as being the same also preserve the algebraic structure in the sense of being algebraic isomorphism. For Banach *-algebras, we then further demand that these morphisms preserve involution.

Theorem 0.47 (Gelfand—Naimark theorem for abelian C^*-algebras) *Every commutative C^* algebra $\mathcal{A}$ (respectively, unital commutative C^* algebra) is isometrically isomorphic to $\mathcal{C}_0(\mathrm{Sp}(\mathcal{A}))$ (respectively, $\mathcal{C}(\mathrm{Sp}(\mathcal{A}))$), with the isomorphism given by the Gelfand transform. Given two commutative C^*-algebras $\mathcal{A}$ and $\mathcal{B}$, there is* *moreover a bijective correspondence between *-homomorphisms $\mathscr{I}$: $\mathcal{A} \to \mathcal{B}$ and continuous functions ϑ : $\mathrm{Sp}(\mathcal{B}) \to \mathrm{Sp}(\mathcal{A})$ given by the formula $\widehat{\mathscr{I}(a)} = \widehat{a} \circ \vartheta$ for all $a \in \mathcal{A}$.*

It is well known that the topological properties of a locally compact or compact space X can be read from the algebraic properties of $\mathcal{C}_0(X)$ or $\mathcal{C}(X)$. Thus, one can treat 'commutative' topology (or at least a part of it) as the study of commutative C^*-algebras. That is why the general theory of C^*-algebras is often called 'noncommutative topology'.

We now turn to the second example. It is obvious that norm-closed *-subalgebras of $B(H)$ become themselves C^*-algebras. C^*-algebras obtained in this way are called *concrete* or *represented*. One prominent example is the (in general non-unital) algebra $K(H)$ of compact operators on H. To go from an abstract to a concrete C^*-algebra, we need a notion of a representation.

Definition 0.48 By a *representation* π of a *-algebra $\mathcal{A}$ on a Hilbert space H, we understand a *-homomorphism from $\mathcal{A}$ into $B(H)$. A representation π is *faithful* if its kernel is $\{0\}$ and *non-degenerate* if $\pi(\mathcal{A})H$ is (norm) dense in H.

Theorem 0.49 (Gelfand–Naimark theorem for general C^*-algebras) *Every abstract C^*-algebra $\mathcal{A}$ is isometrically isomorphic to a concrete one, that is, there exists a (faithful) representation π of $\mathcal{A}$ on some Hilbert space H such that $\|\pi(a)\| = \|a\|$ for all $a \in \mathcal{A}$ and $\pi(\mathcal{A})$ is a C^*-subalgebra of $B(H)$.*

The possibility of switching between abstract and concrete pictures is of fundamental importance. We will often use the possibility of changing the representation so that it suits our needs.

Various classes of elements of C^*-algebras correspond to various classes of bounded operators.

Definition 0.50 Let $\mathcal{A}$ be a C^*-algebra. An element $a \in \mathcal{A}$ is called *self-adjoint* or *Hermitian* if $a = a^*$, *normal* if $a^*a = aa^*$, *unitary* if $aa^* = a^*a = \mathbb{1}$ and *positive* if $a = b^*b$ for some $b \in \mathcal{A}$. The self-adjoint elements of $\mathcal{A}$ are denoted by $\mathcal{A}_h$ and the positive elements by $\mathcal{A}_+$. We write $a \leq b$ for $a, b \in \mathcal{A}_h$ if $b - a \in \mathcal{A}_+$. The set of unitaries in $\mathcal{A}$ is denoted by $\mathbb{U}(\mathcal{A})$.

Remark 0.51 The self-adjoint part $\mathcal{A}_h$ of a C^*-algebra $\mathcal{A}$ becomes an algebra when equipped with the so-called Jordan product $a \circ b = \frac{1}{2}(ab + ba)$. The morphisms on $\mathcal{A}$ which behave well with respect to this structure are the so-called Jordan *-morphisms, namely linear maps $\mathcal{J}$ from one C^*-algebra $\mathcal{A}$ into another $\mathcal{B}$, which preserve both involution and the Jordan product. In other words, $\mathcal{J}(a^*) = \mathcal{J}(a)^*$ and $\mathcal{J}(a \circ b) = \mathcal{J}(a) \circ \mathcal{J}(b)$. It is well known that Jordan *-morphisms on C^*-algebras are also automatically contractive and that they satisfy the following useful identities for all $a, b, c \in \mathcal{A}$:

(1) $\mathcal{J}(aba) = \mathcal{J}(a)\,\mathcal{J}(b)\,\mathcal{J}(a)$;
(2) $\mathcal{J}(abc) + \mathcal{J}(cba) = \mathcal{J}(a)\,\mathcal{J}(b)\,\mathcal{J}(c) + \mathcal{J}(c)\,\mathcal{J}(b)\,\mathcal{J}(a)$;
(3) $[\mathcal{J}(ab) - \mathcal{J}(a)\,\mathcal{J}(b)][\mathcal{J}(ab) - \mathcal{J}(b)\,\mathcal{J}(a)] = 0$.

Definition 0.52 Let $\mathcal{A}$ be a unital C^*-algebra. The *spectrum* $\mathrm{sp}(a)$ of an element $a \in \mathcal{A}$ is the set

$$\{\lambda \in \mathbb{C}: a - \lambda\mathbb{1} \text{ is not invertible in } \mathcal{A}\}.$$

Proposition 0.53 *For any element $a \in \mathcal{A}$, $\mathrm{sp}(a)$ is a compact subset of $\{\lambda \in \mathbb{C}: |\langle\lambda| \leq \|a\|\}$. If a is self-adjoint, then $\mathrm{sp}(a) \subseteq \mathbb{R}$, if a is positive, then $\mathrm{sp}(a) \subseteq \mathbb{R}_+$. If p is a self-adjoint idempotent, then $\mathrm{sp}(p) \subseteq \{0, 1\}$, and if u is unitary, then $\mathrm{sp}(p) \subseteq \{\lambda \in \mathbb{C}: |\langle\lambda| = 1\}$. Furthermore, if $\mathcal{B}$ is a unital C^*-subalgebra of $\mathcal{A}$, then for any $a \in \mathcal{B}$, $\mathrm{sp}_{\mathcal{A}}(a) = \mathrm{sp}_{\mathcal{B}}(a)$.*

We introduce two functional calculi valid for this context. When working with non-normal operators, one may use the so-called holomorphic functional calculus, which is based on Cauchy's integration formula.

Theorem 0.54 (Holomorphic functional calculus) *Let $a \in \mathcal{A}$ be given, let $\mathcal{D}$ be a simply-connected domain containing $\mathrm{sp}(a)$ and Γ a simply closed positively oriented contour inside $\mathcal{D}$ encircling $\mathrm{sp}(a)$. Then, for any function f holomorphic on $\mathcal{D}$,*

$$f(a) = \frac{1}{2\pi i} \int_\Gamma f(z)(z\mathbb{1} - a)^{-1}\, dz$$

is a well-defined element of $\mathcal{A}$. In fact, prescription $f \mapsto f(a)$ yields an algebra homomorphism from the set of functions which are holomorphic in $\mathcal{D}$ to $\mathcal{A}$, which maps the function $\iota : z \mapsto z$ onto a.

When dealing with normal elements of a C^*-algebra, one has access to the more powerful continuous functional calculus.

Theorem 0.55 (Continuous functional calculus) *Let $\mathcal{A}$ be unital and $a \in \mathcal{A}$ normal. There exists a unique $*$-isomorphism $f \mapsto f(a)$ from $C(\mathrm{sp}(a))$ onto the C^*-subalgebra of $\mathcal{A}$ generated by a and $\mathbb{1}$, mapping $\iota : t \mapsto t$ onto a and satisfying $\|f\|_\infty = \|f(a)\|$ for each $f \in C(\mathrm{sp}(a))$.*

Corollary 0.56 *Any $a \in \mathcal{A}$ can be written as a linear combination of four unitaries.*

Corollary 0.57 *If $a \in \mathcal{A}_+$, then there is a unique element $b \in \mathcal{A}_+$, such that $b^2 = a$.*

Definition 0.58 We denote the element in $\mathcal{A}_+$ whose square is $a \in \mathcal{A}_+$ by $a^{1/2}$. For any $a \in \mathcal{A}$, we define $|a| := (a^*a)^{1/2}$ and call it the *absolute value* or *modulus* of a.

We pause to collate some basic properties of the cone $\mathcal{A}_+$.

Proposition 0.59 *Let $\mathcal{A}$ be a C^*-algebra.*

(1) *For any $a \in \mathcal{A}_h$, the elements $a_\pm = \frac{1}{2}(|a| \pm a)$ belong to $\mathcal{A}_+$ and are the unique elements of $\mathcal{A}_+$ satisfying $a = a_+ - a_-$ and $a_+ a_- = 0$. In the case where $\mathcal{A}$ is unital, we have for any $a \in \mathcal{A}_+$ that $a \leq \|a\|\mathbb{1}$.*

(2) *If $0 \leq a \leq b$ for some $a, b \in \mathcal{A}$, then:*
  (1) *$0 \leq a^r \leq b^r$ for any $0 < r \leq 1$;*
  (2) *$\|a\| \leq \|b\|$;*
  (3) *$0 \leq c^*ac \leq c^*bc$ for any $c \in \mathcal{A}$;*
  (4) *in the case where $\mathcal{A}$ is unital we have that $0 \leq (b + \lambda\mathbb{1})^{-1} \leq (a + \lambda\mathbb{1})^{-1}$ for any $\lambda > 0$.*

The above technology now enables us to introduce the important notion of approximate identity.

Theorem 0.60 *Let $\mathcal{L}$ be a left ideal of a C^*-algebra $\mathcal{A}$. Then, there is a net (f_λ) of positive contractive elements of $\mathcal{L}$ such that f_λ increases as λ increases, and $\lim_\lambda \|af_\lambda - a\| = 0$ for all $a \in \mathcal{A}$. A similar claim holds for right-ideals.*

Definition 0.61 Let $\mathfrak{m}$ be a left (respectively, right) ideal of a C^*-algebra $\mathcal{A}$. A net (f_λ) of positive contractive elements of $\mathfrak{m}$ is called a *right* (respectively, *left*) *approximate identity* of $\mathfrak{m}$, if f_λ increases as λ increases, and $\lim_\lambda \|af_\lambda - a\| = 0$ (respectively, $\lim_\lambda \|f_\lambda a - a\| = 0$) for all $a \in \mathcal{A}$.

We are now moving to linear forms on a C^*-algebra $\mathcal{A}$.

Definition 0.62 A linear functional ω on $\mathcal{A}$ is said to be *real* or *Hermitian* if $\omega(a) \in \mathbb{R}$ for all $a \in \mathcal{A}_h$, and *positive* if $\omega(a) \geq 0$ for any $a \in \mathcal{A}_+$. A positive functional of norm 1 is called a *state*. A positive functional ω on $\mathcal{A}$ is called *faithful* if $\omega(a) = 0$ for $a \in \mathcal{A}_+$ implies $a = 0$.

Proposition 0.63 *For any normal element a of a C^*-algebra $\mathcal{A}$, we have that:*

- *a is self-adjoint if and only if $\omega(a) \in \mathbb{R}$ for every state ω on $\mathcal{A}$;*
- *a is positive if and only if $\omega(a) \geq 0$ for every state ω on $\mathcal{A}$.*

Proposition 0.64 *A positive linear functional on a C^*-algebra $\mathcal{A}$ is automatically bounded, i.e., $\omega \in \mathcal{A}_+^*$. If the algebra $\mathcal{A}$ is unital (with unit $\mathbb{1}$), then $\|\omega\| = \omega(\mathbb{1})$.*

Definition 0.65 If $\mathcal{A}$ is a C^*-subalgebra of $B(H)$ and $\xi \in H$, then $\omega_\xi : a \mapsto \langle a\xi, \xi \rangle$ is a positive linear functional on $\mathcal{A}$. If $\|\xi\| = 1$, then ω_ξ is a state, called a *vector state*.

The following definition introduces an important class of linear forms.

Definition 0.66 A functional $\omega \in \mathcal{A}^*$ is called *tracial* if $\omega(ab) = \omega(ba)$ for all $a, b \in \mathcal{A}$.

Notation 0.67 For $\omega \in \mathcal{A}_+^*$, we write $N_\omega := \{a \in \mathcal{A} : \omega(a^*a) = 0\}$.

It is easy to see that N_ω is a left ideal in $\mathcal{A}$. If ω is tracial, then the ideal is two-sided.

Notation 0.68 For $\omega \in \mathcal{A}_+^*$, we denote by η_ω the quotient map $\mathcal{A} \ni a \mapsto a + N_\omega \in \mathcal{A}/N_\omega$. We denote by $\langle \cdot, \cdot \rangle_\omega$ the inner product on $\mathcal{A}/N_\omega$ defined by $\langle \eta_\omega(a), \eta_\omega(b) \rangle_\omega := \omega(b^*a)$.

The Cauchy–Bunyakovsky–Schwarz (*CBS* for short) inequality for the above inner product leads to an important inequality for elements of $\mathcal{A}_+^*$:

$$|\omega(b^*a)|^2 \le \omega(a^*a)\omega(b^*b). \tag{0.3}$$

To construct a faithful representation of an abstract C^*-algebra on a Hilbert space, we use the ingenious *Gelfand–Naimark–Segal construction* (*GNS* for short).

Definition 0.69 (GNS representation) Let H_ω be the Hilbert space completion of $\mathcal{A}/N_\omega$ with the inner product from Notation 0.68. Then, for each $a \in \mathcal{A}$, $\pi_\omega(a) : \eta_\omega(b) \mapsto \eta_\omega(ab)$ extends to a bounded operator on H_ω. The representation π_ω of $\mathcal{A}$ on H_ω obtained in this manner is called the *GNS representation of $\mathcal{A}$ associated with ω*.

Remark 0.70 In the case where the C^*-algebra is unital, it is common practice to denote $\eta_\omega(\mathbb{1})$ by Ω_ω. Each $\eta_\omega(a)$ $(a \in \mathcal{A})$ is then of the form $\pi_\omega(a)\Omega_\omega$. It is then an easy exercise to see that

$$\omega(a) = \langle \pi_\omega(a)\Omega_\omega, \Omega_\omega \rangle \text{ for each } a \in \mathcal{A}.$$

Since $\pi_\omega(\mathcal{A})\Omega_\omega$ is dense in H_ω, the GNS representation is clearly non-degenerate. However, more is true. From the fact that $\omega(a^*a) = \langle \pi_\omega(a^*a)\Omega_\omega, \Omega_\omega \rangle = \|\pi_\omega(a)\Omega_\omega\|^2$ for all $a \in \mathcal{A}$, it is clear that ω is faithful if and only if $\pi_\omega(a)\Omega_\omega \ne 0$ for every non-zero $a \in \mathcal{A}$.

The facts noted above justify the following definition.

Definition 0.71 Let $\{\pi, H\}$ be a representation of a C^*-algebra $\mathcal{A}$. We say that a vector $\xi \in H$ is *cyclic* if $\pi(\mathcal{A})\xi$ is dense in H, and *separating* if for any $a \in \mathcal{A}$, we have that $\pi(a)\xi \ne 0$ whenever $a \ne 0$.

Before moving on from a review of the GNS representation, we record a result pointing to the uniqueness of this representation. This aspect of uniqueness will ultimately find a fuller expression in Haagerup's theory of the standard form of a von Neumann algebra, which we shall briefly encounter in Section 8.5 and then explore further in Section 10.6.

Proposition 0.72 *Let ω be a state on a unital C^*-algebra $\mathcal{A}$ and $\{\pi_0, H_0\}$ a representation of $\mathcal{A}$ which admits a cyclic vector ξ_0 for which $\omega(\cdot) = \langle \pi_0(\cdot)\xi_0, \xi_0 \rangle$. Then, there exists a unitary $u : H_0 \to H_\omega$ such that $u(\xi_0) = \Omega_\omega$ and $u^*\pi_\omega u = \pi_0$.*

Proof. Consider the formal operator $u \colon \pi_0(\mathcal{M})\xi_0 \to \pi_\omega(\mathcal{M})\Omega_\omega$: $\pi_0(a)\xi_0 \mapsto \pi_\omega(a)\Omega_\varphi$. The operator is clearly a linear bijection from $\pi_0(\mathcal{M})\xi_0$ to $\pi_\omega(\mathcal{M})\Omega_\omega$. The cyclicity of both ξ_0 and Ω_ω ensure that the spaces $\pi_0(\mathcal{M})\xi_0$ and $\pi_\omega(\mathcal{M})\Omega_\omega$ are respectively dense in H_0 and H_ω. Thus, u is both densely defined and has dense range. Since we further have that $\|\pi_0(a)\xi_0\|^2 = \omega(a^*a) = \|\pi_\omega(a)\Omega_\omega\|^2$, the action of u is in fact isometric, which in turn ensures that it uniquely extends to a linear isometry from H_0 onto H_ω; that is a unitary. Given $a, b \in \mathcal{M}$, we moreover have that

$$u(\pi_0(a)\pi_0(b)\xi_0) = u(\pi_0(ab)\xi_0) = \pi_\omega(ab)\Omega_\omega$$
$$= \pi_\omega(a)\pi_\omega(b)\Omega_\omega = \pi_\omega(a)u(\pi_0(b)\xi_0).$$

Since $\{\pi_0(b)\xi_0 \colon a \in \mathcal{M}\}$ is dense in H_0, this can only be if $\pi_0(\cdot) = u^*\pi_\omega(\cdot)u$, thereby proving the second claim. $\square$

We close our review of C^*-algebras with a description of the second dual.

Definition 0.73 Let $\mathcal{A}$ be any subset of $B(H)$. The *commutant* $\mathcal{A}'$ of $\mathcal{A}$ is the set $\{a' \in B(H) \colon aa' = a'a \text{ for all } a \in \mathcal{A}\}$. The *centre* $\mathcal{Z}(\mathcal{A})$ of $\mathcal{A}$ is defined as $\mathcal{A} \cap \mathcal{A}'$.

Definition 0.74 A representation $\{\pi, H\}$ of a C^*-algebra $\mathcal{A}$ is a *universal representation* if for any other representation $\{\rho, K\}$, there exists a σ-weakly continuous *-homomorphism $\widetilde{\rho}$ from $\pi(\mathcal{A})''$ onto $\rho(\mathcal{A})''$ such that $\widetilde{\rho} \circ \pi = \rho$.

It is an exercise to see that universal representations are unique up to σ-weakly continuous *-isomorphisms.

Theorem 0.75 *With $\mathfrak{S}(\mathcal{A})$ denoting the set of states (referred to as the* state space*) of a C^*-algebra $\mathcal{A}$, the representation $\{\oplus_{\omega \in \mathfrak{S}(\mathcal{A})} \pi_\omega; \oplus_{\omega \in \mathfrak{S}(\mathcal{A})} H_\omega\}$ is a universal representation of $\mathcal{A}$. For any universal representation $\{\pi, H\}$, there is a unique linear isometry $\widetilde{\pi}$ from the second dual $\mathcal{A}^{**}$ onto $\pi(\mathcal{A})''$ which is a $(\mathcal{A}^{**}, \sigma(\mathcal{A}^{**}, \mathcal{A}^*))$ to $(\pi(\mathcal{A})'', \sigma - \text{weak})$ homeomorphism.*

In view of the above facts, we shall consistently identify $\mathcal{A}^{**}$ with $\pi(\mathcal{A})''$ as described above and refer to it as the *universal enveloping von Neumann algebra* of $\mathcal{A}$.

0.4 Von Neumann algebras

The theory of von Neumann algebras is very rich, and we are dealing here only with its most basic aspects. The reader interested in the theory, in addition to texts and monographs mentioned in the Preface and in Section 0.3, could consult the books of Kaplansky [**Kap68**], Sunder [**Sun87**], Zhu [**Zhu93**] and volume 3 of Takesaki [**Tak03b**].

The presentation in this section is strongly influenced by [**Tak02**].

Definition 0.76 A *-subalgebra $\mathcal{M}$ of $B(H)$ is called a (concrete) *von Neumann algebra* if $\mathcal{M} = \mathcal{M}''$. Note that $\mathbb{1} = \mathbb{1}_H \in \mathcal{M}$. We say that $\mathcal{M}$ *acts* on H. A von Neumann algebra is called a *factor* if the centre of the algebra is trivial, i.e., $\mathcal{Z}(\mathcal{M}) = \mathbb{C}\mathbb{1}_H$.

It is clear that $B(H)$ is a factor von Neumann algebra. In fact, $B(H)' = \mathbb{C}\mathbb{1}_H$ and $(\mathbb{C}\mathbb{1}_H)' = B(H)$.

Definition 0.77 We say that von Neumann algebras $\mathcal{M}_1$ and $\mathcal{M}_2$, acting, respectively, on H_1 and H_2, are *isomorphic* if there exists a *-preserving algebra isomorphism Φ of $\mathcal{M}_1$ onto $\mathcal{M}_2$. It is then automatically norm-preserving and σ-weakly bicontinuous. We denote this fact by $\mathcal{M}_1 \cong \mathcal{M}_2$. If, for some unitary $u : H_1 \to H_2$ (i.e., $u^*u = 1_{H_1}$, $uu^* = 1_{H_2}$) we have $\Phi(a) = uau^*$, we say that Φ is a *spatial isomorphism*, and the algebras are *spatially isomorphic*.

There is also an abstract counterpart to the notion of a von Neumann algebra, introduced by Shôichirô Sakai in [**Sak56**].

Definition 0.78 An (abstract) C^* algebra $\mathcal{M}$ is called an *abstract von Neumann algebra* or a *W^*-algebra* if, as a Banach space, it is the dual of another Banach space.

In this case, the *predual* Banach space is unique (see [**Sak71**, 1.13.3]) and we denote it by $\mathcal{M}_*$. In the present context uniqueness of the predual ensures that $(\mathcal{M}_*)^* \simeq \mathcal{M}$, where $\simeq$ denotes the Banach space isometric isomorphism. From the duality theory for locally convex spaces, we know that $\mathcal{M}_*$ can be identified with the set of $\sigma(\mathcal{M}, \mathcal{M}_*)$-continuous functionals on $\mathcal{M}$, with duality given by $\langle a, \omega \rangle := \omega(a)$.

The following theorem corresponds to the Gelfand–Naimark theorem for general C^*-algebras.

Theorem 0.79 *Every abstract von Neumann algebra $\mathcal{M}$ is isometrically isomorphic to a concrete one, i.e., there exists a (faithful) representation π of $\mathcal{M}$ on some Hilbert space H such that $\|\pi(a)\| = \|a\|$ for all $a \in \mathcal{M}$ and $\pi(\mathcal{M})$ is a von Neumann algebra acting on H.*

Proposition 0.80 *If $\mathcal{M}$ is a von Neumann algebra acting on a Hilbert space H, then the $\sigma(\mathcal{M}, \mathcal{M}_*)$-topology on $\mathcal{M}$ is exactly the σ-weak topology. In other words, the predual $\mathcal{M}_*$ of $\mathcal{M}$ consists of σ-weakly continuous functionals on $\mathcal{M}$.*

Proposition 0.81 *For a functional $\omega \in \mathcal{M}^*$, the following conditions are equivalent.*

(1) *ω is weakly continuous.*
(2) *ω is strongly continuous.*
(3) *ω is strongly* continuous.*

For a functional $\omega \in \mathcal{M}^*$, the following conditions are equivalent.

(4) *ω is σ-weakly continuous.*
(5) *ω is σ-strongly continuous.*
(6) *ω is σ-strongly* continuous.*

Definition 0.82 A functional $\omega \in \mathcal{M}^*$ is called

(1) *completely additive* if $\omega(\sum_{i\in I} a_i) = \sum_{i\in I}\omega(a_i)$ for any family $\{a_i\}_{i\in I}$ of positive operators from $\mathcal{M}$ with

$$\sup_{J\subseteq I,\, J\text{ finite}} \|\sum_{i\in J} a_i\| < \infty;$$

(2) *completely additive on projections* if $\omega(\sum_i p_i) = \sum_i \omega(p_i)$ for any orthogonal family of projections from $\mathcal{M}$.

(3) If in addition ω is a positive functional, we say that it is *normal* if $a_i \nearrow a$ implies $\omega(a_i) \nearrow \omega(a)$ for any increasing net $(a_i)_{i\in I}$ from $\mathcal{M}_+$ with supremum $a \in \mathcal{M}$.

Theorem 0.83 [Bla06, Theorem III.2.1.4], [Tak02, Corollary III.3.11]

For a functional $\omega \in \mathcal{M}^*$, the following conditions are equivalent.

(1) $\omega \in \mathcal{M}_*$.

(2) ω *is σ-weakly (or $\sigma(\mathcal{M},\mathcal{M}_*)$, σ-strongly, σ-strongly*) continuous.*

(3) ω *is completely additive.*

(4) ω *is completely additive on projections.*

(5) $\omega = \sum_{n\in\mathbb{N}} \langle(\cdot)\xi_n, \zeta_n\rangle$ *for some $\xi_n, \zeta_n \in H$ with $\sum_{n\in\mathbb{N}}(\|\xi_n\|^2 + \|\zeta_n\|^2) < \infty$.*

If ω is a state, then any one (and therefore all) of the above conditions is equivalent to ω being normal. In this case, condition (5) above may be reformulated as the claim that $\omega = \sum_{n\in\mathbb{N}} \omega_{\xi_n}$ for some $\xi_n \in H$ with $\sum_{n\in\mathbb{N}} \|\xi_n\|^2 = 1$.

Any element of $\mathcal{M}_*$ may be written as a linear combination of four states.

In view of the above result, even the non-positive elements of $\mathcal{M}_*$ are sometimes also referred to as 'normal' functionals. The earlier description of normal elements of the algebra in terms of their behaviour with respect to states extends to normal states in the von Neumann algebra context.

Although condition (5) above does not make sense for abstract von Neumann algebras, the rest of the result has an abstract counterpart. We record the core of that result as a separate statement.

Theorem 0.84 *Let M be an abstract von Neumann algebra. For any state $\omega \in M_+^*$, the following conditions are equivalent.*

> (1) *$\omega \in M_*$.*
> (2) *ω is σ-weakly (or equivalently weak*) continuous.*
> (3) *ω is completely additive on projections.*
> (4) *ω is normal.*

Given any increasing net (a_γ) of positive elements of some concrete von Neumann algebra $M \subset B(H)$ increasing to $a \in B(H)$, it is clear from Proposition 0.12 that $a \in M$. The same property holds for abstract von Neumann algebras. (See, for example, [**Sak71**, Lemma 1.7.4] and its proof.) When combined with the above analysis, this observation leads to the following elegant conclusion.

Proposition 0.85 *Let M be an abstract von Neumann algebra. Then, any bounded increasing net in M_+ has a supremum, which coincides with its weak* limit. In particular, the weak* continuous states of M determine the order on M.*

The class of normal states is also large enough to effectively identify self-adjoint and positive elements.

Proposition 0.86 *For any normal element a of a von Neumann algebra M, we have that:*

> * *a is self-adjoint if and only if $\omega(a) \in \mathbb{R}$ for every normal state ω;*
> * *a is positive if and only if $\omega(a) \geq 0$ for every normal state ω.*

In the von Neumann algebra context, the second dual serves the purpose of achieving a decomposition of any functional $\omega \in M^*$ into a normal and singular part.

Theorem 0.87 *Let M be a von Neumann algebra. Then, there exists a central projection p in M^{**} such that $pM^{**} \equiv M$ and $p.M^* \equiv M_*$ (where $p.\omega := \omega(\cdot p)$). Any state $\omega \in M^*$ uniquely extends with preservation of the norm to a normal state on M^{**}. Given a state $\omega \in M^*$ and using the same notation for the extension, we define $\omega_n = p.\omega$ and $\omega_s = (\mathbb{1} - p).\omega$, respectively, as the normal and singular parts of ω.*

The following result provides criteria for testing a state for singularity.

Proposition 0.88 [Tak02, Theorem III.3.8] *A state ω of a von Neumann algebra is singular if and only if for every non-zero projection e we can find a non-zero sub-projection e_1 for which we have that $\omega(e_1) = 0$.*

The characterisation of normal states given in Theorem 0.84 now easily yields the following conclusion.

Proposition 0.89 *Any $*$-isomorphism $\mathscr{I}$ from one von Neumann algebra $\mathcal{M}_1$ onto another $\mathcal{M}_2$ is automatically a σ-weak$\,$–$\,\sigma$-weak and σ-strong$\,$–$\,\sigma$-strong homeomorphism.*

Proof. Since both $\mathscr{I}$ and $\mathscr{I}^{-1}$ are contractive, $\mathscr{I}$ is an isometric isomorphism, and hence so is the *Banach adjoint* $\mathscr{I}^*$ of $\mathscr{I}$. For any $a \in \mathcal{M}_1$, we have that $\mathscr{I}(a^*a) = \mathscr{I}(a)^*\mathscr{I}(a)$ since any $b \in \mathcal{M}_1^+$ may be written in the form $b = a^*a$, this shows that $\mathscr{I}$ is an order-isomorphism. But if that is the case, we will for any net $(a_i) \subseteq \mathcal{M}_1$ and $a \in \mathcal{M}_1$ have that $a_i \nearrow a$ if and only if $\mathscr{I}(a_i) \nearrow \mathscr{I}(a)$. But then the preceding theorem will ensure that for any state ω on $\mathcal{M}_2$, we have that $\omega \in (\mathcal{M}_2)_*$ if and only $\omega \circ \mathscr{I} \in (\mathcal{M}_1)_*$. Thus, $\mathscr{I}^*$ restricts to an isometric isomorphism from $(\mathcal{M}_2)_*$ onto $(\mathcal{M}_1)_*$. It is now an exercise to see that $\mathscr{I}$ is the Banach adjoint of this restriction, proving that $\mathscr{I}$ is a σ-weak$\,$–$\,\sigma$-weak homeomorphism.

For the claim regarding the σ-strong topology, we note that by Theorem 0.83, the seminorms determining the σ-strong topology on say $\mathcal{M}_2$ are all of the form $\omega(a^*a)^{1/2}$ where $\omega \in (\mathcal{M}_2)_*^+$ and $a \in \mathcal{M}_2$. Taking into account that $\mathscr{I}^*$ restricts to a linear bijection from $(\mathcal{M}_2)_*^+$ to $(\mathcal{M}_1)_*^+$, the claim now follows from the obvious fact that for any $\omega \in (\mathcal{M}_2)_*^+$, any $a \in \mathcal{M}_1$ and any net $(a_\gamma) \subset \mathcal{M}_1$, we have $\omega \circ \mathscr{I}(|a - a_\gamma|^2) = \omega(|\mathscr{I}(a) - \mathscr{I}(a_\gamma)|^2)$. $\square$

Jordan $*$-morphisms on von Neumann algebras also exhibit somewhat more elegant behaviour than their C^*-algebraic counterparts.

Proposition 0.90 (BR87, Proposition 3.2.2) *Let $\mathscr{J}$ be a Jordan $*$-homomorphism from one von Neumann algebra $\mathcal{M}_1$ into another $\mathcal{M}_2$, and let $\mathcal{B}$ be the σ-weakly closed $*$-subalgebra of $\mathcal{M}_2$*

*generated by $\mathscr{J}(\mathcal{M}_1)$. Then, there exists a projection $e \in \mathcal{B} \cap \mathcal{B}'$ such that $a \mapsto e\,\mathscr{J}(a)$ is a *-homomorphism, and $a \mapsto (\mathbb{1} - e)\,\mathscr{J}(a)$ a *-antihomomorphism.*

Notation 0.91 The following notation is 'inherited' from $B(H)$:

$$\mathcal{M}_h = \mathcal{M} \cap B(H)_h, \qquad\qquad \mathcal{M}_+ = \mathcal{M} \cap B(H)_+,$$
$$\mathbb{P}(\mathcal{M}) = \mathcal{M} \cap \mathbb{P}(B(H)), \qquad \mathbb{U}(\mathcal{M}) = \mathcal{M} \cap \mathbb{U}(B(H)).$$

The following two theorems belong to the most basic set of principles in the whole theory of operator algebras.

Theorem 0.92 (Von Neumann double commutant theorem) *A *-subalgebra $\mathcal{M}$ of $B(H)$ containing $\mathbb{1}_H$ is a von Neumann algebra if and only if it is closed in any of the non-uniform topologies. In particular, if a unital *-subalgebra $\mathcal{A}$ of a von Neumann algebra $\mathcal{M}$ is dense in $\mathcal{M}$ in one of the non-uniform topologies, then it is dense in $\mathcal{M}$ in all non-uniform topologies.*

Theorem 0.93 (Kaplansky's density theorem) *Let $\mathcal{M}$ be a von Neumann algebra and $\mathcal{A} \subseteq \mathcal{M}$ a *-subalgebra, dense in one of the non-uniform topologies. Then, the unit ball of $\mathcal{A}$ (respectively, $\mathcal{A}_h$, $\mathcal{A}_+$) is dense in any of the non-uniform topologies in the unit ball of $\mathcal{M}$ (respectively, $\mathcal{M}_h$, $\mathcal{M}_+$).*

Below, we list facts that follow easily from von Neumann's double commutant theorem.

Proposition 0.94

(1) *For any family of projections $\{p_i\}_{i \in I}$ from $\mathcal{M}$, both $\bigvee_{i \in I} p_i$ and $\bigwedge_{i \in I} p_i$ belong to $\mathcal{M}$.*

(2) *The sum of an orthogonal family $\{p_i\}_{i \in I}$ from $\mathcal{M}$ belongs to $\mathcal{M}$.*

(3) *The initial and final projections of a partial isometry u from $\mathcal{M}$ belong to $\mathcal{M}$.*

(4) *If $(a_i)_{i \in I}$ is an increasing net of positive operators from $\mathcal{M}$ with an upper bound, then its supremum exists and belongs to $\mathcal{M}$.*

(5) *If* $\mathfrak{m}$ *is a σ-weakly closed left (respectively, right) ideal in $\mathcal{M}$, then there exists a projection $p \in \mathcal{M}$ such that $\mathfrak{m} = \mathcal{M}p$ (respectively, $\mathfrak{m} = p\mathcal{M}$). The projection p is the supremum of a right approximate (respectively, left approximate) identity in* $\mathfrak{m}$. *If* $\mathfrak{m}$ *is a two-sided ideal, then p belongs to the centre $\mathcal{Z}(\mathcal{M})$ of $\mathcal{M}$.*

(6) *If $a \in \mathcal{M}$, then $\mathfrak{n}(a), \mathfrak{s}_l(a), \mathfrak{s}_r(a) \in \mathcal{M}$. If $a \in \mathcal{M}_h$, then $\mathfrak{s}(a) \in \mathcal{M}$.*

(7) *The spectral resolution (e_λ) of an operator $a \in \mathcal{M}_h$ consists of projections from* $\mathcal{M}$.

(8) *If $a \in \mathcal{M}_h$ and $f \in \mathcal{B}_b(\mathrm{sp}(a))$, then $f(a) \in \mathcal{M}$. In particular, if $a \in \mathcal{M}_+$, then $a^{1/2} \in \mathcal{M}_+$.*

(9) *If $a \in \mathcal{M}$ has polar decomposition $a = u|a|$, then $u \in \mathcal{M}$ and $|a| \in \mathcal{M}_+$.*

We can now introduce the important concept of a polar decomposition of functionals in $\mathcal{M}_*$. We start with a characterisation of N_ω for $\omega \in \mathcal{M}_*^+$:

Lemma 0.95 *If $\omega \in \mathcal{M}_*^+$, then the left ideal N_ω is σ-weakly closed in* $\mathcal{M}$.

Proof.   Let $(a_i)_{i \in I}$ be a net in $\mathcal{N}_\omega$ that converges σ-weakly to some $a \in \mathcal{M}$. By the CBS inequality (see (0.3)), $\omega(a^* a_i) = 0$ for each $i \in I$. Since by Proposition 0.11(2)(c) $a^* a_i \to a^* a$ σ-weakly, we have, by Proposition 0.80, that $\omega(a^* a) = \lim_{i \in I} \omega(a^* a_i) = 0$ and hence that $a \in N_\omega$. $\square$

Definition 0.96 Let $\omega \in \mathcal{M}_*^+$ be given. Then, by proposition 0.94(5), there exists a projection $p \in \mathcal{M}$ such that $N_\omega = \mathcal{M}p$. We define the *support projection* of ω to be $\mathrm{supp}\,\omega = \mathbb{1} - p$.

It is easy to see that $(\mathrm{supp}\,\omega)^\perp$ is the largest projection p in $\mathcal{M}$ such that $\omega(p) = 0$.

Proposition 0.97 (Polar decomposition of normal functionals) *For any $\omega \in \mathcal{M}_*$, there exists a partial isometry $u \in \mathcal{M}$ and a functional $\rho \in \mathcal{M}_*^+$ related by the equality $\omega(a) = \rho(au)$ for all $a \in \mathcal{M}$, with in addition $\|\omega\| = \|\rho\|$ and $\mathfrak{s}(\rho) = uu^*$. Such a decomposition is* *unique and we write $|\omega|$ for ρ. For the partial isometry in the decomposition, we have $u^*u = \mathbf{s}(\omega^*)$, where the functional ω^* is defined by $\omega^*(a) = \overline{\omega(a^*)}$ for all $a \in \mathcal{M}$.*

Definition 0.98 A projection $p \in \mathcal{M}$ is called *σ-finite* if for any orthogonal family $\{p_i\}_{i \in I}$ of non-zero projections from $\mathcal{M}$ such that $p = \sum_{i \in I} p_i$ we have $\#I \leq \aleph_0$ where $\#I$ denotes the cardinality of I. A von Neumann algebra $\mathcal{M}$ is *σ-finite* or *countably decomposable* if $\mathbb{1}$ is σ-finite.

The theory of GNS representations of von Neumann algebras, and in particular of σ-finite von Neumann algebras, has its own distinctive flavour. Having introduced the notion of σ-finite algebras, we pause to analyse the structure of their GNS representations.

Proposition 0.99 *Let ω be a state on a von Neumann algebra. Then, ω is a normal state if and only if for the GNS representation $\{\pi_\omega, H_\omega\}$ the map π_ω is normal (σ-weak to σ-weak continuous) and $\pi_\omega(\mathcal{M})$ a von Neumann algebra.*

We leave the 'if' part of the above proposition as an exercise. The 'only if' part is the content of [**Bla06**, Proposition III.2.2.3] and [**Tak02**, Proposition III.3.15]. We shall refer to representations of the above type, where the associated *-homomorphism is normal, as *normal representations*. With this proposition as a starting point, we are now ready to investigate σ-finite algebras.

Theorem 0.100 *Let $\mathcal{M}$ be a von Neumann algebra. Then, the following are equivalent.*

    (1) *$\mathcal{M}$ is σ-finite.*
    (2) *$\mathcal{M}$ admits a faithful normal state.*
    (3) *There exists a faithful normal representation $\{\pi_0, H_0\}$ of $\mathcal{M}$ for which $\pi_0(\mathcal{M})$ admits a cyclic and separating vector ξ_0.*

Proof. $(1 \Leftrightarrow 2)$: First, consider the implication $(2 \Rightarrow 1)$. Given a family $\{p_i\}$ of non-zero mutually orthogonal projections and a normal state ω, the normality of the state ensures that $\omega(\sum_i p_i) = \sum_i \omega(p_i)$. If the state is additionally faithful, then each $\omega(p_i)$ is non-zero. The obvious finiteness of the sum $\sum_i \omega(p_i)$ then ensures that the sum must be made up of countably many terms and hence that the family $\{p_i\}$ is countable. It remains to prove the converse. For that, we start by using Zorn's lemma to select a maximal family of normal states with mutually orthogonal supports $\{p_i : i \in I\}$. We claim that $\sum_i p_i = \mathbb{1}$. If not, then $(\mathbb{1} - \sum_i p_i)\mathcal{M}(\mathbb{1} - \sum_i p_i)$ is a non-trivial von Neumann algebra, which must admit a normal state ν. The support q of this state is a subprojection of $(\mathbb{1} - \sum_i p_i)$. We can extend this state to all of $\mathcal{M}$ in the obvious way by simply sending each $a \in \mathcal{M}$ to $\nu((\mathbb{1} - \sum_i p_i)a(\mathbb{1} - \sum_i p_i))$. It is an exercise to see that this extension is still a normal state with the same support projection. But this would contradict the maximality of the family $\{\psi_i\}$. Thus, we must have $\sum_i p_i = \mathbb{1}$ as claimed. Since $\mathcal{M}$ is σ-finite, the collection $\{p_i\}$ must be countable. We may then of course assume that $I = \mathbb{N}$. The state we need is defined by setting $\omega = \sum_n \frac{1}{2^n}\psi_n$.

Now, suppose that $\{a_\gamma\} \in \mathcal{M}_+$ increases to $a \in \mathcal{M}_+$. The fact that all terms in the sum $\omega(a_\gamma) = \sum_n \frac{1}{2^n}\psi_n(a_\gamma)$ are positive means that we may interchange summation and suprema. The normality of the ψ_ns, therefore, ensure that

$$\sup_\gamma \omega(a_\gamma) = \sup_\gamma \sum_i \frac{1}{2^n}\psi_n(a_\gamma) = \sum_n \frac{1}{2^n}\psi_n(a) = \omega(a);$$

that is, that ω is normal. All we still need to do is prove that ω is faithful. So, let $a \in \mathcal{M}_+$ be given with $a \neq 0$. Then, $a^{1/2} \neq 0$. We leave it as an exercise to verify that $a^{1/2} = \sum_n a^{1/2}p_n$. (It follows from the fact that the partial sums converge σ-strongly to both left- and right-hand sides of the equality.) This then ensures that there must be some n_0 such that $a^{1/2}p_{n_0} \neq 0$, or equivalently $p_{n_0}ap_{n_0} \neq 0$. Recalling that ψ_{n_0} is faithful on $p_{n_0}\mathcal{M}p_{n_0}$, we therefore have $0 < \psi_{n_0}(p_{n_0}ap_{n_0}) = \psi_{n_0}(a) \leq \omega(a)$, as was required.

$(2 \Rightarrow 3)$: Given a faithful normal state ω, the normality of $\{\pi_\omega, H_\omega\}$ is ensured by the preceding proposition, with the fact that Ω_ω is separating following from the equality $\|\pi_\omega(a)\Omega_\omega\|^2 = \omega(a^*a)$ combined with the faithfulness of ω.

$(3 \Rightarrow 2)$: Assuming (3) to hold, we may clearly assume ξ_0 to be a unit vector. The implication then follows by first using the normality of π_0 to see that $\omega(\cdot) = \langle \pi_0(\cdot)\xi_0, \xi_0 \rangle$ defines a normal state. Since for any $a \in \mathcal{M}$ we have that $\|\pi_0(a)\xi_0\|^2 = \omega(a^*a)$, the fact that ξ_0 separating clearly ensures that ω is faithful. $\qquad\square$

For general von Neumann algebras, we have the following interesting connection between cyclic and separating vectors.

Proposition 0.101   *Let M be a von Neumann algebra acting on the Hilbert space H. For any vector $\xi \in H$, we have that ξ is cyclic with respect to the action of M if and only if it is separating for M'*

Proof.   First, suppose ξ to be cyclic with respect to the action of M, and let $a' \in M'$ be given with $a'\xi = 0$. We will then for any $a \in M$ have that $a'(a\xi) = a(a'\xi) = 0$. By the cyclicity of ξ, the map a' is therefore 0-valued on a dense subspace of H, whence $a' = 0$. Thus, ξ must be separating for M'.

Now, suppose that ξ is separating for M' and let p' be the orthogonal projection onto the closure $[M\xi]$ of $M\xi$. Since for any $a \in M$ we will by continuity have that $a[M\xi] \subseteq [M\xi]$, it is then clear that $p'a = ap'$ and hence that $p' \in M'$. In view of the fact that $\xi \in [M\xi]$, we have $(\mathbb{1} - p')\xi = O$. Since ξ is separating, this of course ensures $\mathbb{1} - p' = 0$ or equivalently that $[M\xi] = H$. $\qquad\square$

Definition 0.102 For each $a \in M$, there exists a smallest central projection z such that $a = az = za$. It is called the *central support* or *central cover* of a, and it is denoted by $\mathbf{z}(a)$.

Two projections $p, q \in M$ are said to be *centrally orthogonal* if $\mathbf{z}(p)\mathbf{z}(q) = 0$.

The following criterion for central orthogonality is well known.

Proposition 0.103   *For any two projections p and q in a von Neumann algebra M, the following are equivalent.*

(1) *p and q are centrally orthogonal.*
(2) *$pMq \neq \{0\}$.*
(3) *There is no non-zero partial isometry v such that $v^*v \leq p$ and $vv^* \leq q$.*

Let K be a closed linear subspace of H invariant under M, i.e., such that $aK \subseteq K$ for all $a \in M$. We denote by a_K the operator on K obtained by restricting a to K. Note that if $e' \in \mathbb{P}(M')$, then $e'H$ is invariant under M.

Definition 0.104 Let $e \in \mathbb{P}(\mathcal{M})$ and $e' \in \mathbb{P}(\mathcal{M}')$. We put $\mathcal{M}_e :=$ $\{a_{eH} : a \in e\mathcal{M}e\}$ (respectively, $\mathcal{M}_{e'} := \{a_{e'H} : a \in \mathcal{M}\}$). Then, $\mathcal{M}_e$ (respectively, $\mathcal{M}_{e'}$) is a von Neumann algebra called the *reduced von Neumann algebra of* $\mathcal{M}$ *on* $K := eH$ (respectively, the *induced von Neumann algebra of* $\mathcal{M}$ *on* $K := e'H$). In particular, for any $z \in \mathbb{P}(\mathcal{Z}(\mathcal{M}))$, we can form the algebra $\mathcal{M}_z$, often written simply as $\mathcal{M}z$.

Proposition 0.105 *If $e \in \mathcal{M}$, then $(\mathcal{M}_e)' = (\mathcal{M}')_e$.*

Definition 0.106 Let $\{\mathcal{M}_i\}_{i\in I}$ be a family of von Neumann algebras acting on Hilbert spaces H_i. Let $H := \sum_{i\in I}^{\oplus} H_i = \{\{\xi_i\}_{i\in I} : \xi_i \in H_i, \sum_{i\in I} \|\xi_i\|^2 < \infty\}$. For each bounded family $\{a_i\}_{i\in I}$ in $\prod_{i\in I} \mathcal{M}_i$, we define a bounded operator a on H by $a(\{\xi_i\}_{i\in I}) = \{a_i\xi_i\}_{i\in I}$ and denote it by $\sum_{i\in I}^{\oplus} a_i$. The set of such operators is denoted by $\sum_{i\in I}^{\oplus} \mathcal{M}_i$ and called the *direct sum* of $\{\mathcal{M}_i\}_{i\in I}$. If the index set is finite, we use $\oplus$ instead of $\sum^{\oplus}$.

Proposition 0.107 *If $\{z_i\}_{i\in I}$ is any orthogonal family of central projections in $\mathcal{M}$ such that $\sum_{i\in I} z_i = \mathbb{1}$, then $\mathcal{M} \cong \sum_{i\in I}^{\oplus} \mathcal{M}_{z_i}$; if, on the other hand, $\mathcal{M} = \sum_{i\in I}^{\oplus} \mathcal{M}_i$ for a family of von Neumann algebras $\{\mathcal{M}_i\}_{i\in I}$, then there is an orthogonal family $\{z_i\}$ of central projections in $\mathcal{M}$ such that $\sum_{i\in I} z_i = \mathbb{1}$ and $\mathcal{M}_i \cong \mathcal{M}_{z_i}$.*

Definition 0.108 Let $\mathcal{M}_1$ and $\mathcal{M}_2$ be von Neumann algebras acting on H_1 and H_2, respectively. Let $H := H_1 \otimes H_2$. For $a_1 \in \mathcal{M}_1$ and $a_2 \in \mathcal{M}_2$, there is a unique bounded operator a acting on H such that $a(\xi_1 \otimes \xi_2) = a_1(\xi_1) \otimes a_2(\xi_2)$. We denote this operator by $a_1 \otimes a_2$ and call it a *simple tensor*. The von Neumann subalgebra of $B(H)$ generated by simple tensors, i.e., the closure of the *-algebra of finite linear combinations of simple tensors in one of the non-uniform topologies, is called the *tensor product* of $\mathcal{M}_1$ and $\mathcal{M}_2$ and is denoted by $\mathcal{M}_1\overline{\otimes}\mathcal{M}_2$.

Theorem 0.109 *One has $(\mathcal{M}_1\overline{\otimes}\mathcal{M}_2)' = \mathcal{M}_1'\overline{\otimes}\mathcal{M}_2'$. Hence*

$$\mathcal{Z}(\mathcal{M}_1\overline{\otimes}\mathcal{M}_2) = \mathcal{Z}(\mathcal{M}_1)\overline{\otimes}\mathcal{Z}(\mathcal{M}_2).$$

Definition 0.110 We say that projections p and q from $\mathcal{M}$ are *equivalent* and write $p \sim q$, if they are initial and final projections of a partial isometry $u \in \mathcal{M}$, i.e. $p = u^*u$ and $q = uu^*$. If $p \sim r \leq q$ for some $p, q, r \in \mathbb{P}(\mathcal{M})$, then we write $p \precsim q$ and say that p is *dominated* by q. If $p \nsim q$ and $p \precsim q$, we write $p \prec q$ and say that q *strictly dominates* p.

Proposition 0.111 $\sim$ *is an equivalence relation on* $\mathbb{P}(\mathcal{M})$, *and* $\precsim$ *is a partial order on* $\mathbb{P}(\mathcal{M})$. *In particular,* $p \precsim q$ *and* $q \precsim p$ *implies* $p \sim q$ *for* $p, q \in \mathbb{P}(\mathcal{M})$.

Proposition 0.112 *If* $\{p_i\}_{i \in I}$ *and* $\{q_i\}_{i \in I}$ *are two orthogonal families of projections from* $\mathcal{M}$ *such that* $p_i \sim q_i$ *for each* $i \in I$, *then* $\sum_{i \in I} p_i \sim \sum_{i \in I} q_i$. *If, on the other hand,* $p_i \precsim q_i$ *for each* $i \in I$, *then* $p \precsim q$.

Theorem 0.113 (Comparability theorem) *For any* $p, q \in \mathbb{P}(\mathcal{M})$, *there exists a central projection* $z \in \mathcal{M}$ *such that* $pz \precsim qz$ *and* $pz^\perp \succsim qz^\perp$.

Corollary 0.114 *If* $\mathcal{M}$ *is a factor, then for any* $p, q \in \mathbb{P}(\mathcal{M})$, *either* $p \precsim q$ *or* $q \precsim p$. *In other words,* $\precsim$ *determines a total order on* $\mathbb{P}(\mathcal{M})$.

We know from the polar decomposition of $a \in \mathcal{M}$ that $\mathbf{s}_l(a) \sim \mathbf{s}_r(a)$. This leads to the highly useful result of Kaplansky [**Kap68**, p. 81].

Proposition 0.115 (Kaplansky's parallelogram law) *For any* $p, q \in \mathbb{P}(\mathcal{M})$,

$$p \vee q - q \sim p - p \wedge q.$$

Proof.   Note that $\mathfrak{n}(q^\perp p) = p^\perp + p \wedge q$ and $\mathfrak{n}(pq^\perp) = q + p^\perp \wedge q^\perp$. Hence, $\mathbf{s}_r(q^\perp p) = \mathfrak{n}(q^\perp p)^\perp = p - p \wedge q$ and $\mathbf{s}_l(q^\perp p) = \mathbf{s}_r(pq^\perp) = \mathfrak{n}(pq^\perp)^\perp = p \vee q - q$. Thus, $\mathbf{s}_l(q^\perp p) \sim \mathbf{s}_r(q^\perp p)$ yields the result. $\square$

Definition 0.116 A projection $p \in \mathcal{M}$ is called:

(1) *minimal,* if $p \neq 0$ and $0 \neq q \leq p$ implies $q = p$;
(2) *abelian,* if $p \neq 0$ and $\mathcal{M}_p$ is abelian;
(3) *finite,* if $p \sim q \leq p$ implies $q = p$;
(4) *infinite,* if it is not finite;

(5) *properly infinite*, if $p \neq 0$ and pz is infinite for any central projection z such that $pz \neq 0$.

(6) *purely infinite*, if $p \neq 0$ and there is no non-zero finite projection $q \in \mathcal{M}$ with $q \leq p$;

Proposition 0.117

(1) *A non-zero subprojection of a minimal projection is equal to the projection.*

(2) *A subprojection of an abelian projection is an abelian projection.*

(3) *A subprojection of a finite projection is finite.*

(4) *A non-zero subprojection of a purely infinite projection is purely infinite.*

We pause to, respectively, further investigate abelian and finite projections.

Proposition 0.118 *If $z \in \mathcal{Z}(\mathcal{M})$ and $\{e_i\}_{i \in I}$, $\{f_j\}_{j \in J}$ are two orthogonal families of abelian projections from $\mathcal{M}$ such that $\mathbf{z}(e_i) = \mathbf{z}(f_j) = z$ for all $i \in I$, $j \in J$, and $\sum_{i \in I} e_i = \sum_{j \in J} e_i = z$, then the cardinal numbers of I and J are equal. Similarly, if $z \in \mathcal{Z}(\mathcal{M})$ and $\{e_i\}_{i \in I}$, $\{f_j\}_{j \in J}$ are two orthogonal families of equivalent finite projections from $\mathcal{M}$ such that $\mathbf{z}(e_i) = \mathbf{z}(f_j) = z$ for all $i \in I$, $j \in J$, and $\sum_{i \in I} e_i = \sum_{j \in J} e_i = z$, then the cardinal numbers of I and J are equal.*

Proposition 0.119 *Abelian projections are 'minimal' in the following sense: if $p, q \in \mathbb{P}(\mathcal{M})$ with p abelian and $p \leq \mathbf{z}(q)$, then $p \precsim q$. An abelian projection in a factor is a minimal projection.*

Passing to finite projections, we first note the following.

Theorem 0.120 *The set of finite projections in a von Neumann algebra is a modular lattice (cf. Definition 1.3).*

Proposition 0.121 *Let $\{p_i\}$ be a family of centrally orthogonal finite projections in a von Neumann algebra M. Then, $p = \sum_i p_i$ is also finite. In addition, the supremum of any family of finite central projections is again finite.*

Proof. Let each p_i be finite, and let $u \in \mathcal{M}$ be a partial isometry such that $uu^* \leq u^*u = p$. For each i, we then have $(\mathbf{z}(p_i)u)(\mathbf{z}(p_i)u)^* = \mathbf{z}(p_i)uu^* \leq \mathbf{z}(p_i)u^*u = (\mathbf{z}(p_i)u)^*(\mathbf{z}(p_i)u) = \mathbf{z}(p_i)p = p_i$. The finiteness of the p_is then ensures that $\mathbf{z}(p_i)uu^* = p_i$ for each i. It is an exercise to see that $\mathbf{z}(p) = \sum_i \mathbf{z}(p_i)$ from which it then follows that $uu^* = u[\sum_i \mathbf{z}(p_i)]u^* = \sum_i \mathbf{z}(p_i)(uu^*) = \sum_i p_i = p$.

Let $\{z_i\}$ be a family of finite central projections and $\{w_j\}$ a maximal family of mutually orthogonal finite central subprojections of $z = \vee_i z_i$. Clearly, $\sum_j w_j \leq z$. If indeed $z - \sum_j w_j = z(\mathbb{1} - \sum_j w_j) \neq 0$, there has to be at least one z_{i_0} for which $z_{i_0}(\mathbb{1} - \sum_j w_j) \neq 0$. Since $z_{i_0}(\mathbb{1} - \sum_j w_j)$ is then a finite central projection orthogonal to all w_js, this would contradict the maximality of $\{w_j\}$. So, we must have $z = \sum_j w_j$ with the claim then following from the first part of the lemma. $\square$

Proposition 0.122 *Let q and r be subprojections of a finite projection p. If $q \sim r$, then $(p - q) \sim (p - r)$.*

Proof. If $(p-q) \not\sim (p-r)$, then by the comparability theorem there is a nonzero central projection z such that either $z(p-q)$ is equivalent to a proper subprojection s of $z(p-r)$ or vice versa. Suppose that we in fact have $z(p-q)$ equivalent to a proper subprojection s of $z(p-r)$. Then, $p = z(p-q) + zq + (1-z)p$ is equivalent to the proper subprojection $s + zr + (1-z)p$. Thus, p cannot be finite. $\square$

As a last word on finite projections, we note the following very useful decomposition result.

Proposition 0.123 *For any projection $p \in \mathcal{M}$ in a von Neumann algebra, there exists a central projection z such that pz is finite and $p(\mathbb{1} - z)$ properly infinite.*

Proof. It follows from Proposition 0.105 that any central projection in $p\mathcal{M}p$ is of the form pz_0 for some central projection of $\mathcal{M}$. So, in proving this lemma, we may pass to $p\mathcal{M}p$ and assume that $p = \mathbb{1}$. Now, let z be the supremum of all the finite central projections. By Proposition 0.121, z, and therefore also zp, is finite. Finally note that by construction $\mathbb{1} - z$ cannot have any finite central subprojections, and hence, it must be properly infinite. $\square$

Lemma 0.124 (Halving lemma) *Let $p \in \mathbb{P}(\mathcal{M})$.*

(1) *If p is properly infinite, then there is a $q \in \mathbb{P}(\mathcal{M})$ with $q \leq p$ such that $q \sim p - q \sim p$.*

(2) *If p does not have any abelian subprojection, then there is a $q \in \mathbb{P}(\mathcal{M})$ such that $q \sim p - q$.*

Definition 0.125 A von Neumann algebra $\mathcal{M}$ is said to be:

(1) *discrete*, if every non-zero central projection majorises a non-zero abelian projection;

(2) *continuous*, if there are no abelian projections in it;

(3) *finite*, if $\mathbb{1}$ is finite;

(4) *infinite*, if $\mathbb{1}$ is infinite;

(5) *purely infinite*, if $\mathbb{1}$ is purely infinite;

(6) *properly infinite*, if $\mathbb{1}$ is properly infinite;

(7) *semifinite*, if there are no purely infinite projections in its centre.

Definition 0.126 A von Neumann algebra $\mathcal{M}$ is said to be:

(1) *of type I*, if it is discrete;

(2) *of type II*, if there are no non-zero abelian projections and no purely infinite projections in $\mathcal{M}$;

(3) *of type III*, if it is purely infinite;

(4) *of type I_α* with α a cardinal number $\leq \#\mathcal{M}$, if $\mathbb{1}$ is the sum of α abelian projections with central support $\mathbb{1}$;

(5) *of type I_∞*, if it is properly infinite and of type I;

(6) *of type II_1*, if it is finite and of type II;

(7) *of type II_α* with α a cardinal number such that $\aleph_0 \leq \alpha \leq \#\mathcal{M}$, if $\mathbb{1}$ is the sum of α equivalent finite projections with central support $\mathbb{1}$;

(8) *of type II_∞*, if it is properly infinite and of type II.

Theorem 0.127 (Structure of von Neumann algebras)

(1) *Every von Neumann algebra is (isomorphic to) a direct sum of algebras of type I, II_1, II_∞, III (with some summands possibly missing). Each factor von Neumann algebra is of one of the types I, II_1, II_∞, III.*

(2) *Every von Neumann algebra of type I can be uniquely represented as a direct sum of algebras of type I_α, $\alpha \leq \#\mathcal{M}$. If $\mathcal{M}$ is finite, then all the α s in the direct sum are finite. An algebra of type I_α is isomorphic to $\mathcal{K}\,\overline{\otimes}B(H)$, where $\mathcal{K}$ is abelian and H is α-dimensional. A factor of type I is of type I_α for some cardinal α, that is isomorphic to $B(H)$ with H α-dimensional.*

(3) *Every von Neumann algebra $\mathcal{M}$ of type II_∞ can be represented in a unique way as a direct sum of algebras of type II_α, $\aleph_0 \leq \alpha \leq \#\mathcal{M}$. An algebra of type II_α is isomorphic to $\mathcal{N}\,\overline{\otimes}B(H)$, where $\mathcal{N}$ is of type II_1 and H is α-dimensional. If $\mathcal{M}$ is a factor, then $\mathcal{N}$ is also a factor.*

(4) *Every finite-dimensional von Neumann algebra is a direct sum of a finite number of type I_n factors, with $n \in \mathbb{N}$. A finite-dimensional factor is isomorphic, for some $n \in \mathbb{N}$, to the algebra of complex $n \times n$-matrices.*

(5) *Every finite von Neumann algebra is a direct sum of σ-finite ones.*

(6) *Every commutative von Neumann algebra is of type I_1.*

We will make the above structure theorem for von Neumann algebras clearer by explaining points (2) and (3) of the theorem in more detail. To this aim, we introduce a useful notion of a matrix unit (we use Definition IV.1.7 of Takesaki [**Tak02**]).

Definition 0.128 A family $\{u_{i,j}\}_{i,j\in I}$ of elements of a von Neumann algebra $\mathcal{M}$ is called a *matrix unit* in $\mathcal{M}$ if

(1) $u_{i,j}^* = u_{j,i}$;

(2) $u_{i,j}u_{k,\ell} = \delta_{j,k}u_{i,\ell}$;

(3) $\sum_{i\in I} u_{i,i} = \mathbb{1}$,

with summation in the strong topology (observe that by (1) and (2) $u_{i,i} \in \mathbb{P}(\mathcal{M})$).

The simplest example of a matrix unit can be found in the full von Neumann algebra $B(H)$.

Proposition 0.129 *Let $\{e_i\}_{i\in I}$ be a maximal family of minimal projections in $B(H)$. By Structure Theorem 0.127 and Proposition*

0.119, $\sum_{i \in I} e_i = \mathbb{1}$ and all the projections are mutually equivalent. Choose one minimal projection e_{i_0} from the family. Let $\{u_i\}_{i \in I}$ be partial isometries such that $e_{i_0} = u_i^* u_i$ and $e_i = u_i u_i^*$ for all $i \in I$. Put $u_{i,j} := u_i u_j^*$ for $i, i \in I$. Then, $\{u_{i,j}\}_{i,j \in I}$ is a matrix unit. Let $\mathcal{M} = \mathcal{N} \overline{\otimes} B(H)$. Each element $a \in \mathcal{M}$ can be written in the form $a = \sum_{i,j \in I} a_{i,j} \otimes u_{i,j}$, with all $a_{i,j}$ in $\mathcal{N}$, where the sum converges strongly.

Now, we can write every element of $\mathcal{M} = \mathcal{N} \overline{\otimes} B(H)$ as an infinite matrix with entries in $\mathcal{N}$.

Notation 0.130 Let $a \in \mathcal{M} = \mathcal{N} \overline{\otimes} B(H)$. If $a_{i,j}$s are as in a previous proposition, we write $a = (a_{i,j})_{i,j \in I}$.

Proposition 0.131 *Let $a, b \in \mathcal{M} = \mathcal{N} \overline{\otimes} B(H)$. Then, with the above notation, $(ab)_{i,j} = \sum_{k \in I} a_{i,k} b_{k,j}$, with the sum σ-strong*-convergent.*

0.5 Affiliated operators

Operators affiliated with a von Neumann algebra are those unbounded operators whose spectral projections belong to the algebra. They are home to all the classes of operators important for the noncommutative theory.

Lemma 0.132 *Let $\mathcal{M}$ be a von Neumann algebra. For an unbounded operator x on H, the following conditions are equivalent:*

    (1) $u'x = xu'$ *(or, equivalently, $u'^* x u' = x$) for any $u' \in \mathbb{U}(\mathcal{M}')$;*
    (2) $u'x \subseteq xu'$ *for any $u' \in \mathbb{U}(\mathcal{M}')$;*
    (3) $a'x \subseteq xa'$ *for any $a' \in \mathcal{M}'$.*

Proof. (1)$\Rightarrow$(2): is obvious.

(2)$\Rightarrow$(3) We start with representing a' as a linear combination of unitaries: $a' = \sum_{n=1}^{4} \lambda_n u_n'$. Then

$$\mathrm{dom}(xa') = \{\xi \in H : a'\xi \in \mathrm{dom}(x)\}$$

$$= \{\xi \in H : \sum_{n=1}^{4} \lambda_n u_n' \xi \in \mathrm{dom}(x)\}$$

$$\supseteq \bigcap_{n=1}^{4} \{\xi \in H : u_n'\xi \in \mathrm{dom}(x)\} = \bigcap_{n=1}^{4} \mathrm{dom}(xu_n')$$

$$\supseteq \bigcap_{n=1}^{4} \mathrm{dom}(u_n'x) = \mathrm{dom}(x) = \mathrm{dom}(a'x).$$

For $\xi \in \mathrm{dom}(a'x)$, we have

$$a'x\xi = \sum_{n=1}^{4} \lambda_n u_n' x\xi = \sum_{n=1}^{4} \lambda_n x u_n'\xi = x\Big(\sum_{n=1}^{4} \lambda_n u_n'\xi\Big) = xa'\xi,$$

which ends the proof.

(3)$\Rightarrow$(1): Take $u' \in \mathrm{U}(\mathcal{M}')$. By (3), $u'x \subseteq xu'$, so that $\mathrm{dom}(x) = \mathrm{dom}(u'x) \subseteq \mathrm{dom}(xu') = u'^*(\mathrm{dom}(x))$, which yields $u'(\mathrm{dom}(x)) \subseteq \mathrm{dom}(x)$. Using u'^* instead of u' gives $u'^*(\mathrm{dom}(x)) \subseteq \mathrm{dom}(x)$. Hence, $u'(\mathrm{dom}(x)) = \mathrm{dom}(x)$, which yields $\mathrm{dom}(u'x) = \mathrm{dom}(xu')$, so that $u'x = xu'$. $\qquad\square$

Definition 0.133 A (not necessarily bounded, not necessarily densely defined) operator x on H is *affiliated* to the von Neumann algebra $\mathcal{M}$ if it satisfies one of the equivalent conditions of the previous lemma. The set of operators affiliated to $\mathcal{M}$ is denoted by $^{\eta}\mathcal{M}$. The set of closed densely defined operators affiliated with $\mathcal{M}$ is denoted by $\overline{\mathcal{M}}$, the subset of self-adjoint operators by $\overline{\mathcal{M}}_h$, and the subset of $\overline{\mathcal{M}}_h$ consisting of positive operators by $\overline{\mathcal{M}}_+$.

Lemma 0.134

(1) *If $x \in {}^{\eta}\mathcal{M}$ and $x \in B(H)$, then $x \in \mathcal{M}$.*

(2) *If $x, y \in {}^{\eta}\mathcal{M}$ and $\lambda \in \mathbb{C}$, then $x + y \in {}^{\eta}\mathcal{M}$, $xy \in {}^{\eta}\mathcal{M}$ and $\lambda x \in {}^{\eta}\mathcal{M}$.*

(3) *If $x \in {}^{\eta}\mathcal{M}$ is densely defined, then $x^* \in {}^{\eta}\mathcal{M}$.*

(4) *If $x \in {}^{\eta}\mathcal{M}$ is preclosed (closable), then $[x] \in {}^{\eta}\mathcal{M}$.*

(5) *If $x \in {}^{\eta}\mathcal{M}$ is self-adjoint and (e_λ) is the spectral resolution of x, then $e_\lambda \in \mathcal{M}$ for all $\lambda \in \mathbb{R}$. Moreover, $f(x) \in {}^{\eta}\mathcal{M}$ for each $f \in \mathcal{B}([0, \infty))$, and one can choose a core D for all $f(x)$ with $f \in \mathcal{B}([0, \infty))$ (see Theorem 0.34) to be affiliated with $\mathcal{M}$.*

(6) *If x is a closed, densely defined operator on H with polar decomposition $x = u|x|$, then $x \in \overline{\mathcal{M}}$ if and only if $u \in \mathcal{M}$ and $|x| \in \overline{\mathcal{M}}_+$.*

(7) *If $\{p_n\}_{n\in\mathbb{N}}$ is an orthogonal family of projections from $\mathcal{M}$ and $\{\lambda_n\}$ is a family of positive numbers, then $\sum_{n\in\mathbb{N}}\lambda_n p_n \in {}^\eta\mathcal{M}$.*

Proof.   Let $u' \in \mathbf{U}(\mathcal{M}')$.

(1) By definition and von Neumann's double commutant theorem.

(2) Suppose $x, y \in {}^\eta\mathcal{M}$. Since $\operatorname{dom}(u'(x+y)) = \operatorname{dom}(x+y)$ and $\operatorname{dom}(u'x + u'y) = \operatorname{dom}(u'x) \cap \operatorname{dom}(u'y) = \operatorname{dom}(x) \cap \operatorname{dom}(y) = \operatorname{dom}(x+y)$, by Proposition 0.24, we have $u'(x+y) = u'x+u'y = xu'+yu' = (x+y)u'$, so that $x+y \in {}^\eta\mathcal{M}$. Also, $u'xy = xu'y = xyu'$, so that $xy \in {}^\eta\mathcal{M}$. It is obvious that $\lambda x \in {}^\eta\mathcal{M}$ for any $\lambda \in \mathbb{C}$.

(3) Suppose $x \in {}^\eta\mathcal{M}$ be densely defined. Then, by Proposition 0.31(4) and (7),

$$u'x^* \subseteq (xu'^*)^* = (u'^*x)^* = x^*u'$$

which shows that $x^* \in {}^\eta\mathcal{M}$.

(4) Assume that $x \in {}^\eta\mathcal{M}$ is preclosed. Then, u'^*xu' is preclosed and $[x] = [u'^*xu'] = u'^*[x]u'$, so that $[x] \in {}^\eta\mathcal{M}$.

(5) If $x \in {}^\eta\mathcal{M}$ and $u' \in \mathbf{U}(\mathcal{M}')$, then

$$x = u'xu'^* = u'\int_{-\infty}^{\infty}\lambda de_\lambda u'^* = \int_{-\infty}^{\infty}\lambda d(u'e_\lambda u'^*).$$

By the uniqueness of the spectral decomposition, $e_\lambda = u'e_\lambda u'^*$ for all $\lambda \in \mathbb{R}$, so that $e_\lambda \in {}^\eta\mathcal{M}$ and by (1), $e_\lambda \in \mathcal{M}$. This implies $f(x) \in {}^\eta\mathcal{M}$ for each $f \in \mathcal{B}([0,\infty))$. It is clear that a countable union of ranges of spectral projections of x (see Theorem 0.34) is affiliated with $\mathcal{M}$.

(6) Suppose $x \in \overline{\mathcal{M}}$, with polar decomposition $x = u|x|$. We have, for any $u' \in \mathbf{U}(\mathcal{M}')$,

$$x = u'^*xu' = u'^*u|x|u' = u'^*uu'u'^*|x|u'$$

whence $u'^*uu' = u$ and $u'^*|x|u' = |x|$ by the uniqueness of the polar decomposition. Hence, $u \in \mathcal{M}$ and $|x| \in {}^\eta\mathcal{M}$. The other direction is obvious.

(7) It is obvious (cf. Notation 0.40). $\square$

Definition 0.135 A closed subspace K of H is called *affiliated* to $\mathcal{M}$ if the orthogonal projection onto the subspace belongs to $\mathcal{M}$.

We finish the section by introducing order in the set $\overline{\mathcal{M}}_+$. The definition is chosen in such a way that all our definitions of the order relation agree on common domains.

Definition 0.136 Let $x, y \in \overline{\mathcal{M}}_+$. We say that x is *less than or equal to* y and write $x \leq y$ if $\mathrm{dom}(y^{1/2}) \subseteq \mathrm{dom}(x^{1/2})$ and $\|x^{1/2}\xi\| \leq \|y^{1/2}\xi\|$ for each $\xi \in \mathrm{dom}(y^{1/2})$.

Note that for all $x \in \overline{\mathcal{M}}_+$ we have $0 \leq x$, so there is no conflict of notation.

0.6 Generalised positive operators

The presentation of generalised positive operators in this section is based on Haagerup's seminal paper on operator-valued weights [**Haa79b**].

Definition 0.137 A *generalised positive operator affiliated to* $\mathcal{M}$ is a map $m : \mathcal{M}_*^+ \to [0, \infty]$ satisfying:

(1) $m(\omega + \omega') = m(\omega) + m(\omega')$ for all $\omega, \omega' \in \mathcal{M}_*^+$;
(2) $m(\lambda\omega) = \lambda m(\omega)$ for $\lambda \geq 0$, $\omega \in \mathcal{M}_*^+$;
(3) m is lower semicontinuous (in the norm topology on $\mathcal{M}_*^+$).

The collection of these is called the *extended positive part* of $\mathcal{M}$ and is denoted by $\widehat{\mathcal{M}}_+$.

For $m, n \in \widehat{\mathcal{M}}_+, a \in \mathcal{M}$ and $\lambda \geq 0$, we define $m + n, \lambda m$ and $a^* m a$ in $\widehat{\mathcal{M}}_+$ by

$$(m + n)(\omega) := m(\omega) + n(\omega) \text{ for } \omega \in \mathcal{M}_*^+;$$

$$(\lambda m)(\omega) := \lambda m(\omega) \text{ for } \omega \in \mathcal{M}_*^+;$$

$$(a^* m a)(\omega) := m(a\omega a^*) \text{ for } \omega \in \mathcal{M}_*^+,$$

where $a\omega a^* \in \mathcal{M}_*^+$ is defined by $(a\omega a^*)(b) := \omega(a^* b a)$ for each $b \in \mathcal{M}$. If $z \in \mathcal{Z}(\mathcal{M})_+$, we write mz or zm instead of $z^{1/2} m z^{1/2}$.

For $m_1, m_2 \in \widehat{\mathcal{M}}_+$, we write $m_1 \leq m_2$ if $m_1(\omega) \leq m_2(\omega)$ for all $\omega \in \mathcal{M}_*^+$. If $(m_i)_{i \in I}$ is an increasing net in $\widehat{\mathcal{M}}_+$, then we define $m = \sup m_i \in \widehat{\mathcal{M}}_+$ by $m(\omega) := \sup_{i \in I} m_i(\omega)$ for $\omega \in \mathcal{M}_*^+$ and write $m_i \nearrow m$. Similarly, for an arbitrary family $\{m_i\}_{i \in I}$ from $\widehat{\mathcal{M}}_+$, an element $m = \sum_{i \in I} m_i$ is defined by $m(\omega) = \sum_{i \in I} m_i(\omega)$ for $\omega \in \mathcal{M}_*^+$.

Remark 0.138 It is clear that $m+n$, λm and $a^* m a$ are well defined. This is also true of $\sup m_i$ and $\sum m_i$: the additivity of $\sup m_i$ is easy to show and the supremum of lower semicontinuous functions is lower semicontinuous.

Example 0.139 For $a \in \mathcal{M}_+$, define $m_a(\omega) = \omega(a)$, $\omega \in \mathcal{M}_*^+$. It is clear that the map $a \mapsto m_a$ preserves order. Thus, we can view $\mathcal{M}_+$ as a subset of $\widehat{\mathcal{M}}_+$.

Notation 0.140 For $x \in \overline{\mathcal{M}}_+$ with spectral representation $x = \int_0^\infty \lambda de_\lambda$, put $m_x(\omega) = \int_0^\infty \lambda d\omega(e_\lambda)$ for $\omega \in \mathcal{M}_*^+$.

Proposition 0.141 *If $x \in \overline{\mathcal{M}}_+$, then $m_x \in \widehat{\mathcal{M}}_+$. The map $x \mapsto m_x$ is injective.*

Proof. In fact, if $a_n = \int_0^n \lambda de_\lambda$, then $m_x(\omega) = \sup_n \omega(a_n)$; hence, m_x is lower semicontinuous. For $\xi \in H$

$$m_x(\omega_\xi) = \int_0^\infty \lambda d\langle e_\lambda \xi, \xi\rangle = \begin{cases} \|x^{1/2}\xi\|^2 & \xi \in \mathrm{dom}(x^{1/2}) \\ \infty & \xi \notin \mathrm{dom}(x^{1/2}). \end{cases}$$

If $m_x = m_y$, then $\mathrm{dom}(x^{1/2}) = \mathrm{dom}(y^{1/2})$ and $\|x^{1/2}\xi\| = \|y^{1/2}\xi\|$ for $\xi \in \mathrm{dom}(x^{1/2})$. By the uniqueness of the polar decomposition, $x^{1/2} = y^{1/2}$, so that $x = y$. $\qquad\square$

Thus, positive self-adjoint operators affiliated to $\mathcal{M}$ can be viewed as elements of $\widehat{\mathcal{M}}_+$. Note that under this identification, the positive self-adjoint operator $\sum_{n \in \mathbb{N}} \lambda_n p_n$ defined by Notation 0.40 can be regarded as a sum of a series of generalised positive operators $\lambda_n p_n$.

Definition 0.142 A *positive quadratic form on H* is a map $\mathfrak{s} : H \to [0, \infty]$ satisfying:

    (1) $\mathfrak{s}(\lambda\xi) = |\lambda|^2 \mathfrak{s}(\xi)$ for all $\xi \in H$;
    (2) $\mathfrak{s}(\xi + \eta) + \mathfrak{s}(\xi - \eta) = 2\mathfrak{s}(\xi) + 2\mathfrak{s}(\eta)$ for all $\xi, \eta \in H$;

(3) $\mathfrak{s}$ is lower semicontinuous on H.

The form $\mathfrak{s}$ is *affiliated* to $\mathcal{M}$, if

(4) $\mathfrak{s}(u'\xi) = \mathfrak{s}(\xi)$ for any unitary $u' \in \mathcal{M}'$ and any $\xi \in H$.

We define the *domain* of the positive quadratic form $\mathfrak{s}$ by

$$\mathrm{dom}(\mathfrak{s}) := \{\xi \in H : \mathfrak{s}(\xi) < \infty\}$$

and the *null space* of the form by

$$N_\mathfrak{s} := \{\xi \in H : \mathfrak{s}(\xi) = 0\}.$$

The set of positive quadratic forms on H affiliated to $\mathcal{M}$ is denoted by $\mathfrak{S}_\mathcal{M}$.

Theorem 0.143 *There is a one-to-one correspondence between:*

(a) *generalised positive operators m affiliated to $\mathcal{M}$;*

(b) *positive quadratic forms $\mathfrak{s}$ on H affiliated to $\mathcal{M}$;*

(c) *pairs (K, x), where K is a closed subspace of H affiliated to $\mathcal{M}$, and x is a positive self-adjoint operator with domain dense in K, affiliated to $\mathcal{M}$.*

Proof. We will show the existence of three maps: $\widehat{\mathcal{M}}_+ \ni m \mapsto \mathfrak{s}m \in \mathfrak{S}_\mathcal{M}$, $\mathfrak{S}_\mathcal{M} \ni \mathfrak{s} \mapsto (K_\mathfrak{s}, x_\mathfrak{s})$, with $K_\mathfrak{s}$ and $x_\mathfrak{s}$ as in (b) above, and $(K, x) \mapsto m_{(K,x)} \in \widehat{\mathcal{M}}_+$, with K and x as in (c) above. It will be clear from the constructions that each of the mappings is injective and that their composition is the identity mapping on $\widehat{\mathcal{M}}_+$.

(a) $\Rightarrow$ (b). Define $\widehat{\mathcal{M}}_+ \ni m \mapsto \mathfrak{s}m \in \mathfrak{S}$, where $\mathfrak{s}m : H \ni \xi \mapsto m(\omega_\xi)$. Let $m \in \widehat{\mathcal{M}}_+$. To show that $\mathfrak{s}m$ is a positive quadratic form on H affiliated to $\mathcal{M}$, we have to check that it satisfies conditions (1) to (4) of Definition 0.142. We have $\mathfrak{s}m(\lambda\xi) = m(\omega_{\lambda\xi}) = |\lambda|^2 m(\omega_\xi) = |\lambda|^2\mathfrak{s}m(\xi)$, hence (1). Similarly, (2) follows from the equality $\omega_{\xi+\eta} + \omega_{\xi-\eta} = 2(\omega_\xi + \omega_\eta)$. For (3), note first that for any $a \in \mathcal{M}$,

$$(\omega_\xi - \omega_\eta)(a) = \frac{1}{2}\left(\langle a(\xi+\eta), \xi - \eta\rangle + \langle a(\xi-\eta), \xi + \eta\rangle\right)$$

which implies that

$$\|\omega_\xi - \omega_\eta\| \leq \|\xi - \eta\|\|\xi + \eta\|.$$

Hence, if the sequence $(\xi_n)_{n\in\mathbb{N}}$ converges to ξ in the norm of H, then also the sequence $(\omega_{\xi_n})_{n\in\mathbb{N}}$ converges in the norm of $\mathcal{M}_*$ to ω_ξ.

Thus, $\mathfrak{s}m(\xi) = m(\omega_\xi) \leq \limsup m(\omega_{\xi_n}) = \limsup \mathfrak{s}m(\xi_n)$, which yields (3). (4) follows immediately from the equality $\omega_{u'\xi} = \omega_\xi$ for all $u' \in \mathbb{U}(\mathcal{M}')$.

(b) $\Rightarrow$ (c). Let $\mathfrak{s}$ be a positive quadratic form on H affiliated to $\mathcal{M}$. Put $K_\mathfrak{s} = \overline{\mathrm{dom}(\mathfrak{s})}$. Note that by Definition 0.142(1),(2) both $\mathrm{dom}(\mathfrak{s})$ and $N_\mathfrak{s}$ are linear spaces. Using again Definition 0.142(1),(2) and polarisation, we will easily find a sesquilinear form $\langle \cdot, \cdot \rangle_\mathfrak{s}$ such that $\mathfrak{s}(\xi) = \langle \xi, \xi \rangle_\mathfrak{s}$. Let $q_\mathfrak{s} : \mathrm{dom}(\mathfrak{s}) \to \mathrm{dom}(\mathfrak{s})/N_\mathfrak{s}$ be the quotient map, and let L be the completion of $\mathrm{dom}(\mathfrak{s})/N_\mathfrak{s}$ with respect to the inner product induced from $\langle \cdot, \cdot \rangle_\mathfrak{s}$, that is $\langle q_\mathfrak{s}\xi, q_\mathfrak{s}\eta \rangle_L = \langle \xi, \eta \rangle_\mathfrak{s}$ for any $\xi, \eta \in \mathrm{dom}(\mathfrak{s})$. If we denote by $\| \cdot \|_L$ the corresponding Hilbert space norm on L, then $\mathfrak{s}(\xi) = \|q_\mathfrak{s}\xi\|_L^2$ for $\xi \in \mathrm{dom}(q_\mathfrak{s})$ and $\mathfrak{s}(\xi) = \infty$ for $\xi \notin \mathrm{dom}(q_\mathfrak{s})$. The operator $q_\mathfrak{s}$, as an operator from K to L, is closed. To see this, let $(\xi_n)_{n \in \mathbb{N}}$ be a sequence from $\mathrm{dom}(\mathfrak{s})$ converging to some $\xi \in K$ and such that the sequence $(q_\mathfrak{s}\xi_n)_{n \in \mathbb{N}}$ from L converges to some $\eta \in L$. Then, using Definition 0.142(3), we get $\mathfrak{s}(\xi) \leq \sup_{n \in \mathbb{N}} \mathfrak{s}(\xi_n) = \sup_{n \in \mathbb{N}} \|q_\mathfrak{s}\xi_n\|_L^2 < \infty$, so that $\xi \in \mathrm{dom}(q_\mathfrak{s})$. For a fixed $\varepsilon > 0$, there is some n_0 such that, for $n, m \geq n_0$, $\|q_\mathfrak{s}\xi_n - q_\mathfrak{s}\xi_m\|_L \leq \varepsilon$, that is $\mathfrak{s}(\xi_n - \xi_m) \leq \varepsilon^2$. Hence, again by lower semicontinuity of $\mathfrak{s}$, we get $\mathfrak{s}(\xi_n - \xi) \leq \sup_m \mathfrak{s}(\xi_n - \xi_m) \leq \varepsilon^2$, so that $\|q_\mathfrak{s}\xi_n - q_\mathfrak{s}\xi\|_L \leq \varepsilon^2$, that is $q_\mathfrak{s}\xi = \eta$. By a result fully analogous to that of Proposition 0.32, but for operators between different Hilbert spaces (see [**KR83**, Theorem 2.7.8(v)]), we get that $q_\mathfrak{s}^* q_\mathfrak{s}$ is an (obviously positive) self-adjoint operator on K. Put $x_\mathfrak{s} = q_\mathfrak{s}^* q_\mathfrak{s}$. Then $x_\mathfrak{s}^{1/2} = |q_\mathfrak{s}|$ and

$$\|x_\mathfrak{s}^{1/2}\xi\| = \|q_\mathfrak{s}\xi\| \quad \text{for all } \xi \in \mathrm{dom}(x_\mathfrak{s}^{1/2}) = \mathrm{dom}(q_\mathfrak{s}).$$

Since $u'\mathrm{dom}(\mathfrak{s}) \subseteq \mathrm{dom}(\mathfrak{s})$ by Definition 0.142(4), we have $u'K \subseteq K$ for any unitary $u' \in \mathcal{M}'$. Similarly, $x_\mathfrak{s} \in {}^\eta \mathcal{M}$.

(c) $\Rightarrow$ (a). Let K and x be as in (c). Denote by p the projection from H on $H \ominus K$ and let $x = \int_0^\infty \lambda de_\lambda$ with $e_\lambda \nearrow \mathbb{1} - p$ for $\lambda \to \infty$. By Proposition 0.94(7) and Definition 0.135, we have $e_\lambda \in \mathcal{M}$ for all $\lambda \in [0, \infty)$ and $p \in \mathcal{M}$. Moreover, $x^{1/2} = \int_0^\infty \lambda^{1/2} de_\lambda$ and $\mathrm{dom}(x^{1/2}) = \{\xi \in K : \int_0^\infty \lambda d(e_\lambda \xi, \xi) < \infty\}$. Put $m_{(K,x)}(\omega) = \int_0^\infty \lambda d(\omega(e_\lambda)) + \infty \cdot \omega(p)$. For $\omega = \omega_\xi$, we have $m_{(K,x)}(\omega_\xi) = \|x^{1/2}\xi\|^2$ for $\xi \in \mathrm{dom}(x^{1/2})$, $m(\omega_\xi) = \infty$ otherwise. The proof of lower semicontinuity is the same as in Remark 0.138. The uniqueness on ω_ξ is clear; by Theorem 0.83(6), other ωs are sums of ω_ξ. $\square$

Theorem 0.144 *Each $m \in \widehat{\mathcal{M}}_+$ has a unique spectral decomposition of the form*

$$m(\omega) = \int_0^\infty \lambda d\omega(e_\lambda) + \infty \cdot \omega(p), \quad \omega \in \mathcal{M}_*^+ \qquad (0.4)$$

which we write simply as

$$m = \int_0^\infty \lambda de_\lambda + \infty \cdot p$$

where (e_λ) is an increasing, right-continuous family of projections from $\mathcal{M}$ and $p = \mathbb{1} - e_\infty$ with $e_\infty = \lim_{\lambda \to \infty} e_\lambda$. Moreover:

(1) $e_0 = 0$ *if and only if* $m(\omega) > 0$ *for all* $\omega \in \mathcal{M}_*^+ \setminus \{0\}$;
(2) $p = 0$ *if and only if* $\{\omega \in \mathcal{M}_*^+ : m(\omega) < \infty\}$ *is dense in* $\mathcal{M}_*^+$.

Proof. The representation (0.4) is a consequence of Theorem 0.143 and its proof. For the proof of (1) and (2), assume that $\mathcal{M}$ acts in such a Hilbert space H that any state ω is a vector state. By Theorem 0.143, $e_0 = 0$ if and only if x is injective if and only if $m(\omega_\xi) > 0$ for all $\xi \in H \setminus \{0\}$, which proves (1). If $p = 0$, then $H = K$, so that $\mathrm{dom}(x^{1/2}) = \{\xi : m(\omega_\xi) < \infty\}$ is dense in H. Let $\omega \in \mathcal{M}_*^+$, $\omega = \omega_\xi$. Choose $\xi_n \in \mathrm{dom}(x^{1/2})$ so that $\xi_n \to \xi$. Then, $\omega_n = \omega_{\xi_n} \to \omega$ and $m(\omega_n) < \infty$. This shows '$\Rightarrow$' in (2). If $p \neq 0$, the set $\{\omega \in \mathcal{M}_*^+ : \omega(p) = 0\}$, which contains the set $\{\omega \in \mathcal{M}_*^+ : m(\omega) < \infty\}$, is not dense in $\mathcal{M}_*^+$ (as it is closed). $\qquad\square$

Remark 0.145 Any normal *-isomorphism $\mathscr{I}$ from $\mathcal{M}$ onto $\mathcal{N}$ where $\mathcal{M}$ and $\mathcal{N}$ are von Neumann algebras, in a very natural way admits a map from $\widehat{\mathcal{M}}_+$ onto $\widehat{\mathcal{N}}_+$. Given $m \in \widehat{\mathcal{M}}_+$ with

$$m(\omega) = \int_0^\infty \lambda d\omega(e_\lambda) + \infty \cdot \omega(p), \quad \omega \in \mathcal{M}_*^+$$

one simply defines $\mathscr{I}(m)$ by means of the prescription

$$\mathscr{I}(m)(\omega) = \int_0^\infty \lambda d\omega(\mathscr{I}(e_\lambda)) + \infty \cdot \omega(\mathscr{I}(p)), \quad \omega \in \mathcal{N}_*^+.$$

Remark 0.146 Many useful aspects of the Borel functional calculus are also applicable to generalised positive operators. If, for example, we are given an increasing function $f : [0, \infty] \to [0, \infty]$ which satisfies

$f(0) = 0$ and $\lim_{s \nearrow \infty} f(s) = \infty$, and which is continuous on $[0, b_f]$ where $b_f = \sup\{s \in [0, \infty) : f(s) < \infty\}$, then given $m \in \widehat{\mathcal{M}}_+$ with

$$m(\omega) = \int_0^\infty \lambda d\omega(e_\lambda) + \infty \cdot \omega(p), \quad \omega \in \mathcal{M}_*^+$$

we may define $f(m)$ by means of the prescription

$$f(m)(\omega) = \int_0^{b_f} f(\lambda) d\omega(e_\lambda) + \infty \cdot \omega(p + (e_\infty - e_{b_f})), \quad \omega \in \mathcal{M}_*^+.$$

Remark 0.147 If, for some $m \in \widehat{\mathcal{M}}_+$, $m(\omega) < \infty$ for all $\omega \in \mathcal{M}_*^+$, then there is an $a \in \mathcal{M}_+$ such that $m(\omega) = \omega(a)$ for all $\omega \in \mathcal{M}_*^+$. This is an easy consequence of Theorem 0.143.

Proposition 0.148 *Any* $m \in \widehat{\mathcal{M}}_+$ *is a pointwise limit of an increasing sequence of bounded operators from* $\mathcal{M}_+$.

Proof. If $m = \int_0^\infty \lambda de_\lambda + p \cdot \infty$, then $a_n := \int_0^n \lambda de_\lambda + np \nearrow m$. $\qquad\square$

Since we can treat $\overline{\mathcal{M}}_+$ as a subset of $\widehat{\mathcal{M}}_+$, it is important to see whether the inclusion $x \mapsto m_x$ respects the order introduced in Definition 0.136.

Proposition 0.149 *For* $x, y \in \overline{\mathcal{M}}_+$, $x \leq y$ *if and only if* $m_x \leq m_y$. *In other words,* $\leq$ *is the order obtained by restricting the usual order of generalised positive operators.*

Proof. By definition, $x \leq y$ implies $m_x(\omega_\xi) \leq m_y(\omega_\xi)$ for each $\xi \in \text{dom}(y^{1/2})$. Since an arbitrary $\omega \in \mathcal{M}_*$ can be represented as a sum $\omega = \sum_{n \in \mathbb{N}} \omega_{\xi_n}$, and $m_y(\omega_\xi) = \infty$ whenever $\xi \notin \text{dom}(y^{1/2})$, we have $m_x \leq m_y$. Now, if $m_x \leq m_y$ and $\xi \in \text{dom}(y^{1/2})$, then $m_y(\omega_\xi) < \infty$, hence $m_x(\omega_\xi) < \infty$ and $\xi \in \text{dom}(x^{1/2})$. Thus, $x \leq y$. $\qquad\square$

Hence, the order on $\overline{\mathcal{M}}_+$ inherited from $\widehat{\mathcal{M}}_+$ coincides with the order in $\overline{\mathcal{M}}_+$. Moreover, $\overline{\mathcal{M}}_+$ is hereditary with respect to the order of generalised positive operators:

Proposition 0.150 *Let* $m, n \in \widehat{\mathcal{M}}_+$ *with* $m \leq n$. *If* $n = m_y$ *with* $y \in \overline{\mathcal{M}}_+$, *then* $m = m_x$ *for some* $x \in \overline{\mathcal{M}}_+$.

Proof.   Let $m := \int_0^\infty \lambda\, de_\lambda + p \cdot \infty$ and $n := \int_0^\infty \lambda\, df_\lambda + q \cdot \infty$. It is clear that $m \le n$ implies $p \le q$. Since $n = m_y$, we have $q = 0$, hence $p = 0$, which yields the result. $\qquad\square$

We also have the following lemma.

Lemma 0.151   *If $m_i \nearrow m$ in $\widehat{\mathcal{M}}_+$ and $a \in \mathcal{M}$, then $a^* m_i a \nearrow a^* m a$.*

Proof.   Obvious from the definitions. $\qquad\square$

Example 0.152   Let $\mathcal{M} = L^\infty(\mathcal{X}, \mathfrak{S}, \nu)$ where $(\mathcal{X}, \mathfrak{S}, \nu)$ is a localisable measure space. Such measure spaces are of course characterised by the fact that L^∞ appears as the dual of $L^1(\mathcal{X}, \mathfrak{S}, \nu)$ [**Sal20**, Theorem 4.35]. Hence, $\mathcal{M}$ is a W^*-algebra for which $\mathcal{M}_*^+ \equiv L_+^1(\mathcal{X}, \mathfrak{S}, \nu)$. Let $f \colon \mathcal{X} \to [0, \infty]$ be a $\mathfrak{S}$-measurable function. Put

$$m_f(\omega) = \int_\Omega f \cdot \omega \, d\mu \text{ for } \omega \in \mathcal{M}_*^+.$$

Then $m_f \in \widehat{\mathcal{M}}_+$ by Fatou's lemma. Moreover, $m_f = m_g$ implies $f = g$ μ-a.e. On the other hand, if $m \in \widehat{\mathcal{M}}_+$, we can find a sequence $f_k \in L^\infty(\mathcal{X}, \mathfrak{S}, \nu)$ s.t. $\int f_k \omega \, d\mu \nearrow m(\omega)$. Put $f = \sup f_k$. Then $m = m_f$. Hence, $\widehat{\mathcal{M}}_+$ can be identified with the set of equivalence classes of measurable functions $\mathcal{X} \to [0, \infty]$.

We shall ultimately see in Theorem 1.54 that all abelian von Neumann algebras are *-isomorphic to $L^\infty(\mathcal{X}, \mathfrak{S}, \nu)$-spaces of the above type. Hence, the above example clearly shows that the notion of generalised positive operators is a true noncommutative analogue of the notion of equivalence classes of possibly infinite-valued non-negative measurable functions.

PART 1

FOUNDATIONAL EXAMPLES

This part of the book is devoted to two special cases of noncommutative measure theory, already considered to be classical. Namely, in Chapter 1, we deal with the abelian case, while in Chapter 2, we deal with the case of $B(H)$.

These two examples are in a way extreme. On the one hand, we have an abelian algebra, and we expect the theory to be well known and well understood. Nevertheless, the reader may find that both the complete classification of such algebras and the interplay between abelian von Neumann algebras, measure algebras and L^∞-spaces are not something to be taken for granted, and they can expect some surprises. On the other hand, we have the simplest example of a factor, a 'most noncommutative' algebra, one with a trivial centre.

In classical measure theory, we deal with *measures* and *integrals*. After moving to the noncommutative context of operator algebras, we tend to use similar terminology when speaking about functions defined on the algebras that take positive possibly infinite values, are additive and positively homogeneous. The proper counterpart of an *integral* in the von Neumann algebra context is the notion of a weight.

Definition P1.1 A map $\varphi : \mathcal{M}_+ \to [0, \infty]$ is a *weight* on $\mathcal{M}$ if

1. $\varphi(a + b) = \varphi(a) + \varphi(b)$ for all $a, b \in \mathcal{M}_+$;
2. $\varphi(\lambda a) = \lambda \varphi(a)$ for all $a \in \mathcal{M}_+$, $\lambda \geq 0$.

A weight φ is *finite* if $\varphi(\mathbb{1}) < \infty$.

By a small abuse of language, we often treat weights as *measures* on $\mathcal{M}$. It should be noted that a finite weight is a restriction to $\mathcal{M}_+$ of an element of $\mathcal{M}_+^*$. If $\varphi(\mathbb{1}) = 1$, then φ is a restriction of a *state* on $\mathcal{M}$, and in what follows, we disregard the distinction between the two, calling φ a state.

Definition P1.2 A weight φ is *faithful* if $\varphi(a) = 0$ implies that $a = 0$ for $a \in \mathcal{M}_+$.

Definition P1.3 A weight φ is *normal* if for any bounded from above, increasing net $(a_i)_{i \in I}$ from $\mathcal{M}_+$, we have $\varphi(\sup_{i \in I} a_i) = \sup_{i \in I} \varphi(a_i)$.

A serious study of weights will be undertaken in Chapter 4, and for the reader's convenience, Definitions P1.1, P1.2, and P1.3 will be repeated there in a broader context.

Definition P1.4 We call a couple consisting of a von Neumann algebra $\mathcal{M}$ endowed with a faithful normal weight φ a *von Neumann measure algebra*. If, in fact, φ is a (faithful normal) state, we refer to the pair $(\mathcal{M}, \varphi)$ as a *von Neumann probability algebra*.

The choice of calling pairs $(\mathcal{M}, \varphi)$ of the above form von Neumann measure algebras rather than something like *noncommutative measure spaces* is deliberate. The terminology has already been used in [**BGL22**], and its choice will be substantiated in Chapter 1, with Definition 1.15 (p. 69) repeating and extending Definition P1.4.

One can easily construct a more general definition of an *operator measure algebra* and speak about C^*-measure algebras, AW^*-measure algebra, etc. It always makes sense to assume that the weight in question is *faithful*. There are no 'sets of measure zero' in the noncommutative world, and the classical definition of a measure algebra (see Definition 1.10 (p. 67)) is meant to get rid of them. One can imagine a use of non-faithful weights, but we can usually reduce the situation to the case of faithful weights in a straightforward manner. On the other hand, our second assumption, namely that the weight is normal, is natural only in the case of von Neumann algebras. This should already become apparent in Chapter 1.

There is one more natural assumption that has not been mentioned up to now: we usually deal with weights that are finite on a sufficiently rich set of operators. We call such weights *semifinite*. Since the notion is a bit more delicate, we do not give a precise definition here. It will first be carefully introduced in the two 'special' cases, and then in full generality in Chapters 3 and 4.

There is one kind of weight that is of special interest, namely *traces*. The idea is that traces are weights that give equal values to projections *of the same size*. Here, the same size means being equivalent in the Murray–von Neumann sense; see Definition 0.110 (p. 40). So the resemblance to the Lebesgue (or Haar) measure hides deep subtleties (in an abelian algebra, two distinct projections are never equivalent). The reader can regard *size* of the projection as its *dimension*. This point of view turned out to be really fruitful and gave the impetus for development of the theory for Murray and von Neumann. It is elaborated in Section 3.1 (p.131).

Definition P1.5 A *trace* $\tau : \mathcal{M}_+ \to [0, \infty]$ is a *weight* which additionally satisfies

$$\tau(a) = \tau(u^* a u)$$

for each $a \in \mathcal{M}_+$ and all $u \in \mathbb{U}(\mathcal{M})$.

Obviously, all weights on an abelian algebra are traces. The flagship noncommutative example of a trace will be given in Chapter 2, and the full use of general traces will be demonstrated in Chapter 3. The reader who decides to go directly to Part 2 or Part 3 of the book should at least read Definition 1.15 (p. 69), as it repeats and extends Definition P1.4 of a von Neumann measure algebra.

Chapter 1
Abelian von Neumann algebras

The objective of this chapter is to prepare the reader for the future interpretation of non-abelian von Neumann algebras as noncommutative L^∞-spaces. We achieve this by presenting a proof of Maharam's theorem interpreted from a von Neumann algebra perspective, which is in essence a structural classification of abelian von Neumann algebras. We then apply the resultant theory to show that abelian von Neumann algebras may variously be represented as L^∞-spaces living on either localisable, decomposable, or Radon measure spaces. As a consequence, we then show that the categories of abelian von Neumann (measure) algebras and localisable measure algebras are functorially equivalent.

1.1 A measure space primer

We give below a few basic definitions concerning measure-space theory. We assume that the reader knows the rudiments of measure spaces. The material can be found in any good book on measure theory. The one we recommend is Salamon's *Measure and Integration* ([**Sal20**]).

Let X be a set and Σ a σ-algebra of subsets of X. The couple (X, Σ) is called a *measurable space* or just *measure space*. The elements of Σ are said to be Σ-*measurable* (or just *measurable* if the σ-algebra is understood). If (X_1, Σ_1) and (X_2, Σ_2) are two measurable spaces, then a function $f : X_1 \to X_2$ is called (Σ_1, Σ_2)-*measurable* (or simply *measurable* if the σ-algebras are understood) if the pre-image $f^{-1}(B)$ is Σ_1-measurable for each Σ_2-measurable $B \subseteq X_2$.

A *measure* on the measurable space (X, Σ) is any function μ satisfying:

1. σ-**additivity:** for any sequence (A_n) of pairwise disjoint measurable sets,

$$\mu(\bigcup_{n=1}^{\infty} A_n) = \sum_{n=1}^{\infty} \mu(A_n);$$

Noncommutative measures and L^p and Orlicz Spaces, with Applications to Quantum Physics. Stanisław Goldstein and Louis Labuschagne, Oxford University Press. ©Stanisław Goldstein and Louis Labuschagne (2025).
DOI: 10.1093/oso/9780198950202.003.0003

   2. $\mu(\emptyset) = 0$ (or, equivalently, $\mu(A) < \infty$ for some $A \in \Sigma$).

As is usual, we here sometimes write (X, μ) for (X, Σ, μ) when Σ is understood.

A measure space (X, Σ, μ) is called *finite* if $\mu(X) < \infty$. If $\mu(X) = 1$, we call it a *probability space*. We say that a measure space (X, Σ, μ) is *σ-finite* if there is a countable family $\{E_n\}$ of measurable subsets of X such that $X = \bigcup_{n=1}^{\infty} E_n$ and $\mu(E_n) < \infty$ for all n. We call a measure space (X, Σ, μ) *semifinite* if for any $\emptyset \neq A \in \Sigma$, there is $B \in \Sigma$ such that $\emptyset \neq B \subseteq A$ and $0 < \mu(B) < \infty$.

The following definitions might not be known to the reader.

We will say that a measure space (X, Σ, μ) is *decomposable* (sometimes called *strictly localisable*) if we can partition X into measurable μ-finite subsets X_t such that

$$\Sigma = \{E \subset X : E \cap X_t \in \Sigma \text{ for all } t\}$$

and $\mu(E) = \sum_t \mu(E \cap X_t)$ for all $E \in \Sigma$.

We further say that (X, Σ, μ) is a *Dedekind measure space* if, for any collection $\mathcal{E}$ of sets in Σ, there exists a set $H \in \Sigma$ (called the *envelope* of $\mathcal{E}$) satisfying the following requirements.

   1. $\mu(E \backslash H) = 0$ for any $E \in \mathcal{E}$.
   2. If for any other set $G \in \Sigma$ we have that $\mu(E \backslash G) = 0$ for any $E \in \mathcal{E}$, then $\mu(H \backslash G) = 0$.

(In the quotient algebra $\Sigma / \mathcal{Z}(\Sigma)$, the element $[H]$, where H is the envelope of $\mathcal{E}$, corresponds to the lowest upper bound of the set $\{[E] : E \in \mathcal{E}\}$.)

A measure space is *localisable* if it is both semifinite and Dedekind. We will give several characterisations of these later. It is interesting that it is not so easy to find examples of non-localisable measure spaces. This can be seen by noting that any decomposable (strictly localisable) measure space is localisable (which, in turn, encompasses all finite and σ-finite measure spaces). To see this, first note that decomposable measure spaces are trivially semifinite. So, all that remains is to show that decomposable measure spaces are Dedekind. A proof of this fact can be found in [**Fre04**, 211Ld(i)].

We will often 'decompose' a measure space (X, Σ, μ) into pieces with finite measure. With this in mind, we now describe the 'disjoint sum' of measure spaces $(\Omega_i, \Sigma_i, \mu_i)$. Let Ω be the disjoint union of the sets Ω_i. Define a σ-algebra Σ on Ω by $E \in \Sigma$ if and only if $E \cap \Omega_i \in \Sigma_i$ for all i. It is an exercise that Σ is indeed a σ-algebra. Define a measure ν on Σ by $\nu(E) = \sum_i \mu_i(E \cap \Omega_i)$. We can now check that (Ω, Σ, μ) is a measure space. In fact, it is *decomposable* in the sense defined above if $\mu_i(\Omega_i) < \infty$ for each i. It is an exercise from the definition of Σ that a scalar-valued function f on Ω is Σ-measurable if and only if its restriction to Ω_i is measurable on Ω_i for all i (that is, if and only if f is *locally measurable*).

Definition 1.1 By analogy with the above example, given a von Neumann algebra $\mathcal{M}$, a trace τ from $\mathcal{M}_+$ to $[0, \infty]$ is called *semifinite* if for any non-zero projection p in $\mathcal{M}$, we can find a non-zero subprojection q such that $\tau(q) < \infty$. We write *f.n.s.* for *faithful normal semifinite*. We shall say that a *-isomorphism θ between two von Neumann algebras $\mathcal{M}_1$ and $\mathcal{M}_2$ equipped with f.n.s. traces τ_1 and τ_2 is *measure preserving* if $\tau_2 \circ \theta = \tau_1$.

Note that the definitions above are given for arbitrary von Neumann algebras, without checking whether the traces in question really exist. By *trace*, we understand here the functional defined in the introduction to Part 2. Since this chapter is devoted to abelian algebras only, the definitions will be repeated in later chapters.

Example 1.2 Perhaps, the simplest example of a semifinite but non-Dedekind (so non-localisable) measure is the following: Let X be an uncountably infinite set. Let Σ be the set of subsets of X that are either countable or their complement is countable, and let μ be the counting measure. Then, (X, Σ, μ) is a complete semifinite measure space, but it is not Dedekind. Indeed, if E is an uncountable subset with an uncountable complement, then the countable subsets of E have no envelope in Σ. This measure 'has support' X, in the sense that the empty set is the largest set in Σ of measure zero.

Another semifinite non-Dedekind space may be found in [**Fre04**, 216D]. Note that Fubini's theorem is not valid for localisable measure spaces (see e.g. [**Fre04**, 252K and Volume 3]).

1.2 A Boolean and measure algebra primer

We start with fixing terminology on structures we shall need in the sequel.

Definition 1.3 Let $(\mathfrak{L}, \leq)$ be a *partially ordered set*. We say that it is a *lattice* if, for all $a, b \in \mathfrak{L}$, the elements $\sup\{a, b\}$ and $\inf\{a, b\}$ (denoted by $a \vee b$ and $a \wedge b$ and called *join* and *meet*, respectively) exist in $\mathfrak{L}$. A lattice is *complete* (respectively, *sequentially complete*) if all its subsets (respectively, all its countable subsets) have both join and meet.

It will often be convenient to treat lattices as algebraic structures. From this point of view, $(\mathfrak{L}, \vee, \wedge)$ is a *lattice* if $\vee$ and $\wedge$ are binary commutative and associative operations, satisfying, for all $a, b \in \mathfrak{L}$, the *absorption laws*:

$$a \vee (a \wedge b) = a, \ a \wedge (a \vee b) = a$$

(which together imply the *idempotent laws*: $a \vee a = a$ and $a \wedge a = a$). Then, we *define* the partial order $\leq$ on $\mathfrak{L}$ by saying that $a \leq b$ whenever $a \wedge b = a$ (or $a \vee b = b$). We say that the lattice $\mathfrak{L}$ is *bounded* if in $\mathfrak{L}$, there exist elements 0 and 1 such that $0 \leq a \leq 1$ for all $a \in \mathfrak{L}$. From the algebraic point of view, it is the same as claiming that 0 and 1 are unit elements for $\vee$ and $\wedge$, respectively.

Two important properties of lattices are modularity and distributivity. We call a lattice $\mathfrak{L}$ *modular* if, for all $a, b, c \in \mathfrak{L}$, $a \leq c$ implies $a \vee (b \wedge c) = (a \vee b) \wedge c$. We call it *distributive* if, for all $a, b, c \in \mathfrak{L}$, $a \vee (b \wedge c) = (a \vee b) \wedge (a \vee c)$ and (which turns out to follow automatically) $a \wedge (b \vee c) = (a \wedge b) \vee (a \wedge c)$. Obviously, a distributive lattice is also modular.

An *orthocomplementation* on a bounded lattice is a mapping which assigns to each element a an *orthocomplement* $a^{\perp}$, that satisfies complement laws: $a^{\perp} \vee a = 1$ and $a^{\perp} \wedge a = 0$, is involutive: $a^{\perp\perp} = a$ and order-reversing: $a \leq b$ implies $b^{\perp} \leq a^{\perp}$. An *orthocomplemented lattice* or *ortholattice* is a bounded lattice equipped with an orthocomplementation.

In the following, you will find the prototypical examples of lattices:

Example 1.4

- Any algebra Σ of subsets of a set X, with the join $\vee$ being the union $\cup$ and the meet $\wedge$ the intersection $\cap$, is a distributive lattice, which becomes an ortholattice if we define orthcomplementation by $E^{\perp} := X \setminus E$. If Σ is a σ-algebra, then it is, in addition, sequentially complete.
- Let $\mathcal{M}$ be a von Neumann algebra. Then, $\mathbb{P}(\mathcal{M})$ is a complete lattice (see Proposition 0.94). It is modular if $\mathcal{M}$ is finite (see Theorem 0.120) and distributive if $\mathcal{M}$ is commutative.

Definition 1.5 We define a *Boolean ring* $\mathfrak{B}$ to be a ring $(\mathfrak{B}, \boxplus, \boxdot)$ for which we have that $a \boxdot a = a$ for every $a \in \mathfrak{B}$. If the ring admits a multiplicative identity, we shall refer to $\mathfrak{B}$ as a *Boolean algebra*. Elements $a, b \in \mathfrak{B}$ are called disjoint if $a \boxdot b = 0$.

Each Boolean algebra $\mathfrak{B}$ admits a natural ordering, where we say that $a \leq b$ precisely when $a \boxdot b = a$. This is easily seen to be a partial ordering with least element 0 and greatest element 1.

Remark 1.6 Note that the operations $\vee$, $\wedge$ and $a \mapsto a^{\perp}$ defined by

$$a \vee b := (a \boxplus b) \boxplus (a \boxdot b) , \ a \wedge b := a \boxdot b \text{ and } a^{\perp} := 1 \boxplus a$$

turn $\mathfrak{B}$ into an orthocomplemented distributive lattice, and the partial orders of the two structures coincide.

- $\vee$ and $\wedge$ are associative and commutative.
- $a \wedge a = a$ and $a \vee a = a$.
- $a \vee (a \wedge b) = a \wedge (a \vee b)$.
- $(a \vee b)^{\perp} = a^{\perp} \wedge b^{\perp}$ and $(a \wedge b)^{\perp} = a^{\perp} \vee b^{\perp}$.
- $a \wedge (b \vee c) = (a \wedge b) \vee (a \wedge c)$ and $a \vee (b \wedge c) = (a \vee b) \wedge (a \vee c)$.
- $a \vee 0 = a$ and $a \wedge 1 = a$.
- $a \wedge b = 1$ and $a \vee b = 0$ if and only if $b = a^{\perp}$.

Conversely, any object $\mathfrak{B}$ that forms an orthocomplemented distributive lattice with operations $\vee$, $\wedge$ and $\perp$ becomes a Boolean algebra when the operations '$\boxplus$' and '$\boxdot$' are defined by $a \boxplus b := (a \vee b) \wedge ((a \wedge b)^{\perp})$ and $a \boxdot b := a \wedge b$.

The following concepts will form the basis for our further analysis.

Definition 1.7

- We say that a Boolean algebra $\mathfrak{B}$ is *Dedekind complete* (respectively, *Dedekind σ-complete*), if it forms a complete (respectively, sequentially complete) lattice.
- A subset C of a Boolean algebra $\mathfrak{B}$ is called sequentially order-closed (respectively, order-closed) if any increasing sequence in C (respectively, any upward-directed net in C) has a unique least upper bound in $\mathfrak{B}$ that belongs to C. By the same token, a Boolean homomorphism which preserves suprema of increasing nets is said to be order continuous. If it only preserves suprema of increasing sequences, we say that it is σ-order continuous.

In the sequel, we shall often identify Boolean algebras by specifying their lattice operations (where we include, by a slight abuse of language, the orthocomplementation). Our earlier examples of lattices now become natural examples of Boolean algebras.

Example 1.8

- Let Σ be an algebra of subsets of a set X. It forms a Boolean algebra with operations $E \boxplus F := E \triangle F = (E \setminus F) \cup (F \setminus E)$ and $E \boxdot F := E \cap F$. If Σ is a σ-algebra, then it is, in addition, Dedekind σ-complete.
- Let $\mathcal{M}$ be a commutative von Neumann algebra. Then, $\mathbb{P}(\mathcal{M})$ is a Dedekind complete Boolean algebra with operations $p \boxplus q := (p \wedge q^{\perp}) \vee (p^{\perp} \wedge q)$ and $p \boxdot q := p \wedge q = pq$.

For Boolean algebras, we define a Boolean homomorphism to be a ring homomorphism which preserves the multiplicative identity. Similarly, we define an isomorphism, an ideal and a principal ideal by treating a Boolean algebra as a ring with unit.

It is not difficult to see that, by definition, all Boolean algebras are automatically commutative, and to conclude that $a \boxplus a = 0$ for all $a \in \mathfrak{B}$.

Remark 1.9 Note that a subset $\mathcal{I}$ of a Boolean algebra $\mathfrak{B}$ is an ideal of $\mathfrak{B}$ if and only if for all $a, b \in \mathcal{I}$, we have $a \vee b \in \mathcal{I}$, and also that

$a \in \mathcal{I}$ and $b \in \mathfrak{B}$ with $b \leq a$ implies $b \in \mathcal{I}$. If an ideal is sequentially order-closed, we say it is a σ-ideal.

It is an exercise to see that if $\mathcal{I}$ is an ideal of some Boolean algebra $\mathfrak{B}$, then $\mathfrak{B}/\mathcal{I}$ will again be a Boolean algebra. In fact, if $\mathfrak{B}$ is σ-complete and $\mathcal{I}$ a σ-ideal, then $\mathcal{B}/\mathcal{I}$ is also σ-complete.

We are particularly interested in a specific subcategory of Boolean algebras, the so-called measure algebras.

Definition 1.10

- A Dedekind σ-complete Boolean algebra $\mathfrak{A}$ is said to be a *measure algebra* if it admits a function $\overline{\mu} : \mathfrak{A} \to [0, \infty]$ with the properties that $\overline{\mu}(p) = 0$ precisely when $p = 0$ and that for any countable set $\{p_n\}$ of mutually disjoint elements, it holds that $\overline{\mu}(\sup_n p_n) = \sum_{n=1}^{\infty} \overline{\mu}(p_n)$. The quantity $\overline{\mu}$ is referred to as a measure on $\mathfrak{A}$ and is said to be *semifinite* if for every non-zero $p \in \mathfrak{A}$, we can find a non-zero element q with $q \leq p$ such that $\overline{\mu}(q) < \infty$.
- A measure algebra $(\mathfrak{A}, \overline{\mu})$ is said to be *finite* if $\overline{\mu}(\mathbb{1}) < \infty$, and a *probability* algebra if $\overline{\mu}(\mathbb{1}) = 1$.
- Two measure algebras $(\mathfrak{A}_0, \overline{\mu}_0), (\mathfrak{A}_1, \overline{\mu}_1)$ are said to be *isomorphic* if there exists a Boolean isomorphism π from $\mathcal{A}_0$ to $\mathfrak{A}_1$ such that $\overline{\mu}_1 \circ \pi = \overline{\mu}_0$.
- A measure algebra $(\mathfrak{A}, \overline{\mu})$ is said to be *localisable* if $\mathfrak{A}$ is Dedekind complete and $\overline{\mu}$ is semifinite.

Example 1.11 We present an important example to illustrate the above concepts. For any measure space (X, Σ, μ), Σ is a σ-complete Boolean algebra with the lattice operations given by

$$E \wedge F := E \cap F, \quad E \vee F := E \cup F, \quad E^{\perp} := X \setminus E$$

and with units $\mathbb{1} := X, \quad 0 := \emptyset$ for all $E, F \in \Sigma$. The collection $\mathcal{Z}(\Sigma)$ of sets of measure 0 is a σ-ideal of this Boolean algebra, which then ensures that the quotient $\Sigma/\mathcal{Z}(\Sigma)$ is again a σ-complete Boolean algebra with algebra operations given by

$$[E] \vee [F] = [E \cup F], \quad [E] \wedge [F] = [E \cap F], \quad \text{and} \quad [E]^{\perp} = [X \setminus E].$$

The action of μ canonically extends to $\Sigma/\mathcal{Z}(\Sigma)$, by means of the prescription $\overline{\mu}([E]) = \mu(E)$. The pair $(\Sigma/\mathcal{Z}(\Sigma), \overline{\mu})$ is then a measure algebra. From the foregoing discussion, it is clear that $\Sigma/\mathcal{Z}(\Sigma)$ is Dedekind complete whenever (X, Σ, μ) is Dedekind, and that $\overline{\mu}$ is semifinite precisely when μ is. It follows that $(\Sigma/\mathcal{Z}(\Sigma), \overline{\mu})$ is a localisable measure algebra precisely when (X, Σ, μ) is a localisable measure space. For simplicity, we will in the future denote the quotient map $E \to [E]$ by $E \to \overline{E}$.

In view of the importance of the above, we formalise it as a definition.

Definition 1.12 The measure algebra $(\mathfrak{A}, \overline{\mu})$ of a measure space (X, Σ, μ) is defined as the Boolean algebra $\mathfrak{A} = \Sigma/\mathcal{Z}(\Sigma)$, where $\mathcal{Z}(\Sigma)$ are the sets of measure 0, equipped with the measure $\overline{\mu}$, where $\overline{\mu}$ is defined by $\overline{\mu}([E]) = \mu(E)$ for some representative E from $[E] \in \Sigma/\mathcal{Z}(\Sigma)$. We shall refer to such measure algebras as concrete measure algebras.

Example 1.13 A second example of a localisable measure algebra is given by $(\mathbb{P}(\mathcal{M}), \tau\!\restriction\!\mathbb{P}(\mathcal{M}))$, where $\mathcal{M}$ is an abelian von Neumann algebra and τ a faithful normal semifinite trace on $\mathcal{M}$. The projection lattice $\mathbb{P}(\mathcal{M})$ is easily seen to be a Boolean algebra when equipped with the operations $p \boxplus q := p + q - 2pq$ and $p \boxdot q := pq$. Note that then $p \vee q = p + q - pq$ and $p \wedge q = pq$. This Boolean algebra is Dedekind complete (see, for example, Proposition 0.94) and $\tau\!\restriction\!\mathbb{P}(\mathcal{M})$ is semifinite (see, for example, Definition 3.32 and Proposition 3.31). More generally, a set of commuting projections in a von Neumann algebra is a Boolean algebra if it contains 0 and is closed under $\perp$ (i.e., $p \mapsto p^{\perp} = \mathbb{1} - p$), and infima (i.e., products pq), and it is a complete Boolean algebra if, in addition, it is closed under suprema of increasing nets.

The importance of the two examples of localisable algebras presented above stems from the functorial correspondence between these two classes of objects and localisable measure algebras. Establishing this correspondence is one of the main goals of this chapter.

Proposition 1.14 *Let $\mathcal{M}$ be an abelian von Neumann algebra. Then, $\mathcal{M}$ admits a faithful normal semifinite trace.*

Proof. We use Zorn's lemma to select a maximal family $\{\psi_\alpha : \alpha \in A\}$ of normal states with orthogonal supports p_α. Then, we must have that $\mathbb{1} = \vee_\alpha p_\alpha$, since if this was not the case, $(\mathbb{1} - \vee_\alpha p_\alpha)\mathcal{M}$ would be a non-trivial abelian von Neumann algebra. Given a normal state ρ on $(\mathbb{1} - \vee_\alpha p_\alpha)\mathcal{M}$, the function $\psi_0 : a \mapsto \rho((\mathbb{1} - \vee_\alpha p_\alpha)a)$ would then be a normal state on $\mathcal{M}$ with support equal to a subprojection of $(\mathbb{1} - \vee_\alpha p_\alpha)$, which would contradict the maximality of $\{\psi_\alpha : \alpha \in A\}$.

Consider the affine functional $\tau : \mathcal{M}_+ \to [0,1]$ defined by $\tau(a) = \sum_\alpha \psi_\alpha(a)$. Since $\mathbb{1} = \vee_\alpha p_\alpha$, it is clear that we will for any given $a \in \mathcal{M}_+$ have that $a \neq 0$ if and only if $ap_{\alpha_0} \neq 0$ for some α_0, which ensures that $\tau(a) \neq 0$. Thus, τ is faithful. Had a been a non-zero projection ap_{α_0} would be a non-zero subprojection for which $0 < \tau(ap_{\alpha_0}) = \psi_{\alpha_0}(a) < \infty$, ensuring the semifiniteness of τ. In closing, we verify that τ is normal. Given a net $(a_\gamma) \subset \mathcal{M}_+$ increasing to say $a \in \mathcal{M}$, it follows from Propositions 0.12 and 0.81 that $\psi_\alpha(a) = \sup_\gamma \psi_\alpha(a_\gamma)$. It is now an exercise to conclude from this equality that $\tau(a) = \sup_\gamma \tau(a_\gamma)$. $\qquad\square$

With the preceding discussion as background, we are now ready to formally introduce the notion of a von Neumann measure algebra.

Definition 1.15 We call a couple consisting of a von Neumann algebra $\mathcal{M}$ endowed with a faithful normal weight φ a *von Neumann measure algebra*. If φ is a trace, the von Neumann measure algebra will be called *tracial*. If, on the other hand, φ is a (faithful normal) state, we shall refer to the pair $(\mathcal{M}, \varphi)$ as a *von Neumann probability algebra*.

Two von Neumann measure algebras $(\mathcal{M}_1, \varphi_{\mathcal{M}_1})$ and $(\mathcal{M}_2, \varphi_{\mathcal{M}_2})$ are said to be *isomorphic* if there exists a $*$-isomorphism $\Phi : \mathcal{M}_1 \to \mathcal{M}_2$ such that $\varphi_{\mathcal{M}_1} = \varphi_{\mathcal{M}_2} \circ \Phi$ on $(\mathcal{M}_1)_+$.

A *direct sum* of von Neumann measure algebras $(\mathcal{M}_i, \varphi_i)$ is the von Neumann measure algebra $(\mathcal{M}, \varphi_{\mathcal{M}})$ such that $\mathcal{M}$ is the direct sum $\oplus_i \mathcal{M}_i$ of $\mathcal{M}_i$ and $\varphi_{\mathcal{M}}$ is given by $\varphi_{\mathcal{M}}(\oplus_i a_i) = \sum_i \varphi_i(a_i)$.

Proposition 1.16 shows that in the case of abelian von Neumann measure algebras, two such algebras are isomorphic if and only if their projection lattices are isomorphic as measure algebras.

Proposition 1.16 *Let $\mathcal{M}_1$ and $\mathcal{M}_2$ be abelian von Neumann algebras equipped with faithful normal semifinite traces τ_1 and τ_2. Every Boolean homomorphism θ from $\mathbb{P}(\mathcal{M}_1)$ to $\mathbb{P}(\mathcal{M}_2)$ extends to a $*$-homomorphism $\mathscr{I} : \mathcal{M}_1 \to \mathcal{M}_2$. Moreover:*

- *θ is an isomorphism if and only if $\mathscr{I}$ is an isometric isomorphism;*
- *θ is order continuous if and only if $\mathscr{I}$ is normal, in which case θ is also measure preserving if and only if $\tau_1 = \tau_2 \circ \mathscr{I}$.*

Proof. Define a mapping $\mathscr{I}$ on the span of $\mathbb{P}(\mathcal{M}_1)$ by setting $\mathscr{I}(s) = \sum_{i=1}^{n} \lambda_i \theta(e_i)$ for any 'simple function' $s = \sum_{i=1}^{n} \lambda_i e_i$. (Note that when we use the term 'simple function' we understand $e_1, \ldots, e_n$ to be mutually orthogonal projections.) Given another representation $\sum_{j=1}^{m} \alpha_j f_j$ of s, we may pass to a common refinement of the 'sets' $e_1, \ldots, e_n$ and $f_1, \ldots, f_m$ to see that $\mathscr{I}(\sum_{i=1}^{n} \lambda_i e_i) = \mathscr{I}(\sum_{j=1}^{m} \alpha_j f_j)$. So, $\mathscr{I}$ is a well-defined bijective linear map from the simple functions of $\mathcal{M}_1$ to those of $\mathcal{M}_2$. The map $\mathscr{I}$ clearly preserves adjoints and squares of self-adjoint simple functions. Hence, it is a *-isomorphism. For any positive simple function $s = \sum_{i=1}^{n} \lambda_i e_i$, we have that $\|s\|_\infty = \max(\lambda_1, \ldots, \lambda_n)$. This ensures that

$$\|\mathscr{I}(s)\|_\infty = \|\sum_{i=1}^{n} \lambda_i \theta(e_i)\|_\infty \leq \|s\|_\infty \|\sum_{i=1}^{n} \theta(e_i)\|_\infty = \|s\|_\infty.$$

By Theorem 0.79 and Proposition 0.94(7), $\mathrm{span}(\mathbb{P}(\mathcal{M}_1))$ is norm-dense in $\mathcal{M}_1$. Thus, $\mathscr{I}$ extends to a *-homomorphism from $\mathcal{M}_1$ into $\mathcal{M}_2$.

If θ is an isomorphism, then for the induced map on the simple functions, we would similarly have that $\|s\|_\infty \leq \|\mathscr{I}(s)\|_\infty$. Hence, $\mathscr{I}$ is an isometric isomorphism between the simple functions. In this case, the simple functions of $\mathcal{M}_1$ of course map onto those of $\mathcal{M}_2$, and so the norm density of $\mathrm{span}(\mathbb{P}(\mathcal{M}_i))$ in $\mathcal{M}_i$ ($i = 1, 2$) ensures that here $\mathscr{I}$ extends to an isometric $*$-isomorphism from $\mathcal{M}_1$ into $\mathcal{M}_2$.

To see the second bullet, notice that given a set $\{e_i\}_{i \in I}$ of mutually orthogonal projections, $\sum_{i \in I} e_i$ equals $\sup_F \sum_{i \in F} e_i$, where the supremum is taken over finite subsets F of I. If θ is normal, then

$$\mathscr{I}\left(\sum_{i \in I} e_i\right) = \theta\left(\sum_{i \in I} e_i\right) = \sup_F \theta\left(\sum_{i \in F} e_i\right)$$

which equals $\sup_F \sum_{i \in F} \theta(e_i) = \sum_{i \in I} \theta(e_i) = \sum_{i \in F} \mathscr{I}(e_i)$. For a normal state ω on $\mathcal{M}_2$, we have $\omega \circ \mathscr{I}\left(\sum_{i \in I} e_i\right) = \omega\left(\sum_{i \in I} \mathscr{I}(e_i)\right) = \sum_{i \in I} \omega \circ \mathscr{I}(e_i)$. So, $\omega \circ \mathscr{I}$ is a normal functional by Theorem 0.84. Since the dual of $\mathscr{I}$ maps normal states to normal functionals, it is an exercise to see that $\mathscr{I}$ is normal.

To see the final claim, notice that for any non-negative simple function $s = \sum_{i=1}^{n} \lambda_i e_i$ of $\mathcal{M}_1$, we have

$$\tau_1(\mathscr{I}(s)) = \sum_{i=1}^{n} \lambda_i \tau_2(\theta(e_i)) = \sum_{i=1}^{n} \lambda_i \tau_1(e_i) = \tau_1(s).$$

Given any positive $a \in \mathcal{M}_1$, we may select a sequence (s_n) of non-negative simple functions increasing to a. The normality of the traces then ensures that $\tau_1(a) = \tau_2(\mathscr{I}(a))$. $\qquad\square$

Remark 1.17 It is useful to note that a Boolean isomorphism π between Dedekind complete Boolean algebras $\mathfrak{B}_0$ and $\mathfrak{B}_1$ is automatically order bicontinuous. To see this, notice that the fact that both π and π^{-1} preserve order ensures that $(a_\alpha) \subset \mathfrak{B}_0$ is an increasing net if and only if $(\pi(a_\alpha))$ is. Since $a_\alpha \leq \sup_\beta a_\beta$ for each α, we have $\pi(a_\alpha) \leq \pi(\sup_\beta a_\beta)$ and hence that $\sup_\alpha \pi(a_\alpha) \leq \pi(\sup_\beta a_\beta)$. Repeating this argument for π^{-1} instead of π shows that $\sup_\beta a_\beta \leq \pi^{-1}(\sup_\alpha \pi(a_\alpha))$ and hence that $\pi(\sup_\beta a_\beta) \leq \sup_\alpha \pi(a_\alpha)$. We, therefore, have $\sup_\alpha \pi(a_\alpha) = \pi(\sup_\beta a_\beta)$ as required.

Any measure-preserving Boolean isomorphism between measure algebras $(\mathfrak{A}_0, \overline{\mu}_0)$ and $(\mathfrak{A}_1, \overline{\mu}_1)$ clearly preserves semifiniteness. It is not difficult to modify the argument at the end of the preceding remark to show that measure-preserving Boolean isomorphisms preserve Dedekind completeness. We record this observation as a proposition.

Proposition 1.18 *Let $(\mathfrak{A}_0, \overline{\mu}_0)$ and $(\mathfrak{A}_1, \overline{\mu}_1)$ be isomorphic measure algebras. Then, $(\mathfrak{A}_0, \overline{\mu}_0)$ is localisable if and only if $(\mathfrak{A}_1, \overline{\mu}_1)$ is.*

For any point set X, the power set $\mathcal{P}X$ may be realised as a Boolean algebra by the prescriptions $X = 1$, $\emptyset = 0$, $E \boxplus F = E \triangle F$ (where $E \triangle F$ denotes the symmetric difference of E and F), and $E \boxdot F = E \cap F$. A deep fact regarding Boolean algebras, verified by Marshall Stone, is that every Boolean algebra admits a representation as a subalgebra of the power set of some point set Z. We pause to record this fact.

Theorem 1.19 (Stone's theorem). *Let $\mathfrak{B}$ be a Boolean algebra, and let Z be the set of ring homomorphisms from $\mathfrak{B}$ onto the two point Boolean algebra $\mathbb{Z}_2 \equiv \{0, 1\}$. Then, the map $\mathfrak{B} \to \mathcal{P}Z : a \mapsto \widehat{a}$ given by $\widehat{a} = \{z : z \in Z,\, z(a) = 1\}$ is an injective Boolean homomorphism with $\widehat{1}_{\mathfrak{B}} = Z$.*

On the basis of the above theorem, the *Stone space* of a Boolean algebra $\mathfrak{B}$ is defined to be the set $Z(\mathfrak{B}) = Z$ of non-zero ring homomorphisms from $\mathfrak{B}$ to $\mathbb{Z}_2$, with the map described in the above theorem referred to as the *Stone representation* of $\mathfrak{B}$. The Stone space admits a canonical topology, which is elucidated by the following theorem.

Theorem 1.20 (Stone space topology). *Let Z be the Stone space of a Boolean algebra $\mathfrak{B}$, and let*

$$\mathfrak{T} = \{G : G \subseteq Z,\, \text{for every } z \in G,\, \text{there exists } a \in \mathfrak{B}\, \text{such that } z \in \widehat{a} \subseteq G\}.$$

Then $\mathfrak{T}$ is a topology on Z, under which Z is a compact Hausdorff space which is zero-dimensional, that is, the collection $\mathcal{O}$ of clopen subsets of Z is a basis. Moreover, $\mathcal{O} = \{\widehat{a} : a \in \mathfrak{B}\}$.

Proofs of the preceding results may be found in [**Fre04**, 311]. Putting them together, we see that every Boolean algebra is representable as the Boolean algebra of all clopen subsets of a zero-dimensional compact space. (We remind the reader that a zero-dimensional topological space is a space for which each point has a neighbourhood base consisting of clopen sets.)

We may exploit the structure offered by the Stone space to establish the following very important result. We shall content ourselves with merely outlining the proof. Details may be found in [**Fre04**, 314M, 321J].

Theorem 1.21 (Loomis-Sikorski). *Every measure algebra is isomorphic to a concrete measure algebra.*

Proof. See, for example, [**Fre04**, 314M/321J] for more details if needed.

Given a measure algebra $(\mathfrak{A}, \overline{\mu})$, we pass to the Stone space Z, and let $\mathcal{O}$ be the algebra of clopen subsets of Z. This algebra is of course nothing but $\{\widehat{a} : a \in \mathfrak{A}\}$. The collection $\mathcal{I}$ of meagre subsets of Z turns out to be a σ-ideal of the Boolean algebra $\mathcal{P}Z$. Now, let Σ be the algebra

$$\Sigma = \{E \bigtriangleup F : E \in \mathcal{O}, F \in \mathcal{I}\}.$$

It is an exercise to see that Σ is a σ-algebra of subsets of Z which contains the σ-ideal $\mathcal{I}$. The fact that Σ is σ-complete and $\mathcal{I}$ a σ-ideal ensures that $\Sigma/\mathcal{I}$ is σ-complete. We know that $\mathfrak{A}$ and $\mathcal{O}$ are isomorphic as Boolean algebras. The next step is to show that $\mathcal{O}$ is isomorphic to the Boolean algebra $\Sigma/\mathcal{I}$. This is achieved by showing that for every $E \in \Sigma$, there is exactly one $F \in \mathcal{O}$ for which $E \bigtriangleup F \in \mathcal{I}$. The action of this isomorphism then identifies F with $[E]$. It follows that $\mathfrak{A}$ and $\Sigma/\mathcal{I}$ are isomorphic as Boolean algebras. What we still need is a measure on (Z, Σ) for which $\mathcal{I}$ are the sets of measure zero. With θ denoting the isomorphism from $\Sigma/\mathcal{I}$ onto $\mathfrak{A}$, the quantity $\nu(E) = \overline{\mu}(\theta([E]))$, where $E \in \Sigma$, is the measure we seek.

Consider the map $\pi : \Sigma \to \mathfrak{A}$ where we say that $\pi(C) = a$ precisely when $C \bigtriangleup \widehat{a} \in \mathcal{I}$ for some $a \in \mathfrak{A}$. Checking reveals that π is a well-defined surjective homomorphism for which the kernel is precisely $\mathcal{I}$. The σ-completeness of $\mathcal{I}$ then ensures that π is σ-order continuous. This fact, in turn, guarantees that the action of $\nu = \overline{\mu} \circ \pi$ on Σ yields a well-defined measure on (Z, Σ). Hence, (Z, Σ, ν) is a measure space for which π identifies $(\mathfrak{A}, \overline{\mu})$ with the measure algebra of (Z, Σ, ν). $\square$

1.3 A structural description of abelian von Neumann measure algebras

The aim of this section is to demonstrate the possibility of performing a complete structural classification of abelian von Neumann algebras using purely operator-algebraic means.

The presentation is based on Fremlin's exposition of Maharam's theorem (see [**Fre04**, Chapter 33]). Nevertheless, only one key result is used explicitly without being proved (see the remark before Theorem 1.23). For the rest, we either present operator-algebraic versions of the proofs or, in a few cases, give independent proofs of the needed facts.

In this section, $\mathcal{M}$ stands for a von Neumann algebra, and $\mathcal{A}$ for an abelian von Neumann algebra.

It is 'folklore' that the Maharam theorem, which classifies localisable measure algebras, can be seen as a theorem on the classification of abelian von Neumann algebras. But, as seen from Theorem 1.68, the proper operator-algebraic counterpart of the notion of *measure algebra* is that of *von Neumann measure algebra* (see Definition 1.15). Hence, our aim will be to classify up to isomorphism all von Neumann measure algebras $(\mathcal{M}, \varphi)$ for *abelian* $\mathcal{M}$.

A subset G of a Boolean algebra (respectively, complete Boolean algebra) $\mathfrak{B}$ *generates* (respectively, *completely generates*) $\mathfrak{B}$ if $\mathfrak{B}$ is the smallest Boolean subalgebra (respectively, complete Boolean subalgebra) of $\mathfrak{B}$ containing G. We say that a complete Boolean algebra $\mathfrak{B}$ is of *Maharam type* κ if κ is the smallest cardinal of any subset of G completely generating $\mathfrak{B}$. We denote the cardinal by $\kappa(\mathfrak{B})$. A complete Boolean algebra $\mathfrak{B}$ is *Maharam-type-homogeneous* if $\kappa(\mathfrak{B}_a) = \kappa(\mathfrak{B})$ for any $a \in \mathfrak{B}$ (with $\mathfrak{B}_a$ the principal ideal generated by a).

We shall need the following simple lemma.

Lemma 1.22 *If $\mathfrak{B}$ is a Boolean algebra completely generated by a finite set of generators, then it is finite. If, on the other hand, $\mathfrak{B}$ is a complete Boolean algebra of infinite Maharam type, and G completely generates $\mathfrak{B}$, then the Boolean subalgebra $\mathfrak{B}_G$ of $\mathfrak{B}$ generated by G completely generates $\mathfrak{B}$ and $\#\mathfrak{B}_G = \#G$.*

Proof. First note that a finite set $G \subseteq \mathfrak{B}$ generates a finite Boolean algebra (which is then also Dedekind complete). It is an exercise to check that if C_0 is a Boolean subalgebra of $\mathfrak{B}$ and $c \in \mathfrak{B}$, then $C := \{(a \cap c) \cup (b \setminus c) : a, b \in C_0\}$ is a Boolean subalgebra of $\mathfrak{B}$ generated by C_0 and c (see [**Fre04**, 312N]). Starting from $D_0 := \{0, 1\}$ and adding elements one by one, we can show that the Boolean subalgebra generated by a finite G is also finite.

If G is arbitrary, let $\mathfrak{B}_G$ be the Boolean subalgebra of $\mathfrak{B}$ generated by G. For all finite subsets $F \subseteq G$, let $\mathfrak{B}_F$ be the finite Boolean subalgebra of $\mathfrak{B}$ generated by F. For finite $F_1, F_2 \subseteq G$, we have $\mathfrak{B}_{F_1} \cup \mathfrak{B}_{F_2} \subseteq \mathfrak{B}_{F_1 \cup F_2}$, which implies that $\mathfrak{B}_G = \bigcup \{\mathfrak{B}_F : F \subseteq G \text{ finite}\}$. If G is of infinite cardinality κ, then the number of its finite subsets is also κ, and the cardinality of $\mathfrak{B}_G$, being a union of κ finite sets, is at most κ. Since $G \subseteq \mathfrak{B}_G$, we have $\#\mathfrak{B}_G = \kappa$. If G completely generates

$\mathfrak{B}$, then $G \subseteq \mathfrak{B}_G \subseteq \mathfrak{B}$ implies that $\mathfrak{B}_G$ completely generates $\mathfrak{B}$, which ends the proof (cf. the proof of 331G in [**Fre04**]). $\square$

The following is the most difficult step in the proof of the Maharam theorem, and the only one which we shall not prove (see [**Fre04**, 331I]). The proof relies on a rather tricky application of transfinite induction.

Theorem 1.23   *Let $(\mathfrak{A}_1, \mu_1)$ and $(\mathfrak{A}_2, \mu_2)$ be Maharam-type-homogeneous measure algebras of the same Maharam type and with $\mu_1(1) = \mu_2(1) < \infty$. Then, they are isomorphic as measure algebras.*

We call von Neumann algebras without minimal projections *non-atomic*, and those isomorphic to a direct sum of some (cardinal) number of copies of $\mathbb{C}$ *purely atomic*. We will write the latter direct sum as $\mathbb{C}^\kappa$, where κ is the cardinal (usually this is written as $\ell^\infty(\kappa)$). We start the effort of classifying abelian von Neumann measure algebras $(\mathcal{A}, \tau)$ by reducing it to the case of non-atomic abelian von Neumann probability spaces. This step is not strictly necessary, and we could jump directly to κ-homogeneous von Neumann probability spaces. Nevertheless, it makes the proof slightly easier to follow and seems a natural step for an operator algebraist.

Lemma 1.24   *An abelian von Neumann algebra $\mathcal{A}$ is a direct sum of a non-atomic algebra and a purely atomic algebra.*

Proof.   If $p \in \mathbb{P}(\mathcal{A})$ is minimal, then $\mathcal{A}p \simeq \mathbb{C}$. Indeed, the only projections in $\mathcal{A}p$ are 0 and p, and a von Neumann algebra is generated by its projections (see Theorem 0.79 and Proposition 0.94(7)), so that $\mathcal{A}p = \mathbb{C}p$. Let $\{p_i\}_{i \in I}$ be the family of all minimal projections in $\mathcal{A}$. These are mutually orthogonal (since $p_i p_j$ is a projection). Now, let $p = \sum_{i \in I} p_i$. It is then an exercise to see that $\mathcal{A}p \equiv \oplus_{i \in I} \mathcal{A}p_i$. Since, as noted above, each $\mathcal{A}p_i$ is a copy of $\mathbb{C}$, it clearly follows that $\mathcal{A}p \simeq \mathbb{C}^{\#(I)}$, with $\mathcal{A}p^\perp$ by construction containing no minimal projections. $\square$

Proposition 1.25   *An abelian non-atomic von Neumann measure algebra $(\mathcal{A}, \tau_{\mathcal{A}})$ with $\tau_{\mathcal{A}}(\mathbb{1}_{\mathcal{A}}) = \infty$ is isomorphic to a direct sum of (abelian non-atomic) von Neumann probability algebras $(\mathcal{A}_i, \tau_i)$.*

Proof.   Since $\mathcal{A}$ is abelian, any orthogonal family (p_i) of projections in $\mathcal{A}$ summing to $\mathbb{1}_{\mathcal{A}}$ yields a decomposition of $\mathcal{A}$ into a direct sum

$\oplus_i \mathcal{A}_i$ of von Neumann subalgebras, where each $\mathcal{A}_i$ corresponds to the reduced algebra $\mathcal{A}p_i$. We shall find such a family (p_i) with $\tau_{\mathcal{A}}(p_i) = 1$ for every i. Since $\tau = \tau_{\mathcal{A}}$ is semifinite, there is a non-zero projection p in $\mathcal{A}$ such that $\tau(p) < \infty$. Since p is not minimal, let us write it as $q + r$ with $q, r \in \mathbb{P}(\mathcal{A}) \setminus \{0\}$, and denote by p_1 one of q and r chosen so that $\tau(p_1) \leq 2^{-1}\tau(p)$. Further divisions lead to $p_k \in \mathbb{P}(\mathcal{A})$ with $\tau(p_k) \leq 2^{-k}\tau(p)$. Hence, there are non-zero projections in $\mathcal{A}$ with arbitrarily small trace. Consequently, the set $\{e \in \mathbb{P}(\mathcal{A}) : \tau(e) \leq 1\}$ is non-empty, and its maximal element $f \in \mathbb{P}(\mathcal{A})$ satisfies $\tau(f) = 1$. Indeed, by semifiniteness, $1 - f$ dominates a τ-finite projection, and hence, by the above argument, a non-zero projection q in $\mathcal{A}$ with trace $< 1 - \tau(f)$ whenever the latter is non-zero. Thus, $\tau(f + q) < 1$, contradicting the maximality of f.

A small modification of the last argument shows that for any real $t > 0$ and projection p in $\mathcal{A}$ with $\tau(p) \geq t$, there exists a projection $f \leq p$ with $\tau(f) = t$.

Let F be a maximal orthogonal family of non-zero projections of trace 1 in $\mathcal{M}$, necessarily infinite. Let $g = \mathbb{1}_{\mathcal{A}} - \sum_{f \in F} f$. Obviously, $\tau(g) < 1$ by the maximality. Choose a countable subfamily $H = \{h_k\}$ from F. By the last paragraph, we may for all k find a projection $g_k \leq h_k$ such that $\tau(g_k) = \tau(g)$. Now

$$G = \{g + (h_1 - g_1), g_1 + (h_2 - g_2), g_2 + (h_3 - g_3), \dots\}$$

is an orthogonal family of projections of trace 1, and its members are orthogonal to all projections from $F \setminus H$. Moreover, their sum equals $\mathbb{1}_{\mathcal{A}} - \sum_{f \in F \setminus H} f$. Hence, the orthogonal family $J := G \cup (F \setminus H)$ consists of projections of trace 1 summing to $\mathbb{1}_{\mathcal{A}}$. Put $\tau_p := \tau|\mathcal{A}p$ for all $p \in J$. The direct sum of the von Neumann probability spaces $(\mathcal{A}_p, \tau_p)_{p \in J}$ is exactly $(\mathcal{A}, \tau)$. $\qquad\square$

There are several notions which we will need in order to formulate the theorem on isomorphism of noncommutative measure algebras.

Definition 1.26 The *decomposability number* $\mathfrak{d}(\mathcal{M})$ of a von Neumann algebra $\mathcal{M}$ is the maximal cardinality of an orthogonal family of non-zero projections in $\mathcal{M}$. The *decomposability number* $\mathfrak{d}_{\mathcal{M}}(p)$ of a projection p in $\mathcal{M}$ is the decomposability number $\mathfrak{d}(\mathcal{M}_p)$ of the reduced von Neumann algebra $\mathcal{M}_p$. A von Neumann algebra $\mathcal{M}$ (respectively, $p \in \mathbb{P}(\mathcal{M})$) is *countably decomposable* or *σ-finite* if

$\mathfrak{d}(\mathcal{M}) \leq \aleph_0$ (respectively, $\mathfrak{d}_{\mathcal{M}}(p) \leq \aleph_0$). If $(\mathcal{M}, \tau_{\mathcal{M}})$ is a von Neumann measure algebra, with $\tau_{\mathcal{M}}$ a faithful normal semifinite trace on $\mathcal{M}$, then for any $p \in \mathbb{P}(\mathcal{M})$, define the *magnitude* of p by

$$\mathfrak{m}_{\mathcal{M}}(p) = \begin{cases} \tau_{\mathcal{M}}(p), & \text{if } \tau_{\mathcal{M}}(p) < \infty, \\ \mathfrak{d}_M(p), & \text{otherwise.} \end{cases}$$

Note that although the terms *decomposability number* and *magnitude* are defined for von Neumann algebras, in the case that the algebra $\mathcal{A}$ considered is abelian, they have the same meaning (as defined in [**Fre04**]) in the Boolean algebra $\mathbb{P}(\mathcal{A})$.

It is well known that $\mathcal{M}$ is σ-finite if and only if it admits a faithful normal state. The one direction of this is obvious (by applying the state to the sum of the projections). The other direction follows from a routine Zorn's lemma argument of the type we have seen before, on sets of mutually orthogonal projections that are the supports of normal states.

Lemma 1.27 *Let κ be an infinite cardinal. If a von Neumann algebra $\mathcal{M}$ can be written as a direct sum of κ σ-finite algebras, then $\mathfrak{d}(\mathcal{M}) = \kappa$.*

Proof. By assumption, there is an orthogonal family $\{z_j\}_{j<\kappa}$ of σ-finite central projections in $\mathcal{M}$ satisfying $\sum_{j<\kappa} z_j = \mathbb{1}$. By definition, $\kappa \leq \mathfrak{d}(\mathcal{M})$. Let $P := \{p_i\}_{i \in I}$ be any orthogonal family of projections in $\mathcal{M}$. Then, $Q := \{p_i z_j : p_i z_j \neq 0\}$ forms an orthogonal family. For each $j < \kappa$, we have $1 \leq \#\{i \in I : p_i z_j \neq 0\} \leq \aleph_0$ by the σ-finiteness of the z_js. Consequently, $\#P \leq \#Q \leq \kappa \cdot \aleph_0 = \kappa$, so that $\mathfrak{d}(\mathcal{M}) \leq \kappa$. $\square$

Definition 1.28 Let $\mathcal{A}$ be an abelian von Neumann algebra. We denote by $\mathfrak{e}(\mathcal{A})$ the smallest cardinality of a set P of projections in $\mathcal{A}$ such that P *generates* $\mathcal{A}$ as a von Neumann algebra. Since any von Neumann subalgebra contains $\mathbb{1}$, we may as well say that P and $\mathbb{1}$ generate $\mathcal{A}$ as a von Neumann algebra. For this reason, if $\mathcal{A}$ is one-dimensional, then we take $\mathfrak{e}(\mathcal{A}) = 0$.

When we speak about generators, we will always mean projections. There is a good reason for this—by Theorem 1.29, a set of projections P generates an abelian von Neumann algebra $\mathcal{A}$ if and only if it *completely generates* the Boolean algebra $\mathbb{P}(\mathcal{A})$. That is, the smallest order-closed (hence Dedekind complete) Boolean subalgebra of $\mathbb{P}(\mathcal{A})$ containing P equals $\mathbb{P}(\mathcal{A})$. This coincides with Fremlin's notion of *complete* or τ-*generation* of Boolean algebras (see also [**Fre04**, 331E]). Additionally, we will avoid problems connected with the generation of a von Neumann algebra by a finite number of operators (see [**She12**]).

The general (not necessarily σ-finite) case of much of the following theorem is contained in Bade's result ([**Bad54**, Theorem 3.4], see also [**DS88**, Lemma XVII.3.6]). We decided to give it an independent proof that we have not seen in the literature, avoiding Bade's proof and helping keep this section independent of the earlier material.

Theorem 1.29 *Let $\mathcal{A}$ be an abelian non-atomic von Neumann algebra acting (non-degenerately) in a Hilbert space H and let $P \subseteq \mathbb{P}(\mathcal{A})$. Then:*

(a) *If $\mathcal{A}$ is σ-finite and $P \subseteq \mathbb{P}(\mathcal{A})$ generates $\mathcal{A}$ as a von Neumann algebra, then for any $p \in \mathbb{P}(\mathcal{A})$, there is countable set $Q \subseteq P$ such that p is in the complete Boolean subalgebra generated by Q.*

(b) *The Boolean subalgebra of $\mathbb{P}(\mathcal{A})$ generated by P is strongly dense in $\mathbb{P}(\mathcal{A})$.*

(c) *We have $\mathfrak{k}(\mathcal{A}) = \kappa(\mathbb{P}(\mathcal{A}))$.*

Proof. Let $\mathcal{A}$ be σ-finite and assume that P is a subset of $\mathbb{P}(\mathcal{A})$ generating $\mathcal{A}$ as a von Neumann algebra. The first half of Lemma 1.22 implies that P is infinite. Otherwise, $\mathbb{P}(\mathcal{A})$ would be finite, which is impossible for a non-atomic $\mathcal{A}$. By the second half of the same lemma, we may assume that P is a Boolean algebra.

Let A be the set of all linear combinations of elements of P. This is a *-algebra generating $\mathcal{A}$, and hence, by von Neumann's double commutant theorem, A is strongly dense in $\mathcal{A}$ (see Theorem 0.92). Take an arbitrary $p \in \mathbb{P}(\mathcal{A})$. We will show that p is in the smallest complete Boolean algebra $\mathfrak{B}$ containing P. This implies that $\mathfrak{B} = \mathbb{P}(\mathcal{A})$ and $\mathfrak{k}(\mathcal{A}) \geq \kappa(\mathbb{P}(\mathcal{A}))$.

By the Kaplansky density theorem (Theorem 0.93), the positive part of the unit ball of A is strongly dense in the positive part of the unit ball of $\mathcal{A}$. There is a directed set D and a net $(x_d)_{d \in D}$ in $\mathcal{A}_+$ with $\|x_d\| \leq 1$ such that $x_d \to p$ strongly. We can assume that for all d,

$x_d = \sum_{i=1}^{m_d} \gamma_i^{(d)} p_i^{(d)}$, where $m_d \in \mathbb{N}$, $0 \le \gamma_i^{(d)} \le 1$ for $i = 1, \ldots, m_d$, with orthogonal families $\{p_i^{(d)}\}_{i=1}^{m_d} \subset P$.

Choose a vector $\xi \in H, \|\xi\| = 1$ that is separating for $\mathcal{A}$ (see, for example, Theorem 0.100). Let, for each $n \in \mathbb{N}$, $d_n \in D$ be such that $\|x_d \xi - p\xi\| \le \frac{1}{n2^n}$ for all $d \ge d_n$. We can also assume that the sequence d_n is increasing. Put $q_n := \chi_{[1/n,1]}(x_{d_n})$. Note that $q_n = \sum \{p_i^{(d_n)} : \gamma_i^{(d_n)} \ge 1/n\} \in P$ for all n. Let $r_j := \sup_{n \ge j} q_n$ for all $j \in \mathbb{N}$. Then, the family $\{r_j\}$ is decreasing and $p \le r_j$ for all j. In fact, we would for some $j_0 \in \mathbb{N}$ otherwise have $pr_{j_0}^{\perp} \xi \ne 0$, and for all $j \ge j_0$,

$$\|x_{d_j}(pr_{j_0}^{\perp})\xi\| = \|x_{d_j} r_j^{\perp}(pr_{j_0}^{\perp})\xi\| \le \|x_{d_j} q_j^{\perp}(pr_{j_0}^{\perp})\xi\| \le \frac{1}{j}\|q_j^{\perp} pr_{j_0}^{\perp}\xi\| \le \frac{1}{j}.$$

On the other hand,

$$x_{d_j}(pr_{j_0}^{\perp})\xi = (pr_{j_0}^{\perp})x_{d_j}\xi \to (pr_{j_0}^{\perp})p\xi \ne 0$$

(as ξ is separating for $\mathcal{A}$), which yields a contradiction.

Let $r := \inf_{j \in \mathbb{N}} r_j$. Then $p \le r$. For all j, we have

$$\|(r-p)\xi\|^2 \le \|r_j(r-p)\xi\|^2 \le \sum_{n \ge j} \|q_n(r-p)\xi\|^2$$

$$\le \sum_{n \ge j} n^2 \|x_{d_n} q_n(r-p)\xi\|^2 \le \sum_{n \ge j} n^2 \|x_{d_n}(r-p)\xi\|^2$$

$$\le \sum_{n \ge j} n^2 \|x_{d_n}(r-p)\xi - p(r-p)\xi\|^2$$

$$= \sum_{n \ge j} n^2 \|(r-p)(x_{d_n}-p)\xi\|^2$$

$$\le \sum_{n \ge j} \frac{n^2}{n^2 4^n} = \frac{1}{3 \cdot 4^{j-1}}.$$

Since j was arbitrary, we get $(r-p)\xi = 0$, and, as ξ was separating for $\mathcal{A}$, we have $r = p$. Thus, p is in the smallest complete Boolean algebra $\mathfrak{B}_Q$ containing the countable set $Q := \{q_n\}_{n \in \mathbb{N}} \subseteq P$. This ends the proof of (a). Note that p is moreover a strong limit of the decreasing sequence r_j, and r_j is a strong limit of the increasing sequence $\sup_{j \le n \le j+m} q_n$ as $m \to \infty$. This shows the density of the Boolean algebra generated by P in $\mathbb{P}(\mathcal{A})$ and also gives (b).

Item (c) easily follows from (a). $\qquad\square$

Definition 1.30 Let κ be a cardinal. We say that an abelian von Neumann algebra $\mathcal{A}$ is κ-*homogeneous* or *homogeneous of type κ* if for any non-zero projection p in $\mathcal{A}$, we have $\kappa = \mathfrak{k}(\mathcal{A}) = \mathfrak{k}(\mathcal{A}_p)$. We say that a non-zero projection p from $\mathcal{A}$ is κ-homogeneous if $\mathcal{A}p$ is κ-homogeneous. An abelian von Neumann algebra or a projection in such an algebra is called homogeneous if it is κ-homogeneous for some cardinal κ.

Note that since $\mathcal{A}$ is abelian, we can simply use $\mathcal{A}p$ instead of $\mathcal{A}_p$, treating it as a von Neumann algebra with unit p.

The following facts will be useful.

Lemma 1.31 *If a von Neumann algebra $\mathcal{A}$ has no minimal projections, then $\mathfrak{k}(\mathcal{A}) \geq \aleph_0$. Indeed, if $\mathcal{A}$ is homogeneous, then $\mathfrak{k}(\mathcal{A})$ is either 0 (and $\mathcal{A} = \mathbb{C}\mathbb{1}$) or an infinite cardinal.*

Proof. If $\mathfrak{k}(\mathcal{A}) < \aleph_0$, then by Lemma 1.22 and Theorem 1.29(c), $\mathbb{P}(\mathcal{A})$ is finite. Hence, $\mathcal{A}$ has a minimal projection. If $\mathcal{A}$ is homogeneous, then the minimal projection must be $\mathbb{1}$ and $\mathcal{A} = \mathbb{C}\mathbb{1}$. $\qquad\square$

Lemma 1.32 *For any non-zero projection $p \in \mathcal{A}$, there is a non-zero projection $q \leq p$ such that $\mathcal{A}q$ is homogeneous.*

Proof. Let $K = \{\mathfrak{k}(\mathcal{A}r) \colon r \in \mathbb{P}(\mathcal{A}), 0 \neq r \leq p\}$, and let κ be the least member of K. Let $q \in \mathbb{P}(\mathcal{A})$ be such that $\mathfrak{k}(\mathcal{A}q) = \kappa$. If $r \in \mathbb{P}(\mathcal{A}), 0 \neq r \leq q$, then $\mathfrak{k}(\mathcal{A}r) \leq \mathfrak{k}(\mathcal{A}q)$. In fact, if $G \subseteq \mathbb{P}(\mathcal{A}q)$ generates $\mathbb{P}(\mathcal{A}q)$, then the set $\{gr \colon g \in G\}$ generates $\mathcal{A}r$. Hence, $\mathfrak{k}(\mathcal{A}r) = \mathfrak{k}(\mathcal{A}q)$, and $\mathcal{A}q$ is homogeneous. $\qquad\square$

Lemma 1.33 *A (σ-finite, non-atomic) abelian von Neumann algebra is a direct sum of (σ-finite, non-atomic) homogeneous von Neumann algebras.*

Proof. Take a maximal orthogonal family $\{q_i\}$ of projections from $\mathcal{A}$ such that $\mathcal{A}q_i$ is homogeneous. By the previous lemma, $\sum_i q_i = \mathbb{1}$. $\qquad\square$

Definition 1.34 The *homogeneous component of type κ* of an abelian von Neumann algebra $\mathcal{A}$ is the ideal $\mathcal{A}\,e_\kappa$, where e_κ is the supremum of all non-zero projections $e \in \mathcal{A}$ such that $\mathcal{A}e$ is homogeneous of type κ.

Often we will also call e_κ the homogeneous component of type κ. We disregard the trivial components of $\mathcal{A}$, that is, those κ for which there are no non-zero κ-homogeneous projections in $\mathcal{A}$.

Lemma 1.35 *For a cardinal κ, let e_κ denote the homogeneous component of $\mathcal{A}$ of type κ. If $\kappa \neq \kappa'$, then $e_\kappa \perp e_{\kappa'}$. Moreover, $\sum_\kappa e_\kappa = \mathbb{1}$, when the sum is over cardinals.*

Proof. In fact,

$$e_\kappa e_{\kappa'} = \sup\{ee' : e \text{ (respectively, } e') \text{ is homogeneous of type } \kappa$$
$$(\text{respectively, } \kappa')\} = 0.$$

Indeed, if e is homogeneous of type κ and e' is homogeneous of type κ', then $e \perp e'$; otherwise, ee' would be both κ- and κ'-homogeneous, which is impossible. Also taking into account Lemma 1.33, we hence have $\sum_\kappa e_\kappa = \mathbb{1}$. $\qquad\square$

Note that e_κ can, by Lemma 1.31, only be non-zero if κ is infinite or zero. Moreover, e_κ is not necessarily κ-homogeneous (for example, e_0 is not 0-homogeneous if $\mathcal{A}$ has at least two atoms). Nevertheless, we have the following lemma.

Lemma 1.36 *If the homogeneous component of type κ of an abelian σ-finite von Neumann algebra $\mathcal{A}$ is $\mathcal{A}$ itself, then $\mathcal{A}$ is κ-homogeneous.*

Proof. Denote by P the set of κ-homogeneous projections in $\mathcal{A}$. By definition, $\sup P = \mathbb{1}$. Note that if $p \in P$ and $0 \neq q \leq p$, then $q \in P$. By Zorn's lemma, there exists a maximal orthogonal family of κ-homogeneous projections $Q \subset P$. If $e_0 := \sup Q \neq \mathbb{1}$, then $q_0 := pe_0^\perp \neq 0$ for some $p \in P$. Thus, $q_0 \in P$ and q_0 is orthogonal to all projections from Q. This contradicts the maximality of Q. Hence $e_0 = \mathbb{1}$. For each $q \in Q$, let G_q denote a set of cardinality κ generating $\mathcal{A}q$. Since Q is at most countable, and $\bigcup_{q \in Q} G_q$ generates $\mathcal{A}$ as a von Neumann algebra, we have $\mathfrak{k}(\mathcal{A}) \leq \kappa$. If $0 \neq r \in \mathbb{P}(\mathcal{A})$, then $rp \neq 0$ for some $p \in P$; hence, $\kappa = \mathfrak{k}(\mathcal{A}(rp)) \leq \mathfrak{k}(\mathcal{A}r) \leq \mathfrak{k}(A) \leq \kappa$, and $\mathcal{A}$ is indeed κ-homogeneous. $\qquad\square$

The most important part of Maharam's theorem is contained in the following result (formulated here in the language of von Neumann algebras).

Theorem 1.37 *Let $(\mathcal{A}, \tau_{\mathcal{A}})$ and $(\mathcal{B}, \tau_{\mathcal{B}})$ be such that $\tau_{\mathcal{A}}(\mathbb{1}) = \tau_{\mathcal{B}}(\mathbb{1})$ and, for some infinite cardinal κ, both $\mathcal{A}$ and $\mathcal{B}$ are κ-homogeneous. Then $(\mathcal{A}, \tau_{\mathcal{A}})$ and $(\mathcal{B}, \tau_{\mathcal{B}})$ are isomorphic as von Neumann measure algebras.*

Proof. This follows directly from Proposition 1.16 (see also Theorem 1.68) and Theorem 1.23. $\qquad\square$

The measure algebra version of the following result is due to Fremlin (see [**Fre04**, 332J]).

Theorem 1.38 *Let e_κ and f_κ denote the homogeneous components of type κ of algebras $\mathcal{A}$ and $\mathcal{B}$, respectively. Von Neumann measure algebras $(\mathcal{A}, \tau_{\mathcal{A}})$ and $(\mathcal{B}, \tau_{\mathcal{B}})$ are isomorphic if and only if, for every infinite cardinal κ, $\mathfrak{m}_{\mathcal{A}}(e_\kappa) = \mathfrak{m}_{\mathcal{B}}(f_\kappa)$ and, for each $0 < \gamma < \infty$,*

$$\#\{minimal\ p \in \mathbb{P}(\mathcal{A}) \colon \tau_{\mathcal{A}}(p) = \gamma\} = \#\{minimal\ q \in \mathbb{P}(\mathcal{B}) \colon \tau_{\mathcal{B}}(q) = \gamma\}.$$

Proof. ($\Rightarrow$) Let Φ denote the isomorphism. It is clear that $\Phi(e_\kappa) = f_\kappa$, which implies that $\mathfrak{d}_{\mathcal{A}}(e_\kappa) = \mathfrak{d}_{\mathcal{B}}(f_\kappa)$. Since $\tau_A(p) = \tau_B(\Phi(p))$ for all $p \in \mathbb{P}(\mathcal{A})$, we have $\tau_A(e_\kappa) = \tau_B(f_\kappa)$ for all κ. Hence $\mathfrak{m}_{\mathcal{A}}(e_\kappa) = \mathfrak{m}_{\mathcal{B}}(f_\kappa)$. The isomorphism Φ takes minimal projections in $\mathbb{P}(\mathcal{A})$ (atoms of $\mathbb{P}(\mathcal{A})$) into minimal projections in $\mathbb{P}(\mathcal{B})$ (atoms of $\mathbb{P}(\mathcal{B})$), and if $\tau_A(p) = \gamma$, then $\tau_B(\Phi(p)) = \gamma$, which shows the second statement.

($\Leftarrow$) Assume now that for all infinite cardinals κ, $\mathfrak{m}_{\mathcal{A}}(e_\kappa) = \mathfrak{m}_{\mathcal{B}}(f_\kappa)$ and, for each $0 < \gamma < \infty$, $\#\{$minimal $p \in \mathbb{P}(\mathcal{A}) \colon \tau_{\mathcal{A}}(p) = \gamma\} = \#\{$minimal $q \in \mathbb{P}(\mathcal{B}) \colon \tau_{\mathcal{B}}(q) = \gamma\}$. It is clear that the assumptions guarantee that the noncommutative measure algebras $\mathcal{A}$ and $\mathcal{B}$ have the same number λ of minimal projections. Hence, the purely atomic parts of $\mathcal{A}$ and $\mathcal{B}$, equipped with the restrictions of the trace, are isomorphic as von Neumann measure algebras.

If, for some infinite κ, $\mathfrak{m}_{\mathcal{A}}(e_\kappa) = \mathfrak{m}_{\mathcal{B}}(f_\kappa) < \infty$, then the von Neumann algebras $\mathcal{A}e_\kappa$ and $\mathcal{B}f_\kappa$ are σ-finite and constitute their own κ-homogeneous components. By Lemma 1.36, they are both κ-homogeneous as von Neumann algebras; hence, $(\mathcal{A}e_\kappa, \tau_{\mathcal{A}}|\mathcal{A}e_\kappa)$ and $(\mathcal{B}f_\kappa, \tau_{\mathcal{B}}|\mathcal{B}f_\kappa)$ are isomorphic as von Neumann measure algebras by Theorem 1.37.

On the other hand, if $\mathfrak{m}_{\mathcal{A}}(e_\kappa) = \mathfrak{m}_{\mathcal{B}}(f_\kappa) = \lambda$ for some infinite cardinal λ, then by Propositions 1.25 and 1.27, $\mathcal{A}e_\kappa$ and $\mathcal{B}f_\kappa$ are direct sums of

$\lambda = \mathfrak{d}(\mathcal{A}e_\kappa) = \mathfrak{d}(\mathcal{B}f_\kappa)$ noncommutative measure algebras $(\mathcal{A}p_i, \tau_{\mathcal{A}}|\mathcal{A}p_i)$ and $(\mathcal{B}q_i, \tau_{\mathcal{B}}|\mathcal{B}q_i)$, respectively, with $\tau_{\mathcal{A}}(p_i) = \tau_{\mathcal{B}}(q_i) = 1$ for all i. Note that $\mathcal{A}p_i$ and $\mathcal{B}q_i$ are κ-homogeneous components of themselves. Indeed, if, say, p_i is not a κ-component of itself, then by Lemma 1.32, there would be a κ'-homogeneous projection $p \leq p_i$ with $\kappa' \neq \kappa$. But $p \leq e_\kappa$, so Lemma 1.35 yields a contradiction. Again, by Lemma 1.36, $\mathcal{A}p_i$ and $\mathcal{B}q_i$ are κ-homogeneous. By Theorem 1.37, the von Neumann measure algebras $(\mathcal{A}p_i, \tau_{\mathcal{A}}|\mathcal{A}p_i)$ and $(\mathcal{B}q_i, \tau_{\mathcal{B}}|\mathcal{B}q_i)$ are isomorphic. The direct sums of all the isomorphisms produce an isomorphism of $(\mathcal{A}, \tau_{\mathcal{A}})$ and $(\mathcal{B}, \tau_{\mathcal{B}})$. $\qquad\square$

1.3.1 Infinite tensor products

To properly demonstrate the von Neumann algebra version of the probabilistic case of Maharam's theorem, we will use various notions of infinite tensor products: of Hilbert spaces, of C^*-algebras, and of von Neumann algebras. To avoid misunderstanding, we take some time to carefully describe the tensor products. For further background or definitions if needed, we direct the reader to, for example, [**Bla06**, III.3.1] and [**KR83**, Chapter 11]. In the definitions, the index set is arbitrary. For our purposes, it is most convenient to have an infinite cardinal κ as the set of indices. We freely use the notation introduced here in the proofs that follow. We also use the notation '$\Subset$' to denote a finite subset.

We start with a family $\{(H_i, \xi_i)\}_{i \in K}$ of Hilbert spaces H_i with a distinguished unit vector ξ_i. For $F \Subset K$, we put $H_F = \bigotimes_{i \in F} H_i$ and $\xi_F = \bigotimes_{i \in F} \xi_i$. For $F, G \Subset K$ with $F \subseteq G$, we have a unique isometric linear mapping $h_{GF} : H_F \to H_G$ such that $\bigotimes_{i \in F} \eta_i \mapsto \bigotimes_{i \in G} \zeta_i$, where $\zeta_i = \eta_i$ for $i \in F$ and $\zeta_i = \xi_i$ for $i \in G \setminus F$. Then, $(\{H_F\}_{F \Subset K}, \{h_{GF}\}_{F \subseteq G \Subset K})$ forms an inductive system of Hilbert spaces. Note that $\bigcup h_F(H_F)$ is dense in the Hilbert space inductive limit H_K, and there exists a unique unit vector $\xi_K \in H_K$ such that $h_F \xi_F = \xi_K$ for all $F \Subset K$. We say that the pair (H_K, ξ_K) is the infinite tensor product of $\{(H_i, \xi_i)\}_{i \in K}$.

We now define infinite tensor products of C^*-algebras [**KR83**, §11.4]. Consider a family $\{A_i\}_{i \in K}$ of unital C^*-algebras. For $F \Subset K$, we put $A_F = \bigotimes_{i \in F} A_i$, where the tensor product used is the minimal one. For $F, G \Subset K$ with $F \subseteq G$, we have a unique

$*$-isomorphism $\theta_{GF} : A_F \to A_G$ such that $\theta_{GF}(a_F) = a_F \otimes \mathbb{1}_{G \backslash F}$. Then, $(\{A_F\}_{F \Subset K}, \{\theta_{GF}\}_{F \subseteq G \Subset K})$ forms an inductive system of C^*-algebras. We denote by $(A_K, \{\theta_F\}_{F \Subset K})$ the inductive limit of the system. Note that $\bigcup_{F \Subset K} \theta_F(A_F)$ is uniformly dense in A_K. We say that the pair A_K is the infinite tensor product of $\{A_i\}_{i \in K}$ and write $A_K = \bigotimes_{i \in K} A_i$.

Finally, we will consider infinite tensor products of von Neumann algebras, or, more precisely, of von Neumann probability algebras. We start with a family $(\mathcal{M}_i, \varphi_i)_{i \in K}$, where $\mathcal{M}_i$ are (necessarily σ-finite) von Neumann algebras and φ_i are faithful normal states on $\mathcal{M}_i$. First, we treat $\mathcal{M}_i$ as unital C^* algebras and define their C^*-algebraic infinite tensor product M_K, as we did above. There is a unique state φ_K on M_K such that $\varphi_F(\bigotimes_{i \in F} a_i) = \prod_{i \in F} \varphi_i(a_i)$ for any $F \Subset K$ and $a_i \in \mathcal{M}_i$ for $i \in F$ (see [**KR83**, Proposition 11.4.6]). We call the state φ_K the *product state* and denote it by $\bigotimes_{i \in K} \varphi_i$. Having the algebra M_K and the state φ_K on it, we can build the GNS representation $(\pi_{\varphi_K}, H_{\varphi_K}, \xi_{\varphi_K})$ of M_K and define $\mathcal{M}_K := \pi_{\varphi_K}(M_K)''$ acting on H_{φ_K} with a cyclic and separating vector ξ_{φ_K}. Let $\overline{\varphi}_K$ be the faithful normal state on $\mathcal{M}_K$ given by the vector state ξ_{φ_K}. Then, we call the pair $(\mathcal{M}_K, \overline{\varphi}_K)$ the *infinite tensor product of von Neumann probability algebras* $(\mathcal{M}_i, \varphi_i)$ and denote it by $\overline{\bigotimes}_{i \in K}(\mathcal{M}_i, \varphi_i)$.

Now assume that for each $\mathcal{M}_i$ acting on a Hilbert space H_i, there exists a cyclic and separating vector ξ_i (see Theorem 0.100). For each $i \in K$, let φ_i be the vector state corresponding to ξ_i. Let π_i denote the (identity) representation of $\mathcal{M}_i$ on H_i. For all $F \Subset K$, we form the tensor product $\pi_F := \bigotimes_{i \in F} \pi_i$ in an obvious way. We easily check (see [**KR83**, 11.5.30]) that there is a unique representation π_K (denoted by $\bigotimes_{i \in K} \pi_i$) of M_K on H_K such that $\pi_K(\theta_F(a_F))h_F = h_F \pi_F(a_F)$ for all $a_F \in A_F$, for all $F \Subset K$. Moreover, the vector ξ_K is cyclic for π_K, and the vector state ω_{ξ_K} is, in fact, the product state $\bigotimes_{i \in K} \omega_{\xi_i}$. Hence, $\varphi_K = \omega_{\xi_K}$ and, by Proposition 0.72, π_K is unitarily equivalent to the GNS representation π_{φ_K}. That means that $\mathcal{M}_K$ is generated by $\pi_K(M_K)$ and that $\overline{\varphi}_K$ is the (extension of the) vector state ξ_K. The vector ξ_K is not only cyclic, but also separating. This follows, for example, from the fact that the vectors ξ_i are all separating, hence cyclic for the algebras $\mathcal{M}_i'$. Hence, $\overline{\bigotimes}_{i \in K}(\mathcal{M}_i', \omega_{\xi_i})$, when represented in H_K, has a cyclic vector ξ_K. But $\overline{\bigotimes}_{i \in K}(\mathcal{M}'_i, \omega_{\xi_i}) = \mathcal{M}'_K$. (This is just an extension of Theorem 0.109 in accordance with the above ideas. See [**Bla06**, III.3.1.2] for details.) Hence, ξ_K is separating for $\mathcal{M}_K$. This means that the representation π_K is in fact faithful, so that we can identify M_K as a subset of $\mathcal{M}_K$.

We use the definitions above in the following special case. In the sequel, we will use the space $\mathbb{C}^2$ in three different guises: as a Hilbert space, as a C^*-algebra, and as a von Neumann algebra. We fix an infinite cardinal κ, take $K := \kappa$, and then, for all $i < \kappa$, put $H_i := H$ and $\xi_i := \xi$, where $H = \mathbb{C}^2$ and $\xi = (\frac{1}{\sqrt{2}}, \frac{1}{\sqrt{2}})$. We write $(H_\kappa, \xi_\kappa) = (H, \xi)^{\otimes \kappa}$. Next, $A_i := A$ for all $i < \kappa$, where $A = \mathbb{C}^2$. We denote by A_κ the C^*-algebraic tensor product of κ copies of A. Finally, $\mathcal{A} := \mathbb{C}^2$, τ is the normalised trace on $\mathcal{A}$, and $\mathcal{M}_i := A$, $\tau_i := \tau$ for all $i < \kappa$, with $(\mathcal{A}_\kappa, \tau_\kappa)$ being the infinite tensor product of κ copies of the von Neumann measure algebra $(\mathcal{A}, \tau)$. It follows from the above that ξ is both cyclic and separating for the action of $\mathcal{A}$. The latter implies, in particular, that ξ_κ is separating and cyclic for $\mathcal{A}_\kappa$, and τ_κ is a faithful normal tracial state on $\mathcal{A}_\kappa$.

The following proposition gives much important information about our 'building blocks' for the classification of abelian von Neumann algebras. The proof of its last part closely follows the corresponding part in [**Fre04**, 331K].

Proposition 1.39 *For an infinite cardinal κ, the von Neumann algebra $\mathcal{A}_\kappa$ is abelian, σ-finite, non-atomic and κ-homogeneous.*

Proof. It is obvious that the C^*-algebra A_κ is abelian, and so the continuity of multiplication on bounded subsets of $\mathcal{A}_\kappa$ together with the Kaplansky density theorem (see, for example, Proposition 0.11 (2) and Theorem 0.93) implies that $\mathcal{A}_\kappa$ is abelian. It is σ-finite, indeed, it admits a faithful normal tracial state τ_κ, and a cyclic and separating vector ξ_κ.

Let p_j, with $j < \kappa$, be projections of the form $((1,0)) \otimes \mathbb{1}_{\kappa \setminus \{j\}}$ (that is the 'elementary tensor' which is $\mathbb{1}_2$ in each component except the jth, which is $(1,0) \in \mathbb{C}^2$). For all $F \Subset \kappa$, the set $P := \{p_j\}_{j \in F}$ generates $\mathcal{A}_F \overline{\otimes} \mathbb{C}\mathbb{1}_{\kappa \setminus F}$. Since $\bigcup_{F \Subset \kappa} \theta_F(A_F)$ is uniformly dense in A_κ, it is strongly dense in $\mathcal{A}_\kappa$. So the family $\{p_j\}_{j < \kappa}$ generates $\mathcal{A}_\kappa$. Hence, $\mathfrak{e}(\mathcal{A}_\kappa) \leq \kappa$. If $p \in \mathbb{P}(\mathcal{A}_\kappa)$ is non-zero, then by Theorem 1.29(b), with P as above, there is a countable subset J of κ such that for some non-zero $q \in \mathbb{P}(\mathcal{A}_J)$, we have $p = q \overline{\otimes} \mathbb{1}_{\kappa \setminus J}$.

Suppose now that $p \in \mathbb{P}(\mathcal{A}_\kappa)$ is minimal. For each $j < \kappa$, we have $0 \neq p = pp_j + pp_j^\perp$; hence (at least) one of pp_j and $pp_j^\perp$ is non-zero. Let q_j be either p_j or $p_j^\perp$, with the provision that $pq_j \neq 0$. Since p is minimal and $pq_j \leq p$, we have $p \leq q_j$ for all $j < \kappa$. Let $F \Subset \kappa$ be such that $\tau_\kappa(p) > 1/2^{\#F}$, and let $q_F = \prod_{j \in F} q_j$. Then

$$\tau_\kappa(q_F) = 1/2^{\#F} < \tau_\kappa(p) \leq \tau_\kappa(q_F),$$

a contradiction. Hence, there are no minimal projections in $\mathcal{A}_\kappa$.

We check the equality $\mathfrak{k}(\mathcal{A}_\kappa) = \mathfrak{k}(\mathcal{A}_\kappa)p$ for any non-zero $p \in \mathbb{P}(\mathcal{A}_\kappa)$. Indeed, if G is a set of generators for $\mathcal{A}_\kappa$, then $\{gp \colon g \in G\}$ is a set of generators for $\mathcal{A}_\kappa p$, and hence, the inequality $\mathfrak{k}(\mathcal{A}_\kappa) \geq \mathfrak{k}(\mathcal{A}_\kappa)p$ holds. In the other direction, note that minimal projections occur in neither $\mathcal{A}_\kappa$ nor $\mathcal{A}_\kappa p$. By Lemma 1.31, we have $\aleph_0 \leq \mathfrak{k}(\mathcal{A}_\kappa p) \leq \mathfrak{k}(\mathcal{A}_\kappa) \leq \kappa$. Consequently, if $\kappa = \aleph_0$, then $\mathfrak{k}(\mathcal{A}_\kappa p) = \mathfrak{k}(\mathcal{A}_\kappa)$. Otherwise, by Theorem 1.29(b), $p = q \overline{\otimes} \mathbb{1}_{\kappa \setminus J}$ for some countable set J and $q \in \mathcal{A}_J$, so that $\mathcal{A}_\kappa p = \mathcal{A}_J q \overline{\otimes} \mathcal{A}_{\kappa \setminus J}$. If G generates $\mathcal{A}_\kappa p$, then $\{g \mathbb{1}_{\kappa \setminus J} \colon g \in G\}$ generates $\mathcal{A}_{\kappa \setminus J}$, so that the isomorphism of $\mathcal{A}_\kappa$ and $\mathcal{A}_{\kappa \setminus J}$ implies that $\mathfrak{k}(\mathcal{A}_\kappa) = \mathfrak{k}(\mathcal{A}_{\kappa \setminus J}) \leq \mathfrak{k}(\mathcal{A}_\kappa p)$, and again $\mathfrak{k}(\mathcal{A}_\kappa p) = \mathfrak{k}(\mathcal{A}_\kappa)$. Hence, once we prove that $\mathfrak{k}(\mathcal{A}_\kappa) \geq \kappa$, the proof of the proposition will be finished.

First observe that for any sequence $(i_k)_{k \in \mathbb{N}}$ of (pairwise) distinct indices from κ, we have $\bigwedge_{k \in \mathbb{N}} p_{i_k} = 0$ and $\bigvee_{k \in \mathbb{N}} p_{i_k} = \mathbb{1}$. In fact, the normality of τ_κ on the unit ball implies that

$$\tau_\kappa\Big(\bigwedge_{k \in \mathbb{N}} p_{i_k}\Big) = \lim_{n \to \infty} \tau_\kappa\Big(\bigwedge_{k \leq n} p_{i_k}\Big) = \lim_{n \to \infty} 2^{-n} = 0$$

and we get the same result if we replace each p_{i_k} with $p_{i_k}^\perp$. Now, for a non-zero $q_0 \in \mathbb{P}(\mathcal{A}_\kappa)$ and $\varepsilon > 0$, put

$$U(p, \varepsilon) := \{q \in \mathbb{P}(\mathcal{A}_\kappa) \colon \|(q - p)\xi_\kappa\| \leq \varepsilon\}.$$

There is $\delta > 0$ such that

$$\#\{i < \kappa \colon p_i \in U(p, \delta)\} < \infty.$$

Otherwise, there would be an infinite sequence of distinct indices i_k such that $p_{i_k} \in U(p, 2^{-k})$, and we would obtain

$$1 = \tau_\kappa(\mathbb{1}) = \tau_\kappa(p) + \tau_\kappa(p^\perp)$$

$$= \tau_\kappa(p(\bigwedge_{k\in\mathbb{N}} p_{i_k})^\perp) + \tau_\kappa((\bigvee_{k\in\mathbb{N}} p_{i_k})p^\perp)$$

$$\leq \sum_{k\in\mathbb{N}} \tau_\kappa(pp_{i_k}^\perp) + \sum_{k\in\mathbb{N}} \tau_\kappa(p_{i_k}p^\perp)$$

$$= \sum_{k\in\mathbb{N}} \tau_\kappa(p(p - p_{i_k})) + \sum_{k\in\mathbb{N}} \tau_\kappa(p_{i_k}(p_{i_k} - p))$$

$$= \sum_{k\in\mathbb{N}} \|p(p - p_{i_k})\xi_\kappa\|^2 + \sum_{k\in\mathbb{N}} \|p_{i_k}(p_{i_k} - p)\xi_\kappa\|^2$$

$$\leq \sum_{k\in\mathbb{N}} \|(p - p_{i_k})\xi_\kappa\|^2 + \sum_{k\in\mathbb{N}} \|(p_{i_k} - p)\xi_\kappa\|^2$$

$$\leq 2\sum_{k\in\mathbb{N}} 4^{-k} = \frac{2}{3} < 1,$$

a contradiction.

Let now $Q \subseteq \mathbb{P}(\mathcal{A}_\kappa)$ be such that Q generates $\mathcal{A}_\kappa$ as a von Neumann algebra and that $\#Q = \mathfrak{k}(\mathcal{A}_\kappa)$. By Lemma 1.22, we can assume that Q is a Boolean algebra. For any non-zero $p \in \mathbb{P}(\mathcal{A}_\kappa)$, there exists $\delta > 0$ such that the set $\{j < \kappa : p_j \in U(p, \delta)\}$ is finite. Take $n \in \mathbb{N}$ such that $2^{-n+1} \leq \delta$. By Theorem 1.29(b), there is a projection $q \in Q$ such that $q \in U(p, 2^{-n})$. If $\|(r - q)\xi_\kappa\| \leq 2^{-n}$, then

$$\|(r - p)\xi_\kappa\| \leq 2^{-n} + \|(p - q)\xi_\kappa\| \leq 2^{-n+1} \leq \delta.$$

So $U(q, 2^{-n}) \subseteq U(p, \delta)$. Hence, the number of indices $j < \kappa$ such that $p_j \in U(q, 2^{-n})$ is finite. Moreover, $p \in U(q, 2^{-n})$. Let $\mathcal{U}$ be the family of sets of the form $U = U(q, 2^{-n})$ with $q \in Q$ and $n \in \mathbb{N}$, for which the set of indices $J_U := \{j < \kappa : p_j \in U\}$ is finite. Then, $\mathcal{U}$ covers $\mathbb{P}(\mathcal{A}_\kappa) \supseteq P$, where $P := \{p_i\}_{i<\kappa}$. Since $\sup_{U\in\mathcal{U}} J_U \leq \aleph_0$, we have

$$\kappa = \#\kappa \leq \#\mathcal{U} \cdot \aleph_0 = \#\mathcal{U} \leq \mathfrak{k}(\mathcal{A}_\kappa) \cdot \aleph_0 = \mathfrak{k}(\mathcal{A}_\kappa)$$

which ends the proof. $\qquad\qquad\square$

Corollary 1.40 *Let $\mathcal{A}$ be an abelian, non-atomic and κ-homogeneous von Neumann algebra, and let τ_A be a faithful normal trace on $\mathcal{A}$*

with $\tau_{\mathcal{A}}(\mathbb{1}) = 1$. Then, $(\mathcal{A}, \tau_{\mathcal{A}})$, as a von Neumann measure algebra, is isomorphic to $(\mathcal{A}_\kappa, \tau_\kappa)$.

Proof.   This follows directly from Theorems 1.39 and 1.37. $\square$

The following theorem (together with Theorem 1.38) gives a complete classification, up to isomorphism, of abelian von Neumann measure algebras (c.f. [**Fre04**, 332C]).

Theorem 1.41 (Maharam's theorem for von Neumann measure algebras).   *Let $(\mathcal{A}, \tau_{\mathcal{A}})$ be an abelian von Neumann measure algebra. For $\kappa = 0$, we put $\mathcal{A}_0 := \mathbb{C}$ and let τ_0 be the normalised trace on $\mathcal{A}_0$. Then, there are families $(\kappa_i)_{i \in I}$ and $(\gamma_i)_{i \in I}$ such that each κ_i is either 0 or an infinite cardinal, and each γ_i is a strictly positive real number, and such that*

$$(\mathcal{A}, \tau_{\mathcal{A}}) \simeq \oplus_{i \in I} (\mathcal{A}_{\kappa_i}, \gamma_i \, \tau_{\kappa_i}).$$

as von Neumann measure algebras.

Proof.   Let P be the collection of all minimal projections in $\mathcal{A}$, and let $q := \sup P$. By the proof of Lemma 1.24, $\mathcal{A}q$ is purely atomic and isomorphic to $\mathbb{C}^\lambda$ with $\lambda := \#P$, and $\mathcal{A}q^\perp$ non-atomic. If $\tau_{\mathcal{A}}(q^\perp) = \infty$, then we use Proposition 1.25 to decompose $q^\perp$ into a sum of non-zero finite-trace projections. By Lemma 1.33, we can find an orthogonal family $\{z_j\}_{j \in J}$ of homogeneous finite-trace projections with $\sum_{j \in J} z_j = q^\perp$. Put $I := P \cup J$ and $z_p := p$ for $p \in P$. Then, $\{z_i\}_{i \in I}$ forms an orthogonal family of homogeneous finite-trace projections in $\mathcal{A}$ with $\sum_{i \in I} z_i = \mathbb{1}$.

Put $\gamma_i := \tau_{\mathcal{A}}(z_i)$ for each $i \in I$. If $i \in P$, then clearly

$$(\mathcal{A}z_i, \tau_{\mathcal{A}}|\mathcal{A}z_i) = (\mathcal{A}p, \tau_{\mathcal{A}}|\mathcal{A}p) \simeq (\mathbb{C}, \gamma_i \tau_0).$$

For each $i \in J$, there is, by Theorem 1.40, an infinite cardinal κ_i such that

$$\left(\mathcal{A}z_i, \frac{1}{\gamma_i} \tau_{\mathcal{A}}|\mathcal{A}z_i\right) \simeq (\mathcal{A}_{\kappa_i}, \tau_{\kappa_i}).$$

Putting $\kappa_i := 0$ for each $i \in P$, we have for each $i \in I$ an isomorphism Φ_i between $(\mathcal{A}z_i, \tau_{\mathcal{A}}|\mathcal{A}z_i)$ and $(\mathcal{A}_{\kappa_i}, \gamma_i \tau_{\kappa_i})$. The direct sum of those isomorphisms yields the desired isomorphism of $(\mathcal{A}, \tau_{\mathcal{A}})$ with $\oplus_{i \in I} (\mathcal{A}_{\kappa_i}, \gamma_i \tau_{\kappa_i})$. $\square$

Since every abelian von Neumann algebra has a faithful normal semifinite trace (see Proposition 1.14), we have the following corollary.

Corollary 1.42 *An abelian von Neumann algebra is isomorphic to a direct sum of $\mathbb{C}^\lambda$ for some cardinal λ and a direct sum of $\mathcal{A}_\kappa$ for some infinite cardinals κ.*

Remark 1.43 Together, Theorems 1.38 and 1.41 have a myriad of consequences and applications, just as in the measure algebra case in [**Fre04**]. Many of the latter applications may be found in Chapter 33 in that reference, for example, after the statements of the analogous results and in the *Notes* sections therein. We will not take the time here to spell out a sample of these, but the interested reader could look in [**Fre04**] for details in that parallel case. Note that Theorem 1.38 may be used to refine Theorem 1.41 in various ways, using additional information about the magnitudes $\mathfrak{m}_{\mathcal{A}}(e_\kappa)$ and Lemma 1.27. For example, one may 'clean up' the 'summand clumps' associated with each cardinal. For each distinct infinite cardinal κ, there are, up to measure isomorphism, two cases: one may either 1) take all of the associated scaling factors γ_i to be 1 or 2) replace all associated γ_i by a single $\gamma \neq 1$. (See the relationship between $\mathfrak{m}_{\mathcal{A}}(e_\kappa)$ and γ_i in [**Fre04**, 332K(a)].) In particular, an abelian von Neumann probability algebra is isomorphic to a finite or countable direct sum $\oplus_n (\mathcal{A}_{\kappa_n}, \gamma_n)$, with the cardinals κ_n infinite and distinct if non-zero, and with $\sum_n \gamma_n = 1$. If, in addition, the predual is separable (or under some equivalent 'countability' condition), then κ_n is 0 or $\aleph_0$ for all n.

We remark that there is another famous and similar sounding theorem of Kuratowski for Polish spaces up to Borel space isomorphism [**Sri98**]. In particular, any two uncountable standard Borel spaces are Borel isomorphic.

1.4 A Radon measure space primer

Let X be a set with a locally compact Hausdorff topology $\mathfrak{T}$. The Borel σ-algebra $\mathscr{B}(X)$ on X is the σ-algebra generated by $\mathfrak{T}$. A Borel measure μ on X is said to be *outer regular on a measurable set E* if

$$\mu(E) = \inf\{\mu(U)\colon U \in \mathfrak{T},\, E \subseteq U\}$$

and *inner regular on* E if on the other hand

$$\mu(E) = \sup\{\mu(C)\colon C \subseteq E,\, C \text{ is compact}\}.$$

We simply say that μ is *outer regular* (respectively, *inner regular*) if it is outer (respectively, inner) regular on every measurable set.

Radon measures on such X are a very special category of Borel measures on X, constructed from positive linear functionals on $C_c(X)$—the continuous functions of compact support. There are, however, several different definitions of Radon measures in the literature, all justified by some particular factor an author may need. Hence, we pause to give some background regarding the issues that gave rise to some of these various versions. We shall start with the general process of constructing a Radon measure.

Let X be as before and I a positive linear functional on $C_c(X)$. By the Riesz representation theorem, there exists a Borel measure μ for which we have $I(f) = \int f\, d\mu$ for each $f \in C_c(X)$. The measure μ is usually constructed in a real variables course by first defining the set function μ^* on open subsets by means of the prescription

$$\mu^*(U) = \sup\{I(f)\colon f \in C_c(X), 0 \le f \le \chi_U, \operatorname{supp}(f) \subseteq U\}$$

and then extending to all subsets of X by the prescription

$$\mu^*(A) = \inf\{\mu^*(U)\colon U \text{ open and } A \subseteq U\}. \tag{1.1}$$

This quantity turns out to be an outer measure for which the μ^*-measurable sets include the Borel σ-algebra. Hence, μ^* restricts to a measure μ on the Borel σ-algebra $\mathscr{B}(X)$ that is finite on compact sets, for which we have $I(f) = \int f\, d\mu$ for all $f \in C_c(X)$, and which is outer regular on all elements of $\mathscr{B}(X)$, but inner regular on open sets only. For some authors, it is this measure μ which is a Radon measure. However, it does have some shortcomings, namely that the measure may not be inner regular on all Borel sets (see [**Sal20**, Exercise 3.23]), and indeed, it may not be semifinite.

Recall from, for example, [**Fol16**, Exercise 1.15] that an arbitrary measure ν on a measure space (X, Σ) may be written as $\nu = \nu_{sf} + \nu_\infty$, where ν_{sf} is semifinite and ν_∞ assumes only the values 0 and ∞. In fact

$$\nu_{sf}(E) = \sup\{\nu(F) : F \in \Sigma, \nu(F) < \infty,\ F \subset E\}$$

and will be referred to below as the *semifinite part* of ν. All measurable sets with $\nu(E) < \infty$ are trivially semifinite, and so $\nu(E) = \nu_{sf}(E)$ for such sets.

Many authors insist on Radon measures being inner regular on all measurable sets. It is clear that any inner regular measure on $(X, \mathscr{B}(X))$ that is finite valued on compact sets is semifinite. Thus, semifiniteness presents itself as the possible criterion needed to ensure universal inner regularity. If the measure constructed as described immediately after Equation (1.1) is semifinite, then it is indeed universally inner regular. If, however, this measure is not semifinite, further modification is necessary. In such a case, the 'semifinite part' of the measure μ defined above, which we shall denote by μ_0, is indeed inner regular on all Borel sets and may alternatively be realised by the prescription

$$\mu_0(E) = \sup\{\mu(C)\colon C \subseteq E,\ C \text{ is compact}\}.$$

(See [**Coh13**, Exercise 5, §7.2] for this fact.) This measure μ_0 moreover also satisfies $I(f) = \int f\,d\mu_0$ for all $f \in C_c(X)$ and, hence, may also be regarded as a measure fulfilling the objective of the Riesz representation theorem (see, for example, [**Sal20**] for an outstanding account of this and of the issues involved). It is such inner regular Borel measures that many other authors refer to as Radon measures. However, the benefit of ensuring inner regularity (via passage to μ_0) is not without cost. Specifically, the measure μ_0 can only be outer regular if it agrees with μ [**Coh13**, Exercise 5(b), §7.2]. Thus, ensuring inner regularity often means sacrificing outer regularity. Using the fact that its semifinite part is inner regular, there is one serendipitous observation we may make regarding μ, and that is that μ is in fact inner regular on sets of finite μ-measure (and more generally on all sets of σ-finite μ-measure). This comes from the fact that μ agrees with its semifinite part on such sets.

Our ultimate interest as far as Radon measures are concerned is to establish good connections with decomposable measure spaces, and with abelian von Neumann algebras. For this, further structure is needed–the Borel σ-algebra is not large enough, in general. To add such a structure, we enlarge $\mathscr{B}(X)$ to, respectively, the μ^* and μ_0^*-measurable sets. We write μ_0^* for the outer measure induced by μ_0 on the power set $\mathcal{P}(X)$, and $\mathcal{M}_{\mu_0^*}$ for the μ_0^*-measurable sets. That is, $\mu_0^*(E)$ is the infimum of $\mu_0(B)$ over Borel sets B containing E. We shall retain the notation μ_0^* for the restriction of μ_0^* to $\mathcal{M}_{\mu_0^*}$. This is a measure that extends μ_0 (by Caratheodory's theory); in particular, its domain contains all Borel sets. The μ^*-measurable sets will similarly be denoted by $\mathcal{M}_{\mu^*}$, and the restriction of μ^* to $\mathcal{M}_{\mu^*}$ by μ^*.

There is one more approach to Radon measures we shall describe, namely the Bourbaki approach. The enlargement of $(X, \mathcal{B}(X), \mu)$ to $(X, \mathcal{M}_{\mu^*}, \overline{\mu})$ will also here prove to be useful. The reader may find details of the description that follows on [**Coh13**, pages 215 and 216]. As before, we start with a locally compact Hausdorff space X and a positive linear functional I on $C_c(X)$, which Bourbaki calls a Radon measure. With $\mathcal{I}_+(X)$ denoting the set of all $[0, \infty]$-valued lower semicontinuous functions on X, Bourbaki then uses the prescription

$$I^*(f) = \sup\{I(g) \colon g \in C_c(X), 0 \le g \le f\}$$

to first extend I to a sublinear functional on $\mathcal{I}_+(X)$, after which he then uses the formula

$$I^*(f) = \inf\{I^*(h) \colon h \in \mathcal{I}_+(X), f \le h\}$$

to further extend I to a sublinear functional on the cone of all $[0, \infty]$-valued functions on X, for which we also have that $I^*(\alpha f) = \alpha I^*(f)$ for all $0 \le \alpha < \infty$. It follows that the set $\mathcal{F}$ of functions $f : X \to \mathbb{R}$ for which $I^*(|f|) < \infty$ is a real vector space and that $N : \mathcal{F} \to \mathbb{R} : f \mapsto I^*(|f|)$ is a seminorm on $\mathcal{F}$. Bourbaki then defines $L^1(X, I)$ as the closure of $C_c(X)$ in $(\mathcal{F}, N)$. When checking that I is N-continuous on $C_c(X)$, one can use this continuity to extend I to all of $L^1(X, I)$. The functions in $L^1(X, I)$ are said to be *I-integrable*, and the extension of I to $L^1(X, I)$ referred to as *the integral*, with $I(f)$ then denoted as $\int f \, dI$. A function $f : X \to \mathbb{R}$ is said to be *I-measurable* if for each compact subset K of X and each $\epsilon > 0$, there is a compact subset K_ϵ of K for which $f : K_\epsilon \to \mathbb{R}$ is continuous and $I^*(\chi_{K - K_\epsilon}) < \epsilon$. A subset A of X is, respectively, called *I*-integrable and *I*-measurable, whenever the function χ_A is *I*-integrable or, respectively, *I*-measurable. For our purposes, the following theorem is now crucial. It clearly shows that the Bourbaki picture is compatible with that described by $(X, \mathcal{M}_{\mu^*}, \overline{\mu})$.

Theorem 1.44 ([Coh13, Theorem 7.5.5])   *Let I be a positive linear functional on $C_c(X)$ (a positive Radon measure in Bourbaki terminology). We then have that:*

1. *a subset $A \subseteq$ is I-measurable if and only if it is μ^*-measurable;*
2. *a function $f : X \to \mathbb{R}$ is I-measurable if and only if it belongs to $\mathcal{M}_{\mu^*}$;*
3. *$L^1(X, I) = L^1(X, \mathcal{M}_{\mu^*}, \overline{\mu}, \mathbb{R})$;*
4. *$\int f \, dI = \int f \, d\overline{\mu}$ for each $f \in L^1(X, I)$.*

We shall ultimately see that of the two measure spaces $(X, \mathcal{M}_{\mu_0^*}, \mu_0^*)$ and $(X, \mathcal{M}_{\mu^*}, \mu^*)$, it is the framework afforded by $(X, \mathcal{M}_{\mu_0^*}, \mu_0^*)$ that is well suited to the description of abelian von Neumann algebras. In particular, as shown by the following theorem, it is effectively localisable. See [**BGL22**, Theorem 6.2] for a proof of this fact.

Theorem 1.45   *The measure space $(X, \mathcal{M}_{\mu_0^*}, \mu_0^*)$ is decomposable. Specifically, there exists a disjoint family $\mathscr{C}$ of compact sets for which following holds.*

a) *$\mu_0(K) > 0$ for each $K \in \mathscr{C}$.*
b) *For any open set U and any $K \in \mathscr{C}$, we have that $\mu_0(U \cap K) > 0$ whenever $U \cap K \neq \emptyset$.*
c) *For any compact set K_0 we have that $K_0 \cap K \neq \emptyset$ for at most countably many sets K in $\mathscr{C}$.*
d) *For any $E \in \mathcal{M}_{\mu_0^*}$ we have that $\mu_0^*(E) = \sum_{K \in \mathscr{C}} \mu_0^*(E \cap K)$.*
e) *A subset E of X belongs to $\mathcal{M}_{\mu_0^*}$ if and only if for each $K \in \mathscr{C}$, $E \cap K \in \mathcal{M}_{\mu_0^*}$.*

There is a version of this result that holds for $(X, \mathcal{M}_{\mu^*}, \mu^*)$, but with the trade-off that while (c) will hold for all sets $E \in \mathcal{M}_{\mu^*}$ with $\mu^*(E) < \infty$, (d) will only hold for sets $E \in \mathcal{M}_{\mu^*}$ for which $\mu^*(E) < \infty$. For details, see [**Coh13**, Proposition 7.5.3].

With Theorem 1.45 as a background, the framework below now emerges as a formalisation of the structure of $(X, \mathcal{M}_{\mu_0^*}, \mu_0^*)$. This formalised framework is what Fremlin calls a Radon measure space, and we shall follow him in embracing this definition.

Definition 1.46 A *Radon measure space* is a quadruple $(X, \mathfrak{T}, \mathscr{B}, \mu)$ for which we have that:

(i) $(X, \mathscr{B}, \mu)$ is a complete measure space;

(ii) if $E \subseteq X$ and $E \cap F \in \mathscr{B}$ for all $F \in \mathscr{B}$ with $\mu(F) < \infty$, then $E \in \mathscr{B}$;

(iii) $\mathfrak{T}$ is a Hausdorff topology on X;

(iv) $\mathfrak{T} \subset \mathscr{B}$;

(v) for every $E \in \mathscr{B}$, we have $\mu(E) = \sup\{\mu(C) \colon C \subseteq E, C \text{ is compact}\}$;

(vi) for every x, there is an open neighbourhood U of x with $\mu(U) < \infty$.

Since it follows from (vi) that compact sets are μ-finite, μ induces a positive functional on the continuous functions of compact support. Conversely, if X is locally compact, any such functional comes from integration on the Radon measure space $(X, \mathcal{M}_{\mu_0^*}, \mu_0^*)$ above. See [**Fre04**, Volume 4] for more on Radon measure spaces in this sense (§436 there treats the Riesz representation theorem).

A minor modification of the proof of Theorem 1.45 (see [**BGL22**, Theorem 6.2]) now yields the following fact.

Theorem 1.47 ([Fre89, Theorem 1.10]) *Any Radon measure space* $(X, \mathfrak{T}, \mathscr{B}, \mu)$ *is decomposable in the sense that there exists a disjoint family* $\mathscr{C}$ *of compact sets for which following holds.*

a) $\mu(K) > 0$ *for each* $K \in \mathscr{C}$.

b) *For any open set* U *and any* $K \in \mathscr{C}$, *we have that* $\mu(U \cap K) > 0$ *whenever* $U \cap K \neq \emptyset$.

c) *For any compact set* K_0 *we have that* $K_0 \cap K \neq \emptyset$ *for at most countably many sets* K *in* $\mathscr{C}$.

d) *For any* $E \in \mathscr{B}$ *we have that* $\mu(E) = \sum_{K \in \mathscr{C}} \mu(E \cap K)$.

e) *A subset* E *of* X *belongs to* $\mathscr{B}$ *if and only if for each* $K \in \mathscr{C}$, $E \cap K \in \mathscr{B}$.

1.5 A measure-theoretic description of localisable measure algebras

We pass to the task of characterising localisable measure algebras. This will affirm the status of Radon measure spaces as a class of well-described measure spaces unifying the concepts of 'localisable' and 'decomposable'.

Theorem 1.48 *For a measure algebra $(\mathfrak{A}, \overline{\mu})$, the following are equivalent.*

> (1.) $(\mathfrak{A}, \overline{\mu})$ *is localisable;*
> (2.) $(\mathfrak{A}, \overline{\mu})$ *is isomorphic to the measure algebra of a localisable measure space.*
> (3.) $(\mathfrak{A}, \overline{\mu})$ *is isomorphic to the measure algebra of a decomposable measure space.*
> (4.) $(\mathfrak{A}, \overline{\mu})$ *is isomorphic to the measure algebra of a Radon measure space $(X, \mathfrak{T}, \mathscr{B}, \mu)$.*

Proof. By Theorem 1.47, the measure algebra of any Radon measure space is isomorphic to the measure algebra of a decomposable measure space. (With $\mathscr{C}$ denoting the collection of compact sets guaranteed by Theorem 1.47, the isomorphism in question is realised by the prescription $[F] \to \oplus_{K \in \mathcal{C}}[F \cap K]$. The content of Theorem 1.47 ensures that this map is indeed a measure algebra isomorphism.) We leave the verification of these details to the reader. Any decomposable (strictly localisable) measure space is, of course, localisable. The implications $(4) \Rightarrow (3) \Rightarrow (2)$ are, therefore, immediate. For any concrete measure algebra, it is a straightforward exercise to see that the measure algebra is localisable if and only if the underlying measure space is localisable. The equivalence $(2) \Leftrightarrow (1)$ is, therefore, a consequence of Theorem 1.21 and Proposition 1.18.

Therefore, it remains to prove the implication $(2) \Rightarrow (4)$. The measure algebra of a finite measure space (Ω, Σ, μ) is isomorphic to the measure algebra of a compact Radon measure space in the sense of Definition 1.46. Indeed, by the Loomis–Sikorski theorem (Theorem 1.21) and its proof, up to measure algebra isomorphism, we may take Ω to be a Stone space, and Σ to consist of $C \triangle F$ for C clopen and F meagre. Also, such F are μ-null. Since μ is finite, items (ii) and (vi) in Definition 1.46 are trivial, as is (iii). Since any open set U equals $\bar{U} \triangle F$ for the meagre set $F = \bar{U} \setminus U$, and since $\mu(C \triangle F) = \mu(C)$, we also have items (iv) and (v) in that definition. Items (ii)–(vi) are easily seen to remain true for the completion of μ, and hence, we may assume that (i) also holds. Thus, we have a compact (finite) Radon measure space in the sense of Definition 1.46.

Let (X, Σ, μ) be a localisable measure space, and let (p_i) be a maximal set of mutually orthogonal non-zero elements of $\Sigma/\mathcal{Z}(\Sigma)$ with

$\bar{\mu}(p_i) < \infty$ for each i. Such a maximal set exists by Zorn's lemma. Let $p = \vee_i p_i$. If $p \neq 1$, then, by semifiniteness, there exists an element $q \in \Sigma/\mathcal{Z}(\Sigma)$ with $\bar{\nu}(q) < \infty$ such that $q \leq 1 - p$. This contradiction shows that $p = 1$. For each i, let E_i be a μ-finite set in Σ corresponding to p_i. The measure space $\oplus_i(E_i, \Sigma \cap E_i, \mu_{E_i})$ is clearly decomposable. Let $P \in \Sigma$ be the envelope of the collection $\{E_i\}$. It is then an exercise to see that $[E_i] \leq [P]$ for each i and hence that $[P] = 1$

We claim that the measure algebra $\Sigma/\mathcal{Z}(\Sigma)$ is isomorphic to the measure algebra $\oplus_i(\Sigma \cap E_i)/\mathcal{Z}(\Sigma \cap E_i)$ with the isomorphism implemented by the prescription $\theta : [F] \to \oplus_i[F \cap E_i]$. By the argument in the second paragraph, each $(\Sigma \cap E_i)/\mathcal{Z}(\Sigma \cap E_i)$ is isomorphic to the measure algebra of a compact Radon measure space. Hence, the implication will follow if, in addition to the first claim, we can also show that the 'disjoint sum' of a collection of compact Radon measure spaces is again a Radon measure spaces. We proceed to verify these two claims. The fact that $\theta : [F] \to \oplus_i[F \cap E_i]$ is a Boolean homomorphism trivially follows from the fact that each $\theta_i : [F] \to [F \cap E_i]$ is a Boolean homomorphism. The injectivity of this map follows from the fact that $1 = \oplus_i p_i = \oplus_i[E_i]$. Thus, if $K \in \Sigma$ is a set with a non-zero measure, the fact that $[K] = \oplus_i[K]p_i$ ensures that $[K]p_i = [K \cap E_i]$ must be non-zero for some i. In other words, $[K] \neq 0$ implies $\theta([K])$. To see that θ is also surjective, notice that given a collection of sets $F_i \in \Sigma \cap E_i$, it is a straightforward exercise to verify that the envelope $H \in \Sigma$ of the collection $\{F_i\}$ satisfies $[H \cap E_i] = [F_i]$ for each i and hence that $\theta([H]) = \oplus_i[F_i]$.

We finally show that θ is measure preserving. For any $F \in \Sigma$ and any finite collection $\mathcal{F}$ of is, the fact that the E_is differ by sets of measure 0 ensures that $\sum_{i \in \mathcal{F}} \mu(F \cap E_i) = \mu(\cup_{i \in \mathcal{F}}(F \cap E_i)) = \mu(F \cap (\cup_{i \in \mathcal{F}})) \leq \mu(F)$. Since $\sum_i \mu(F \cap E_i)$ is the limit of such 'finite' partial sums, we must have $\sum_i \mu(F \cap E_i) \leq \mu(F)$. So, if $\sum_i \mu(F \cap E_i) = \infty$, we have equality. If, on the other hand, $\sum_i \mu(F \cap E_i) < \infty$, then only countably many of the $\mu(F \cap E_i)$s are non-zero. Let $\mathcal{C}$ be the collection of is, for which the $\mu(F \cap E_i)$s are non-zero. We claim that $F\backslash(\cup_{i \in \mathcal{C}}(F \cap E_i))$ has measure 0. If this is not the case, semifiniteness of the measure space ensures that there then exists a subset K of $F\backslash(\cup_{i \in \mathcal{C}}(F \cap E_i))$ with finite positive measure. The injectivity of the map θ then ensures that we must have $[K \cap E_j] \neq 0$ for some j, where, by construction, this j necessarily does not belong to $\mathscr{C}$. But since $[K \cap E_j] \leq [F \cap E_j]$, the element $[F \cap E_j]$ must then also be non-zero; equivalently, $\mu(F \cap E_j) \neq 0$— a clear contradiction. Thus, as claimed, $\mu(F) = \cup_{i \in \mathscr{C}}(F \cap E_i)$. Since the sets E_i differ by sets of measure 0, we can 'disjointify' these sets by an inductive process, to obtain a related set $\{F_i : i \in \mathscr{C}\}$ of mutually disjoint sets for which $\mu(F \cap E_i) = \mu(f \cap F_i)$ for each $i \in \mathscr{C}$. The countable additivity of the measures then ensures that $\mu(F) = \mu(\cup_{i \in \mathscr{C}}(F \cap E_i)) = \mu(\cup_{i \in \mathscr{C}}(F \cap F_i) = \sum_{i \in \mathscr{C}} \mu(F \cap F_i) = \sum_{i \in \mathscr{C}} \mu(F \cap E_i) = \sum_{i \in \mathscr{C}} \mu(F \cap E_i)$. So, in all cases, we have that $\mu(F) = \sum_{i \in \mathscr{C}} \mu(F \cap E_i)$, or equivalently that $\bar{\mu}([F]) = \mu(F) = \sum_{i \in \mathscr{C}} \mu(F \cap E_i) = \sum_{i \in \mathscr{C}} \bar{\mu}_{E_i}([F \cap E_i]) = (\oplus_i \bar{\mu}_{E_i})(\theta[F])$.

We now show that the measure algebra $\oplus_i (\Sigma \cap E_i)/\mathcal{Z}(\Sigma \cap E_i)$ is isomorphic to the measure algebra of a Radon measure space. We noted earlier that each $(\Sigma \cap E_i)/\mathcal{Z}(\Sigma \cap E_i)$ is isomorphic to the measure algebra of a finite (compact) Radon measure space. So, we may assume that we are dealing with a measure space (Ω, Σ_0, μ), which is the 'disjoint sum' of the compact Radon measure spaces $(\Omega_i, \Sigma_i, \mu_i)$. This is clearly a complete measure space: for every subset E of a ν-null set in Σ_0, each $E \cap \Omega_i$ will be measurable and null since μ_i is complete. So $E \in \Sigma_0$. We give Ω the disjoint sum topology (where a set is open if and only if its intersection with every Ω_i is open). This topology is locally compact and is easily seen to be contained in Σ_0. Items (ii) and (vi) in Definition 1.46 are clear if we take $F = \Omega_i$ there. In (vi), we assume that $x \in \Omega_i$. For (ii), if $E \cap \Omega_i \in \Sigma$ for all i, it follows by definition that $E \in \Sigma$. The validity of (iii) and (iv) in Definition 1.46 is clear. It is an exercise to see that (v) holds too, using the fact that the subsets Ω_i form an open covering of any compact set and hence have a finite subcover. Thus, (Ω, Σ_0, μ) is a Radon measure space.

We show that the obvious isomorphism from $\Omega/\mathcal{Z}(\Omega)$ to $\oplus_i \Omega_i/\mathcal{Z}(\Omega_i)$ is measure preserving. Given $q_i \in \Omega_i/\mathcal{Z}(\Omega_i)$ and $r \in \Omega/\mathcal{Z}(\Omega)$ of the form $r = \oplus_i q_i$, where $q_i \in \Omega_i/\mathcal{Z}(\Omega_i)$ for each i. Assuming that the q_is correspond to sets F_i, r will then correspond to $F = \cup_i F_i$, from which it is then clear that $\bar{\mu}(r) = \mu(F) = \sum_i \mu_i(F_i) = \sum_i \bar{\mu}_i(q_i)$. Hence, the obvious isomorphism from $\Omega/\mathcal{Z}(\Omega)$ to $\oplus_i \Omega_i/\mathcal{Z}(\Omega_i)$ is a measure algebra isomorphism. $\qquad\square$

Note that the Radon measure space constructed in the preceding proof is decomposable by construction.

1.6 From abelian von Neumann algebras to L^∞ spaces

Corollary 1.49 *The pair $(\mathcal{A}_\kappa, \tau_\kappa)$ is isomorphic, as a von Neumann measure algebra, to $(L^\infty(\Omega_\kappa, \Sigma_\kappa, \nu_\kappa), \sigma_\kappa)$, where σ_κ is given by $\sigma_\kappa(a) = \int_{\Omega_\kappa} a\, d\nu_\kappa$ for $a \in L^\infty(\Omega_\kappa, \Sigma_\kappa, \nu_\kappa)$.*

Proof. Since $(\mathcal{A}_\kappa, \tau_\kappa)$ was obtained from A_κ via the GNS construction with φ_κ, and since $\varphi_\kappa = \sigma_\kappa \circ \theta$ by the last result, it is enough to check that the GNS construction with σ_κ leads from $\mathcal{C}(\Omega_\kappa)$ to $L^\infty(\Omega_\kappa, \Sigma_\kappa, \nu_\kappa)$ acting by multiplication on $L^2(\Omega_\kappa, \Sigma_\kappa, \nu_\kappa)$. To see this, we first note that H_σ, obtained as the completion of $\eta_{\sigma_\kappa}(\mathcal{C}(\Omega_\kappa))$ with the inner product $\langle \eta_{\sigma_\kappa}(f), \eta_{\sigma_\kappa}(g) \rangle_{\sigma_\kappa} := \int_{\Omega_\kappa} f\bar{g}\, dz\nu_\kappa$, coincides with $L^2(\Omega_\kappa, \nu_\kappa)$. In fact, this follows immediately from the possibility of approximation of every square-integrable function by a continuous function in L^2-norm (see, for example, [**Rud74**, Theorem 3.14]).

For the sake of clarity, we will, in the ensuing analysis, suppress the subscript κ. Now, let π_σ be the GNS representation of $\mathcal{C}(\Omega)$ on $L^2(\Omega, \nu)$. Recall that in the GNS representation, as realised above, the constant function 1 is a cyclic vector. So, given $f, g \in \mathcal{C}(\Omega)$, we have $\pi_\sigma(f)g = \pi_\sigma(f)\pi_\sigma(g)1 = \pi_\sigma(fg)1 = fg.1 = fg$. The density of $\mathcal{C}(\Omega)$ in $L^2(\Omega, \nu)$, now ensures that for any $f \in \mathcal{C}(\Omega)$ and any $g \in L^2(\Omega, \nu)$, we have $\pi_\sigma(f)g = M_f g$, where M_f is the multiplication operator with symbol f. This clearly shows that $\pi_\sigma(\mathcal{C}(\Omega)) \subset \{M_f : f \in L^\infty(\Omega, \nu)\}$. We then trivially also have $\{M_f : f \in L^\infty(\Omega, \nu)\} \subset \pi_\sigma(\mathcal{C}(\Omega))'$. If we can prove that the converse inclusion also holds, the result will follow from the von Neumann double commutant theorem.

So, let $T \in \pi_\sigma(\mathcal{C}(\Omega))'$ be given and set $f = T(1) \in L^2(\Omega, \nu)$. For any $g \in L^\infty(\Omega, \nu)$, we have that

$$T(g) = T \circ M_g(1) = M_g \circ T(1) = M_g(f) = fg = M_f(g).$$

For f and g as above, we also have $\|fg\|_2 = \|T(g)\|_2 \leq \|T\| \cdot \|g\|_2$. When combined with the Hölder inequality, the density of $L^\infty(\Omega, \nu)$ in $L^2(\Omega, \nu)$, therefore, ensures that, in fact, $f \in L^\infty(\Omega, \nu)$ with $\|f\|_\infty = \|T\|$. It now clearly follows that $T = M_f$.

The correspondence between the traces σ_κ and τ_κ should be clear if we take into account the correspondence between ν_κ and μ_κ above. $\square$

Corollary 1.50 *For $\kappa = \aleph_0$, the pair $(\mathcal{A}_\kappa, \tau_\kappa)$ is isomorphic, as a von Neumann measure algebra, to $(L^\infty([0,1], \Lambda, \lambda), \text{where } \lambda \text{ is}$

the Lebesgue measure on $[0,1]$ with the σ-algebra Λ of Lebesgue-measurable subsets of $[0,1]$.

Proof. The complete proof can be found in [**Fre04**, 254K]. The main idea is the (well-known) construction of a measure-preserving bijection between Ω_κ and $[0,1]$, by first putting $\Phi(a) = \sum_{i=0}^{\infty} 2^{-i-1} a_i$ for $a = (a_i)_{i<\kappa}$ and then mending it in a countable number of points to make it a bijection. $\qquad\qquad\square$

The energetic reader may show, using associativity of the product of cardinals, that the last corollary holds for $\kappa > \aleph_0$ with Lebesgue measure λ replaced by its power λ^κ.

If the Radon–Nikodym theorem is to form part of our analysis, we shall need a version of that theorem up to the task of dealing with the most general L^∞-spaces that may be realised as abelian von Neumann algebras. We consider measures ν on (X, Σ), with $\nu \ll \mu$ denoting *absolute continuity*: $\nu(E) = 0$ whenever $\mu(E) = 0$. As we learn in a real variable course, the complex or signed measure case of the Radon–Nikodym theorem follows immediately from the positive measure case, using either the Hahn decomposition or the variation measure $|\nu|$.

Given measures μ and ν on a σ-algebra Σ, we say that ν is μ-*semifinite* if whenever $\nu(E) > 0$, then E has a μ-finite measurable subset F with $\nu(F) > 0$. This condition was considered in some of the Radon–Nikodym theorems in [**Fre04**, Volume II]. It is automatic if μ is finite and coincides with semifiniteness if $\mu = \nu$. We say that ν is *strongly μ-semifinite*, if whenever $\nu(E) > 0$, the set E has a μ-finite measurable subset F with $0 < \nu(F) < \infty$. One may check that ν is strongly μ-semifinite if and only if ν is μ-semifinite and semifinite.

Example 1.51 If λ is Lebesgue measure and μ is counting measure on the Lebesgue σ-algebra of $[0,1]$ then λ satisfies $\lambda \ll \mu$ but λ is not μ-semifinite.

Given a non-negative measurable function f, we will below write μ_f for $\mu_f(E) = \int_E f \, d\mu$.

Theorem 1.52 [**Fre04**, 234O]   *(Radon–Nikodym for localizable measures) Fix a localisable measure space (X, Σ, μ) and a positive* *measure ν on (X, Σ). Set $\mathcal{M} = L^\infty(X, \Sigma, \mu)$. Then, there exists measurable $h : X \to [0, \infty)$ with $\nu(E) = \int_E h \, d\mu$ for all $E \in \Sigma$ if and only if $\nu \ll \mu$ and ν is strongly μ-semifinite (that is, whenever $\nu(E) > 0$, then E has a μ-finite measurable subset F with $0 < \nu(F) < \infty$).*

We point out that using a quite different proof, two additional equivalent conditions were added to the above theorem in [**BGL22**, Theorem 4.4]. In addition, further extensions and generalisations of this result were also obtained in [**BGL22**]. These results are then arguably the most general cycle of Radon–Nikodym theorems for 'classical' measure spaces in print. We, in particular, note that when for a localisable measure space $L^\infty((X, \Sigma, \mu)$ is identified with the algebra of multiplication operators on $L^2(X, \Sigma, \mu)$ with symbols in $L^\infty(X, \Sigma, \mu)$, the Radon–Nikodym derivative described above is precisely the Pedersen–Takesaki Radon–Nikodym derivative (an object that will be formally introduced in Definition 4.27). Of course, to make this claim sound, we do need to know that the algebra of multiplication operators with symbols in $L^\infty(X, \Sigma, \mu)$ is a von Neumann algebra (a fact proved later on in this section).

Given a measure space (X, Σ, μ), we often write τ_μ for the functional $f \mapsto \int_X f \, d\mu$ on $L^\infty_+(X, \Sigma, \mu)$. We shall say τ_μ is *faithful* if for any $f \in L^\infty_+(X, \Sigma, \mu)$, $f = 0$ precisely when $\tau_\mu(f)$, and *normal* if for any increasing net (f_α) in $L^\infty_+(X, \Sigma, \mu)$ with lowest upper bound $f \in L^\infty(X, \Sigma, \mu)$, we have $\sup_\alpha \tau_\mu(f_\alpha) = \tau_\mu(f)$. Semifiniteness of the underlying measure space may also be expressed at the level of $L^\infty(X, \Sigma, \mu)$ as the statement that for any non-zero projection $p \in L^\infty(X, \Sigma, \mu)$, we can find a non-zero subprojection $q \in L^\infty(X, \Sigma, \mu)$ with $\tau_\mu(q) < \infty$.

Remark 1.53 One may also, for example, apply Theorems 1.38 and 1.41 to the question of characterising $*$-isomorphisms $\pi : L^\infty(X, \mathcal{A}, \mu) \to L^\infty(Y, \mathcal{B}, \nu)$, for localisable measure spaces. These are all induced by a measure algebra isomorphism between the measure algebras of $(X, \mathcal{A}, \mu)$ and $(Y, \mathcal{B}, \nu_h)$ (see, for example, Proposition 1.16), where ν_h is as in the Radon–Nikodym theorem with $h : Y \to (0, \infty)$ $\mathcal{B}$-measurable. Such a measure algebra isomorphism may clearly be described in terms of Theorems 1.38 and

1.41. Note that h is the Radon–Nikodym derivative of the measure $\nu'(E) = \bar{\mu}(\pi^{-1}(\chi_E))$ with respect to ν (see the discussion above Theorem 1.69 for further details). The Maharam decompositions of ν and ν' may differ very slightly. The homogeneous components are the same, but note, for example, that $\oplus_{n=1}^{\infty} \mathcal{A}_\kappa$ is $*$-isomorphic to $\mathcal{A}_\kappa$ for infinite κ.

When combined with Proposition 1.16, Theorem 1.48 yields the following serendipitous conclusion.

Theorem 1.54 *For any von Neumann algebra $\mathcal{M}$ equipped with a faithful normal semifinite trace τ, the following are equivalent.*

> (1.) $\mathcal{M}$ *is abelian.*
>
> (2.) $\mathcal{M}$ *is $*$-isomorphic to $L^\infty(X, \Sigma, \mu)$ where (X, Σ, μ) is a localisable measure space.*
>
> (3.) $\mathcal{M}$ *is $*$-isomorphic to $L^\infty(X, \Sigma, \mu)$ where (X, Σ, μ) is a decomposable measure space.*
>
> (4.) $\mathcal{M}$ *is $*$-isomorphic to $L^\infty(X, \mathscr{B}, \mu)$ where $(X, \mathfrak{T}, \mathscr{B}, \mu)$ is a Radon measure space.*

In each of the cases (2)–(4), the $*$-isomorphism $\mathscr{I}$ can be chosen so as to satisfy $\tau = \int_X \mathscr{I}(\cdot)\, d\mu$.

Proof. Any Radon measure space is 'decomposable' (see Theorem 1.47), and any decomposable measure space is localisable. Thus, $(4)\Rightarrow(3)\Rightarrow(2)\Rightarrow(1)$ holds. The implication $(1)\Rightarrow(4)$ is a consequence of applying Proposition 1.16 to the fact that by Theorem 1.48, $(\mathbb{P}(\mathcal{M}), \tau)$ is isomorphic via a measure-preserving map θ to the concrete measure algebra of a Radon measure space (X, μ). $\square$

1.7 From L^∞ spaces to abelian von Neumann algebras

For a measure space (X, Σ, μ), it is elementary to verify that $\mathcal{M} = L^\infty(X, \Sigma, \mu)$ is a commutative C^*-algebra. Realised as a C^*-algebra, the projections in $L^\infty(X, \Sigma, \mu)$ correspond to the a.e. equivalence classes $[\chi_E]$ of the characteristic function of sets $E \in \Sigma$. The *self-adjoint part* $\mathcal{M}_{\mathrm{sa}}$ corresponds to the real-valued elements of $\mathcal{M}$.

1.7.1 L^∞-spaces on localisable measure spaces

Proposition 1.55 *Suppose that (X, Σ, μ) is localisable, and $\mathcal{M} = L^\infty(X, \Sigma, \mu)$. Then, integration with respect to μ is a faithful normal semifinite trace on $\mathcal{M}$.*

Proof. By Theorem 1.48 and Proposition 1.16, we may clearly assume that $(X, \Sigma, \mu_\tau) = \oplus_i (X_i, \Sigma \cap X_i, \nu_i)$, where each ν_i is a finite measure. The semifiniteness and faithfulness of the trace induced by the integral is easy to see. Given any non-zero element f of $L^\infty(X, \Sigma, \mu)$, we may use the semifiniteness of the measure space to select a measurable set $E \subset \operatorname{supp}(f)$ with finite positive measure. Then, $f\chi_E$ is majorised by f with $\int f\chi_E \, d\mu \leq \|f\|_\infty \mu(E) < \infty$, thereby proving the semifiniteness of the trace induced by the integral. It remains to prove the normality. So, suppose that we are given $\{f_\alpha\} \subset L^\infty(X, \Sigma, \mu_\tau)$ and $f \in L^\infty(X, \Sigma, \mu_\tau)$ such that $M_{f_\alpha} \nearrow M_f$. For each i, $M_{f_\alpha} M_{\chi_{X_i}} = M_{f_\alpha \chi_{X_i}}$ will then increase to $M_f M_{\chi_{X_i}} = M_{f\chi_{X_i}}$ as α increases. This ensures that $M_{f_\alpha \chi_{X_i}}$ converges σ-strong* to $M_{f\chi_{X_i}}$ and hence also σ-weakly. So, by trace duality $\tau(M_{f_\alpha \chi_{X_i}} . \mathbb{1}) \to \tau(M_{f\chi_{X_i}} . \mathbb{1})$ as α increases. In other words, for each i, we have $\sup_\alpha \int_{X_i} f_\alpha \, d\mu = \int_{X_i} f \, d\mu$. On taking note of the fact that for each $g \in L^\infty(X, \Sigma, \mu)$, we have that $\int_X g \, d\mu \sum_i \int_{X_i} g\chi_{X_i} \, d\nu_i$, the claim will follow if we are able to show that $\sup_\alpha \sum_i \int_{X_i} f_\alpha \chi_{X_i} \, d\nu_i = \sum_i \sup_\alpha \int_{X_i} f_\alpha \chi_{X_i} \, d\nu_i$. We leave this as an exercise. $\qquad\square$

We pause to note the following useful fact before proving the main theorem of this section.

Lemma 1.56 *Let (X, μ) and (Y, ν) be measure spaces with $L^\infty(Y, \nu) = L^1(Y, \nu)^*$. If there is a *-isomorphism θ from $L^\infty(X, \mu)$ onto $L^\infty(Y, \nu)$ for which $\tau_\nu \circ \theta = \tau_\mu$, then also $L^\infty(X, \mu) = L^1(X, \mu)^*$.*

Proof. The measurable simple functions in L^p are norm-dense in L^p. By the equality above, the mapping θ restricts to an isometry θ' on the space of measurable simple integrable functions. Hence, $L^1(X, \mu) \cong L^1(Y, \nu)$ isometrically via a (unique) continuous map ι extending θ'. Taking the dual of ι, we get a chain of isometries

$$L^\infty(X, \mu) \overset{\theta}{\cong} L^\infty(Y, \nu) \cong L^1(Y, \nu)^* \overset{\iota^*}{\cong} L^1(X, \mu)^*.$$

One may check that the composition ρ of these isomorphisms is the canonical map $L^\infty(X, \mu) \to L^1(X, \mu)^*$ (Indeed, by linearity and density, it suffices to test this on $\rho(q)(p)$, where q (respectively, p) is the characteristic function of a measurable (respectively, μ-finite measurable) set (viewing $p \in L^1(X, \mu)$). $\qquad\qquad\qquad\qquad\qquad\qquad\square$

It is well known that for a measure space (X, Σ, μ), the measure μ is semifinite if and only if the canonical representation $\pi : f \mapsto M_f$ of $L^\infty(X, \Sigma, \mu)$ as multiplication operators on $L^2(X, \mu)$ is faithful. Indeed, if it is faithful and $\mu(E) = \infty$, then there must exist a function $g \in L^2$, which we may take to be the characteristic function of a μ-finite set F, such that $\chi_E\, g = \chi_E\, \chi_F = \chi_{E \cap F} \neq 0$. So, $0 < \mu(E \cap F) < \infty$, ensuring the semifiniteness of μ. Conversely, suppose that μ is semifinite and let $\epsilon > 0$ be given. Then, $|f| \geq \|f\|_\infty - \epsilon$ on a non-null set $E \in \Sigma$. If $F \subset E$ with $0 < \mu(F) < \infty$, then $\|f \chi_F\|_2^2 \geq (\|f\|_\infty - \epsilon)^2 \|\chi_F\|_2^2$. Thus, the canonical $*$-representation of $\mathcal{M}$ on $L^2 = L^2(X, \mu)$ is an isometric copy of $\mathcal{M}$.

Theorem 1.57   *Let (X, Σ, μ) be a measure space, and let $\mathcal{M} = L^\infty(X, \Sigma, \mu)$. Then, the following are equivalent.*

 (i) $\mathcal{M} = L^1(X, \Sigma, \mu)^*$.

 (ii) *(X, Σ, μ) is localisable (semifinite and Dedekind).*

 (iii) *μ is semifinite and $\mathcal{M}$ is an abstract von Neumann algebra.*

 (iv) *The canonical representation of $\mathcal{M}$ as multiplication operators on $L^2(X, \mu)$ is faithful and has range which is a m.a.s.a. (that is, a maximal abelian subalgebra of $B(L^2(X, \mu))$).*

 (v) *The canonical representation of $\mathcal{M}$ as multiplication operators on $L^2(X, \mu)$ is faithful and has range which is a von Neumann algebra.*

 (vi) *μ is semifinite and every strongly μ-semifinite positive measure ν on (X, Σ) with $\nu \ll \mu$ admits a μ-measurable function $h : X \to [0, \infty)$ on X such that $\nu = \mu_h$.*

Proof.   The equivalence (i)$\Leftrightarrow$(ii) is a well-known fact (see, for example, [**Sal20**, Theorem 4.35(ii)]), with (ii) $\Rightarrow$(vi) following from Theorem 1.52. The implication (v)$\Rightarrow$(iii) is clear. To see that (iii)$\Rightarrow$(ii) holds, notice that the order properties of an (abstract) von Neumann algebra ensure that $\Sigma/\mathcal{Z}(\Sigma)$ is Dedekind and hence localisable. But (X, Σ, μ)

is then also localisable. (We leave this verification as an exercise.) Next, if (iv) holds, then the fact that $\pi(\mathcal{M})$ is a m.a.s.a. ensures that $\pi(\mathcal{M}) = \pi(\mathcal{M})''$, at which point we may invoke the von Neumann double commutant theorem to see that (v) holds.

Next, we show that (ii) implies (iv). Supposing that (X, Σ, μ) is localisable, it then follows from Theorem 1.48 that $(\Sigma/\mathcal{Z}(\Sigma), \bar{\mu})$ is isomorphic to the measure algebra $(\oplus_\alpha(\Sigma_\alpha/\mathcal{Z}(\Sigma_\alpha), \bar{\mu}_\alpha))$ of a disjoint sum $\oplus_\alpha(X_\alpha, \Sigma_\alpha, \mu_\alpha)$ of finite measure spaces. In addition, by Proposition 1.16, the measure-preserving Boolean isomorphism identifying these two algebras extends to a trace-preserving *-isomorphism from $L^\infty(X, \Sigma, \mu)$ to $L^\infty(\oplus_\alpha(X_\alpha, \Sigma_\alpha, \mu_\alpha))$. The isomorphism induces a well-defined affine and order-preserving isomorphism U from the lattice of characteristic functions in $L^\infty(\oplus_\alpha(X_\alpha, \Sigma_\alpha, \mu_\alpha))$ to those in $L^\infty(X, \Sigma, \mu)$, with $\chi_E \equiv \chi_{[E]}$ mapping onto $\chi_{\theta[E]}$. Now, let $\sum_{k=1}^n \alpha_n \chi_{E_n}$ be a simple function in $L^2(\oplus_\alpha(X_\alpha, \Sigma_\alpha, \mu_\alpha))$ with the E_n's disjoint. Then, $\|\sum_{k=1}^n \alpha_n \chi_{E_n}\|_2^2 = \sum_{k=1}^n |\alpha_n|^2 \mu_\theta(E_n) = \sum_{k=1}^n |\alpha_n|^2 \mu(\theta([E_n])) = \|\sum_{k=1}^n \alpha_n \chi_{\theta([E_n])}\|_2^2$. Thus, U extends to a unitary from $L^2(\oplus_\alpha(X_\alpha, \Sigma_\alpha, \mu_\alpha))$ to $L^2(X, \Sigma, \mu)$. This fact then clearly shows that the multiplication algebra of $L^\infty(X, \Sigma, \mu)$ is unitarily equivalent to the multiplication algebra of $L^\infty(\oplus_\alpha(X_\alpha, \Sigma_\alpha, \mu_\alpha))$, and hence, for all practical purposes, we may replace (X, Σ, μ) with $\oplus_\alpha(X_\alpha, \Sigma_\alpha, \mu_\alpha)$.

The next step is to observe that the argument in the second half of the proof of Corollary 1.49 shows that for each α, the algebra of multiplication operators on $L^2(X_\alpha, \Sigma_\alpha, \mu_\alpha)$ with symbols in $L^\infty(X_\alpha, \Sigma_\alpha, \mu_\alpha)$ is a m.a.s.a. Since, for each $p \in \{2, \infty\}$, we have $L^p(\oplus_\alpha(X_\alpha, \Sigma_\alpha, \mu_\alpha)) = \oplus_\alpha L^p(X_\alpha, \Sigma_\alpha, \mu_\alpha)$, the same is, therefore, true of $L^\infty(\oplus_\alpha(X_\alpha, \Sigma_\alpha, \mu_\alpha))$.

The theorem will now follow if we are able to prove that (vi)$\Rightarrow$(i). We will only use the case of (vi) where $\nu \leq \mu$. It is classical (see, e.g. [**Sal20**, Theorem 4.33(iv)]) that for semifinite measures, the canonical map $L^\infty \to (L^1)^*$ is an isometric injection. It suffices to show that any $\varphi \in \mathrm{ball}(L^1(\mu)^*)_+$ is 'integration' with respect to the measure $h\,d\mu$, for some $h \in L^\infty(\mu)$. Define $\nu'(E) = \varphi(\chi_E)$ for μ-finite $E \in \Sigma$. This is countably additive on the measurable μ-finite sets, by the matching part in the standard proof that $L^\infty = (L^1)^*$ for finite measures (for example, see the first paragraph of the proof of [**Fol16**, Theorem 6.15]). Define $\nu(E) = \sup \nu'(F)$, where the supremum is over μ-finite measurable $F \subset E$. This is easily checked to be a measure with

$$\nu(E) = \sup \nu'(F) = \sup \varphi(\chi_F) \leq \sup \|\chi_F\|_1 \leq \mu(E).$$

Moreover, if $\nu(E) > 0$, then there exists μ-finite measurable $F \subset E$ with $0 < \nu'(F) = \nu(F) < \infty$. So, ν is strongly μ-semifinite. By (vi), there exists measurable $h : X \to [0, \infty)$ on X with $\nu = \mu_h$. Let D be a set on which $h > 1$. By semifiniteness, we may assume that $\mu(D) < \infty$. Since $0 \leq \int_D (h - 1)\, d\mu = \nu(D) - \mu(D) \leq 0$, we see that $h = 1$ μ-a.e. on D. So, $\mu(D) = 0$, and we may assume that $h : X \to [0, 1]$ and $h \in L^\infty(\mu)$. We have

$$\varphi(f) = \int_X f\, d\nu = \int_X fh\, d\mu$$

when f is the characteristic function of a measurable μ-finite set, hence also for simple functions which are linear combinations of such characteristic functions. Now, let $f \in L^1(\mu)_+$, and (s_n) be a sequence of the latter simple functions with $s_n \geq 0$ and $s_n \nearrow f$. It follows from Lebesgue's monotone convergence theorem that $\int_X (f - s_n)\, d\mu \to 0$, so that $\varphi(s_n) \to \varphi(f)$, and $\varphi(s_n) = \int_X s_n h\, d\mu \to \int_X fh\, d\mu$. Thus, $\varphi(f) = \int_X fh\, d\mu$ for all $f \in L^1(\mu)$. $\square$

1.7.2 L^∞-spaces on Dedekind measure spaces

Theorem 1.57 characterises those semifinite measure spaces (X, Σ, μ) for which the space $L^\infty(X, \mu)$ is a von Neumann algebra with predual $L^1(X, \mu)$. With some more refined technology and a deeper understanding of abelian von Neumann algebras at our disposal, we are now ready to consider the related problem of characterising measure spaces (X, Σ, μ) for which $L^\infty(X, \mu)$ is a von Neumann algebra with no assumption on the identity of a predual and no assumption of semifiniteness. The achieved analysis will also demonstrate the extent to which the normality of τ_μ will ensure that the ambient measure space is localisable.

Dedekind measure spaces as defined earlier may, of course, not be semifinite. However, it is an interesting fact that they may be decomposed into a semifinite (and so localisable) part and an 'infinite only' part.

Lemma 1.58 *Suppose that (X, Σ, μ) is a Dedekind measure space and $\mathcal{M} = L^\infty(X, \Sigma, \mu)$. Then, there exists $S \in \Sigma$ such that $(S, \mu{\restriction}S)$*

is semifinite and the projection lattice of $L^\infty(S, \mu{\restriction}S)$ is Dedekind complete, and such that $\mu(E) = \infty$ for any measurable subset E of $X \setminus S$ with $\mu(E) > 0$.

Theorem 1.59 *Let μ be a measure on a measurable space (X, Σ), and set $\mathcal{M} = L^\infty(X, \Sigma, \mu)$. The following are equivalent.*

(i) *$\mathcal{M}$ is an abstract von Neumann algebra.*

(ii) *$\mu \cong \nu$ for a localisable measure ν on (X, Σ). The measure ν can be chosen to agree with μ on the 'semifinite part' of μ.*

Proof. (i) $\Rightarrow$ (ii) Let $(S, \mu{\restriction}S)$ be the semifinite part of μ as in Lemma 1.58 and the lines after it. Let $e \in \mathbb{P}(\mathcal{M})$ correspond to S, and set $\mathcal{N} = \mathcal{M}e$. Then, $\mathcal{N} \cong L^\infty(S, \mu{\restriction}S)$. Let $\varphi_0 = \varphi{\restriction}\mathcal{N}_+ = \int_S \cdot\, d(\mu{\restriction}S)$. Then $(S, \mu{\restriction}S)$ is localisable by Lemma 1.58.

The 'antisemifinite part' $\mathcal{M}(1 - e)$ is a commutative von Neumann algebra, so by Theorem 1.54, there exists a $*$-isomorphism $\pi : \mathcal{M}(1 - e) \to L^\infty(\Omega, \Sigma_0, \nu')$, for a localisable measure space (Ω, Σ_0, ν'). The projections in $\mathcal{M}(1 - e)$, which correspond to the Σ-measurable subsets of $X \setminus S$, are taken by π to the (equivalence class of) characteristic functions of sets in Σ_0. Define a measure ν on $X \setminus S$ by $\nu(F) = \int_\Omega \pi(\chi_F)\, d\nu'$ for $F \in \Sigma, F \subset X \setminus S$. Since ν' is semifinite, so is ν. Therefore, ν is a localisable measure.

(ii) $\Rightarrow$ (i) The equivalence of the measures ensures that $L^\infty(X, \Sigma, \nu) = L^\infty(X, \Sigma, \mu)$, with $L^\infty(X, \Sigma, \nu)$ being an abstract von Neumann algebra by Theorem 1.57. $\qquad\square$

We recall some interesting facts, a couple of which will be used in the background in the remainder of this section. It is an interesting exercise that a measure space (X, Σ, μ) is Dedekind if and only if $L^\infty_{\mathbb{R}}(X, \Sigma, \mu)$ is boundedly complete (i.e. every bounded subset has a supremum). To see the one direction, suppose that (x_t) is an increasing net of projections in $\mathcal{M}$ with supremum x. Then, $x \leq 1$, and so for each t, we have $x_t = x x_t$. Therefore, $(x + 1)x = \sup_t (x + 1)x_t = \sup_t 2x_t = 2x$. That is, x is a projection. A proof of the other direction may be found in the details of the proof of, for example, [**Fre04**, 363M].

As mentioned earlier, $L^\infty_{\mathbb{R}}(X, \Sigma, \mu) = C(\Omega, \mathbb{R})$ for a compact space Ω, the spectrum, or maximal ideal space of $L^\infty(X, \mu)$, which in this case is called the *Stone space*. This equality encodes a very elegant connection between Dedekind measure spaces and Stonean spaces, which we pause to record as a proposition. We remind the reader that a compact Hausdorff space is said to be a *Stonean space*; the closure of any open set is open. We shall need some additional background on Stonean spaces in the form of the lemmata below, both due to M.H. Stone [**Sto49**]. The proofs we present are those recorded in [**Sak71**, Propositions 1.3.1 and 1.3.2].

Lemma 1.60   *Let Ω be a Stonean space. Then, every element $a \in C(\Omega)$ can be uniformly approximated by finite linear combinations of projections.*

Proof.   Given $a \in C(\Omega)$, we may clearly assume that a is a positive function on Ω. Given $\epsilon > 0$, let $0 = \lambda_0 < \lambda_1 < \lambda_2 < \cdots < \lambda_n = \|a\| + 1$ be a partitioning of the interval $[0, \|a\| + 1]$ into subintervals of length less than ϵ (that is $\lambda_{i+1} - \lambda_i < \epsilon$ ($i = 1, 2, \ldots, n - 1$)). Define inductively the clopen sets $G_1, G_2, \ldots, G_n$ by $G_1 = \overline{\{t \colon a(t) < \lambda_1, t \in \Omega\}}$ and

$$G_i = \overline{\left\{ t \colon a(t) < \lambda_i, t \in \Omega - \bigcup_{j=1}^{i-1} G_j \right\}} \quad (i = 2, 3, \ldots, n).$$

It is easy to check that the G_is are mutually disjoint. Since they are all clopen, the characteristic functions $p_i = \chi_{G_i}$ turn out to be continuous projections. Moreover, by construction

$$\left| a(t) - \sum_{i=1}^{n} \lambda_i x_i(t) \right| < \epsilon \quad t \in \Omega;$$

that is, $\left\| a - \sum_{i=1}^{n} \lambda_i x_i \right\|_\infty < \epsilon$. $\square$

It is known that for any compact Hausdorff space Ω, the space $C(\Omega, \mathbb{R})$ is boundedly complete if and only if Ω is Stonean. We shall prove only the implication that we need. For a proof of the reverse implication, see, for example, [**DDLS16**, Theorem 2.3.3]. For interest's sake, we remark that Ω is Stonean if and only if $C(\Omega, \mathbb{R})$ is an injective Banach space (see, for example, [**Fre04**, 363R]), and it is not hard to show that the latter is equivalent to $C(\Omega)$ being injective as a complex Banach space. In this case, $C(\Omega)$ is called a (commutative) *AW*-algebra*.

Lemma 1.61 *Let Ω be a compact Hausdorff space. Suppose that every bounded upwardly directed set $\{f_\alpha\}$ of real-valued, non-negative functions in $C(\Omega)$ has a least upper bound in $C(\Omega)$. Then, Ω is Stonean.*

Proof. Let G be an open set in Ω, and let (f_α) be the set of all continuous functions on Ω such that $0 \le f_\alpha \le 1$, with $\mathrm{supp}(f_\alpha) \subset G$. Then, $\{f_\alpha\} \subset C(\Omega)$ is a bounded upwardly directed set. Set $f = \sup_\alpha f_\alpha$ in $C(\Omega)$ with the function $g \colon \Omega \to [0,1]$ defined pointwise by $g(t) = \sup_\alpha f_\alpha(t)$ for all $t \in \Omega$. We then clearly have that $f \le 1$. Since $f_\alpha \le f$ for each α, we must also have that $g(t) \le f(t)$ for all $t \in \Omega$. But by Urysohn's lemma, there must for each $t \in G$ exist an f_α such that $f_\alpha(t) = 1$. Hence, $g(t) = 1$ on G, and so the same must be true of f on G. By continuity, we in fact have that $f(t) = 1$ on $\overline{G}$. Suppose that there exists a $t_0 \in \Omega - \overline{G}$ such that $f(t_0) > 0$. Now, use Urysohn's lemma to select a positive h in $C(\Omega)$ such that $h = 1$ on $\overline{G}$ and $h(t_0) = 0$. Then, $f_\alpha \le f \wedge h < f$, a contradiction. We, therefore, have that $f^{-1}(0) = \Omega - \overline{G}$ showing that $\Omega - \overline{G}$ is closed and $\overline{G}$ therefore open. $\qquad\square$

Proposition 1.62 *Let (X, Σ, μ) be a Dedekind measure space. Then, the maximal ideal space Ω of $L^\infty_{\mathbb{R}}(X, \Sigma, \mu)$ is a Stonean space. Conversely, if Ω a Stonean space, then there is a Dedekind measure space (X, Σ, μ) with $L^\infty_{\mathbb{R}}(X, \Sigma, \mu) \cong C(\Omega, \mathbb{R})$.*

Proof. Since (as noted earlier) $L^\infty_{\mathbb{R}}(X, \Sigma, \mu)$ is boundedly complete, the one direction follows by applying Lemma 1.61 to the fact that $L^\infty_{\mathbb{R}}(X, \Sigma, \mu) = C(\Omega, \mathbb{R})$. For the converse, we define a measure on the projection lattice of $C(\Omega, \mathbb{R})$ by setting $\bar\mu(p) = \infty$ for any non-zero projection p. Theorem 1.21 then gives a Dedekind measure space (X, Σ, μ). But by Lemma 1.60, the fact that Ω is Stonean ensures that the projections in $C(\Omega)$ are densely spanning. Hence, the proof of Proposition 1.16 shows that $L^\infty(X, \Sigma, \mu) \cong C(\Omega)$ $*$-isomorphically. $\qquad\square$

Putting the above facts together, a measure space is Dedekind if and only if the spectrum (or Stone space) Ω above is Stonean, or equivalently $C(\Omega)$ or its isomorphic copy $L^\infty(X, \mu)$ has the above properties.

Example 1.63 It is evident from the equivalence of (i) and (ii) in Theorem 1.57 that $\mathcal{M} = L^\infty(X, \mu)$ is a von Neumann algebra if (X, μ) is Dedekind and integration τ_μ with respect to μ is a semifinite normal weight on $\mathcal{M}$. However, if μ is not semifinite, so that ψ_μ is allowed to be infinite on 'large pieces', then $\mathcal{M}$ need not be a von Neumann algebra. Indeed, we give a counterexample to the question of whether $L^\infty(X, \mu)$ is a von Neumann algebra if and only if we have that both the measure space (X, μ) is Dedekind and integration ψ_μ is a normal weight. To do this, one may take the canonical example of a commutative AW^*-algebra, which is not a von Neumann algebra, and assign it the trivial 'always infinite' weight.

More specifically, consider the *Dixmier algebra D*, the quotient of the bounded Borel functions on $[0, 1]$ by the closed ideal of bounded Borel functions vanishing on the complement of a meagre Borel set [**Dix51**]. Being a commutative C^*-algebra, $D = C(Y)$ for a compact space Y. It can be shown that D_{sa} is boundedly complete (see, for example, [**Dix51**, **KR86**]). Hence, by the facts summarised below Theorem 1.59, Y is Stonean, and there is a Dedekind measure space (X, Σ, μ) with $L^\infty(X, \Sigma, \mu) \cong C(Y)$ $*$-isomorphically. However, $L^\infty(Y, \Sigma, \mu)$ is not a von Neumann algebra, since it is well known that $C(Y)$ is not a von Neumann algebra (see, for example, [**KR86**]). To see that the integral φ with respect to μ is normal, note that by the lines just below Theorem 1.59, μ only takes values 0 and ∞. For any non-negative measurable simple function $s = \sum_{k=1}^{n} t_k \chi_{E_k}$ with $t_k > 0$, we see that $\int s\, d\mu$ is either 0 or ∞. So, the integral of any non-negative measurable function is either 0 or ∞. Suppose that (f_t) is an increasing bounded non-negative net in $L^\infty(Y, \Sigma, \mu)$. Thus, $(\varphi(f_t))$ is an increasing net in $\{0, \infty\}$. If $\varphi(f_t) = 0$ for all t, then $f_t = 0$ μ-a.e., and so $\sup_t f_t$ is 0. If $\varphi(f_t) = \infty$ for some t, then $\sup_t \varphi(f_t) = \varphi(\sup_t f_t)$. It is now clear that φ is normal. Clearly, φ is also completely additive.

1.8 A categorical description of abelian von Neumann measure algebras

Remarkable as the preceding results may be, the insights gleaned from Propositions 1.16 yield an even more remarkable conclusion, namely that the class of abelian von Neumann measure algebras and the class of localisable measure algebras are basically the same thing. To fully appreciate this statement, we need some concepts from category theory.

Definition 1.64 A category $\mathcal{C}$ consists of the following three mathematical entities:

- A class $\mathbf{ob}(C)$, whose elements are called *objects*.
- A class $\mathbf{hom}(C)$, whose elements are called morphisms. Morphisms represent the natural 'transitions' from one object to another. Each morphism f has a source object $s(f)$ and target object $t(f)$. A morphism f from $s(f) = a$ to $t(f) = b$ would be written as $f : a \to b$. The class of all morphisms from a to b is denoted by $\mathbf{hom}_C(a, b)$ or simply $\mathbf{hom}(a, b)$ if $\mathcal{C}$ is understood.
- A binary operation $\circ$ governing composition of morphisms. Specifically, for any three objects a, b, and c, we have that $\circ : (\mathbf{hom}(b, c), \mathbf{hom}(a, b)) \to \mathbf{hom}(a, c)$, where the composition of $f : a \to b$ and $g : b \to c$ is denoted by $g \circ f$. Composition must satisfy the *associativity axiom* in that for $f : a \to b$, $g : b \to c$, and $h : c \to d$, we must have $h \circ (g \circ f) = (h \circ g)f$, and the *identity axiom* meaning that for every object x, there must exist a morphism $1_x : x \to x$, such that for every morphism $f : a \to b$, we have that $1_b \circ f = f = f \circ 1_a$. (it follows from the above that there is exactly one identity morphism for every object.)

Next, we define the notion of a functor, where functors are the means of 'comparing' two given categories. In particular, when we say that two categories are 'functorially equivalent', we basically say that the two categories are from a categorical point of view 'the same thing'.

Definition 1.65 Let $\mathcal{C}$ and $\mathcal{D}$ be categories.

A functor F from $\mathcal{C}$ to $\mathcal{D}$ is a mapping acting on both the objects and morphisms, which

- associates each object a in $\mathcal{C}$ to an object $F(a)$ in $\mathcal{C}$;
- associates each morphism $f : a \to b$ in $\mathcal{C}$ to a morphism $F(f) : F(a) \to F(b)$ in such a way that the following conditions are satisfied:
  - For every object $a \in \mathcal{C}$, we have that $F(1_a) = 1_{F(a)}$,
  - and $F(g \circ f) = F(g) \circ F(f)$ for any two morphisms $f : a \to b$ and $g : b \to c$ in $\mathcal{C}$.

The categories $\mathcal{C}$ and $\mathcal{D}$ are said to be *(functorially) equivalent* if there exist functors $F : \mathcal{C} \to \mathcal{D}$ and $G : \mathcal{D} \to \mathcal{C}$, and two natural isomorphisms $\epsilon : FG \to \mathrm{id}_{\mathcal{D}}$ and $\eta : \mathrm{id}_{\mathcal{C}} \to GF$. Here, $\mathrm{id}_{\mathcal{C}}$ and $\mathrm{id}_{\mathcal{D}}$ denote identity functors on $\mathcal{C}$ and $\mathcal{D}$, assigning each object and morphism to itself.

Remark 1.66 When trying to prove the (functorial) equivalence of two categories, it is not always that easy to construct a pair (F, G) of functors satisfying the criteria in the preceding definition. Fortunately, the existence of a single functor F from $\mathcal{C}$ to $\mathcal{D}$, which satisfies certain additional regularity properties, is known to be sufficient to guarantee functorial equivalence. Specifically, the categories $\mathcal{C}$ and $\mathcal{D}$ are functorially equivalent if there exists a functor F from $\mathcal{C}$ to $\mathcal{D}$ that is

- *full*, meaning that for any two objects a and b of $\mathcal{C}$, the map from $\mathbf{hom}_{\mathcal{C}}(a, b)$ to $\mathbf{hom}_{\mathcal{D}}(F(a), F(b))$ induced by F is surjective;
- *faithful*, meaning that for any two objects a and b of $\mathcal{C}$, the map from $\mathbf{hom}_{\mathcal{C}}(a, b)$ to $\mathbf{hom}_{\mathcal{D}}(F(a), F(b))$ induced by F is injective;
- *essentially surjective*, meaning that each object d in $\mathcal{D}$ is isomorphic to an object of the form $F(c)$, where $c \in \mathcal{C}$.

We are now finally ready to give a precise mathematical meaning to the claim that *'the class of abelian von Neumann measure algebras and the class of localisable measure algebras are basically the same thing'*. The first step in doing so is to introduce the appropriate categories.

Definition 1.67 The category **AvNMA** is defined to be the collection of semifinite abelian von Neumann measure algebras $(\mathcal{M}, \tau)$, with the natural morphisms between two such object $(\mathcal{M}_1, \tau_1)$ and $(\mathcal{M}_2, \tau_2)$ from **AvNMA** defined to be normal *-homomorphisms $\mathscr{I} : \mathcal{M}_1 \to \mathcal{M}_2$ for which we have $\tau_2 \circ \mathscr{I} = \tau_1$ (see Definition 1.15). Next, the category **LMA** is defined the collection of localisable measure algebras with the natural morphisms in this case being the measure-preserving order continuous Boolean homomorphisms.

Theorem 1.68 *The categories* **AvNMA** *of semifinite abelian von Neumann measure algebras equipped with faithful normal semifinite traces, and* **LMA** *of localisable measure algebras, are equivalent.*

Proof. Let F be the mapping that sends each object $(\mathcal{M},\tau)$ in **AvNMA** to $(\mathbb{P}(\mathcal{M}),\tau{\upharpoonright}\mathbb{P}(\mathcal{M}))$ in **LMA** and sends a morphism $\mathscr{I} : (\mathcal{M}_1,\tau_1) \to (\mathcal{M}_2,\tau_2)$ to $F(\mathscr{I}) = \mathscr{I}{\upharpoonright}\mathbb{P}(\mathcal{M}_1)$. It is now not difficult to conclude from Proposition 1.16 that F sends $\mathscr{I} : (\mathcal{M}_1,\tau_1) \to (\mathcal{M}_2,\tau_2)$ to $F(\mathscr{I}) : F(\mathcal{M}_1,\tau_1) \to F(\mathcal{M}_2,\tau_2)$ in such a way that

- $F(1_{(\mathcal{M},\tau)}) = 1_{F(\mathcal{M},\tau)}$;
- and $F(\mathscr{I} \circ \mathscr{J}) = F(\mathscr{I}) \circ F(\mathscr{J})$ for morphisms $\mathscr{J} : (\mathcal{M}_1,\tau_1) \to (\mathcal{M}_2,\tau_2)$ and $\mathscr{I} : (\mathcal{M}_2,\tau_2) \to (\mathcal{M}_3,\tau_3)$.

In other words, F is a functor.

It is clear from Proposition 1.16 that F, in fact, induces a bijection from $\mathbf{hom}((\mathcal{M}_1,\tau_1),(\mathcal{M}_2,\tau_2))$ to $\mathbf{hom}((\mathbb{P}(\mathcal{M}_1),\tau_1),(\mathbb{P}(\mathcal{M}_2),\tau_2))$. Hence, F is both full and faithful.

In closing, we note two facts. First, by Theorem 1.48, any localisable measure algebra is isomorphic to $(\Sigma/\mathcal{Z}(\Sigma),\bar{\mu})$ for some localisable measure space (X,Σ,μ), and second, for such a measure space, the pair $(L^{\infty}(X,\Sigma,\mu), \int_X \cdot\, d\mu)$ will then be, by Theorem 1.57 and Corollary 1.55, an abelian von Neumann measure algebra. Since the projection lattice of $L^{\infty}(X,\Sigma,\mu)$ is a copy of $\Sigma/\mathcal{Z}(\Sigma)$, it follows that F is essentially surjective.

This then suffices to ensure that the categories **AvNMA** and **LMA** are functorially equivalent. $\qquad\qquad\square$

There is a version of the last result that does not require weights or measures on the algebras. Note that the semifinite measures on a Dedekind complete Boolean algebra $\mathcal{B}$ are all 'equivalent' in some sense and constitute a group. To see this, fix one such semifinite measure $\bar{\mu}$. Then, $(\mathcal{B},\bar{\mu})$ is isomorphic to a concrete measure algebra of a localisable measure space (X,Σ,μ), with $\bar{\mu}$ corresponding to the restriction of the weight $\omega(f) = \int_X f\, d\mu$, by Theorem 1.21. Let $\mathcal{M}$ be the associated (abstract) von Neumann algebra, i.e., $L^{\infty}(X,\Sigma,\mu)$. Semifinite measures on $\mathcal{B}$ correspond bijectively to faithful normal semifinite weights on $\mathcal{M}_+$, as may be seen again by Theorem 1.21, by thinking of $\mathcal{M}$ as $C(Z)$, where Z is the Stone space. By Radon–Nikodym Theorem 1.52, such weights on $\mathcal{M}_+$ are all equivalent to ω, and equal to ω_h for (μ-a.e. unique) strictly positive measurable functions $h : X \to (0,\infty)$.

The (μ-a.e. equivalence classes of the) latter functions on X clearly form a group.

It is, therefore, natural to define a category **LBA** consisting of the Dedekind-complete Boolean algebras that admit some (unspecified) localisable measure, together with the order continuous morphisms of Boolean algebras. Let **AvNA** be the category of abelian von Neumann algebras and normal $*$-homomorphisms.

Theorem 1.69 **LBA** *and* **AvNA** *are equivalent categories.*

This follows from the proof of Theorem 1.68. See also [**Pav20**]; the above shows that **LBA** is equivalent to the categories mentioned in the main theorem of that paper.

Remark 1.70 The projection lattice of a direct sum $(\oplus_i \mathcal{M}_i, \oplus_i \tau_i)$ of abelian von Neumann measure algebras is clearly of the form $(\prod_i \mathbb{P}(\mathcal{M}_i), \tau)$, where here τ has the action $\tau((a_i)_{i\in I}) = \sum_i \tau(a_i)$ (see [**Fre04**], 322L). Moreover, if each $(\mathcal{M}_i, \tau_i)$ is isomorphic to say $(L^\infty(X_i, \Sigma_i, \mu_i), \int \cdot \, d\mu_i)$, then $(\oplus_i \mathcal{M}_i, \oplus_i \tau_i)$ will be isomorphic to

$$(L^\infty(\oplus_i(X_i, \Sigma_i, \mu_i)), \int \cdot \, d(\oplus_i \mu_i)) = (\oplus_i L^\infty(X_i, \Sigma_i, \mu_i), \oplus_i \int \cdot \, d\mu_i).$$

The action of the functor will then associate each $(\mathbb{P}(\mathcal{M}_i), \tau_i)$ with $(\Sigma_i/\mathcal{Z}(\Sigma_i), \overline{\mu_i})$. Thus, the functor takes direct sums of von Neumann measure algebras to *simple products* of localisable measure algebras, which correspond to disjoint sums of measure spaces.

Chapter 2
The Schatten–von Neumann classes

The Schatten–von Neumann classes were the first well-developed infinite-dimensional example of noncommutative L^p and Orlicz spaces and, in a very real sense, therefore represents the genesis of the theory described in the rest of the book. In order to set the scene for what follows, we provide full details of this construction. The approach we follow is strongly based on that of Diestel, Jarchow and Tonge [**DJT95**]. Almost all the constructs that play such an important role in developing the theory of L^p-spaces for semifinite algebras already appear here in embryonic form. This example is, therefore, a helpful indicator of the kinds of constructs that need to be developed if we wish to posit a matching theory for general semifinite von Neumann algebras.

2.1 Definition and basic properties
of the trace on $B(H)$

Definition 2.1 For a fixed orthonormal basis $(v_i) \subset H$, we define the trace Tr on $B(H)_+$ by $\mathrm{Tr}(f) = \sum_i \langle fv_i, v_i \rangle$.

We list some basic facts about the trace Tr.

Theorem 2.2

(1) Tr *is additive and homogeneous on* $B(H)_+$.
(2) $\mathrm{Tr}(f^*f) = \mathrm{Tr}(ff^*)$ *for any* $f \in B(H)$.
(3) Tr *is independent of the choice of orthonormal basis* (v_i).
(4) *For any projection* $e \in B(H)$, $\mathrm{Tr}(e) = \mathrm{rank}(e)$.
(5) *If* $0 \leq a \leq b$ *then* $\mathrm{Tr}(a) \leq \mathrm{Tr}(b)$. *In particular, if* $(b_\alpha) \subset B(H)_+$ *increases to* $b \in B(H)$, *then* $\sup_\alpha \mathrm{Tr}(b_\alpha) = \mathrm{Tr}(b)$.

Proof. Property (1) is obvious. To see that (2) holds, let an orthonormal basis (ξ_i) be given and notice that we may use Parseval's identity and Fubini's theorem to see that

Noncommutative measures and L^p and Orlicz Spaces, with Applications to Quantum Physics. Stanisław Goldstein and Louis Labuschagne, Oxford University Press. © Stanisław Goldstein and Louis Labuschagne (2025).
DOI: 10.1093/oso/9780198950202.003.0004

$$\sum_i \langle f^* f \xi_i, \xi_i \rangle = \sum_i \langle f \xi_i, f \xi_i \rangle = \sum_i \sum_j \langle f \xi_i, \xi_j \rangle \overline{\langle f \xi_i, \xi_j \rangle}$$

$$= \sum_j \sum_i \langle \xi_i, f^* \xi_j \rangle \langle f^* \xi_j, \xi_i \rangle = \sum_j \sum_i \langle f^* \xi_j, \xi_i \rangle \overline{\langle f \xi_j, \xi_i \rangle} = \sum_j \langle f^* f \xi_j, \xi_j \rangle.$$

Given two orthonormal bases (ξ_i) and (η_i), we may of course select a unitary $u \in B(H)$ such that $u \eta_i = \xi_i$ for all i. Since by (2) $\mathrm{Tr}(u^* f u) = \mathrm{Tr}(f^{1/2} u^* u f^{1/2}) = \mathrm{Tr}(f)$ for any $f \in B(H)_+$, property (3) therefore follows from (2). Property (4) is by definition. To see this, let (ζ_j) be an orthonormal basis of $e(H)$, and then select an orthonormal basis (ξ_i) of H containing (ζ_j). Then $\mathrm{Tr}(e) = \sum_i \langle e \xi_i, \xi_i \rangle = \sum_j \langle \zeta_j, \zeta_j \rangle = \mathrm{rank}(e)$.

Finally, notice that if $0 \leq a \leq b$, then $\mathrm{Tr}(a) \leq \mathrm{Tr}(b)$ almost by definition. Therefore, it remains to prove that if $(b_\alpha) \subset B(H)_+$ increases to $b \in B(H)$, then $\sup_\alpha \mathrm{Tr}(b_\alpha) = \mathrm{Tr}(b)$. From what we just verified, it is clear that $\sup_\alpha \mathrm{Tr}(b_\alpha) \leq \mathrm{Tr}(b)$. So, we just need to show that we also have $\mathrm{Tr}(b) \leq \sup_\alpha \mathrm{Tr}(b_\alpha)$. Let $(\xi_i) \subset H$ be an orthonormal basis. Since vector states are normal, it is clear that $\sup_\alpha \langle b_\alpha \xi_i, \xi_i \rangle = \langle b \xi_i, \xi_i \rangle$ for each i. From this, it follows that we will for any finite subset $\mathcal{F}$ of the index set $\mathcal{I}$ have $\sum_{i \in \mathcal{F}} \langle b \xi_i, \xi_i \rangle = \sup_\alpha \sum_{i \in \mathcal{F}} \langle b_\alpha \xi_i, \xi_i \rangle \leq \sup_\alpha \mathrm{Tr}(b_\alpha)$. The claim now follows by noticing that $\mathrm{Tr}(b)$ is the limit of the net $(\sum_{i \in \mathcal{F}} \langle b \xi_i, \xi_i \rangle)_{\mathcal{F}}$. $\qquad\square$

2.2 Definition and basic properties of the Schatten–von Neumann classes

The starting point for our analysis is the spectral theorem for compact operators (see, for example, [**DJT95**, Theorem 4.1] for a proof).

Theorem 2.3 (Spectral theorem for compact operators) *An operator $f \in B(H)$ is compact if and only if it admits a representation of the form*

$$f = \sum_{n=1}^{\infty} \alpha_n \langle \cdot, x_n \rangle y_n$$

where (x_n) and (y_n) are orthonormal sequences, and (α_n) is a null sequence of scalars.

Definition 2.4 Let $1 \leq p < \infty$. We define the pth Schatten class $\mathscr{S}_p(H)$ to be the set of all compact operators a for which the sequence (α_n) mentioned above belongs to $\ell_p(\mathbb{N})$. The norm ς_p on $\mathscr{S}_p$ is given by $\varsigma_p(a) = (\sum_{n=1}^{\infty} |\alpha_n|^p)^{1/p}$.

Remark 2.5 Given $1 \leq p < q < \infty$, it follows from known facts about ℓ_p that $\mathscr{S}_p(H) \subset \mathscr{S}_q(H)$ with $\varsigma_q \leq \varsigma_p$.

We pass to showing that the norm is well defined and that it is indeed a norm. As a first step, we note the following.

Corollary 2.6 *Given a compact operator f, the representation in the spectral theorem may be chosen so that (α_n) is a decreasing sequence of non-negative scalars.*

Proof. Suppose that f can be written in the form $f = \sum_{n=1}^{\infty} \alpha_n \langle \cdot, x_n \rangle y_n$ where all the α_ns are non-zero. It is then an exercise to see that $|f| = \sum_{n=1}^{\infty} |\alpha_n| \langle \cdot, y_n \rangle y_n$. We may clearly reorder the terms so that $|\alpha_n|$ decreases as n increases. Let $f = v|f|$ be the polar decomposition of f. The closure of the range of $|f|$ is the closure of $\mathrm{span}\{y_n : n \in \mathbb{N}\}$. So, the initial projection of v is just $\sum_{n=1}^{\infty} \langle \cdot, y_n \rangle y_n$, ensuring that (vy_n) is again an orthonormal system. The representation we seek is then $f = v|f| = \sum_{n=1}^{\infty} |\alpha_n| \langle \cdot, y_n \rangle vy_n$. $\qquad\square$

We show that the scalars (α_n), as described in the preceding corollary, are uniquely determined. We shall do this by means of the so-called *approximation numbers* of operators in $B(H)$. For any $f \in B(H)$, the nth *approximation number* of f is defined to be $a_n(f) = \inf\{\|f - g\| : g \in B(H), \dim(g(H)) < n\}$. Note that $a_1(f)$ is just $\|f\|$. We clearly have that f is a limit of finite-rank operators (and hence compact) if and only if $a_n(f) \to 0$ as $n \to \infty$.

Proposition 2.7 *Let f be a compact operator represented as in the above corollary. Then $\alpha_n = a_n(f)$.*

Proof. Let f be given as in the hypothesis, and for a fixed n, set $f_n = \sum_{k=1}^{n-1} \alpha_k \langle \cdot, x_k \rangle y_k$. The operator f_n clearly has rank less than n. For any $x \in H$, we have that

$$\|(f - f_n)(x)\| \;=\; \|\sum_{k=n}^{\infty} \alpha_k \langle x, x_k \rangle y_k\|$$

$$\leq \; (\sum_{k=n}^{\infty} \alpha_k^2 |\langle x, x_k \rangle|^2)^{1/2}$$

$$\leq \; \alpha_n (\sum_{k=n}^{\infty} |\langle x, x_k \rangle|^2)^{1/2}$$

$$\leq \; \alpha_n \|x\|.$$

Therefore, it is clear that $a_n(f) \leq \|f - f_n\| \leq \alpha_n$.

For the converse inequality, let $g \in B(H)$ be any operator of rank less than n. Then, g cannot be injective on the n-dimensional subspace $\mathrm{span}\{x_k : 1 \leq k \leq n\}$. Therefore, we may select a unit vector of the form $x_0 = \sum_{k=1}^{n} \beta_k x_k$ in the kernel of g. Since then $f(x_0) = \sum_{k=1}^{n} \alpha_k \beta_k y_k$, it follows that

$$a_n(f) \;\geq\; \|f - g\|$$

$$\geq\; \|f(x_0)\|$$

$$=\; (\sum_{k=1}^{n} \alpha_k^2 |\beta_k|^2)$$

$$\geq\; \alpha_n (\sum_{k=1}^{n} |\beta_k|^2)$$

$$=\; \alpha_n \|x_0\|^2 = \alpha_n. \qquad \square$$

Having established the well-definiteness of ς_p, we proceed to showing that $\mathscr{S}_p$ is a linear space. This is achieved by (for now) showing that ς_p is a quasi-norm. Once trace duality has been established, the fact that ς_p is in fact a norm will be clear.

Lemma 2.8 *For any $f, g \in B(H)$, we have that*

$$a_n(f) = a_n(f^*) = a_n(|f|)$$

and that

$$a_{n+m-1}(f + g) \leq a_n(f) + a_m(g) \text{ and } a_{n+m-1}(fg) \leq a_n(f) . a_m(g)$$

for all $n, m \in \mathbb{N}$.

Proof.   It is an exercise to see that an operator $u \in B(H)$ has rank k if and only if the same is true of u^*. The fact that $a_n(f) = a_n(f^*)$ then follows from this observation and the fact that $\|f - u\| = \|f^* - u^*\|$. Next, let $f = v|f|$ be the polar decomposition of f. If $u \in B(H)$ has rank less than n, then so does vu and v^*u. Since then $a_n(f) \le \|f - vu\| \le \||f| - u\|$ and $a_n(|f|) \le \||f| - v^*u\| \le \|f - u\|$, it is clear that both $a_n(f) \le a_n(|f|)$ and $a_n(|f|) \le a_n(f)$ must hold.

Let $u_0, u_1 \in B(H)$ be given with $\mathrm{rank}(u_0) < n$ and $\mathrm{rank}(u_1) < m$. Then, it is clear that $a_{n+m-1}(f+g) \le \|(f+g)-(u_0+u_1)\| \le \|f-u_0\| + \|g - u_1\|$. Taking the infimum over u_0 and u_1 leads to the conclusion that

$$a_{n+m-1}(f + g) \le a_n(f) + a_m(g).$$

With u_0 and u_1 as before, it also follows that $w = u_0 g + (f - u_0)u_1$ has rank less than $n + m - 1$, which in this case ensures that $a_{n+m-1}(fg) \le \|fg - w\| = \|(f - u_0)(g - u_1)\| \le \|f - u_0\|.\|g - u_1\|$. Taking the infimum over u_0 and u_1 now leads to the conclusion that

$$a_{n+m-1}(fg) \le a_n(f).a_m(g). \qquad \square$$

Proposition 2.9 *Let $1 \le p < \infty$ be given. For any $f, g \in \mathscr{S}_p(H)$, we have $\varsigma_p(f + g) \le 2^{1/p}(\varsigma_p(f) + \varsigma_p(g))$. Therefore, the quantity ς_p is a quasi-norm.*

Proof. The fact that for any scalar λ we have that $\varsigma_p(\lambda f) = |\lambda|\varsigma_p(f)$ is clear. It is, therefore, only the weighted subadditivity that needs attention. An application of Lemma 2.8 shows that

$$
\begin{aligned}
\varsigma_p(f + g) &= \left(\sum_{n=1}^{\infty} a_n(f + g)^p\right)^{1/p} \\
&= \left(\sum_{n=1}^{\infty}(a_{2n-1}(f + g)^p + a_{2n}(f + g)^p)\right)^{1/p} \\
&\le 2^{1/p}\left(\sum_{n=1}^{\infty} a_{2n-1}(f + g)^p\right)^{1/p} \\
&\le 2^{1/p}\left(\sum_{n=1}^{\infty}(a_n(f) + a_n(g))^p\right)^{1/p} \\
&\le 2^{1/p}\left(\left(\sum_{n=1}^{\infty} a_n(f)^p\right)^{1/p} + \left(\sum_{n=1}^{\infty} a_n(g)^p\right)^{1/p}\right) \\
&= 2^{1/p}(\varsigma_p(f) + \varsigma_p(g))
\end{aligned}
$$

where the final inequality follows from the standard Minkowski inequality for sequences. $\qquad \square$

We pause to draw some immediate conclusions before going on to elucidate a tracial approach to the Schatten classes.

Remark 2.10 Notice that for any $f, g \in B(H)$, we will by Lemma 2.8 have that $a_n(fg) \leq a_1(f)a_n(g) = \|f\|a_n(g)$ and similarly that $a_n(fg) \leq a_n(f)\|g\|$ for all n. If further $f \in \mathscr{S}_p$, these facts ensure that $fg, gf \in \mathscr{S}_p$ with $\varsigma_p(fg), \varsigma(gf) \leq \|g\|\varsigma(f)_p$.

The fact that $a_n(f) = a_n(f^*)$ for each $1 \leq p < \infty$, each $f \in \mathscr{S}_p$ and each n similarly ensures that $\mathscr{S}_p$ is a self-adjoint subspace of $B(H)$ for which $\varsigma_p(f) = \varsigma_p(f^*)$. If $f \in \mathscr{S}_p$, the real and imaginary parts of f will, therefore, still belong to $\mathscr{S}_p$. Since also $a_n(f) = a_n(|f|)$ for each n, $|f|$ will be in $\mathscr{S}_p$ if f is. So, any self-adjoint element f of $\mathscr{S}_p$ can be written as the difference of two positive elements of $\mathscr{S}_p$ (namely $|f| \pm f$). Thus, each $\mathscr{S}_p$ is the span of the positive cone of $\mathscr{S}_p$.

For each $1 \leq p < \infty$, the finite-rank operators will moreover be dense in $\mathscr{S}_p$. To see this, let $f \in \mathscr{S}_p$ be given in the form $f = \sum_{n=1}^{\infty} a_n(f)\langle \cdot, x_n \rangle y_n$, and let f_m be the finite-rank operator $f_m = \sum_{n=1}^{m} a_n(f)\langle \cdot, x_n \rangle y_n$. Since, by assumption, $\varsigma_p(f) = (\sum_{n=1}^{\infty} a_n(f)^p)^{1/p} < \infty$, it clearly follows that $\varsigma_p(f - f_m) = \varsigma_p(\sum_{n=m+1}^{\infty} a_n(f)\langle \cdot, x_n \rangle y_n) = (\sum_{n=m+1}^{\infty} a_n(f)^p)^{1/p} \to 0$ as $m \to \infty$.

2.3 A tracial approach to the Schatten–von Neumann classes

Given $f \in \mathscr{S}_p(H)$ with $f = \sum_{n=1}^{\infty} a_n(f)\langle \cdot, x_n \rangle y_n$, it is then an exercise to see that $|f|^p = \sum_{n=1}^{\infty} a_n(f)^p \langle \cdot, y_n \rangle y_n$ with $\mathrm{Tr}(|f|^p) = \sum_{n=1}^{\infty} a_n(f)^p = \varsigma_p(f)^p$ (whence $\mathrm{Tr}(|f|^p) < \infty$). To see this, simply compute Tr using an orthonormal basis which contains (y_n). If, conversely, $\mathrm{Tr}(|f|^p) < \infty$, then for any $\epsilon > 0$, we will have $\mathrm{Tr}(\chi_{(\epsilon,\infty)}(|f|)) \leq \frac{1}{\epsilon^p}\mathrm{Tr}(|f|^p) < \infty$, proving that $\chi_{(\epsilon,\infty)}(|f|)$ (and hence also $\chi_{(\epsilon,\infty)}(|f|)|f|$) is finite rank. Since $\| |f| - \chi_{(\epsilon,\infty)}(|f|)|f| \| \leq \epsilon$, $|f|$ is the limit of finite-rank operators and f is, therefore, compact. If f is of the form $f = \sum_{n=1}^{\infty} a_n(f)\langle \cdot, x_n \rangle y_n$, we will have $\sum_{n=1}^{\infty} a_n(f)^p = \mathrm{Tr}(|f|^p) < \infty$, showing that $f \in \mathscr{S}_p(H)$.

We formulate the conclusions of the above discussion as a proposition.

Proposition 2.11 *The space $\mathscr{S}_p(H)$ corresponds exactly to the set $\{f \in B(H): \mathrm{Tr}(|f|^p) < \infty\}$ with ς_p given by $\varsigma_p = \mathrm{Tr}(|\cdot|^p)^{1/p}$.*

We have already noted that $\mathscr{S}_p$ is a 'self-adjoint' subspace of $B(H)$, which then allows one to decompose each $f \in \mathscr{S}_p$ into a real and imaginary part inside $\mathscr{S}_p$. We may now make sharper statements about the further decomposition of a self-adjoint element $f \in \mathscr{S}_p$ into the difference of two positive elements. Specifically, considering $f_+ = f\chi_{[0,\infty)}(f)$ and $f_- = -f\chi_{(-\infty,0)}(f)$, the fact that $f_\pm^p \leq |f|^p$ ensures that each of f_+ and f_- will also belong to $\mathscr{S}_p$.

With Tr and $B(H)$ playing the roles of some sort of noncommutative integral and concomitant space of measurable operators, the pth Schatten class can, from a structural point of view, therefore be regarded as a noncommutative version of ℓ_p-space, with compact operators corresponding to c_0 and $B(H)$ to ℓ_∞. In view of these facts, we may equivalently denote $\mathscr{S}_p(H)$ by $L^p(B(H), \mathrm{Tr})$, with $\|\cdot\|_p$ denoting the (quasi-)norm $\mathrm{Tr}(|\cdot|^p)^{1/p}$, and similarly write $L^\infty(B(H), \mathrm{Tr})$ for $B(H)$. *Warning*: In the literature, $\mathscr{S}_\infty(H)$ is more commonly identified with the compact operators $\mathcal{K}(H)$, which is some sort of noncommutative c_0 space. Therefore, the reader should be careful not to confuse $\mathscr{S}_\infty(H) = \mathcal{K}(H)$ with $L^\infty(B(H), \mathrm{Tr}) = B(H)$.

The description of $\mathscr{S}_p$ afforded by the preceding proposition now enables us to prove the completeness of the Schatten classes.

Proposition 2.12   *For each $1 \leq p < \infty$, the space $(\mathscr{S}_p(H), \varsigma_p)$ is complete.*

Proof.   Let $(f_n) \subset \mathscr{S}_p$ be a Cauchy sequence, and let (v_i) be an orthonormal basis for H. It then follows that

$$\varsigma_p(f_n - f_m) = \left(\sum_{k \geq 1} a_k (f_n - f_m)^p \right)^{1/p} \geq a_1(f_n - f_m) = \|f_n - f_m\|.$$

So, (f_n) is also Cauchy in $\mathcal{K}(H)$ and, hence, must converge to some compact operator f. Given $\epsilon > 0$, select $n_0 \in \mathbb{N}$ such that $\varsigma_p(f_n - f_m) < \epsilon$ for all $n, m \geq n_0$. Let $\mathcal{F}$ be a finite subset of the index set of the orthonormal basis (v_i). For any $n \geq n_0$, we will then have that

$$\sum_{i \in \mathcal{F}} \langle |(f - f_n)|^p v_i, v_i \rangle = \lim_{m \to \infty} \sum_{i \in \mathcal{F}} \langle |(f_m - f_n)|^p v_i, v_i \rangle$$

$$\leq \limsup_{m \to \infty} \mathrm{Tr}(|f_m - f_n|^p) = \limsup_{m \to \infty} \varsigma_p(f_m - f_n)^p \leq \epsilon^p.$$

Since $\mathrm{Tr}(|f - f_n|^p)$ is the limit of the net $(\sum_{i \in \mathcal{F}} \langle |(f - f_n)|^p v_i, v_i \rangle)_{\mathcal{F}}$, it follows that $\mathrm{Tr}(|f - f_n|^p) \leq \epsilon^p$. By the preceding proposition, this, on the one hand, shows that $f - f_n$ (and therefore also f) is in $\mathscr{S}_p$, and, on the other that $\varsigma_p(f - f_n) \leq \epsilon$ for all $n \geq n_0$. Thus, $f_n \to f$ in $\mathscr{S}_p$, showing that $\mathscr{S}_p$ is complete. $\qquad\square$

The spaces $\mathscr{S}_1$ and $\mathscr{S}_2$ (also, respectively, referred to as the Trace Class and Hilbert–Schmidt operators) have some particularly elegant properties which we now investigate.

Theorem 2.13 *The trace* Tr *canonically extends to a linear functional on $\mathscr{S}_1$ for which we have that*

- $|\mathrm{Tr}(f)| \leq \mathrm{Tr}(|f|)$;
- $\mathrm{Tr}(f^*) = \overline{\mathrm{Tr}(f)}$;
- $\|f\| \leq \mathrm{Tr}(|f|)$;
- $\mathrm{Tr}(xf) = \mathrm{Tr}(fx)$

for all $f \in \mathscr{S}_1(H)$ and all $x \in B(H)$.

Proof. It is fairly easy to check that Tr induces an affine functional on the positive cone of $\mathscr{S}_1$. It clearly follows from the discussion preceding this theorem that the positive cone of $\mathscr{S}_1$ spans all of $\mathscr{S}_1$. This affine functional on $\mathscr{S}_1(H)_+$, therefore, extends to a linear functional $\mathscr{S}_1(H) \to \mathbb{C} : f \mapsto \mathrm{Tr}(f)$ on all of $\mathscr{S}_1$.

If $f \in \mathscr{S}_1$, then, of course, $\mathrm{Tr}(|f|) = \varsigma_1(f) = \sum_{n \geq 1} a_n(f)$, with the operator f admitting an orthonormal system representation of the form $f = \sum_{n=1}^{\infty} a_n(f)\langle \cdot, x_n \rangle y_n$. For convenience sake, we may use an orthonormal basis (ξ_i) that contains (x_n) to compute the trace. The quantity $\mathrm{Tr}(f) = \sum_i \langle f\xi_i, \xi_i \rangle$ then corresponds to the absolutely convergent series $\sum_{n=1}^{\infty} a_n(f)\langle y_n, x_n \rangle$. Using this representation of $\mathrm{Tr}(f)$, it is now clear that

$$|\mathrm{Tr}(f)| = |\sum_{n=1}^{\infty} a_n(f)\langle y_n, x_n \rangle| \leq \sum_{n=1}^{\infty} a_n(f) = \mathrm{Tr}(|f|).$$

In terms of the given orthonormal system representation of f, we further have that $f^* = \sum_{n=1}^{\infty} a_n(f)\langle \cdot, y_n \rangle x_n$ and hence that

$$\overline{\mathrm{Tr}(f)} = \sum_{n=1}^{\infty} a_n(f)\overline{\langle y_n, x_n \rangle} = \sum_{n=1}^{\infty} a_n(f)\langle x_n, y_n \rangle = \mathrm{Tr}(f^*).$$

The third claim follows from the dual facts that $\|f\| = \sup\{\langle |f|\xi, \xi\rangle : \xi \in H, \|\xi\| = 1\}$ and that we may use any orthonormal basis to compute $\mathrm{Tr}(|f|)$. To see the final claim, we recall that any operator in $B(H)$ can be written as a linear combination of four unitaries. So, in principle, we only need to prove this fact in the case where x is a unitary. However, given a unitary $u \in B(H)$, the fact that Tr is independent of the orthonormal basis used ensures that

$$\mathrm{Tr}(uf) = \sum_i \langle uf\xi_i, \xi_i\rangle = \sum_i \langle ufu\xi_i, u\xi_i\rangle = \sum_i \langle fu\xi_i, \xi_i\rangle = \mathrm{Tr}(fu). \quad \square$$

Theorem 2.14 *The space $(\mathscr{S}_2(H), \varsigma_2)$ is a Hilbert space. For any orthonormal basis (v_i) of H, the canonical inner product on $\mathscr{S}_2(H)$ (which yields the norm ς_2) is given by $\langle f, g\rangle_{\mathscr{S}_2} = \sum_i \langle fv_i, gv_i\rangle = \mathrm{Tr}(g^*f)$ for all $fg \in \mathscr{S}_2(H)$.*

Proof. We first make the very obvious observation that by the definition of the (quasi)-norms ς_1 and ς_2, $f \in \mathscr{S}_2$ if and only if $f^*f \in \mathscr{S}_1$, with $\varsigma_2(f)^2 = \mathrm{Tr}(f^*f) = \varsigma_1(f^*f)$. When combined with the polarisation identity $fg = \frac{1}{4}\sum_{k=0}^3 i^k(g + i^k f^*)^*(g + i^k f^*)$, it follows that for any $f, g \in \mathscr{S}_2$, the product fg belongs to $\mathscr{S}_1$. (Taking into account the fact that $\mathrm{Tr}(x^*x) = \mathrm{Tr}(xx^*)$ for any $x \in B(H)$, we even have $\mathrm{Tr}(fg) = \mathrm{Tr}(gf)$ for all $f, g \in \mathscr{S}_2$.) For any $f, g \in \mathscr{S}_2$, Theorem 2.13 then ensures that $\mathrm{Tr}(g^*f) = \overline{\mathrm{Tr}(f^*g)}$. Let (v_i) be an orthonormal basis of H. Since for any $f, g \in \mathscr{S}_2$, we have $\mathrm{Tr}(g^*f) = \sum_i \langle g^*fv_i, v_i\rangle = \sum_i \langle fv_i, gv_i\rangle$, it follows that the prescription $\langle \cdot, \cdot\rangle_{\mathscr{S}_2} : \mathscr{S}_2(H) \times \mathscr{S}_2(H) \to \mathbb{C} : (f, g) \mapsto \sum_i \langle fv_i, gv_i\rangle$ yields a well-defined inner product on $\mathscr{S}_2(H)$, for which the corresponding norm is precisely $(\sum_i \|fv_i\|^2)^{1/2} = (\sum_i (\langle f^*fv_i, v_i\rangle)^2)^{1/2} = \mathrm{Tr}(|f|^2)^{1/2} = \varsigma_2(f)$. Thus, $\mathscr{S}_2(H)$ is, in fact, a Hilbert space. $\quad \square$

Remark 2.15 It is worth noting that $\mathscr{S}_2(H)$ may also be realised as the Hilbert space tensor product $H \otimes \overline{H}$ of H with the conjugate Hilbert space $\overline{H}$. To see this, note that any rank 1 operator f may be written in the form $f = \langle \cdot, y\rangle x$. Each such operator $\langle \cdot, y\rangle x$ is identified with the simple tensor $x \otimes \overline{y}$. As the span of operators of the form $\langle \cdot, y\rangle x$ (the finite-rank operators) is dense in $\mathscr{S}_2(H)$, so too is the span of the simple tensors in $H \otimes \overline{H}$. Therefore, the claim follows from the fact that the induced linear map on the span of the rank 1 operators is, in fact, an isometry. To see this, notice, for example, that given two rank 1 operators $\langle \cdot, y \rangle x$ and $\langle \cdot, b \rangle a$ and an orthonormal basis (v_i), we will have

$$
\begin{aligned}
\langle \langle \cdot, y \rangle x, \langle \cdot, b \rangle a \rangle_{\mathscr{S}_2}
&= \operatorname{Tr}((\langle \cdot, b \rangle a)^* \circ (\langle \cdot, y \rangle x)) \\
&= \operatorname{Tr}((\langle \cdot, a \rangle b) \circ (\langle \cdot, y \rangle x)) \\
&= \langle x, a \rangle \operatorname{Tr}(\langle \cdot, y \rangle b) \\
&= \langle x, a \rangle \sum_i \langle v_i, y \rangle \langle b, v_i \rangle \\
&= \langle x, a \rangle \cdot \langle b, y \rangle \\
&= \langle x, a \rangle \cdot \langle \overline{y}, \overline{b} \rangle \\
&= \langle x \otimes \overline{y}, a \otimes \overline{b} \rangle .
\end{aligned}
$$

2.4 Trace duality

To prove the promised trace duality, we need a Hölder-type inequality for the spaces $\mathscr{S}_p$. For this, we shall need the following lemma, which is interesting in its own right. The lemma exploits the rich structure of $\mathscr{S}_2(H)$ noted above.

Lemma 2.16 *Let $v \in B(H)$ be given. Then, $T_v : \mathscr{S}_2(H) \to \mathscr{S}_2(H) :$ $f \mapsto vfv^*$ is a bounded linear operator. Moreover, for any $1 \le r < \infty$, T_v belongs to $\mathscr{S}_r(\mathscr{S}_2(H))$ whenever v belongs to $\mathscr{S}_r(H)$, in which case $\varsigma_r(T_v) = \varsigma_r(v)^2$. (The converse implication also holds (see* **[DJT95,** *Theorem 6.2]), but we shall not need this.)*

Proof. The first claim easily follows from the first observation in Remark 2.10. For the second assertion, suppose that v admits and orthonormal system representation of the form $v = \sum_{n=1}^{\infty} \alpha_n \langle \cdot, x_n \rangle y_n = \sum_{n=1}^{\infty} \alpha_n y_n \otimes \overline{x_n}$, where $\alpha_n \ge 0$ for each n. For each $f \in \mathscr{S}_2(H)$, direct computation shows that

$$
T_v(f) = vfv^* = \sum_{m,n \ge 1} \alpha_n \alpha_m \langle f y_m, y_n \rangle \langle \cdot, x_m \rangle x_n,
$$

which may be rewritten as

$$
T_v(f) = \sum_{m,n \ge 1} \alpha_n \alpha_m \langle f, y_n \otimes \overline{y_m} \rangle (x_n \otimes \overline{x_m}).
$$

This is clearly an orthonormal system representation of T_v, from which it now follows that

$$\varsigma_r(T_v)^r = \sum_{m,n \geq 1} \alpha_n^r \alpha_m^r = \left(\sum_{n \geq 1} \alpha_n^r\right)^2 = \varsigma_r(v)^{2r}.$$

The claim clearly follows. $\qquad\qquad\square$

Theorem 2.17 (Hölder's inequality) *Let $1 \leq p, q, r < \infty$ be given, with $\frac{1}{r} = \frac{1}{p} + \frac{1}{q}$. For any $f \in \mathscr{S}_p(H)$ and $g \in \mathscr{S}_q(H)$, we will then have $fg \in \mathscr{S}_r(H)$ with $\varsigma_r(fg) \leq \varsigma_p(f)\varsigma_q(g)$.*

As noted in Remark 2.10, the result still holds if $q = 1$ with $B(H)$ substituted for $\mathscr{S}_q(H)$.

Proof.   We may suppose that $p \leq q$. Using Lemma 2.8, it follows that

$$
\begin{aligned}
\left(\sum_{n=1}^{\infty} a_n(fg)^r\right)^{1/r} &= \left(\sum_{n=1}^{\infty} (a_{2n}(fg)^r + a_{2n-1}(fg)^r)\right)^{1/r} \\
&\leq 2^{1/r}\left(\sum_{n=1}^{\infty} a_{2n-1}(fg)^r\right)^{1/r} \\
&\leq 2^{1/r}\left(\sum_{n=1}^{\infty} a_n(f)^r a_n(g)^r\right)^{1/r} \\
&\leq 2^{1/r}\varsigma_p(f)\varsigma_q(g).
\end{aligned}
$$

This shows that $fg \in \mathscr{S}_r(H)$ with $\varsigma_r(fg) \leq 2^{1/r}\varsigma_p(f)\varsigma_q(g)$. We now apply what we have just verified to the operators described in the lemma to see that

$$\varsigma_r(fg) = \varsigma_r(T_{fg})^{1/2} \leq 2^{1/(2r)}\varsigma_p(T_f)^{1/2}\varsigma_q(T_g)^{1/2} = 2^{1/(2r)}\varsigma_p(f)\varsigma_q(g).$$

Carrying on inductively shows that $\varsigma_r(fg) \leq 2^{1/(2^n r)}\varsigma_p(f)\varsigma_q(g)$ for each n, which yields the required inequality. $\qquad\square$

Corollary 2.18   *Let $1 \leq p, q < \infty$ be given with $1 = \frac{1}{p} + \frac{1}{q}$. For any $f \in \mathscr{S}_p(H)$ and $g \in \mathscr{S}_q(H)$, we will then have that $\mathrm{Tr}(fg) = \mathrm{Tr}(gf)$.*

Proof.   The fact that $\mathscr{S}_p . \mathscr{S}_q \subset \mathscr{S}_1$ ensures that Tr has a well-defined action on $\mathscr{S}_p . \mathscr{S}_q$. In the case where both f and g are finite rank, the result follows from Theorem 2.13. The general case now follows by using the fact that the finite-rank operators are dense in both $\mathscr{S}_p$ and $\mathscr{S}_q$. $\qquad\qquad\square$

Theorem 2.19 Trace duality

(1) *Let $1 < p, q < \infty$ be given, with $1 = \frac{1}{p} + \frac{1}{q}$. Then, the prescription $\psi : \mathscr{S}_q(H) \to (\mathscr{S}_p(H))^*$ obtained by setting $\psi(f)(g) = \mathrm{Tr}(fg)$ for all $f \in \mathscr{S}_q(H)$ and $g \in \mathscr{S}_p(H)$ is a linear isometry from $\mathscr{S}_q(H)$ to $(\mathscr{S}_p(H))^*$.*

(2) *Similarly, $B(H)$ is linearly isometric to $(\mathscr{S}_1(H))^*$, and $(\mathcal{K}(H))^*$ (the dual of the compact operators) to $\mathscr{S}_1(H)$.*

It is clear from Theorem 2.19 that the pair $(\mathscr{S}_p, \varsigma_p)$ will for any $1 \leq p < \infty$ be a Banach space.

Proof.   We first prove the second set of claims. In view of the similarity of the proofs, we shall here only show that $(\mathcal{K}(H))^*$ (the dual of the compact operators) is isomorphic to $\mathscr{S}_1(H)$. Notice that $\psi(f)(x) \leq \|f\|_1 \|x\|$ for any $f \in \mathscr{S}_1(H)$ and any $x \in \mathcal{K}(H)$. So, $f \mapsto \psi(f)$ is a contractive embedding of $\mathscr{S}_1(H)$ into $(\mathcal{K}(H))^*$.

If, on the other hand, we are given some $\varphi \in (\mathcal{K}(H))^*$, then $(\xi, \eta) \mapsto \varphi(\xi \otimes \overline{\eta})$ defines a sesquilinear form, which is bounded by $\|\varphi\|$ since $|\varphi(\xi \otimes \overline{\eta})| \leq \|\varphi\|.\mathrm{Tr}(|\xi \otimes \overline{\eta}|) = \|\varphi\|.\|\xi\|.\|\eta\|$. Hence, there must exist some $f \in B(H)$ such that $\varphi(\xi \otimes \overline{\eta}) = \langle f\xi, \eta \rangle$ for all $\xi, \eta \in H$. This latter formula clearly extends to the span of operators of the form $\xi \otimes \overline{\eta}$, namely the finite-rank operators.

Let $f = v|f|$ be the polar decomposition of f. For any finite orthonormal system $\eta_1, \ldots, \eta_k$, Parseval's inequality ensures that $g = \sum_{i=1}^{k} \eta_i \otimes \overline{\eta_i}$ is a norm 1 finite-rank operator. We, therefore, have that

$$\sum_{i=1}^{k} \langle |f|\eta_i, \eta_i \rangle = \sum_{i=1}^{k} \langle f\eta_i, v\eta_i \rangle = \varphi\left(\sum_{i=1}^{k} \eta_i \otimes \overline{v\eta_i}\right)$$

$$= \varphi(gv^*) \leq \|\varphi\|\|gv^*\| \leq \|\varphi\|.$$

On taking the supremum over all finite orthonormal system, it follows that

$$\mathrm{Tr}(|f|) \leq \|\varphi\| < \infty.$$

So, we in fact have that $f \in \mathscr{S}_1(H)$. Since, for any rank 1 operator, we have $\varphi(\xi \otimes \overline{\eta}) = \langle a\xi, \eta \rangle = \mathrm{Tr}(f(\xi \otimes \overline{\eta})) = \psi(f)(\xi \otimes \overline{\eta})$ and since the span of the rank 1 operators (the finite-rank operators) is dense in $\mathcal{K}(H)$, $\psi(f)$ and φ will agree on all of $\mathcal{K}(H)$. So, ψ is surjective. Notice that the previous displayed inequality also shows that $\|f\|_1 \leq \|\varphi\| = \|\psi(f)\|$, which, by the first part of the proof, ensures that ψ is an isometry.

Now suppose that we are given $1 < p, q < \infty$ with $1 = \frac{1}{p} + \frac{1}{q}$. Theorem 2.17 clearly shows that $\psi(f) = \mathrm{Tr}(f \cdot)$ will, for any $f \in \mathscr{S}_q$, be an element of $(\mathscr{S}_p)^*$ with $\|\psi(f)\| \leq \varsigma_q(f)$. So, as before, $f \mapsto \psi(f)$ is a contractive embedding of $\mathscr{S}_q(H)$ into $(\mathscr{S}_p(H))^*$. Let $f = \sum_{n \geq} a_n(f)\langle \cdot, x_n \rangle y_n$ be an orthonormal system representation of f. The fact that ψ is in fact an isometry follows on noticing that

$$
\varsigma_q(f) = \left(\sum_{n \geq 1} a_n(f)^q \right)^{1/q} = \sup\{ |\sum_{n \geq 1} a_n(f)\alpha_n| : \sum_{n \geq 1} |\alpha_n|^p \leq 1 \}
$$

$$
= \sup\{ |\mathrm{Tr}(fg)| : \sum_{n \geq 1} \alpha_n \langle \cdot, y_n \rangle x_n, \sum_{n \geq 1} |\alpha_n|^p \leq 1 \}
$$

$$
\leq \sup\{ |\mathrm{Tr}(fg)| : g \in \mathscr{S}_q, \varsigma_q(g) \leq 1 \}
$$

$$
= \|\psi\|.
$$

If, on the other hand, we are given some $\varphi \in (\mathscr{S}_p(H))^*$, it is once again an interesting exercise to see that $(\xi, \eta) \mapsto \varphi(\xi \otimes \overline{\eta})$ defines a bounded sesquilinear form and hence that there must exist some $f \in B(H)$ such that $\varphi(\xi \otimes \overline{\eta}) = \langle f\xi, \eta \rangle$ for all $\xi, \eta \in H$. Where the proof of this part differs from the previous one is in the method used to show that $f \in \varsigma_q(H)$.

Let $f = v|f|$ be the polar decomposition of a. It is then an exercise to see that $\varphi_{v^*} : g \mapsto \varphi(v^* g)$ is still a bounded linear functional on $\mathscr{S}_p$, but now with $\varphi_{v^*}(\xi \otimes \overline{\eta}) = \langle |f^*|\xi, \eta \rangle$ for all $\xi, \eta \in H$. To see this, recall that $v|f|v^* = |f^*|$. Suppose that there is some $\epsilon > 0$ for which the spectral projection $\chi_{(\epsilon, \infty)}(|f^*|)$ is infinite dimensional. We may then select an orthonormal system (x_n) within $\chi_{(\epsilon, \infty)}(|f^*|)(H)$ and a sequence $(\alpha_n) \in \ell_p \backslash \ell_1$, with $\alpha_n \geq 0$ for all n. Then, the operator $g = \sum_{n \geq 1} \alpha_n \langle \cdot, x_n \rangle x_n = \sum_{n \geq 1} \alpha_n x_n \otimes \overline{x_n}$ belongs to $\mathscr{S}_p$, which ensures that $\varphi_{v^*}(g)$ is a well-defined scalar. However, by writing g_m for $\sum_{n=1}^{m} \alpha_n x_n \otimes \overline{x_n}$, we will, by continuity, have $\varphi_{v^*}(g) = \lim_{m \to \infty} \varphi_{v^*}(g_m) = \lim_{m \to \infty} \sum_{n=1}^{m} \alpha_n \langle |f^*| x_n, x_n \rangle \geq \epsilon \lim_{m \to \infty} \sum_{n=1}^{m} \alpha_n = \infty$. This clearly cannot be, which then shows that each $\chi_{(\epsilon, \infty)}(|f^*|)$ is of finite rank and $|f^*|$, therefore, is compact. Let the orthonormal system representation of $|f^*|$ be $|f^*| = \sum_{n \geq 1} a_n(|f^*|)\langle \cdot, x_n \rangle x_n$. For any $(\alpha_n) \in \ell_p(\mathbb{N})$ with each $\alpha_n \geq 0$, we will then see that $g = \sum_{n \geq 1} \alpha_n \langle \cdot, x_n \rangle x_n$ belongs to $\mathscr{S}_p$. Careful checking now shows that we have $\sum_{n=1}^{\infty} a_n(|f^*|)\alpha_n = \lim_{m \to \infty} \sum_{n=1}^{m} a_n(|f^*|)\alpha_n = \lim_{m \to \infty} \sum_{n=1}^{m} \alpha_n \langle |f^*| x_n, x_n \rangle = \lim_{m \to \infty} \varphi_{v^*}(g_m) = \varphi_{v^*}(g) < \infty$. This suffices to show that $(a_n(|f^*|)) \in \ell_q$ and hence that $|f^*| \in \mathscr{S}_q$ (equivalently $f \in \mathscr{S}_q$). But then $\psi(f)$ is a bounded linear functional, which agrees with φ on a dense subspace of $\mathscr{S}_p$ (the finite-rank operators) and hence on all of $\mathscr{S}_p$. $\qquad\square$

PART 2

TRACIAL CASE

This part of the book deals mainly with tracial von Neumann measure algebras, that is, couples consisting of a semifinite von Neumann algebra $\mathcal{M}$ and a faithful normal trace. Even readers interested only in the general L^p and/or Orlicz spaces need to learn the core components of the material placed in this part.

We start with Chapter 3 on measure and integral in the tracial case. Section 3.1 (p. 131) on the dimension function, which corresponds to a noncommutative measure, describes the results obtained by Murray and von Neumann in the 1930s. The section is not compulsory reading, but is extremely useful in building intuition for the subsequent material on traces, and also gives insight into the geometry underlying the classification of von Neumann algebras into the various types. However, the serious student should master the core material from each of the other sections in this chapter. We recommend reading the whole of Section 3.2 (p. 143) on traces and at least those parts of Section 3.3 (p. 162) on measurability that relate to τ-measurability and, if one is interested in the densities with respect to a trace, also the material on local measurability. The rest of Chapter 3 must be mastered in its entirety, with the possible exception of Section 3.6 (p. 182) on τ-local convergence in measure.

Chapter 4 contains material on weights and their densities with respect to a trace. The reader can omit the material dealing with special notions of semifiniteness of weights (points (2)–(4) in Definition 4.9 (p. 198) and from Lemma 4.12 (p. 200) to Remark 4.15 (p. 201)), but should read the rest of Section 4.1 (p. 196), together with the whole Section 4.2 (p. 203) and Section 4.3 (p. 205) up to Proposition 4.30 (p. 212). Theorem 4.32 (p. 213) which indicates which weights have densities that are locally measurable operators (as opposed to generalised positive operators) is nevertheless worth looking at.

We recommend reading Chapter 5 in its entirety, as it is the basic tool used in the construction of L^p and Orlicz spaces. We advise against omitting Sections 6.1 (p. 245) and 6.2 (p. 266) of Chapter 6, even for those interested only in the general case. Together, Chapters 1 and 6 and Section 10.1 (p. 413) outline a continuous line of development from classical L^p and Orlicz spaces to the versions of these spaces associated with the most general von Neumann algebras. Section 6.3 (p. 276) on Orlicz spaces should be read by all. Although the material may be new to a large part of the audience, Orlicz space technology will ultimately prove to be a crucial tool in developing some of the more advanced theories, like, for example, the theory on complex interpolation in Chapter 11. Anyone interested in interpolation and/or Banach function spaces should read Chapter 7.

Chapter 3
Noncommutative measure theory—tracial case

In this chapter, we deal with various aspects of noncommutative measure and integration theory in the tracial case. We start with the notion of a dimension function, which is in effect a *measure* on a factor, and then move to a *measure* and *integral* on an arbitrary semifinite von Neumann algebra in the form of a faithful normal semifinite trace. In the rest of the chapter, we discuss various notions of measurability, including a foray into classes of localised measurability, and their algebraic, topological, and order properties.

3.1 The dimension function

This section is devoted to a noncommutative counterpart of a *measure*, already introduced by Murray and von Neumann (see [**MvN36**, Part II]). It should be obvious from Chapter 1 that in the von Neumann algebra in question, the place of sets from classical measure theory is taken by projections. In fact, as we learnt in Chapter 1, classes of measurable sets from a measure space (X, Σ, μ) will modulo sets of zero measure become projections in $L^\infty(X, \Sigma, \mu)$. It should also be clear from Chapter 2 that the standard trace Tr on $B(H)$, when restricted to $\mathbb{P}(B(H))$, maps projections p to the dimension of their ranges $\dim pH$ and, moreover, enjoys all the properties of a measure. Note that the ranges of equivalent projections in $B(H)$ have the same dimension. This substantiates the following definition.

Definition 3.1 Let $\mathcal{M}$ be a von Neumann factor. A mapping $D\colon \mathbb{P}(M) \to [0, \infty]$ is called a *dimension function* if it satisfies:

(1) $D(p + q) = D(p) + D(q)$ for $p \perp q$;
(2) $D(p) = D(q)$ if $p \sim q$;

Noncommutative measures and L^p and Orlicz Spaces, with Applications to Quantum Physics. Stanisław Goldstein and Louis Labuschagne, Oxford University Press. © Stanisław Goldstein and Louis Labuschagne (2025).
DOI: 10.1093/oso/9780198950202.003.0005

(3) $D(p) < \infty$ if and only if p is finite;

(4) $D(p) > 0$ for $p \neq 0$.

The conditions of the definition require some comment. The first one, additivity on orthogonal projections, is obvious. The second constitutes the essence of the definition: we want projections *of equal size* to have the same measure. We see that the assumption of factoriality of $\mathcal{M}$ is here indispensable. First, two projections in a factor are always comparable (see Corollary 0.114), which makes the set $\mathbb{P}(\mathcal{M})/\sim$ totally ordered. For non-factorial algebras, the situation is different. If, for example, $\mathcal{M}$ is abelian, then $p \precsim q$ is equivalent to $p \leq q$ and $p \sim q$ to $p = q$. Therefore, $\mathbb{P}(\mathcal{M})/\sim$ can be identified with $\mathbb{P}(\mathcal{M})$, which is not totally ordered except in trivial cases. Condition (3) guarantees that the function is not identically infinite and together with (4) implies that $D(p) = 0$ if and only if $p = 0$.

It is easy to show, again using Corollary 0.114, that if $D(p) = D(q)$ and p is finite, then q is also finite and necessarily $p \sim q$. Hence, one is tempted to replace 'if' in Definition 3.1(2) with 'if and only if'. This would also cause condition (4) to be superfluous. Unfortunately, it does not work even for the standard dimension function (restriction of the standard trace) on $B(H)$. Indeed, if H can accommodate an uncountable family of non-zero mutually orthogonal projections, then there are in $B(H)$ inequivalent projections of infinite dimension. The problem vanishes if we restrict ourselves to σ-finite algebras. Otherwise, the way out is to admit infinite cardinals as values of our dimension function. This is indeed what Tomiyama has done in [**Tom58**]. In the following definition, we use the notion of *decomposability number* (see Definition 1.26). It is also useful to take note of the notion of *magnitude* in that definition.

Definition 3.2 A *generalised dimension function* (or a *magnitude function*) is a function $\bar{D} : \mathbb{P}(\mathcal{M}) \to \mathcal{R}_+ \cup \{\kappa : \aleph_0 \leq \kappa \leq \mathfrak{d}(M)\}$ (i.e., we form the union of $\mathcal{R}_+$ with a set of infinite cardinals $\leq \mathfrak{d}(M)$), satisfying:

(1) $\bar{D}(p + q) = \bar{D}(p) + \bar{D}(q)$ for $p \perp q$;

(2) $\bar{D}(p) = \bar{D}(q)$ if and only if $p \sim q$;

(3) $\bar{D}(p) < \infty$ if and only if p is finite.

Example 3.3 Assume that H is a finite- or infinite-dimensional Hilbert space and $\mathcal{M} = B(H)$ acts in H. Then, $\mathbb{P}(\mathcal{M}) \ni p \mapsto \dim pH$ is a dimension function (or a generalised dimension function, depending on the definition of dim). The proof of this fact is left to the reader.

This is obviously the prototype of the dimension function we consider in this section.

The following proofs imitate the proofs from Sunder [**Sun87**] and Tomiyama [**Tom58**]. We give them here for completeness.

The following theorem shows that the dimension function exists on each von Neumann factor, is unique up to a positive constant, and enjoys a number of additional properties:

Theorem 3.4 *For each von Neumann factor $\mathcal{M}$, there is a unique, up to a positive constant c, dimension function $D : \mathbb{P}(\mathcal{M}) \to [0, \infty]$. The function is completely additive and if, for an increasing sequence of projections (p_n), we have $p_n \nearrow p$, then $D(p_n) \nearrow D(p)$. Moreover, the set $\Delta := D(\mathbb{P}(\mathcal{M}))$ is determined solely by the type of the factor. Namely, we have:*

(1) $\Delta = \{c, 2c, \ldots, nc\}$ *for $\mathcal{M}$ of type $I_n, n < \infty$;*
(2) $\Delta = \{c, 2c, \ldots, \infty\}$ *for $\mathcal{M}$ of type I_∞;*
(3) $\Delta = [0, c]$ *for $\mathcal{M}$ of type II_1;*
(4) $\Delta = [0, \infty]$ *for $\mathcal{M}$ of type II_∞;*
(5) $\Delta = \{0, \infty\}$ *for $\mathcal{M}$ of type III.*

We start with the following lemma, which has a clear number-theoretic counterpart.

Lemma 3.5 (Division with remainder) *Let $\mathcal{M}$ be a von Neumann factor, and let $p, q \in \mathbb{P}(\mathcal{M})$, with $q \neq 0$. There is an orthogonal family $\{p_i\}_{i \in I}$, with all $p_i \in \mathbb{P}(\mathcal{M})$, and a projection $r \in \mathcal{M}$ such that:*

(1) $p_i \sim q$ *for all $i \in I$;*
(2) $r \prec q$;
(3) $p = \sum_{i \in I} p_i + r.$

If I is infinite, we can take $r = 0$.

Proof. If $p \prec q$, put $r := p$ and $I := \emptyset$. Otherwise, $q \precsim p$, in which case there exists a projection $p' \sim q$ with $p' \leq p$. By Zorn's lemma, we may select a maximal orthogonal family $\{p_i\}_{i \in I}$ such that $p_i \sim q, p_i \leq p$ for each $i \in I$. Put $r = p - \sum_{i \in I} p_i$. By maximality, $r \prec q$. Fix an i_0 in I. We have $r \prec p \sim p_{i_0}$. If I is infinite, there is a bijection between I and $I \backslash \{i_0\}$. By Proposition 0.112, we then have

$$p = \sum_{i \in I} p_i + r \sim \sum_{i \in I \backslash \{i_0\}} p_i + r \precsim \sum_{i \in I} p_i \precsim p. \tag{3.1}$$

Hence, $p \sim \sum_{i \in I} p_i$. Let u be a partial isometry satisfying $u^*u = p$ and $uu^* = \sum_{i \in I} p_i$. Put $p'_i = u^*p_i u$ for each $i \in I$. Then $p = \sum_{i \in I} p'_i$ and $p'_i \sim p$ for each $i \in I$. $\square$

It is clear that the decomposition is not unique. However, it is true that the cardinality of the index set I does not depend on the decomposition. In investigating this question, we treat the finite and infinite cases separately.

Lemma 3.6 *Let $\mathcal{M}$ be a von Neumann factor and let $p, q \in \mathbb{P}(\mathcal{M})$, with p finite and $q \neq 0$. If $p = \sum_{i \in I} p_i + r = \sum_{j \in J} p'_j + r'$, where $p_i \sim p'_j \sim q$ for each $i \in I, j \in J$ and $r \prec q, r' \prec q$, then $\#I = \#J < \infty$.*

Proof. It is clear that both I and J must be finite: If I is infinite, then $\sum_{i \in I} p_i \sim \sum_{i \in I \backslash \{i_0\}} p_i$, as in (3.1). So, $\sum_{i \in I} p_i$, and hence also p, cannot be finite.

Suppose now that $\#I < \#J$. Let $i \mapsto i'$ be an injective embedding of I into J, and let $j_0 \in J$ be different from any i'. Since $r \prec q \sim p'_{j_0}$, we can choose $r_0 \in \mathbb{P}(\mathcal{M})$ satisfying $r_0 \sim r$ and $r_0 < p'_{j_0}$, where the last inequality is an abbreviation for $r_0 \leq p'_{j_0}$ and $r_0 \neq p'_{j_0}$. We have (with the same convention)

$$p = \sum_{i \in I} p_i + r \sim \sum_{i \in I} p'_{i'} + r_0 < \sum_{i \in I} p'_{i'} + p'_{j_0} \leq \sum_{j \in J} p'_j \leq p$$

which is impossible. Hence, $\#I \geq \#J$ and, by symmetry, $\#J \geq \#I$. $\square$

To facilitate the reading of the proofs, we will call the number $\#I$ from the above lemma the *integral quotient* of p and q and denote it by $\left\lfloor \frac{p}{q} \right\rfloor$. The integral quotient obviously depends only on the equivalence class of q. We will write statements like $kq \precsim p \prec (k+1)q$ for $k := \left\lfloor \frac{p}{q} \right\rfloor$, with the meaning that kq represents a sum of the form $\sum_{i \in I} p_i$ with $\#I = k$, as in Lemma 3.5 (see also Lemma 3.6). We shall need the following technical lemmas.

Lemma 3.7 *Let $p, q, r \in \mathbb{P}(\mathcal{M})$ be finite and non-zero. Then*

$$\left\lfloor \frac{p}{q} \right\rfloor \left\lfloor \frac{q}{r} \right\rfloor \leq \left\lfloor \frac{p}{r} \right\rfloor < \left(\left\lfloor \frac{p}{q} \right\rfloor + 1 \right) \left(\left\lfloor \frac{q}{r} \right\rfloor + 1 \right). \tag{3.2}$$

Furthermore, if $p \perp q$, then

$$\left\lfloor \frac{p}{r} \right\rfloor + \left\lfloor \frac{q}{r} \right\rfloor \leq \left\lfloor \frac{p+q}{r} \right\rfloor < \left\lfloor \frac{p}{r} \right\rfloor + \left\lfloor \frac{q}{r} \right\rfloor + 2. \tag{3.3}$$

Proof.   Using our notation,

$$\left\lfloor \frac{p}{q} \right\rfloor \left\lfloor \frac{q}{r} \right\rfloor r \precsim \left\lfloor \frac{p}{q} \right\rfloor q \precsim p.$$

So, the left-hand-side inequality in (3.2) follows from the fact that $\left\lfloor \frac{p}{r} \right\rfloor$ is the largest integer k such that $kr \precsim p$. The left inequality in (3.3) is obtained in the same way.

Assume $\left\lfloor \frac{p}{r} \right\rfloor \geq \left(\left\lfloor \frac{p}{q} \right\rfloor + 1 \right) \left(\left\lfloor \frac{q}{r} \right\rfloor + 1 \right)$. Since $q \prec \left(\left\lfloor \frac{q}{r} \right\rfloor + 1 \right) r$, we have

$$\left(\left\lfloor \frac{p}{q} \right\rfloor + 1 \right) q \precsim \left(\left\lfloor \frac{p}{q} \right\rfloor + 1 \right) \left(\left\lfloor \frac{q}{r} \right\rfloor + 1 \right) r \precsim \left\lfloor \frac{p}{r} \right\rfloor r \precsim p$$

which is impossible. This confirms the right-hand-side inequality in (3.2). For the right-hand-side inequality in (3.3), suppose that

$$\left\lfloor \frac{p+q}{r} \right\rfloor \geq \left(\left\lfloor \frac{p}{r} \right\rfloor + \left\lfloor \frac{q}{r} \right\rfloor + 2 \right).$$

Then, we obtain

$$p + q \precsim \left\lfloor \frac{p+q}{r} \right\rfloor r \geq \left(\left\lfloor \frac{p}{r} \right\rfloor + \left\lfloor \frac{q}{r} \right\rfloor + 2 \right) r = \left(\left\lfloor \frac{p}{r} \right\rfloor + 1 \right) r + \left(\left\lfloor \frac{q}{r} \right\rfloor + 1 \right) r \succ p + q,$$

which by Theorem 0.120 is impossible. Therefore, $\left\lfloor \frac{p+q}{r} \right\rfloor < \left\lfloor \frac{p}{r} \right\rfloor + \left\lfloor \frac{q}{r} \right\rfloor + 2$, which ends the proof. $\square$

A simple lemma will clarify the situation in the infinite case.

Lemma 3.8 *Let $\mathcal{M}$ be a von Neumann algebra. Suppose that for an infinite orthogonal family $\{p_i\}_{i \in I}$ of σ-finite projections from $\mathcal{M}$ and for an arbitrary family $\{q_j\}_{j \in J}$ of projections from $\mathcal{M}$, we have $\sum_{i \in I} p_i = \sum_{j \in J} q_j$. Then $\#J \leq \#I$. In particular, if J is also infinite and $q_j s$ are σ-finite, then $\#I = \#J$.*

Proof. Let $p := \sum_{i \in I} p_i$. Since the theorem can be proved in the reduced algebra $\mathcal{M}_p$, we may assume that $p = \mathbb{1}_{\mathcal{M}}$. Moreover, by assumption, the $\mathcal{M}_{p_i}$s are all σ-finite, and hence, each of them admits a faithful normal state φ_i. Let $J_i := \{j \in J : p_i q_j p_i \neq 0\}$. Note that $J = \bigcup_{i \in I} J_i$. Indeed, if for some $j_0 \in J$ we had $p_i q_{j_0} p_i = 0$ for all $i \in I$, then $p_i \perp q_{j_0}$ for all $i \in I$, whence $q_{j_0} = 0$, which is impossible (as orthogonal families consist of non-zero projections). Now, $p_i = \sum_{j \in J} p_i q_j p_i$, so that $\varphi_i(p_i) = \sum_{j \in J_i} \varphi_i(p_i q_j p_i) < \infty$. Therefore, $\#J_i \leq \aleph_0$ for each $i \in I$, and $\#J \leq \aleph_0 \cdot \#I = \#I$. $\qquad\square$

The only 'hard' case of Theorem 3.4 is that of a type II factor. Indeed, in factors of type I, any finite non-zero projection p majorises a minimal projection e, so that it is, in fact, an integral multiple of a minimal projection. Thus, the dimension function can be defined on p as $\lfloor \frac{p}{e} \rfloor$, even if e is not an actual subprojection of p. In the type III case, there are no non-zero finite projections, so that the dimension function can only have values 0 and ∞. In the type II case, we need a special 'rapidly decreasing' sequence of projections to clarify the structure.

Lemma 3.9 *In any factor $\mathcal{M}$ of type II, there is a sequence of finite projections $e_0, e_1, e_2, \ldots$ such that $\left\lfloor \frac{e_n}{e_{n+1}} \right\rfloor \geq 2$ for all n.*

A sequence satisfying the condition of the lemma is called a *fundamental sequence* for the factor $\mathcal{M}$.

Proof. Note that no projection in $\mathcal{M}$ is minimal. Let e_0 be any non-zero finite projection from $\mathcal{M}$. Once a non-zero finite projection e_n is identified, we can find in $\mathcal{M}$ a non-zero projection $f \leq e_n$ such that $f \neq e_n$, $f \prec e_n$ and $\left\lfloor \frac{e_n}{f} \right\rfloor > 0$. If $\left\lfloor \frac{e_n}{f} \right\rfloor \geq 2$, we put $e_{n+1} := f$. Otherwise, we have $\left\lfloor \frac{e_n}{f} \right\rfloor = 1$ and then $e_n = f + r$ with $r \prec f$. Therefore, $q \in \mathbb{P}(\mathcal{M})$

satisfies $r \sim q$ and $q \leq f$. Thus, $r+q \precsim e_n$ and, consequently, $\lfloor \frac{e_n}{r} \rfloor \geq 2$. In this case, we put $e_{n+1} := r$. It is clear that e_{n+1} constructed in this manner is non-zero and finite. $\qquad\square$

The following technical lemma shows the possibility of defining a 'real' quotient in the type II case.

Lemma 3.10 *Let $\{e_n\}_{n \in \mathbb{N}}$ be a fundamental sequence for a type II factor $\mathcal{M}$. Let $p, q, r, e \in \mathbb{P}(\mathcal{M})$ be finite and non-zero. Then, $\lfloor \frac{p}{e_n} \rfloor \neq 0$ for sufficiently large n, $\lfloor \frac{p}{e_n} \rfloor \nearrow \infty$ and the limit $\left[\frac{p}{q} \right] :=$ $\lim_{n \to \infty} \left(\lfloor \frac{p}{e_n} \rfloor / \lfloor \frac{q}{e_n} \rfloor \right)$ exists, with $0 < \left[\frac{p}{q} \right] < \infty$. Moreover,*

(1) *if $p \sim q$, then $\left[\frac{p}{r} \right] = \left[\frac{q}{r} \right]$;*

(2) $\left[\frac{p}{p} \right] = 1$;

(3) $\left[\frac{p}{q} \right] = \left[\frac{q}{p} \right]^{-1}$;

(4) $\left[\frac{p}{q} \right] = \left[\frac{p}{r} \right] \left[\frac{r}{q} \right]$

(5) *if $p \perp q$, then $\left[\frac{p+q}{r} \right] = \left[\frac{p}{r} \right] + \left[\frac{q}{r} \right]$;*

(6) *if $p \precsim q$, then $\left[\frac{p}{r} \right] \leq \left[\frac{q}{r} \right]$.*

We call the limit α from the above lemma the *real quotient* of p and e and denote it by $\left[\frac{p}{e} \right]$. It is obvious that the real quotient depends only on the equivalence class of e, that is, $\left[\frac{p}{e} \right] = \left[\frac{p}{f} \right]$ for $f \in \mathbb{P}(\mathcal{M})$, $f \sim e$. It will become clear from the proof that the real quotient also does not depend on the choice of the fundamental sequence.

Proof of the lemma. Note that $\lfloor \frac{p}{e_n} \rfloor = 0$ implies that $p \prec e_n$, so $\lfloor \frac{e_n}{p} \rfloor \geq 1$. If $\lfloor \frac{p}{e_n} \rfloor = 0$ for all n, then $\lfloor \frac{e_0}{p} \rfloor \geq \lfloor \frac{e_0}{e_n} \rfloor \lfloor \frac{e_n}{p} \rfloor \geq 2^n$ for all n, which is impossible. Hence, $\lfloor \frac{p}{e_{n_0}} \rfloor \geq 1$ for some $n_0 \in \mathbb{N}$, and for all $n \geq n_0$ and $k \in \mathbb{N}$, we then have $\lfloor \frac{p}{e_{n+k}} \rfloor \geq \lfloor \frac{p}{e_n} \rfloor \lfloor \frac{e_n}{e_{n+k}} \rfloor \geq 2^k$. Consequently, $\lfloor \frac{p}{e_n} \rfloor \nearrow \infty$ when $n \to \infty$.

For n such that $\lfloor \frac{p}{e_n} \rfloor \neq 0$, we have, by inequalities (3.2), that

$$\left\lfloor \frac{p}{e_{n+k}} \right\rfloor \leq \left(\left\lfloor \frac{p}{e_n} \right\rfloor + 1 \right) \left(\left\lfloor \frac{e_n}{e_{n+k}} \right\rfloor + 1 \right) \qquad (3.4)$$

and for n such that $\left\lfloor \frac{e}{e_n} \right\rfloor \neq 0$, we have that

$$\left\lfloor \frac{e}{e_{n+k}} \right\rfloor \geq \left\lfloor \frac{e}{e_n} \right\rfloor \left\lfloor \frac{e_n}{e_{n+k}} \right\rfloor . \tag{3.5}$$

By (3.4) and (3.5),

$$\frac{\left\lfloor \frac{p}{e_{n+k}} \right\rfloor}{\left\lfloor \frac{e}{e_{n+k}} \right\rfloor} \leq \frac{\left\lfloor \frac{p}{e_n} \right\rfloor + 1}{\left\lfloor \frac{e}{e_n} \right\rfloor} \frac{\left\lfloor \frac{e_n}{e_{n+k}} \right\rfloor + 1}{\left\lfloor \frac{e_n}{e_{n+k}} \right\rfloor} . \tag{3.6}$$

Now put $\alpha_n := \left\lfloor \frac{p}{e_n} \right\rfloor / \left\lfloor \frac{e}{e_n} \right\rfloor$. Note that (for n large enough) $0 < \alpha_n < \infty$. By (3.6), $\limsup_\ell \alpha_\ell \leq \alpha_n + 1/\left\lfloor \frac{e}{e_n} \right\rfloor$, whence $\limsup_\ell \alpha_\ell \leq \liminf_n \alpha_n$ and $\left[\frac{p}{e} \right] := \lim_n \alpha_n$ exists and is finite. Note that the existence and finiteness of $\left[\frac{e}{p} \right]$ implies that $\left[\frac{p}{e} \right] \neq 0$. Statements (1)–(4) and (6) are immediate by definition, and (5) follows directly from (3.3).

To see that the real quotient does not depend on the choice of a fundamental sequence, start with two fundamental sequences, $\{e_n\}$ and $\{f_n\}$, and form from them a third fundamental sequence that consists of infinitely many of both the e_ns and f_ns. The 'mixed' sequence must then lead to the same limit as each of the two sequences we started with. $\qquad\square$

Proof of Theorem 3.4.

Step 1 *Assume that $\mathcal{M}$ is a factor of type III.*

Define the dimension function D by $D(0) := 0$ and $D(p) = \infty$ for $p \neq 0$, so that the range of D is $\Delta = \{0, \infty\}$. Since there are no non-zero finite projections in $\mathcal{M}$, it is clear that D is the unique function satisfying conditions (1)–(4) of Definition 3.1.

Step 2 *Assume that $\mathcal{M}$ is a factor of type I_n, with finite or infinite n.*

Fix a minimal projection $e \in \mathcal{M}$ and choose any $c > 0$. Define $D_c(p) := c \left\lfloor \frac{p}{e} \right\rfloor$ if p is finite, $D_c(p) := \infty$ if p is infinite. Since e is minimal, Lemma 3.5 then ensures that $p = \left\lfloor \frac{p}{e} \right\rfloor e$ for a finite p. As noted after Lemma 3.6, $D_c(p)$ does not depend on the choice of e. It is clear that D_c satisfies conditions (1)–(4) of Definition 3.1. If D is any function satisfying conditions (1)–(4) of Definition 3.1, we may let $c := D(e)$ where e is any minimal projection of $\mathcal{M}$. Since e is minimal, it is finite and consequently c is also finite. Clearly, $D = D_c$. Note that the range Δ of D_c is exactly the set $\{c, 2c, \ldots, nc\}$ if n is finite and $\{c, 2c, \ldots, \infty\}$ otherwise.

Step 3 *If $\mathcal{M}$ is a factor of type II, it admits a dimension function, which is unique up to a positive scalar factor.*

Fix a non-zero finite projection $r \in \mathcal{M}$. Define, for $p \in \mathbb{P}(\mathcal{M})$,

$$D_r(p) = \begin{cases} 0, & \text{if } p = 0 \\ \left[\frac{p}{r}\right], & \text{if p is non-zero and finite} \\ \infty, & \text{otherwise.} \end{cases} \tag{3.7}$$

Conditions (1)–(4) of Definition 3.1 are clearly satisfied, so that D_r is a dimension function. Assume now that D is another dimension function on $\mathcal{M}$. Then, D and D_r are both 0 on the zero projection, and both are infinite on any infinite projection. Let us see how they act on non-zero finite projections. For any finite non-zero $p, q \in \mathbb{P}(\mathcal{M})$, we have

$$\left\lfloor \frac{p}{q} \right\rfloor D(q) \le D(p) \le \left(\left\lfloor \frac{p}{q} \right\rfloor + 1 \right) D(q).$$

Hence, for any fundamental sequence $\{e_n\}_{n \in \mathbb{N}}$,

$$\left\lfloor \frac{p}{e_n} \right\rfloor / \left(\left\lfloor \frac{r}{e_n} \right\rfloor + 1 \right) \le D(p)/D(r) \le \left(\left\lfloor \frac{p}{e_n} \right\rfloor + 1 \right) / \left\lfloor \frac{r}{e_n} \right\rfloor.$$

When n increases without bound, both the left- and right-hand sides of the above inequality tend to the real quotient $\left[\frac{p}{r}\right] = D_r(p)$. Hence,

$$D(p) = D(r)D_r(p)$$

so that D is a (non-zero, finite) multiple of D_r.

Step 4 *Suppose that D is a dimension function and assume that p and q are finite projections. We have $D(p) \le D(q)$ if and only if $p \precsim q$, and $D(p) < D(q)$ if and only if $p \prec q$.*

In fact, if $p \sim q$, then $D(p) = D(q)$ by Definition 3.1(2). If $p \prec q$, then there is a projection r such that $p \sim r \le q$ and $r \ne q$. By Definition 3.1(2),(3), and (4), $D(p) < D(q)$. By symmetry, if $q \prec p$, then $D(q) < D(p)$. Since exactly one of these three cases must occur, the inequalities between the values of the dimension function imply the corresponding order relations between projections.

Step 5 *If $p \in \mathbb{P}(\mathcal{M})$ and $\{p_i\}_{i \in \mathbb{N}}$ is a family of non-zero projections in $\mathcal{M}$ such that $\sum_{i \in \mathbb{N}} D(p_i) < D(p)$, then there exists an orthogonal family $\{q_i\}_{i \in \mathbb{N}}$ of subprojections of p such that $q_i \sim p_i$ for each $i \in \mathbb{N}$.*

We prove this by induction. By Step 3, $p_1 \prec p$; hence, $p_1 \sim q_1 \leq p$ and $D(q_1) < D(q)$. Suppose that we have already defined n non-zero mutually orthogonal projections $q_1, q_2, \ldots, q_n$ such that $q_i \sim p_i$ for each $i = 1, 2, \ldots, n$ and $\sum_{i=1}^{n} D(q_i) < D(p)$. Then, by the assumption and the finite additivity of D, $D(p - \sum_{i=1}^{n} q_i) > D(p_{n+1})$. Hence, again by Step 3, $p_{n+1} \prec p - \sum_{i=1}^{n} q_i$ and so there exists a (non-zero) projection q_{n+1} such that $p_{n+1} \sim q_{n+1} \leq p$ and $\sum_{i=1}^{n+1} D(q_i) < D(p)$.

Step 6 *Dimension functions are completely additive*

We have to show that for any orthogonal family $\{p_i\}_{i\in I}$ of non-zero projections in $\mathcal{M}$, we have

$$D\left(\sum_{i\in I} p_i\right) = \sum_{i\in I} D(p_i). \tag{3.8}$$

We can suppose that all projections p_i are finite and I is at most countable; otherwise, both sides of (3.8) are infinite. In particular, we do not need to consider factors of type III. Also, (3.8) follows by definition if I is finite, and hence so do the results for factors of type I_n with finite n, where there are no countably infinite orthogonal families of non-zero projections. Thus, we may assume that I is countably infinite. In a factor of type I_∞, values of D on the projections are bounded below by a value of D on a minimal projection, so that again both sides of (3.8) are infinite. We are left with factors of type II, where a fundamental sequence delivers non-zero projections r with arbitrarily small $D(r)$.

For any $K \subseteq I$, we will write p_K for $\sum_{i\in K} p_i$. By the finite additivity and monotonicity of D (see Step 3), we get the inequality $\sum_{i\in I} D(p_i) \leq D(p_I)$. Assume that $\sum_{i\in I} D(p_i) < D(p_I)$. Choose non-zero $q \in \mathbb{P}(\mathcal{M})$ so that $\sum_{i\in I} D(p_i) + D(q) < D(p_I)$. Now choose a finite $J \subset I$ so that $\sum_{i\in I\setminus J} D(p_i) < D(q)$. Use Step 5 to find an orthogonal family $\{q_i\}_{i\in I\setminus J}$ such that $q_i \sim p_i$ for each $i \in I \setminus J$ and $q_i \leq q$. By Proposition 0.112,

$$p_{I\setminus J} = \sum_{i\in I\setminus J} p_i \sim \sum_{i\in I\setminus J} q_i \leq q.$$

In particular, $D(p_{I\setminus J})$ and $D(p_I)$ are finite and

$$D(p_I) = D(p_J) + D(p_{I\setminus J}) \leq D(p_J) + D(q)$$

$$< D(p_J) + D(p_I) - \sum_{i\in I} D(p_i) = D(p_I) - \sum_{i\in I\setminus J} D(p_i) < D(p_I)$$

which is impossible. Consequently, (3.8) is satisfied.

Step 7 *Dimension functions preserve suprema of increasing sequences.*

In fact, suppose that $(p_n)_{n\in\mathbb{N}}$ is an increasing sequence of projections. Put $p_0 := 0$. Then, $p = \sum_{n=1}^{\infty}(p_n - p_{n-1})$. Therefore, by Step 6,

$$D(p) = \sum_{n=1}^{\infty} D((p_n) - D(p_{n-1})) = \sup_{n} D(p_n).$$

Step 8 *If $\mathcal{M}$ is a factor of type II, then the range Δ of a dimension function D is the interval $[0, c]$, where $c := D(\mathbb{1})$. (Hence c is infinite and $\Delta = [0, \infty]$ for a factor of type II_∞.)*

By monotonicity (Step 4), we have $\Delta \subseteq [0, c]$. Combining Steps 4–6 we see that whenever $\alpha, \beta \in \Delta$ with $\alpha < \beta$, then $\beta - \alpha \in \Delta$, and whenever $\alpha_1, \alpha_2, \cdots \in \Delta$ with $\alpha_1 + \alpha_2 + \cdots \leq c$, then $\alpha_1 + \alpha_2 + \cdots \in \Delta$. In particular, if $\alpha \in \Delta$, and $k \in \mathbb{N}$ is such that $k\alpha \leq c$, then $k\alpha \in \Delta$. Since Δ contains arbitrarily small numbers (for example, values of D on a fundamental sequence), Δ is dense in $[0, c]$. If $\alpha \in (0, c)$ and $\alpha_n \nearrow \alpha$, $\alpha_n \in \Delta, \alpha_0 = 0$, then $\alpha = \sum_{n=1}^{\infty}(\alpha_n - \alpha_{n-1}) \in \Delta$, so that $\Delta = [0, c]$.

This ends the proof of the theorem. $\qquad\square$

Theorem 3.4 does not discriminate between infinite projections. If we are interested in a dimension function that assigns different values to inequivalent projections, we should use the generalised dimension function. We formulate a theorem analogous to Theorem 3.4 on the existence and properties of the generalised dimension function by exchanging a value ∞ with a cardinal segment from $\aleph_0$ up to $\mathfrak{d}(\mathcal{M})$, the decomposability number of $\mathcal{M}$.

Theorem 3.11 *Each von Neumann factor $\mathcal{M}$ admits a generalised dimension function $\bar{D} : \mathbb{P}(\mathcal{M}) \to [0, \infty) \cup [\aleph_0, \mathfrak{d}(\mathcal{M})]$, where $[\aleph_0, \mathfrak{d}(\mathcal{M})]$ consists of cardinals, which is unique up to a positive constant factor c. The function $\bar{D}$ is completely additive and if, for an increasing sequence of projections (p_n), we have $p_n \nearrow p$, then $\bar{D}(p_n) \nearrow \bar{D}(p)$. Moreover, the set $\bar{\Delta} := \bar{D}(\mathbb{P}(\mathcal{M}))$ is determined solely by the type of the factor. Namely, we have:*

(1) $\bar{\Delta} = \{c, 2c, \ldots, nc\}$ *for $\mathcal{M}$ of type $I_n, n < \infty$;*
(2) $\bar{\Delta} = \{c, 2c, \ldots\} \cup [\aleph_0, \mathfrak{d}(\mathcal{M})]$ *for $\mathcal{M}$ of type I_∞;*
(3) $\bar{\Delta} = [0, c]$ *for $\mathcal{M}$ of type II_1;*

(4) $\bar{\Delta} = [0, \infty) \cup [\aleph_0, \mathfrak{d}(\mathcal{M})]$ *for $\mathcal{M}$ of type II_∞;*
(5) $\bar{\Delta} = \{0\} \cup [\aleph_0, \mathfrak{d}(\mathcal{M})]$ *for $\mathcal{M}$ of type III.*

We shall not give the proof of the theorem as it does not differ substantially from the proof of Theorem 3.4. We will just comment on the differences. In the present case, the value of $\bar{D}$ on an infinite projection is exactly the decomposability number of the projection. By Lemma 3.8, if we represent an infinite projection as a sum of an infinite orthogonal family of σ-finite projections, the cardinality of the family will be equal to the decomposability number of the projection. This clearly implies complete additivity of the generalised dimension function. It is also clear that inequivalent infinite projections in a factor have different decomposability numbers, and that any cardinal between $\aleph_0$ and $\mathfrak{d}(\mathcal{M})$ is a decomposability number of a projection from the algebra.

There are several natural questions connected with the material just presented. Here is the list.

(1) Both the dimension function and the generalised dimension function preserve suprema of increasing sequences of projections. Do they also preserve suprema of an increasing directed family of projections? In other words, are they *normal*?

The answer is an easy 'no' for the generalised dimension function (in infinite non-σ-finite factors): if a directed family p_i of σ-finite projections increases to a non-σ-finite projection p, then $\sup \bar{D} = \aleph_0 < \mathfrak{d}(p)$. On the other hand, dimension functions are normal, although the direct proof of this fact is by no means easy. Fortunately, normality of dimension functions, which is trivial for factors of type III, follows immediately from the existence of faithful normal traces on semifinite algebras (see Theorem 3.38 and the remark following it).

(2) Can we define dimension functions on *non-factorial* von Neumann algebras?

The answer is 'yes', although the functions are no longer number- (or cardinal-) valued. An interested reader should consult the papers of Tomiyama ([**Tom58**]) and Sherman ([**She07**]).

(3) Dimension functions correspond more or less to measures. What about integrals? Can we *extend* dimension functions from the lattice of projections to the whole algebra?

The answer is again 'yes', and the objects we obtain for finite algebras, namely centre-valued traces, are instrumental for the further development of noncommutative measure and integral.

The theory is developed in Section 3.2, and the more difficult case of general semifinite algebras can be found in Tomiyama [**Tom58**], Sherman [**She07**] and Takesaki [**Tak02**, Section V.2]).

3.2 Traces

The existence of centre-valued traces on finite von Neumann algebras and semifinite traces on semifinite algebra is a non-trivial matter. In this section, we shall give full proofs of both these facts.

3.2.1 Centre-valued traces

There are two basic approaches to proving the existence of centre-valued traces. The first uses the device of monic projections and the other the Ryll-Nardzewski fixed point theorem. We shall prefer an updated version of the latter. The basic approach here is to show that every normal functional on $\mathcal{Z}(\mathcal{M})$ admits a unique extension to a tracial normal functional τ_η on $\mathcal{M}$, and then to show that this process in fact realises a linear map from $\mathcal{Z}(\mathcal{M})_*$ to $\mathcal{M}_*$ for which the adjoint is a trace-like map that characterises finite von Neumann algebras, namely a map of the following type:

Definition 3.12 [**SZ79**, 7.11, 7.12] A *(canonical) centre-valued* (or *central) trace* on $\mathcal{M}$ is a linear bounded map $\mathscr{T}$ from $\mathcal{M}$ onto $\mathcal{Z}(\mathcal{M})$ such that:

(1) $\mathscr{T}(ab) = \mathscr{T}(ba)$ for all $a, b \in \mathcal{M}$;
(2) $\mathscr{T}(c) = c$ for $c \in \mathcal{Z}(\mathcal{M})$.

In proving the existence of centre-valued traces, our first order of business is to gain a deeper understanding of what weak compactness means for subsets of $\mathcal{M}_*$.

Lemma 3.13 *Let $\mathcal{M}$ be a von Neumann algebra, $q \in \mathcal{M}$ a projection, and (e_n) an increasing sequence of projections in $\mathcal{M}$ such that $e_n \preceq q$ for every n. Then $\vee_{n \geq 1} e_n \preceq q$.*

Proof.   The fact that (e_n) is increasing ensures that the sequence given by $p_1 = e_1$ and $p_n = e_n - e_{n-1}$ for all $n \geq 2$ is a sequence of mutually orthogonal projections with $\sum_n p_n = \vee_{n\geq 1} e_n$. We construct a mutually orthogonal sequence of projections (q_n) such that $p_n \sim q_n \leq q$ for each n. By Proposition 0.123, it suffices to separately consider the cases where q is, respectively, properly infinite and finite. First, assume q to be properly infinite. By repeated application of the Halving Lemma (Lemma 0.124), we may construct a sequence (q_n) of mutually orthogonal projections such that $q_n \leq q$ and $q_n \sim q$ for each n. Given that then $p_n \preceq q \sim q_n$ for each n, an application of Proposition 0.112 then shows that $\vee_{n\geq 1} e_n = \sum_n p_n \preceq \sum_n q_n \leq q$, as required.

We proceed to the case where q is finite. By hypothesis, $p_1 = e_1 \preceq q$. So, there must exist a projection q_1 with $p_1 \sim q_1 \leq q$. Now, suppose that $\{q_1, q_2, \ldots, q_n\}$ have been found. Then, it follows from Proposition 0.112 that $e_n = \sum_{1\leq k\leq n} p_k \sim \sum_{1\leq k\leq n} q_k \leq q$. Since, by hypothesis, $e_{n+1} = \sum_{1\leq k\leq n+1} p_k \preceq q$, there must exist a partial isometry w such that $w^*w = \sum_{1\leq k\leq n+1} p_k$ and $ww^* \leq q$. Now, it is an exercise to see that $\left(\sum_{1\leq k\leq n} q_k\right) \sim \left(\sum_{1\leq k\leq n} p_k\right) \sim w\sum_{1\leq k\leq n} p_k w^*$ with $w\sum_{1\leq k\leq n} p_k w^* \leq w\sum_{1\leq k\leq n+1} p_k w^* = ww^* \leq q$. The finiteness of q now enables us to conclude from Proposition 0.122 that $q - \left(\sum_{1\leq k\leq n} q_k\right) \sim q - w\sum_{1\leq k\leq n} p_k w^*$. Since $p_{n+1} \sim wp_{n+1}w^* = ww^* - w\sum_{1\leq k\leq n} p_k w^* \leq q - w\sum_{1\leq k\leq n} p_k w^* \sim q - \left(\sum_{1\leq k\leq n} q_k\right)$, there must exist a projection $q_{n+1} \leq q - \left(\sum_{1\leq k\leq n} q_k\right)$ with $p_{n+1} \sim q_{n+1}$. The claim now follows by induction. $\square$

Lemma 3.14 *Let $\mathcal{M}$ be a finite von Neumann algebra with (p_n) a sequence of mutually orthogonal projections in $\mathcal{M}$. If (q_n) is another sequence of projections for which $p_n \sim q_n$ for each n, then $q_n \to 0$ in the σ-strong* topology.*

Proof. We claim that given projections e_1, e_2 and f_1, f_2 with $e_i \preceq f_i$ for $i = 1, 2$ and $f_1 \perp f_2$, we then have $e_1 \vee e_2$. To see this, note that we may use Kaplansky's formula to conclude that $e_1 \vee e_2 - e_2 \sim e_1 - e_1 \wedge e_2 \leq e_1 \preceq f_1$. The claim then follows from Proposition 0.112. An inductive application of this fact now shows that we will have $\vee_{m\geq k\geq n} q_k \preceq \sum_{m\geq k\geq n} p_k \leq \sum_{k\geq n} p_k$ for any $n < m$. Lemma 3.13 then informs us that, in fact,

$$\bigvee_{k\geq n} q_k \preceq \sum_{k\geq n} p_k.$$

For ease of notation, we write e_n for $\vee_{k \geq n} q_k$, and e for $\wedge_n e_n$. The fact that $e_n \preceq \sum_{k \geq n} p_k$ ensures, of course, that there is some subprojection r_n of $\sum_{k \geq n} p_k$ with $e_n \sim r_n$. Proposition 0.122 then ensures that

$$\mathbb{1} - \sum_{k \geq n} p_k \leq \mathbb{1} - r_n \sim \mathbb{1} - e_n \leq \mathbb{1} - e.$$

Noticing that the projections $\sum_{k \geq n} p_k$ decrease to 0, we may then apply Lemma 3.13 to conclude that

$$\mathbb{1} = \mathbb{1} - \bigwedge_{n \geq k} \sum_{k \geq n} p_k = \bigvee_{n \geq 1} \left(\mathbb{1} - \sum_{k \geq n} p_k\right) \preceq \mathbb{1} - e \leq \mathbb{1}.$$

The finiteness of $\mathbb{1}$ then ensures that $\wedge_n e_n = e = 0$, or equivalently $\mathbb{1} = \vee_n(\mathbb{1} - e_n\} = \vee_n(\mathbb{1} - \vee_{k \geq n} q_k)$. Since $(e_n) = (\vee_{k \geq n} q_k)$ is a decreasing sequence of projections, this can only happen if $(\mathbb{1} - \vee_{k \geq n} q_k)$ increases to $\mathbb{1}$ (implying convergence in the σ-strong* topology). So, $(\vee_{k \geq n} q_k)$ converges σ-strong* to 0, and since $q_n \leq \vee_{k \geq n} q_k$ for each n, so does (q_n). $\qquad\qquad\square$

The required description of weak compactness is provided by the following lemma (due to Akemann). As Akemann showed, the converse implication also holds (see [**Tak02**, Theorem III.5.4]), but we shall not need this.

Lemma 3.15 *Let $\mathcal{M}_*$ be the predual of a von Neumann algebra $\mathcal{M}$, and let $\mathcal{F}$ be a norm-bounded subset of $\mathcal{M}_*$. Then, $\mathcal{F}$ will be relatively weakly compact if for any sequence (e_n) of projections decreasing to 0, we have $\psi(e_n) \to 0$ uniformly for $\psi \in \mathcal{F}$.*

Proof. Assume that the stated condition is satisfied. We know that each $\psi \in \mathcal{F}$ is completely additive on projections. We claim that the stated condition ensures that $\mathcal{F}$ is uniformly completely additive on projections in the sense that for any family $\{e_i\} \subset \mathcal{M}$ of mutually disjoint projections, the series $\sum_i \psi(e_i)$ will converge to $\psi(\sum_i e_i)$ uniformly for $\psi \in \mathcal{F}$. That means that for any $\epsilon > 0$, there exists a finite subset J_ϵ of the index set I such that for all finite subsets J of $I - J_\epsilon$ and all $\psi \in \mathcal{F}$, we have $|\sum_{i \in J} \psi(e_i)| < \epsilon$. So, if this failed, we would be able to find a $\delta > 0$ such that for any finite subset J of I, there is an infinite set of ψs in $\mathcal{F}$ for which $|\sum_{i \in J_0} \psi(e_i)| > \delta$ for some finite subset J_0 of $I - J$. Proceeding inductively, we can then select a sequence

$(\psi_n) \subset \mathcal{F}$ and a sequence of finite subsets $J_n \subset I$ such that for each n, we have $|\sum_{i \in J_n} \psi_n(e_i)| > \delta$, and $i > j$ for any $i \in J_{n+1}$ and $j \in J_n$. The sequence $q_n = \sum_{i \in J_n}$ is then a sequence of mutually orthogonal projections for which $|\psi_n(q_n)| \geq \delta$ for each n. This is in clear violation of the condition stated in the hypothesis. Therefore, that condition must ensure that $\mathcal{F}$ is 'uniformly completely additive on projections'.

Next, observe that the norm boundedness of $\mathcal{F}$ ensures that as a subset of $\mathcal{M}^*$, it is relatively weak* compact. Recall that the weak topology on $\mathcal{M}_*$ is just the subspace topology inherited from the weak* topology on $\mathcal{M}^*$. We know that the weak* closure $\overline{\mathcal{F}}^{w*}$ in $\mathcal{M}^*$ is weak* compact. If we can show that it is actually contained in $\mathcal{M}_*$, the claim will follow. Let $\psi \in \overline{\mathcal{F}}^{w*}$ be given. There must then exist a net $(\psi_j) \subset \mathcal{F}$, which in $\mathcal{M}^*$ is weak* convergent to ψ. Since each ψ_j is completely additive on projections, we will for any mutually orthogonal family $\{e_i\}$ of projections and any j have that $\lim_i \psi_i(e_j) = \psi(e_j)$ and $\sum_i \psi_j(e_i) = \psi_j(\sum_i e_i)$, and also that $\lim_j \psi_j(\sum_i e_i) = \psi(\sum_i e_i)$. If we combine this with the fact that $\sum_i \psi_j(e_i)$ converges to $\psi_j(\sum_i e_i)$ uniformly in j, then it follows that $\sum_i \psi(e_i)$ converges to $\psi(\sum_i e_i)$. Thus, ψ is completely additive on projections and therefore must belong to $\mathcal{M}_*$. $\qquad\square$

Equipped with the above lemma, we are now able to prove the following result.

Lemma 3.16 *Let $\mathcal{M}$ be a von Neumann algebra and $\psi \in \mathcal{M}_*$ a normal state. Then, $K_\varphi = \{\varphi(u \cdot u^*) : u \in \mathcal{U}(\mathcal{M})\}$ is relatively weakly compact.*

Proof. Let $K_\varphi = \{\varphi(u \cdot u^*) : u \in \mathcal{U}(\mathcal{M})\}$. Suppose that K_φ is not relatively weakly compact. By the lemma, there must then exist a sequence (e_n) of mutually orthogonal projections for which $\{\psi(e_n) : \psi \in K_\varphi\}$ does not converge uniformly to 0. This means that there must exist some $\delta > 0$, a subsequence (e_{n_k}) of (e_n), and unitaries (u_k) such that $\varphi(u_k e_{n_k} u_k^*) \geq \delta$ for all k. Since we clearly have $u_k e_{n_k} u_k^* \sim e_{n_k}$ for each k, Lemma 6.4.14 then informs us that $(u_k e_{n_k} u_k^*)$ is strongly convergent to 0. But if that is the case, we must have $\varphi(u_k e_{n_k} u_k^*) \to 0$ as $k \to \infty$. This is a clear contradiction, proving the claim. $\qquad\square$

At this point, we pass to an analysis of the existence and normality of tracial functionals. Existence is proven by invoking the Ryll–Nardzewski fixed point theorem and applying it to the above fact.

Theorem 3.17 Ryll–Nardzewski fixed-point theorem   *Let Q be a nonempty, weakly compact, convex subset of a locally convex Hausdorff linear topological space E, and let S be a noncontracting semigroup of weakly continuous affine maps of Q into itself. Then, there is a common fixed point of S in Q.*

Lemma 3.18   *Let $\mathcal{M}$ be a finite von Neumann algebra. Then, every normal functional $\eta \in \mathcal{Z}(\mathcal{M})_*$ admits a unique normal extension to a tracial functional $\tau_\eta \in \mathcal{M}_*$. This extension is moreover norm preserving. If, in addition, η is a positive functional, then so is τ_η.*

Proof.   Given $\eta \in \mathcal{Z}(\mathcal{M})_*$, one first applies the Hahn–Banach theorem to extend η to a normal linear functional φ_η on $\mathcal{M}$. The next step is to apply the Ryll–Nardzewski fixed-point theorem to the closed convex hull of K_φ (which is weakly compact) to obtain an extension τ_η of η to a normal tracial functional on $\mathcal{M}$. So, η admits at least one tracial extension which can be chosen to be normal.

We proceed to prove the claim regarding the preservation of the norm. First, consider the case where τ_η is normal. Let $\tau_\eta = v.|\tau_\eta|$ then be the polar form of a normal tracial extension τ_η of η. Notice that then $\tau_\eta(a) = \tau_\eta(uau^*) = |\tau_\eta|(uau^*v) = (u^*.|\tau_\eta|.u)(a(u^*vu))$ for any $a \in \mathcal{M}$ and any $u \in \mathcal{U}(\mathcal{M})$. By the uniqueness of the polar decomposition of τ_η, we must have that $|\tau_\eta| = (u^*.|\tau_\eta|.u)$ and that $v = u^*vu$ for all $u \in \mathcal{U}(\mathcal{M})$. So, $|\tau_\eta|$ is a tracial functional and $v \in \mathcal{Z}(\mathcal{M})$. Since $v \in \mathcal{Z}(\mathcal{M})$, the expression $v.|\tau_\eta|/_{\mathcal{Z}(\mathcal{M})}$ is, in fact, the polar form of η. Therefore, we have $\|\tau_\eta\| = \tau_\eta(v^*) = \eta(v^*) = \|\eta\|$.

With the preservation of the norm verified, the claim regarding positive functionals is now an easy consequence of Proposition 0.64.

We finally consider the question of the uniqueness of the extension. Suppose that ν is another extension of η for which we have $u^*.\nu.u = \nu$ for every $u \in \mathcal{U}(\mathcal{M})$. Then, $\nu - \tau_\eta$ is an extension of the 0 functional on $\mathcal{Z}(\mathcal{M})$ for which we have $u^*.(\nu - \tau_\eta).u = (\nu - \tau_\eta)$ for all $u \in \mathcal{U}(\mathcal{M})$. But we saw above that all such extensions are norm-preserving. So, we must have $\|\nu - \tau_\eta\| = 0$.   □

The next step is a description of the normality of tracial functionals.

Proposition 3.19 ([Tak02, Proposition V.1.34]) *For any mutually orthogonal family $\{e_j : j \in J\}$ of equivalent projections in a von Neumann algebra, there exist a nonzero central projection z and a family $\{p_i : i \in I\}$ of mutually orthogonal projections with $J \subset I$, such that $p_j \sim e_j z$ for each $j \in J$ and $p_0 = z_0 - \sum_{i \in I} p_i \prec p_k$ for each $k \in I$.*

Proof. Given a mutually orthogonal family $\{e_j : j \in J\}$, we use Zorn's lemma to select a maximal family $\{\widetilde{e}_i : i \in I\}$ of equivalent mutually orthogonal projections containing $\{e_j : j \in J\}$. We then set $e_0 = \mathbb{1} - \sum_i \widetilde{e}_i$ and use the comparability theorem (Theorem 0.113) to select a central projection z so that for some $k \in I$, we have $ze_0 \preceq \widetilde{e}_k$ and $(\mathbb{1} - z)\widetilde{e}_k \preceq (\mathbb{1} - z)e_0$. The equivalence of the $\widetilde{e}_i$s ensures that this holds for all $k \in I$. If $ze_0 \sim z\widetilde{e}_k$, then we would have $\widetilde{e}_k \preceq e_0$, in which case e_0 would admit a subprojection equivalent to each of the $\widetilde{e}_k$s. But this would contradict the maximality of $\{\widetilde{e}_i : i \in I\}$, and therefore, we must have $ze_0 \prec z\widetilde{e}_k$ for each k. Setting $p_k = z\widetilde{e}_k$ and $p_0 = ze_0$, we have $z - \sum_i p_i = pe_0 \prec p_k$ for each k, as claimed. $\qquad\square$

Lemma 3.20 ([Tak02, Proposition V.2.5]) *Let $\mathcal{M}$ be a von Neumann algebra and $\omega \in \mathcal{M}_*$ be a tracial state. Then, ω is normal if and only if the restriction $\omega{\upharpoonright}\mathcal{Z}(\mathcal{M})$ to the centre is normal.*

Proof. Recall that ω canonically extends to $\mathcal{M}^{**}$. The weak* density of $\mathcal{M}$ in $\mathcal{M}^{**}$ alongside the weak* continuity of the extension ensures that it is again a tracial state. The normal and singular parts (ω_n and ω_s) of ω are then defined as the restrictions of $p\omega$ and $(\mathbb{1} - p)\omega$ to $\mathcal{M}$ where p is as in Theorem 0.87. The centrality of p ensures that both are again positive tracial functionals. This clearly shows that we need only show that singular positive tracial functionals that are normal on the centre must be 0 for the theorem to hold. So, let ω be singular and, by contradiction, assume that ω is non-zero but normal on $\mathcal{Z}(\mathcal{M})$.

To start with, we show that $\mathbf{z}(\omega{\upharpoonright}\mathcal{Z}(\mathcal{M}))$ is finite. To see this, recall that, by Proposition 0.123, $\mathbf{z}(\omega{\upharpoonright}\mathcal{Z}(\mathcal{M}))$ may be written as an orthogonal sum $\mathbf{z}(\omega{\upharpoonright}\mathcal{Z}(\mathcal{M})) = z_{pi} + z_{fin}$ of a properly infinite z_{pi} and finite projection z_{fin}. If now the properly infinite part was non-zero, we would be able to find a non-zero proper subprojection p of z_{pi} for which we have $p \sim (z_{pi} - p) \sim z_{pi}$ (see Lemma 0.124). The tracial property then ensures that $\omega(p) = \omega(z_{pi} - p) = \omega(z_{pi})$, which in turn implies that $\omega(z_{pi}) = 2\omega(z_{pi})$ and hence that $\omega(z_{pi}) = 0$. But this contradicts the fact that ω is faithful on $\mathbf{z}(\omega{\restriction}\mathcal{Z}(\mathcal{M}))\mathcal{Z}(\mathcal{M})$.

The fact that $\omega(\mathbb{1}) \neq 0$ clearly implies that $\mathbf{z}(\omega{\restriction}\mathcal{Z}(\mathcal{M})) \neq 0$. By Proposition 0.88, there exists a non-zero subprojection e of $\mathbf{z}(\omega{\restriction}\mathcal{Z}(\mathcal{M}))$ such that $\omega(e) = 0$. When applying Proposition 3.19 to $\mathbf{z}(\omega{\restriction}\mathcal{Z}(\mathcal{M}))\mathcal{M}$, there must then exist a non-zero central subprojection z_0 of $\mathbf{z}(\omega{\restriction}\mathcal{Z}(\mathcal{M}))$ and a family $\{p_i\}$ of mutually orthogonal subprojections of $\mathbf{z}(\omega{\restriction}\mathcal{Z}(\mathcal{M}))$, each of which is equivalent to $e\mathbf{z}(\omega{\restriction}\mathcal{Z}(\mathcal{M}))$, with in addition $p_0 = z_0 - \sum_i p_i \prec p_j$ for each j. We then have $\omega(p_i) = \omega(ez_0) \leq \omega(e) = 0$ and $\omega(p_0) \leq \omega(f_i) = 0$. The finiteness of $\mathbf{z}(\omega{\restriction}\mathcal{Z}(\mathcal{M}))$ of course ensures that $\sum_i p_i$ is also finite. This, in turn, guarantees that the index set I is finite. (If I were infinite, it would admit an infinite proper subset I_0 of I, which is bijective to I. By Proposition 0.112, the term $\sum_{i \in I_0} p_i$ would correspond to a proper but equivalent subprojection of $\sum_{i \in I} p_i$. But this would contradict the finiteness of $\sum_{i \in I} p_i$.) Taking I to be $\{1, 2, \ldots, n\}$, the equalities verified above then show that $\omega(z_0) = \sum_{i=1}^{n} \omega(p_i) = 0$. Once again, this contradicts the fact that ω is faithful on $\mathbf{z}(\omega{\restriction}\mathcal{Z}(\mathcal{M}))\mathcal{Z}(\mathcal{M})$. Therefore, we must have $\omega = 0$ as required. $\qquad\square$

We are now finally ready to prove our main theorem, namely the following.

Theorem 3.21

(a) *For a centre-valued trace $\mathscr{T}$ on a finite von Neumann algebra, the following are equivalent:*
- *$\mathscr{T}$ is contractive;*
- *$\mathscr{T}$ is positive;*
- *$\mathscr{T}$ is normal.*

(b) *A von Neumann algebra $\mathcal{M}$ is finite if and only if there exists a unique normal (and hence positive and contractive) centre-valued trace $\mathscr{T}$ on $\mathcal{M}$. The centre-valued trace has the following additional properties.*

(1) *$\mathscr{T}(ca) = c\mathscr{T}(a)$ for $c \in \mathcal{Z}(\mathcal{M})$, $a \in \mathcal{M}$.*

(2) *If $a \in \mathcal{M}_+$ and $\mathscr{T}(a) = 0$, then $a = 0$ (i.e., $\mathscr{T}$ is faithful).*

(3) *For each $a \in \mathcal{M}$, $\mathscr{T}(a)$ is in the norm-closed convex hull of the set $\{uau^* : u \in \mathcal{U}(\mathcal{M})\}$.*

Proof. First consider part (a). Since $\mathscr{T}$ will by definition preserve $\mathbb{1}$, Proposition 0.64 ensures that whenever $\mathscr{T}$ is contractive, its dual will map states to states. Therefore, $\mathscr{T}$ is positive. Next, suppose that $\mathscr{T}$ is positive. Again, since by definition $\mathscr{T}$ preserves the identity, $\nu \circ \mathscr{T}$ will be a state whenever ν is a state on $\mathcal{Z}(\mathcal{M})$. Then, it clearly follows from Lemma 3.20 that $\nu \mapsto \nu \circ \mathscr{T}$ will send normal states to normal states, which in turn suffices to prove that $\mathscr{T}$ is normal. We pass to part (b). The contractivity of normal centre-valued traces, claimed in part (a), will be obtained as yet another additional property alongside those mentioned in part (b).

So, let $\mathcal{M}$ be a finite von Neumann algebra. The uniqueness criterion in Lemma 3.18 ensures that the prescription $\eta \mapsto \tau_\eta$ defines a linear map $\mathscr{S} \colon \mathcal{Z}(\mathcal{M})_* \to \mathcal{M}_*$, with the preservation of the norm ensuring that $\mathscr{S}$ is contractive. Now, we set $\mathscr{T} = \mathscr{S}^*$. Since for any $a, b \in \mathcal{M}$ and any normal functional η on $\mathcal{Z}(\mathcal{M})$, we have $\eta(\mathscr{T}(ab)) = \tau_\eta(ab) = \tau_\eta(ba) = \eta(\mathscr{T}(ba))$, the fact that $\mathscr{T}(ab) = \mathscr{T}(ba)$ is clear. This proves the existence.

We go on to prove the uniqueness. Let $\mathscr{T}'$ be an arbitrary normal centre-valued trace on $\mathcal{M}$. For any functional $\eta \in \mathcal{Z}(\mathcal{M})$, the properties described in Definition 3.12 ensure that $\eta \circ \mathscr{T}'$ is then an extension of η to a normal tracial functional on $\mathcal{M}$. By the uniqueness criterion of Lemma 3.18, this functional must agree with the normal functional τ_η. For any $\eta \in \mathcal{M}_*$, we therefore have $\eta \circ \mathscr{T}' = \eta \circ \mathscr{T}$. This shows that the predual of $\mathscr{T}'$ agrees with $\mathcal{S}$, which, of course, ensures that $\mathcal{T}' = \mathcal{S}^* = \mathcal{T}$. With the uniqueness established, the contractivity on the predual of $\mathscr{T}$ (and hence also of $\mathscr{T}$) now follows from Lemma 3.18.

Property (1) follows from the uniqueness criterion in Lemma 3.18, and the fact that for any normal state ω on $\mathcal{Z}(\mathcal{M})$, the functionals $c.(\omega \circ \mathscr{T})$ and $(c.\omega) \circ \mathscr{T}$ are both tracial extensions of $c.\omega$.

To see that (2) holds, let $x \in \mathcal{M}_+$ be given with $x \neq 0$, and let η be a normal state on $\mathbf{z}(s(x))\mathcal{Z}(\mathcal{M})$. The support projection p of η is a central projection (element of $\mathcal{Z}(\mathcal{M})$) that satisfies $p \leq \mathbf{z}(s(x))$. Now let $q = s(\tau_\eta)$. For any $u \in \mathcal{U}(\mathcal{M})$ and any $a \in \mathcal{M}$, we have $\tau_\eta((uq^\perp u^*)a^*a(uq^\perp u^*)) = \tau_\eta(q^\perp u^* a^* a u q^\perp) = \tau_\eta(q^\perp (au)^*(au)q^\perp) = 0$. Thus, $uq^\perp u^* \leq q^\perp$ for all $u \in \mathcal{U}(\mathcal{M})$. But then also $q^\perp \leq uq^\perp u^*$ for all $u \in \mathcal{U}(\mathcal{M})$, whence $uqu^* = q$ for all $u \in \mathcal{U}(\mathcal{M})$. Thus $q \in \mathcal{Z}(\mathcal{M})$. Using the centrality of q and the properties already verified, easy checking now shows that $\mathcal{Z}(\mathcal{M})q^{\perp} \subseteq \{c \in \mathcal{Z}(\mathcal{M}) \colon \eta(c^*c)\} = \mathcal{Z}(\mathcal{M})p^{\perp}$ and that $\mathcal{M}p^{\perp} \subseteq \{a \in \mathcal{M} \colon \tau_{\eta}(a^*a)\} = \mathcal{M}q^{\perp}$. This can only be if $p = q$. Now, if $xp = 0$, then $\mathbf{z}(s(x)) - p$ rather than $\mathbf{z}(s(x))$ would be the central carrier of $s(x)$. So, we must have $xp \neq 0$. But then we must have $\eta(\mathcal{T}(x)) = \tau_{\eta}(x) \neq 0$ and hence $\mathcal{T}(x) \neq 0$.

Claim (3) follows from the Hahn–Banach separation theorem applied to the construction of $\mathcal{T}(a)$. For each normal functional φ, the value $\varphi(\mathcal{T}(a))$ is a limit of terms of the form $\varphi(b_{\gamma})$ where $(b_{\gamma}) \subset \overline{\mathrm{co}}\{\varphi(u \cdot u^*) \colon u \in \mathcal{U}(\mathcal{M})\}$. So, there is no normal functional separating a from this convex set.

To see that the existence of a centre-valued trace forces a von Neumann algebra to be finite, note that if $\mathcal{M}$ is a general von Neumann algebra that has a faithful normal centre-valued trace $\mathcal{T}$, and if v is an isometry (i.e., $v^*v = \mathbb{1}$), then $\mathbb{1} = \mathcal{T}(v^*v) = \mathcal{T}(vv^*)$, which ensures that $\mathcal{T}(\mathbb{1} - vv^*) = 0$. So, $\mathbb{1} = vv^*$ by the faithfulness of $\mathcal{T}$, and hence, $\mathcal{M}$ is finite. $\qquad\square$

The next corollary states that the centre-valued trace restricted to projections constitutes a (centre-valued) dimension function.

Corollary 3.22   *Let $\mathcal{M}$ be a finite von Neumann algebra with the centre-valued trace $\mathcal{T}$. Then, for $e, f \in \mathbb{P}(\mathcal{M})$, we have $e \sim f$ (respectively, $e \precsim f$) if and only if $\mathcal{T}(e) = \mathcal{T}(f)$ (respectively, $\mathcal{T}(e) \leq \mathcal{T}(f)$).*

Proof.   It follows from conditions (1) and (2) of Definition 3.12 and (3) of Theorem 3.21 that $e \sim f$ (respectively, $e \precsim f$) implies $\mathcal{T}(e) = \mathcal{T}(f)$ (respectively, $\mathcal{T}(e) \leq \mathcal{T}(f)$). Let now $e, f \in \mathbb{P}(\mathcal{M})$ and assume that $\mathcal{T}(e) \leq \mathcal{T}(f)$. By the Comparability Theorem 0.113, there is a central projection $z \in \mathcal{M}$ such that $ez \precsim fz$ and $fz^{\perp} \precsim ez^{\perp}$. From the assumption, $\mathcal{T}(ez^{\perp}) = \mathcal{T}(e)z^{\perp} \leq \mathcal{T}(f)z^{\perp} = \mathcal{T}(fz^{\perp})$. On the other hand, from $fz^{\perp} \precsim ez^{\perp}$ it follows that $\mathcal{T}(ez^{\perp}) \geq \mathcal{T}(fz^{\perp})$. Thus, $\mathcal{T}(ez^{\perp}) = \mathcal{T}(fz^{\perp})$. If $ez^{\perp} \not\sim fz^{\perp}$, then $ez^{\perp} \sim g \leq fz^{\perp}$ and $g \neq fz^{\perp}$. This implies that $\mathcal{T}(ez^{\perp}) = \mathcal{T}(g) \leq \mathcal{T}(fz^{\perp})$ and $\mathcal{T}(g) \neq \mathcal{T}(fz^{\perp})$, by the faithfulness of $\mathcal{T}$ (see Theorem 3.21(5)), a contradiction. Hence, $ez^{\perp} \sim fz^{\perp}$ and $e \precsim f$. If $\mathcal{T}(e) = \mathcal{T}(f)$, then $e \precsim f$ and $f \precsim e$, and by Proposition 0.111, $e \sim f$. $\qquad\square$

Lemma 3.23 *Let $\mathcal{M}$ be a finite von Neumann algebra, and let $\mathcal{T}$ be the centre-valued trace on $\mathcal{M}$. Then, $\mathcal{T}(p \vee q) \leq \mathcal{T}(p) + \mathcal{T}(q)$ for all $p, q \in \mathbb{P}(\mathcal{M})$.*

Proof. By Kaplansky's parallelogram law (Proposition 0.115), $\mathcal{T}(p \vee q - p) = \mathcal{T}(q - p \wedge q)$, so that $\mathcal{T}(p \vee q) - \mathcal{T}(p) = \mathcal{T}(q) - \mathcal{T}(p \wedge q)$. Hence, $\mathcal{T}(p \vee q) = \mathcal{T}(p) + \mathcal{T}(q) - \mathcal{T}(p \wedge q) \leq \mathcal{T}(p) + \mathcal{T}(q)$. $\qquad\square$

3.2.2 Traces: definition and basic properties

Definition 3.24 A functional $\tau \colon \mathcal{M}_+ \to [0, \infty]$ is called a *trace* on $\mathcal{M}$ if it satisfies the following conditions.

(1) $\tau(a + b) = \tau(a) + \tau(b)$ for all $a, b \in \mathcal{M}_+$ (additivity).
(2) $\tau(\lambda a) = \lambda \tau(a)$ for all $a \in \mathcal{M}_+$ and $\lambda \geq 0$, with the provision that $0 \cdot \infty = 0$ (homogeneity).
(3) $\tau(a^* a) = \tau(a a^*)$ for all $a \in \mathcal{M}$.

Definition 3.25 We say that a trace τ is:

(1) *normal* if for any $a \in \mathcal{M}_+$ and any net (a_i) with $a_i \in \mathcal{M}_+$ such that $a_i \nearrow a$, we have $\tau(a_i) \nearrow \tau(a)$;
(2) *faithful* if $\tau(a) = 0$ for some $a \in \mathcal{M}_+$ implies $a = 0$;
(3) *semifinite* if the linear span of the set $\{a \in \mathcal{M}_+ : \tau(a) < \infty\}$ is σ-weakly dense in $\mathcal{M}$.

Definition 3.26 For a trace τ, we introduce the following standard notation:

$$N_\tau = \{a \in \mathcal{M} : \tau(a^* a) = 0\}$$

$$\mathfrak{p}_\tau = \{a \in \mathcal{M}_+ : \tau(a) < \infty\}$$

$$\mathfrak{n}_\tau = \{a \in \mathcal{M} : a^* a \in \mathfrak{p}_\tau\}$$

$$\mathfrak{m}_\tau = \text{linear span of } \mathfrak{p}_\tau.$$

Proposition 3.27 *Let τ by a trace on $\mathcal{M}$. Then:*

(1) $\mathfrak{p}_\tau$ *is a hereditary subcone of* $\mathcal{M}_+$;

(2) N_τ *and* $\mathfrak{n}_\tau$ *are two-sided ideals in* $\mathcal{M}$;

(3) $\mathfrak{m}_\tau$ *is the linear span of* $\mathfrak{n}_\tau^*\mathfrak{n}_\tau$ *and* $\mathfrak{m}_\tau \subseteq \mathfrak{n}_\tau \cap \mathfrak{n}_\tau^*$;

(4) $\mathfrak{m}_\tau$ *is a* *-*subalgebra of* $\mathcal{M}$;

(5) $\tau\!\restriction\!\mathfrak{p}_\tau$ *extends to a positive linear form on* $\mathfrak{m}_\tau$ *(denoted also by* τ*) and* $\mathfrak{m}_\tau \cap \mathcal{M}_+ = \mathfrak{p}_\tau$.

Proof.   Claim (1) is obvious.

(2) Let $a, b \in \mathcal{M}$. Then

$$(a+b)^*(a+b) = 2(a^*a + b^*b) - (a-b)^*(a-b) \leq 2(a^*a + b^*b)$$

which shows that both N_τ and $\mathfrak{n}_\tau$ are linear spaces. Moreover,

$$\tau(((ba)^*(ba)) = \tau(a^*b^*ba) \leq \|b\|\tau(a^*a)$$

so that both N_τ and $\mathfrak{n}_\tau$ are left ideals. But condition (3) in Definition 3.24 ensures that N_τ and $\mathfrak{n}_\tau$ are, in fact, two-sided ideals.

(3) If $a \in \mathfrak{p}_\tau$, then $a^{1/2} \in \mathfrak{n}_\tau$, so that $a = (a^*)^{1/2}a^{1/2} \in \mathfrak{n}_\tau^*\mathfrak{n}_\tau$ and $\mathfrak{p}_\tau \subseteq \mathfrak{n}_\tau^*\mathfrak{n}_\tau$. On the other hand, if $a, b \in \mathfrak{n}_\tau$, then the polarisation identity

$$b^*a = \frac{1}{4}\sum_{n=0}^{3} i^n (a + i^n b)^*(a + i^n b)$$

shows that $b^*a \in \mathfrak{m}_\tau$ and consequently $\mathfrak{n}_\tau^*\mathfrak{n}_\tau \subseteq \mathfrak{m}_\tau$. Finally, since $\mathfrak{n}_\tau$ is a two-sided ideal, $\mathfrak{p}_\tau \subseteq \mathfrak{n}_\tau^*\mathfrak{n}_\tau \subseteq \mathfrak{n}_\tau \cap \mathfrak{n}_\tau^*$, which ends the proof of (3).

(4) It is obvious that $\mathfrak{m}_\tau$ is a linear space. That $ab \in \mathfrak{m}_\tau$ whenever $a, b \in \mathfrak{m}_\tau$ follows easily from (3) and the fact that $\mathfrak{n}_\tau$ is a left ideal. Finally, $a^* \in \mathfrak{m}_\tau$ for $a \in \mathfrak{m}_\tau$ follows immediately from (3).

(5) If $a = a_1 - a_2 + ia_3 - ia_4 = b_1 - b_2 + ib_3 - ib_4 \in \mathfrak{m}_\tau$ with $a_i, b_i \in \mathfrak{p}_\tau$ for $i = 1, \ldots, 4$, we have $a_1 - a_2 = b_1 - b_2$ and $a_3 - a_4 = b_3 - b_4$, which yields $\tau(a_1) - \tau(a_2) = \tau(b_1) - \tau(b_2)$ and $\tau(a_3) - \tau(a_4) = \tau(b_3) - \tau(b_4)$. Therefore, the number $\tau(a_1) - \tau(a_2) + i\tau(a_3) - i\tau(a_4)$ does not depend on the representation of a as a linear combination of four elements of $\mathfrak{p}_\tau$ and we define $\tau(a)$ as that number. τ extended in this manner is clearly linear and Hermitian. If $a \in \mathfrak{m}_\tau \cap \mathcal{M}_+$, then $0 \leq a = a_1 - a_2 \leq a_1 \in \mathfrak{p}_\tau$, so that $a \in \mathfrak{p}_\tau$ and τ is positive on $\mathfrak{m}_\tau$. $\square$

Here is a list of easy properties of the trace.

Proposition 3.28 *Let τ be a trace on $\mathcal{M}$. Then:*

(1) $\tau(u^*au) = \tau(a)$ *for any $a \in \mathcal{M}_+$ and $u \in \mathbb{U}(\mathcal{M})$;*

(2) *if $e, f \in \mathbb{P}(\mathcal{M})$ and $e \sim f$ (respectively, $e \precsim f$), then $\tau(e) = \tau(f)$ (respectively, $\tau(e) \leq \tau(f)$);*

(3) *if τ is faithful, $\mathcal{M}$ is a factor and $\tau(e) = \tau(f)$ (respectively, $\tau(e) \leq \tau(f)$, $\tau(e) < \tau(f)$) for some $e, f \in \mathbb{P}(\mathcal{M})$, then $e \sim f$ (respectively, $e \precsim f$, $e \prec f$).*

(4) *if τ is faithful and $\tau(e) < \infty$ for some $e \in \mathbb{P}(\mathcal{M})$, then e is finite;*

Proof. (1) We have $\tau(u^*au) = \tau((a^{1/2}u)(a^{1/2}u)^*) = \tau(a)$.

(2) $e \sim f$ means $e = u^*u$, $f = uu^*$ for some $u \in \mathcal{M}$; hence, $\tau(e) = \tau(f)$. The other part follows by Definition 0.110 and the monotonicity of the trace.

(3) Since $\mathcal{M}$ is a factor, $e \nsim f$, implies that $e \prec f$ or $f \prec e$. But then $\tau(e) < \tau(f)$ or $\tau(f) < \tau(e)$, a contradiction. The other parts follow as in (2).

(4) Let $f \in \mathbb{P}(\mathcal{M})$ be such that $f \leq e$ and $f \sim e$. Then, $\tau(e - f) = 0$, hence $e = f$. This implies the finiteness of e. $\qquad\square$

Proposition 3.29 *If τ is a normal trace on $\mathcal{M}$, then there is a unique central projection $z \in \mathcal{M}$ such that τ is faithful on $(\mathcal{M}z)_+$ and (the extension of) τ is zero on $\mathcal{M}z^{\perp}$.*

Proof. By Kaplansky's law 0.115 and Proposition 3.28(2), we have $\tau(p \vee q) + \tau(p \wedge q) = \tau(p) + \tau(q)$ for any $p, q \in \mathbb{P}(\mathcal{M})$. Hence, $p, q \in N_\tau$ implies $p \vee q \in N_\tau$. Hence, the family of projections in N_τ is upward directed. Since τ is normal, $\tau(\sup \mathbb{P}(N_\tau)) = 0$. Put $z := \sup \mathbb{P}(N_\tau)^{\perp}$. Then, $z^{\perp}$ is the largest projection that is annihilated by τ, from which it is clear that τ is faithful on $(z\mathcal{M}z)_+$ and (the extension of) τ is zero on $z^{\perp}\mathcal{M}z^{\perp}$. It can now easily be checked that $\tau(uz^{\perp}u^*) = \tau(z^{\perp}) = 0$, so that $uz^{\perp}u^* \leq z^{\perp}$, for all $u \in \mathbb{U}(\mathcal{M})$. But this means that $uz^{\perp}u^* = z^{\perp}$, for all $u \in \mathbb{U}(\mathcal{M})$, and hence that $z \in \mathcal{Z}(\mathcal{M})$. $\qquad\square$

Definition 3.30 The projection z from the proposition above is called the *support (projection)* of τ and is denoted by $\operatorname{supp}\tau$. The orthogonal complement of $\operatorname{supp}\tau$ is called the *null projection* of τ and is denoted by $e_0(\tau)$.

Proposition 3.31 *Let τ be a faithful normal trace on a von Neumann algebra $\mathcal{M}$. The following conditions are equivalent.*

(1) *$\mathfrak{p}_\tau$ generates $\mathcal{M}$ as a von Neumann algebra (equivalently, the $*$-algebra $\mathfrak{m}_\tau$ is dense in $\mathcal{M}$ in any of the following topologies: weak, σ-weak, strong, σ-strong, strong*, σ-strong*).*

(2) *There exists an orthogonal family $\{e_i\}_{i\in I}$ such that $e_i \in \mathbb{P}(\mathcal{M}) \cap \mathfrak{p}_\tau$ for all $i \in I$, and $\sum_{i\in I} e_i = \mathbb{1}$.*

(3) *For each non-zero $e \in \mathbb{P}(\mathcal{M})$, there exists $f \in \mathbb{P}(\mathcal{M})$ such that $0 \neq f \leq e$ and $\tau(f) < \infty$.*

(4) *There exists a family $\{\omega_i\}_{i\in I}$ with $\omega_i \in \mathcal{M}_*^+$ for all $i \in I$, with pairwise orthogonal supports, such that $\sum_{i\in I} \operatorname{supp} \omega_i = \mathbb{1}$ and $\tau = \sum_{i\in I} \omega_i$ (pointwise).*

Proof.   (3) $\Rightarrow$ (2): Choose $\{e_i\}$ to be a maximal family of mutually orthogonal non-zero projections of finite trace (use Zorn's lemma and condition (3) to show its existence).

(2) $\Rightarrow$ (1): Let $\{e_i\}_{i\in I}$ be the family of condition (2). For finite $J \subseteq I$, put $f_J = \sum_{i\in J} e_i$. We will show that for $a \in \mathcal{M}$, $f_J a f_J \in \mathfrak{m}_\tau$ and $f_J a f_J \to a$ strongly, which shows (1) by Theorem 0.92. In fact, for any $i,j \in I$, we have $e_i, e_j \in \mathfrak{n}_\tau \cap \mathfrak{n}_\tau^*$ and $\mathfrak{n}_\tau$ is a left ideal in $\mathcal{M}$, so that $e_i a e_j \in \mathfrak{n}_\tau \mathfrak{n}_\tau^* \subseteq \mathfrak{m}_\tau$. Now, $f_J \to 1$ strongly and, for any $\xi \in H$,

$$\|(f_J a f_J - a)\xi\| \leq \|(f_J a(1 - f_J))\xi\| + \|(1 - f_J)a\xi\|$$

$$\leq \|a\|\|(1 - f_J)\xi\| + \|(1 - f_J)a\xi\| \to 0$$

which ends the proof.

(1) $\Rightarrow$ (3): Let $e \in \mathcal{M}$. By the Kaplansky density theorem 0.93, there is a net $(a_i)_{i\in I}$, with $a_i \in \mathcal{M}_+, \|a_i\| \leq 1$ such that $a_i \to e$ strongly and $\tau(a_i) < \infty$ for all i. Then, $\tau(e a_i e) = \tau(a_i^{1/2} e a_i^{1/2}) \leq \tau(a_i) < \infty$ for all i. Since $e a_i e \to e$ strongly, there must exist an i_0 such that $e a_{i_0} e \neq 0$. Hence, for some $\epsilon > 0$, $0 \neq \epsilon \chi_{[\epsilon,\infty)}(e a_{i_0} e) \leq e a_{i_0} e \leq e$. Put $f = \chi_{[\epsilon,\infty)}(e a_{i_0} e)$. Then $0 \neq f \leq (1/\epsilon)e$, so that $f \leq e$ and $\tau(f) < \infty$.

(2) $\Rightarrow$ (4): Let $\{e_i\}$ be an orthogonal family of projections from $\mathfrak{p}_\tau$ with $\sum e_i = \mathbb{1}$. Then, as in (2) $\Rightarrow$ (1), for all i,j and $a \in \mathcal{M}_+$, we have $e_j a e_i \in \mathfrak{m}_\tau$ and $\tau(a) = \sum_{i,j} \tau(e_j a e_i) = \sum_i \tau(e_i a e_i)$. Put $\omega_i = e_i \tau e_i$. Then $\operatorname{supp} \omega_i = e_i$ and $\tau = \sum \omega_i$.

(4) $\Rightarrow$ (2): Choose $e_i = \operatorname{supp} \omega_i$. Then, $e_i \in \mathbb{P}(\mathcal{M}) \cap \mathfrak{p}_\tau$ for all $i \in I$, and $\sum_{i\in I} e_i = \mathbb{1}$, so that (2) holds. $\square$

Definition 3.32 We say that a faithful normal trace τ is *semifinite* if it satisfies one of the equivalent conditions of Proposition 3.31. We write *f.n.s.* instead of *faithful normal semifinite*.

Corollary 3.33 *A semifinite normal trace τ is σ-weakly lower semicontinuous.*

Proof. Proposition 3.31(4) shows that τ is a sum of σ-weakly continuous functionals from $\mathcal{M}_*^+$, which implies that it is σ-weakly lower semicontinuous. $\square$

Proposition 3.34 *If τ is a normal trace on $\mathcal{M}$, then there exists a unique central projection $z \in \mathcal{M}$ such that τ is semifinite on $\mathcal{M}z$ and $\tau(a) = \infty$ for every non-zero $a \in \mathcal{M}_+ z^{\perp}$.*

Proof. Let $\bar{\mathfrak{n}}$ be the σ-weak closure of $\mathfrak{n}_\tau$. By Proposition 3.27(2), $\bar{\mathfrak{n}}$ is then also a two-sided ideal in $\mathcal{M}$, which by Proposition 0.94(5) ensures that there is a central projection $z \in \mathcal{M}$ such that $\bar{\mathfrak{n}} = \mathcal{M}z$. Notice that then

$$\mathrm{span}(\mathfrak{p}_\tau) = \mathfrak{m}_\tau = \mathrm{span}(\mathfrak{n}_\tau^* \mathfrak{n}_\tau) \subset \mathrm{span}(\bar{\mathfrak{n}}^* \bar{\mathfrak{n}}) = \mathcal{M}z.$$

So, $\overline{\mathrm{span}(\mathfrak{p}_\tau)} \subset \mathcal{M}z$. But we conversely also have that $\mathcal{M}z = \mathrm{span}(\bar{\mathfrak{n}}^* \bar{\mathfrak{n}}) \subset \overline{\mathrm{span}(\mathfrak{p}_\tau)}$. To see this, let $a, b \in \bar{\mathfrak{n}}$ be given and select nets $(a_\gamma), (b_\rho)$ in $\mathfrak{n}_\tau$ converging to a and b respectively. For each fixed ρ, the net $(b_\rho^* a_\gamma)_\gamma \subset \mathrm{span}(\mathfrak{n}_\tau^* \mathfrak{n}_\tau) = \mathrm{span}(\mathfrak{p}_\tau)$ converges to $b_\rho^* a$. Therefore, the net $(b_\rho^* a)$ belongs to $\overline{\mathrm{span}(\mathfrak{p}_\tau)}$. But then the limit $b^* a$ also belongs to $\overline{\mathrm{span}(\mathfrak{p}_\tau)}$. This then ensures that $\mathcal{M}z = \mathrm{span}(\bar{\mathfrak{n}}^* \bar{\mathfrak{n}}) \subset \overline{\mathrm{span}(\mathfrak{p}_\tau)}$ and hence $\overline{\mathrm{span}(\mathfrak{p}_\tau)} = \mathcal{M}z$, thereby proving the first claim.

If, on the other hand, there existed a non-zero $a \in \mathcal{M}_+ z^{\perp}$, then also $a^{1/2} \in \mathcal{M}z^{\perp}$, so that $a^{1/2} \notin \mathfrak{n}_\tau$ and $\tau(a) = \infty$. $\square$

Definition 3.35 The projection z from the previous proposition is called the *semifinite projection* of τ and is denoted by $e_\infty(\tau)$.

Obviously, $e_0(\tau) \leq e_\infty(\tau)$ (cf. Definition 3.30).

3.2.3 Existence of traces

Lemma 3.36 ([Ped18, Lemma 2]) *Let p be a projection in $\mathcal{M}$. If $p\mathcal{M}p$ admits a normal tracial state τ, then there exists a normal semifinite trace on $\mathbf{z}(p)\mathcal{M}$.*

Proof. Let $\mathbf{z}(p)$ be the central support of p. On applying the negation of Proposition 0.103 to the pair $\mathbf{z}(p)$ and p, it is clear that there exists a nonzero subprojection q of $\mathbf{z}(p)$ for which we have that $q \preceq p$. A standard Zorn's lemma argument now shows that there exists a maximal mutually orthogonal family of subprojections $\{q_i\}$ of $\mathbf{z}(p)$ such that $\sum_i q_i \leq \mathbf{z}(p)$, with $q_i \preceq p$ for each i. We then claim that $\sum_i q_i = \mathbf{z}(p)$. To see that this is so, notice that if not, the fact that $0 \neq \mathbf{z}(p) - \sum_i q_i \leq \mathbf{z}(p)$ ensures that the central carrier of $\mathbf{z}(p) - \sum_i q_i$ is a subprojection of $\mathbf{z}(p)$. So again by Proposition 0.103, there exists a non-zero subprojection q_0 of $\mathbf{z}(p) - \sum_i q_i$ for which we have that $q_0 \preceq p$. But this contradicts the maximality of $\{q_i\}$, establishing the claim.

Stated in terms of partial isometries, the above fact means that there exists a family $\{v_i\}$ of partial isometries such that $\sum_i v_i^* v_i = \mathbf{z}(p)$, with $v_i v_i^* \leq p$ for each i.

For any $a \in \mathbf{z}(p)\mathcal{M}_+$, we now define $\widetilde{\tau}(a)$ by

$$\widetilde{\tau}(a) = \sum_i \tau(v_i a v_i^*).$$

Since all terms are positive, summation and supremum can be interchanged, and hence, the quantity $\widetilde{\tau}$ is clearly normal and additive on $\mathbf{z}(p)\mathcal{M}_+$. To see that $\widetilde{\tau}$ is a trace, notice that for any $a \in \mathbf{z}(p)\mathcal{M}$, we have that

$$\widetilde{\tau}(a^*a) = \sum_j \widetilde{\tau}(a^* v_j^* v_j a) = \sum_{i,j} \tau(v_i a^* v_j^* v_j a v_i^*)$$

$$= \sum_{i,j} \tau(v_j a v_i^* v_i a^* v_j^*) = \sum_i \widetilde{\tau}(a v_i^* v_i a^*) = \widetilde{\tau}(aa^*).$$

It remains to show that $\widetilde{\tau}$ is semifinite. To see this, note that with F ranging over the finite subsets of the index set I, the net of projections $e_F = \sum_{i \in F} v_i^* v_i$ increases to $\mathbf{z}(p)$ and hence is σ-strong* convergent to $\mathbf{z}(p)$. For any finite subset F of I and any $a \in \mathcal{M}_+$, we have that

$$\widetilde{\tau}(e_F a e_F) = \sum_{i \in I} \tau(v_i e_F a e_F v_i^*) = \sum_{i \in F} \tau(v_i e_F a e_F v_i^*) < \infty.$$

The fact that $(e_F a e_F)$ is σ-strong* convergent to $\mathbf{z}(p)a$, now ensures that $\widetilde{\tau}$ is semifinite. $\qquad\square$

Lemma 3.37 *The sum of semifinite normal traces with orthogonal supports is again a semifinite normal trace.*

Proof. Let $\{\tau_i\}$ be a family of normal semifinite traces with orthogonal supports $\{z_i\}$, and define τ by $\tau(a) = \sum_i \tau_i(a)$ for all $a \in \mathcal{M}_+$. The traciality of each τ_i ensures that τ is tracial. Normality of τ follows from the fact that when all terms are positive, suprema and summation may be interchanged. We leave it as an exercise to show that the support of τ is just $z = \sum_i z_i$. We then clearly have that $\mathfrak{n}_{\tau_i} z_i \subset \mathfrak{n}_\tau$ for each i. Taking σ-weak closures, we have that $\mathcal{M} z_i \subset \overline{\mathfrak{n}_\tau}^{w*}$ for each i. It is, moreover, trivial that $\mathcal{M}(\mathbb{1} - z) \subset \overline{\mathfrak{n}_\tau}^{w*}$. Since $\mathbb{1} = (\mathbb{1} - z) + \sum_i z_i$, these facts ensure that $\mathcal{M} \subset \overline{\mathfrak{n}_\tau}^{w*}$. $\qquad\square$

Theorem 3.38 *A von Neumann algebra $\mathcal{M}$ is semifinite if and only if it admits a faithful normal semifinite trace τ.*

Proof. We first prove the 'if' part. Assume that $\mathcal{M}$ admits a faithful normal semifinite trace τ. We need to show that there are no purely infinite projections in the centre of $\mathcal{M}$. Let $0 \neq z$ be non-zero projection the centre of $\mathcal{M}$. By the definition of semifiniteness for a trace and Proposition 3.31, there is a non-zero projection $q \in \mathcal{M}$ such that $q \leq z$ and $\tau(q) < \infty$. Now suppose that $q \sim q_0$ with $q_0 \leq q$. There must then exist a partial isometry $v \in \mathcal{M}$ so that $q = v^*v$ and $q_0 = vv^*$. Since $0 \leq \tau(q - q_0) = \tau(v^*v - vv^*) = \tau(v^*v - v^*v) = 0$, the faithfulness of τ ensures that $q = q_0$. Thus, q is finite, and $\mathcal{M}$ is therefore semifinite.

For the 'only if' part, we first show that $\mathcal{M}$ admits normal semifinite traces. So, let p be a non-zero finite projection. Then, $p\mathcal{M}p$ is a finite von Neumann algebra that must admit a centre-valued trace $\mathscr{T}$. For any normal state ψ on $\mathcal{Z}(\mathcal{M})$, $\psi \circ \mathscr{T}$ will then be a normal tracial state on $p\mathcal{M}p$. By Lemma 3.36, this trace extends to an f.n.s. trace on $\mathbf{z}(p)\mathcal{M}$ and hence to a semifinite normal trace on $\mathcal{M}$.

Now, select a maximal family $\{\tau_\alpha\}$ of semifinite traces with mutually orthogonal supports $\{z_\alpha\}$. By the preceding lemmata, $\sum_\alpha \tau_\alpha$ is semifinite and normal. We claim that $\sum_\alpha z_\alpha = \mathbb{1}$. If not, then by selecting a finite subprojection p of $(\mathbb{1} - \sum_\alpha z_\alpha)$ and applying the same argument as above, we could find a non-trivial semifinite normal trace on $\mathbf{z}(p)\mathcal{M}$. The support of this trace is clearly a central subprojection of $\mathbf{z}(p)$, which in turn is a central subprojection of $(\mathbb{1} - \sum_\alpha z_\alpha)$. Therefore, this clearly contradicts the maximality of $\{\tau_\alpha\}$, which then proves the claim. $\qquad\qquad\square$

In closing, we show that if $\mathcal{M}$ is a factor, the converse of Proposition 3.28(4) also holds. Alongside this fact, the foregoing analysis shows that for semifinite factors, the restriction of an f.n.s. trace to the projection lattice of such an algebra is a dimension function. Given that dimension functions are unique up to positive scalar multiples, the normality of f.n.s. traces therefore implies the normality of dimension functions. (Here, one needs to use the fact that dimension functions on type III algebras can trivially be seen to be normal.)

Proposition 3.39 *Let τ be a trace on a factor von Neumann algebra $\mathcal{M}$. If τ is f.n.s. and $e \in \mathbb{P}(\mathcal{M})$ is finite, then $\tau(e) < \infty$.*

Proof.   Let $\tau(e) = \infty$. By Proposition 3.31, there exists a projection f with $f \leq e$ and $\tau(f) < \tau(e)$. Let $\{e_i\}_{i \in I}$ be a maximal orthogonal family of non-zero equivalent projections in $\mathcal{M}$ with $\tau(e_i) < \infty$ and $e' := \sum_{i \in I} e_i \leq e$. This family must be infinite since if it were finite, then given i_0 we would have $\tau(e_{i_0}) < \infty = \tau(e - e')$, whence $e_{i_0} \prec e - e'$ by Proposition 3.28(3). Thus, there would then exist some projection e_0 equivalent to all the other e_i's with $e_0 \leq e - \sum_i e_i \leq e$, clearly contradicting the maximality. By maximality, $e - e' \prec e_i$ (for all i). Choose a specific $i_0 \in I$. Then, $e - e' \prec e_{i_0}$ and, by Proposition 0.112, $\sum_{i \in I, i \neq i_0} e_i \sim e'$. Hence,

$$e = (e - e') + e' \prec e_{i_0} + \sum_{i \in I, i \neq i_0} e_i = e' \leq e.$$

This is impossible if e is finite, hence the result. $\qquad\qquad\square$

3.2.4 Traces and finite von Neumann algebras

Definition 3.40 A family $\{\tau_i\}_{i \in I}$ of non-zero traces on $\mathcal{M}$ is called *sufficient* if for any non-zero $a \in \mathcal{M}_+$, there is $i_0 \in I$ such that $\tau_{i_0}(a) \neq 0$.

Theorem 3.41 *An algebra $\mathcal{M}$ is finite if and only if it possesses a sufficient family of finite normal traces. An algebra $\mathcal{M}$ is finite and σ-finite if and only if it admits a faithful normal tracial state.*

Proof. '$\Rightarrow$' Assume $\mathcal{M}$ is finite. Let $\{\mu_i\}_{i \in I}$ be a maximal family of non-zero elements of $\mathcal{Z}(\mathcal{M})_*^+$ with mutually orthogonal supports. It is clear that $\sum_{i \in I} \operatorname{supp} \mu_i = \mathbb{1}$. Let $\tau_i := \mu_i \circ \mathscr{T}$, where $\mathscr{T}$ is the centre-valued trace on $\mathcal{M}$. Then, $\{\tau_i\}_{i \in I}$ is a sufficient family of finite normal traces on $\mathcal{M}$. If $\mathcal{M}$ is σ-finite, then $\mathcal{Z}(\mathcal{M})$ is σ-finite. So, by the discussion preceding Lemma 1.27, there is a faithful normal state μ on it, so that $\tau := \mu \circ \mathscr{T}$ is a tracial state on $\mathcal{M}$.

'$\Leftarrow$' Let $0 \neq p \in \mathbb{P}(\mathcal{M})$ be such that $p \sim \mathbb{1}$. Let $\{\tau_i\}_{i \in I}$ be a sufficient family of finite normal traces on $\mathcal{M}$. Put $z_i := \operatorname{supp} \tau_i$. Then, $\tau_i(pz_i) = \tau_i(z_i)$ for each $i \in I$, hence $pz_i = z_i$ for each $i \in I$. Thus, $p = \mathbb{1}$, which implies that $\mathbb{1}$ is finite. Therefore, $\mathcal{M}$ is finite. If $\mathcal{M}$ admits a faithful normal tracial state, then it must obviously be σ-finite. $\qquad\square$

Proposition 3.42 *If $\mathcal{M}$ is finite and τ is a normal semifinite trace on $\mathcal{M}$, then the restriction of τ to the centre of $\mathcal{M}$ is semifinite.*

Proof. Let τ be a normal semifinite trace on $\mathcal{M}$ and let $a \in \mathcal{M}_+$. By Proposition 3.21(6),

$$\mathscr{T}(a) = \text{norm}-\lim \sum_{k=1}^{k_n} \lambda_k^{(n)} u_k^{(n)*} a u_k^{(n)}$$

for some unitaries $u_k^{(n)}$ and non-negative numbers $\lambda_k^{(n)}$ with $\sum_{k=1}^{k_n} \lambda_k^{(n)} = 1$. Since τ is σ-weakly (so also norm) lower semicontinuous (see Corollary 3.33), we have

$$(\tau \circ \mathscr{T})(a) \leq \liminf_n \sum_{k=1}^{k_n} \lambda_k^{(n)} \tau(u_k^{(n)*} a u_k^{(n)}) = \tau(a).$$

We are ready to prove that τ is semifinite on the centre of $\mathcal{M}$. Let $0 \neq p \in \mathbb{P}(\mathcal{Z}(\mathcal{M}))$. By the semifiniteness of τ on $\mathcal{M}$, there is a non-zero $q \in \mathbb{P}(\mathcal{M})$ such that $q \leq p$. Now $\mathcal{T}(q) \leq \mathcal{T}(p) = p$, and for some $\epsilon > 0$ the spectral projection $r := \chi_{(\epsilon,\infty)}(\mathcal{T}(q)) \neq 0$. We have $\epsilon r \leq r\mathcal{T}(q) \leq \mathcal{T}(q)$, so that $\tau(r) \leq (1/\epsilon)\tau(\mathcal{T}(q)) \leq (1/\epsilon)\tau(q) < \infty$. This means that τ is semifinite. $\qquad\square$

Corollary 3.43 *Any normal semifinite trace τ on a finite von Neumann algebra $\mathcal{M}$ is of the form $\mu \circ \mathcal{T}$, where μ is a normal semifinite trace on $\mathcal{Z}(\mathcal{M})$.*

Proof.   It is clear that if μ is a normal semifinite trace on $\mathcal{Z}(\mathcal{M})$, then $\mu \circ \mathcal{T}$ is a normal semifinite trace on $\mathcal{M}$ (just note that $\mathcal{T}$ acts as the identity on $\mathcal{Z}(\mathcal{M})$). If τ is a normal semifinite trace on $\mathcal{M}$, then by the proposition its restriction μ to $\mathcal{Z}(\mathcal{M})_+$ is semifinite. Hence, it is a sum of a family $\{\mu_i\}$ with $\mu_i \in \mathcal{Z}(\mathcal{M})_*^+$ with pairwise orthogonal supports. By Lemma 3.18, $\mu_i \circ \mathcal{T} = \mu_i$ on $\mathcal{Z}(\mathcal{M})$ for all i, and by uniqueness in the same lemma, $\tau\!\restriction\!(\mathrm{supp}\,(\mu_i)\mathcal{M}) = \mu_i \circ \mathcal{T}$. Therefore $\tau = \mu \circ \mathcal{T}$. $\quad\square$

Definition 3.44 Let τ be a faithful normal semifinite trace on $\mathcal{M}$. We say that τ is *bounded away from* 0 if

$$\inf\{\tau(p)\colon p \in \mathbb{P}(\mathcal{M}), p \neq 0\} > 0. \tag{3.9}$$

Lemma 3.45 *An f.n.s. trace τ on $\mathcal{M}$ can be bounded away from 0 only if $\mathcal{M}$ is of type I and the centre of $\mathcal{M}$ is purely atomic. If τ is not bounded away from zero, then we can find an orthogonal sequence (p_n) of non-zero projections from $\mathcal{M}$ such that $\tau(p_n) \to 0$.*

Proof.   By Theorem 3.38, $\mathcal{M}$ must be semifinite. If $\mathcal{M}$ contains a continuous summand, we may assume that $\mathcal{M}$ is actually continuous. Let $p \in \mathbb{P}(\mathcal{M})$ in $\mathcal{M}$ be such that $0 < \tau(p) < \infty$. Using the Halving Lemma 0.124(2), we can construct an orthogonal sequence (p_n) of projections from $\mathcal{M}$ starting with p and such that $\tau(p_n) \to 0$; in particular, τ is not bounded away from 0.

Now, assume that $\mathcal{M}$ is discrete and contains a direct summand of type I_α of the form $\mathcal{K}\overline{\otimes}B(H)$ with non-atomic K and α-dimensional H. Let r be any minimal projection in the $B(H)$. For a projection p in $\mathcal{K}$, any subprojection of $p \otimes r$ must be of the form $q \otimes r$ with $q \leq p$

being a projection in $\mathcal{K}$. Hence, the formula $\mu(a) := \tau(a \otimes r)$ defines a semifinite trace on $\mathcal{K}$. We can now start with any non-zero projection p in $\mathcal{K}$ of finite trace, and divide it into two non-zero projections, of which at least one will have trace $\leq \tau(p)/2$. It is clear that we can continue the process, producing an orthogonal sequence (p_n) of projections from $\mathcal{K}$ such that $\tau(p_n \otimes r) = \mu(p_n) \to 0$. We have produced orthogonal sequence $(p_n \otimes r)$ in $\mathcal{M}$ with $\tau(p_n \otimes r) \to 0$; in particular, τ is not bounded away from 0.

Finally, let $\mathcal{M}$ be an infinite direct sum $\sum_{i \in I}^{\oplus} \mathcal{F}_i$ of factors of type I_α. Let r_i be a minimal projection in $\mathcal{F}_i$. If τ is not bounded away from 0, there is a sequence (p_n) of (different) projections from the family $\{r_i\}$ such that $\tau(p_n) \to 0$. The sequence is clearly orthogonal. $\qquad\square$

Lemma 3.46 *If the algebra $\mathcal{M}$ is infinite-dimensional, then there exists an infinite orthogonal family (p_n) of non-zero projections from $\mathcal{M}$.*

Proof. This follows from Lemma 3.45 if τ is not bounded away from zero. If τ is bounded away from zero, then $\mathcal{M}$ is an infinite direct sum of factors of type I_α, and the units of the factors form the desired family. $\qquad\square$

3.3 Measurability

In this section, we deal with the important concept of measurable operators. The various classes of measurable operators will have the structure of a topological *-algebra and will contain all the noncommutative function spaces we are going to consider.

The number of different operator classes we introduce here may seem overwhelming. Some of them are, however, given here only for the sake of completeness. Nevertheless, all contributed to a growth in the understanding of what measurability means for operator algebras. The differences between various classes of measurable operators disappear if we deal with finite traces or if we restrict our attention to factors.

Definition 3.47 Let $\mathcal{M}$ be a semifinite von Neumann algebra endowed with a faithful normal semifinite trace τ. Then:

(1) $F(\mathcal{M}, \tau)$ denotes the set $\{a \in \mathcal{M} : \tau(|a|) < \infty\}$; its elements are called τ-*finite*.

(2) $K(\mathcal{M}, \tau)$ denotes the set

$$\{a \in \overline{\mathcal{M}} : \tau(\chi_{(\epsilon, \infty)}(|a|)) < \infty \text{ for each } \epsilon > 0\};$$

its elements are called τ-*compact*.

(3) $S(\mathcal{M}, \tau)$ denotes the set

$$\{a \in \overline{\mathcal{M}} : \tau(\chi_{(\epsilon, \infty)}(|a|)) < \infty \text{ for some } \epsilon > 0\};$$

its elements are called τ-*measurable*.

(4) $S(\mathcal{M})$ denotes the set

$$\{a \in \overline{\mathcal{M}} : \chi_{(\epsilon, \infty)}(|a|) \text{ is finite for some } \epsilon > 0\};$$

its elements are called *(Segal) measurable*.

(5) $LS(\mathcal{M}, \tau)$ denotes the set $\{a \in \overline{\mathcal{M}} :$ there exists a sequence $z_n \in \mathcal{Z}(\mathcal{M})$ such that $z_n \nearrow \mathbb{1}$ and $az_n \in S(\mathcal{M}, \tau)$ for each $n\}$; its elements are called *locally τ-measurable*.

(6) $LS(\mathcal{M})$ denotes the set $\{a \in \overline{\mathcal{M}} :$ there exists a sequence $z_n \in \mathcal{Z}(\mathcal{M})$ such that $z_n \nearrow \mathbb{1}$ and $az_n \in S(\mathcal{M})$ for each $n\}$; its elements are called *locally (Segal) measurable*.

Of the sets defined above, the most important for our purposes is that of τ-measurable operators. Whenever it does not lead to confusion, we will denote it by $\widetilde{\mathcal{M}}$. The notion was introduced by Nelson [**Nel74**]. It turned out to be most useful for the modern approach to noncommutative L^p-spaces. In particular, the class is large enough to serve the purpose of a formalism for the theory of noncommutative L^p- and Orlicz spaces, both in the semifinite and the general case. The class of measurable operators was introduced much earlier by Segal [**Seg53**]. Although not used much in the sequel, its obvious advantage is its lack of dependence on the trace. Many results are proved for both τ measurable and measurable operators, since the proofs are essentially the same, *mutatis mutandis*. The definition of locally measurable operators was given by Sankaran [**San59**], with the class of locally τ-measurable operators appearing in a paper by Cecchini [**Cec78**]. The importance of the class of locally measurable operators will be explained in Section 3.4.

τ-compact operators were introduced by Fack and Kosaki [**FK86**], and investigated thoroughly by Ströh and West [**SW93**]. An inquisitive reader will certainly notice the lack of operators compact with respect to a von Neumann algebra. This class was introduced and explored by Kaftal [**Kaf77**, **Kaf78**] and further investigated in type III factors by Halpern and Kaftal [**HK86**, **HK87**]. It is not described here as it does not fit the scheme we are using.

We know that the sum of two unbounded densely defined operators can happen to have a domain consisting only of 0. Thus, to have a non-trivial algebraic structure for our operators, we have to assume that their domains are 'large enough', so that intersections of the domains are again dense in H. Below, readers will find the proper definitions.

Definition 3.48 Let τ be a faithful normal semifinite trace on $\mathcal{M}$. A subspace D of H is called τ-*dense* (respectively, *strongly dense*) in H if there is a sequence (p_n) of projections from $\mathcal{M}$ such that $p_n H \subseteq D, p_n \nearrow \mathbb{1}$ and, for each $n, \tau(p_n^\perp)$ is finite (respectively, $p_n^\perp$ is finite). A sequence (p_n) from the definition is called a *determining* sequence for D.

Remark 3.49 It is clear that if (p_n) is a determining sequence for a τ-dense (respectively, strongly dense) subspace of H, then $\tau(p_n^\perp) \to 0$ (respectively, $p_n \to 0$ strongly) as $n \to \infty$. Furthermore, a τ-dense subspace is strongly dense, and a strongly dense subspace is dense in H.

Proposition 3.50

(1) *The intersection of a countable number of τ-dense subspaces of H is τ-dense in H.*

(2) *The intersection of a finite number of strongly dense subspaces of H is strongly dense in H.*

Proof. (1) Let $\{D_k\}$ be a countable family of τ-dense subspaces of H. Let $D := \bigcap D_k$ and let $(p_n^{(k)})$ be a determining sequence for D_k, for each k. By passing to a subsequence, we can assume that for each n

and k, $\tau(p_n^{(k)\perp}) < 1/2^{n+k}$. Put $q_n = \bigwedge_{k=1}^{\infty} p_n^{(k)}$. Then

$$\tau(q_n^{\perp}) = \tau\left(\bigvee_{k=1}^{\infty} p_n^{(k)\perp}\right) \leq \sum_{k=1}^{\infty} \tau(p_n^{(k)\perp}) \leq \sum_{k=1}^{\infty} 1/2^{k+n} = 1/2^n \to 0$$

as $n \to \infty$. Obviously, q_n is increasing, and the faithfulness of τ ensures that $q_n \nearrow 1$. It is also clear that, for each n, $q_n H \subseteq D$. Thus, (q_n) is a determining sequence for D, and D is τ-dense.

(2) Assume now that D_1 and D_2 are strongly dense subspaces of H, with determining sequences (p_n) and (q_n). We will show that (r_n), where $r_n := p_n \wedge q_n$, is a determining sequence for $D := D_1 \cap D_2$. Note first that $r_1^{\perp} = p_1^{\perp} \vee q_1^{\perp}$ is a finite projection in $\mathcal{M}$, and that for each n, $r_n^{\perp} \leq r_1^{\perp}$, so that $p_n, q_n, r_n^{\perp} \in \mathcal{M}_{r_1^{\perp}}$. Let $\mathscr{T}$ be the centre-valued trace on $\mathcal{M}_{r_1^{\perp}}$. Then, $\mathscr{T}(p_n^{\perp}), \mathscr{T}(q_n^{\perp}) \searrow 0$ and, by Lemma 3.23,

$$\mathscr{T}(r_n^{\perp}) = \mathscr{T}(p_n^{\perp} \vee q_n^{\perp}) \leq \mathscr{T}(p_n^{\perp}) + \mathscr{T}(q_n^{\perp}) \searrow 0.$$

Together with the obvious fact that $r_n^{\perp}$ is decreasing, the faithfulness of $\mathscr{T}$ yields $r_n^{\perp} \searrow 0$. It is clear that $r_n H \subseteq D$, which means that r_n is a determining sequence for D. $\qquad\square$

Definition 3.51 A preclosed (closable) operator $a \in {}^{\eta}\mathcal{M}$ is called *τ-premeasurable* (respectively, *premeasurable*) if it has a τ-dense (respectively, strongly dense) domain.

We need the following simple lemma.

Lemma 3.52

(1) *Let $p, q \in \mathbb{P}(\mathcal{M})$. If $p \wedge q = 0$, then $p \preceq q^{\perp}$.*

(2) *Let D be strongly dense and let (p_n) be a determining sequence for D. If $q \in \mathbb{P}(\mathcal{M})$ is such that $q \wedge p_n = 0$ for each n, then $q = 0$.*

(3) *Let D be strongly dense and let (p_n) be a determining sequence for D. If $q, r \in \mathbb{P}(\mathcal{M})$ is such that $q \wedge p_n = r \wedge p_n$ for each n, then $q = r$.*

Proof. (1) By Kaplansky's parallelogram law (see Proposition 0.115),

$$p = p - p \wedge q \sim p \vee q - q \leq q^{\perp}.$$

(2) By (1), $q \preceq p_n^{\perp} \searrow 0$. Hence, $\mathscr{T}(q) \leq \mathscr{T}(p_n) \to 0$ yields $\mathscr{T}(q) = 0$ and, by the faithfulness of $\mathscr{T}$, $q = 0$.

(3) Put $p = q - q \wedge r$. By assumption, for each n, we have $q \wedge p_n = (q \wedge r) \wedge p_n$, which implies that $p \wedge p_n = 0$. By (2), $p = 0$, so that $q = q \wedge r$. Similarly, $r = q \wedge r$, which yields $q = r$. $\square$

Lemma 3.53 *Let $a \in \overline{\mathcal{M}}$. If $\xi \in \chi_{[0,\epsilon]}(|a|)H$, then $\|a\xi\| \leq \epsilon\|\xi\|$. If, on the other hand, $0 \neq \xi \in \chi_{(\epsilon,\infty)}(|a|)H$, then $\|a\xi\| > \epsilon\|\xi\|$.*

Proof. In the first case, $\|a\xi\| = \||a|\xi\| = \||a|\chi_{[0,\epsilon]}(|a|)\xi\| \leq \epsilon\|\xi\|$. For the second part, by the right-continuity of the spectral decomposition there is an $\epsilon' > \epsilon$ such that $\xi \in \chi_{[\epsilon',\infty)}(|a|)$. Let $\int_{-\infty}^{\infty} \lambda de_\lambda$ be the spectral decomposition of $|a|$. Then

$$\|a\xi\|^2 = \||a|\xi\|^2 = \int_{-\infty}^{\infty} \lambda^2 d(\langle e_\lambda \xi, \xi \rangle)$$

$$\geq \int_{[\epsilon',\infty)} \lambda^2 d(\langle e_\lambda \xi, \xi \rangle) \geq \int_{[\epsilon',\infty)} \epsilon'^2 d(\langle e_\lambda \xi, \xi \rangle)$$

$$= \epsilon'^2 \|\xi\|^2 > \epsilon^2 \|\xi\|^2.$$

$\square$

Lemma 3.54

(1) *Let $a \in {}^\eta\mathcal{M}$ be preclosed and let $p \in \mathbb{P}(\mathcal{M})$ be such that $pH \subseteq \mathrm{dom}(a)$. Then $ap \in \mathcal{M}$. In particular, if (p_n) is a determining sequence for the domain $\mathrm{dom}(a)$ of a τ-premeasurable (respectively, premeasurable) operator a, then ap_n is bounded for each n.*

(2) *Let $a \in \overline{\mathcal{M}}$. Assume that $p \in \mathbb{P}(\mathcal{M})$ is such that $pH \subseteq \mathrm{dom}(a)$. Then, $\chi_{(\|ap\|,\infty)}(|a|) \preceq p^{\perp}$.*

Proof. (1) If a is preclosed and $pH \subseteq \mathrm{dom}(a)$, then ap is closed and everywhere defined, hence bounded by the closed graph theorem.

(2) The operator ap is closed and everywhere defined, hence bounded. We have $ap \in {}^\eta\mathcal{M}$, so that $ap \in \mathcal{M}$. Put $\epsilon := \|ap\|$. By Lemma 3.53, $p \wedge \chi_{(\epsilon,\infty)}(|a|) = 0$. By Lemma 3.52(1), $\chi_{(\epsilon,\infty)}(|a|) \preceq p^{\perp}$ $\square$

Proposition 3.55 *Let $a \in \overline{\mathcal{M}}$. The following conditions are equivalent.*

(1) *a is τ-measurable (respectively, a is measurable).*

(2) *$|a|$ is τ-measurable (respectively, $|a|$ is measurable).*

(3) *The domain of a is τ-dense (respectively, strongly dense) in H.*

(4) *There is a projection $p \in \mathcal{M}$ such that $pH \subseteq \mathrm{dom}(a)$ and $\tau(p^{\perp})$ (respectively, $p^{\perp}$) is finite.*

Proof. $(1){\Rightarrow}(2)$ is true by definition.

$(2){\Rightarrow}(3)$: Let $a \in \widetilde{\mathcal{M}} = S(\mathcal{M}, \tau)$ (respectively, $a \in S(\mathcal{M})$), and let $\epsilon > 0$ be such that $\tau(\chi_{(\epsilon,\infty)}(|a|))$ (respectively, $\chi_{(\epsilon,\infty)}(|a|)$) is finite. Choose an increasing sequence $\epsilon_n \to \infty$ with $\epsilon_1 := \epsilon$. Then, the sequence $p_n := \chi_{(\epsilon_n,\infty)}(|a|)$ satisfies (3).

$(3){\Rightarrow}(4)$ is obvious.

$(4){\Rightarrow}(1)$: This is a direct consequence of Lemma 3.54(2). $\square$

Corollary 3.56 *The closure $[a]$ of a τ-premeasurable (respectively, premeasurable) operator a is τ-measurable (respectively, measurable).*

Lemma 3.57 *Let $a, b \in {}^{\eta}\mathcal{M}$, and let $p, q \in \mathbb{P}(\mathcal{M})$ be such that $pH \subseteq \mathrm{dom}(a)$, $qH \subseteq \mathrm{dom}(b)$ and $ap, bq \in \mathcal{M}$. Put $e := \mathrm{n}(p^{\perp}bq)$. Then, $e^{\perp} \precsim p^{\perp}$, $p^{\perp}bqe = 0$ and $e \wedge q \subseteq \mathrm{dom}(ab)$.*

Proof.   We have $e^{\perp} = \mathrm{s}_r(p^{\perp}bq) \sim \mathrm{s}_l(p^{\perp}bq) \leq p^{\perp}$. The equality $p^{\perp}bqe = 0$ is evident from the definition of e. Thus, $\xi \in eH$ implies $p^{\perp}bq\xi = 0$, so that $bq\xi = pbq\xi \subseteq pH \subseteq \mathrm{dom}(a)$. Hence, $eH \subseteq \mathrm{dom}(abq)$ and $(e \wedge q)H \subseteq \mathrm{dom}(ab)$. $\square$

Lemma 3.58 *If $a, b \in {}^{\eta}\mathcal{M}$ are τ-premeasurable (respectively, premeasurable), then $\mathrm{dom}(ab)$ is τ-dense (respectively, strongly dense).*

Proof.   Let (p_n) be a determining sequence for $\mathrm{dom}(a)$ and let (q_n) be a determining sequence for $\mathrm{dom}(b)$. Put $e_n := \mathrm{n}(p_n^{\perp}bq_n)$. By Lemma 3.57, for each n, $\tau(e_n^{\perp})$ (respectively, $e_n^{\perp}$) is finite. Put $f_n := e_n \wedge q_n$. Again by Lemma 3.57, $f_n \subseteq \mathrm{dom}(ab)$ for each n. Clearly, $\tau(f_n^{\perp})$ (respectively, $f_n^{\perp}$) is finite. We shall show that $f_n \nearrow 1$. First, by Lemma 3.57,

$$p_{n+1}^{\perp}bq_{n+1}f_n = p_{n+1}^{\perp}bq_{n+1}q_ne_nf_n = p_{n+1}^{\perp}p_nbq_nf_n = 0.$$

Hence, $f_n \leq \mathrm{m}(p_{n+1}^{\perp} b q_{n+1}) = e_{n+1}$. We also have $f_n \leq q_n \leq q_{n+1}$, hence $f_n \leq f_{n+1}$. Suppose $f_n \nearrow f$. Using the normality of the centre-valued trace $\mathscr{T}$ on the finite algebra $f_1^{\perp} \mathcal{M} f_1^{\perp}$, we get $\mathscr{T}(e_n^{\perp}) \leq \mathscr{T}(p_n^{\perp}) \searrow 0$ and $\mathscr{T}(q_n^{\perp}) \searrow 0$, so that by Lemma 3.23, $\mathscr{T}(f_n^{\perp}) \searrow \mathscr{T}(f^{\perp}) = 0$ and $f = \mathbb{1}$. Hence, ab has a τ-dense (respectively, strongly dense) domain. $\qquad\square$

Corollary 3.59 *If $a \in \widetilde{\mathcal{M}} = S(\mathcal{M}, \tau)$ (respectively, $a \in S(\mathcal{M})$) and $b \in \mathcal{M}$, then $a + b, ab \in \widetilde{\mathcal{M}} = S(\mathcal{M}, \tau)$ (respectively, $a + b, ab \in S(\mathcal{M})$).*

Proof. Since a is closed, both $a + b$ and ab are closed. Note that $\mathrm{dom}(a + b) = \mathrm{dom}(a)$; hence, the measurability of $a + b$ follows from Proposition 3.55, (1)$\Leftrightarrow$(3). By Lemma 3.58, ab has a strongly dense domain, which implies its measurability (see Proposition 3.55, (1)$\Leftrightarrow$(3)). $\qquad\square$

An important property of τ-measurable (respectively, measurable) operators stated in the next proposition is their rigidity: if they agree on a τ-dense (respectively, strongly dense) domain, they are equal. This property will be further generalised in Proposition 3.71.

Proposition 3.60

(1) *If $a, b \in {}^{\eta}\mathcal{M}$ are premeasurable, D is a strongly dense (in particular, a τ-dense) subspace of $\mathrm{dom}(a) \cap \mathrm{dom}(b)$ and $a{\restriction}D = b{\restriction}D$, then $[a] = [b]$.*

(2) *If $a \in {}^{\eta}\mathcal{M}$ is premeasurable and D is a strongly dense (in particular, a τ-dense) subspace of $\mathrm{dom}(a)$, then D is a core for $[a]$.*

(3) *If $a \in {}^{\eta}\mathcal{M}$ is premeasurable and (p_n) is a determining sequence for $\mathrm{dom}(a)$, then $D_0 := \bigcup_n p_n H$ is a core for $[a]$.*

Proof. (1) We may assume that $a, b \in S(\mathcal{M})$. Let $\mathcal{N}$ be the von Neumann algebra $\mathcal{M} \otimes B(\mathbb{C}^2)$ acting in $H \oplus H$. Denote by p_a and p_b the projections onto the graphs $\mathcal{G}(a), \mathcal{G}(b) \subseteq H \oplus H$ of a and b, respectively. Then, $\mathcal{G}(a)$ and $\mathcal{G}(b)$ are invariant under all elements of $\mathcal{N}' = \mathcal{M}' \otimes \mathbb{1}_{\mathbb{C}^2}$, hence $p_a, p_b \in \mathcal{N}$. Let (p_n) be a determining sequence for D and let

$q_n := p_n \otimes \mathbb{1}_{\mathbb{C}^2}$ for each n. One easily checks that (q_n) is a determining sequence for a strongly dense subspace of $H \oplus H$. Moreover,

$$\mathcal{G}(a) \cap (q_n H \oplus q_n H) = \{(\xi, a\xi) : \xi \in q_n H, a\xi \in q_n H\}$$
$$= \{(\xi, b\xi) : \xi \in q_n H, b\xi \in q_n H\}$$
$$= \mathcal{G}(b) \cap (q_n H \oplus q_n H).$$

Hence, $p_a \wedge q_n = p_b \wedge q_n$ for each n, and by Lemma 3.52(3), $p_a = p_b$, so that $[a] = [b]$.

(2) It is clear that $a{\restriction}D$ is premeasurable and that $[a{\restriction}D]{\restriction}D = [a]{\restriction}D$. By (1), $[a{\restriction}D] = [a]$, hence also $[[a]{\restriction}D] = [a]$, so that D is a core for $[a]$.

(3) follows immediately from (2). $\qquad\qquad\qquad\qquad\qquad\qquad\qquad\square$

Corollary 3.61 $a \in LS(\mathcal{M})$ *(respectively, $a \in LS(\mathcal{M}, \tau)$) if and only if $|a| \in LS(\mathcal{M})$ (respectively, $|a| \in LS(\mathcal{M}, \tau)$)*

Proof. Obvious by definition. $\qquad\qquad\qquad\qquad\qquad\qquad\qquad\qquad\square$

Corollary 3.62 *If $a \in \widetilde{\mathcal{M}} = S(\mathcal{M}, \tau)$ (respectively, $a \in S(\mathcal{M})$, $a \in LS(\mathcal{M}, \tau)$, $a \in LS(\mathcal{M})$), then $a^* \in \widetilde{\mathcal{M}} = S(\mathcal{M}, \tau)$ (respectively, $a^* \in S(\mathcal{M})$, $a^* \in LS(\mathcal{M}, \tau)$, $a^* \in LS(\mathcal{M})$)*

Proof. Let $a \in \overline{\mathcal{M}}$ have polar decomposition $a = u|a|$. Then, $a^* = |a|u^*$, and the statements follow from Proposition 3.55, $(1) \Rightarrow (2)$ and Corollaries 3.59 and 3.61. $\qquad\qquad\qquad\qquad\qquad\qquad\square$

The relation between various classes of measurable operators is elucidated in Propositions 3.63–3.65 and Theorem 3.66. Most of the implications have appeared in the literature in one form or another, although some seem new. Diagram 3.10 should facilitate the reader's orientation. Readers with a deeper interest in measurability in von Neumann algebras will find the survey of Muratov and Chilin [**MC16**] invaluable, where they report on their own work on the subject and cite many other sources. In particular, they investigate the dependence of the class $\widetilde{\mathcal{M}} = S(\mathcal{M}, \tau)$ on the trace τ. It is worth noting that the class $LS(\mathcal{M}, \tau)$ of locally τ-measurable operators does not depend on the choice of τ (see [**Cec78**]).

Proposition 3.63 *Let τ be a faithful normal semifinite trace on $\mathcal{M}$. We have*

$$
\begin{array}{ccccccc}
 & & \mathcal{M} & & & & S(\mathcal{M}) \\
 & \nearrow & & \nwarrow & \nearrow & & \nwarrow \\
F(\mathcal{M},\tau) & & & S(\mathcal{M},\tau) & & & LS(\mathcal{M}) \subseteq \overline{\mathcal{M}} \\
 & \nwarrow & & \nearrow & \nwarrow & & \nearrow \\
 & & K(\mathcal{M},\tau) & & & LS(\mathcal{M},\tau) &
\end{array}
$$

$$(3.10)$$

Proof. The inclusion $F(\mathcal{M},\tau) \subseteq K(\mathcal{M},\tau)$ follows from Chebyshev's inequality:

$$
\epsilon \chi_{(\epsilon,\infty)}(|a|) \leq |a| \chi_{(\epsilon,\infty)}(|a|) \leq |a|.
$$

Once one observes that projections from $\mathcal{M}$ that have a finite trace are necessarily finite, all the other inclusions become completely obvious. $\square$

We are going to describe the dependence of the classes on dimensionality, factoriality and finiteness of $\mathcal{M}$, and on properties of τ, such as being finite or bounded away from zero.

Proposition 3.64 *Let τ be a faithful normal semifinite trace on $\mathcal{M}$.*

(1) *If $\mathcal{M}$ is a factor, then*

$$
S(\mathcal{M},\tau) = S(\mathcal{M}) = LS(\mathcal{M},\tau) = LS(\mathcal{M}).
$$

(2) *If $\mathcal{M}$ is finite, then*

$$
S(\mathcal{M}) = LS(\mathcal{M}) = \overline{\mathcal{M}}.
$$

(3) *The trace τ is bounded away from zero if and only if $K(\mathcal{M},\tau) \subseteq \mathcal{M}$, and if and only if $\mathcal{M} = S(\mathcal{M},\tau)$.*

Proof. (1) The centre of a factor is trivial, and hence, there is no difference between local and non-local measurability. Moreover, in a factor, the trace of a projection is finite if and only if the projection is finite; hence, there is no difference between versions with and without τ.

(2) If $\mathcal{M}$ is finite, then every densely defined operator affiliated with $\mathcal{M}$ is measurable, i.e., $S(\mathcal{M}) = \overline{\mathcal{M}}$.

(3) If τ is bounded away from zero, then no unbounded operator $a \in \overline{\mathcal{M}}$ is τ-measurable. In fact, assume that $a \in \overline{\mathcal{M}}$ is not bounded and let $e_n = \chi_{[n,n+1)}(|a|)$. Then, infinitely many e_ns are non-zero, which shows that a is not τ-measurable. Hence, $\mathcal{M} = S(\mathcal{M}, \tau)$, which implies that $K(\mathcal{M}, \tau) \subseteq \mathcal{M}$. If τ is not bounded away from zero, then the algebra $\mathcal{M}$ is infinite-dimensional, and there is, by Lemma 3.46, an orthogonal sequence of non-zero projections (p_n) such that $\tau(p_n) \to 0$. We may assume that $\sum_{n=1}^{\infty} \tau(p_n) < \infty$. Then, the operator $\sum_{n=1}^{\infty} np_n$ belongs to $K(\mathcal{M}, \tau)$ and is unbounded. Therefore, $K(\mathcal{M}, \tau) \not\subseteq \mathcal{M}$, and consequently $\mathcal{M} \neq S(\mathcal{M}, \tau)$. $\qquad\square$

Proposition 3.65

    (1) *If $\mathcal{M}$ is finite-dimensional (so that the trace τ is necessarily finite), then*

$$F(\mathcal{M}, \tau) = \mathcal{M} = K(\mathcal{M}, \tau) = S(\mathcal{M}, \tau)$$
$$= S(\mathcal{M}) = LS(\mathcal{M}, \tau) = LS(\mathcal{M}) = \overline{\mathcal{M}}.$$

    (2) *If $\mathcal{M}$ is infinite-dimensional, but the trace τ is finite, then*

$$F(\mathcal{M}, \tau) = \mathcal{M} \subsetneq K(\mathcal{M}, \tau) = S(\mathcal{M}, \tau)$$

$$= S(\mathcal{M}) = LS(\mathcal{M}, \tau) = LS(\mathcal{M}) = \overline{\mathcal{M}}.$$

    (3) *If the trace τ is infinite, and $\mathcal{M}$ is of type I with finite-dimensional centre, then*

$$F(\mathcal{M}, \tau) \subsetneq K(\mathcal{M}, \tau) \subsetneq \mathcal{M} = S(\mathcal{M}, \tau)$$

$$= S(\mathcal{M}) = LS(\mathcal{M}, \tau) = LS(\mathcal{M}) \subsetneq \overline{\mathcal{M}}.$$

Proof.   (1) is obvious since in a finite-dimensional algebra $F(\mathcal{M}, \tau) = \mathcal{M} = \overline{\mathcal{M}}$.

(2) If the trace τ is finite, then $F(\mathcal{M}, \tau) = \mathcal{M}$ and $K(\mathcal{M}, \tau) = \overline{\mathcal{M}}$. For τ to be bounded away from zero, $\mathcal{M}$ would have to be a direct sum of factors of type I (see Lemma 3.45). But τ is finite, so none of these factors can be infinite-dimensional. Hence, $\mathcal{M}$ is an infinite direct sum

of factors of type I_n with $n < \infty$. Since τ is finite, it cannot be bounded away from zero. By Proposition 3.64(3), $\mathcal{M} \neq S(\mathcal{M}, \tau)$.

(3) If $\mathcal{M}$ is of type I with finite-dimensional centre, then $\mathcal{M}$ is a finite direct sum of factors of type I, with at least one of them infinite. Thus, τ is bounded away from zero, and by Proposition 3.64, (1) and (3), $\mathcal{M} = S(\mathcal{M}, \tau) = S(\mathcal{M}) = LS(\mathcal{M}, \tau) = LS(\mathcal{M})$. Hence, $K(\mathcal{M}, \tau) \subseteq \mathcal{M}$, but $\tau(\mathbb{1}) = \infty$, so that $\mathbb{1} \notin K(\mathcal{M}, \tau)$ and $K(\mathcal{M}, \tau) \neq \mathcal{M}$. Let now (p_n) be an infinite orthogonal sequence of projections from $\mathcal{M}$. Then, the operator $\sum_{n=1}^{\infty}(1/n)p_n$ belongs to $K(\mathcal{M}, \tau)$, but not to $F(\mathcal{M}, \tau)$. Obviously, $\mathcal{M} \neq \overline{\mathcal{M}}$. $\qquad\square$

Theorem 3.66 clarifies which inclusions in diagram 3.10 of Proposition 3.63 are, in fact, equalities.

Theorem 3.66

(1) $F(\mathcal{M}, \tau) = \mathcal{M}$ *iff τ is finite.*

(2) $\mathcal{M} = S(\mathcal{M}, \tau)$ *iff τ is bounded away from 0.*

(3) $F(\mathcal{M}, \tau) = K(\mathcal{M}, \tau)$ *iff $\mathcal{M}$ is finite-dimensional.*

(4) $K(\mathcal{M}, \tau) = S(\mathcal{M}, \tau)$ *iff τ is finite.*

(5) $S(\mathcal{M}, \tau) = S(\mathcal{M})$ *iff finite projections from $\mathcal{M}$ have finite trace.*

(6) $S(\mathcal{M}) = LS(\mathcal{M})$ *iff the centre of the properly infinite part of $\mathcal{M}$ is finite-dimensional.*

(7) $S(\mathcal{M}, \tau) = LS(\mathcal{M}, \tau)$ *iff the restriction of τ to the finite part of $\mathcal{M}$ is finite, and the centre of the properly infinite part of $\mathcal{M}$ is finite-dimensional.*

(8) $LS(\mathcal{M}, \tau) = LS(\mathcal{M})$ *iff the centre of the type II part of $\mathcal{M}$ is σ-finite.*

(9) $LS(\mathcal{M}) = \overline{\mathcal{M}}$ *iff $\mathcal{M}$ is finite.*

Proof. (1) is clear from the definition.

(2) This is part of Proposition 3.64(3).

(3) '$\Leftarrow$' is obvious since in finite-dimensional algebras $F(\mathcal{M}, \tau) = \mathcal{M} = \overline{\mathcal{M}}$.

'$\Rightarrow$' If $\mathcal{M}$ is infinite-dimensional, then as in the proof of Proposition 3.65(3), we can construct the operator a such that $a \in K(\mathcal{M}, \tau)$ and $a \notin F(\mathcal{M}, \tau)$.

(4) '$\Leftarrow$' If τ is finite, then clearly $K(\mathcal{M}, \tau) = \overline{\mathcal{M}}$, hence the result.

'$\Rightarrow$' If τ is infinite, then $\mathbb{1} \in S(\mathcal{M}, \tau)$ and $\mathbb{1} \notin K(\mathcal{M}, \tau)$.

(5) '$\Leftarrow$' is clear from the definitions.

'$\Rightarrow$' Suppose that $p \in \mathbb{P}(\mathcal{M})$ is such that p is finite and $\tau(p) = \infty$. Take a maximal orthogonal family of non-zero subprojections of p of finite trace (it exists and sums up to p, by Proposition 3.31). It is clear that we may then select a countable subfamily $\{p_k\}$ such that $\sum_{k=1}^{\infty} \tau(p_k) = \infty$. Let $a := \sum_{k=1}^{\infty} k p_k$. Then, $a \in S(\mathcal{M})$, since $\chi_{(0,\infty)}(a) = \sum_{k=1}^{\infty} p_k \leq p$, which is finite, but for each natural $n, \chi_{(n,\infty)}(a) = \sum_{k=n+1}^{\infty} p_k$ with $\tau(\sum_{k=n+1}^{\infty} p_k) = \infty$. Hence, $a \in S(\mathcal{M})$, but $a \notin S(\mathcal{M}, \tau)$.

(6) '$\Leftarrow$' If the algebra $\mathcal{M}$ is a direct sum of a finite von Neumann algebra and a finite number of infinite factors, then the equality $S(\mathcal{M}) = LS(\mathcal{M})$ follows from (1) and (2) of Proposition 3.64.

'$\Rightarrow$' We assume that $S(\mathcal{M}) = LS(\mathcal{M})$ and that the centre of the properly infinite part of $\mathcal{M}$ is infinite-dimensional. We construct an operator a such that $a \in LS(\mathcal{M})$ and $a \notin S(\mathcal{M})$. There is no loss of generality in assuming additionally that $\mathcal{M}$ is properly infinite. Let (z_k) be an orthogonal sequence of non-zero central projections from $\mathcal{M}$ such that $\sum_{k=1}^{\infty} z_k = \mathbb{1}$. Define $a := \sum_{k=1}^{\infty} k z_k$. If $w_n = \sum_{k=1}^{n} z_k$, then $w_n \nearrow \mathbb{1}$ and $w_n^{\perp}$ is infinite (as a central projection in a properly infinite algebra). Note that $a w_n \in \mathcal{M} \subseteq S(\mathcal{M})$, so that $a \in LS(\mathcal{M})$. On the other hand, for any $\epsilon > 0$, the spectral projection $\chi_{(\epsilon,\infty)}$ contains some $w_n^{\perp}$, so it must be infinite. Hence, $a \notin S(\mathcal{M})$, which ends the proof.

(7) '$\Leftarrow$' Let $z \in \mathcal{Z}(\mathcal{M})$ be such that $z\mathcal{M}$ is finite and $(\mathbb{1} - z)\mathcal{M}$ is properly infinite. It is clear that $S(z\mathcal{M}, \tau) = zS(\mathcal{M}, \tau)$ and $S((\mathbb{1} - z)\mathcal{M}, \tau) = (\mathbb{1} - z)S(\mathcal{M}, \tau)$. Take $a \in LS(\mathcal{M}, \tau)$. Then, $a \in LS(\mathcal{M})$, so that by (6) $a \in S(\mathcal{M})$. Note that $za \in S(z\mathcal{M}, \tau)$ by (5) applied to $z\mathcal{M}$, and that $(\mathbb{1} - z)a \in S((\mathbb{1} - z)\mathcal{M}, \tau)$ by Proposition 3.64(1). Hence, $a \in S(\mathcal{M}, \tau)$.

'$\Rightarrow$' Assume $S(\mathcal{M}, \tau) = LS(\mathcal{M}, \tau)$. It is enough to show that if either the restriction of τ to the finite part of $\mathcal{M}$ is infinite or the centre of the properly infinite part of $\mathcal{M}$ is infinite-dimensional, we get a contradiction. In both cases, there is an orthogonal sequence of non-zero projections $z_k \in \mathcal{Z}(\mathcal{M})$ such that for each $n \in \mathbb{N}$, $\sum_{k=n}^{\infty} \tau(z_k) = \infty$. Indeed, if the restriction of τ to the finite part of $\mathcal{M}$ is infinite, we can choose z_k so that $\tau(z_k) < \infty$ for each k, but $\sum_{k=1}^{\infty} \tau(z_k) = \infty$, as in (5)'$\Rightarrow$'. If, on the other hand, there is an orthogonal sequence of non-zero projections in the centre of the properly infinite part of $\mathcal{M}$, then all the projections z_k are infinite, so that $\tau(z_k) = \infty$ for each k.

Put $w_n = \sum_{k=1}^{n} z_k$ and $w = \sup_n w_n$. Then, $v_n := w_n + (\mathbb{1} - w) \nearrow \mathbb{1}$. Let $a \in {}^{\eta}\mathcal{M}$ be the operator $\sum_{k=1}^{\infty} k z_k$. For each n, we then have $v_n a = \sum_{k=1}^{n} k z_k \in \mathcal{M} \subseteq S(\mathcal{M}, \tau)$, so that $a \in LS(\mathcal{M}, \tau)$, but $a \notin S(\mathcal{M}, \tau)$ (cf. the proof of (5)).

(8) '$\Leftarrow$' Suppose first that $\mathcal{M}$ is of type I_n with $n < \infty$. Take $a \in LS(\mathcal{M})$. By (7), $a \in S(\mathcal{M})$. Let $\{f_k\}$ be an orthogonal family of abelian projections from $\mathcal{M}$ such that $\sum_{k=1}^{n} f_k = \mathbb{1}$ (which implies that $\mathbf{z}(f_k) = \mathbb{1}$ for each k), and denote by $v_{i,k}$ partial isometries from $\mathcal{M}$ such that $v_{i,k}^* v_{i,k} = f_k$ and $v_{i,k} v_{i,k}^* = f_i$. By Lemma 3.59, the operators $a f_k \in S(\mathcal{M})$, and by Proposition 3.55 the $|a f_k|$s are also measurable. Let $e_{k,m} := \chi_{(m,\infty)}(|a f_k|)$ and $z_{k,m} := \mathbf{z}(e_{k,m})$. Note that $\mathfrak{n}(|a f_k|) = \mathfrak{n}(a f_k) \geq f_k^{\perp}$, so that $e_{k,m} \leq \mathbf{s}_l(|a f_k|) \leq f_k$. Since $f_k \mathcal{M} f_k = \mathcal{Z}(\mathcal{M}) f_k$ (see Proposition 0.105), we have $e_{k,m} = z_{k,m} f_k$. We check that for a fixed k, $z_{k,m} \searrow 0$ as $m \to \infty$. In fact,

$$z_{k,m} = \sum_{j=1}^{n} f_j z_{k,m} = \sum_{j=1}^{n} v_{j,k} f_k v_{j,k}^* z_{k,m} = \sum_{j=1}^{n} v_{j,k} e_{k,m} v_{j,k}^*$$

and so $e_{k,m} \searrow 0$ as $m \to \infty$ implies $z_{k,m} \searrow 0$ as $m \to \infty$.

Put $z_m := z_{1,m}^{\perp} z_{2,m}^{\perp} \cdots z_{n,m}^{\perp}$. Clearly, $z_m \nearrow \mathbb{1}$ as $m \to \infty$. Since $z_{k,m}^{\perp} f_k = e_{k,m}^{\perp} f_k$, we have that

$$f_k z_m = f_k e_{k,m}^{\perp} w_k$$

where $w_k := z_{1,m}^{\perp} \cdots z_{k-1,m}^{\perp} z_{k+1,m}^{\perp} \cdots z_{n,m}^{\perp}$. Hence, the closed operators $a f_k z_m = a f_k e_{k,m}^{\perp} w_k$ are bounded with $\|a f_k z_m\| \leq m$. Therefore, the closed operators $a z_m = a(\sum_{k=1}^{n} f_k z_m) \supseteq \sum_{k=1}^{n} a f_k z_m$ (see Proposition 0.24) are also bounded, and consequently $a \in LS(\mathcal{M}, \tau)$.

Assume now that $\mathcal{M}$ is a σ-finite algebra of type II_1, and let $a \in LS(\mathcal{M})$. Again, by (7), $a \in S(M)$. If τ is finite, there is nothing to prove (see Proposition 3.64), so assume that τ is infinite. By Lemma 3.42, the restriction of τ to $\mathcal{M}$ is semifinite. By Proposition 3.31(2), there is an orthogonal family of non-zero projections $\{w_i\}_{i \in I}$ in the centre $\mathcal{Z}(\mathcal{M})$ such that $\tau(w_i) < \infty$ and $\sum_{i \in I} w_i = \mathbb{1}$. Since the algebra is σ-finite, I is countable. Put $z_n := \sum_{k=1}^{n} w_k$. Then, $z_n \nearrow \mathbb{1}$ as $n \to \infty$, and $a z_n \in S(\mathcal{M} z_n, \tau)$, by (5). Obviously, $a z_n$ is also τ-measurable in $\mathcal{M}$ and hence $a \in LS(\mathcal{M}, \tau)$.

Now, let $a \in S(\mathcal{M})$ and $\mathcal{M}$ be properly infinite. Let $p := \chi_{[0,\lambda]}(|a|)$ be such that $ap \in \mathcal{M}$ and $p^{\perp}$ is finite. Using Lemma 3.59 and the inclusion $a = a(p + p^\perp) \supseteq ap + ap^\perp$, we easily get $a = ap + ap^\perp$, with $ap^\perp \in S(M)$. If the centre of type II part of $\mathcal{M}$ is σ-finite, the same is true of $\mathcal{M}_{p^\perp}$. If this is the case, $ap^\perp \in LS(\mathcal{M}, \tau)$ by the previous part of the proof. Again, by Lemma 3.59, $a \in LS(\mathcal{M}, \tau)$.

It is clear from the above proof that whenever $a \in S(\mathcal{M})$ and $\mathcal{M}$ is either of type I_n (with an arbitrary n) or of type II with a σ-finite centre, one can find a sequence of central projections z_n such that az_n is bounded and $z_n \nearrow \mathbb{1}$. If now $a \in LS(\mathcal{M})$, then there is a sequence (w_k) of central projections such that $aw_k \in S(\mathcal{M})$ and $w_k \nearrow \mathbb{1}$. Let $v_n^{(k)} \in \mathcal{Z}(\mathcal{M})$ be such that $a(w_k - w_{k-1})v_n^{(k)} \in \mathcal{M}$ and $v_n^{(k)} \nearrow w_k - w_{k-1}$ (with $w_0 = 0$) as $n \to \infty$. Put $z_n^{(k)} = \sum_{j=1}^k v_n^{(k)}$. It is routine to check that $z_k^k \nearrow \mathbb{1}$, with the boundedness of az_k^k implying that $a \in LS(\mathcal{M}, \tau)$.

'$\Rightarrow$' Now, assume that $\mathcal{M}$ is a non-σ-finite von Neumann algebra of type II. Let $p \in \mathbb{P}(\mathcal{M})$ be finite with central support $\mathbb{1}$; this can easily be done using an exhaustion argument on recalling that the sum of centrally orthogonal finite projections is again finite. Let now (p_k) be an orthogonal sequence of non-zero subprojections of p with central support $\mathbb{1}$ such that $\sum_{k=1}^\infty p_k = p$; this could be done by consecutive halving of projections (see Lemma 0.124)(2), starting with p. Note that $p_k z \neq 0$ for all non-zero $z \in \mathcal{Z}(\mathcal{M})$. Let $a := \sum_{k=1}^\infty k p_k$. Then $a \in S(\mathcal{M})$, since $\chi_{(n,\infty)}(|a|) = \sum_{k=n+1}^\infty p_k \leq p$ is finite. Let (z_n) be a sequence of central projections such that $z_n \nearrow \mathbb{1}$. For sufficiently large n, the projection z_n is a sum of an uncountable number of orthogonal central projections, say z_α (otherwise, $(z_n - z_{n-1})$ with $z_0 = 0$ would be a countable orthogonal sequence of σ-finite central projections with sum $\mathbb{1}$). For such n, az_n cannot be τ-measurable, since $\tau(\chi_{(n,\infty)}(|az_n|)) = \tau(\sum_{k=n+1}^\infty p_k)$ and $\tau(p_{n+1}) = \sum_\alpha \tau(p_{n+1}z) = \infty$, as an uncountable sum of positive numbers. Hence, $a \notin LS(\mathcal{M}, \tau)$, which ends the proof.

(9) '$\Leftarrow$' This is part of Proposition 3.64.

'$\Rightarrow$' Suppose that $\mathcal{M}$ is not finite. We are going to build a closed and densely defined operator a that is not locally measurable. Let z be the non-zero central projection for which $\mathcal{M}z$ is the properly infinite part of $\mathcal{M}$. If $a \notin LS(\mathcal{M}z, \tau)$, then $a \notin LS(\mathcal{M}, \tau)$, and so we can assume that $z = \mathbb{1}$. Choose an orthogonal sequence of properly infinite projections (p_k) from $\mathcal{M}$ with central support $\mathbb{1}$ (for example, by consecutive halving, starting with $\mathbb{1}$), and let $a := \sum_{k=1}^\infty k p_k$. Then, $a \in \overline{\mathcal{M}}$, but $a \notin LS(\mathcal{M})$, since $\mathbf{z}(a) = \mathbb{1}$ and for any central projection z the projection $\chi_{(n,\infty)}(|az|) = \sum_{k=n+1}^\infty p_k z)$ is infinite. $\qquad\square$

3.4 Algebraic properties of measurable operators

In this section, we show (in Theorem 3.70) that the space $LS(\mathcal{M})$ of locally measurable operators forms, with strong sum and strong product (see Theorem 3.68), a *-algebra, and that $LS(M, \tau), S(\mathcal{M})$ and $\widetilde{\mathcal{M}} = S(\mathcal{M}, \tau)$ are *-subalgebras. The importance of locally measurable operators stems from the fact that as far as measurability of operators is concerned, they in a sense form the largest class worth considering. If we want to include an operator $x \in \overline{\mathcal{M}}$ in such a class, we should clearly require that ax is closable for any $a \in \mathcal{M}$, so that the strong product of a and x can be defined. As shown by Yeadon in his paper [**Yea75**] that generalises a previous result of Dixon [**Dix71**], this implies that x is, in fact, locally measurable. Another important application of the notion will be found in Section 4.3.

Lemma 3.67 *If $a, b \in {}^{\eta}\mathcal{M}$ are τ-premeasurable (respectively, premeasurable), then $a + b$ and ab are τ-premeasurable (respectively, premeasurable).*

Proof. By Proposition 3.50 and Lemma 3.58, both $a + b$ and ab have τ-dense (respectively, strongly dense) domains, and so it is enough to show that they are closable. By Corollary 3.62, both $[a]^*$ and $[b]^*$ are measurable, hence $\mathrm{dom}([a]^* + [b]^*)$ and $\mathrm{dom}([b]^*[a]^*)$ are dense and the operators $([a]^* + [b]^*)^*$ and $([b]^*[a]^*)^*$ exist and are closed. Consequently, both

$$a + b \subseteq [a] + [b] \subseteq ([a]^* + [b]^*)^*$$

and

$$ab \subseteq [a][b] \subseteq ([b]^*[a]^*)^*$$

are closable. $\qquad\square$

Theorem 3.68 *The space $\widetilde{\mathcal{M}} = S(\mathcal{M}, \tau)$ (respectively, $S(\mathcal{M})$) with the operations $(a, b) \mapsto [a + b]$ of strong sum, $(a, b) \mapsto [ab]$ of strong product, together with the operation of multiplication by (complex) scalar $(\lambda, a) \mapsto \lambda a$ and the adjoint operation $*$ forms a $*$-algebra.*

Proof. If $a, b, c \in {}^{\eta}\mathcal{M}$ are τ-premeasurable (respectively, premeasurable), then by Lemma 3.67, $(a + b) + c$, $a + (b + c)$, $(ab)c$, $a(bc)$, $(a + b)c$,

$ac + bc$, $c(a + b)$, and $ca + cb$ are all τ-premeasurable (respectively, premeasurable), and by Proposition 3.60, one has

$$[[a + b] + c] = [a + [b + c]], \qquad [[ab]c] = [a[bc]],$$
$$[[a + b]c] = [[ac] + [bc]], \qquad [c[a + b]] = [[ca] + [cb]],$$
$$[a + b]^* = [a^* + b^*], \qquad [ab]^* = [b^*a^*].$$

It is obvious that λa is τ-premeasurable (respectively, premeasurable) if a is τ-premeasurable (respectively, premeasurable), for any $\lambda \in \mathbb{C}$. Similarly, $\lambda a \in \widetilde{\mathcal{M}} = S(\mathcal{M}, \tau)$ (respectively, $\lambda a \in S(\mathcal{M})$) for $a \in \widetilde{\mathcal{M}} = S(\mathcal{M}, \tau)$ (respectively, $a \in S(\mathcal{M})$), for any $\lambda \in \mathbb{C}$. The result follows. $\square$

Notation 3.69 In the sequel, we will denote the operations of strong sum and strong product by $\bar{+}$ and $\bar{\cdot}$, respectively. That is, $a\bar{+}b := [a + b]$ and $a\bar{\cdot}b := [ab]$. Similarly, we write $a\bar{-}b$ for $[a - b]$. We shall simply write $a + b$ for the strong sum and ab for the strong product whenever the meaning is obvious from the context and does not lead to confusion.

Theorem 3.70 *If $a, b \in LS(\mathcal{M}, \tau)$ (respectively, $LS(\mathcal{M})$), then $a + b, ab$ are both closable, $[a + b], [ab] \in LS(\mathcal{M}, \tau)$ (respectively, $LS(\mathcal{M})$), and the space $LS(\mathcal{M}, \tau)$ (respectively, $LS(\mathcal{M})$) with the operations of strong sum, strong product, multiplication by (complex) scalar $(\lambda, a) \mapsto \lambda a$ and the adjoint operation $*$ forms a $*$-algebra.*

Proof. If $a, b \in LS(\mathcal{M}, \tau)$ (respectively, $LS(\mathcal{M})$), then we can find one sequence (z_n) of central projections with $z_n \nearrow \mathbb{1}$ such that $az_n, bz_n \in S(\mathcal{M}, \tau)$ (respectively, $S(\mathcal{M})$). Since $(a + b)z_n = az_n + bz_n$, $a\bar{+}b \in LS(\mathcal{M}, \tau)$ (respectively, $LS(\mathcal{M})$). For the product, note that $az_n bz_n \in S(\mathcal{M}, \tau)$ (respectively, $S(\mathcal{M})$), and that $z_n b \subseteq bz_n$ by Lemma 0.132, which implies that $az_n bz_n \subseteq abz_n$. By Proposition 3.60, $az_n bz_n = abz_n$. This means that $abz_n \in S(\mathcal{M}, \tau)$ (respectively, $abz_n \in S(\mathcal{M})$) and $ab \in LS(\mathcal{M}, \tau)$ (respectively, $ab \in LS(\mathcal{M})$). Finally, $a \in \overline{\mathcal{M}}$ implies that $z_n a$ is densely defined and $z_n a \subseteq az_n$. Hence, $(az_n)^*$ exists and $(az_n)^* \subseteq (z_n a)^* = a^* z_n$. By Proposition 3.60, $a^* z_n = (az_n)^* \in S(\mathcal{M}, \tau)$ (respectively, $a^* z_n \in S(\mathcal{M})$) and $a \in LS(\mathcal{M}, \tau)$ (respectively, $a \in LS(\mathcal{M})$). $\square$

The following proposition strengthens the conclusions of Proposition 3.60.

Proposition 3.71 *Let* $a, b \in {}^{\eta}\mathcal{M}$ *be* τ-*premeasurable (respectively, premeasurable), and let* $x \in {}^{\eta}\mathcal{M}$. *Then:*

(1) *If* D *is a dense subspace of* $\mathrm{dom}(a) \cap \mathrm{dom}(b)$ *and* $a{\upharpoonright}D = b{\upharpoonright}D$, *then* $[a] = [b]$; *in particular, if* $a, b \in \widetilde{\mathcal{M}} = S(\mathcal{M}, \tau)$ *(respectively,* $a, b \in S(\mathcal{M})$*) agree on a dense subspace, then* $a = b$.

(2) *If* x *is closable and* $a \subseteq x$, *then* $[x] = [a]$.

(3) *If* x *is densely defined and* $x \subseteq a$, *then* $[x] = [a]$.

Proof. (1) $[a]\bar{-}[b]$ is τ-measurable (respectively, measurable) and $[a]\bar{-}[b] \supseteq 0{\upharpoonright}D$, so $([a]\bar{-}[b])^* \subseteq 0$. On the other hand, $([a]\bar{-}[b])^*$ is τ-measurable (respectively, measurable), so it must be 0, which implies that $[a] = [b]$ by Theorem 3.70.

(2) The assumptions ensure that x is τ-premeasurable (respectively, premeasurable), so we get the result from (1).

(3) follows from (2) by taking adjoints. $\qquad\square$

3.5 Topological properties of measurable operators

In this section, we will introduce the so-called *measure* topology (respectively, the topology of *convergence in measure*) in the *-algebra $\widetilde{\mathcal{M}}$ and show that it turns $\widetilde{\mathcal{M}}$ into a complete topological *-algebra.

One can obtain similar results for other classes of measurability, but that would require the introduction of another topology, that of local convergence in measure (see [**Yea73**]), which will not be used in the sequel.

Notation 3.72 We denote by $\mathcal{N}(\epsilon, \delta)$ the set $\{a \in \widetilde{\mathcal{M}}$: there exists a projection $p = p_{\epsilon, \delta} \in \mathcal{M}$ such that $pH \subseteq \mathrm{dom}(a)$, $\|ap\| \leq \epsilon$ and $\tau(p^{\perp}) \leq \delta\}$.

Lemma 3.73 *Let* $a \in \widetilde{\mathcal{M}}$. *Then:*

(1) $a \in \mathcal{N}(\epsilon, \delta)$ *iff* $\tau(\chi_{(\epsilon, \infty)}(|a|)) \leq \delta$. *In particular,* $a \in \mathcal{N}(\epsilon, \delta)$ *iff* $|a| \in \mathcal{N}(\epsilon, \delta)$.

(2) *For all $a \in \widetilde{\mathcal{M}}$ and $\delta > 0$, there is an $\epsilon > 0$ such that $a \in \mathcal{N}(\epsilon, \delta)$.*

Proof. (1) '$\Leftarrow$' Put $p := \chi_{(\epsilon,\infty)}(|a|)$ in the definition of $\mathcal{N}(\epsilon, \delta)$. (1) '$\Rightarrow$' This follows directly from Lemma 3.54(2).

(2) follows immediately from (1), the definition of τ-measurability and the normality of the trace. $\qquad\qquad\qquad\qquad\qquad\qquad\qquad\qquad\square$

Lemma 3.74 *For all $\epsilon, \epsilon_1, \epsilon_2, \delta, \delta_1, \delta_2 > 0$, and $\lambda \in \mathbb{C}$, we have:*

(1) $\mathcal{N}(\epsilon, \delta)^* = \mathcal{N}(\epsilon, \delta);$
(2) $\mathcal{N}(|\lambda|\epsilon, \delta) = \lambda\mathcal{N}(\epsilon, \delta);$
(3) *if $\epsilon_1 \le \epsilon_2$ and $\delta_1 \le \delta_2$, then $\mathcal{N}(\epsilon_1, \delta_1) \subseteq \mathcal{N}(\epsilon_2, \delta_2);$*
(4) $\mathcal{N}(\epsilon_1, \delta_1) \cap \mathcal{N}(\epsilon_2, \delta_2) \supseteq \mathcal{N}(\min(\epsilon_1, \epsilon_2), \min(\delta_1, \delta_2));$
(5) $\mathcal{N}(\epsilon_1, \delta_1) \overline{+} \mathcal{N}(\epsilon_2, \delta_2) \subseteq \mathcal{N}(\epsilon_1 + \epsilon_2, \delta_1 + \delta_2);$
(6) $\mathcal{N}(\epsilon_1, \delta_1) \overline{\cdot} \mathcal{N}(\epsilon_2, \delta_2) \subseteq \mathcal{N}(\epsilon_1\epsilon_2, \delta_1 + \delta_2).$

Proof. (1) Assume that $a \in \widetilde{\mathcal{M}}$, and let $a = u|a|$ be its polar decomposition. Then, u maps $\mathbf{s}(|a|)$ isometrically onto $\mathbf{s}(|a^*|)$ with in addition $|a^*| = u|a|u^*$. For any $\xi \in H$, we have

$$\langle |a^*|\xi, \xi \rangle = \int_{(0,\infty)} \lambda d\langle e_\lambda(|a^*|)\xi, \xi \rangle$$

and

$$\langle |a^*|\xi, \xi \rangle = \langle |a|u^*\xi, u^*\xi \rangle = \int_{(0,\infty)} \lambda d\langle e_\lambda(|a|)u^*\xi, u^*\xi \rangle$$
$$= \int_{(0,\infty)} \lambda d\langle (ue_\lambda(|a|)u^*)\xi, \xi \rangle.$$

By the uniqueness of the spectral decomposition, $e_\lambda(|a^*|) = ue_\lambda(|a|)u^*$ for each $\lambda \ge 0$. Hence, $e_{(\epsilon,\infty)}(|a^*|) = ue_{(\epsilon,\infty)}(|a|)u^*$ and

$$\tau(\chi_{(\epsilon,\infty)}(|a^*|)) = \tau(u\chi_{(\epsilon,\infty)}(|a|)u^*) = \tau(u^*u\chi_{(\epsilon,\infty)}(|a|))$$
$$= \tau(\mathbf{s}(|a|)\chi_{(\epsilon,\infty)}(|a|)) = \tau(\chi_{(\epsilon,\infty)}(|a|))$$

for each $\epsilon > 0$, which implies (1).

(2) and (3) are obvious, and (4) follows immediately from (3).

(5) If $a \in \mathcal{N}(\epsilon_1, \delta_1)$ and $b \in \mathcal{N}(\epsilon_2, \delta_2)$, then there exist projections $p, q \in \mathcal{M}$ such that $pH \subseteq \text{dom}(a)$, $qH \subseteq \text{dom}(b)$, $\|ap\| \le \epsilon_1$, $\|bq\| \le \epsilon_2$,

and $\tau(p^\perp) \le \delta_1$, $\tau(q^\perp) \le \delta_2$. Put $r := p \wedge q$. Then, $rH \subseteq \mathrm{dom}(a\overline{+}b)$, $\|(a\overline{+}b)r\| = \|ar + br\| \le \epsilon_1 + \epsilon_2$ and $\tau(r^\perp) \le \delta_1 + \delta_2$, which implies (5).

(6) Assume that $a \in \mathscr{N}(\epsilon_1, \delta_1)$ and $b \in \mathscr{N}(\epsilon_2, \delta_2)$. Choose projections $p, q \in \mathcal{M}$ such that $pH \subseteq \mathrm{dom}(a)$, $qH \subseteq \mathrm{dom}(b)$, $\|ap\| \le \epsilon_1$, $\|bq\| \le \epsilon_2$, and $\tau(p^\perp) \le \delta_1$, $\tau(q^\perp) \le \delta_2$. Let $e := \mathrm{n}(p^\perp bq)$ and $f := e \wedge q$. Then, by Lemma 3.57, $fH \subseteq \mathrm{dom}(ab)$ and $pbqe = bqe$, so that $abf = abqef = apbqef = apbqf$. Thus $\|(a\overline{\cdot}b)f\| = \|abf\| \le \epsilon_1\epsilon_2$. Similarly, again by Lemma 3.57, $\tau(f^\perp) \le \tau(e^\perp) + \tau(q^\perp) \le \tau(p^\perp) + \tau(q^\perp) \le \delta_1 + \delta_2$, which ends the proof. $\qquad\square$

Proposition 3.75 *The sets $\mathscr{N}(\epsilon, \delta)$ form a basis of neighbourhoods of 0 for a vector space topology on $\widetilde{\mathcal{M}}$, called the measure topology.*

Proof. By definition, a set $A \subseteq \widetilde{\mathcal{M}}$ is open in the measure topology if, for each $a \in A$, there are $\epsilon, \delta > 0$ such that $a \overline{+} \mathscr{N}(\epsilon, \delta) \subseteq A$. It follows from (4) and (5) of Lemma 3.74 that for each $a \in \widetilde{\mathcal{M}}$ the sets $a + \mathscr{N}(\epsilon, \delta)$ form a basis of neighbourhoods of a for a translation-invariant topology on $\widetilde{\mathcal{M}}$. The neighbourhoods $\mathscr{N}(\epsilon, \delta)$ are balanced (or circled) by (2) and (3) of the lemma, and absorbing by additionally using Lemma 3.73(2). Together with Lemma 3.74, this yields the result (see, for example, [**SW99**, 1.2]). $\qquad\square$

Theorem 3.76 *The algebra $\widetilde{\mathcal{M}}$ of τ-measurable operators endowed with the measure topology is a complete metrisable topological *-algebra in which $\mathcal{M}$ is dense.*

Proof. The joint continuity of the multiplication easily follows from Lemma 3.74 (6) and Lemma 3.73 (2). The continuity of the * operation follows from Lemma 3.74(1). Note that $\{\mathscr{N}(1/n, 1/n) : n \in \mathbb{N}\}$ forms a countable base of neighbourhoods at 0 for the measure topology. The topology is Hausdorff, since $\bigcap_{\epsilon, \delta > 0} \mathscr{N}(\epsilon, \delta) = \{0\}$. In fact, if for a fixed ϵ we have $a \in \mathscr{N}(\epsilon, \delta)$ for each $\delta > 0$, then by Lemma 3.73(1), $\tau(\chi_{(\epsilon, \infty)}(|a|)) = 0$. Since this is true for all $\epsilon > 0$, we get $a = 0$. The countability of the base at 0 of a Hausdorff vector topology implies its metrisability; see [**SW99**, 6.1].

Let us show that $\mathcal{M}$ is dense in $\widetilde{\mathcal{M}}$ in the measure topology. Let $a \in \widetilde{\mathcal{M}}$ and let (p_n) be a determining sequence for $\mathrm{dom}(a)$. The normality of τ implies $\tau(p_n^\perp) \to 0$. Take any $\epsilon, \delta > 0$, and choose n_0 so that $\tau(p_n^\perp) \le \delta$ for $n \ge n_0$. Then, $a - ap_n \in \mathcal{N}(\epsilon, \delta)$ for $n \ge n_0$. In fact, it is enough to take $p_{\epsilon,\delta} := p_n$ in the definition of $\mathcal{N}(\epsilon, \delta)$ (see Notation 3.72).

To show that $\widetilde{\mathcal{M}}$ is complete in the measure topology, it is enough to show that every sequence in $\widetilde{\mathcal{M}}$ which is Cauchy in measure converges in measure to an element of $\widetilde{\mathcal{M}}$. This follows directly from the metrisability of $\widetilde{\mathcal{M}}$ (or, even simpler, the existence of a countable base of neighbourhoods of 0 in $\widetilde{\mathcal{M}}$). We can assume that the Cauchy sequence is taken from $\mathcal{M}$. In fact, for a sequence (a_n) from $\widetilde{\mathcal{M}}$, we can pick a sequence (a_n') from $\mathcal{M}$ in such a way that $a_n - a_n' \in \mathcal{N}(1/n, 1/n)$ for all $n \in \mathbb{N}$. Then, (a_n') is Cauchy, and its limit (if any) is also the limit of (a_n). By passing to a subsequence, we can also assume that $a_n - a_{n+1} \in \mathcal{N}(1/2^{n+1}, 1/2^{n+1})$. So let (a_n) be a Cauchy sequence from $\mathcal{M}$ satisfying the above condition. Then, by Lemma 3.74, also $a_n^* - a_{n+1}^* \in \mathcal{N}(1/2^{n+1}, 1/2^{n+1})$. Choose projections p_n' and p_n'' from $\mathcal{M}$ so that $\|(a_{n+1} - a_n)p_n'\| \le 1/2^{n+1}$, $\|(a_{n+1}^* - a_n^*)p_n''\| \le 1/2^{n+1}$, $\tau(p_n'^\perp) \le 1/2^{n+1}$ and $\tau(p_n''^\perp) \le 1/2^{n+1}$. Put $p_n := p_n' \wedge p_n''$. Then $\|(a_{n+1} - a_n)p_n\| \le 1/2^{n+1}$ and $\tau(p_n^\perp) \le 1/2^n$. For each $n \in \mathbb{N}$, put $q_n = \bigwedge_{k=n+1}^\infty p_k$, so that $\tau(q_n^\perp) \le \sum_{k=n+1}^\infty 1/2^k = 1/2^n$.

We calculate, for $m \ge n+1$ and $\ell \in \mathbb{N}$,

$$\|(a_{m+\ell} - a_m)q_n\| \le \sum_{k=m}^{m+\ell-1} \|(a_{k+1} - a_k)q_n\| \le \sum_{k=m}^{m+\ell-1} \|(a_{k+1} - a_k)p_k\|$$

$$\le \sum_{k=m}^{m+\ell-1} 1/2^{k+1} \le 1/2^m$$

and similarly

$$\|(a_{m+\ell}^* - a_m^*)q_n\| \le 1/2^m.$$

Therefore, if $\xi \in q_n H$, then both $(a_n \xi)$ and $(a_n^* \xi)$ are Cauchy sequences in H. Thus, we can define operators a_0 and b_0 with domain $D = \bigcup_{n \in \mathbb{N}} q_n H$ by $a_0 \xi := \lim_{m \to \infty} a_m \xi$ and $b_0 \xi := \lim_{m \to \infty} a_m^* \xi$. Clearly $a_0, b_0 \in {}^\eta \mathcal{M}$ with their domain D τ-dense in H. Moreover, for all $\xi, \eta \in D$, we have $\langle a_0 \xi, \eta \rangle = \lim_{m \to \infty} \langle a_m \xi, \eta \rangle = \lim_{m \to \infty} \langle \xi, a_m^* \eta \rangle = \langle \xi, b_0 \eta \rangle$, so that $a_0 \subseteq b_0^*$ and a_0 is premeasurable. Put $a = [a_0]$, so that $a \in \widetilde{\mathcal{M}}$.

We will now show that (a_n) actually converges in measure to a. Take any $\epsilon, \delta > 0$, and let n_0 be such that $\epsilon \geq 1/2^{n_0+1}$ and $\delta \geq 1/2^{n_0}$. We claim that $a - a_m \in \mathcal{N}(\epsilon, \delta)$ for $m \geq n_0 + 1$. In fact, it is enough to take $p := q_{n_0}$ in the definition of $\mathcal{N}(\epsilon, \delta)$. We have $\tau(q_{n_0}^{\perp}) \leq 1/2^{n_0} \leq \delta$. Moreover,

$$
\begin{aligned}
\|(a - a_m)q_{n_0}\| &= \sup_{\xi \in H, \|\xi\| \leq 1} \|(a - a_m)q_{n_0}\xi\| \\
&\leq \sup_{\xi \in H, \|\xi\| \leq 1} \limsup_{\ell \to \infty} \|(a_{m+\ell} - a_m)q_{n_0}\xi\| \\
&\leq \sup_{\xi \in H, \|\xi\| \leq 1} \limsup_{\ell \to \infty} \|(a_{m+\ell} - a_m)q_{n_0}\|\|\xi\| \\
&\leq \sup_{\xi \in H, \|\xi\| \leq 1} \limsup_{\ell \to \infty} (1/2^m)\|\xi\| \leq \epsilon.
\end{aligned}
$$

This ends the proof of the theorem. $\qquad\qquad\qquad\qquad\qquad\qquad\qquad\square$

3.6 τ-local convergence in measure

In addition to the topologies on the spaces $LS(\mathcal{M}, \tau)$ and $LS(\mathcal{M})$, there are other topologies on $S(\mathcal{M}, \tau) = \widetilde{\mathcal{M}}$ sometimes also referred to in the literature as (weak) local convergence in measure. These topologies should, however, not be confused with either of the topologies on $LS(\mathcal{M}, \tau)$ or $LS(\mathcal{M})$. They do not correspond to yet more classes of measurable operators. They are rather coarser topologies on $S(\mathcal{M}, \tau)$, which are sometimes useful in establishing noncommutative versions of classical results. Given their utility for the class currently under investigation, we shall briefly describe these topologies, starting by introducing the most basic building blocks. For $\epsilon, \delta > 0$ and a projection $p \in \mathcal{M}$ with finite trace, we define the sets

$$
\mathcal{V}(\epsilon, \delta, p) = \{x \in \widetilde{\mathcal{M}} : \exists q \in \mathbb{P}(\mathcal{M})(q \leq p, \|xq\| \leq \epsilon\delta \text{ and } \tau(p - q) \leq \delta)\}
$$

and

$$
\mathcal{W}(\epsilon, \delta, p) = \{x \in \widetilde{\mathcal{M}} : \exists q \in \mathbb{P}(\mathcal{M})(q \leq p, \|qxq\| \leq \epsilon\delta \text{ and } \tau(p - q \leq \delta)\}.
$$

It is not difficult to modify the earlier arguments to show that both of the classes $\mathfrak{B}_{lc} = \{\mathcal{V}(\epsilon, \delta, p)\}_{\epsilon, \delta, p}$ and $\mathfrak{B}_{blc} = \{\mathcal{W}(\epsilon, \delta, p)\}_{\epsilon, \delta, p}$ satisfy the following criteria.

- $\cap \mathfrak{B} = \{0\}$.
- For any $U_1, U_2 \in \mathfrak{B}$ there exists $U_3 \in \mathfrak{B}$ such that $U_3 \subset U_1 \cap U_2$.
- Each set $U \in \mathfrak{B}$ is circled and absorbent.
- For any $U \in \mathfrak{B}$ there exists $\widetilde{U} \in \mathfrak{B}$ such that $\widetilde{U} + \widetilde{U} \subset U$.

Linear Hausdorff topologies are, of course, characterised by neighbourhood bases of 0 satisfying the above criteria [**Jar81**, Theorem 2.2.5]. Hence, both classes generate linear Hausdorff topologies on $\widetilde{\mathcal{M}}$.

Definition 3.77 The linear topologies generated by the bases of neighbourhoods of zero given by $\mathfrak{B}_{lc} = \{\mathscr{V}(\epsilon, \delta, p)\}_{\epsilon, \delta, p}$ and $\mathfrak{B}_{blc} = \{\mathscr{W}(\epsilon, \delta, p)\}_{\epsilon, \delta, p}$ are respectively defined to be the topologies τ_{lc} of τ-local convergence in measure, and τ_{blc} of bilateral τ-local convergence in measure.

The topologies of τ-local and bilateral τ-local convergence in measure were first introduced by Bikchentaev in [**Bik06**, **Bik04**]. Both were discussed at some length in [**Bik06**], with Bikchentaev using the term *weak τ-local convergence* for what we prefer to call *bilateral τ-local convergence*. The utility of especially bilateral τ-local convergence in measure for noncommutative Banach function spaces was first demonstrated by Dodds, Dodds and de Pagter in their 1993 paper [**DDdP89**] where this topology was introduced under the name of local convergence in measure. (Readers who wish to see a brief description of what is meant by the term 'noncommutative Banach function space', may jump forward to Subsection 7.2.1.) Since then, several further applications of this topology have been made, including the following.

- [**DDdP89**, Theorem 5.13] shows that a properly symmetric noncommutative Banach function space is contained in its Köthe bidual if and only if its closed unit ball is closed with respect to τ_{blc}.
- Both [**DDD$^+$95**] and [**DDS97**] are devoted to the Kadec–Klee property for symmetric Banach function spaces. The lasting impression left by the analysis in these papers is that one can only properly understand the noncommutative Kadec–Klee property in terms of the τ_{blc} topology. Consider, for example, [**DDD$^+$95**, Theorem 2.7] and [**DDS97**, Theorems 2.6, 2.8, 3.1 and 3.2, and Corollary 2.7].

- In [**DDST05**], the authors make the case that a proper noncommutative Yosida–Hewitt theorem should be understood in terms of the τ_{blc} topology. Here, Theorem 3.5 is the primary theorem, but Corollaries 4.4 and 4.5 are also very interesting. They respectively show that for a symmetric Banach function space, the unit ball is closed in τ_{blc} topology iff such a space has the Fatou property iff the unit ball is complete in the τ_{blc} topology.

Lemma 3.78 ([DDST05], Lemma 3.3) *For $\mathcal{N}(\epsilon, \delta)$ and $\mathcal{U}(\epsilon, \delta) = \{x \in \widetilde{\mathcal{M}} \colon \exists p \in \mathbb{P}(\mathcal{M})(\|pxp\| \leq \epsilon\delta \text{ and } \tau(p^{\perp}) \leq \delta)\}$, we have that*

$$\mathcal{N}(\epsilon, \delta) \subseteq \mathcal{U}(\epsilon, \delta) \subseteq \mathcal{N}(\epsilon, 2\delta).$$

If for some fixed projection p of finite trace we apply the above lemma to the collection of sets $\{\mathcal{W}(\epsilon, \delta, p) \cap p\widetilde{\mathcal{M}}p \colon \epsilon > 0, \delta > 0\}$, it is clear that this collection will produce the same topology on $p\widetilde{\mathcal{M}}p$ as the collection $\{\mathcal{N}_{p\mathcal{M}p}(\epsilon, \delta) \colon \epsilon > 0, \delta > 0\}$. (Here, $\{\mathcal{N}_{p\mathcal{M}p}(\epsilon, \delta) \colon \epsilon > 0, \delta > 0\}$ denotes the canonical neighbourhood base at 0 for the topology of convergence in measure on $p\widetilde{\mathcal{M}}p$ realised by $\tau_{|p\mathcal{M}p}$.) The topology τ_{blc} is, therefore, nothing more than the topology corresponding to convergence in measure in every compression $p\widetilde{\mathcal{M}}p$ of $\widetilde{\mathcal{M}}$ by a projection of finite trace.

Proof. It is clear that $\mathcal{N}(\epsilon, \delta) \subseteq \mathcal{U}(\epsilon, \delta)$, and so it only remains to prove that $\mathcal{U}(\epsilon, \delta) \subseteq \mathcal{N}(\epsilon, 2\delta)$. Let $x \in \widetilde{\mathcal{M}}$ be given, and let q_x be the orthogonal projection on the kernel of $q^{\perp}x$ and observe that

$$q_x^{\perp} = \mathbf{s}_l((q^{\perp}x)^*) \sim \mathbf{s}_l(q^{\perp}x) \leq q^{\perp}.$$

Consequently, $xq_x = qxq_x$ and $q_x^{\perp} \preceq q^{\perp}$. We obtain

$$x(q_x \wedge q) = qx(q_x \wedge q) = qxq(q_x \wedge q)$$

so that

$$\|x(q_x \wedge q)\|_{\infty} \leq \|qxq\|_{\infty} \leq \epsilon$$

and

$$\tau((q_x \wedge q)^{\perp}) = \tau(q_x^{\perp} \vee q^{\perp}) \leq \tau(q_x^{\perp}) + \tau(q^{\perp}) \leq 2\delta$$

which suffices to complete the proof. $\square$

We have seen that the topologies τ_{lc} and τ_{blc} are linear topologies with, especially, τ_{blc} proving to be a very useful tool in the theory of noncommutative Banach function spaces. One question we still need to answer is whether these topologies are also 'algebraic' in the same sense as the the topology of convergence. The answer seems to be 'yes, but not quite'. The fact of the matter is that multiplication proves to be separately continuous in both topologies but, in general, not jointly continuous. This fact goes some way towards explaining why these topologies do not also correspond to distinct classes of measurable operators. The claim regarding separate continuity is recorded in the following theorem.

Theorem 3.79 ([Bik04, Theorem 1]) *For a fixed y, both the mappings $x \mapsto xy$ and $x \mapsto yx$ are τ_{lc} and τ_{blc} continuous.*

As noted below, there are conditions under which joint continuity will hold on general semifinite von Neumann algebras, but as is recorded in the final theorem, it is precisely the class of finite von Neumann algebras for which multiplication is always jointly continuous.

Proposition 3.80 ([Bik06, Lemma 3.10]) *Suppose that $x, y, x_i,$ $y_i \in \widetilde{\mathcal{M}}$ for all $i \in I$.*

(1) *If $x_i \to x$ in measure and $y_i \to y$ τ-locally in measure, then $x_i y_i \to xy$ τ-locally in measure.*

(2) *Suppose that $(x_i)_{i \in I}$ is bounded in the topology of convergence in measure. Then, the following also holds.*

    (1) *If both $x_i \to x$ and $y_i \to y$ converge τ-locally in measure, then $x_i y_i \to xy$ τ-locally in measure.*

    (2) *If $x_i \to x$ bilaterally τ-locally in measure and $y_i \to y$ τ-locally in measure, then $x_i y_i \to xy$ bilaterally τ-locally in measure.*

Theorem 3.81 ([Bik06, Theorem 4.1]) *For a von Neumann algebra $\mathcal{M}$ with a faithful normal semifinite trace τ, the following conditions are equivalent.*

(1) *$\mathcal{M}$ is a finite von Neumann algebra.*

(2) *The topologies τ_{lc} and τ_{blc} agree.*

(3) *Multiplication on $\widetilde{\mathcal{M}}$ is jointly continuous in the topology τ_{lc}.*
(4) *Multiplication on $\widetilde{\mathcal{M}}$ is jointly continuous in the topology τ_{blc}.*
(5) *Involution on $\widetilde{\mathcal{M}}$ is continuous in the topology τ_{lc}.*

3.7 Order properties of measurable operators

Up to now, we have defined the order $\leq$ in a C^*-algebra, in the set of bounded and unbounded operators acting on a Hilbert space H, and another order $\widehat{\leq}$ in the set of generalised positive operators $\widehat{\mathcal{M}}_+$, associated with a von Neumann algebra $\mathcal{M}$. For semifinite von Neumann algebras and measurable operators associated with them, the natural order is given by the following.

Notation 3.82 For $a, b \in S(\mathcal{M})_h$, we write $a\,\overline{\leq}\,b$ if $b\overline{-}a \in S(\mathcal{M})_+$.

Lemma 3.83 *For $a, b \in S(\mathcal{M}, \tau)_+$ (or $a, b \in S(\mathcal{M})_+$), we have $a \leq b$ iff $a\,\overline{\leq}\,b$.*

Proof. We have $\mathrm{dom}(x) \subseteq \mathrm{dom}(x^{1/2})$ for any $x \in \overline{\mathcal{M}}_+$. If $a \leq b$, then $\langle a\xi, \xi \rangle \leq \langle b\xi, \xi \rangle$ for all $\xi \in \mathrm{dom}(b^{1/2}) \supseteq \mathrm{dom}(a) \cap \mathrm{dom}(b)$, so that $a\,\overline{\leq}\,b$. To prove the reverse implication, note that by Propositions 3.50 and 3.60(2), $\mathrm{dom}(a) \cap \mathrm{dom}(b)$ is a core for $b^{1/2}$. Choose $\xi \in \mathrm{dom}(b^{1/2})$. Let $\xi_n \in \mathrm{dom}(a) \cap \mathrm{dom}(b)$ be such that $\xi_n \to \xi$ and $b^{1/2}\xi_n \to b^{1/2}\xi$. Then, $\langle a(\xi_n - \xi_m), \xi_n - \xi_m \rangle \leq \langle b(\xi_n - \xi_m), \xi_n - \xi_m \rangle$ for all n, m, so that $\|a^{1/2}\xi_n - a^{1/2}\xi_m\| \leq \|b^{1/2}\xi_n - b^{1/2}\xi_m\|$ and $(a^{1/2}\xi_n)$ is a Cauchy sequence. Since $a^{1/2}$ is closed, we have $\xi \in \mathrm{dom}(a^{1/2})$. The inequality $\|a^{1/2}\xi\| \leq \|b^{1/2}\xi\|$ holds for all $\xi \in \mathrm{dom}(a) \cap \mathrm{dom}(b)$, and therefore, since $\mathrm{dom}(a) \cap \mathrm{dom}(b)$ is the core of $b^{1/2}$, the inequality must hold for $\xi \in \mathrm{dom}(b^{1/2})$ as well. $\qquad\square$

Corollary 3.84 *The inclusions $\widetilde{\mathcal{M}}_+ = S(\mathcal{M}, \tau)_+ \hookrightarrow S(\mathcal{M})_+ \hookrightarrow \overline{\mathcal{M}}_+ \hookrightarrow \widehat{\mathcal{M}}_+$ are order-preserving.*

It should be noted that one of the consequences of Lemma 3.83 is that in order to prove that $a \leq b$ for (τ-) measurable a, b, it suffices to show that $\langle a\xi, \xi \rangle \leq \langle b\xi, \xi \rangle$ for $\xi \in \mathrm{dom}(a) \cap \mathrm{dom}(b)$.

We will now show that all the classes of measurable operators that we have considered are hereditary with respect to the order.

Proposition 3.85 *If $b \in \widetilde{\mathcal{M}}_+ = S(\mathcal{M},\tau)_+$ (respectively, $b \in S(\mathcal{M})_+$, $b \in LS(\mathcal{M},\tau)_+$, $b \in LS(\mathcal{M})_+$) and $a \in \overline{\mathcal{M}}_+$ is such that $a \leq b$, then $a \in \widetilde{\mathcal{M}}_+ = S(\mathcal{M},\tau)_+$ (respectively, $a \in S(\mathcal{M})_+$, $a \in LS(\mathcal{M},\tau)_+$, $a \in LS(\mathcal{M})_+$).*

Proof. Since $b \in \widetilde{\mathcal{M}}_+$ (respectively, $b \in S(\mathcal{M})_+$), the domain of $b^{1/2}$ is τ-dense (respectively, strongly dense), so that the domain of $a^{1/2}$ is τ-dense (respectively, strongly dense). Consequently, $a = (a^{1/2})^2$ is τ-measurable (respectively, measurable). If $b \in LS(\mathcal{M},\tau)_+$ (respectively, $b \in LS(\mathcal{M})_+$), then for some sequence (z_n) of central projections from $\mathcal{M}$ increasing to $\mathbb{1}$, $bz_n \in S(\mathcal{M},\tau)_+$ (respectively, $bz_n \in S(\mathcal{M})_+$). The first part of the proof shows that $az_n \in S(\mathcal{M},\tau)_+$ (respectively, $az_n \in S(\mathcal{M})_+$), so that $a \in LS(\mathcal{M},\tau)_+$ (respectively, $a \in LS(\mathcal{M})_+$). $\qquad\square$

Proposition 3.86 *Let $b \in \widetilde{\mathcal{M}}_+$ (respectively, $b \in S(\mathcal{M})_+$) and let (a_i) be an increasing net in $\widetilde{\mathcal{M}}_+$ (respectively, $S(\mathcal{M})_+$) with $a_i \leq b$ for all i. Then, $a := \sup_i a_i$ exists in $\widetilde{\mathcal{M}}$ (respectively, $S(\mathcal{M})$). Moreover, we have $\mathrm{dom}(a^{1/2}) = \{\xi\colon \sup_i \|a_i^{1/2}\xi\| < \infty\}$ and $\|a^{1/2}\xi\| = \sup_i \|a_i^{1/2}\xi\|$ for $\xi \in \mathrm{dom}(a^{1/2})$.*

Proof. Let $m \in \widehat{\mathcal{M}}_+$ be the supremum of m_{a_i}. Then, $m \leq m_b$, and so by Proposition 0.150, there is an $a \in \widetilde{\mathcal{M}}_+$ (respectively, $a \in S(\mathcal{M})_+$) such that $m = m_a$. Evidently, $a = \sup_i a_i$. We have $\mathrm{dom}(a) = \{\xi \in H\colon m_a(\omega_\xi) < \infty\} = \{\xi \in H\colon \sup_i m_{a_i}(\omega_\xi) < \infty\} = \{\xi \in H\colon \sup_i \|a_i^{1/2}\xi\| < \infty\}$. The equality $\|a^{1/2}\xi\| = \sup_i \|a_i^{1/2}\xi\|$ follows from $\|a^{1/2}\xi\| = m_a(\omega_\xi) = \sup_i m_{a_i}(\omega_\xi) = \sup_i \|a_i^{1/2}\xi\|$. $\qquad\square$

Lemma 3.87 *If $a,b \in LS(\mathcal{M})_+$ with $a \leq b$ and $d \in LS(\mathcal{M})$, then $d^{*\cdot}a^{\cdot}d \leq d^{*\cdot}b^{\cdot}d$.*

Proof. We need to check that

$$\mathrm{dom}(|b^{1/2\cdot}d|) \subseteq \mathrm{dom}(|a^{1/2\cdot}d|) \quad\text{and}\quad \||b^{1/2\cdot}d|\xi\| \leq \||b^{1/2\cdot}d|\xi\|$$

for all $\xi \in \mathrm{dom}(|b^{1/2\cdot}d|)$. Since $\mathrm{dom}(b^{1/2}) \subseteq \mathrm{dom}(a^{1/2})$, we have that

$$\mathrm{dom}(b^{1/2}d) \subseteq \mathrm{dom}(a^{1/2}d) \subseteq \mathrm{dom}(|a^{1/2\cdot}d|)$$

and
$$\|b^{1/2}d\xi\| \leq \|a^{1/2}d\xi\| \leq \||b^{1/2\,\overline{\cdot}}d|\xi\|$$

for $\xi \in \mathrm{dom}(b^{1/2}d)$. If (ξ_n) is a sequence from $\mathrm{dom}(b^{1/2}d)$ convergent to $\xi \in \mathrm{dom}(b^{1/2\,\overline{\cdot}}d)$ and $b^{1/2}d\xi_n \to (b^{1/2\,\overline{\cdot}}d)\xi$, then $(a^{1/2}d\xi_n)$ is Cauchy in H. Thus, since $(a^{1/2\,\overline{\cdot}}d)$ is closed, we have $\xi \in \mathrm{dom}(a^{1/2\,\overline{\cdot}}d)$. The required norm inequality now easily follows from the convergence of $(a^{1/2}d\xi_n)$ to $(a^{1/2\,\overline{\cdot}}d)\xi$. $\qquad\square$

Proposition 3.88 *If we have* $a_i, a \in \widetilde{\mathcal{M}}_+$ *(respectively,* $a_i, a \in S(\mathcal{M})_+$*) with* $a_i \nearrow a$*, then for all* $b \in \widetilde{\mathcal{M}}$ *(respectively,* $b \in S(\mathcal{M})$*),* $b^{*\,\overline{\cdot}}a_i\overline{\cdot}b \nearrow b^{*\,\overline{\cdot}}a\overline{\cdot}b$.

Proof.   It follows from Lemma 3.87 that $b^{*\,\overline{\cdot}}a_i\overline{\cdot}b \nearrow d$ for some $d \in \widetilde{\mathcal{M}}_+$ (respectively, $d \in S(\mathcal{M})_+$), and by Proposition 3.86, we have $d \leq b^{*\,\overline{\cdot}}a\overline{\cdot}b$,

$$\|(a_i^{1/2\,\overline{\cdot}}b)\xi\| = \|(b^{*\,\overline{\cdot}}a_i\overline{\cdot}b)^{1/2}\xi\| \nearrow \|d^{1/2}\xi\|\text{for all } \xi \in D(d^{1/2})$$

and

$$\|a_i^{1/2}b\xi\| \nearrow \|a^{1/2}b\xi\|\text{for all } \xi \in \mathrm{dom}(a^{1/2}b) \subseteq \mathrm{dom}((b^{*\,\overline{\cdot}}a\overline{\cdot}b)^{1/2})$$

$$\subseteq D(d^{1/2}).$$

Note that for all i and all $\xi \in \mathrm{dom}(a^{1/2}b)$, $\mathrm{dom}(a^{1/2}b) \subseteq \mathrm{dom}(a_i^{1/2}b)$ and

$$\|(b^{*\,\overline{\cdot}}a\overline{\cdot}b)^{1/2}\xi\| = \||a_i^{1/2\,\overline{\cdot}}b|\xi\| = \|a_i^{1/2}b\xi\|.$$

Hence, $\|d^{1/2}\xi\| = \|a^{1/2\,\overline{\cdot}}b\xi\|$ for all $\xi \in \mathrm{dom}(a^{1/2}b)$, which implies $\langle d\xi, \xi\rangle = \langle(b^{*\,\overline{\cdot}}a\overline{\cdot}b)\xi, \xi\rangle$ for all $\xi \in \mathrm{dom}(b^*ab)$. This means that the τ-measurable (respectively, measurable) operators $b^{*\,\overline{\cdot}}a\overline{\cdot}b$ and d are equal on a τ-dense (respectively, strongly dense) domain. By Proposition 3.71(1), $b^{*\,\overline{\cdot}}a\overline{\cdot}b = d$. $\qquad\square$

As an easy corollary, we get consistency of notation when an operator is treated as both a measurable and a generalised operator.

Corollary 3.89 *If* $d \in \mathcal{M}$ *and* $a \in \widetilde{\mathcal{M}}_+$ *(respectively,* $a \in S(\mathcal{M})_+$*), then*

$$d^* m_a d = m_{d^{*\,\overline{\cdot}}a\overline{\cdot}d}.$$

Proof. Let (a_n) be a sequence from $\mathcal{M}$ such that $a_n \nearrow a$. Then $d^* m_{a_n} d \nearrow d^* m_a d$ and $d^* a_n d \nearrow d^{*-}\bar{a}\bar{d}$. We have $d^* m_{a_n} d = m_{d^* a_n d} \nearrow m_{d^{*-}\bar{a}\bar{d}}$, and so the result follows from Lemma 0.151. $\qquad\square$

Proposition 3.90 ([Sch86, 2.2D]) *If $a, b \in \widetilde{\mathcal{M}}_+$ (respectively, $a, b \in S(\mathcal{M})_+$) with $a \leq b$, then there exists an operator $d \in \mathcal{M}_+$ with $d \leq \mathbf{s}(b)$ such that $a = b^{1/2} d b^{1/2}$.*

Proof. Assume first that $\mathbf{s}(b) = \mathbb{1}$. By assumption, $\mathrm{dom}(b^{1/2}) \subseteq \mathrm{dom}(a^{1/2})$. By definition (see Definition 0.23(7)), we have the equality $\mathrm{dom}(b^{1/2}) = b^{-1/2}\mathrm{dom}(b^{-1/2})$. For $\xi \in \mathrm{dom}(b^{1/2})$ put $x(b^{1/2}\xi) := a^{1/2}\xi$. Then, $x = a^{1/2}b^{-1/2}$ satisfies $\|x\xi\| \leq 1$ for all ξ in a dense subspace $\mathrm{dom}(x) = \mathrm{dom}(b^{-1/2})$. Thus, $[x]$ is bounded. Since $x \in {}^\eta\mathcal{M}$ (see Lemma 0.134), we have $[x] \in \mathcal{M}$. Thus, $[x]b^{1/2} = a^{1/2}$ on a τ-dense (respectively, strongly dense) domain $\mathrm{dom}(b^{1/2})$, and hence, they are equal (see Proposition 3.60). By Proposition 0.31(7), $([x]b^{1/2})^* = b^{1/2}[x]^*$, so that, with $d := [x]^*[x] \in \mathcal{M}_+$, we get $d \leq \mathbb{1}$ and $a = b^{1/2} d b^{1/2}$.

If $\mathbf{s}(b)$ is not equal to $\mathbb{1}$, we can repeat the proof on the Hilbert subspace $\mathbf{s}(b)H$, and note that a must vanish on the orthogonal complement of the subspace. $\qquad\square$

Lemma 3.91 *Let $a \in \widetilde{\mathcal{M}}_+$ be given with $a \in \mathcal{N}(\epsilon, \delta)$. Then $a^{1/2} \in \mathcal{N}(\sqrt{\epsilon}, \delta)$*

Proof. Note that if $a \in \mathcal{N}(\epsilon, \delta)$, then of course $\tau(\chi_{(\epsilon,\infty)}(a)) \leq \delta$. Since by the Borel functional calculus $\chi_{(\epsilon,\infty)}(a) = \chi_{(\sqrt{\epsilon},\infty)}(a^{1/2})$, we, in fact, have $a^{1/2} \in \mathcal{N}(\sqrt{\epsilon}, \delta)$ $\qquad\square$

Lemma 3.92 *Let $a, b \in \widetilde{\mathcal{M}}_+$ with $a \leq b$. If $b \in \mathcal{N}(\epsilon, \delta)$, then also $a \in \mathcal{N}(\epsilon, \delta)$.*

Proof. Assume that $b \in \mathcal{N}(\epsilon, \delta)$. If $0 \neq \xi \in \chi_{(\epsilon,\infty)}(a)H$, then by Lemmas 3.53 and 3.91, $\|b^{1/2}\xi\| \geq \|a^{1/2}\xi\| > \epsilon^{1/2}\|\xi\|$. If, on the other hand, $\xi \in \chi_{[0,\epsilon]}(b) = \chi_{[0,\epsilon^{1/2}]}(b^{1/2})$, then $\|b^{1/2}\xi\| \leq \epsilon^{1/2}\|\xi\|$. Hence, $\chi_{(\epsilon,\infty)}(a) \wedge \chi_{[0,\epsilon]}(b)) = 0$, which by Lemma 3.52(1) yields $\tau(\chi_{(\epsilon,\infty)}(b)) \leq \tau(\chi_{(\epsilon,\infty)}(a)) \leq \delta$, so that $b \in \mathcal{N}(\epsilon, \delta)$. $\qquad\square$

Proposition 3.93 *The positive cone $\widetilde{\mathcal{M}}_+$ is closed in measure in $\widetilde{\mathcal{M}}$.*

Proof. Since $\widetilde{\mathcal{M}}$ is metrisable in the measure topology, it is enough to show that if a sequence (a_n) from $\widetilde{\mathcal{M}}_+$ converges to a, then $a \in \widetilde{\mathcal{M}}_+$. Since the adjoint operation is continuous, we must have $a \in \widetilde{\mathcal{M}}_h$. Assume that $a \notin \widetilde{\mathcal{M}}_+$. Then $a = \int_{\mathbb{R}} \lambda de_\lambda$ with $\int_{(-\infty,-\epsilon]} \lambda de_\lambda \neq 0$ for some $\epsilon > 0$. Put $p := \chi_{(-\infty,-\epsilon]}(a)$. Since $pa_n p \to pap$, we can take pap in place of a and assume that $a \leq -\epsilon \mathbb{1}$. Now put $\delta := \min(1, \tau(1))$ If $b \in \widetilde{\mathcal{M}}_+$ belongs to $a + N(\epsilon/2, \delta/2)$, then by Lemma 3.73(1),

$$\delta/2 \geq \tau(\chi_{(\epsilon/2,\infty)}(|b - a|)) = \tau(\chi_{(\epsilon/2,\infty)}(b - a)) \geq \tau(\chi_{(\epsilon/2,\infty)}(b + \epsilon\mathbb{1}))$$
$$= \tau(1)$$

which is impossible. Therefore, we cannot have $a_n \to a$, and consequently the assumption that a is not positive leads to a contradiction. $\qquad\qquad\square$

Lemma 3.94 *Let (a_n) be a sequence in $\widetilde{\mathcal{M}}_+$ that converges to 0 in measure. Then, there exists a subsequence (a_{n_k}) of (a_n) and some $b \in \widetilde{\mathcal{M}}_+$ such that $2^k a_{n_k} \leq b$ for all $k \in \mathbb{N}$.*

Proof. Suppose that (a_n) is a sequence in $\widetilde{\mathcal{M}}_+$ converging to 0 in the topology of convergence in measure. Observe that the collection of sets $U_n = \mathcal{N}(2^{-n}, 2^{-n})$ $(n \in \mathbb{N})$ constitutes a countable neighbourhood base at 0 for the topology of convergence in measure on $\widetilde{\mathcal{M}}$. The convergence of (a_n) to 0 in measure now ensures that we can select natural numbers n_k such that $a_n \in 2^{-k}U_k$ for each $n \geq n_k$. Since for each k we clearly have that $2^{-(k+1)}U_{k+1} \subseteq 2^{-k}U_k$, we may select the n_ks to be increasing. The subsequence we seek is then (a_{n_k}). We proceed to show that this sequence satisfies the hypothesis of the lemma. First, note that by construction $2^k x_{n_k} \in U_k$ for each $k \in \mathbb{N}$. Now consider the sequence $w_m = \sum_{k=1}^m 2^k x_{n_k}$ $(m \in \mathbb{N})$. Let $\mathcal{N}$ be an arbitrary neighbourhood of 0 in the measure topology. For N large enough, we will then have $U_m \subseteq \mathcal{N}$ for each $m \geq N$. Observe that by Lemma 3.74(5), we have

$$w_{m+n} - w_m = \sum_{k=m+1}^{m+n} 2^k x_{n_k}$$
$$\in \sum_{k=m+1}^{m+n} \mathcal{N}(2^{-k}, 2^{-k})$$

$$\subseteq \;\; \mathcal{N}\Big(\sum_{k=m+1}^{m+n} 2^{-k}, \; \sum_{k=m+1}^{m+n} 2^{-k} \Big)$$

$$\subseteq \;\; \mathcal{N}(2^{-m}, 2^{-m}) = U_m$$

$$\subseteq \;\; \mathcal{N}$$

whenever $m \geq N$. This clearly shows that (w_n) is a Cauchy sequence in $\widetilde{\mathcal{M}}$. By the completeness of $\widetilde{\mathcal{M}}$, (w_n) must converge to some $b \in \widetilde{\mathcal{M}}$. But since (w_m) is by construction an increasing sequence in $\widetilde{\mathcal{M}}_+$, we not only have that $b \geq 0$ (by the closedness of $\widetilde{\mathcal{M}}_+$ in the topology of convergence in measure), but also that $w_m \leq b$ for all m. This in particular also ensures that $2^k x_{n_k} \leq b$ for all $k \in \mathbb{N}$, as required. $\qquad\square$

Corollary 3.95 *Let $\mathcal{M}$ and $\mathcal{N}$ be semifinite von Neumann algebras endowed with f.n.s. traces $\tau_\mathcal{M}, \tau_\mathcal{N}$. Assume that $T : \widetilde{\mathcal{M}} \to \widetilde{\mathcal{N}}$ is a positive map, and let $(a_n) \subseteq \widetilde{\mathcal{M}}_+$ be a sequence converging to 0 in $\tau_\mathcal{M}$-measure. Then, there exists a subsequence (a_{n_k}) of (a_n) such that $T(a_{n_k}) \to 0$ in $\tau_\mathcal{N}$-measure.*

Proof. By Lemma 3.94, there exists a subsequence (a_{n_k}) of (a_n) and some $b \in \widetilde{\mathcal{M}}_+$ such that $2^k a_{n_k} \leq b$ for all $k \in \mathbb{N}$. The positivity of T then ensures that $0 \leq T(a_{n_k}) \leq 2^{-k}T(b)$ for all $k \in \mathbb{N}$. The sequence $(2^{-k}T(b))$ trivially converges to 0 in $\tau_\mathcal{N}$-measure. By Lemma 3.92, the same must then be true of $(T(a_{n_k}))$. $\qquad\square$

3.8 Jordan morphisms on $\widetilde{\mathcal{M}}$

Proposition 3.96 *Let $\mathcal{M}_1$ and $\mathcal{M}_2$ be semifinite von Neumann algebras with f.n.s. traces τ_1 and τ_2, respectively, and let $\mathscr{J} : \widetilde{\mathcal{M}}_1 \to \widetilde{\mathcal{M}}_2$ be a Jordan *-morphism (see Remark 0.51). Then, $\mathscr{J}$ is positivity preserving. Moreover $\mathscr{J}$ maps $\mathcal{M}_1$ into $\mathcal{M}_2$ in a uniformly continuous manner.*

Proof. Since $\mathscr{J}$ preserves squares of self-adjoint elements of $\widetilde{\mathcal{M}}_1$, it easily follows from Lemma 3.91 that it is positivity preserving with $\mathscr{J}(\mathbb{1})$ a projection commuting with all elements of $\mathscr{J}(\widetilde{\mathcal{M}}_1)$ since

$$\mathscr{J}(a) = \mathscr{J}(\mathbb{1})\,\mathscr{J}(a)\,\mathscr{J}(\mathbb{1}) \text{ for all } a \in \widetilde{\mathcal{M}}_1.$$

For any $a \in \mathcal{M}_1^+$, we of course have $0 \leq a \leq \|a\|\mathbb{1}$. It, therefore, follows from what we have just noted that then $0 \leq \mathscr{J}(a) \leq \|a\|\mathscr{J}(\mathbb{1})$.

Therefore, $\mathscr{J}(a) \in \mathcal{M}_2^+$ with $\|\mathscr{J}(a)\| \leq \|a\| < \infty$ whenever $a \in \mathcal{M}_1^+$. We then clearly have that $\mathscr{J}(\mathcal{M}_1) \subseteq \mathcal{M}_2$ and also that the stated continuity claim holds. $\qquad\square$

Corollary 3.97 *Let $\mathcal{M}_1$ and $\mathcal{M}_2$ be semifinite von Neumann algebras with f.n.s. traces τ_1 and τ_2, respectively, and let $\mathscr{J} : \widetilde{\mathcal{M}}_1 \to \widetilde{\mathcal{M}}_2$ be a Jordan *-morphism. Then, $\mathscr{J}$ maps projections in $\mathcal{M}_1$ onto projections in $\mathcal{M}_2$.*

Proof. Easy consequence of the above proposition. $\qquad\square$

Proposition 3.98 *Let $\mathcal{M}_1$ and $\mathcal{M}_2$ be semifinite von Neumann algebras with f.n.s. traces τ_1 and τ_2, respectively, and let $\mathscr{J} : \widetilde{\mathcal{M}}_1 \to \widetilde{\mathcal{M}}_2$ be a Jordan *-morphism. Then, $\mathscr{J}$ is continuous in measure if and only if $\mathscr{J}$ is the continuous (in measure) extension of a Jordan *-morphism $\mathscr{J}_0 : \mathcal{M}_1 \to \mathcal{M}_2$ for which $\tau_2 \circ \mathscr{J}_0$ is $\epsilon - \delta$ absolutely continuous with respect to τ_1, in the sense that for every $\epsilon > 0$, there exists a $\delta > 0$ such that $\tau_2(\mathscr{J}_0(e)) \leq \epsilon$ whenever $e \in \mathcal{M}_1$ is a projection satisfying $\tau_1(e) \leq \delta$.*

Proof. Suppose that $\mathscr{J}$ is continuous with respect to the topology of convergence in measure. We know from Proposition 3.96 that $\mathscr{J}$ maps $\mathcal{M}_1$ into $\mathcal{M}_2$ in a uniformly continuous manner with norm $\|\mathscr{J}\|$. So, we only need to verify the claim about absolute continuity. Given a basic neighbourhood of 0 of the form

$$\mathscr{N}_2(\epsilon_2, \epsilon) = \{a \in \widetilde{\mathcal{M}}_2 : \tau_2(\chi_{(\epsilon_2, \infty)}(|a|)) \leq \epsilon\},$$

there exists $\epsilon_1, \delta > 0$ so that $\mathscr{J}$ maps $\mathscr{N}_1(\epsilon_1, \delta) \cap \mathcal{M}_1$, where

$$\mathscr{N}_1(\epsilon_1, \delta) = \{a \in \widetilde{\mathcal{M}}_1 : \tau_1(\chi_{(\epsilon_1, \infty)}(|a|)) \leq \delta\}$$

into $\mathscr{N}_2(\epsilon_2, \epsilon)$. So, given a non-zero projection $e \in \mathcal{M}_1$ with $\tau_1(e) \leq \delta$ it is clear that $(1 + \epsilon_2)e \in \mathscr{N}_1(\epsilon_1, \delta)$ and hence that $(1 + \epsilon_2)\mathscr{J}(e) \in \mathscr{N}_2(\epsilon_2, \epsilon)$. Since $\mathscr{J}(e)$ is again a projection, this is sufficient to imply $\tau_2(\mathscr{J}(e)) \leq \epsilon$.

Conversely, suppose that we are given a Jordan morphism $\mathscr{J}_0 : \mathcal{M}_1 \to \mathcal{M}_2$ such that $\tau_2 \circ \mathscr{J}_0$ is $\epsilon - \delta$ absolutely continuous with respect to τ_1. As a positive map, $\mathscr{J}_0$ is of course uniformly continuous with norm say $\|\mathscr{J}_0\|$. Let $0 < \tilde{\epsilon}, \tilde{\delta}$ be given and select $\delta > 0$ so that

$\tau_2(\mathscr{J}_0(e)) < \tilde{\delta}$ whenever $e \in \mathcal{M}_1$ is a projection with $\tau_1(e) < 2\delta$. We show that then

$$\mathscr{J}_0(\mathscr{N}_1(\epsilon, \delta) \cap \mathcal{M}_1) \subseteq \mathscr{N}_2(\tilde{\epsilon}, \tilde{\delta})$$

whenever $\sqrt{2\|\mathscr{J}_0\|}\epsilon \leq \tilde{\epsilon}$. This will be sufficient to establish the continuity of $\mathscr{J}_0$ (at 0). Thus, let $a \in \mathscr{N}_1(\epsilon, \delta) \cap \mathcal{M}_1$ be given. Then, of course, $a^* \in \mathscr{N}_1(\epsilon, \delta)$. Consequently, we may find projections $e, f \in \mathcal{M}_1$ with

$$\tau_1(e), \tau_1(f) < \delta \quad \text{and} \quad \|a(\mathbb{1} - e)\|, \|a^*(\mathbb{1} - f)\| \leq \epsilon.$$

Now let $g = e \vee f$. Then, since τ_1 is a trace, $\tau_1(g) \leq \tau_1(e) + \tau_1(f) < 2\delta$ and hence $\tau_2(\mathscr{J}_0(g)) < \tilde{\delta}$ by the τ_1-absolute continuity of $\tau_2 \circ \mathscr{J}_0$. It remains to show that $\|\mathscr{J}_0(a)(\mathbb{1} - \mathscr{J}_0(g))\| \leq \sqrt{2\|\mathscr{J}_0\|}\epsilon$. Since $\mathbb{1} - g = \mathbb{1} - (e \vee f) = (\mathbb{1} - e) \wedge (\mathbb{1} - f)$ is majorised by both $\mathbb{1} - e$ and $\mathbb{1} - f$, it follows that $\|a(\mathbb{1} - g)\| = \|a(\mathbb{1} - e)(\mathbb{1} - g)\| \leq \|a(\mathbb{1} - e)\| \leq \epsilon$ and similarly that $\|a^*(\mathbb{1} - g)\| \leq \epsilon$.

We pause to note that $\mathscr{J}(xyx) = \mathscr{J}(x)\mathscr{J}(y)\mathscr{J}(x)$ for all $x, y \in \mathcal{M}$. This can be seen by first noting that one may, by direct computation, conclude from the equality $\mathscr{J}(x + y)^3 = \mathscr{J}((x + y)^3)$ that $\mathscr{J}(xyx + yxy) = \mathscr{J}(x)\mathscr{J}(y)\mathscr{J}(x) + \mathscr{J}(y)\mathscr{J}(x)\mathscr{J}(y)$, and similarly from $\mathscr{J}(x - y)^3 = \mathscr{J}((x - y)^3)$ that $\mathscr{J}(xyx - yxy) = \mathscr{J}(x)\mathscr{J}(y)\mathscr{J}(x) - \mathscr{J}(y)\mathscr{J}(x)\mathscr{J}(y)$. When combined with positivity, this identity then enables us to conclude that

$$
\begin{aligned}
|\mathscr{J}_0(a)(\mathbb{1} - \mathscr{J}_0(g))|^2 &= (\mathbb{1} - \mathscr{J}_0(g))|\mathscr{J}_0(a)|^2(\mathbb{1} - \mathscr{J}_0(g)) \\
&\leq (\mathbb{1} - \mathscr{J}_0(g))(|\mathscr{J}_0(a)|^2 + |\mathscr{J}_0(a^*)|^2) \\
&\quad (\mathbb{1} - \mathscr{J}_0(g)) \\
&= (\mathbb{1} - \mathscr{J}_0(g))(\mathscr{J}_0(|a|^2 + |a^*|^2))(\mathbb{1} - \mathscr{J}_0(g)) \\
&= \mathscr{J}_0((\mathbb{1} - g)(|a|^2 + |a^*|^2)(\mathbb{1} - g)) \\
&= \mathscr{J}_0(|a(\mathbb{1} - g)|^2 + |a^*(\mathbb{1} - g)|^2).
\end{aligned}
$$

The fact that both $a(\mathbb{1} - g)$ and $a^*(\mathbb{1} - g)$ are bounded will, when combined with the above inequality, ensure that $|\mathscr{J}_0(a)(\mathbb{1} - \mathscr{J}_0(g))|^2$ is bounded, and hence also $\mathscr{J}_0(a)(\mathbb{1} - \mathscr{J}_0(g))$. In fact, it follows from the above that

$$
\begin{aligned}
\|\mathscr{J}_0(a)(\mathbb{1} - \mathscr{J}_0(g))\|^2 \;&=\; \||\mathscr{J}_0(a)(\mathbb{1} - \mathscr{J}_0(g))|^2\| \\
&\leq\; \|\mathscr{J}_0(|a(\mathbb{1} - g)|^2 + |a^*(\mathbb{1} - g)|^2)\| \\
&\leq\; \|\mathscr{J}_0\|(\||a(\mathbb{1} - g)|^2\| + \||a^*(\mathbb{1} - g)|^2\|) \\
&\leq\; 2\|\mathscr{J}_0\|\epsilon.
\end{aligned}
$$

Since both $\widetilde{\mathcal{M}}_1$ and $\widetilde{\mathcal{M}}_2$ are complete linear metric spaces with $\mathcal{M}_1$ dense in $\widetilde{\mathcal{M}}_1$, $\mathscr{J}_0$ then allows for a continuous extension of $\mathscr{J}_0$ to all of $\widetilde{\mathcal{M}}_1$. It is an exercise to see that the extension is still a Jordan *-morphism. $\qquad\square$

In closing, we present an automatic continuity result of Weigt [**Wei09**], effectively showing that all Jordan morphisms $\mathscr{J} : \widetilde{\mathcal{M}}_1 \to \widetilde{\mathcal{M}}_2$ are continuous extensions of Jordan morphisms $\mathscr{J}_0 : \mathcal{M}_1 \to \mathcal{M}_2$.

Definition 3.99 For a linear map $T : \mathcal{A} \to \mathcal{B}$ between linear metric spaces $\mathcal{A}$ and $\mathcal{B}$, we define the *separating space* $\mathcal{S}(T, \mathcal{B})$ to be

$$
\mathcal{S}(T, \mathcal{B}) = \{b \in \mathcal{B} : x_n \to 0 \text{ and } T(x_n) \to b \text{ for some sequence } (x_n) \subseteq \mathcal{A}\}.
$$

It can be easily verified that $\mathcal{S}(T, \mathcal{B})$ is a vector subspace of $\mathcal{B}$. The following version of Closed Graph Theorem not closed graph theorem is valid in this context:

Theorem 3.100 ([KN63, p.101]) *Let $T : \mathcal{A} \to \mathcal{B}$ be a linear map between complete linear metric spaces $\mathcal{A}$ and $\mathcal{B}$. Then, T is continuous whenever $\mathcal{S}(T, \mathcal{B}) = \{0\}$.*

We finally arrive at the promised automatic continuity result. Alongside Proposition 3.98, this then provides a complete depiction of the nature of Jordan *-morphisms from $\widetilde{\mathcal{M}}_1$ to $\widetilde{\mathcal{M}}_2$.

Theorem 3.101 *Every Jordan *-morphism $\widetilde{\mathscr{J}} : \widetilde{\mathcal{M}}_1 \to \widetilde{\mathcal{M}}_2$ is automatically continuous with respect to the topologies of convergence in measure and is, therefore, the continuous extension of a Jordan *-morphism $\mathscr{J}$ from $\mathcal{M}_1$ to $\mathcal{M}_2$ for which $\tau_2 \circ \mathscr{J}$ is $\epsilon - \delta$ absolutely continuous with respect to τ_1.*

Proof.   Once the first claim has been established, the second will trivially follow from Propositions 3.96 and 3.98. Therefore, we only need to prove the first claim. The Closed Graph Theorem not closed graph theorem is known to hold for complete linear metric spaces [**KN63**, p 101]. Having established earlier that $\widetilde{\mathcal{M}}_1$ and $\widetilde{\mathcal{M}}_2$ belong to this category of spaces, the proof of automatic continuity consists of nothing more than showing that $\widetilde{\mathscr{I}}$ fulfils the prerequisites of this particular flavour of the Closed Graph Theorem not closed graph theorem. This amounts to showing that $\mathcal{S}(\widetilde{\mathscr{I}}, \widetilde{\mathcal{M}}_2) = \{0\}$ where $\mathcal{S}(\widetilde{\mathscr{I}}, \widetilde{\mathcal{M}}_2)$ is the set consisting of all elements $b \in \widetilde{\mathcal{M}}_2$, for which we can find a sequence $(a_n) \subseteq \widetilde{\mathcal{M}}_1$ such that $a_n \to 0$ and $\widetilde{\mathscr{I}}(a_n) \to b$ in measure. So, let $b \in \mathcal{S}(\widetilde{\mathscr{I}}, \widetilde{\mathcal{M}}_2)$ be given and let $(a_n) \subseteq \widetilde{\mathcal{M}}_1$ be a sequence such that $a_n \to 0$ and $\widetilde{\mathscr{I}}(a_n) \to b$ in measure. We will show that $b = 0$. Notice that we now have that $\mathrm{Re}(a_n) = \frac{1}{2}(a_n + a_n^*) \to 0$ and $\widetilde{\mathscr{I}}(\mathrm{Re}(a_n)) = \frac{1}{2}(\widetilde{\mathscr{I}}(a_n) + \widetilde{\mathscr{I}}(a_n)^*) \to \mathrm{Re}(b)$, and similarly that $\mathrm{Im}(a_n) \to 0$ and $\widetilde{\mathscr{I}}(\mathrm{Im}(a_n)) \to \mathrm{Im}(b)$. Thus, for the task of verifying that $b = 0$, we may without loss of generality assume that $b = b^*$, and $a_n = a_n^*$ for all $n \in \mathbf{N}$. By the continuity of multiplication and the fact that $\widetilde{\mathscr{I}}$ preserves squares of self-adjoint elements, we have that $|a_n|^2 = a_n^2 \to 0$ and $\widetilde{\mathscr{I}}(a_n^2) = \widetilde{\mathscr{I}}(a_n)^2 \to |b|^2 = b^2$ in measure. But, by Corollary 3.95, there exists a subsequence (a_{n_k}) of (a_n) such that $\widetilde{\mathscr{I}}(a_{n_k}^2) \to 0$ in measure. For this subsequence, we surely still have $\widetilde{\mathscr{I}}(a_{n_k}^2) \to |b|^2$. Thus, the uniqueness of the limit ensures that $|b|^2 = 0$, or equivalently that $b = 0$, as required. $\square$

Chapter 4
Weights and densities

In this chapter, we provide further foundational material by putting in place a noncommutative theory of measures for the non-tracial case. The basic strategy employed here is to regard a weight on a von Neumann algebra as a noncommutative version of an integral with respect to some measure. We start by presenting the basic theory of normal weights (including a comparison of the different notions of semifiniteness) before addressing the issue of existence of faithful normal semifinite (f.n.s.) weights and then showing that such weights extend to the class of generalised positive operators. That then forms the foundation for proving a noncommutative Radon–Nikodym theorem for semifinite normal weights on semifinite algebras.

4.1 Weights

Let $\mathcal{M}$ be a von Neumann algebra acting on a Hilbert space H.

Definition 4.1 A map $\varphi : \mathcal{M}_+ \to [0, \infty]$ is a *weight* on $\mathcal{M}$ if:

(1) $\varphi(a + b) = \varphi(a) + \varphi(b)$ for all $a, b \in \mathcal{M}_+$;
(2) $\varphi(\lambda a) = \lambda \varphi(a)$ for all $a \in \mathcal{M}_+$, $\lambda \geq 0$.

A weight φ is *finite* if $\varphi(\mathbb{1}) < \infty$.

Notation 4.2 For a weight φ on $\mathcal{M}$, we have

$$N_\varphi := \{a \in \mathcal{M} : \varphi(a^*a) = 0\}$$

$$\mathfrak{p}_\varphi := \{a \in \mathcal{M}_+ : \varphi(a) < \infty\}$$

$$\mathfrak{n}_\varphi := \{a \in \mathcal{M} : \varphi(a^*a) < \infty\}$$

$$\mathfrak{m}_\varphi := \text{linear span of } \mathfrak{p}_\varphi.$$

Noncommutative measures and L^p and Orlicz Spaces, with Applications to Quantum Physics. Stanisław Goldstein and Louis Labuschagne, Oxford University Press. © Stanisław Goldstein and Louis Labuschagne (2025).
DOI: 10.1093/oso/9780198950202.003.0006

Proposition 4.3

(1) $\mathfrak{p}_\varphi$ *is a hereditary subcone of* $\mathcal{M}_+$.

(2) N_φ *and* $\mathfrak{n}_\varphi$ *are left ideals in* $\mathcal{M}$.

(3) $\mathfrak{m}_\varphi$ *is the linear span of* $\mathfrak{n}_\varphi^* \mathfrak{n}_\varphi$ *and* $\mathfrak{m}_\varphi \subseteq \mathfrak{n}_\varphi \cap \mathfrak{n}_\varphi^*$.

(4) $\mathfrak{m}_\varphi$ *is a* *-subalgebra of* $\mathcal{M}$.

(5) $\varphi{\upharpoonright}\mathfrak{p}_\varphi$ *extends to a positive linear form on* $\mathfrak{m}_\varphi$ *(denoted also by* φ*) and* $\mathfrak{m}_\varphi \cap \mathcal{M}_+ = \mathfrak{p}_\varphi$.

Proof. The proof is almost identical to the proof of Proposition 3.31, the only difference being that both N_φ and $\mathfrak{n}_\varphi$ are only left ideals. This is, in fact, the only place where the tracial property (see Definition 3.24(3)) is used in the proof of Proposition 3.31. $\qquad\square$

Theorem 4.4 ([Haa75a])   *For any weight* φ *on* $\mathcal{M}$*, the following conditions are equivalent.*

(1) *If* $(a_i)_{i \in I} \subset \mathcal{M}_+$ *is an increasing net bounded from above, then* $\varphi(\sup_{i \in I} a_i) = \sup_{i \in I} \varphi(a_i)$.

(2) φ *is completely additive; that is, whenever* $\{a_i\}_{i \in I}$ *is a family from* $\mathcal{M}_+$ *such that* $\sum_{i \in I} a_i$ *exists in the strong topology of* $\mathcal{M}$*, we have* $\varphi(\sum_{i \in I} a_i) = \sum_{i \in I} \varphi(a_i)$.

(3) φ *is* σ*-weakly lower semicontinuous (i.e., for all* $\lambda \geq 0$ *the set* $\{a \in \mathcal{M}_+ : \varphi(a) \leq \lambda\}$ *is* σ*-weakly closed).*

(4) *For some* $F \subseteq \mathcal{M}_*^+$ $\varphi(a) = \sup_{\omega \in F} \omega(a)$ *for all* $a \in \mathcal{M}_+$.

(5) $\varphi(a) = \sup\{\omega(a) : \omega \in \mathcal{M}_*^+, \omega \leq \varphi\}$ *for all* $a \in \mathcal{M}_+$.

(6) *For a family* $\{\varphi_i\}_{i \in I} \subset \mathcal{M}_*^+$*,* $\varphi(a) = \sum_{i \in I} \varphi_i(a)$ *for all* $a \in \mathcal{M}_+$.

Definition 4.5 A weight φ is *normal* if it satisfies any of (1)–(6). φ is *faithful* if $\varphi(a) = 0$ implies $a = 0$ for $a \in \mathcal{M}_+$.

Example 4.6 ([Haa75a]) Complete additivity on projections does not imply normality. Indeed, let $\mathcal{M} = l^\infty$ with $\varphi((\alpha_n)_{n \in \mathbb{N}}) = \sum_{n \in \mathbb{N}} \alpha_n$ if $\#\{n : \alpha_n \neq 0\} < \infty$ and $\varphi((\alpha_n)_{n \in \mathbb{N}}) = \infty$ otherwise. Then φ is not normal: for $a = (\alpha_n) := (2^{-n})$ and $a^{(n)} = (\alpha_1, \dots \alpha_n, 0, \dots)$, we have $\varphi(a^{(n)}) = 1 - 2^{-n}$ and $\varphi(a) = \infty$, so that $a^{(n)} \nearrow a$ but $\varphi(a^{(n)})$ does not converge to $\varphi(a)$. On the other hand, condition $\varphi(\sum_{i \in I} p_i) = \sum_{i \in I} \varphi(p_i)$ is satisfied for any orthogonal family of projections from $\mathcal{M}$.

Proposition 4.7 *If φ is a normal weight on $\mathcal{M}$, then there exists a unique projection $p \in \mathcal{M}$ such that φ is faithful on $\mathcal{M}_{p^\perp}$ and (the extension of) φ equals zero on $\mathcal{M}_p$.*

Proof. Note that, according to Theorem 4.4(5), we have

$$N_\varphi = \bigcap_{\omega \in \mathcal{M}_*, \, \omega \leq \varphi} N_\omega.$$

Since all N_ωs are σ-weakly closed (see Lemma 0.95), N_φ is also σ-weakly closed. By Propositions 0.94(5) and 4.3(2), there is a projection $p \in \mathcal{M}$ such that $N_\varphi = \mathcal{M}p$. If $\varphi(a) = 0$ for some $0 \neq a \in p^\perp \mathcal{M}_+ p^\perp$, then $a^{1/2} \in N_\varphi$, so that $a \in p\mathcal{M}p$—a contradiction. We leave the proof of uniqueness as an exercise. $\qquad\square$

Definition 4.8 For a normal weight φ, the projection p from the proposition above is called the *null projection* of φ and is denoted by $e_0(\varphi)$. The orthogonal complement of $e_0(\varphi)$ is called the *support projection* of φ and is denoted by $\operatorname{supp}\varphi$.

It is clear that φ is faithful if and only if $e_0(\varphi) = 0$.

The definition of semifiniteness of a weight might not seem the most natural at first. Below, we introduce versions of the notion that correspond to various conditions in Proposition 3.31.

Definition 4.9 A normal weight φ on $\mathcal{M}$ is called:

  (1) *semifinite* if $\mathfrak{p}_\varphi$ generates $\mathcal{M}$ as a von Neumann algebra;
  (2) *orthogonally semifinite* if there is an orthogonal family $\{e_i\}_{i \in I}$ of projections from $\mathcal{M}$ such that $\sum_{i \in I} e_i = \mathbb{1}$ and for all i, $\varphi(e_i) < \infty$;
  (3) *strictly semifinite* if $\varphi = \sum_{i \in I} \varphi_i$ for a family of functionals from $\mathcal{M}_*^+$ with orthogonal supports;
  (4) *strongly semifinite* if for every non-zero projection e from $\mathcal{M}$, there is a non-zero subprojection f of e such that $\varphi(f) < \infty$; in other words, if the set of projections of finite weight is order-dense in the set of all projections in $\mathcal{M}$.

Strictly semifinite weights were introduced by Combes in [**Com71**]. Strongly semifinite weights were defined by Trunov [**Tru78**], who called them *locally finite*, and Gardner [**Gar79**], more or less at the same time. They were called *densely semifinite* in [**GP15**]. We decided to use the name *strongly semifinite*, since the paper of Trunov is in Russian and not easily accessible. The name *orthogonally semifinite* was coined in [**GP15**], but main results on such weights had been obtained earlier in [**STS02**] and [**HKZ91**]. This notion may, in fact, become redundant—it follows from the paper [**HKZ91**] of Halpern, Kaftal and Zsidó that for σ-finite algebras, there is no difference between semifinite and orthogonally semifinite.

In sharp contrast to the case of traces, the existence of faithful normal (strictly) semifinite weights is a fairly simple matter to prove.

Lemma 4.10   *The sum of normal states with orthogonal supports is a semifinite normal weight.*

Proof.   Let $\{\psi_i\}$ be a family of normal states with orthogonal supports p_i, and define φ on $\mathcal{M}_+$ by $\varphi(a) = \sum_i \psi_i(a)$ for all $a \in \mathcal{M}_+$. Since all terms are positive, we may interchange summation and suprema. It is, therefore, clear that the normality of each ψ_i ensures the normality of φ.

To see that φ is semifinite, write p for $\sum_{i \in I} p_i$ and note that with F ranging over the finite subsets of the index set I, the net of projections $e_F = (\sum_{i \in F} p_i) + (\mathbb{1} - p)$ increases to $\mathbb{1}$ and hence is σ-strong* convergent to $\mathbb{1}$. For any finite subset F of I and any $a \in \mathcal{M}_+$, we have that

$$\varphi(e_F a e_F) = \sum_{i \in I} \psi_i(e_F a e_F) = \sum_{i \in F} \psi_i(p_i a p_i) = \sum_{i \in F} \psi_i(a) < \infty.$$

The fact that $(e_F a e_F)$ is σ-strong* convergent to a now ensures that φ is semifinite. $\square$

Proposition 4.11   *Each von Neumann algebra admits a faithful normal (strictly) semifinite weight.*

Proof.   The proof closely follows that of Theorem 0.100. Select a maximal family $\{\psi_i\}$ of normal states with mutually orthogonal supports $\{p_i\}$, and define φ to be $\varphi = \sum_i \psi_i$. A virtual repeat of the earlier argument now shows that $\sum_i p_i = \mathbb{1}$. Taking the lemma into account, all we still need to do is prove that φ is faithful. This again follows a very similar path to corresponding part of the proof of Theorem 0.100. $\qquad\square$

The former description of the various types of weights will not be used in future chapters. Nevertheless, the reader may benefit from a thorough understanding of those special families of weights. The importance of the notion of locally measurable operators in the sense of Sankaran [**San59**] was mentioned at the beginning of Section 3.4. In Section 4.3, we shall see the importance of Radon–Nikodym-type theorems. In particular, we are going to show that each normal weight φ on a semifinite algebra $\mathcal{M}$ with a f.n.s. trace τ possesses a *density* with respect to the trace (see Theorems 4.25 and 4.26). The properties of the weights correspond to appropriate properties of the densities. Theorem 4.32 states that the property of a weight, which corresponds to the local measurability of its density, is exactly strong semifiniteness.

Lemma 4.12 *A weight is strongly semifinite if and only if it is orthogonally semifinite on each reduced von Neumann algebra $\mathcal{M}_e$ with $e \in \mathbb{P}(\mathcal{M})$.*

Proof. '$\Rightarrow$' Let $\{e_i\}$ be a maximal family of mutually orthogonal non-zero subprojections of e of finite weight (use Zorn's lemma and strong semifiniteness to show its existence). If $\sum e_i < e$, then we could enlarge the family by adding to it a non-zero subprojection of $e - \sum e_i$. Hence, $\sum e_i = e$ and $\varphi{\restriction}\mathcal{M}_e$ is orthogonally semifinite.

'$\Leftarrow$' Let $0 \neq e \in \mathbb{P}(\mathcal{M})$. If φ is orthogonally semifinite on $\mathcal{M}_e$, then $e = \sum e_i$ with $e_i \in \mathbb{P}(\mathcal{M})$, $\varphi(e_i) < \infty$, and at least one of the e_is must be non-zero. $\qquad\square$

Theorem 4.13

    (1) *Every finite weight is both strongly and strictly semifinite.*

    (2) *Every strongly semifinite or strictly semifinite weight is orthogonally semifinite.*

    (3) *Every orthogonally semifinite weight is semifinite.*

Proof. (1) is obvious by definition.

(2) If φ is strictly semifinite, then $\varphi = \sum \varphi_i$ with $\varphi_i \in \mathcal{M}_*^+$. It is enough to take $e_i := \operatorname{supp} \varphi_i$. If φ is strongly semifinite, then it is orthogonally semifinite by Lemma 4.12.

(3) Assume that $\{e_i\}_{i\in I}$ is the family from the definition of orthogonal semifiniteness. For $J \subseteq I$, J finite, put $f_J = \sum_{i \in J} e_i$. We are going to show that for any $a \in \mathcal{M}$, $f_J a f_J \in \mathfrak{m}_\varphi$ and $f_J a f_J \to a$ strongly, which shows that φ is semifinite. In fact, for any $i, j \in I$, we have $e_i, e_j \in \mathfrak{n}_\varphi \cap \mathfrak{n}_\varphi^*$ and $\mathfrak{n}_\varphi$ is a left ideal, so that $e_i a e_j \in \mathfrak{n}_\varphi^* \mathfrak{n}_\varphi \subseteq \mathfrak{m}_\varphi$. Now, $f_J \to \mathbb{1}$ strongly and, for any $\xi \in H$,

$$\|(f_J a f_J - a)\xi\| \leq \|f_J a (1 - f_J)\xi\| + \|(1 - f_J)a\xi\|$$
$$\leq \|a\|\|(1 - f_J)\xi\| + \|(1 - f_J)a\xi\| \to 0$$

which ends the proof of (3). $\qquad\qquad\qquad\qquad\qquad\qquad\qquad\square$

Hence, for an arbitrary algebra and a normal weight, we have

$$\begin{array}{ccccc}
 & & \text{strictly semifinite} & & \\
 & \nearrow & & \searrow & \\
\text{finite} & & & & \text{orthogonally semifinite} \to \text{semifinite} \\
 & \searrow & & \nearrow & \\
 & & \text{strongly semifinite} & &
\end{array}$$

Proposition 4.14 *Let φ be a faithful orthogonally semifinite weight on $\mathcal{M}$. Then, there exists a countable family $\{e_n\}_{n\in\mathbb{N}}$ such that $e_n \in \mathbb{P}(\mathcal{M})$ and $\varphi(e_n) < \infty$ for each $n \in \mathbb{N}$, with $\sum_{n\in\mathbb{N}} e_n = \mathbb{1}$, if and only if the algebra $\mathcal{M}$ is σ-finite.*

Proof. '$\Rightarrow$' Let $\{f_i\}_{i\in I}$ such that for all $i \in I$, $0 \neq f_i \in \mathbb{P}(\mathcal{M})$, $\sum_{i\in I} f_i = \mathbb{1}$ and $\#I > \aleph_0$. Suppose that there is a countable family $\{e_n\}$ of mutually orthogonal projections from $\mathcal{M}$ such that $\varphi(e_n) < \infty$ for all n and $\sum e_n = \mathbb{1}$. As in the proof of part (3) '$\Rightarrow$' of Theorem 4.13, $e_n f_i e_n \in \mathfrak{m}_\varphi$ for any i, n. By normality, $\sum_{i\in I} \varphi(e_n f_i e_n) = \varphi(e_n (\sum_{i\in I} f_i)e_n) = \varphi(e_n) < \infty$, so that for each n, there is a set I_n such that $\#I \backslash I_n \leq \aleph_0$ and $\varphi(e_n f_i e_n) = 0$ for $i \in I_n$. Thus, the intersection of I_ns is non-empty and there is $i_0 \in I$ such that $\varphi(e_n f_{i_0} e_n) = 0$ for all n. We have $|\varphi(e_n f_{i_0} e_m)| \leq \varphi(e_n f_{i_0} e_n)^{1/2} \varphi(e_m)^{1/2} = 0$ for all n, m. Hence, $\varphi(f_{i_0}) = \sum_{n,m} \varphi(e_n f_{i_0} e_m) = 0$, which contradicts the faithfulness of φ.

'$\Leftarrow$' is obvious from the definitions of σ-finiteness of the algebra. $\quad\square$

Remark 4.15 For a normal trace $\varphi = \tau$ on $\mathcal{M}$, the four notions of semifiniteness from Definition 4.9 are equivalent. This is exactly the content of Proposition 3.31.

Proposition 4.16 *If φ is a normal weight on $\mathcal{M}$, then there exists a unique projection $q \in \mathcal{M}$ such that the weight $q\varphi q := \varphi(q \cdot q)$ is semifinite with $\varphi(a) = \infty$ for every $a \in \mathcal{M}_+$ with $(\mathbb{1} - q)a(\mathbb{1} - q) \neq 0$.*

Proof. By Propositions 0.94(5) and 4.3(2), there is a projection $q \in \mathcal{M}$ such that $\overline{\mathfrak{n}_\varphi}^{\sigma - \mathrm{w}} = \mathcal{M}q$. The same argument as was used in the first part of the proof of Proposition 3.34 can now be used to show that $\overline{\mathrm{span}(\mathfrak{p}_\tau)} = q\mathcal{M}q$. On setting $q.\varphi.q = \varphi(q \cdot q)$, it is now an exercise to see that $\mathfrak{n}_{q\varphi q} = \mathfrak{n}_\varphi \oplus \mathcal{M}(\mathbb{1} - q)$ and hence that $\overline{\mathfrak{n}_{q\varphi q}}^{\sigma - \mathrm{w}} = \mathcal{M}$. This, in turn, ensures that $\overline{\mathrm{span}(\mathfrak{p}_{q\varphi q})} = \overline{\mathrm{span}(\mathfrak{n}_{q\varphi q}^* \mathfrak{n}_{q\varphi q})} = \mathcal{M}$, and hence that $q\varphi q$ is semifinite.

If, on the other hand, $(\mathbb{1} - q)a(\mathbb{1} - q) \neq 0$ for some $a \in \mathcal{M}_+$, then $a^{1/2}$ clearly does not belong to $\overline{\mathfrak{n}_\varphi}^{\sigma - \mathrm{w}} = \mathcal{M}q$, which ensures that $\varphi(a) = \infty$ as required. We leave the proof of uniqueness as an exercise. $\square$

Definition 4.17 For a normal weight φ, the orthogonal projection q from Proposition 4.16 is called the *semifinite projection* and is denoted by $e_\infty(\varphi)$.

It is clear that φ is semifinite if and only if $e_\infty(\varphi) = \mathbb{1}$. Obviously, $e_0(\varphi) \leq e_\infty(\varphi)$.

Notation 4.18 For a weight on $\mathcal{M}$, *f.n.s.* abbreviates faithful normal semifinite.

Definition 4.19 *Representation (π_φ, H_φ) induced by φ is defined as follows:* $H_\varphi = (\mathfrak{n}_\varphi / N_\varphi)^\sim$, *where the completion is with respect to the scalar product given by* $\langle \eta_\varphi(a), \eta_\varphi(b) \rangle_\varphi = \varphi(b^* a)$, *where* $\eta_\varphi : \mathfrak{n}_\varphi \to \mathfrak{n}_\varphi / N_\varphi$ *is the quotient map, and* $\pi_\varphi \colon \mathcal{M} \to B(H_\varphi)$ *is given by* $\pi_\varphi(a)\eta_\varphi(b) = \eta_\varphi(ab)$ *for* $a \in \mathcal{M}, b \in \mathfrak{n}_\varphi$.

Note that the representation π_φ is normal if φ is normal, faithful if π_φ is faithful and non-degenerate if π_φ is semifinite.

Proposition 4.20 *If φ is f.n.s., then $\pi_\varphi \colon \mathcal{M} \to B(H_\varphi)$ is a $*$-isomorphism of $\mathcal{M}$ onto $\pi_\varphi(M)$.*

4.2 Extensions of weights and traces

Theorem 4.21 *Any normal weight φ on $\mathcal{M}$ has a unique extension to $\widehat{\mathcal{M}}_+$ (denoted also by φ) such that:*

(1) $\varphi(\lambda m) = \lambda\varphi(m)$;

(2) $\varphi(m + n) = \varphi(m) + \varphi(n)$;

(3) *if $m_i \nearrow m$, then $\varphi(m_i) \nearrow \varphi(m)$.*

Proof. Take any normal weight φ on $\mathcal{M}$ and any $m \in \widehat{\mathcal{M}}_+$. Then, by Theorems 0.144 and 4.4, $\varphi = \sum_{i \in I} \omega_i$ with $\omega_i \in \mathcal{M}_*^+$, and $m = \int_0^\infty \lambda de_\lambda + \infty \cdot p$, so that

$$\varphi(m) := \lim_{n \to \infty} \sum_{i \in I} \omega_i(x_n) = \sum_{i \in I} \omega_i(m) \text{ for } x_n = \int_0^n \lambda de_\lambda + np.$$

(1)–(3) are easily checked; if φ' is another extension of φ to $\widehat{\mathcal{M}}_+$, then

$$\varphi'(m) = \sum_{i \in I} \omega_i(m) = \varphi(m).$$

$\square$

Notation 4.22 Let $\mathcal{M}$ be semifinite and let τ be an f.n.s. trace on $\mathcal{M}$. For $a, b \in \mathcal{M}_+$,

$$a \cdot b := a^{1/2}ba^{1/2}.$$

Theorem 4.23 *The map $(a, b) \mapsto \tau(a \cdot b)$ has a unique extension $(m, n) \mapsto \tau(m \,\widehat{\cdot}\, n)$ to $\widehat{\mathcal{M}}_+ \times \widehat{\mathcal{M}}_+$ such that:*

(1) $\tau(m \,\widehat{\cdot}\, n) = \tau(n \,\widehat{\cdot}\, m)$ *for $m, n \in \widehat{\mathcal{M}}_+$;*

(2) τ *is homogeneous and additive with respect to both m and n;*

(3) *if $m_\alpha \nearrow m$ and $n_\beta \nearrow n$, then $\tau(m_\alpha \,\widehat{\cdot}\, n_\beta) \nearrow \tau(m \,\widehat{\cdot}\, n)$. This extension satisfies*

(4) $\tau((ama^*) \,\widehat{\cdot}\, n) = \tau(m \,\widehat{\cdot}\, (a^*na))$ *for $m, n \in \widehat{\mathcal{M}}_+, a \in \mathcal{M}$.*

Proof. For $m \in \widehat{\mathcal{M}}_+$ as in Theorem 0.144, put $m_n := \int_0^n \lambda de_\lambda + np$. For $m, n \in \widehat{\mathcal{M}}_+$, put $\tau(m \,\widehat{\cdot}\, n) := \sup_{n,k} \tau(m_n \cdot n_k)$. Evidently, τ satisfies (1). For $m \in \widehat{\mathcal{M}}_+$, put $\tau_m(a) := \tau(m \,\widehat{\cdot}\, a)$ for $a \in \mathcal{M}_+$.

Since $\tau_m(a) = \sup_n \tau(m_n^{1/2} a m_n^{1/2})$, it follows that τ_m is a normal weight. By Theorem 4.21, the extension of τ_m to $\widehat{\mathcal{M}}_+$ is given by

$$\tau_m(a) = \sup_k \tau_m(a_k) = \sup_{n,k} \tau(m_n \cdot a_k) = \tau(m \,\widehat{\cdot}\, a).$$

By Theorem 4.21, $n \mapsto \tau(m \,\widehat{\cdot}\, n)$ is homogeneous, additive and normal. As $\tau(m \,\widehat{\cdot}\, n) = \tau(n \,\widehat{\cdot}\, m)$, (2) and (3) are satisfied. Uniqueness follows from the uniqueness in Theorem 4.21. If $m, n \in L^2(\mathcal{M}, \tau)$, then $\tau((ama^*) \,\widehat{\cdot}\, n) = \tau(m \,\widehat{\cdot}\, (a^* na))$. Any $m \in \mathcal{M}_+$ is the limit of an increasing sequence from $L^2(\mathcal{M}, \tau)$. (To see this, let $e_n \in \mathcal{M}$ be such that $e_n \nearrow \mathbb{1}$ and $\tau(e_n) < \infty$, and put $m_n = m^{1/2} e_n m^{1/2} \nearrow m$. Then, $\tau(m_n^2) = \tau(e_n m e_n m e_n) \le \|m\|^2 \tau(e_n) < \infty$.) But this then implies the validity of the formula for $m, n \in \mathcal{M}_+$. By Proposition 0.148, we have (4). $\qquad\square$

Proposition 4.24 *If $a, b \in LS(\mathcal{M})_+$, then*

$$\tau(m_a \,\widehat{\cdot}\, m_b) = \tau(a^{1/2} \,\overline{\cdot}\, b \,\overline{\cdot}\, a^{1/2}) = \tau(b^{1/2} \,\overline{\cdot}\, a \,\overline{\cdot}\, b^{1/2}).$$

Proof.   Let $u \in \mathcal{M}$ be such that $u^* u = \mathbf{s}_r(a^{1/2} \,\overline{\cdot}\, b^{1/2})$. Using the Spectral Theorem 0.33, we can form an increasing sequence (d_n) from $\mathcal{M}_+$ such that $d_n \nearrow b^{1/2} \,\overline{\cdot}\, a \,\overline{\cdot}\, b^{1/2}$ and that $\mathbf{s}(d_n) \le \mathbf{s}_r(b^{1/2} \,\overline{\cdot}\, a \,\overline{\cdot}\, b^{1/2}) = \mathbf{s}_r(|a^{1/2} \,\overline{\cdot}\, b^{1/2}|^2) = \mathbf{s}_r(|a^{1/2} \,\overline{\cdot}\, b^{1/2}|) = \mathbf{s}_r(a^{1/2} \,\overline{\cdot}\, b^{1/2})$. Now, $u d_n u^* \nearrow u(b^{1/2} \,\overline{\cdot}\, a \,\overline{\cdot}\, b^{1/2}) u^*$ by Proposition 3.88. We have, by the last statement of Theorem 0.37, that

$$\begin{aligned}
\tau(a^{1/2} \,\overline{\cdot}\, b \,\overline{\cdot}\, a^{1/2}) &= \tau((a^{1/2} \,\overline{\cdot}\, b^{1/2})(a^{1/2} \,\overline{\cdot}\, b^{1/2})^*) \\
&= \tau(u(a^{1/2} \,\overline{\cdot}\, b^{1/2})^* (a^{1/2} \,\overline{\cdot}\, b^{1/2}) u^*) \\
&= \tau(u(b^{1/2} \,\overline{\cdot}\, a \,\overline{\cdot}\, b^{1/2}) u^*) = \lim_{n \to \infty} \tau(u d_n u^*) \\
&= \lim_{n \to \infty} \tau(d_n^{1/2} u^* u d_n^{1/2}) = \lim_{n \to \infty} \tau(d_n) \\
&= \tau(b^{1/2} \,\overline{\cdot}\, a \,\overline{\cdot}\, b^{1/2}).
\end{aligned}$$

Next, let (a_n) and (b_m) be increasing sequences from $\mathcal{M}_+$ such that $a_n \nearrow a$ and $b_m \nearrow b$. Then, by the first part of the proof and Proposition 3.88,

$$\tau(m_a \widehat{\cdot} m_b) = \lim_{m,n\to\infty} \tau(a_n \cdot b_m)$$

$$= \lim_{m\to\infty} \lim_{n\to\infty} \tau(b_m^{1/2} a_n b_m^{1/2})$$

$$= \lim_{m\to\infty} \tau(b_m^{1/2} a b_m^{1/2})$$

$$= \lim_{m\to\infty} \tau(a^{1/2} \widehat{\cdot} b_m \widehat{\cdot} a^{1/2})$$

$$= \tau(a^{1/2} \widehat{\cdot} b \widehat{\cdot} a^{1/2})$$

which ends the proof. $\qquad\square$

From now on, we shall write $\tau(x \cdot y)$ instead of $\tau(x \widehat{\cdot} y)$ for $x, y \in \widehat{\mathcal{M}}_+$, and $\tau(a \cdot b)$ instead of $\tau(b^{1/2} \widehat{\cdot} a \widehat{\cdot} b^{1/2})$ for $a, b \in LS(\mathcal{M})_+$. Note that now $\tau(x \cdot \mathbb{1}) = \tau(x)$, as expected.

4.3 Density of weights with respect to a trace

The next result is a version of the Radon–Nikodym theorem that is most useful in applications. It is an easy corollary from the already powerful Pedersen–Takesaki Radon–Nikodym theorem for two weights [**PT73**] (further generalised by [**Vae01b**]), a difficult theorem using the full apparatus of modular theory. We felt that the tracial version of the theorem should be proved early on in an 'elementary' way.

Theorem 4.25 *Let $\mathcal{M}$ be a semifinite algebra equipped with an f.n.s. trace τ. Then, for any normal state ω, there is a τ-measurable operator g_ω such that $\omega(a) = \tau(g_\omega \cdot a)$ for all $a \in \mathcal{M}_+$.*

Proof. We claim that the space of functionals $\{\tau(f\cdot) : f \in \mathfrak{m}_\tau\}$ is norm dense in $\mathcal{M}_*$. To see this, suppose that we are given $a \in \mathcal{M}$ such that $\tau(fa) = 0$ for all $f \in \mathfrak{m}_\tau$. So, for any f of the form $f = xx^*u^*$ where u is the partial isometry in the polar decomposition $a = u|a|$ and $x \in \mathfrak{n}_\tau$, we have that $0 = \tau(fa) = \tau(x^*|a|x)$. In other words, $x^*|a|x = 0$ for all $x \in \mathfrak{n}_\tau$, which can only be the case if $a = 0$. This is enough to ensure the norm density of $\{\tau(f\cdot) : f \in \mathfrak{m}_\tau\}$ in $\mathcal{M}_*$.

Therefore, we may select $(g_n) \subset \mathfrak{m}_\tau$ so that $(\tau(g_n\cdot))$ converges to ω in norm. Since $(\tau(g_n^*\cdot))$ will then also converge to ω, we may replace each g_n with its real part if necessary, and assume that $g_n = g_n^*$ for each n.

Now, let $e_n = \chi_{[0,\|g_n\|]}(g_n)$ for every n. By passing to a subsequence, if necessary, we may assume that (e_n) converges σ-weakly to some e_0. Since, for every $x \in \mathcal{M}$, we have that

$$|\tau(g_n e_n x) - \omega(e_0 x)| \leq |\tau(g_n e_n x) - \omega(e_n x)| + |\omega(e_n x) - \omega(e_0 x)| \quad (4.1)$$

$$\leq \|\tau(g_n \cdot) - \omega\| \cdot \|x\| + |\omega(e_n x) - \omega(e_0 x)|,$$

it clearly follows that $(\tau(g_n e_n \cdot))$ converges weakly to $\omega(e_0 \cdot)$ in $\mathcal{M}_*$. But that ensures that $\tau(g_n(\mathbb{1} - e_n)) \to \omega(\mathbb{1} - e_0)$. Since, by construction, $\tau(g_n(\mathbb{1} - e_n)) \leq 0$ for every n with $\omega(\mathbb{1} - e_0) \geq 0$, this can only be the case if in fact $|\tau(g_n(\mathbb{1} - e_n))| \to 0$. But, by construction, each $a \mapsto \tau(g_n(e_n - \mathbb{1})a)$ is a positive functional with norm $\tau(g_n(e_n - \mathbb{1})) = |\tau(g_n(\mathbb{1} - e_n))|$. So in fact $(\tau(g_n(\mathbb{1} - e_n) \cdot))$ converges to 0 in norm in $\mathcal{M}_*$, ensuring that $(\tau(g_n e_n \cdot))$ actually converges in norm to ω. We have, therefore, shown that we may assume that $(g_n) \subset \mathfrak{m}_\tau^+$. It is a fairly straightforward exercise to see that $\tau(|g_n - g_m|) \leq \|\tau((g_n - g_m) \cdot)\|$ for each pair (m, n). Given $\epsilon > 0$ and selecting $N \in \mathbb{N}$ so that $\|\tau((g_n - g_m) \cdot)\| \leq \epsilon^2$ for every $n, m \geq N$, we may now use the Borel functional calculus to see that

$$\tau(\chi_{(\epsilon,\infty)}(|g_n - g_m|)) \leq \epsilon^{-1}\tau(|g_n - g_m|) \leq \epsilon^{-1}\|\tau((g_n - g_m) \cdot)\| \leq \epsilon$$

for every $n, m \geq N$. In other words, $(g_n - g_m) \in \mathcal{N}(\epsilon, \epsilon)$ for all $n, m \geq N$. So, (g_n) is Cauchy in $\widetilde{\mathcal{M}}$ and must therefore converge in measure to some $g \in \widetilde{\mathcal{M}}_+$.

Now, let $m \in \mathbb{N}$ be given and let $p_m = \chi_{[0,m]}(g)$. We clearly have $g_n p_m \to g p_m$ in measure, and hence, given $\epsilon > 0$, there exists some $N \in \mathbb{N}$ such that $(g p_m - g_n p_m) \in \mathcal{N}(\epsilon, \epsilon)$ for all $n \geq N$. This, in turn, ensures that we can find projections $(q_n)_{n \in \mathbb{N}}$ so that $\|(g p_m - g_n p_m)q_n\| \leq \epsilon$ and $\tau(\mathbb{1} - q_n) \leq \epsilon$ for all $n \geq N$. For any $x \in \mathfrak{m}_\tau$, we then have

$$|\tau((g p_m - g_n p_m)q_n x)| \leq \|(g p_m - g_n p_m)q_n\|\tau(|x|) \leq \epsilon\tau(|x|)$$

$$\text{for all } n \geq N.$$

Thus, $\lim_{n \to \infty} \tau((g p_m - g_n p_m)q_n x) = 0$. However, we also have

$$|\tau(g p_m(\mathbb{1} - q_n)x))| = |\tau(x p_m g(\mathbb{1} - q_n)))| \leq \|x\|\tau(g(\mathbb{1} - q_n)) \to 0$$

as $n \to \infty$, which ensures that $\lim_{n \to \infty} \tau(g_n p_m q_n x) = \tau(g p_m x)$. By passing to a subsequence, if necessary, we may assume that (q_n) converges σ-weakly to some q_0 as $n \to \infty$. For any $y \in \mathfrak{n}_\tau$, we then have that $\tau(y^*(1 - q_n)y) \to \tau(y^*(1 - q_0)y)$ as $m \to \infty$ as $n \to \infty$. But we also have that $|\tau(y^*(1 - q_n)y)| \leq \|y\|^2\tau(1 - q_n)$ and hence that $\tau(y^*(1 - q_n)y) \to 0$ as $n \to \infty$. The faithfulness of τ now guarantees that $y^*(1 - q_0)y = 0$ for all $y \in \mathfrak{n}_\tau$, and hence that $q_0 = 1$. It is clear that $\tau(g_np_m\cdot) \to \omega(p_m\cdot)$ in norm. So, arguing as in equation (4.1), it follows that $\tau(g_np_mq_n\cdot) \to \omega(p_m\cdot)$ weakly in $\mathcal{M}_*$. Thus, for $x \in \mathfrak{m}_\tau$, we have as before that $\lim_{n\to\infty}\tau(g_np_mq_nx) \to \omega(e_mx)$ and hence that $\tau(gp_mx) = \omega(p_mx)$. That ensures that for any $x \in \mathfrak{m}_\tau^+$ we have that $\tau(g.x) = \sup_{m\geq 1}\tau((gp_m).x) = \lim_{m\geq 1}\omega(gp_mx) = \omega(x)$, which is enough to ensure that $g = g_\omega$. $\qquad\square$

Theorem 4.26 *Let $\mathcal{M}$ be semifinite and let τ be an f.n.s. trace on $\mathcal{M}$.*

(1) *The map $m \mapsto \tau_m$, where $\tau_m(a) := \tau(m \cdot a)$ for $a \in \mathcal{M}_+$, is a homogeneous and additive bijection of $\widehat{\mathcal{M}}_+$ onto the set of normal weights on $\mathcal{M}$. Moreover,*

$$\tau_{ama^*} = a^*\tau_m a$$

$$m \leq n \text{ iff } \tau_m \leq \tau_n$$

$$m_i \nearrow m \text{ iff } \tau_{m_i} \nearrow \tau_m.$$

(2) *Let $m = \int \lambda de_\lambda + \infty \cdot p \in \widehat{\mathcal{M}}_+$. Then, $e_0(\tau_m) = e_0$ and $e_\infty(\tau_m) = p^\perp$, so that τ_m is faithful iff $e_0 = 0$ and τ_m is semifinite iff $p = 0$.*

Proof. (1) All, except bijectivity, follows from Theorem 4.23. Surjectivity follows from Theorem 4.25: If $\omega \in \mathcal{M}_*^+$, then $\omega = \tau(m\cdot)$ for some $m \in \widehat{\mathcal{M}}_+$. Let φ be an arbitrary normal weight. Then, by Theorem 4.4(4), $\varphi = \sum \omega_{i\in I}$ for some $\omega_i \in \mathcal{M}_*^+$. Now, $\omega_i = \tau(m_i\cdot)$ so that $\varphi = \tau(m\cdot)$, where $m = \sum_{i\in I} m_i \in \widehat{\mathcal{M}}_+$. For injectivity, assume that $\tau_m = \tau_n$ for some $m, n \in \widehat{\mathcal{M}}_+$. Then, $\tau_m(a) = \tau_n(a)$ for all $a \in \mathcal{M}_+$, so that $\tau_a(m) = \tau_a(n)$ for all $a \in \mathcal{M}_+$. By Theorem 4.25, $\omega(m) = \omega(n)$ for all $\omega \in \mathcal{M}_*^+$, which implies $m = n$. (Similarly, if $\tau_m \leq \tau_n$, then $m \leq n$; if $\tau_{m_i} \nearrow \tau_m$, then $m_i \nearrow m$.)

(2) $e_0(\tau_m) = e_0$: Put $m_n = \int_0^n \lambda de_\lambda + np$. Then $\tau_{m_n} \nearrow \tau_m$. For a projection $q \in \mathcal{M}$, we have $\tau_m(q) = 0$ iff for all n one has $\tau_{m_n}(q) = 0$ iff $\tau(qm_nq) = 0$ for all n iff $\mathbf{s}(m_n) \perp q$ for all $n \in \mathbb{N}$ iff $q \leq e_0$. Hence, $e_0(\tau_m) = e_0$.

$e_\infty(\tau m) = p^\perp$: Define $e_n := e_{[0,n]}(m)$, and let $k_j \in \mathcal{M}$, $j \in J$ be such that $k_j \in \mathfrak{n}_\tau$ and $k_j \to \mathbb{1}$ σ-strongly. Then, $k_j e_n \in \mathfrak{n}_{\tau m}$ for all $j \in J$, $n \in \mathbb{N}$, so that $e_n = \lim_{j \in J} k_j e_n$ belongs to the σ-strong closure of $\mathfrak{n}_{\tau m}$, which implies that $p^\perp$ is in the σ-strong closure of $\mathfrak{n}_{\tau m}$. Thus, there is a net $(d_i)_{i \in I}$ such that $d_i \in \mathfrak{n}_{\tau m}$ for all $i \in I$ and $d_i \to p^\perp$ σ-strongly. By Kaplansky's density theorem 0.93, we can choose the d_i's so that $\|d_i\| \le 1$. For any $a \in \mathcal{M}$, $d_i a d_i \in \mathfrak{n}_{\tau m}$, so that $d_i a d_i \to p^\perp a p^\perp$ σ-strongly. Hence, $p^\perp \le e_\infty(\tau m)$.

Put now $q := e_\infty(\tau m) - p^\perp$. Again, there exists a net $(d_i)_{i \in I}$ from $\mathfrak{n}_{\tau m}$ such that $d_i \to q$ σ-strongly. Since $m \ge qmq$ (in $\widehat{\mathcal{M}}_+$), we have $\tau m(q d_i^* d_i q) = \tau(m \cdot q d_i^* d_i q) = \tau(qmq \cdot d_i^* d_i) \le \tau(m \cdot d_i^* d_i) = \tau m(d_i^* d_i) < \infty$. On the other hand, if $q \ne 0$, then for sufficiently large $i \in I$, one has $d_i q \ne 0$; hence,

$$\tau m(q d_i^* d_i q) = \tau(m \cdot q d_i^* d_i q) = \lim_{m \to \infty} \tau(m_m \cdot q d_i^* d_i q)$$

$$= \lim_{m \to \infty} \tau(q m_m q \cdot q d_i^* d_i q) = \lim_{m \to \infty} m\tau(q d_i^* d_i q) = \infty.$$

Thus, $q = 0$ and $e_\infty(\tau m) = p^\perp$. $\qquad\square$

For the rest of this chapter, we fix a faithful normal weight φ on $\mathcal{M}$.

Definition 4.27 If $m \in \widehat{\mathcal{M}}_+$ is such that $\varphi = \tau m$, then m is called the *Radon–Nikodym derivative of φ with respect to τ*, and we write $d\varphi/d\tau := m$.

At this point, we need to sound a warning. Theorem 4.26 is a theorem about elements of $\widehat{\mathcal{M}}_+$ that can be regarded, without loss of generality, as positive quadratic forms. Hence, when claiming that $\tau_{ama^*} = a^*\tau_m a$ and that $\tau_{m_1+m_2} = \tau_{m_1} + \tau_{m_2}$, the product ama^* and the sum $m_1 + m_2$ is actually a statement about quadratic forms, not operators, and should therefore be interpreted in the sense of the definitions made at the start of Section 0.6. It therefore behoves us to see how in the case where h, h_1 and h_2 are self-adjoint positive operators affiliated to $\mathcal{M}$, the operator versions of these statements compare to the quadratic form versions. For this, we need the technology of form sums. We will here only summarise the essentials. Readers who wish to see a more complete treatment are referred to [**Tar13**]. Each of h_1 and h_2 induces closed sesquilinear forms $\mathfrak{t}_i(\xi, \zeta) = \langle h_i^{1/2}\xi, h_i^{1/2}\zeta \rangle$ $(i = 1, 2)$ on their respective domains, where ξ and ζ are vectors in the underlying Hilbert space. When speaking of representing such operators as elements of $\widehat{\mathcal{M}}_+$, it is actually these sesquilinear forms that we have in mind. These forms have an obvious action on vector states $\rho_{\xi,\xi} : \mathcal{M} \to \mathbb{R} : a \mapsto \langle a\xi, \xi \rangle$ with such a $\rho_{\xi,\xi}$ being mapped to $\mathfrak{t}_i(\xi, \xi)$ if $\xi \in \mathrm{dom}(h_i)$, and ∞ if not. This action may then in a natural way be extended to an action on the cone of positive normal functionals. The sum $\mathfrak{t}_1 + \mathfrak{t}_2$ is again a sesquilinear form, and in the particular case where $\mathrm{dom}(h_1^{1/2}) \cap \mathrm{dom}(h_2^{1/2})$ is dense, there exists a unique densely defined positive self-adjoint operator g characterised by the requirements that $\mathrm{dom}(g) = \mathrm{dom}(h_1^{1/2}) \cap \mathrm{dom}(h_2^{1/2})$, and that $\langle g\xi, \zeta \rangle = \langle h_1^{1/2}\xi, h_1^{1/2}\zeta \rangle + \langle h_2^{1/2}\xi, h_2^{1/2}\zeta \rangle$ for all $\xi \in \mathrm{dom}(g)$ and $\zeta \in \mathrm{dom}(h_1^{1/2}) \cap \mathrm{dom}(h_2^{1/2})$. This operator g is called the form sum of h_1 and h_2, and will be denoted by $g = h_1 \mathbin{\widehat{+}} h_2$. The natural domain of $h_1 \mathbin{\widehat{+}} h_2$ is moreover precisely $\mathrm{dom}(h_1^{1/2}) \cap \mathrm{dom}(h_2^{1/2})$ (see [**Tar13**, Theorem 3.5]). Thus, in the case where the terms we add in Theorem 4.26 correspond to positive self-adjoint operators h_1 and h_2 with $\mathrm{dom}(h_1^{1/2}) \cap \mathrm{dom}(h_2^{1/2})$ dense, the sum will correspond to the operator $h_1 \mathbin{\widehat{+}} h_2$. Even in the case where $\mathrm{dom}(h_1^{1/2}) \cap \mathrm{dom}(h_2^{1/2})$ is not dense, it is still possible to give expression to $h_1 \mathbin{\widehat{+}} h_2$, but in that case, $h_1 \mathbin{\widehat{+}} h_2$ turns out to be a linear relation (a multi-valued map), rather than an operator. Similar comments apply to the expression $a^* \tau_h a$ in Theorem 4.26, where we actually have what may be called a *form product*. Given a bounded operator a and a closed positive definite sesquilinear form $\mathfrak{t}$, we may define an associated sesquilinear form $a.\mathfrak{t}.a^*$ by the formal prescription $a.\mathfrak{t}.a^*(\xi, \zeta) = \mathfrak{t}(a^*\xi, a^*\zeta)$. If $\mathfrak{t}$ then corresponds to the densely defined self-adjoint positive operator h, we may write $a \mathbin{\widehat{\cdot}} h \mathbin{\widehat{\cdot}} a^*$ for the unique, possibly not densely defined, self-adjoint positive operator corresponding to the closed positive definite sesquilinear form $a.\mathfrak{t}.a^*$. The relevance of this for Theorem 4.23 lies in the fact that if $a, h \in \mathcal{M}_+$, then $a^{1/2} \tau_h a^{1/2}$ actually equals $\tau_{a \cdot h}$.

Proposition 4.28

(1) *If h_1 and h_2 are self-adjoint positive operators for which $\mathrm{dom}(h_1^{1/2}) \cap \mathrm{dom}(h_2^{1/2})$ is dense in H and $h_1 + h_2$ is essentially self-adjoint, then the unique self-adjoint extension of $h_1 + h_2$ is precisely $h_1 \mathbin{\widehat{+}} h_2$.*

(2) *If h is a self-adjoint positive operator and a is a bounded operator such that $\mathrm{dom}(h^{1/2}a^*)$ is dense and aha^* essentially*

> *self-adjoint, then the unique self-adjoint extension of aha^* is precisely $a \cdot h \cdot a^*$.*

Proof.   Let h_1 and h_2 be self-adjoint positive operators affiliated to $\mathcal{M}$ for which $\mathrm{dom}(h_1^{1/2}) \cap \mathrm{dom}(h_2^{1/2})$ is dense in H and $h_1 + h_2$ is essentially self-adjoint. Now, let $\xi \in \mathrm{dom}(h_1) \cap \mathrm{dom}(h_2)$ and $\zeta \in \mathrm{dom}(h_1 \widehat{+} h_2) = \mathrm{dom}(h_1^{1/2}) \cap \mathrm{dom}(h_2^{1/2})$. By the Borel functional calculus, we will then also have that $\xi \in \mathrm{dom}(h_1^{1/2}) \cap \mathrm{dom}(h_2^{1/2}) = \mathrm{dom}(h_1 \widehat{+} h_2)$. It then clearly follows that

$$
\begin{aligned}
\langle (h_1 + h_2)\xi, \zeta \rangle &= \langle h_1\xi, \zeta \rangle + \langle h_2\xi, \zeta \rangle \\
&= \langle h_1^{1/2}\xi, h_1^{1/2}\zeta \rangle + \langle h_2^{1/2}\xi, h_2^{1/2}\zeta \rangle \\
&= \langle (h_1 \widehat{+} h_2)^{1/2}\xi, (h_1 \widehat{+} h_2)^{1/2}\zeta \rangle \\
&= \langle \xi, (h_1 \widehat{+} h_2)\zeta \rangle.
\end{aligned}
$$

It follows that $h_1 + h_2 \subseteq (h_1 \widehat{+} h_2)^* = h_1 \widehat{+} h_2$. But since, by assumption, $h_1 + h_2$ is essentially self-adjoint, it must have a unique self-adjoint extension. This extension must then clearly be $h_1 \widehat{+} h_2$.

Next, let h be a self-adjoint positive operator and a be a bounded operator for which $\mathrm{dom}(h^{1/2}a^*)$ is dense and aha^* essentially self-adjoint. Let h correspond to the sesquilinear form $\mathfrak{t}(\xi, \zeta) = \langle h^{1/2}\xi, h^{1/2}\zeta \rangle$. Then, by definition, $a\widehat{\cdot}h\widehat{\cdot}a^*$ will correspond to the sesquilinear form $a.\mathfrak{t}.a^*(\xi, \zeta) = \langle h^{1/2}a^*\xi, h^{1/2}a^*\zeta \rangle$. In other words, $a\widehat{\cdot}h\widehat{\cdot}a^* = |h^{1/2}a^*|^2$, in which case $\mathrm{dom}(a\widehat{\cdot}h\widehat{\cdot}a^*) = \mathrm{dom}(h^{1/2}a^*)$. Now, let $\xi \in \mathrm{dom}(aha^*) = \mathrm{dom}(ha^*)$ and $\zeta \in \mathrm{dom}(a\widehat{\cdot}h\widehat{\cdot}a^*) = \mathrm{dom}(h^{1/2}a^*)$ be given. Observe that we will then also have that $\xi \in \mathrm{dom}(h^{1/2}a^*)$. Consequently,

$$
\langle aha^*\xi, \zeta \rangle = \langle ha^*\xi, a^*\zeta \rangle = \langle h^{1/2}a^*\xi, h^{1/2}a^*\zeta \rangle = \langle \xi, |h^{1/2}a^*|^2\zeta \rangle.
$$

This proves that $aha^* \subseteq (|h^{1/2}a^*|^2)^* = (a\widehat{\cdot}h\widehat{\cdot}a^*)^* = a\widehat{\cdot}h\widehat{\cdot}a^*$. Thus, $a\widehat{\cdot}h\widehat{\cdot}a^*$ is a self-adjoint extension of aha^*. But since aha^* is essentially self-adjoint, that extension must be unique and must, therefore, agree with $a\widehat{\cdot}h\widehat{\cdot}a^*$. $\qquad\square$

The following proposition is useful when comparing different traces on the same von Neumann algebra.

Proposition 4.29 *Let τ be an f.n.s. trace on a semifinite von Neumann algebra $\mathcal{M}$. Given any other semifinite normal trace τ' on $\mathcal{M}$, we have that:*

(1) *$\tau + \tau'$ is an f.n.s. trace;*

(2) *there exists a central element $0 \leq a \leq \mathbb{1}$ such that $\tau(x) = (\tau + \tau')(a \cdot x)$ and $\tau'(x) = (\tau + \tau')((\mathbb{1} - a) \cdot x)$ for all $x \in \mathcal{M}_+$, with $\mathbf{s}(a) = \mathbb{1}$ and $\mathbf{s}(\mathbb{1} - a) = \operatorname{supp} \tau'$.*

Proof.   The functional $\tau_0 := \tau + \tau'$ is clearly a faithful normal trace. Now, observe that $\mathfrak{m}_\tau \cdot \mathfrak{m}_{\tau'} \subseteq \mathfrak{m}_\tau \cap \mathfrak{m}_{\tau'} \subseteq \mathfrak{m}_{\tau_0}$ with $\mathfrak{m}_\tau \cdot \mathfrak{m}_{\tau'}$ σ-weakly dense in $\mathcal{M}$. Thus, $\mathfrak{m}_{\tau_0}$ is σ-weakly dense, and hence, τ_0 is semifinite. Set $a = \frac{d\tau}{d\tau_0} \in \widehat{\mathcal{M}}_+$ (see Definition 4.27). Since $\tau \leq \tau_0$, the Radon–Nikodym Theorem for f.n.s. traces (see Theorem 4.26) yields $0 \leq a \leq \mathbb{1}$. The inequality $\tau \leq \tau_0$ implies that $\mathfrak{n}_{\tau_0} \subseteq \mathfrak{n}_\tau$. Using this fact, a suitable polarisation identity now shows that for all $x, y \in \mathfrak{n}_{\tau_0}$ and any $u \in \mathcal{M}$, we have that

$$\langle \pi_{\tau_0}(u)\pi_{\tau_0}(a)\eta(x), \eta(y) \rangle_{\tau_0} = \langle \pi_{\tau_0}(a)\eta(x), \eta(u^* y) \rangle_{\tau_0}$$

$$= \tau(y^* u x) = \langle \pi_{\tau_0}(a)\eta(ux), \eta(y) \rangle_{\tau_0} = \langle \pi_{\tau_0}(au)\eta(x), \eta(x) \rangle_{\tau_0}.$$

This clearly ensures that $\pi_{\tau_0}(a) \in \pi_{\tau_0}(\mathcal{M})'$ and hence that $a \in \mathcal{M}'$, since the representation π_{τ_0} is faithful.

We have thus far proven that

$$\tau_0(ay^* x) = \tau(y^* x) \text{ for every } x, y \in \mathfrak{n}_{\tau_0}. \tag{4.2}$$

For any $b \in \mathcal{M}_+$, we may select a net $(b_i)_{i \in I}$ in $\mathfrak{m}_{\tau_0}$ increasing to b. When combined with what we have just noted, the normality of the traces involved now shows that $\tau(b) = \sup_{i \in I} \tau(b_i) = \sup_{i \in I} \tau_0(a \cdot b_i) = \tau_0(a \cdot b)$. This proves that $\tau = (\tau + \tau')(a \cdot \cdot)$. The fact that $\mathbf{s}(a) = \mathbb{1}$ is a trivial consequence of the statement regarding supports in the preceding theorem.

Note that it follows trivially from equation (4.2) that $\tau_0((\mathbb{1} - a) \cdot y^* x) = \tau'(y^* x)$ for every $x, y \in \mathfrak{n}_{\tau_0}$. Thus, a similar proof to the one above shows that $\tau'(\cdot) = (\tau + \tau')((\mathbb{1} - a) \cdot \cdot)$. The element $\mathbb{1} - a$ clearly plays the role of a Radon–Nikodym derivative. The uniqueness of the Radon–Nikodym derivative therefore ensures that $\mathbb{1} - a = \frac{d\tau'}{d\tau_0}$. The claim about the supports, therefore, follows from the preceding theorem. $\square$

Proposition 4.30 *Let τ be an f.n.s. trace on $\mathcal{M}$. For any normal trace τ' on $\mathcal{M}$, we have $\frac{d\tau'}{d\tau} \in \widehat{\mathcal{Z}(\mathcal{M})}_+$.*

Proof. First, let τ' be semifinite, and let $\tau_0 := \tau + \tau'$ and $a := \frac{d\tau_0}{d\tau}$. We claim that $a^{-1}(\mathbb{1} - a) = \frac{d\tau'}{d\tau}$, which by the fact that $a \in \mathcal{Z}(\mathcal{M})$ will then show that $\frac{d\tau'}{d\tau} \in \overline{\mathcal{Z}(\mathcal{M})}$ in this case (cf. Proposition 0.26(5)). To see this, note that for any $b \in \mathcal{M}_+$, we have

$$\tau([a^{-1}(\mathbb{1} - a)] \cdot b) = \tau_0(a \cdot b^{1/2}[a^{-1}(\mathbb{1} - a)]b^{1/2})$$

$$= \tau_0((\mathbb{1} - a) \cdot b) = \tau'(b).$$

Let τ' be an arbitrary normal trace. Then, $e_\infty = e_\infty(\tau') \in \overline{\mathcal{Z}(\mathcal{M})}$. Since $e_\infty \tau' e_\infty$ is semifinite, we know from the first part of the proof that $\frac{d(e_\infty \tau' e_\infty)}{d\tau} \in \mathcal{Z}(\mathcal{M})$. For simplicity of notation, we will write h_∞ for $\frac{d(e_\infty \tau' e_\infty)}{d\tau}$.

Suppose that $h_\infty = \int \lambda de_\lambda$. Since for any $b \in \mathcal{M}e_\infty^\perp$ we have $\tau(h_\infty b) = (e_\infty \tau' e_\infty)(b) = 0$, it is clear that $\mathbf{s}(h_\infty) \leq e_\infty$. The prescription $h_{\tau'} = \int \lambda de_\lambda + \infty \cdot (\mathbb{1} - e_\infty)$ will therefore yield a well-defined element of $\widehat{\mathcal{Z}(\mathcal{M})}_+$. It is now an exercise to see that for any $b \in \mathcal{M}_+$, we have that $\tau(h_{\tau'} \cdot b) = \tau(h_\infty \cdot b) + \tau([\infty \cdot (\mathbb{1} - e_\infty)] \cdot b) = (e_\infty \tau' e_\infty)(b) + \tau([\infty \cdot (\mathbb{1} - e_\infty)] \cdot b) = \tau'(b)$. It is therefore clear that $\frac{d\tau'}{d\tau} = h_{\tau'}$, which proves the claim. $\qquad\qquad\square$

We finish the section by showing how the density of a weight with respect to a trace depends on the properties of the weight. Theorem 4.32 has been obtained by Trunov [**Tru82**], using modular theory. Again, the proof presented here is elementary and is based on the proof from [**GP15**].

Lemma 4.31 *Let $\mathcal{M}$ be a semifinite properly infinite von Neumann algebra endowed with an f.n.s. trace τ, and let $h \in \overline{\mathcal{M}}_+$. Assume that there exist an increasing sequence of natural numbers $1 = n_1 < n_2 < \ldots$ and $e \in \mathbb{P}(\mathcal{M})$ such that $\chi_{[n_k, n_{k+1})}(h) \succsim e$ for each k. Then, the weight τ_h is not strongly semifinite.*

Proof. First, find projections $e_k \sim e$ such that

$$e_k \leq \chi_{[n_k, n_{k+1})}(h) \text{for all } k \in \mathbb{N}.$$

Choose $u_k \in \mathcal{M}$ so that $u_k^* u_k = e_1, u_k u_k^* = e_k$. Let (α_k) be a sequence of positive reals such that $\sum_{k=1}^{\infty} \alpha_k^2 = 1$ and $\sum_{k=1}^{\infty} \alpha_k^2 n_k = \infty$; this is easy, using $n_k \geq k$. For any $\epsilon > 0$, there is an m_0 such that for each $m \geq m_0$

$$\| \sum_{k=m}^{\infty} \alpha_k u_k \|^2 = \| \sum_{k=m}^{\infty} \alpha_k^2 u_k^* u_k \| = \| \sum_{k=m}^{\infty} \alpha_k^2 e_1 \| \leq \epsilon.$$

Hence, $v := \sum_{k=1}^{\infty} \alpha_k u_k$ exists (with convergence in norm topology) and belongs to $\mathcal{M}$. Put $p := vv^*$. Since $v^* v = e_1$, p is a non-zero projection. We will show that if $0 \neq q \leq p$, then $\varphi(q) = \infty$. Put $g := v^* q v$. Then, $g \leq v^* v = e_1$ and $q = vgv^*$. From the spectral theorem, we easily get $h \geq \sum_{m=1}^{\infty} n_m e_m$. We calculate

$$\varphi(q) = \tau(h \cdot q) = \tau(h \cdot pqp) = \tau(h \cdot vv^* q vv^*)$$

$$= \tau(h \cdot vgv^*) = \tau(v^* h v \cdot g) = \tau((\sum_{j=1}^{\infty} \alpha_j u_j^*) h (\sum_{k=1}^{\infty} \alpha_k u_k^*) \cdot g)$$

$$\geq \tau((\sum_{j=1}^{\infty} \alpha_j u_j^*)(\sum_{m=1}^{\infty} n_m e_m)(\sum_{k=1}^{\infty} \alpha_k u_k^*) \cdot g)$$

$$\geq \tau((\sum_{j,k,m=1}^{\infty} \alpha_j \alpha_k n_m u_j^* e_m u_k) \cdot g) = \tau((\sum_{m=1}^{\infty} \alpha_m^2 n_m u_m^* e_m u_m) \cdot g)$$

$$= \sum_{m=1}^{\infty} \alpha_m^2 n_m \tau(e_1 \cdot g) = \sum_{m=1}^{\infty} \alpha_m^2 n_m \tau(g) = \infty. \qquad \square$$

Theorem 4.32 *Let $\mathcal{M}$ be a semifinite von Neumann algebra endowed with an f.n.s. trace τ. Assume that, φ is an f.n.s. weight on $\mathcal{M}$. Then, φ is strongly semifinite if and only if $d\varphi/d\tau$ is locally measurable.*

Proof. '$\Rightarrow$' Assume that $h := d\varphi/d\tau$ is not locally measurable. We will show that τ_h is not strongly semifinite. By assumption, there exists a (largest) projection $z \in \mathcal{Z}(\mathcal{M})$ such that for any projection $z' \in \mathcal{Z}(\mathcal{M})$, if $0 \neq z' \leq z$, then hz' is not measurable. We can assume that $z = \mathbb{1}$. In fact, if φ is strongly semifinite, then it is strongly semifinite on each $\mathcal{M}z$, with $z \in \mathbb{P}(\mathcal{Z}(\mathcal{M}))$.

By the structure theorem for semifinite von Neumann algebras (Theorem 0.127), we have $\mathcal{M} = \sum_{\alpha}^{\oplus} \mathcal{M}z_\alpha$ with $z_\alpha \in \mathbb{P}(\mathcal{Z}(\mathcal{M}))$, $\sum_{\alpha} z_\alpha = \mathbb{1}$, where α runs over infinite cardinals $\leq \#\mathcal{M}$ (with some of the z_α possibly zero), and with each $\mathcal{M}z_\alpha \simeq \mathcal{N}_\alpha \overline{\otimes} \mathcal{F}_\alpha$ where $\mathcal{N}_\alpha$ is finite and $\mathcal{F}_\alpha$ a factor of type I_α. As above, we can assume that

$\mathcal{M} = \mathcal{N} \bar{\otimes} \mathcal{F}$, where $\mathcal{N}$ is finite and $\mathcal{F}$ is an infinite-dimensional type I factor. Note that $\mathcal{Z}(\mathcal{M}) = \mathcal{Z}(\mathcal{N}) \otimes \mathbb{C}\mathbb{1}_{\mathcal{F}}$. Let $\tau_{\mathcal{N}}$ be a normalised trace on $\mathcal{N}$ (i.e. $\tau(\mathbb{1}_{\mathcal{N}}) = 1$) and Tr a normalised trace on $\mathcal{F}$ (i.e., the trace of any minimal projection from $\mathcal{F}$ is 1). Let f be a minimal projection from $\mathcal{F}$ and $e := \mathbb{1}_{\mathcal{N}} \otimes f$. It is clear that e is finite.

Fix $n \in \mathbb{N}$. Then, for each $m \geq n$, there is a largest central projection $w_m = u_m \otimes \mathbb{1}_{\mathcal{F}}$ such that

$$\chi_{[n,m)}(h)w_m \succsim ew_m.$$

As m increases, w_m has to increase to some w. Suppose $w \neq 1$. Then, for each $m \geq n$,

$$\chi_{[n,m)}(h)w^{\perp} \precsim ew^{\perp}.$$

By Lemma 3.13,

$$\chi_{[n,\infty)}(h)w^{\perp} = \bigvee_{m=n}^{\infty} \chi_{[n,m)}(h)w^{\perp} \precsim ew^{\perp}$$

so that $\chi_{[n,\infty)}(h)w^{\perp}$ is finite. This implies that $hw^{\perp}$ is measurable, which contradicts the first paragraph of the proof. Therefore $w = 1$. This means that we can successively choose $1 = n_1 < n_2 < \cdots$ and projections $z_k = z_{\mathcal{N}}^{(k)} \otimes \mathbb{1}_{\mathcal{F}}$ in such a way that

$$\chi_{[n_k,n_{k+1})}(h)z_k \succsim ez_k \text{ with } \tau_{\mathcal{N}}(z_{\mathcal{N}}^{(k)}) \geq 1 - 1/2^{k+1}.$$

Now, we put $z_{\mathcal{N}} = \bigwedge_{k=1}^{\infty} z_{\mathcal{N}}^{(k)}$ and $z = z_{\mathcal{N}} \otimes \mathbb{1}_{\mathcal{F}}$. We have $\tau_{\mathcal{N}}(z_{\mathcal{N}}) \geq 1/2$, whence $z \neq 0$. Moreover,

$$\chi_{[n_k,n_{k+1})}(h)z_k \succsim ez$$

which implies that

$$\chi_{[n_k,n_{k+1})}(h)z = (\chi_{[n_k,n_{k+1})}(h)z_k)z \succsim ez.$$

Now, as in the first paragraph of the proof, we can assume that $z = 1$ and then use Lemma 4.31 to show that φ is not strongly semifinite.

'$\Leftarrow$' Let h be locally measurable. Choose non-zero $p \in \mathbb{P}(\mathcal{M})$. Since τ is strongly semifinite (see Remark 4.15), there is a non-zero $q \in \mathbb{P}(\mathcal{M})$ with $q \leq p$ and $\tau(q) < \infty$. From the definition of local measurability of h, we get the existence of a $z \in \mathbb{P}(\mathcal{Z}(\mathcal{M}))$ such that $hz \in S(\mathcal{M})$

and $qz \neq 0$. Put $e_n := \chi_{[0,n]}(hz)$. Note that $e_n^{\perp}$ is finite for sufficiently large n. If $qz \precsim e_n^{\perp}$ for each n, then applying the centre-valued trace $\mathscr{T}$ (on the reduced algebra $\mathcal{M}_{e_1^{\perp}}$) to both sides, we get $qz = 0$, a contradiction. Hence, for some n_0, $r := qz \wedge \chi_{[0,n_0]}(hz) \neq 0$ (see Lemma 3.52). Obviously, $r \leq q \leq p$. Moreover, using Theorem 4.23(4) and Proposition 4.24,

$$\varphi(r) = \tau(h \cdot (qz \wedge e_{n_0})) = \tau(hz \cdot (qz \wedge e_{n_0})) \leq n_0 \tau(qz \wedge e_{n_0})$$

$$\leq n_0 \tau(q) < \infty. \qquad \square$$

Proposition 4.33 *If $\mathcal{M}$ is finite, then any semifinite weight φ on $\mathcal{M}$ is both strictly and strongly semifinite.*

Proof. Let τ be a normal faithful semifinite trace on $\mathcal{M}$. By Proposition 3.64, $d\varphi/d\tau \in \overline{\mathcal{M}} = LS(\mathcal{M})$; hence, by Theorem 4.32, φ is strongly semifinite.

Assume now additionally that $\mathcal{M}$ is σ-finite, and choose τ to be finite. Put $e_n = \chi_{[n,n+1)}(h)$. Then, $\sum_{n=0}^{\infty} e_n = \mathbb{1}$, and $\varphi(e_n) = \tau(he_n) \leq (n+1)\tau(e_n) < \infty$. Thus, $e_n \varphi e_n$ are positive normal functionals on $\mathcal{M}$ with mutually orthogonal supports. We shall show that $\varphi = \sum e_n \varphi e_n$. First of all, for a fixed $a \in \mathcal{M}$, $e_n a e_n \leq \|a\| e_n$, so that $e_n a e_n \in \mathfrak{m}_{\varphi}^{+}$. Hence, $a^{1/2} e_n \in \mathfrak{n}_{\varphi}$ for all n, so that $e_m a e_n \in \mathfrak{m}_{\varphi}$ for all m, n. If $m \neq n$,

$$\varphi(e_m a x e_n) = \tau(h^{1/2} e_m a e_n h^{1/2}) = \tau(a^{1/2} e_n h^{1/2} h^{1/2} e_m a^{1/2})$$

$$= \tau(a^{1/2} e_n h^{1/2} e_n e_m h^{1/2} e_m a^{1/2}) = 0.$$

Consequently, $\varphi(a) = \sum (e_n \varphi e_n)(a)$, which shows that φ is strictly semifinite.

Finally, let $\mathcal{M}$ be an arbitrary finite von Neumann algebra. Then, there exists a family $\{z_i\}$ of non-zero central projections in $\mathcal{M}$ such that $\sum z_i = \mathbb{1}$ and each of the reduced von Neumann algebras $\mathcal{M}_{z_i}$ is σ-finite. Note that $\mathfrak{p}_{\varphi \restriction (\mathcal{M}_{z_i})_+} = \mathfrak{p}_{\varphi} z_i$. Hence, for any i, $\varphi \restriction (\mathcal{M}_{z_i})_+$ is semifinite and thus also strictly semifinite. It follows directly from the definition of strict semifiniteness that φ is also strictly semifinite. $\qquad \square$

Chapter 5
Basic theory of decreasing rearrangements

In this chapter, we develop a noncommutative theory of decreasing rearrangements, closely mimicking the classical theory. This will prove to be an indispensable tool for developing a theory of L^p and Orlicz spaces for not just tracial von Neumann algebras, but also general von Neumann algebras. These decreasing arrangements are constructed from the spectral distribution of the modulus of a given operator and will, as in the classical case, similarly enable us to reduce many general questions to the setting of the Borel σ-algebra on $(0, \infty)$.

5.1 Distributions and reduction to subalgebras

We start by introducing the concept of decreasing rearrangements for τ-measurable operators.

Definition 5.1 Given $f \in \widetilde{\mathcal{M}}$, we define the so-called decreasing rearrangement of f to be the function $\mathbf{m}_f : [0, \infty) \to [0, \infty] : t \mapsto \mathbf{m}_f(t)$ given by $\mathbf{m}_f(t) = \inf\{\|fe\|_\infty : e \in \mathbb{P}(\mathcal{M}), \tau(\mathbb{1} - e) \leq t\}$, and the distribution function of f by $[0, \infty) \to [0, \infty] : s \mapsto d_f(s)$, where $d_f(s) = \tau(\chi_{(s, \infty)}(|f|))$.

We note that what we call the decreasing rearrangement of $f \in \mathcal{M}$ is referred to as the *generalised singular value function* of f by many authors. However, to emphasise the analogy of the theory developed with the classical theory of decreasing rearrangements and Banach function spaces, we prefer the term *decreasing rearrangement*. We collate the basic properties of distributions of elements of $\widetilde{\mathcal{M}}$ and then pause to illustrate an interesting application of this theory, before proceeding with the development of the theory of decreasing rearrangements. For this task, we need the following lemma. A stronger version of this

Noncommutative measures and L^p and Orlicz Spaces, with Applications to Quantum Physics. Stanisław Goldstein and Louis Labuschagne, Oxford University Press. © Stanisław Goldstein and Louis Labuschagne (2025).
DOI: 10.1093/oso/9780198950202.003.0007

fact was first proved for bounded operators by Akemann, Anderson and Pedersen [**AAP82**]. However, the validity of the weaker version below extends to the class of τ-measurable operators. The proof we present is due to Kosaki.

Lemma 5.2 ([Kos84b])   *Let $f, g \in \widetilde{\mathcal{M}}$ be given. Then:*

- *there exists a partial isometry $w \in \mathcal{M}$ so that $\mathrm{Re}(f)_+ \leq w|f|w^*$, where $\mathrm{Re}(f)_+$ is the positive part of $\mathrm{Re}(f)$;*
- *there exist partial isometries $u, v \in \mathcal{M}$ such that $|f + g| \leq u|f|u^* + v|g|v^*$.*

Proof.   We prove the first claim. Let $f = u|f|$ be the polar decomposition of f with e the support projection of $\mathrm{Re}(f)_+$. Now, set $a = \frac{1}{2}e(\mathbb{1} + u)$, and let $a|f|^{1/2} = w|a|f|^{1/2}|$ be the polar decomposition of $a|f|^{1/2}$. Each of e, u, a, and w of course belongs to $\mathcal{M}$, with $a^*a = \frac{1}{4}(\mathbb{1} + u^*)e(\mathbb{1} + u) \leq \mathbb{1}$. In addition, we note that

$$a|f|a^* = w|a|f|^{1/2}|^2 w^* = w|f|^{1/2}a^*a|f|^{1/2}w^* \leq w|f|w^*.$$

Here, we made use of the fact that if $g = u_g|g|$ is the polar decomposition of g, then $gg^* = u_g g^* g u_g^*$, from which it follows that $|g^*| = u_g|g|u_g^*$. In view of the fact that $\mathrm{Re}(f)_+ = e\mathrm{Re}(f)e = \frac{1}{2}e(|f|u^* + u|f|)e$, we therefore have that

$$
\begin{aligned}
w|f|w^* - \mathrm{Re}(f)_+ \; &\geq \; a|f|a^* - \frac{1}{2}e(|f|u^* + u|f|)e \\[2mm]
&= \; \frac{1}{4}e(\mathbb{1} + u)|f|(\mathbb{1} + u^*)e - \frac{1}{2}e(|f|u^* + u|f|)e \\[2mm]
&= \; \frac{1}{4}e[(\mathbb{1} + u)|f|(\mathbb{1} + u^*) - 2(|f|u^* + u|f|)]e \\[2mm]
&= \; \frac{1}{4}e(\mathbb{1} - u)|f|(\mathbb{1} - u^*)e \\[2mm]
&\geq \; 0
\end{aligned}
$$

as required.

We proceed to proving the second claim. Let $(f + g) = v|f + g|$ be the polar decomposition of $(f + g)$. Given that $|f + g| = v^*(f + g) = (f^* + g^*)v$, we have that

$$|f + g| = \frac{1}{2}(v^*(f + g) + (f^* + g^*)v)$$
$$= \mathrm{Re}(v^* f) + \mathrm{Re}(v^* g)$$
$$\leq \mathrm{Re}(v^* f)_+ + \mathrm{Re}(v^* g)_+.$$

By the first part, we may now select partial isometries w_1 and w_2 in $\mathcal{M}$ so that $\mathrm{Re}(v^* f)_+ \leq w_1 |v^* f| w_1^* \leq w_1 |f| w_1^*$ and $\mathrm{Re}(v^* g)_+ \leq w_2 |v^* g| w_2^* \leq w_2 |g| w_2^*$. Therefore,

$$|f + g| \leq \mathrm{Re}(v^* f)_+ + \mathrm{Re}(v^* g)_+ \leq w_1 |f| w_1^* + w_2 |g| w_2^*$$

as required. $\qquad\square$

We will shortly see that there is a close link between decreasing rearrangements and the distribution function. We therefore pause to investigate the properties of the distribution function.

Proposition 5.3 *For any $f \in \widetilde{\mathcal{M}}$, the function $d : \mathbb{R} \to [0, \infty] : s \mapsto d_f(s)$ is non-increasing and right-continuous. This function satisfies the following properties.*

(i) *For any $f \in \widetilde{\mathcal{M}}$ and any non-zero scalar α, we have that $d_f(s) = d_{f^*}(s) = d_{|f|}(s)$ and that $d_{\alpha f}(s) = d_f(s/|\alpha|)$ for all $s \geq 0$.*

(ii) *Given $a, b \in \widetilde{\mathcal{M}}$ with $a \geq b \geq 0$, we then have that $d_a(s) \geq d_b(s)$ for any $s \geq 0$. Moreover, if $(a_\alpha) \subseteq \widetilde{\mathcal{M}}_+$ is a net increasing to a in $\widetilde{\mathcal{M}}$, then $\sup_\alpha d_{a_\alpha}(s) = d_a(s)$ for any $s \geq 0$.*

(iii) *For any $t, s \geq 0$ and any $a, b \in \widetilde{\mathcal{M}}$, $d_{a+b}(t+s) \leq d_a(t) + d_b(s)$.*

(iv) *For any $f \in \widetilde{\mathcal{M}}$ and any $a, b \in \mathcal{M}$, $d_{afb} \leq d_{\|a\|\|b\|f}$.*

(v) *For any $f \in \widetilde{\mathcal{M}}$, $d_f(t) \to 0$ as $t \to \infty$.*

Proof. Assume that $\mathcal{M}$ acts on the Hilbert space H. It is clear from the definition that $\mathbb{R} \to [0, \infty] : s \mapsto d_s(f)$ is non-increasing. Now, let $s_n, s \in (0, \infty)$ be given such that $s_n \searrow s$. From the spectral theorem, it then follows that $\chi_{(s_n, \infty)}(|f|) \nearrow \chi_{(s, \infty)}(|f|)$ strongly as $n \to \infty$. The normality of the trace τ then ensures that $d_{s_n}(f) \nearrow d_s(f)$ as $n \to \infty$. Thus, $\mathbb{R} \to [0, \infty] : s \mapsto d_s(f)$ is right-continuous.

Claim (i): The fact that $d_f = d_{|f|}$ is clear from the definition. The further fact that $\chi_{(s, \infty)}(|\alpha f|) = \chi_{(s/|\alpha|, \infty)}(|f|)$, and hence $d_{\alpha f}(s) = d_f(s/|\alpha|)$ for any $s \geq 0$, follows from the Borel functional calculus.

Next, let $f = u|f|$ be the polar decomposition of $|f|$. As noted in the previous proof, we then have $|f^*| = u|f|u^*$. By the uniqueness of the spectral decomposition, we must then have that $\chi_{(s,\infty)}(|f^*|) = u\chi_{(s,\infty)}(|f|)u^*$ for all $s > 0$. Consequently, $d_{f^*}(s) = \tau(\chi_{(s,\infty)}(|f^*|)) = \tau(u\chi_{(s,\infty)}(|f|)u^*) = \tau(\chi_{(s,\infty)}(|f|)) = d_f(s)$ for all $s > 0$. This then proves the claim.

Claim (ii): First suppose that $a \geq b \geq 0$. If $\chi_{[0,s]}(a)(H) \cap \chi_{(s,\infty)}(b)(H)$ contained non-zero elements, then for any $\xi \in \chi_{[0,s]}(a)(H) \cap \chi_{(s,\infty)}(b)(H)$ with $\|\xi\| = 1$, it would follow that $\langle b\xi, \xi \rangle > s$, and that $\langle a\xi, \xi \rangle \leq s$. But this is impossible since the inequality $a \geq b$ demands that $\langle b\xi, \xi \rangle \leq \langle a\xi, \xi \rangle$. Thus, we must have that $\chi_{[0,s]}(a)(H) \cap \chi_{(s,\infty)}(b)(H) = \{0\}$. Equivalently, $\chi_{[0,s]}(a) \wedge \chi_{(s,\infty)}(b) = 0$. This then enables us to conclude that

$$
\begin{aligned}
\chi_{(s,\infty)}(b) \;&=\; \chi_{(s,\infty)}(b) - \chi_{[0,s]}(a) \wedge \chi_{(s,\infty)}(b) \\
&\sim\; \chi_{[0,s]}(a) \vee \chi_{(s,\infty)}(b) - \chi_{[0,s]}(a) \\
&\leq\; \mathbb{1} - \chi_{[0,s]}(a) \\
&=\; \chi_{(s,\infty)}(a).
\end{aligned}
$$

Consequently, $d_b(s) = \tau(\chi_{(s,\infty)}(b)) \leq \tau(\chi_{(s,\infty)}(a)) \leq d_a(s)$ for all $s \geq 0$.

Next, suppose that $(a_\alpha) \subseteq \widetilde{\mathcal{M}}_+$ is a net increasing to a in $\widetilde{\mathcal{M}}$. From what we have already proved, it is clear that $d_a(s) \geq \sup_\alpha d_{a_\alpha}(s)$.

To prove the converse inequality, we first show that

$$
\chi_{(s,\infty)}(a) \wedge (\wedge_\alpha \chi_{[0,s]}(a_\alpha)) = 0.
$$

If $\chi_{(s,\infty)}(a)(H) \cap (\cap_\alpha \chi_{[0,s]}(a_\alpha)(H))$ contained non-zero elements, then for any $\xi \in \chi_{(s,\infty)}(a)(H) \cap (\cap_\alpha \chi_{[0,s]}(a_\alpha)(H))$ with $\|\xi\| = 1$, it would follow that $\|a\xi\|^2 = \langle a^2\xi, \xi \rangle > s^2$, with in addition $\|a_\alpha\xi\|^2 = \langle a_\alpha^2\xi, \xi \rangle \leq s^2$. But since $a_\alpha \nearrow a$ strongly, we then have that $\|a\xi\| = \lim_\alpha \|a_\alpha\xi\| \leq s$, which is a clear contradiction. Thus, $\chi_{(s,\infty)}(a) \wedge (\wedge_\alpha \chi_{[0,s]}(a_\alpha)) = 0$ as claimed. But in that case, we have that

$$
\begin{aligned}
\chi_{(s,\infty)}(a) \;&=\; \chi_{(s,\infty)}(a) - \chi_{(s,\infty)}(a) \wedge (\wedge_\alpha \chi_{[0,s]}(a_\alpha)) \\
&\sim\; \chi_{(s,\infty)}(a) \vee (\wedge_\alpha \chi_{[0,s]}(a_\alpha)) - \wedge_\alpha \chi_{[0,s]}(a_\alpha) \\
&\leq\; \mathbb{1} - \wedge_\alpha \chi_{[0,s]}(a_\alpha) \\
&=\; \vee_\alpha \chi_{(s,\infty)}(a_\alpha).
\end{aligned}
$$

The normality of the trace now ensures that

$$d_a(s) = \tau(\chi_{(s,\infty)}(a)) \leq \tau(\vee_\alpha \chi_{(s,\infty)}(a_\alpha)) = \sup_\alpha \tau(\chi_{(s,\infty)}(a_\alpha))$$

$$= \sup_\alpha d_{a_\alpha}(s).$$

Claim (iii): Given $a, b \in \widetilde{\mathcal{M}}$, we use the preceding lemma to select partial isometries $u, v \in \mathcal{M}$ such that $|a + b| \leq u|a|u^* + v|b|v^*$. Then, of course $d_{a+b} \leq d_{u|a|u^*+v|b|v^*}$. Let $t, s > 0$ be given. For any norm one element ξ of $\chi_{(t+s,\infty)}(u|a|u^* + v|b|v^*)(H)$, we then have that $\|(u|a|u^* + v|b|v^*)(\xi)\| > t + s$. But then at least one of the inequalities $\|u|a|u^*(\xi)\| > t$ or $\|v|b|v^*(\xi)\| > s$ must hold, or else we will by the triangle inequality have that $\|(u|a|u^* + v|b|v^*)(\xi)\| \leq t + s$. In other words, $\xi \notin \chi_{[0,t]}(u|a|u^*)(H) \cap \chi_{[0,s]}(v|b|v^*)(H)$, or equivalently $\{0\} = \chi_{(t+s,\infty)}(u|a|u^* + v|b|v^*)(H) \cap \chi_{[0,t]}(u|a|u^*)(H) \cap \chi_{[0,s]}(v|b|v^*)$ (H). So, by definition

$$\chi_{(t+s,\infty)}(u|a|u^* + v|b|v^*) \wedge \chi_{[0,t]}(u|a|u^*) \wedge \chi_{[0,s]}(v|b|v^*) = 0.$$

This, in turn, ensures that

$$\chi_{(t+s,\infty)}(u|a|u^* + v|b|v^*)$$
$$= \chi_{(t+s,\infty)}(u|a|u^* + v|b|v^*)$$
$$\qquad - \chi_{(t+s,\infty)}(u|a|u^* + v|b|v^*) \wedge \chi_{[0,t]}(u|a|u^*) \wedge \chi_{[0,s]}(v|b|v^*)$$
$$\sim \chi_{(t+s,\infty)}(u|a|u^* + v|b|v^*) \vee (\chi_{[0,t]}(u|a|u^*) \wedge \chi_{[0,s]}(v|b|v^*))$$
$$\qquad - (\chi_{[0,t]}(u|a|u^*) \wedge \chi_{[0,s]}(v|b|v^*))$$
$$\leq \mathbb{1} - (\chi_{[0,t]}(u|a|u^*) \wedge \chi_{[0,s]}(v|b|v^*))$$
$$= \chi_{(t,\infty)}(u|a|u^*) \vee \chi_{(s,\infty)}(v|b|v^*).$$

But then

$$
\begin{aligned}
d_{a+b}(t+s) &\leq d_{u|a|u^*+v|b|v^*}(t+s) \\
&= \tau(\chi_{(t+s,\infty)}(u|a|u^* + v|b|v^*)) \\
&\leq \tau(\chi_{(t,\infty)}(u|a|u^*) \vee \chi_{(s,\infty)}(v|b|v^*)) \\
&\leq \tau(\chi_{(t,\infty)}(u|a|u^*)) + \tau(\chi_{(s,\infty)}(v|b|v^*)) \\
&= d_{u|a|u^*}(t) + d_{v|b|v^*}(s).
\end{aligned}
$$

Recall that, for example, $|u|a|u^*|^2 \leq |au^*|^2$ and hence that $|u|a|u^*| \leq |au^*|$. Similarly $|ua^*| \leq |a^*|$. We may therefore use part (i) to see that

$d_{u|a|u^*} \leq d_{au^*} = d_{ua^*} \leq d_{a^*} = d_a$ and similarly that $d_{v|b|v*} \leq d_b$. The claim now follows.

Claim (iv): Given $f \in \widetilde{\mathcal{M}}$ and $a \in \mathcal{M}$, we may use parts (i) and (ii) to see that $d_{af} = d_{|af|} = d_{|\,|a|f|} \leq d_{\|a\|\,|f|} = d_{\|a\|f}$. From this inequality and part (i), it then follows that $d_{fa} = d_{a^*f^*} \leq d_{\|a\|f^*} = d_{\|a\|f}$.

Claim (v): To see this, note that the properties of the spectral resolution of $|f|$ ensure that $\chi_{(t,\infty)}(|f|) \searrow 0$ as $t \to \infty$. Since for some $t_0 > 0$, $d_f(t_0) = \tau(\chi_{(t_0,\infty)}(|f|)) < \infty$, the normality of the trace now ensures that $\lim_{t\to\infty} d_f(t) = \lim_{t\to\infty} \tau(\chi_{(t,\infty)}(|f|)) = 0$. $\qquad\square$

The properties of the distribution function allow us to realise $\widetilde{\mathcal{M}}$ as an F-normed space.

Definition 5.4 Let X be a complex vector space.

- A functional $\|\cdot\| : X \to [0,\infty)$ is called an F-norm on X if it satisfies the following criteria for $x, y \in X$ and $\alpha \in \mathbb{C}$:
  - $\|0\| = 0$ and if $\|x\| = 0$, then $x = 0$.
  - $\|e^{it}x\| = \|x\|$ for any $t \in \mathbb{R}$.
  - $\|x + y\| \leq \|x\| + \|y\|$.
  - Given sequences $(x_k) \subseteq X$ and $(\alpha_k) \subseteq \mathbb{C}$ such that $\|x - x_k\| \to 0$ and $|\alpha - \alpha_k| \to 0$ as $k \to \infty$, we will then have that $\|\alpha x - \alpha_k x_k\| \to 0$ as $k \to \infty$.
- A functional $\rho : X \to [0,\infty]$ is called a semimodular (alt. modular) on X if it satisfies the following criteria for $x, y \in X$:

  (a) $\rho(0) = 0$ and $x = 0$ whenever $\rho(\epsilon x) = 0$ for all $\epsilon > 0$ (alt. $x = 0$ whenever $\rho(x) = 0$).

  (b) $\rho(e^{it}x) = \rho(x)$ for any $t \in \mathbb{R}$.

  (c) For any $\epsilon \in [0,1]$, we have that $\rho(\epsilon x + (1-\epsilon)y) \leq \rho(x) + \rho(y)$.

  We say that ρ is a convex modular if instead of (c), we have $\rho(\epsilon x + (1-\epsilon)y) \leq \epsilon\rho(x) + (1-\epsilon)\rho(y)$ for any $\epsilon \in [0,1]$ and any $x, y \in X$.

  If the (semi)modular on X satisfies $\lim_{\epsilon\to 0} \rho(\epsilon x) = 0$ for each $x \in X$, X is said to be a modular space.

Remark 5.5 There is a very close link between F-normed spaces and modular spaces. Any F-norm on a vector space X for which $\epsilon \mapsto \|\epsilon x\|$

is for every x a non-decreasing function of ϵ on $[0, \infty)$ is a modular. Conversely, if X is equipped with a semimodular with respect to which it is a modular space, then the prescription $\|x\| = \inf\{\epsilon > 0 : \rho(\epsilon^{-1}x) \leq \epsilon\}$ yields an F-norm on X with the property that for any sequence $(x_k) \subseteq X$, the claim that $\lim_{k\to\infty} \|x - x_k \cdot \| = 0$ is equivalent to the claim that $\lim_{k\to\infty} \rho(\epsilon(x - x_k)) = 0$ for each $\epsilon > 0$. The interested reader may find proofs of these facts in [**Mus83**, section I.1].

With the above as background, we are now able to make the following conclusion.

Theorem 5.6 *The functional $\rho : \widetilde{\mathcal{M}} \to [0, \infty]$ defined by $\rho(f) = d_f(1)$, is a semimodular on $\widetilde{\mathcal{M}}$ with respect to which $\widetilde{\mathcal{M}}$ is a modular space. Moreover, the prescription $\|f\| = \inf\{\epsilon > 0 : d_f(\epsilon) \leq \epsilon\}$ defines an F-norm on $\widetilde{\mathcal{M}}$. The topology induced by this F-norm is exactly the topology of convergence in measure.*

Proof. We clearly have $\rho(0) = d_0(1) = \tau(\chi_{(1,\infty)}(0)) = 0$. In addition, if for some $f \in \widetilde{\mathcal{M}}$ we have that $0 = \rho(\epsilon f) = d_{\epsilon f}(1) = d_f(\epsilon^{-1}) = \tau(\chi_{(\epsilon^{-1},\infty)}(|f|))$ for all $\epsilon > 0$ (hence $\chi_{(\epsilon^{-1},\infty)}(|f|) = 0$ for all $\epsilon > 0$), we must clearly have that $|f| = 0$ (and hence $f = 0$).

It also follows fairly immediately from Proposition 5.3 that $\rho(e^{it}x) = \rho(x)$ for any $t \in \mathbb{R}$, and that for any $\epsilon \in (0,1)$ and any $f, g \in \widetilde{\mathcal{M}}$, we have $\rho(\epsilon f + (1 - \epsilon)g) = d_{(\epsilon f + (1-\epsilon)g)}(1) \leq d_{\epsilon f}(\epsilon) + d_{(1-\epsilon)g}(1 - \epsilon) = d_f(1) + d_g(1) = \rho(f) + \rho(g)$. Thus, ρ is a semimodular as claimed.

Next, note that, by definition, any element f of $\widetilde{\mathcal{M}}$ must satisfy the property that $\lim_{\gamma\to\infty} d_f(\gamma) = 0$. But since for $\gamma > 0$ we have that $d_f(\gamma) = d_{\gamma^{-1}f}(1)$, this corresponds to the claim that $\lim_{\epsilon\to 0} \rho(\epsilon f) = 0$ for each $f \in \widetilde{\mathcal{M}}$. Thus, $\widetilde{\mathcal{M}}$ is a modular space. The fact that the prescription $\|f\| = \inf\{\epsilon > 0 : d_f(\epsilon) \leq \epsilon\}$ defines an F-norm on $\widetilde{\mathcal{M}}$, now follows from the preceding remark. It therefore remains to show that the F-norm topology agrees with the topology of convergence in measure. Both topologies are metric topologies, and hence, the equivalence will follow if we show that convergence of sequences in the one is equivalent to convergence of sequences in the other. In this regard, recall that convergence of a sequence $(f_n) \subseteq \widetilde{\mathcal{M}}$ to $f \in \widetilde{\mathcal{M}}$ in the F-norm topology is equivalent to the claim that $\lim_{n\to\infty} \rho(\epsilon(f - f_n)) = 0$ for each $\epsilon > 0$. In other words, given $\epsilon > 0$ and some $\delta > 0$, there must exist $N \in \mathbb{N}$ so that $\tau(\chi_{(\epsilon^{-1},\infty)}(|f - f_n|)) = d_{f-f_n}(\epsilon^{-1}) = d_{\epsilon(f-f_n)}(1) = \rho(\epsilon(f - f_n)) \leq \delta$ for all $n \geq N$. But this is exactly convergence in measure. Hence, the result follows. $\qquad\square$

The following result establishes the link between the distribution function and decreasing rearrangements, hinted at earlier.

Proposition 5.7 *For any $f \in \widetilde{\mathcal{M}}$, we have $\mathbf{m}_f(t) = \inf\{s \geq 0 : d_f(s) \leq t\}$. Moreover, the infimum is attained and $d_f(\mathbf{m}_f(t)) \leq t$ for all $t \geq 0$.*

Proof. Assume that $\mathcal{M}$ acts on the Hilbert space H. The second claim follows from the right-continuity of $s \mapsto d_f(s)$. For the first claim let a be the point where the infimum is attained. Then, the inequality $d_f(a) \leq t$ amounts to the claim that $\tau(\mathbb{1} - \chi_{[0,a]}(|f|)) \leq t$. In view of the fact that $\|f\chi_{[0,a]}(|f|)\|_\infty = \| |f|\chi_{[0,a]}(|f|)\|_\infty \leq a$, it clearly follows that $\mathbf{m}_f(t) \leq a$. It remains to prove the converse.

Let $\epsilon > 0$ be given and select $e \in \mathbb{P}(\mathcal{M})$ such that $\tau(\mathbb{1} - e) \leq t$ with $\|fe\|_\infty < \mathbf{m}_f(t) + \epsilon$. For the sake of simplicity, we will write α_ϵ for $\mathbf{m}_f(t) + \epsilon$. Our task is, of course, to show that $a \leq \alpha_\epsilon$.

For any $\xi \in e(H) \cap \chi_{(\alpha_\epsilon,\infty)}(|f|)(H)$ with $\|\xi\| = 1$, the fact that $\chi_{(\alpha_\epsilon,\infty)}(|f|) = \chi_{(\alpha_\epsilon^2,\infty)}(f^*f)$ ensures that $\langle f^*f\xi, \xi \rangle \geq \alpha_\epsilon^2$. But, on the other hand, the fact that $\|fe\| < \alpha_\epsilon$ ensures that $\langle f^*f\xi, \xi \rangle < \alpha_\epsilon$. This cannot be, and hence, we must have that $e \wedge \chi_{(\alpha_\epsilon,\infty)}(|f|) = 0$. But in that case

$$\begin{aligned}
\chi_{(\alpha_\epsilon,\infty)}(|f|) &= \chi_{(\alpha_\epsilon,\infty)}(|f|) - e \wedge \chi_{(\alpha_\epsilon,\infty)}(|f|) \\
&\sim e \vee \chi_{(\alpha_\epsilon,\infty)}(|f|) - e \\
&\leq \mathbb{1} - e.
\end{aligned}$$

Consequently, $d_f(\alpha_\epsilon) = \tau(\chi_{(\alpha_\epsilon,\infty)}(|f|)) \leq \tau(\mathbb{1} - e) \leq t$, which amounts to the claim that $a \leq \alpha_\epsilon = \mathbf{m}_f(t) + \epsilon$, as required. $\qquad\square$

The following remark captures an important consequence of the above result—the fact that decreasing rearrangements are, in some sense, commutatively realised.

Remark 5.8

- Let $f \in \widetilde{\mathcal{M}}$ be given. For any von Neumann subalgebra $\mathcal{M}_0$ of $\mathcal{M}$ containing both f and the spectral projections of $|f|$, the above result actually shows that $\mathbf{m}_f(t) = \inf\{\|fe\|_\infty : e \in \mathbb{P}(\mathcal{M}_0), \tau(\mathbb{1} - e) \le t\}$
- The preceding proposition also shows that the quantity is a faithful extension of the classical concept of a decreasing rearrangement. To see this, observe that in the case where $\mathcal{M} = L^\infty(X, \Sigma, \nu)$ with $\tau = \int \cdot \, d\nu$, the formula in the preceding proposition corresponds to the claim that $\mathbf{m}_f(t) = \inf\{s > 0 : \nu(\{x \in X : |f(x)| > s\}) \le t$.

We pause to present a particularly elegant application of the above facts, which proves to be a useful tool in lifting the classical theory to the noncommutative context.

Proposition 5.9 *Let $\mathcal{M}$ be a von Neumann algebra with no minimal projections. Then, the following holds.*

(a) *For any non-zero projection $e \in \mathcal{M}$, any maximal abelian von Neumann subalgebra $\mathcal{M}_0$ of $e\mathcal{M}e$ also has no minimal projections.*

(b) *Whenever $e\mathcal{M}e$ admits a faithful normal state ω, then any maximal abelian subalgebra $\mathcal{M}_0$ of $e\mathcal{M}e$ corresponds to a classical $L^\infty(\Omega, \Sigma, \nu_\omega)$, where $(\Omega, \Sigma, \nu_\omega)$ is a nonatomic probability space, with the measure ν_ω defined by $\nu_\omega(E) = \omega(\chi_E)$ for each $E \in \Sigma$. In particular, if $\mathcal{M}$ is a finite algebra equipped with a finite faithful normal trace τ, then given any $f \in \widetilde{\mathcal{M}}_+$, we may select a maximal abelian subalgebra $\mathcal{M}_0$ containing all the spectral projections of f. The element f corresponds to a Borel function on $(\Omega, \Sigma, \nu_\tau)$, and the classical decreasing rearrangement of f as Borel function corresponds exactly to the decreasing rearrangement of f as an element of $\widetilde{\mathcal{M}}_+$.*

(c) *Suppose that $\mathcal{M}$ admits a faithful normal semifinite trace τ. For any non-zero projection $e \in \mathcal{M}$, $e\mathcal{M}e$ admits an abelian subalgebra $\mathcal{M}_0$, which has no minimal projections, and on which the restriction of τ is still semifinite. The algebra $\mathcal{M}_0$ corresponds to a classical $L^\infty(\Omega, \Sigma, \nu_\tau)$, where $(\Omega, \Sigma, \nu_\tau)$ is* *a nonatomic measure space, with the measure ν_τ defined by $\nu_\tau(E) = \tau(\chi_E)$ for each $E \in \Sigma$.*

Proof. *Part(a):* If $\mathcal{M}$ has no minimal projections, then the same is true of $e\mathcal{M}e$. Hence, for simplicity, we may replace $\mathcal{M}$ with $e\mathcal{M}e$ throughout. Let $\mathcal{M}_0$ be a commutative von Neumann subalgebra of $\mathcal{M}$. Suppose that e_0 is a minimal projection in $\mathcal{M}_0$. By hypothesis, there must exist a projection $f_0 \in \mathcal{M} \setminus \mathcal{M}_0$ with $0 < f_0 < e_0$. Now, given any other projection e in $\mathcal{M}_0$, we have by commutativity that $e_0 e \in \mathcal{M}_0$ is a subprojection of e_0. So, by minimality,

$$\text{either} \quad e_0 e = 0 \quad (\text{i.e., } e_0 \perp e) \quad \text{or} \quad e_0 e = e_0 \quad (\text{i.e., } e_0 \leq e).$$

Thus, since $f_0 < e_0$, we also have that

$$\text{either} \quad f_0 \perp e \quad \text{or} \quad f_0 < e$$

for any projection e in $\mathcal{M}_0$. But this means that f_0 commutes with all the projections in $\mathcal{M}_0$. Since the span of these projections is dense in $\mathcal{M}_0$, f_0 commutes with $\mathcal{M}_0$. Therefore, $\mathcal{M}_0$ cannot be maximal abelian, since $\{f_0, \mathcal{M}_0\}$ generates a commutative subalgebra that is strictly larger than $\mathcal{M}_0$.

Part(b): The first part of (b) follows from the fact that any commutative von Neumann subalgebra $\mathcal{M}_0$ will correspond to some $L^\infty(\Omega, \Sigma, \rho)$ (see Theorem 1.54). In particular, given a faithful normal state ω on $\mathcal{M}$, it is an exercise to show that the restriction of ω to $\mathcal{M}_0 = L^\infty(\Omega, \Sigma, \rho)$ defines a probability measure $\nu_\omega = \nu$ on (Ω, Σ) (with the same sets of measure zero as ν) by means of the prescription $\nu(E) = \omega(\chi_E) \ E \in \Sigma$. Replacing ρ by ν if necessary, all that remains is to note that the subalgebra $\mathcal{M}_0 = L^\infty(\Omega, \Sigma, \nu)$ has no minimal projections precisely when (Ω, Σ, ν) is nonatomic.

The final part of (b) follows by combining what we have just observed with the preceding remarks and the Borel functional calculus.

Part (c): To prove (c), we first note that the semifiniteness of the trace τ on $\mathcal{M}$ ensures that each projection f in $\mathcal{M}$, admits a subprojection with finite trace. To see this, recall that the semifiniteness of τ ensures that there exists $0 \leq x \leq f$ with $0 < \tau(x) < \infty$. For $\epsilon > 0$ small enough, $e_\epsilon = \chi_{(\epsilon,\infty)}(x)$ will be a non-zero projection with $\epsilon e_\epsilon \leq x \leq f$. But then $\tau(e_\epsilon) \leq \epsilon^{-1}\tau(x) < \infty$, with in addition $e_\epsilon = \lim_{n\to\infty} \epsilon^{2^{-n}} e_\epsilon \leq f$. Here, we used the fact that the square root preserves order. Given a projection e, we may now apply Zorn's lemma to obtain a maximal family $\{e_\alpha\}$ of mutually orthogonal subprojections of e, each with a finite trace. It must then hold that $e = \sum_\alpha e_\alpha$, since if $e - \sum_\alpha e_\alpha \neq 0$, we would be able to select a subprojection e_0 of $e - \sum_\alpha e_\alpha$ with finite trace, which would contradict the maximality of $\{e_\alpha\}$. For each α, we then select a maximal abelian subalgebra $\mathcal{M}_\alpha$ of $e_\alpha \mathcal{M} e_\alpha$. By part (a), there are no minimal projections in any of the $\mathcal{M}_\alpha$s. The algebra we seek is then given by $\mathcal{M}_0 = \oplus_\alpha \mathcal{M}_\alpha$. Since τ is finite on each $\mathcal{M}_\alpha$, it is an exercise to see that the restriction of τ to $\mathcal{M}_0$, is semifinite. The final part of (c) may now be proven using the same sort of argument as in part (b). $\square$

5.2 Algebraic properties of decreasing rearrangements

As before $\mathcal{M}$ is a von Neumann algebra equipped with a faithful normal semifinite trace τ. To start off with, we present yet another way of realising the decreasing rearrangements described earlier.

Lemma 5.10 *For each $t \geq 0$, we let R_t be the set of all τ-measurable operators f satisfying $\tau(\mathbf{s}(|f|)) \leq t$, where $\mathbf{s}(|f|)$ is the support projection of $|f|$. For any $g \in \widetilde{\mathcal{M}}$, we then have that*

$$\mathbf{m}_g(t) = \inf\{\|g - f\| : f \in R_t\}.$$

Proof. Let $t \geq 0$ be given and let $g = u|g|$ be the polar decomposition of g. With $\lambda \mapsto e_\lambda$ denoting the spectral resolution of $|g|$, we set

$$f = u \int_{(\alpha, \infty)} \lambda \, de_\lambda$$

where $\alpha = \mathbf{m}_g(t)$. Then, by construction $\|g - f\| \leq \alpha = \mathbf{m}_g(t)$, with $\tau(\mathbf{s}(|f|)) = d_g(\alpha) = d_g(\mathbf{m}_g(t)) \leq t$ by Proposition 5.7. Therefore, $\inf\{\|g - f\| : f \in R_t\} \leq \mathbf{m}_g(t)$.

For the converse, let $f \in R_t$ be given and set $e = \mathbb{1} - \mathbf{s}(|f|)$. Then $\|ge\| = \|(g - f)e\| \leq \|g - f\|$. In view of the fact that $\tau(\mathbb{1} - e) \leq t$, it follows that $\mathbf{m}_g(t) \leq \|ge\| \leq \|g - f\|$ and hence $\mathbf{m}_g(t) \leq \inf\{\|g - f\| : f \in R_t\}$. $\square$

Corollary 5.11 *For any $f \in \widetilde{\mathcal{M}}$ and any projection $e \in \mathcal{M}$, we have that $\mathbf{m}_{fe}(t) = 0$ whenever $t \geq \tau(e)$.*

The following proposition collates several important properties of decreasing rearrangements. The reader is encouraged to take careful note of these properties, as they will be used repeatedly without comment from here on.

Proposition 5.12   *Let f, a and b be τ-measurable operators.*

(i) *The map $\mathbf{m}_f : (0,\infty) \to \mathbb{R} : t \mapsto \mathbf{m}_f(t)$ is non-increasing and right-continuous. Moreover, $\lim_{t\searrow 0} \mathbf{m}_f(t) = \|f\|_\infty$.*

(ii) *For any $t > 0$ and any $\alpha \in \mathbb{C}$, $\mathbf{m}_f(t) = \mathbf{m}_{f^*}(t) = \mathbf{m}_{|f|}(t)$, and $\mathbf{m}_{\alpha f}(t) = |\alpha|\mathbf{m}_f(t)$.*

(iii) *If $a \geq b \geq 0$, then $\mathbf{m}_a(t) \geq \mathbf{m}_b(t)$ for any $t > 0$. Moreover, if $(a_\alpha) \subseteq \widetilde{\mathcal{M}}_+$ is a net increasing to a in $\widetilde{\mathcal{M}}$, then $\sup_\alpha \mathbf{m}_{a_\alpha}(t) = \mathbf{m}_a(t)$.*

(iv) *For any $s, t > 0$, $\mathbf{m}_{a+b}(t+s) \leq \mathbf{m}_a(t) + \mathbf{m}_b(s)$.*

(v) *For any $s, t > 0$, $\mathbf{m}_{ab}(t+s) \leq \mathbf{m}_a(t)\mathbf{m}_b(s)$. We have, in particular, $\mathbf{m}_{afb}(t) \leq \|a\|\|b\|\mathbf{m}_f(t)$.*

(vi) *Let $\Phi : [0,\infty) \to [0,\infty]$ be a non-decreasing function which is continuous on $[0, b_\Phi]$ where $b_\Phi = \sup\{t \in [0,\infty): \Phi(t) < \infty\}$. If $\Phi(|f|)$ is again τ-measurable, then $\Phi(\mathbf{m}_f(t)) = \mathbf{m}_{\Phi(|f|)}(t)$ for any $t \geq 0$.*

Proof.   *Claim (i):* The fact that $\mathbf{m}_f$ is non-increasing follows by definition. Next, suppose that at some point $t_0 > 0$, $\mathbf{m}_f$ is not right-continuous. There must then exist some $\alpha > 0$ so that $\mathbf{m}_f(t_0) > \alpha \geq \mathbf{m}_f(t_0 + \epsilon)$ for all $\epsilon > 0$. We may then combine the fact that $s \mapsto d_f(s)$ is non-increasing with Proposition 5.7, to conclude from this that $d_f(\alpha) \leq d_f(\mathbf{m}_f(t_0 + \epsilon)) \leq t_0 + \epsilon$ for any $\epsilon > 0$, that is $d_f(\alpha) \leq t_0$. But if that were the case, then by Proposition 5.7, we should have that $\mathbf{m}_f(t_0) \leq \alpha$ which is a clear contradiction. Hence, the claim regarding right-continuity holds.

It remains to prove the claim regarding the left limit at 0. It is clear from the definition that $\|f\| \geq \mathbf{m}_f(t)$ for all $t > 0$. Suppose that for all $\epsilon > 0$, we have that $\|f\| > \alpha \geq \mathbf{m}_f(\epsilon)$. A similar argument to the one used above then leads to the conclusion that $d_f(\alpha) \leq \epsilon$ for any $\epsilon > 0$ and hence that $d_f(\alpha) = 0$. But if that were the case, we ought to have that $\|f\| \leq \alpha$ – a clear contradiction. Hence, the remaining claim follows.

Claim (ii): The fact that $\mathbf{m}_f(t) = \mathbf{m}_{|f|}(t) = \mathbf{m}_{f^*}(t)$ for all $t > 0$ is a clear consequence of part (i) of Proposition 5.3 considered alongside Proposition 5.7.

Claim (iii): First suppose that $a \geq b \geq 0$. A simple application of Proposition 5.7 to part (ii) of Proposition 5.3 then yields the first conclusion.

Next, suppose that $(a_\alpha) \subseteq \widetilde{\mathcal{M}}_+$ is a net increasing to a in $\widetilde{\mathcal{M}}$. Given that, by Proposition 5.3, $d_a(s) = \sup_\alpha d_{a_\alpha}(s)$ for all $s \geq 0$, the claim similarly follows from Proposition 5.7.

Claim (iv): Let $\epsilon > 0$ and $s, t > 0$ be given. By Lemma 5.10, we may find $a_0, b_0 \in \widetilde{\mathcal{M}}$ such that

$$\|a - a_0\| \leq \mathbf{m}_a(t) + \epsilon, \quad \tau(\mathbf{s}(\|a_0\|)) \leq t$$
$$\|b - b_0\| \leq \mathbf{m}_b(s) + \epsilon, \quad \tau(\mathbf{s}(\|b_0\|)) \leq s.$$

We have that

$$\|(a + b) - (a_0 + b_0)\| \leq \|a - a_0\| + \|b - b_0\| \leq \mathbf{m}_a(t) + \mathbf{m}_b(s) + 2\epsilon.$$

In addition, we also have that

$$\begin{aligned}
\tau(\mathbf{s}(|a + b|)) &= \tau(\mathbf{s}(|a|) \vee \mathbf{s}(|b|)) \\
&\leq \tau(\mathbf{s}(|a|)) + \tau(\mathbf{s}(|b|)) \\
&\leq t + s.
\end{aligned}$$

It follows that $\mathbf{m}_{a+b}(t + s) \leq \mathbf{m}_a(t) + \mathbf{m}_b(s) + 2\epsilon$. Given that ϵ was arbitrary, it is clear that $\mathbf{m}_{a+b}(t + s) \leq \mathbf{m}_a(t) + \mathbf{m}_b(s)$ as required.

Claim (v): Let $\epsilon > 0$ and $s, t > 0$ be given, and select a_0 and b_0 as in the proof of the previous claim. Now, set $c = (a - a_0)b_0 + a_0 b$. We then have that

$$\begin{aligned}
\|ab - c\| &= \|ab - (a - a_0)b_0 - a_0 b\| \\
&= \|(a - a_0)(b - b_0)\| \\
&\leq \|(a - a_0)\| . \|(b - b_0)\| \\
&\leq (\mathbf{m}_a(t) + \epsilon)(\mathbf{m}_b(s) + \epsilon).
\end{aligned}$$

Moreover, we also have that

- $\mathbf{s}(|c|) \leq \mathbf{s}(|(a - a_0)b_0|) \vee \mathbf{s}(|a_0 b|)$;
- $\mathbf{s}(|(a - a_0)b_0|) \leq \mathbf{s}(|b_0|)$;
- $\mathbf{s}(|b^* a_0^*|) \leq \mathbf{s}(|a_0^*|)$ together with $\mathbf{s}(|a_0 b|) \sim \mathbf{s}(|b^* a_0^*|)$ and $\mathbf{s}(|a_0^*|) \sim \mathbf{s}(|a_0|)$.

It follows from this that $\tau(\mathbf{s}(|c|)) \leq \tau(\mathbf{s}(|b_0|)) + \tau(\mathbf{s}(|a_0|)) \leq t + s$. Therefore, $\mathbf{m}_{ab}(t+s) \leq (\mathbf{m}_a(t) + \epsilon)(\mathbf{m}_b(s) + \epsilon)$, whence $\mathbf{m}_{ab}(t+s) \leq \mathbf{m}_a(t).\mathbf{m}_b(s)$.

Claim (vi): We may replace $\mathcal{M}$ by a maximal abelian von Neumann subalgebra $\mathcal{M}_0$ to which both $|f|$ and $\Phi(|f|)$ are affiliated (see Remark 5.8). Let e be any projection in this subalgebra. Now, notice that $\mathrm{sp}(|f|) \subseteq [0, \infty)$.

First, suppose that Φ is bounded on $\mathrm{sp}(|f|e)$. (By the Borel functional calculus, $\Phi(|f|e)$ will then, of course, be bounded.) Since $\lim_{u \to \infty} \Phi(u) = \infty$, we must then have that $\mathrm{sp}(|f|e)$ itself is a bounded subset of $[0, \infty)$. Thus, $|f|e$ must be bounded. By the spectral theory for positive elements, we now have that $\| \, |f|e\| = \max\{\lambda \colon \lambda \in \mathrm{sp}(|f|e)\}$. Since Φ is increasing and non-negative on $[0, \infty)$, the Borel functional calculus also ensures that

$$\Phi(\| \, |f|e\|) = \max\{\Phi(\lambda) \colon \lambda \in \mathrm{sp}(|f|e)\} = \|\Phi(|f|e)\|.$$

If Φ is not bounded on $\mathrm{sp}(|f|e)$, then

$$\|\Phi(|f|e)\| = \sup\{\Phi(\lambda) \colon \lambda \in \mathrm{sp}(|f|e)\} = \infty.$$

We then proceed to show that $\Phi(\| \, |f|e\|) = \infty$. If $\mathrm{sp}(|f|e)$ was an unbounded subset of $[0, \infty)$, we would already have $\| \, |f|e\| = \infty$, and therefore, $\Phi(\| \, |f|e\|) = \infty$ as required. Thus, let $\mathrm{sp}(|f|e)$ be a bounded subset of $[0, \infty)$. As noted previously, this forces $\| \, |f|e\| = \max\{\lambda \colon \lambda \in \mathrm{sp}(|f|e)\}$. Since Φ is increasing on $[0, \infty]$ with $\Phi(0) = 0$, we must have $\Phi(\| \, |f|e\|) \geq \Phi(\lambda) \geq 0$ for any $\lambda \in \mathrm{sp}(|f|e)$. The fact that Φ is unbounded on $\mathrm{sp}(|f|e)$ therefore forces $\Phi(\| \, |f|e\|) = \infty$ as required.

The above observations clearly show that

$$\inf\{\Phi(\| \, |f|e\|) \colon e \in \mathbb{P}(\mathcal{M}_0) \text{ with } \tau(\mathbb{1} - e) \leq t\}$$

$$= \inf\{\|\Phi(|f|e)\| \colon e \in \mathbb{P}(\mathcal{M}_0) \text{ with } \tau(\mathbb{1} - e) \leq t\}.$$

But since Φ is increasing and continuous on $[0, b_\Phi]$, we also have that

$$\inf\{\Phi(\| \, |f|e\|) \colon e \in \mathbb{P}(\mathcal{M}_0) \text{ with } \tau(\mathbb{1} - e) \leq t\}$$

$$= \Phi(\inf\{\| \, |f|e\| \colon e \in \mathbb{P}(\mathcal{M}_0) \text{ with } \tau(\mathbb{1} - e) \leq t\}).$$

When combined with the observation in Remark 5.8, this yields the conclusion that $\mathbf{m}_{\Phi(|f|)}(t) = \Phi(\mathbf{m}_{|f|}(t)) = \Phi(\mathbf{m}_f(t))$. $\qquad \square$

5.3 Decreasing rearrangements and the trace

We saw in Theorem 4.23 that the trace may be extended to the extended positive cone $\widehat{\mathcal{M}}_+$ of $\mathcal{M}$ by means of the following prescription: Given $m \in \widehat{\mathcal{M}}_+$ with spectral resolution $m = \int_0^\infty \lambda\, de_\lambda + p.\infty$, we define the trace of m as $\tau(m) = \sup_{n \in \mathbb{N}} \tau(m_n)$, where the m_ns are the bounded elements of $\mathcal{M}$ given by $m_n = \int_0^n \lambda\, de_\lambda + p.n$. Our goal in this section is to show that on $\widetilde{\mathcal{M}}$, this extension exhibits a more subtle behaviour. The first step in realising this goal is the following proposition.

Proposition 5.13 *For any $f \in \widetilde{\mathcal{M}}_+$,*

$$\tau(f) = \int_0^\infty \mathbf{m}_f(t)\, dt.$$

The above proposition provides us with a powerful alternative for defining the trace on $\widetilde{\mathcal{M}}_+$. On considering this result alongside, for example, Proposition 5.12, we may then fairly directly deduce highly refined properties of the trace on $\widetilde{\mathcal{M}}_+$ from matching properties of the decreasing rearrangement. We will briefly address this issue in Proposition 5.17 and then revisit this topic in Chapter 6.

Proof. Let $f \in \widetilde{\mathcal{M}}_+$ be given. For each $n \in \mathbb{N}$, let $f_n = f\chi_{[0,n]}(f)$. By definition, we then have that $\tau(f) = \sup_{n \in \mathbb{N}} \tau(f_n)$. It is clear from the Borel functional calculus that the f_ns increase to f. So, by part (iii) of Proposition 5.12 considered alongside the monotone convergence theorem, we also have that $\int_0^\infty \mathbf{m}_f(t)\, dt = \sup_{n \in \mathbb{N}} \int_0^\infty \mathbf{m}_{f_n}(t)\, dt$. It is, therefore, clear that the proposition will follow for $\widetilde{\mathcal{M}}_+$, if we are able to prove it for $\mathcal{M}_+$. Hence, assume that $f \in \mathcal{M}_+$.

It is well known that any positive measurable function may be written as the increasing limit of a sequence of positive simple functions. If we combine this fact with the Borel functional calculus, it is clear that f can be written as an increasing limit of a sequence of operators (f_N) of the form $\sum_{k=1}^n \alpha_k e_k$, where the α_ks are positive real numbers, and the e_ks mutually orthogonal projections. By the normality of the trace on $\mathcal{M}_+$, $\tau(f) = \sup_N \tau(f_N)$. Once again considering part (iii) of Proposition 5.12 alongside the monotone convergence theorem, we see that $\int_0^\infty \mathbf{m}_f(t)\, dt = \sup_N \int_0^\infty \mathbf{m}_{f_N}(t)\, dt$. Therefore, the result will follow for $\mathcal{M}_+$ if we can verify its validity in the case where $f = \sum_{k=1}^n \alpha_k e_k$. We proceed with the verification of this case. We may assume, without loss of generality, that $\alpha_1 < \alpha_2 < \cdots < \alpha_n$. It then follows from the Borel functional calculus that

$$\chi_{(s,\infty)}(f) = \begin{cases} 0, & \text{if } s \geq \alpha_n \\ \sum_{k=m+1}^n e_k, & \text{if } \alpha_m \leq s < \alpha_{m+1} \\ \sum_{k=1}^n e_k, & \text{if } s < \alpha_1 \end{cases}$$

and hence that

$$d_f(s) = \begin{cases} 0, & \text{if } s \geq \alpha_n \\ \sum_{k=m+1}^n \tau(e_k), & \text{if } \alpha_m \leq s < \alpha_{m+1} \\ \sum_{k=1}^n \tau(e_k), & \text{if } s < \alpha_1 \end{cases}$$

Now, consider the case where each $\tau(e_k)$ is finite. It then follows from Proposition 5.7 that

$$\mathbf{m}_f(t) = \begin{cases} \alpha_n, & \text{if } \tau(e_n) > t \geq 0 \\ \alpha_m, & \text{if } \sum_{k=m}^n \tau(e_k) > t \geq \sum_{k=m+1}^n \tau(e_k) \,. \\ 0, & \text{if } t \geq \sum_{k=1}^n \tau(e_k) \end{cases}$$

It is then clear that $\int_0^\infty \mathbf{m}_f(t)\, dt = \sum_{k=1}^n \alpha_k \tau(e_k) = \tau(f)$ as required.

Now, suppose that some of the $\tau(e_k)$s are infinite and let m_0 be the largest index for which $\tau(e_{m_0}) = \infty$. Suppose, for the sake of the argument, that $n > m_0 \geq 1$. (The proof for the case where $m_0 = n$ is similar.) In this case, we, of course, have $\sum_{k=m}^n \tau(e_k) = \infty$ whenever $m \leq m_0$, and hence, the formula for $\mathbf{m}_f(t)$ becomes

$$\mathbf{m}_f(t) = \begin{cases} \alpha_n, & \text{if } \tau(e_n) > t \geq 0 \\ \alpha_m, & \text{if } \sum_{k=m}^n \tau(e_k) > t \geq \sum_{k=m+1}^n \tau(e_k), \quad m > m_0 \,. \\ \alpha_{m_0}, & \text{if } \infty > t \geq \sum_{k=m_0+1}^n \tau(e_k) \end{cases}$$

But then $\int_0^\infty \mathbf{m}_f(t)\, dt = \infty$, which accords with the fact that $\tau(f) = \sum_{k=1}^n \alpha_k \tau(e_k) = \infty$. $\qquad\square$

Corollary 5.14 *Let $\Phi : [0, \infty) \to [0, \infty]$ be a non-decreasing function which is continuous on $[0, b_\Phi]$, where $b_\Phi = \sup\{t \in [0, \infty): \Phi(t) < \infty\}$, and let $f \in \widetilde{\mathcal{M}}$ be given such that $\Phi(|f|)$ is again τ-measurable. Then $\tau(\Phi(|f|)) = \int_0^\infty \Phi(\mathbf{m}_t(|f|)) \, \mathrm{d}t$.*

Proof.   Apply Proposition 5.13 to part (vi) of Proposition 5.12.   □

We briefly comment on the use of the above corollary.

Remark 5.15 Let $\mathcal{M}$ be a semifinite algebra with a faithful normal semifinite trace τ. Any element a of $\widehat{\mathcal{M}}_+$ is, of course, of the form $a = \int_0^\infty \lambda de(\lambda) + \infty.p$ for some spectral resolution $e(\lambda)$ and projection p. For the extension of the trace to $\widehat{\mathcal{M}}_+$, we will by the discussion at the start of this section clearly have that $\tau(a) = \infty$ if $p \neq 0$. So, if $\tau(a) < \infty$, we must have that $p = 0$ and hence that a is in fact an operator affiliated to $\mathcal{M}$. But more is true in this case. For any $n, m \in \mathbb{N}$ with $n < m$, we will then have that

$$\tau(e_{(n,m]}) \leq \frac{1}{n}\tau(ae_{(n,m]}) \leq \frac{1}{n}\tau(a) < \infty.$$

If now we let $m \to \infty$, we get $\tau(e_{(n,\infty)}) \leq \frac{1}{n}\tau(a) < \infty$. Thus, if $\tau(a) < \infty$, a must in fact correspond to a τ-measurable element of $\widetilde{\mathcal{M}}$. Therefore, given $a \in \widehat{\mathcal{M}}_+$, it follows that $\tau(a) < \infty$ if and only if a corresponds to a positive element of $\widetilde{\mathcal{M}}$ with finite trace.

Given an element a of $\widetilde{\mathcal{M}}$ and a function $\Phi : [0, \infty) \to [0, \infty]$ of the type described in the preceding result, $\Phi(|a|)$ will, in general, not be a member of $\widetilde{\mathcal{M}}$ (unless of course $\mathrm{sp}(|a|) \subseteq [0, b_\Phi)$). However, we are able to give meaning to $\Phi(|a|)$ as an element of $\widehat{\mathcal{M}}_+$. If we apply the observation made previously to this setting, then given $a \in \widetilde{\mathcal{M}}$, it follows that $\tau(\Phi(|a|)) < \infty$ if and only if $\Phi(|a|)$ corresponds to a positive element of $\widetilde{\mathcal{M}}$ with finite trace.

Corollary 5.16 *Let $f, g \in \widetilde{\mathcal{M}}_+$ be given. Then, the following are equivalent.*

    (i) $\mathbf{m}_f(t) \leq \mathbf{m}_g(t)$ *for all $t > 0$.*
    (ii) $d_f(s) \leq d_g(s)$ *for all $s > 0$.*
    (iii) *For any non-decreasing function $\Phi : [0, \infty) \to [0, \infty]$, which is continuous on $[0, b_\Phi]$ where $b_\Phi = \sup\{t \in [0, \infty): \Phi(t) < \infty\}$, we have that $\tau(\Phi(f)) \leq \tau(\Phi(g))$.*

Proof. The implication (ii)$\Rightarrow$(i) is a direct consequence of Proposition 5.7, with (i)$\Rightarrow$(iii) following from the preceding Corollary 5.14 and Remark 5.15. To prove the implication (iii)$\Rightarrow$(ii), we first select a sequence (v_n) of continuous functions on $[0,\infty)$, which increase monotonically to $\chi_{(s,\infty)}$. (In fact, given any $\epsilon > 0$, we may select this sequence so that each v_n agrees with $\chi_{(s,\infty)}$ on $[0,s] \cup [s+\epsilon,\infty)$.) Then, by the Borel functional calculus, $v_n(f) \nearrow \chi_{(s,\infty)}(f)$ and $v_n(g) \nearrow \chi_{(s,\infty)}(g)$ as $n \to \infty$. The normality of the trace, therefore, ensures that $\tau(v_n(f)) \nearrow d_f(s)$ and $\tau(v_n(g)) \nearrow d_g(s)$ as $n \to \infty$. The validity of (ii) now follows from the fact that, by assumption, $\tau(v_n(f)) \leq \tau(v_n(g))$ for every n. $\square$

As promised, we now pause to briefly present a (fairly standard) sampling of the basic properties of the trace on $\widetilde{\mathcal{M}}_+$. The interested reader may easily use the techniques demonstrated in the following proof to deduce further properties.

Proposition 5.17 *The extension of the trace τ to $\widetilde{\mathcal{M}}_+$ is faithful, additive, and positive homogeneous. For any $g \in \widetilde{\mathcal{M}}$, we have $\tau(g^*g) = \tau(gg^*)$. In addition, τ is normal on $\widetilde{\mathcal{M}}_+$, in the sense that if $(f_\alpha) \subseteq \widetilde{\mathcal{M}}_+$ is a net increasing to $f \in \widetilde{\mathcal{M}}_+$, then $\tau(f) = \sup_\alpha \tau(f_\alpha)$.*

Proof. The proof of each of these properties relies on a repeated use of Proposition 5.13. To see that τ is faithful, observe that if $0 = \tau(f) = \int_0^\infty \mathbf{m}_f(s)\,ds$, then $\mathbf{m}_f = 0$ almost everywhere. But since $\mathbf{m}_f$ is nonincreasing, we have $\mathbf{m}_f(s) = 0$ for all $s \geq s_0$ whenever $\mathbf{m}_f(s_0) = 0$. Thus, $\mathbf{m}_f = 0$ on $(0,\infty)$, with the right-continuity of $\mathbf{m}_f$ now ensuring that also $\|f\|_\infty = \lim_{s \searrow 0} \mathbf{m}_f(s) = 0$.

The positive homogeneity of the trace is a direct consequence of part (ii) of Proposition 5.12.

The fact that $\tau(g^*g) = \tau(gg^*)$ for any $g \in \widetilde{\mathcal{M}}$, is similarly an easy consequence of the fact that part (ii) of Proposition 5.12 ensures that $\mathbf{m}_{g^*g} = \mathbf{m}_{|g|}^2 = \mathbf{m}_{|g^*|}^2 = \mathbf{m}_{gg^*}$.

Next, let $(f_\alpha) \subseteq \widetilde{\mathcal{M}}_+$ be a net increasing to $f \in \widetilde{\mathcal{M}}_+$. For each $n \in \mathbb{N}$, let $e_n = \chi_{[0,n]}(f)$. We know from Proposition 3.88 that $(e_n f_\alpha e_n)$ then increases to $f e_n$. But $f e_n$ belongs to $\mathcal{M}_+$, and hence, the same must be true of $\{e_n f_\alpha e_n\}$. It then follows from the normality of the trace on $\mathcal{M}_+$ that $\tau(f e_n) = \sup_\alpha \tau(e_n f_\alpha e_n)$. We next use part (v) of Proposition 5.12 to see that $\tau(e_n f_\alpha e_n) = \int_0^\infty \mathbf{m}_{e_n f_\alpha e_n}(s)\,ds \leq \int_0^\infty \mathbf{m}_{f_\alpha}(s)\,ds = \tau(f_\alpha)$ for each n and each α. But then $\tau(f e_n) \leq \sup_\alpha \tau(f_\alpha)$. By the definition of $\tau(f)$, we then have that $\tau(f) \le \sup_\alpha \tau(f_\alpha)$. But by part (iii) of Proposition 5.12, we must have that $\tau(f_\alpha) \le \tau(f)$ for each α. Hence, we must have that $\tau(f) = \sup_\alpha \tau(f_\alpha)$.

We now prove the additivity of the extension of the trace. Given $f, g \in \widetilde{\mathcal{M}}_+$, the Borel functional calculus ensures that the sequences $(f_n), (g_n) \subseteq \mathcal{M}_+$ defined by $f_n = f\chi_{[0,n]}(f)$ and $g_n = g\chi_{[0,n]}(g)$, respectively, increase to f and g in $\widetilde{\mathcal{M}}$. Then, $(f_n + g_n)$ will increase to $f + g$. So, by what has been proved, we have $\tau(f + g) = \sup_{n \in \mathbb{N}} \tau(f_n + g_n) = \sup_{n \in \mathbb{N}} (\tau(f_n) + \tau(g_n)) = \tau(f) + \tau(g)$. $\qquad\square$

Lemma 5.18 *Let $f \in \widetilde{\mathcal{M}}$ be given. Given any $\alpha > 0$ with $\alpha = d_f(s)$ for some $s > 0$, let $s_\alpha = \inf\{s : d_f(s) \le \alpha\}$. Denoting $f\chi_{(s_\alpha,\infty)}(|f|)$ by f_α, we then have that $\mathbf{m}_{f_\alpha} = \mathbf{m}_f \chi_{[0,\alpha]}$.*

Proof. Let α, s_α, and f_α be as in the hypothesis. From the right-continuity of $s \mapsto d_f(s)$, we have that $d_f(s_\alpha) = \alpha$, with by construction $d_f(s) > \alpha$ whenever $s < s_\alpha$. It is clear that in computing the decreasing rearrangements, we may pass to the abelian von Neumann subalgebra generated by the spectral projections of f. This then allows us access to the Borel functional calculus. Using this calculus, it is now an easy exercise to show that

$$\chi_{(s,\infty)}(f_\alpha) = \begin{cases} \chi_{(s,\infty)}(f), & \text{if } s \ge s_\alpha \\ \chi_{(s_\alpha,\infty)}(f), & \text{if } 0 < s < s_\alpha \end{cases}.$$

Therefore

$$d_{f_\alpha}(s) = \begin{cases} d_f(s), & \text{if } s \ge s_\alpha \\ d_f(s_\alpha), & \text{if } 0 < s < s_\alpha \end{cases}.$$

From these formulae, it now clearly follows that if $t < \alpha$, then $\mathbf{m}_t(f_\alpha) = \inf\{s : d_{f_\alpha}(s) \le t\} = \inf\{s : d_f(s) \le t\} = \mathbf{m}_t(f)$, and similarly that if $t \ge \alpha$, then $\mathbf{m}_t(f_\alpha) = \inf\{s : d_{f_\alpha}(s) \le t\} = 0$. $\qquad\square$

The following result presents a very elegant way of realising what is sometimes called the 'second' decreasing rearrangement in the context of 'non-atomic' von Neumann algebras.

Proposition 5.19 *Let $\mathcal{M}$ have no minimal projections. For any $f \in \widetilde{\mathcal{M}}$ and any $t > 0$, we then have that $\int_0^t \mathbf{m}_f(s)\, ds = \sup\{\tau(e|f|e) : e \in \mathbb{P}(\mathcal{M}), \tau(e) \le t\}$*

Proof.   We may clearly assume that $f \geq 0$. For any $e \in \mathbb{P}(\mathcal{M})$ with $\tau(e) \leq t$, we may next apply Corollary 5.11 and Proposition 5.13 to see that $\tau(efe) = \int_0^t \mathbf{m}_{efe}(s)\, ds \leq \int_0^t \mathbf{m}_f(s)\, ds$. Clearly $\int_0^t \mathbf{m}_f(s)\, ds \geq \sup\{\tau(e|f|e) \colon e \in \mathbb{P}(\mathcal{M}), \tau(e) \leq t\}$.

To prove the converse inequality, we will consider two cases. First, let $s_0 = \inf\{s \geq 0 \colon d_f(s) < \infty\}$, and $\alpha_0 = d_f(s_0)$.

Case 1 $(t \geq \alpha_0)$: In this case, we clearly have $\alpha_0 < \infty$. Since $d_f(s) = \infty$ for all $s < s_0$, it is now also clear that

$$\mathbf{m}_f(v) = \inf\{s > 0 \colon d_f(s) \leq v\} = s_0 \text{ for all } v \geq \alpha_0.$$

It then follows that

$$\int_0^t \mathbf{m}_f(v)\, dv = \int_0^{\alpha_0} \mathbf{m}_f(v)\, dv + (t - \alpha_0)s_0.$$

If, in fact, $t = \alpha_0$, then for $e_0 = \chi_{(s_0,\infty)}(f)$, it follows from the lemma that

$$\tau(fe_0) = \int_0^\infty \mathbf{m}_v(fe_0)\, dv = \int_0^{\alpha_0} \mathbf{m}_v(f)\, dv.$$

Therefore, in this case, $\int_0^t \mathbf{m}_f(s)\, ds \leq \sup\{\tau(e|f|e) \colon e \in \mathbb{P}(\mathcal{M}), \tau(e) \leq t\}$ as required.

Next, consider the case where $t > \alpha_0$. Let $\epsilon > 0$ be given. By construction, $\tau(\chi_{(s_0-\epsilon,\infty)}(f)) = d_f(s_0 - \epsilon) = \infty$. Since, by assumption, $\tau(\chi_{(s_0,\infty)}(f)) = \alpha_0 < \infty$, we must therefore have $\tau(\chi_{(s_0-\epsilon,s_0]}(f)) = \infty$. The fact that $\mathcal{M}$ has no minimal projections now ensures that there must exist a subprojection e_t of $\chi_{(s_0-\epsilon,s_0]}(f)$ with $\tau(e_t) = (t - \alpha_0)$. The easiest way to see this is to use Proposition 5.9 to select an abelian subalgebra of $\chi_{(s_0-\epsilon,s_0]}(f)\mathcal{M}\chi_{(s_0-\epsilon,s_0]}(f)$ with no minimal projections. Proposition 5.9 then informs us that this abelian subalgebra corresponds to a classical L^∞-space living on a nonatomic measure space. The existence of a subprojection e_t for which $\tau(e_t) = (t - \alpha_0)$, therefore, follows from classical measure theory. By construction, e_0 and e_t are mutually orthogonal, with $e_0 + e_t$ therefore a projection with $\tau(e_0 + e_t) = t$, and with

$$\begin{aligned}
\tau(f(e_0 + e_t)) &= \tau(fe_0) + \tau(fe_t) \\
&\geq \int_0^{\alpha_0} \mathbf{m}_f(v)\,dv + (s_0 - \epsilon)\tau(e_t) \\
&= \int_0^{\alpha_0} \mathbf{m}_f(v)\,dv + (s_0 - \epsilon)(t - \alpha_0) \\
&= \int_0^t \mathbf{m}_f(v)\,dv - \epsilon(t - \alpha_0).
\end{aligned}$$

Since $\epsilon > 0$ was arbitrary, it is clear that here too

$$\int_0^t \mathbf{m}_f(s)\,ds \leq \sup\{\tau(|f|e) \colon e \in \mathbb{P}(\mathcal{M}), \tau(e) \leq t\}$$

as required.

Case 2 ($t < \alpha_0$): Let $s_t = \inf\{s > 0 \colon d_f(s) \leq t\}$, and set $\alpha_t = d_f(s_t)$. By the right-continuity of $s \mapsto d_f(s)$, we must have $\alpha_t = d_f(s_t) \leq t$, with $d_f(s) > t$ for all $s < s_t$ by construction.

If, in fact, $t = \alpha_t$, then we may set $e_1 = \chi_{(s_t, \infty)}(f)$ and argue as before to see $\tau(fe_1) = \int_0^t \mathbf{m}_v(f)\,dv$, from which it then follows that $\int_0^t \mathbf{m}_f(s)\,ds \leq \sup\{\tau(|f|e) \colon e \in \mathbb{P}(\mathcal{M}), \tau(e) \leq t\}$ as required.

Next, consider the case where $t > \alpha_t$. Despite the similarity of this proof to the earlier case, we will, for the sake of the reader, provide relevant details. Since, for all $s < s_t$, we have by construction that $d_f(s) > t$, with $d_f(s) \leq \alpha_t$ otherwise, it is clear that for any $\alpha_t < v \leq t$, we must have $\mathbf{m}_f(v) = \inf\{s \colon d_f(s) \leq v\} = s_t$. Therefore, $\int_0^t \mathbf{m}_f(v)\,dv = \int_0^{\alpha_t} \mathbf{m}_f(v)\,dv + (t - \alpha_t)s_t$. Let $\epsilon > 0$ be given. By construction, $\tau(\chi_{(s_t - \epsilon, \infty)}(f)) = d_f(t) > t$, and $\tau(\chi_{(s_t, \infty)}(f)) = \alpha_t$. Hence $\tau(\chi_{(s_t - \epsilon, s_t]}(f)) > t - \alpha_t$. As before, the fact that $\mathcal{M}$ has no minimal projections ensures that there must exist a subprojection e_t of $\chi_{(s_t - \epsilon, s_t]}(f)$ with $\tau(e_t) = (t - \alpha_t)$. Let $e_1 = \chi_{(s_t, \infty)}(f)$. Arguing as in the former case, it now follows that $e_1 + e_t$ is a projection with $\tau(e_1 + e_t) = t$, and with $\tau(f(e_1 + e_t)) \geq \int_0^{\alpha_0} \mathbf{m}_v(f)\,dv + (s_0 - \epsilon)(t - \alpha_0)$. Since $\epsilon > 0$ was arbitrary, it is clear that $\int_0^t \mathbf{m}_f(s)\,ds \leq \sup\{\tau(e|f|e) \colon e \in \mathbb{P}(\mathcal{M}), \tau(e) \leq t\}$ as required. $\qquad\square$

The applicability of the above proposition ranges wider than just von Neumann algebras with no minimal projections. This follows from the following observation.

Proposition 5.20 *Let $\mathcal{M}$ be a semifinite von Neumann algebra equipped with an f.n.s. trace τ. Then, the trace $\tau_\infty = \tau \otimes \int_{[0,1]} (\cdot)\, d\mathrm{m}(t)$ (where m is Lebesgue measure) is an f.n.s. trace on the von Neumann algebra tensor product $\mathcal{M}\overline{\otimes}L^\infty[0,1]$. The algebra $\mathcal{M}\overline{\otimes}L^\infty[0,1]$ has no minimal projections, and if we use the trace τ_∞ to define a topology of convergence in measure on $\mathcal{M}\overline{\otimes}L^\infty[0,1]$, the $*$-isomorphism $\mathscr{I}_\infty : \mathcal{M} \to \mathcal{M}\overline{\otimes}L^\infty[0,1] : a \mapsto a \otimes 1$ extends to a homeomorphism between $\widetilde{\mathcal{M}}$, and the closure of $\mathscr{I}_\infty(\mathcal{M})$ in the topology of convergence in measure on $\mathcal{M}\overline{\otimes}L^\infty[0,1]$. Specifically for any $a \in \widetilde{\mathcal{M}}$, the simple tensor $a \otimes 1$ will then be τ_∞-measurable, with $\mathbf{m}_a(t) = \mathbf{m}_{a\otimes 1}(t)$ for any $t \geq 0$.*

We pause to point out that, as far as $\widetilde{\mathcal{M}} \otimes L^\infty[0,1]$ is concerned, one may certainly make sense of the algebraic tensor product. Simple tensors of the form $a \otimes 1$, where $a \in \widetilde{\mathcal{M}}$ belong to this algebraic tensor product. However, since $\widetilde{\mathcal{M}}$ is in general not locally convex, there is no natural metric or topological structure with which we can equip this tensor product. The action of τ_∞ on simple tensors of the form $a \otimes 1$ is, therefore, derived from the fact that these tensors can also be realised as elements of $\widetilde{\mathcal{M}\overline{\otimes}L^\infty}[0,1]$. To see this notice that given $(a_n) \subset \mathcal{M}$ such that $a_n \to a$ in $\widetilde{\mathcal{M}}$, we will then have $(a_n \otimes 1) \to (a \otimes 1)$ in $\widetilde{\mathcal{M}\overline{\otimes}L^\infty}[0,1]$. There is a theory of tensor products of locally convex algebras of unbounded operators (see, for example, [**FIW14**]), but these algebras are typically not all that closely related to $\widetilde{\mathcal{M}}$.

Proof. We leave the claim that τ_∞ is a faithful normal semifinite trace as an exercise. The claim that $\mathscr{I}_\infty$ extends to a homeomorphism between $\widetilde{\mathcal{M}}$, and the closure of $\mathscr{I}_\infty(\mathcal{M})$ in the topology of convergence in measure, is a direct consequence of Proposition 3.98 applied to the fact that $\tau_\infty \circ \mathscr{I}_\infty = \tau$. Given $a \in \widetilde{\mathcal{M}}$, the fact that $\mathbf{m}_a(t) = \mathbf{m}_{a\otimes 1}(t)$ for all $t \geq 0$ will easily follow if we are able to show that $d_a(t) = d_{a\otimes 1}(t)$ for all $t \geq 0$. To see this, one may check that $|a \otimes 1| = |a| \otimes 1$, and that $p(|a \otimes 1|) = p(|a|) \otimes 1$ for any polynomial in one real variable. Now, use the Stone–Weierstrass theorem to see that $f(|a \otimes 1|) = f(|a|) \otimes 1$ holds for non-negative continuous functions f on $[0,\infty)$ and then use that fact to deduce that this equality also holds for non-negative Borel functions on $[0,\infty)$. Given $t \geq 0$, it then trivially follows that $d_{a\otimes 1}(t) = \tau_\infty(\chi_{(t,\infty)}(|a \otimes 1|)) = \tau_\infty(\chi_{(t,\infty)}(|a| \otimes 1)) = \tau_\infty((\chi_{(t,\infty)}(|a|)) \otimes 1) = \tau(\chi_{(t,\infty)}(|a|)) = d_a(t)$. It remains to prove that

$\mathcal{M}\overline{\otimes}L^{\infty}[0,1]$ has no minimal projections. So, let $p \in \mathbb{P}(\mathcal{M}\overline{\otimes}L^{\infty}[0,1])$ be given and consider the set $\mathcal{P} = \{e \in \mathbb{P}(L^{\infty}[0,1]) \colon p \leq 1 \otimes e\}$. We need to find a non-zero subprojection of p, which is distinct from p. Let $e_m = \wedge\{e \colon e \in \mathcal{P}\}$. The measure space $([0,1], \mathscr{B}([0,1]))$ is, of course, non-atomic under Lebesgue measure, and hence, we may find a non-zero projection $e_0 \in \mathbb{P}(L^{\infty}[0,1])$ for which we have $e_0 \neq e_m$ and $e_0 \leq e_m$. It is not difficult to see that $1 \otimes e_0$ is a central element of $\mathcal{M}\overline{\otimes}L^{\infty}[0,1]$. Hence, $(1 \otimes e_0)p$ is a subprojection of p. Note that we must have that $(1 \otimes e_0)p \neq p$, since equality would ensure that $p \leq (1 \otimes e_0)$, which would contradict the minimality of e_m. If now we can show that $(1 \otimes e_0)p \neq 0$, this projection would be the subprojection of p we seek. Assume, by way of contradiction, that $(1 \otimes e_0)p = 0$. This ensures that $p \leq (1 \otimes 1) - (1 \otimes e_0) = 1 \otimes (1 - e_0)$, and hence that $(1 - e_0) \in \mathcal{P}$. The minimality of e_m then ensures that $e_m \leq (1 - e_0)$. Since by assumption $e_0 \leq e_m$, the only way both statements can be true is when $e_0 = 0$, which contradicts the assumption that $e_0 \neq 0$. Thus, as required, $(1 \otimes e_0)p \neq 0$. $\qquad\square$

The next result and the one that follows it will prove to be very useful when we get to the analysis of Orlicz spaces.

Theorem 5.21 ([HLP29]) *Let f and g be non-negative Borel measurable functions on $[0,\infty)$ that are finite almost everywhere. Then, the following are equivalent.*

> (i) *$\int_0^t f(s)\,ds \leq \int_0^t g(s)\,ds$ for all $t > 0$.*
>
> (ii) *Any non-negative, non-decreasing, convex function $\Psi \colon [0,\infty) \to [0,\infty]$, which is neither identically zero nor infinite valued on $(0,\infty)$, and which is continuous on $[0, b_{\Psi}]$, satisfies the inequality $\int_0^t \Psi(f(s))\,ds \leq \int_0^t \Psi(g(s))\,ds$ for all $t > 0$. (Note that Ψ may be infinite-valued at b_{Ψ}.)*

Proof.   Hardy, Littlewood and Polya [**HLP29**] (see also [**HLP88**, §249]) proved that on a bounded interval $[0,a]$, one has $\int_0^t f(s)\,ds \leq \int_0^t g(s)\,ds$ for all $t \in [0,a]$ if and only if $\int_0^a \Psi(f(s))\,ds \leq \int_0^a \Psi(g(s))\,ds$ for any convex and continuous function Ψ on $[0,a]$. (They actually used a different criterion to (i) above, but their criterion can be seen to be equivalent to the one stated here; see [**BS88**, Chapter 2, Exercise 1].) However, for our purposes, we need to know that the claims in [**HLP29**] hold for functions on the half-line. The thrust of the proof will then be the verification of this extension.

First, assume that Ψ is finite on all of $[0, \infty)$. For any fixed $r > 0$, criterion (i) of course holds for any $0 < t \leq r$. So, by the Hardy–Littlewood–Polya result applied to the interval $[0, r]$, we have $\int_0^r \Psi(f(s)) \, ds \leq \int_0^r \Psi(g(s)) \, ds$. Since $r > 0$ was arbitrary, we are done. It now remains to show that (ii) still holds if Ψ is infinite-valued on part of $(0, \infty)$. We may therefore pass to the case where $b_\Psi < \infty$. Observe that on replacing Ψ with $t \mapsto \Psi(t) - \Psi(0)$, we may, without loss of generality, assume that $\Psi(0) = 0$. Since Ψ is convex, we know that it is left-differentiable at all points of $(0, b_\Psi)$. In fact, this left derivative ψ turns out to be a left-continuous non-negative non-decreasing function on $[0, b_\Phi)$. On defining ψ to be infinite valued on (b_Ψ, ∞), it follows that $\Phi(t) = \int_0^t \psi(s) \, ds$ for all $t > 0$ (see [**NP06**, §1.3 and 1.6]). Now, select a sequence $(t_n) \subseteq (0, b_\Psi)$ increasing to b_Ψ, and define ψ_n by

$$
\psi_n(t) = \begin{cases}
\psi(t), & \text{for all } 0 \leq t \leq t_n \\
\psi(t_n), & \text{for all } t_n \leq t \leq b_\Psi \\
\exp(n(t - b_\Psi))\psi(t_n), & \text{for all } t > b_\Psi
\end{cases}.
$$

On setting $\Psi_n(t) = \int_0^t \psi_n(s) \, ds$, it is an exercise to see that each Ψ_n is finite-valued on all of $(0, \infty)$. From what we have already proved, we then have that $\int_0^t \Psi_n(f(s)) \, ds \leq \int_0^t \Psi_n(g(s)) \, ds$ for all $t > 0$ and all n. Now, observe that for any fixed $t > 0$, the sequence $(\psi_n(t))$ will by construction monotonically increase to $\psi(t)$. We may, therefore, use the monotone convergence theorem to see that $\lim_{n \to \infty} \Psi_n(t) = \lim_{n \to \infty} \int_0^t \psi_n(s) \, ds = \int_0^t \psi(s) \, ds = \Psi(t)$. Since, by construction, the sequence (Ψ_n) is itself increasing, we may now once again use the monotone convergence theorem to see that $\int_0^t \Psi(f(s)) \, ds = \lim_{n \to \infty} \int_0^t \Psi_n(f(s)) \, ds \leq \lim_{n \to \infty} \int_0^t \Psi_n(g(s)) \, ds = \int_0^t \Psi(g(s)) \, ds$ for any $t > 0$. Thus, (ii) holds in general. $\qquad\square$

5.4 Integral inequalities and monotone convergence theorem

The proof of the following result illustrates how the reduction described in Proposition 5.20 may be used in practice.

Theorem 5.22 *Let $a, b \in \widetilde{\mathcal{M}}$ be given and let Φ be a convex function of the type described in the preceding theorem. Then, the following holds.*

- $\int_0^t \mathbf{m}_{a+b}(s)\, ds \leq \int_0^t \mathbf{m}_a(s) + \int_0^t \mathbf{m}_b(s)$ *for any $t > 0$.*

- $\int_0^t \Phi(\mathbf{m}_{a+b}(s))\, ds \leq \int_0^t \Phi(\mathbf{m}_a(s) + \mathbf{m}_b(s))\, ds$ *for all $t \in (0, \infty]$.*

Proof. We observe that for $t \in (0, \infty)$, the second claim will follow from the first on setting $f(s) = \mathbf{m}_{a+b}(s)\chi_{[0,t]}(s)$ and $g(s) = (\mathbf{m}_a(s) + \mathbf{m}_b(s))\chi_{[0,t]}(s)$ in the preceding theorem. The case $t = \infty$ will then follow by letting $t \to \infty$. It, therefore, remains to prove the first claim. It is clear from Proposition 5.20 that we may assume that $\mathcal{M}$ has no minimal projections. This observation then gives us access to Proposition 5.19. With this fact in mind, let $e \in \mathcal{M}$ be a projection with $\tau(e) \leq t$. Now, use Lemma 5.2 to select partial isometries u and v so that $|f + g| \leq u|f|u^* + v|g|v^*$.

We may now apply Proposition 5.19 to see that

$$
\begin{aligned}
\tau(e|f + g|e) &\leq \tau(eu|f|u^*e) + \tau(ev|g|v^*e) \\
&\leq \int_0^t \mathbf{m}_{u|f|u^*}(s)\, ds + \int_0^t \mathbf{m}_{v|g|v^*}(s)\, ds \\
&\leq \int_0^t \mathbf{m}_f(s)\, ds + \int_0^t \mathbf{m}_g(s)\, ds.
\end{aligned}
$$

To conclude the proof, we once again use Proposition 5.19 to see that taking the supremum over all projections e with $\tau(e) \leq t$ yields the required conclusion. $\square$

We are now ready to describe the topology of convergence in measure in terms of decreasing rearrangements.

Proposition 5.23 *The basic neighbourhoods of 0 of the topology of convergence in measure on $\widetilde{\mathcal{M}}$ are given by $\mathcal{N}(\epsilon, \delta) = \{g \in \widetilde{\mathcal{M}} : \mathbf{m}_g(\delta) \leq \epsilon\}$ where $\epsilon, \delta > 0$. Hence, given a net $(f_\alpha) \subseteq \widetilde{\mathcal{M}}$ and $f \in \widetilde{\mathcal{M}}$, we have $f_\alpha \to f$ in the topology of convergence in measure if and only if $\mathbf{m}_{f - f_\alpha}(t) \to 0$ for every $t > 0$.*

Proof. If $g \in \mathcal{N}(\epsilon, \delta)$, then by definition there exists a projection $e \in \mathcal{M}$ such that $\tau(\mathbb{1} - e) \leq \delta$ and $\|fe\| \leq \epsilon$. But then $\mathbf{m}_g(\delta) \leq \epsilon$ by Definition 5.1.

Conversely, suppose that $\mathbf{m}_g(\delta) \leq \epsilon$, and consider the projection $e = \chi_{[0,\rho_\delta]}(|g|)$, where $\rho_\delta = \mathbf{m}_g(\delta)$. We then clearly have that $\|ge\| = \|, |g|e\| \leq \mathbf{m}_g(\delta) \leq \epsilon$. It, moreover, follows from Proposition 5.7 that the projection e satisfies $\tau(\mathbb{1} - e) = \tau(\chi_{(\rho_\delta,\infty)}(|g|)) = d_g(\mathbf{m}_g(\delta)) \leq \delta$. But then $g \in \mathcal{N}(\epsilon, \delta)$ as required. $\qquad\square$

Proposition 5.24 *Let $f \in \widetilde{\mathcal{M}}$ be given. Then, the following are equivalent.*

(i) $d_f(s) < \infty$ *for all $s > 0$.*

(ii) $\mathbf{m}_f(t) \to 0$ *as $t \to \infty$.*

(iii) *There exists a sequence $(f_n) \subseteq \widetilde{\mathcal{M}}$ with $\tau(\mathbf{s}(f_n)) < \infty$ for each n, converging to f in the topology of convergence in measure.*

It is clear from the above that condition (i) above is just an alternative criterion for membership of $K(\mathcal{M}, \tau)$. This fact will prove to be useful in the subsequent analysis.

Proof. **(i)** $\Rightarrow$ **(iii)** : Let $f \in \widetilde{\mathcal{M}}$ be given with $d_f(s) < \infty$ for all $s > 0$, and let $|f| = \int_0^\infty \lambda \, de_\lambda(|f|)$ be the spectral resolution of $|f|$. If $f = u|f|$ is the polar decomposition of f, then clearly $f = u \int_0^\infty \lambda \, de_\lambda(|f|)$. Now set $f_n = u \int_{1/n}^n \lambda \, de_\lambda(|f|)$ for each $n \in \mathbb{N}$. For each n, the left support projection of f_n is just $\chi_{[1/n,n]}(|f|)$. By assumption, $\tau(\chi_{(s,\infty)}(|f|)) < \infty$ for any $0 < s < 1/n$ and hence $\tau(\chi_{[1/n,n]}(|f|)) < \infty$ as required.

Next, set $e_n = \chi_{(n,\infty)}(|f|)$. The τ-measurability of f ensures that $\tau(e_n) = d_f(n) \to 0$ as $n \to \infty$. In addition

$$\|(f - f_n)(\mathbb{1} - e_n)\| = \left\| u \int_0^{1/n} \lambda \, de_\lambda(|f|) \right\| \leq \frac{1}{n}.$$

Hence, as required, (f_n) converges to f in measure.

(iii) $\Rightarrow$ **(ii)** : Assume that *(iii)* holds. Given $\epsilon > 0$, we may then apply Proposition 5.23 to select n_0 so that $\mathbf{m}_{f-f_n}(1) < \epsilon$ for all $n \geq n_0$. Since by Corollary 5.11, $\mathbf{m}_{f_{n_0}}(t) = 0$ for all $t > \tau(\mathbf{s}(f_{n_0}))$, it now follows that

$$\mathbf{m}_f(t) \leq \mathbf{m}_{f_{n_0}}(t - 1) + \mathbf{m}_{f-f_{n_0}}(1) \leq \epsilon \text{ for all } t > \tau(\mathbf{s}(f_{n_0})) + 1.$$

Thus, (ii) follows.

(ii) $\Rightarrow$ **(i)** : Suppose that (ii) holds, and let $\epsilon > 0$ be given. Select $t_0 > 0$ such that $\mathbf{m}_f(t_0) \leq \epsilon$. Using the fact that $s \mapsto d_f(s)$ is non-increasing, we may then conclude from Proposition 5.7 that $d_f(\epsilon) \leq d_f(\mathbf{m}_f(t_0)) \leq t_0 < \infty$. $\qquad\square$

We take some time to briefly investigate the order properties of convergence in measure.

Lemma 5.25 *Let $(f_\alpha) \subset \widetilde{\mathcal{M}}_+$ be a net with $\tau(f_\alpha) < \infty$ for each α which decreases to 0 in $\widetilde{\mathcal{M}}$. Then, the net (f_α) converges to 0 in measure.*

Proof. We may clearly assume that there exists some $f \in \mathcal{M}_+$ with finite trace such that $f_\alpha \leq f$ for all α. The net $(f - f_\alpha)$ will then increase to f. By the normality of the trace on $\widetilde{\mathcal{M}}_+$ (Proposition 5.17), we will then have that $\tau(f) = \sup_\alpha \tau(f - f_\alpha) = \tau(f) - \inf_\alpha \tau(f_\alpha)$, or equivalently that $\tau(f_\alpha) \searrow 0$. For any fixed $t > 0$, the fact that $s \to \mathbf{m}_{f_\alpha}(s)$ is non-increasing now ensures that $\mathbf{m}_{f_\alpha}(t) \leq t^{-1} \int_0^t \mathbf{m}_{f_\alpha}(s)\,ds \leq t^{-1} \int_0^\infty \mathbf{m}_{f_\alpha}(s)\,ds = t^{-1}\tau(f_\alpha)$ and hence that $\lim_\alpha \mathbf{m}_{f_\alpha}(t) = 0$. The claim, therefore, follows from Proposition 5.23. $\qquad\square$

Proposition 5.26 ([DdPS23, Proposition II.6.3]) *Let $(f_\alpha) \subset \widetilde{\mathcal{M}}_+$ be a net decreasing to 0 for which there exists some $f \in K(\mathcal{M}, \tau)$ with $f \geq f_\alpha$ for every α. Then, the net (f_α) converges to 0 in measure.*

Proof. Let $\epsilon, \delta > 0$ be given. We need to show that for some α_0, we will have that $f_\alpha \in \mathcal{N}(\epsilon, \delta)$ whenever $\alpha \geq \alpha_0$. By Proposition 5.3, there exists $\lambda > \epsilon/4$ such that $\tau(\chi_{(\lambda,\infty)}(f)) \leq \delta/2$. Now, let $p = \chi_{(\epsilon/4,\lambda]}(f)$. Since $f \in K(\mathcal{M}, \tau)$, we know that $\tau(p) \leq \tau(\chi_{(\epsilon/4,\infty)}(f)) = d_f(\epsilon/4) < \infty$ and hence that $\tau(fp) \leq \lambda\tau(p) < \infty$. Note that by Proposition 3.88, $fp = pfp \geq pf_\alpha p$ for each α with $pf_\alpha p \searrow 0$. Since by Propositions 5.12 and 5.24 $fp \in K(\mathcal{M}, \tau)$, it now follows from Lemma 5.25 that $(pf_\alpha p)$ converges to 0 in measure. So, there exists some α_0 such that $pf_\alpha p \in \mathcal{N}(\epsilon/4, \delta/2)$ whenever $\alpha \geq \alpha_0$.

Notice that by the choice of p, $\|a(1-p)\chi_{[0,\lambda]}\|_\infty = \|a\chi_{[0,\epsilon/4]}\|_\infty \leq \epsilon/4$ with $\tau(\chi_{(\lambda,\infty)}(f)) \leq \delta/2$. Thus, $a(1-p) \in \mathcal{N}(\epsilon/4, \delta/2)$. Since by Lemma 3.87 $(\mathbb{1} - p)f_\alpha(\mathbb{1} - p) \leq (\mathbb{1} - p)f(\mathbb{1} - p)$ for each α, Lemma 3.92 in turn ensures that $((\mathbb{1} - p)f_\alpha(\mathbb{1} - p)) \subset \mathcal{N}(\epsilon/4, \delta/2)$ for all α.

By Lemma 3.74, $pf_\alpha p + (\mathbb{1} - p)f_\alpha(\mathbb{1} - p) \in \mathcal{N}(\epsilon/2, \delta)$ for each $\alpha \geq \alpha_0$. If therefore we can show that $f_\alpha \leq 2(pf_\alpha p + (\mathbb{1} - p)f_\alpha(\mathbb{1} - p))$ for each α, the claim will then follow from Lemma 3.74. Therefore, it remains to prove the above inequality. For a fixed α, we have $(p - (\mathbb{1} - p))f_\alpha(p - (\mathbb{1} - p)) \geq 0$. This inequality may, in turn, be rephrased as $pf_\alpha(\mathbb{1}-p)+(\mathbb{1}-p)f_\alpha p \leq pf_\alpha p+(\mathbb{1}-p)f_\alpha(\mathbb{1}-p)$. The claim now follows by adding $pf_\alpha p + (\mathbb{1} - p)f_\alpha(\mathbb{1} - p)$ to both sides of the inequality. $\square$

Proposition 5.27 ([DdPS23, Proposition II.6.4])   *Let $(f_\alpha) \subset \widetilde{\mathcal{M}}_+$ be a net decreasing to 0. For any $x \in K(\mathcal{M}, \tau)$, the nets (x^*f_α) and $(f_\alpha x)$ will then both converge to 0 in measure.*

Proof.   We may clearly assume that there exists some $f \in \widetilde{\mathcal{M}}$ such that $f_\alpha \leq f$ for all α. Then, of course, $f_\alpha^{1/2} \leq f^{1/2}$ for each α. By Propositions 5.12 and 5.24, $x^*fx \in K(\mathcal{M}, \tau)$. Since, by Proposition 3.88, $x^*f_\alpha x \searrow 0$ with $x^*f_\alpha x \leq x^*fx$, it follows from Proposition 5.26 that $(x^*f_\alpha x)$ converges to 0 in measure. By Proposition 5.23, this is equivalent to $\mathbf{m}_{x^*f_\alpha x}(t) \to 0$ for every $t > 0$. Since for each α and each $t > 0$ $\mathbf{m}_{x^*f_\alpha x}(t) = \mathbf{m}_{f_\alpha^{1/2}x}(t)^2$, we also have $\mathbf{m}_{f_\alpha^{1/2}x}(t) \to 0$ for every $t > 0$. For every $t > 0$, Proposition 5.12 informs us that $\mathbf{m}_{f_\alpha x}(t) \leq \mathbf{m}_{f_\alpha^{1/2}}(t/2)\mathbf{m}_{f_\alpha^{1/2}x}(t/2) \leq \mathbf{m}_{f^{1/2}}(t/2)\mathbf{m}_{f_\alpha^{1/2}x}(t/2)$. This clearly ensures that $\mathbf{m}_{f_\alpha x}(t) \to 0$ for every $t > 0$. Since by Proposition 5.23 this is equivalent to the fact that (xf_α) converges to 0 in measure, the claim now follows. $\square$

In closing, we shall prove a noncommutative monotone convergence theorem. The following useful Fatou-like lemma was proved in **[Kos84b]**.

Lemma 5.28   *Let $(f_n) \subseteq \widetilde{\mathcal{M}}$ be a sequence converging to $f \in \widetilde{\mathcal{M}}$ in the topology of convergence in measure. Then:*

(i) $\mathbf{m}_f(t) \leq \liminf_{n \to \infty} \mathbf{m}_{f_n}(t)$ *for all $t > 0$;*

(ii) $\mathbf{m}_f(t) = \lim_{n \to \infty} \mathbf{m}_{f_n}(t)$ *if either $s \mapsto \mathbf{m}_f(s)$ is continuous at $s = t$, or $\mathbf{m}_{f_n} \leq \mathbf{m}_f$ for all n.*

Proof.   (i): Let $\epsilon > 0$ be given. On observing that $\mathbf{m}_f(t + \epsilon) \leq \mathbf{m}_{f_n}(t)+\mathbf{m}_{f-f_n}(\epsilon)$, we may apply Proposition 5.23 to see that $\mathbf{m}_f(t+\epsilon) \leq \liminf_{n \to \infty} \mathbf{m}_{f_n}(t)$. Since this holds for any $\epsilon > 0$, it follows from the right-continuity of $t \mapsto \mathbf{m}_f(t)$ that $\mathbf{m}_f(t) \leq \liminf_{n \to \infty} \mathbf{m}_{f_n}(t)$ as required.

(ii): Given $t > 0$ select $\epsilon > 0$ with $\epsilon < t$. Arguing as before, we first note that $\mathbf{m}_{f_n}(t) \leq \mathbf{m}_f(t-\epsilon) + \mathbf{m}_{f_n-f}(\epsilon)$ and then use Proposition 5.23 to conclude from this that $\limsup_{n\to\infty} \mathbf{m}_{f_n}(t) \leq \mathbf{m}_f(t-\epsilon)$. If indeed $s \mapsto \mathbf{m}_f(s)$ is continuous at $s = t$, letting ϵ decrease to 0 yields the conclusion that $\limsup_{n\to\infty} \mathbf{m}_{f_n}(t) \leq \mathbf{m}_f(t)$. On combining this inequality with what we have proved in part (i), it follows that $\lim_{n\to\infty} \mathbf{m}_{f_n}(t)$ exists and equals $\mathbf{m}_f(t)$. The second claim easily follows from part (i). $\qquad\square$

The above Fatou lemma yields the following analogue of the monotone convergence theorem.

Theorem 5.29   *Let $(f_n) \subseteq \widetilde{\mathcal{M}}$ be a sequence of positive operators converging to $f \in \widetilde{\mathcal{M}}$ in the topology of convergence in measure. Then:*

(i) *(Fatou's lemma)* $\tau(f) \leq \liminf_{n\to\infty} \tau(f_n)$;
(ii) *(Monotone convergence)* $\tau(f) = \lim_{n\to\infty} \tau(f_n)$ *whenever* $\mathbf{m}_{f_n} \leq \mathbf{m}_f$ *for all n.*

Proof.   (i): We may use the above lemma alongside the usual Fatou's lemma to conclude that

$$
\begin{aligned}
\tau(f) &= \int_0^\infty \mathbf{m}_f(t)\, dt \\
&\leq \int_0^\infty \liminf_{n\to\infty} \mathbf{m}_{f_n}(t)\, dt \\
&\leq \liminf_{n\to\infty} \int_0^\infty \mathbf{m}_{f_n}(t)\, dt \\
&= \liminf_{n\to\infty} \tau(f_n).
\end{aligned}
$$

(ii): The fact that $\tau(f_n) = \int_0^\infty \mathbf{m}_{f_n}(t)\, dt \leq \int_0^\infty \mathbf{m}_f(t)\, dt = \tau(f)$ for each $n \in \mathbb{N}$, ensures that $\limsup_{n\to\infty} \tau(f_n) \leq \tau(f)$. Considered alongside part (i), this inequality in turn ensures that $\lim_{n\to\infty} \tau(f_n)$ exists and equals $\tau(f)$. $\qquad\square$

Chapter 6
L^p and Orlicz spaces in the tracial case

We will here use the technology of decreasing rearrangement to develop a comprehensive theory of L^p-spaces for tracial semifinite algebras, before finally indicating how that theory may be extended to allow for a theory of noncommutative Orlicz spaces. Readers wishing for a more 'standard' streamlined introduction to L^p-spaces only may stop reading at the end of Section 6.2. However, readers wishing to ultimately master Haagerup L^p-spaces are strongly encouraged to engage the theory of Orlicz spaces as Orlicz technology will prove to be an essential tool in developing that theory. In support of this claim, we note that there is no significant reduction in effort to be had by proving the foundational results in Chapter 10 for L^p-spaces only. On the other hand, the results in Chapter 11 would have been out of reach of the Haagerup approach without Orlicz spaces, since the superspace within which that interpolation takes place is precisely such a noncommutative Orlicz space.

6.1 L^p-spaces for von Neumann algebras with a trace

6.1.1 Definition and (p-)normability

We start with the following definition.

Definition 6.1 Given $0 < p < \infty$, we define the space $L^p(\mathcal{M}, \tau)$ to be the set of all $f \in \widetilde{\mathcal{M}}$ satisfying $\tau(|f|^p) < \infty$. The (p-)norm on such an L^p is defined to be $\|f\|_p = \tau(|f|^p)^{1/p}$. In the case $p = \infty$, we define $L^\infty(\mathcal{M}, \tau)$ to be $\mathcal{M}$ itself.

Our first task is to show that each $L^p(\mathcal{M}, \tau)$ is a well-defined Banach space when $p \geq 1$, and a complete p-normed space when $0 < p < 1$.

Noncommutative measures and L^p and Orlicz Spaces, with Applications to Quantum Physics. Stanisław Goldstein and Louis Labuschagne, Oxford University Press. © Stanisław Goldstein and Louis Labuschagne (2025). DOI: 10.1093/oso/9780198950202.003.0008

For the case $p = \infty$, there is of course nothing to prove. We start our analysis by proving a very general noncommutative version of Hölder's inequality. Here, the Weyl-type inequality, established in part (ii) of the following theorem, proves to be crucial. Given $f \in \widetilde{\mathcal{M}}$, we formally define the quantity $\Delta_t(f)$ by

$$\Delta_t(f) = \exp\left(\int_0^t \log(\mathbf{m}_f(s))\, ds\right), \qquad t > 0.$$

In order to ensure the well-definiteness of this quantity, we will restrict our analysis of this quantity to the class of all $f \in \widetilde{\mathcal{M}}$ for which there exists some $\alpha > 0$ and $C > 0$ for which $\mathbf{m}_f(t) < Ct^{-\alpha}$ for all $t > 0$. We write $\mathscr{L}(\widetilde{\mathcal{M}})$ for this class. It is an interesting exercise to use Proposition 5.12 to prove that this class is closed under finite sums and products. In addition, any $f \in \widetilde{\mathcal{M}}$ for which $\tau(|f|^p) < \infty$ for some $0 < p < \infty$ is in this class. To see this, observe that for such an f, we have that

$$\mathbf{m}_f(t) = (\mathbf{m}_{|f|^p}(t))^{1/p} \leq \left(\frac{1}{t}\int_0^t \mathbf{m}_{|f|^p}(s)\, ds\right)^{1/p}$$

$$\leq t^{-1/p}\left(\int_0^\infty \mathbf{m}_{|f|^p}(s)\, ds\right)^{1/p} = t^{-1/p}\tau(|f|^p)^{1/p}$$

for all $t > 0$.

Theorem 6.2   *Let $f, g \in \widetilde{\mathcal{M}}$ be given.*

(i) *Let $p, q, r > 0$ be given. If $\frac{1}{r} = \frac{1}{p} + \frac{1}{q}$, then $\|fg\|_r \leq \|f\|_p\|g\|_q$.*

(ii) *If in fact $f, g \in \mathscr{L}(\widetilde{\mathcal{M}})$, we have that $\Delta_t(fg) \leq \Delta_t(f)\Delta_t(g)$ for any $t > 0$.*

(iii) *If $F:[0,\infty) \to \mathbb{R}$ is a function for which $F \circ \exp$ is continuous, convex and non-decreasing, we have that*

$$\int_0^t F(\mathbf{m}_{fg}(s))\, ds \leq \int_0^t F(\mathbf{m}_f(s)\mathbf{m}_g(s))\, ds$$

for any $f, g \in \mathscr{L}(\widetilde{\mathcal{M}})$.

We will first prove (ii) for the case where $f, g \in \mathcal{M}$ and then show how the general version of (ii) may be deduced from this case, before finally deducing the general versions of (i) and (iii) from (ii). We will need the following technical result regarding the holomorphic functional calculus.

Lemma 6.3 ([FK52, Lemma 2]) *Let $\mathcal{M}$ be a von Neumann algebra equipped with a finite faithful normal trace. Let F be an analytic function defined on some domain Λ in the complex plane, bounded by a curve Γ, and let $t \mapsto x(t)$ $(0 \le t \le 1)$ be a norm-differentiable family of invertible operators in $\mathcal{M}$, such that the spectrum of each $x(t)$ lies in Λ. Then, $F(x(t))$ is differentiable with respect to t on $(0,1)$, and*

$$\tau\left(\frac{d}{dt}F(x(t))\right) = \tau\left(F'(x(t)) \cdot x'(t)\right).$$

Proof. Recall that by definition $F[x(t)] = \frac{1}{2\pi i}\int_\Gamma F(s)(s\mathbb{1} - x(t))^{-1}\,ds$ for each $0 \le t \le 1$. For any $s \in \rho(x(t_1)) \cap \rho(x(t_0))$, the second resolvent equation informs us that

$$(s\mathbb{1} - x(t_1))^{-1} - (s\mathbb{1} - x(t_0))^{-1}$$
$$= (s\mathbb{1} - x(t_1))^{-1}(x(t_1) - x(t_0))(s\mathbb{1} - x(t_0))^{-1}.$$

We may then use this formula to see that

$$\frac{1}{h}[F(x(t+h)) - F(x(t))]$$
$$= \frac{1}{2\pi i}\int_\Gamma F(s)\tfrac{1}{h}[(s\mathbb{1} - x(t+h))^{-1} - (s\mathbb{1} - x(t))^{-1}]\,ds$$
$$= \frac{1}{2\pi i}\int_\Gamma F(s)(s\mathbb{1} - x(t+h))^{-1}\tfrac{1}{h}[x(t+h) - x(t)](s\mathbb{1} - x(t))^{-1}\,ds.$$

Letting $h \to 0$ and taking account of the fact $(s,t) \mapsto (s\mathbb{1} - X(t))^{-1}$ is uniformly continuous in the uniform topology on the product space $\Gamma \times [0,1]$ now yields the conclusion that

$$\frac{d}{dt}F(x(t)) = \frac{1}{2\pi i}\int_\Gamma F(s)(s\mathbb{1} - x(t))^{-1}x'(t)(s\mathbb{1} - x(t))^{-1}\,ds.$$

Notice that one may use the first resolvent equation stating that for any fixed t, we have

$$(s_1 \mathbb{1} - x(t))^{-1} - (s_0 \mathbb{1} - x(t))^{-1} = (s_0 - s_1)(s_1 \mathbb{1} - x(t))^{-1}$$
$$(s_0 \mathbb{1} - x(t))^{-1}$$

to see that $\frac{d}{ds}(s\mathbb{1} - x(t))^{-1} = -(s\mathbb{1} - x(t))^{-2}$. Taking this into account, we may now use integration by parts to see that we alternatively have

$$F'(x(t)) = \frac{1}{2\pi i} \int_\Gamma F'(s)(s\mathbb{1} - x(t))^{-1}\, ds = \frac{1}{2\pi i} \int_\Gamma F(s)(s\mathbb{1} - x(t))^{-2}\, ds.$$

Since we are dealing with a finite trace (which is continuous with respect to the norm topology on $\mathcal{M}$), it follows that when applying τ to $F'(x(t))x'(t)$ and $\frac{d}{dt}F(x(t))$, respectively, we may do so under the integral sign. This then yields the required conclusion that

$$\begin{aligned}
\tau\left(F'(x(t))x'(t)\right) &= \frac{1}{2\pi i} \int_\Gamma F(s)\tau((s\mathbb{1} - x(t))^{-2}x'(t))\, ds \\
&= \frac{1}{2\pi i} \int_\Gamma F(s)\tau((s\mathbb{1} - x(t))^{-1}x'(t)(s\mathbb{1} - x(t))^{-1})\, ds \\
&= \tau\left(\frac{d}{dt}F(x(t))\right).
\end{aligned}$$

$\square$

We gather some technical facts regarding rearrangements, before starting the actual proof of Theorem 6.2.

Lemma 6.4 *Let $f \in \widetilde{\mathcal{M}}$ be given.*

(i) *If $f \geq 0$, then $\mathbf{m}_{f+\epsilon}(t) = \mathbf{m}_f(t) + \epsilon$ for any $t < \tau(\mathbb{1})$ and any $\epsilon > 0$.*

(ii) *Given a projection $e \in \mathcal{M}$ with $\tau(e) < \infty$, we have that $\mathbf{m}^e_{|fe|} = \mathbf{m}_{|fe|}$, where $\mathbf{m}^e$ denotes the decreasing rearrangement of $|fe|$ with respect to the von Neumann algebra $e\mathcal{M}e$.*

Proof. *Part (i):* Given $s \geq 0$, it follows from the Borel functional calculus that $\chi_{(s+\epsilon,\infty)}(f + \epsilon\mathbb{1}) = \chi_{(s,\infty)}(f)$. It then clearly follows that $d_f(s) = d_{(f+\epsilon\mathbb{1})}(s+\epsilon)$ for all $s > 0$, with $d_{(f+\epsilon\mathbb{1})}(r) = \tau(\mathbb{1})$ if $0 < r < \epsilon$. Therefore, in the case where $0 < t < \tau(\mathbb{1})$, we will have that

$$
\begin{aligned}
\mathbf{m}_f(t) &= \inf\{s > 0 : d_f(s) \le t\} \\
&= \inf\{s > 0 : d_{(f+\epsilon\mathbb{1})}(s+\epsilon) \le t\} \\
&= \inf\{r > \epsilon : d_{(f+\epsilon\mathbb{1})}(r) \le t\} - \epsilon \qquad (\text{set } r = s + \epsilon) \\
&= \inf\{r > 0 : d_{(f+\epsilon\mathbb{1})}(r) \le t\} - \epsilon \\
&= \mathbf{m}_{(f+\epsilon\mathbb{1})}(t) - \epsilon.
\end{aligned}
$$

Part (ii): Observe that $|fe|$ commutes with e, and that $|fe|$ $(\mathbb{1} - e) = 0$. If now we apply the Borel functional calculus to the commutative von Neumann algebra generated by e and the spectral projections of $|fe|$, it follows that $\chi_{(s,\infty)}(|fe|) \le e$ for all $s > 0$. In other words, each such $\chi_{(s,\infty)}(|fe|)$ belongs to $e\mathcal{M}e$. We therefore clearly have that $d^e_{|fe|}(s) = d_{|fe|}(s)$ for all $s > 0$, where $d^e_{|fe|}$ denotes the distribution computed with respect to the compression $e\mathcal{M}e$. This fact suffices to prove the claim. $\qquad\square$

Lemma 6.5 *Let $f \in \mathscr{L}(\widetilde{\mathcal{M}})$ be given with $f \ge 0$. If $\mathcal{M}$ has no minimal projections, we may select an abelian von Neumann subalgebra $\mathcal{M}_1$ of $\mathcal{M}$ so that:*

- *$\mathcal{M}_1$ contains all the spectral projections of f;*
- *τ is semifinite on $\mathcal{M}_1$;*
- *$\mathcal{M}_1$ also has no minimal projections.*

Proof. Let $f \in \mathscr{L}(\widetilde{\mathcal{M}})$ be given with $f \ge 0$. By definition of the class $\mathscr{L}(\widetilde{\mathcal{M}})$, we can find $C > 0$ and $\alpha > 0$ so that $\mathbf{m}_f(t) < Ct^{-\alpha}$ for all $t > 0$. But then each $\chi_{(s,\infty)}(f)$ $(s > 0)$ must have a finite trace by Proposition 5.24. Write e_0 for $\chi_{(0,\infty)}(f)$. The trace τ must therefore be semifinite on the commutative subalgebra of $e_0\mathcal{M}e_0$ containing these spectral projections. It is a fairly straightforward exercise to see that the trace will still be semifinite on any randomly selected maximal abelian subalgebra $\mathcal{M}_0$ of $e_0\mathcal{M}e_0$ containing the former commutative subalgebra. (This follows from the fact that for any non-zero subprojection e of e_0, there must be some $s > 0$ for which the subprojection $e \wedge \chi_{(s,\infty)}(f)$ of e is non-zero.) In addition, $\mathcal{M}_0$ does not have minimal projections according to Proposition 5.9.

Finally, observe that we may use part (c) of Proposition 5.9, to select an abelian subalgebra $\mathcal{M}_c$ of $(\mathbb{1} - e_0)\mathcal{M}(\mathbb{1} - e_0)$ on which τ is still semifinite. The algebra we then seek is $\mathcal{M}_1 = \mathcal{M}_0 \oplus \mathcal{M}_c$. $\qquad\square$

Proof of Theorem 6.2   The first fact we note; is that we may, without loss of generality, assume that $\mathcal{M}$ has no minimal projections (see Proposition 5.20.)

Phase 1 of the proof of part (ii) (bounded case): Let $f, g \in \mathcal{M}$ be given. *We first consider the case where $f, g \geq 0$, and $\tau(\mathbb{1}) < \infty$.* For each $t \in [0, 1]$, set

$$x(t) = \exp(t \log(g + \epsilon\mathbb{1}))^{\frac{.}{.}}(f + \epsilon\mathbb{1})^{2\frac{.}{.}}\exp(t \log(g + \epsilon\mathbb{1})),$$

where $\epsilon > 0$ is arbitrary. We may then apply Lemma 6.3 to see that

$$\tau\left(\frac{d}{dt}\log(x(t))\right) = \tau(x(t)^{-1}x'(t))$$
$$= \tau(x(t)^{-1}(\log(g + \epsilon\mathbb{1})^{\frac{.}{.}}x(t) + x(t)^{\frac{.}{.}}\log(g + \epsilon\mathbb{1})))$$
$$= 2\tau(\log(g + \epsilon\mathbb{1})).$$

But then

$$2\tau(\log(g + \epsilon\mathbb{1})) = \int_0^1 \tau\left(\frac{d}{dt}\log(x(t))\right) dt$$
$$= \tau(\log(x(1))) - \tau(\log(x(0)))$$
$$= \tau(\log((g + \epsilon\mathbb{1})(f + \epsilon\mathbb{1})^2(g + \epsilon\mathbb{1})))$$
$$\quad - \tau(\log((f + \epsilon\mathbb{1})^2)).$$

We may rewrite this last equality as

$$\tau(\log(|(f + \epsilon\mathbb{1})(g + \epsilon\mathbb{1})|)) = \tau(\log(f + \epsilon\mathbb{1})) + \tau(\log(g + \epsilon\mathbb{1})).$$

Since by assumption $\tau(\mathbb{1}) < \infty$, we may use log-rules and Corollary 5.14 to see that

$$\tau(\log(f + \epsilon\mathbb{1})) = \tau(\log((f/\epsilon) + \mathbb{1})) + \log(\epsilon)\tau(\mathbb{1})$$
$$= \int_0^{\tau(\mathbb{1})} \log(\mathbf{m}_{(f/\epsilon)}(s) + 1) \, ds + \log(\epsilon)\tau(\mathbb{1})$$
$$= \int_0^{\tau(\mathbb{1})} \log(\mathbf{m}_f(s) + \epsilon) \, ds.$$

A similar claim obviously also holds for $\tau(\log(g + \epsilon\mathbb{1}))$. If we combine this with what we have just shown, it then follows that for any $\epsilon > 0$, we have that

$$\tau(\log(|(f + \epsilon\mathbb{1})(g + \epsilon\mathbb{1})|)) = \int_0^{\tau(\mathbb{1})} (\log(\mathbf{m}_f(s) + \epsilon) + \log(\mathbf{m}_g(s) + \epsilon)) \, ds$$

Now, we pass to the general case where $\tau(\mathbb{1}) = \infty$ is allowed. Let $\epsilon > 0$ be given. Notice that $|(f + \epsilon\mathbb{1})(g + \epsilon\mathbb{1})|^2 = (g + \epsilon\mathbb{1})(f + \epsilon\mathbb{1})^2(g + \epsilon\mathbb{1}) \geq \epsilon^2(g + \epsilon\mathbb{1})^2 \geq \epsilon^4\mathbb{1}$. Thus, $|(f + \epsilon\mathbb{1})(g + \epsilon\mathbb{1})|$ is of the form $p + \epsilon^2\mathbb{1}$ for some $p \in \mathcal{M}_+$. It then follows from Proposition 5.19 and Lemma 6.4 that

$$\sup\{\tau(e(\log|(f + \epsilon\mathbb{1})(g + \epsilon\mathbb{1})|)e) : e \in \mathbb{P}(\mathcal{M}), \tau(e) = t\}$$
$$= \sup\{\tau(e(\log|(p + \epsilon^2\mathbb{1})|)e) : e \in \mathbb{P}(\mathcal{M}), \tau(e) = t\}$$
$$= \sup\{\tau(e(\log|(p/\epsilon^2 + \mathbb{1})|)e) : e \in \mathbb{P}(\mathcal{M}), \tau(e) = t\} + t\log(\epsilon^2)$$
$$= \int_0^t [\log(\mathbf{m}_p(s)/\epsilon^2 + 1)] \, ds + t\log(\epsilon^2)$$
$$= \int_0^t \log(\mathbf{m}_p(s) + \epsilon^2) \, ds$$
$$= \int_0^t \log(\mathbf{m}_{(f+\epsilon\mathbb{1})(g+\epsilon\mathbb{1})}(s)) \, ds.$$

It is a straightforward exercise to see that $|(f + \epsilon\mathbb{1})(g + \epsilon\mathbb{1})|^2 - \epsilon^4 = (g + \epsilon\mathbb{1})(f + \epsilon\mathbb{1})^2(g + \epsilon\mathbb{1}) - \epsilon^4$ is a linear combination of finite products of elements of $\mathscr{L}(\widetilde{\mathcal{M}})$ and hence itself an element of $\mathscr{L}(\widetilde{\mathcal{M}})$. Thus, by Remark 5.8 and Lemma 6.5, there exists an abelian subalgebra $\mathcal{M}_1$ of $\mathcal{M}$ containing all the spectral projections, on which τ is still semifinite, and which contains no minimal projections. In terms of the formula for computing $\int_0^t \log(\mathbf{m}_{(f+\epsilon\mathbb{1})(g+\epsilon\mathbb{1})}(s)) \, ds$ that we have just verified, this means that we may restrict our attention to projections e with $\tau(e) = t$, which commute with $|(f + \epsilon\mathbb{1})(g + \epsilon\mathbb{1})|$. Let e be such a projection. Then, of course, $e(\log|(f + \epsilon\mathbb{1})(g + \epsilon\mathbb{1})|)e = e(\log|(f + \epsilon\mathbb{1})(g + \epsilon\mathbb{1})e|)e$. Now, by construction, $\mathbf{s}(|(f + \epsilon\mathbb{1})(g + \epsilon\mathbb{1})e|) \leq e$. If $(f + \epsilon\mathbb{1})(g + \epsilon\mathbb{1})e = u|(f + \epsilon\mathbb{1})(g + \epsilon\mathbb{1})e|$ is the polar decomposition of $(f + \epsilon\mathbb{1})(g + \epsilon\mathbb{1})e$, this means that

$$|(f + \epsilon\mathbb{1})(g + \epsilon\mathbb{1})e| = e|(f + \epsilon\mathbb{1})(g + \epsilon\mathbb{1})e| = eu^*(f + \epsilon\mathbb{1})(g + \epsilon\mathbb{1})e.$$

Now, let $(g + \epsilon\mathbb{1})e = v|(g + \epsilon\mathbb{1})e|$ be the polar decomposition of $(g + \epsilon\mathbb{1})e$. On similarly using the fact that $\mathbf{s}(|(g + \epsilon\mathbb{1})e|) \leq e$, we may rewrite the above equality in the following way:

$$|(f + \epsilon\mathbb{1})(g + \epsilon\mathbb{1})e| = eu^*(f + \epsilon\mathbb{1})(g + \epsilon\mathbb{1})e$$
$$= e[u^*(f + \epsilon\mathbb{1})v]e|(g + \epsilon\mathbb{1})e|.$$

An application of the technology we developed for the case $\tau(\mathbb{1}) < \infty$ to the compression $e\mathcal{M}e$, now shows that

$$\tau(e(\log |(f + \epsilon\mathbb{1})(g + \epsilon\mathbb{1})|)e)$$
$$= \tau(e(\log |(f + \epsilon\mathbb{1})(g + \epsilon\mathbb{1})e|)e)$$
$$= \tau(e(\log |e[u^*(f + \epsilon\mathbb{1})v]e|(g + \epsilon\mathbb{1})e|)e)$$
$$= \tau(e(\log |e[u^*(f + \epsilon\mathbb{1})v]e|)e) + \tau(e(\log |(g + \epsilon\mathbb{1})e|)e)$$
$$= \int_0^{\tau(e)} \log(\mathbf{m}^e_{e[u^*(f+\epsilon\mathbb{1})v]e}(s))\, ds + \int_0^{\tau(e)} \log(\mathbf{m}^e_{|(g+\epsilon\mathbb{1})e|}(s)\, ds.$$

By Lemma 6.4, this in turn ensures that

$$\tau(e(\log |(f + \epsilon\mathbb{1})(g + \epsilon\mathbb{1})|)e)$$
$$= \int_0^t \log(\mathbf{m}_{e[u^*(f+\epsilon\mathbb{1})v]e}(s))\, ds + \int_0^t \log(\mathbf{m}_{|(g+\epsilon\mathbb{1})e|}(s))\, ds$$
$$\leq \int_0^t \log(\mathbf{m}_f(s) + \epsilon))\, ds + \int_0^t \log(\mathbf{m}_g(s) + \epsilon)\, ds.$$

If we now take the supremum over all projections e commuting with $|(f + \epsilon\mathbb{1})(g + \epsilon\mathbb{1})|$ for which $\tau(e) = t$, it follows that

$$\int_0^t \log \mathbf{m}_{(f+\epsilon\mathbb{1})(g+\epsilon\mathbb{1})}(s)\, ds \leq \int_0^t \log(\mathbf{m}_f(s) + \epsilon))\, ds$$
$$+ \int_0^t \log(\mathbf{m}_g(s) + \epsilon)\, ds.$$

Observe that for any two bounded positive elements f and g of $\mathcal{M}$, we may use log-rules to write the inequality proven above as the claim that

$$\int_0^t \log \mathbf{m}_{(\epsilon^{-1}f+\mathbb{1})(\epsilon^{-1}g+\mathbb{1})}(s)\, ds$$
$$\leq \int_0^t \log(\epsilon^{-1}\mathbf{m}_f(s) + 1))\, ds + \int_0^t \log(\epsilon^{-1}\mathbf{m}_g(s) + 1)\, ds$$

for all $\epsilon > 0$. Given $r > 0$, we may then apply Theorem 5.21 to the fact that $t^r \circ \exp$ is convex and increasing on $\mathbb{R}$, to conclude that

$$\int_0^t \mathbf{m}^r_{(\epsilon^{-1}f+\mathbb{1})(\epsilon^{-1}g+\mathbb{1})}(s)\, ds \le \int_0^t (\epsilon^{-1}\mathbf{m}_f(s)+1)^r(\epsilon^{-1}\mathbf{m}_g(s)+1)^r\, ds$$

or equivalently that

$$\int_0^t \mathbf{m}^r_{(f+\epsilon\mathbb{1})(g+\epsilon\mathbb{1})}(s)\, ds \le \int_0^t (\mathbf{m}_f(s)+\epsilon)^r(\mathbf{m}_g(s)+\epsilon)^r\, ds$$

for all $\epsilon > 0$. Observe that $\mathbf{m}_{(f+\epsilon\mathbb{1})(g+\epsilon\mathbb{1})} \ge \mathbf{m}_{f(g+\epsilon\mathbb{1})}$ since $|(f+\epsilon\mathbb{1})(g+\epsilon\mathbb{1})| \ge |f(g+\epsilon\mathbb{1})|$. Similarly, $\mathbf{m}_{(g+\epsilon\mathbb{1})f} \ge \mathbf{m}_{gf}$. So since $(gf)^* = fg$ and $[(g+\epsilon\mathbb{1})f]^* = f(g+\epsilon\mathbb{1})$, we have that $\mathbf{m}_{f(g+\epsilon\mathbb{1})} \ge \mathbf{m}_{fg}$ and hence that $\mathbf{m}_{(f+\epsilon\mathbb{1})(g+\epsilon\mathbb{1})} \ge \mathbf{m}_{fg}$. It therefore follows that

$$\int_0^t \mathbf{m}^r_{fg}(s)\, ds \le \int_0^t (\mathbf{m}_f(s)+\epsilon)^r(\mathbf{m}_g(s)+\epsilon)^r\, ds$$

for all $\epsilon > 0$ and hence that

$$\int_0^t \mathbf{m}^r_{fg}(s)\, ds \le \int_0^t (\mathbf{m}_f(s)\mathbf{m}_g(s))^r\, ds.$$

For general possibly non-positive elements f and g of $\mathcal{M}$, we observe that $|fg| = |fu|g|\,| = |\,|fu|.|g|\,|$, where $g = u|g|$ is the polar decomposition of g. Using what we have just verified, it will then follow that $\int_0^t \mathbf{m}^r_{fg}(s)\, ds = \int_0^t \log \mathbf{m}^r_{|fu|\,|g|}(s)\, ds \le \int_0^t (\mathbf{m}_{|fu|}(s)\mathbf{m}_{|g|}(s))^r\, ds \le \int_0^t (\mathbf{m}_f(s)\mathbf{m}_g(s))^r\, ds.$

Phase 2 of the proof of part (ii) (general case): We proceed with using what we have just verified to prove the general version of (ii). We now modify the technique used in the proof of Proposition 5.24 to show that claim (ii) follows from the inequality proved above for bounded elements. Let $f, g \in \mathscr{L}(\mathcal{M})$ be given and let $|f| = \int_0^\infty \lambda\, de_\lambda(|f|)$ be the spectral resolution of $|f|$. If $f = u|f|$ is the polar decomposition of f, then clearly $f = u\int_0^\infty \lambda\, de_\lambda(|f|)$. Now, set $f_n = u\int_0^n \lambda\, de_\lambda(|f|)$ for each $n \in \mathbf{N}$. Each f_n will clearly be bounded, with $|f_n| \le |f|$ for each n (so also $\mathbf{m}_{f_n} \le \mathbf{m}_f$). We proceed to show that $f_n \to f$ in measure as $n \to \infty$. For each n, the left support projection of f_n is just $\chi_{[0,n]}(|f|)$. Writing $e_n = \chi_{(n,\infty)}(|f|)$, the τ-measurability of f ensures that $\tau(e_n) = d_{|f|}(n) \to 0$ as $n \to \infty$. Furthermore, by construction, $\|(f - f_n)(\mathbb{1} - e_n)\| = 0$. Therefore, as required, (f_n) converges to f in measure. We may similarly construct a sequence $(g_n) \subseteq \mathcal{M}$ such that $g_n \to g$ in measure with $\mathbf{m}_{g_n} \le \mathbf{m}_g$ for all n. Of course then also $f_n g_m \to f g_m$ in measure as $n \to \infty$, with $\mathbf{m}_{f_n g_m} \le \mathbf{m}_{f g_m}$ for all n and m. It is now an easy exercise to see that we may then use Theorem 5.29 to conclude that

$$\int_0^t \mathbf{m}^r_{fg_m}(s)\,ds = \lim_{n\to\infty}\int_0^t \mathbf{m}^r_{f_n g_m}(s)\,ds$$

$$\leq \lim_{n\to\infty}\int_0^t \mathbf{m}^r_{f_n}(s)\mathbf{m}^r_{g_m}(s)\,ds$$

$$= \int_0^t \mathbf{m}^r_f(s)\mathbf{m}^r_{g_m}(s)\,ds.$$

Again by Theorem 5.29, taking the limit as $m \to \infty$ yields $\int_0^t \mathbf{m}^r_{fg}(s)\,ds \leq \int_0^t \mathbf{m}^r_f(s)\mathbf{m}^r_g(s)\,ds$. Using the known fact that

$$\exp\left(t^{-1}\int_0^t \log|w(s)|\,ds\right) = \lim_{r\searrow 0}\left[t^{-1}\int_0^t |w(s)|^r\,ds\right]^{1/r}$$

if $\int_0^t |w(s)|^r\,ds < \infty$ for some $r > 0$ (see, for example, exercise 5 of [**Rud74**, Chapter 3] for this), it now follows from the above that

$$\Delta_t(fg) = \exp\left(\int_0^t \log(\mathbf{m}_{fg}(s))\,ds\right)$$

$$\leq \exp\left(\int_0^t \log(\mathbf{m}_f(s)\mathbf{m}_g(s))\,ds\right)$$

$$= \Delta_t(f)\Delta_t(g)$$

as required.

Proof of parts (i) and (iii): By part (ii), we know that $\int_0^t \log(\mathbf{m}_{fg}(s))\,ds \leq \int_0^t \log(\mathbf{m}_f(s)\mathbf{m}_g(s))\,ds$. Thus, to see that (iii) holds, we simply need to consider this fact along with Theorem 5.21. Finally, let $p, q, r > 0$ be given. Hölder's inequality clearly holds if one of $\|f\|_p$ or $\|g\|_q$ is 0. On assuming both to be non-zero, the inequality will then hold similarly if one of $\|f\|_p$ or $\|g\|_q$ is infinite. Hence, we may assume that both are finite. But then $f, g \in \mathscr{L}(\widetilde{\mathcal{M}})$. It then follows from part (iii) that $(\int_0^t \mathbf{m}^r_{fg}(s)\,ds)^{1/r} \leq (\int_0^t (\mathbf{m}_f(s)\mathbf{m}_g(s))^r\,ds)^{1/r}$ for any $t > 0$. On letting $t \to \infty$ and applying the classical Hölder inequality, we have that

$$\|fg\|_r = \left(\int_0^\infty \mathbf{m}^r_{fg}(s)\,ds\right)^{1/r} \leq \left(\int_0^\infty \mathbf{m}^p_f(s)\,ds\right)^{1/p}\cdot\left(\int_0^\infty \mathbf{m}^q_g(s)\,ds\right)^{1/q}$$

$$= \|f\|_p \cdot \|g\|_q$$

as required. $\qquad\square$

The proof of a noncommutative Minkowski inequality is much more straightforward than was the case for Hölder's inequality.

Theorem 6.6 *Let $f, g \in \widetilde{\mathcal{M}}$ be given. For any $1 \leq p < \infty$, we then have $\|f + g\|_p \leq \|f\|_p + \|g\|_p$.*

Proof.   For this result, the hard work was done in Theorem 5.22. To see that the stated claim holds, simply apply that theorem to the function $t \mapsto t^p$, to see that $\int_0^t \mathbf{m}_{f+g}(s)^p \, ds \leq \int_0^t (\mathbf{m}_f(s) + \mathbf{m}_g(s))^p \, ds$ for any $t > 0$. If now we let $t \to \infty$ and apply the usual Minkowski inequality, we will have that

$$\left(\int_0^\infty \mathbf{m}_{f+g}(s)^p \, ds \right)^{1/p} \leq \left(\int_0^\infty (\mathbf{m}_f(s) + \mathbf{m}_g(s))^p \, ds \right)^{1/p}$$

$$\leq \left(\int_0^\infty \mathbf{m}_f(s)^p \, ds \right)^{1/p} + \left(\int_0^\infty \mathbf{m}_g(s)^p \, ds \right)^{1/p}.$$

By Proposition 5.13, this boils down to the statement that $(\tau(|f + g|^p)^{1/p} \leq \tau(|f|^p)^{1/p} + \tau(|g|^p)^{1/p}$. $\square$

We next start the process of proving that $\|\cdot\|_p$ is a p-norm if $0 < p < 1$. This requires some careful preparation.

Lemma 6.7 *Let f be a positive element of $\widetilde{\mathcal{M}}$, $u \in \mathcal{M}$ a contraction and $F : [0, \infty) \to [0, \infty]$ a positive non-decreasing function with $F(0) = 0$, which is continuous on the interval $[0, b_F]$ where $b_F = \sup\{t > 0 : F(t) < \infty\}$. Then:*

- *if F is convex we have that $F \circ \mathbf{m}_{ufu^*} = \mathbf{m}_{F(ufu^*)} \leq \mathbf{m}_{uF(f)u^*}$ whenever $F(f)$ is again τ-measurable;*
- *$F \circ \mathbf{m}_{ufu^*} = \mathbf{m}_{F(ufu^*)} \geq \mathbf{m}_{uF(f)u^*}$ if F is concave.*

Note that it is only in the convex case that b_F can be finite.

The proof relies on the classical fact that a *convex* function of the type described above is of the form $F(t) = \sup_{\gamma \in \Gamma}(c_\gamma t + d_\gamma)$ for some collection of affine linear functions where $d_\gamma \leq 0$ for each γ, with a *concave* function of the above type, of the form $F(t) = \inf_{\gamma \in \Gamma}(c_\gamma t + d_\gamma)$ where in this case $d_\gamma \geq 0$ for each γ.

Proof.   We prove only the first claim. The second can be proven by appropriately modifying the current proof. Hence, let F be a convex function of the type described above, and let $f \in \widetilde{\mathcal{M}}$ be given with $F(f)$ again τ-measurable. Suppose that $\mathcal{M}$ acts on the Hilbert space H. For each unit vector $\xi \in H$ and each $\gamma \in \Gamma$, we then have that

$$
\begin{aligned}
\langle uF(f)u^*\xi, \xi \rangle &\geq \langle u(c_\gamma f + d_\gamma)u^*\xi, \xi \rangle \\
&= \langle c_\gamma uf u^*\xi, \xi \rangle + d_\gamma \|u^*\xi\|^2 \\
&\geq \langle c_\gamma uf u^*\xi, \xi \rangle + d_\gamma \\
&= \langle (c_\gamma uf u^* + d_\gamma \mathbb{1})\xi, \xi \rangle
\end{aligned}
$$

since $d_\gamma \leq 0$. Now take the supremum over γ to see that $\langle uF(f)u^*\xi, \xi \rangle \geq \langle F(uf u^*)\xi, \xi \rangle$ for each $\xi \in H$. This, in particular, shows that if $uF(f)u^*$ is bounded on a subspace of H, then so is $F(uf u^*)$. Since $uF(f)u^*$ is τ-measurable, the same must therefore be true of $F(uf u^*)$. The fact that $F \circ \mathbf{m}_{uf u^*} = \mathbf{m}_{F(uf u^*)}$ therefore follows from earlier results (Proposition 5.12).

Now observe that for any $g \in \widetilde{\mathcal{M}}$, the formula for the decreasing rearrangement in Definition 5.1 may be expressed in the form

$$
\mathbf{m}_g(t) = \inf_{e \in \mathbb{P}(\mathcal{M}), \tau(\mathbb{1}-e) \leq t} \left[\sup_{\xi \in e(H), \|\xi\|=1} \|g\xi\| \right].
$$

In the case where $g \geq 0$, this formula may be rewritten as

$$
\mathbf{m}_{\sqrt{g}}(t) = \inf_{e \in \mathbb{P}(\mathcal{M}), \tau(\mathbb{1}-e) \leq t} \left[\sup_{\xi \in e(H), \|\xi\|=1} \sqrt{\langle g\xi, \xi \rangle} \right].
$$

If we apply this fact to the above inequality, it is clear that we must then have that $\mathbf{m}_{\sqrt{uF(f)u^*}} \geq \mathbf{m}_{\sqrt{F(uf u^*)}}$. Squaring both sides now yields the fact that $\mathbf{m}_{uF(f)u^*} \geq \mathbf{m}_{F(uf u^*)}$, as required. Given that $F \circ \mathbf{m}_{uf u^*} = \mathbf{m}_{F(uf u^*)}$, this proves the claim. $\qquad\square$

We are now finally ready to prove that in the case $0 < p < 1$, the $L^p(\mathcal{M}, \tau)$-spaces are p-normed spaces. This fact follows from part (i) of the following theorem applied to the function $t \mapsto t^p$.

Proposition 6.8 *Let $f, g \in \widetilde{\mathcal{M}}$ be given. Also let $F : [0, \infty) \to [0, \infty]$ be a positive non-decreasing function with $F(0) = 0$ which is continuous on the interval $[0, b_F]$ where $b_F = \sup\{t > 0 : F(t) < \infty\}$, and let $a, b \in \mathcal{M}$ be given with $a^*a + b^*b \leq \mathbb{1}$.*

  (i) *Let F be concave. If f and g are positive, then*
  $$\mathbf{m}_{a^* F(f)\, a + b^* F(f) b} \leq \mathbf{m}_{F(a^* f a + b^* f b)}, \text{ so that } \tau(a^* F(f)a) +$$

$\tau(b^*F(g)b) \leq \tau(F(a^*fa + b^*gb))$. *For general f and g, we have $\tau(F(|f + g|)) \leq \tau(F(|f|)) + \tau(F(|g|))$.*

(ii) *Let F be convex. If f and g are positive, then $\mathbf{m}_{a^*F(f)a+b^*F(f)b} \geq \mathbf{m}_{F(a^*fa+b^*fb)}$ whenever $F(f)$ and $F(g)$ are again τ-measurable, whence $\tau(a^*F(f)a) + \tau(b^*F(g)b) \leq \tau(F(a^*fa + b^*gb))$. If f and g are positive, we have $\tau(F(f + g)) \geq \tau(F(f)) + \tau(F(g))$.*

Proof. Both proofs run along similar lines, and hence, we will only prove part (i). To prove the first claim of part (i), we pass to the von Neumann algebra $\mathcal{M}\overline{\otimes}M_2(\mathbb{C})$ of 2×2 matrices of elements of $\mathcal{M}$. In this algebra, positive elements are of the form $\left[\begin{smallmatrix} a & c \\ c^* & b \end{smallmatrix}\right]$, where $a, b \in \mathcal{M}_+$. It is a semifinite algebra, with the prescription $\widetilde{\tau}\left(\left[\begin{smallmatrix} a & c \\ c^* & b \end{smallmatrix}\right]\right) = \tau(a + b)$ defining a faithful normal semifinite trace. Given $a, b \in \mathcal{M}$ with $a^*a + b^*b \leq \mathbb{1}$, it is now an easy exercise to see that $\left[\begin{smallmatrix} a & 0 \\ b & 0 \end{smallmatrix}\right]$ is a contraction. Therefore, for any two positive elements f and g of $\widetilde{\mathcal{M}}$, we may apply Lemma 6.7 to the pair $\left[\begin{smallmatrix} a & 0 \\ b & 0 \end{smallmatrix}\right]$ and $\left[\begin{smallmatrix} f & 0 \\ 0 & g \end{smallmatrix}\right]$, to see that

$$\mathbf{m}\left(\left[\begin{array}{cc} a^*F(f)a + b^*F(g)b & 0 \\ 0 & 0 \end{array}\right]\right) \leq \mathbf{m}\left(\left[\begin{array}{cc} F(a^*fa + b^*gb) & 0 \\ 0 & 0 \end{array}\right]\right).$$

These decreasing rearrangements are, of course, with respect to $\widetilde{\tau}$. However, for any $h \in \widetilde{\mathcal{M}}_+$, when writing $m(h)$ for $\left[\begin{smallmatrix} h & 0 \\ 0 & 0 \end{smallmatrix}\right]$, it is easy to see that

$$\begin{aligned} d_{m(h)}(s) &= \widetilde{\tau}\left(\chi_{(s,\infty)}\left(\left[\begin{array}{cc} h & 0 \\ 0 & 0 \end{array}\right]\right)\right) \\ &= \widetilde{\tau}\left[\begin{array}{cc} \chi_{(s,\infty)}(h) & 0 \\ 0 & 0 \end{array}\right] \\ &= \tau(\chi_{(s,\infty)}(h)) \\ &= d_h(s). \end{aligned}$$

Thus, by Proposition 5.7, the above inequality may be rephrased as

$$\mathbf{m}_{a^*F(f)a+b^*F(g)b} \leq \mathbf{m}_{F(a^*fa+b^*gb)}.$$

We will prove the second claim in two stages. First, suppose that $f, g \in \widetilde{\mathcal{M}}$ are positive. It then follows from Proposition 3.90 that there exist contractions $c_f, c_g \in \mathcal{M}_+$ supported on $\mathbf{s}(f+g)$, so that $(f + g)^{1/2}c_f(f + g)^{1/2} = f$ and $(f + g)^{1/2}c_g(f + g)^{1/2} = g$. Since $(f + g)^{1/2}(c_f + c_g)(f + g)^{1/2} = f + g$, we must have that $c_f + c_g = \mathbf{s}(f + g)$.

Since $|c_f^{1/2}(f+g)^{1/2}| = f^{1/2}$ and $|c_g^{1/2}(f+g)^{1/2}| = g^{1/2}$, we may select partial isometries u_f and u_g so that for $a = u_f c_f^{1/2}$ and $b = u_g c_g^{1/2}$, we have that $a(f+g)^{1/2} = f^{1/2}$ and $b(f+g)^{1/2} = g^{1/2}$. By construction, a and b satisfy $a^*a + b^*b = \mathbf{s}(f+g)$. We may therefore apply what we have just proven to see that

$$
\begin{aligned}
\tau(F(f)) + \tau(F(g)) &\geq \tau(F(a(f+g)a^*)) + \tau(F(b(f+g)b^*)) \\
&\geq \tau(aF(f+g)a^*) + \tau(bF(f+g)b^*) \\
&= \tau(F(f+g)^{1/2}a^*aF(f+g)^{1/2}) + \\
&\qquad \tau(F(f+g)^{1/2}b^*bF(f+g)^{1/2}) \\
&= \tau(F(f+g)^{1/2}\mathbf{s}(f+g)F(f+g)^{1/2}) \\
&= \tau(F(f+g)).
\end{aligned}
$$

To see the validity of the last equality above, observe that on any interval of the form $[0, \beta]$ where $\beta < \infty$, the fact that $F(0) = 0$ ensures that we may use the Stone–Weierstrass theorem to uniformly approximate $\sqrt{F}$ with polynomials with no constant term on that interval. For any such polynomial p, we have that $p(f+g) = p((f+g)\mathbf{s}(f+g)) = p(f+g)\mathbf{s}(f+g)$ on the subspace $\chi_{[0,\beta]}(f+g)(H)$. But then by the continuous functional calculus, $F(f+g)^{1/2}\mathbf{s}(f+g) = F(f+g)^{1/2}$ on the subspace $\chi_{[0,\beta]}(f+g)(H)$. Hence, we have that $F(f+g)\mathbf{s}(f+g) = F(f+g)$ as τ-measurable operators.

For general $f, g \in \widetilde{\mathcal{M}}$, we use Lemma 5.2 to select partial isometries $u, v \in \mathcal{M}$ so that $|f+g| \leq u|f|u^* + v|g|v^*$. We may then use what we have just proven to see that

$$
\begin{aligned}
\tau(F(|f+g|)) &= \tau(F(u|f|u^* + v|g|v^*)) \\
&\leq \tau(F(u|f|u^*)) + \tau(F(v|g|v^*)) \\
&= \int_0^\infty \mathbf{m}_{F(u|f|u^*)}(s)\,ds + \int_0^\infty \mathbf{m}_{F(v|g|v^*)}(s)\,ds \\
&= \int_0^\infty F(\mathbf{m}_{u|f|u^*}(s))\,ds + \int_0^\infty F(\mathbf{m}_{v|g|v^*}(s))\,ds \\
&\leq \int_0^\infty F(\mathbf{m}_f(s))\,ds + \int_0^\infty F(\mathbf{m}_g(s))\,ds \\
&\leq \int_0^\infty \mathbf{m}_{F(f)}(s)\,ds + \int_0^\infty \mathbf{m}_{F(g)}(s)\,ds \\
&\leq \tau(F(f)) + \tau(F(g)). \qquad \square
\end{aligned}
$$

We pause to note the following easy corollary before concluding this subsection with a preliminary investigation of the action of the trace on $L^1(\mathcal{M},\tau)$ and $L^2(\mathcal{M},\tau)$, and of the bounded elements of $L^p(\mathcal{M},\tau)$ spaces.

Theorem 6.9 *Let $f,g \in \widetilde{\mathcal{M}}$ be given. For any non-decreasing continuous concave function F on $[0,\infty)$ with $F(0) = 0$, we have*

$$\int_0^t F(\mathbf{m}_{f+g}(s))\, ds \leq \int_0^t F(\mathbf{m}_f(s))\, ds + \int_0^t F(\mathbf{m}_g(s))\, ds$$

for all $t > 0$.

Proof.   As we have done many times before, we apply Proposition 5.20 to reduce the proof to the case where $\mathcal{M}$ does not have minimal projections, thus gaining access to Proposition 5.19. We first select partial isometries u and v in $\mathcal{M}$ so that $|f+g| \leq u|f|u^* + v|g|v^*$. Next observe that on any interval of the form $[0,\beta]$ where $\beta < \infty$, the fact that $F(0) = 0$ will as before ensure that we may use the Stone–Weierstrass theorem to uniformly approximate F with polynomials without constant term. For any such polynomial p, we have that $p(e|f+g|e) = ep(|f+g|)e$ on the subspace $\chi_{[0,\beta]}(|f+g|)(H)$. But then by the continuous functional calculus, $eF(|f+g|)e = F(e|f+g|e)$ on the subspace $\chi_{[0,\beta]}(|f+g|)(H)$. Hence, we must have that $eF(|f+g|)e = F(e|f+g|e)$ as τ-measurable operators. Given any projection $e \in \mathcal{M}$ with $\tau(e) = t$, we may then apply part (i) of the preceding theorem and Corollary 5.11 to see that

$$
\begin{aligned}
\tau(eF(|f+g|)e) &= \tau(F(e|f+g|e)) \\
&\leq \tau(F(eu|f|u^*e + ev|g|v^*e)) \\
&\leq \tau(F(eu|f|u^*e)) + \tau(F(ev|g|v^*e)) \\
&= \int_0^\infty F(\mathbf{m}_{eu|f|u^*e}(s))\, ds + \int_0^\infty F(\mathbf{m}_{ev|g|v^*e}(s))\, ds \\
&= \int_0^t F(\mathbf{m}_{eu|f|u^*e}(s))\, ds + \int_0^t F(\mathbf{m}_{ev|g|v^*e}(s))\, ds \\
&\leq \int_0^t F(\mathbf{m}_{|f|}(s))\, ds + \int_0^t F(\mathbf{m}_{|g|}(s))\, ds \\
&= \int_0^t F(\mathbf{m}_f(s))\, ds + \int_0^t F(\mathbf{m}_g(s))\, ds.
\end{aligned}
$$

By Proposition 5.19, taking the supremum over all projections e with $\tau(e) = t$ yields the conclusion. $\qquad\square$

Remark 6.10   Given $p > 0$, it easily follows from Propositions 5.12 and 5.13 that $\|f\|_p = \|\,|f|\,\|_p = \|f^*\|_p$ for any $f \in L^p(\mathcal{M}, \tau)$. In addition, for any $f \in L^p(\mathcal{M}, \tau)$, in case $p \geq 1$, it will follow from Theorem 5.22 and what we have just noted that $\tau(|\mathrm{Re}(f)|^p)^{1/p} \leq \frac{1}{2}(\|f\|_p + \|f^*\|_p) = \|f\|_p$. In case $1 > p > 0$, we may replace Theorem 5.22 with Proposition 6.8 to obtain the conclusion that $\tau(|\mathrm{Re}(f)|^p) \leq 2^{-p}\tau(|f + f^*|^p) \leq 2^{-p}(\tau(|f|^p) + \tau(|f^*|^p\|_p) = 2^{1-p}\tau(|f|^p)$. A similar conclusion clearly holds for $\mathrm{Im}(f)$.

Now, let $f \in L^p(\mathcal{M}, \tau)$ be self-adjoint, and consider the elements $f_+ = f\chi_{[0,\infty)}(f)$ and $f_- = -f\chi_{(-\infty,0)}(f)$. Both f_+ and f_- are positive by construction. Moreover, since for any $t > 0$, $\mathbf{m}_{f_\pm} \leq \mathbf{m}_f$, it is clear from Proposition 5.13 that $\tau((f_\pm)^p) \leq \tau(|f|^p)$ and hence that $f_\pm \in L^p(\mathcal{M}, \tau)$. In fact, since by construction $f_+f_- = 0$, we have $|f|^p = f_+^p + f_-^p$, and hence that $\|f\|_p^p = \|f_+\|_p^p + \|f_-\|_p^p$.

Proposition 6.11

- *The canonical trace τ on $\mathcal{M}_+$ extends to a linear functional on $L^1(\mathcal{M}, \tau)$. In its action on $L^1(\mathcal{M}, \tau)$, we have that $\overline{\tau(a)} = \tau(a^*)$ and $|\tau(a)| \leq \tau(|a|)$ for any $a \in L^1(\mathcal{M}, \tau)$.*
- *The space $L^2(\mathcal{M}, \tau)$ is an inner product space with the inner product given by $\langle a, b \rangle = \tau(b^*a)$ for all $a, b \in L^2(\mathcal{M}, \tau)$.*

Proof.   We start by proving the first two claims regarding $L^1(\mathcal{M}, \tau)$. Given any self-adjoint element a of $L^1(\mathcal{M}, \tau)$, it is clear that $a = a_+ - a_-$ where $a_+ = a\chi_{[0,\infty)}(a)$ and $a_- = -a\chi_{(\infty,0)}(a)$. Both a_+ and a_- are positive by construction, with $a_\pm \in L^1(\mathcal{M}, \tau)$ by Remark 6.10. Now, let p_0 and p_1 be any other pair of positive elements of $L^1(\mathcal{M}, \tau)$ for which $a = p_0 - p_1$. Since then $a_+ + p_1 = p_0 + a_-$, an application of the trace reveals that $\tau(a_+) + \tau(p_1) = \tau(a_+ + p_1) = \tau(p_0 + a_-) = \tau(p_0) + \tau(a_-)$ and hence that $\tau(a_+) - \tau(a_-) = \tau(p_0) - \tau(p_1)$. Thus, for any self-adjoint element $a \in L^1(M, \tau)$, we may uniquely define $\tau(a)$ to be $\tau(p_0) - \tau(p_1)$ where p_0 and p_1 are any two positive elements of $L^1(\mathcal{M}, \tau)$ for which $a = p_0 - p_1$.

Now, let b be another self-adjoint element of $L^1(\mathcal{M}, \tau)$ of the form $b = q_0 - q_1$ where q_0 and q_1 are positive elements of $L^1(\mathcal{M}, \tau)$.

Then, by definition, $\tau(a+b) = \tau(p_0 + q_0) - \tau(p_1 + q_1) = [\tau(p_0) + \tau(q_0)] - [\tau(p_1) + \tau(q_1)] = [\tau(p_0) - \tau(p_1)] + [\tau(q_0) - \tau(q_1)] = \tau(a) + \tau(b)$. We can similarly use the fact that $\tau(\alpha f) = \alpha\tau(f)$ for any $\alpha \geq 0$ and any $f \in \widetilde{\mathcal{M}}_+$, to show that for any a in the self-adjoint portion $L^1(\mathcal{M}, \tau)_h$ of $L^1(\mathcal{M}, \tau)$, we have $\tau(\alpha a) = \alpha\tau(a)$ for any $\alpha \in \mathbb{R}$. Thus, τ is real linear on $L^1(\mathcal{M}, \tau)_h$.

Given a general element $a \in L^1(\mathcal{M}, \tau)$, it follows from Remark 6.10 that $\mathrm{Re}(a), \mathrm{Im}(a) \in L^1(\mathcal{M}, \tau)$. We therefore have that

$$L^1(\mathcal{M}, \tau) = L^1(\mathcal{M}, \tau)_h + iL^1(\mathcal{M}, \tau)_h.$$

On defining $\tau(a)$ to be $\tau(a) = \tau(\mathrm{Re}(a)) + i\tau(\mathrm{Im}(a))$, it is now an exercise to see that this definition ensures that τ is complex linear on $L^1(\mathcal{M}, \tau)$. From this definition, it is now clear that

$$\overline{\tau(a)} = \overline{\tau(\mathrm{Re}(a)) + i\tau(\mathrm{Im}(a))} = \tau(\mathrm{Re}(a)) - i\tau(\mathrm{Im}(a)) = \tau(a^*).$$

We proceed to proving the claims regarding $L^2(\mathcal{M}, \tau)$. It easily follows from Theorem 6.2 that b^*a will belong to $L^1(\mathcal{M}, \tau)$ whenever $a, b \in L^2(\mathcal{M}, \tau)$. Thus, $\langle a, b \rangle = \tau(b^*a)$ is well defined. It follows from what we have already proven that $\overline{\langle a, b \rangle} = \overline{\tau(b^*a)} = \tau((b^*a)^*) = \tau(a^*b) = \langle b, a \rangle$. All other properties of the inner product are easy to verify, including the fact that $\langle a, a \rangle = \tau(|a|^2) = \|a\|_2^2$.

To prove the final claim regarding $L^1(\mathcal{M}, \tau)$, we will make use of the Cauchy–Schwarz inequality for the inner product on $L^2(\mathcal{M}, \tau)$. Given any $x \in L^1(\mathcal{M}, \tau)$, we may let $a = |x|^{1/2}u^*$ and $b = |x|^{1/2}$, where $x = u|x|$ is the polar form of x. It then clearly follows that $\|b\|_2 = \tau(|x|)^{1/2}$ and that $\|a\|_2 = \|a^*\|_2 = \tau(|x|^{1/2}u^*u|x|^{1/2})^{1/2} = \tau(|x|)^{1/2}$. Thus, we have $a, b \in L^2(\mathcal{M}, \tau)$, with $|\tau(x)| = |\tau(u|x|)| = |\tau(a^*b)| = |\langle b, a \rangle| \leq \|b\|_2\|a\|_2 = \tau(|x|)$, as required. $\qquad\square$

Proposition 6.12 *For any $0 < p < 1$, $(L^p \cap L^\infty)(\mathcal{M}, \tau)$ is dense in $L^p(\mathcal{M}, \tau)$. In the case $1 \leq p < \infty$, $(L^1 \cap L^\infty)(\mathcal{M}, \tau)$ is a norm-dense subspace of $L^p(\mathcal{M}, \tau)$.*

Proof. We first prove that $(L^p \cap L^\infty)(\mathcal{M}, \tau)$ is dense in $L^p(\mathcal{M}, \tau)$ for any $0 < p < \infty$. To see this, it is enough to show that any positive element of $L^p(\mathcal{M}, \tau)$ is in the closure of $(L^p \cap L^\infty)(\mathcal{M}, \tau)$. Given $f \in L^p(\mathcal{M}, \tau)$ with $f \geq 0$, it follows from the normality of the trace that $\tau(f^p) = \lim_{n\to\infty} \tau(f^p\chi_{[0,n]}(f^p)) = \sup_{n\in\mathbb{N}} \tau(f^p\chi_{[0,n]}(f^p))$. Therefore,

$\lim_{n\to\infty} \tau((f - f\chi_{[0,n]}(f^p))^p) = \lim_{n\to\infty} \tau(f^p\chi_{(n,\infty)}(f^p)) = 0$. Since $\|f\chi_{[0,n]}(f^p)\|_\infty \leq n^{1/p}$, we have that $(f\chi_{[0,n]}(f^p)) \subseteq (L^p \cap L^\infty)(\mathcal{M},\tau)$ and hence we are done with the first part.

We now move to the case $1 \leq p < \infty$. In the case where $\tau(\chi_{(0,\infty)}(f^p)) < \infty$, the sequence (f_n) constructed above actually lies in $(L^1 \cap L^\infty)(\mathcal{M},\tau)$. To see this, note that in this case $\tau(f_n) = \tau(f_n\chi_{(0,\infty)}(f^p)) \leq \|f_n\|_\infty \tau(\chi_{(0,\infty)}(f^p)) < \infty$.

Next consider the case $\tau(\chi_{(0,\infty)}(f^p)) = \infty$. From Proposition 5.24, it follows that $d_{f^p}(s) < \infty$ for each $s > 0$. By the Borel functional calculus, we have that $f^p\chi_{(1/n,\infty)}(f^p)$ increases to f^p. So, by the normality of the trace, $\lim_{n\to\infty} \tau(f^p\chi_{[0,1/n]}(f^p)) = \tau(f^p) - \lim_{n\to\infty} \tau(f^p\chi_{(1/n,\infty)}(f^p)) = 0$. It follows that the sequence $f\chi_{[0,1/n]}(f^p)$ converges to 0 in L^p-norm. By what we have already shown, the sequence $\widetilde{f}_n = f\chi_{[0,n]}(f^p) - f\chi_{[0,1/n]}(f^p) = f\chi_{(1/n,n]}(f^p)$ ($n \in \mathbb{N}$) must then converge to f in L^p-norm. It remains to show that $(f\chi_{(1/n,n]}(f^p)) \subseteq (L^1 \cap L^\infty)(\mathcal{M},\tau)$. We clearly have $\|f\chi_{(1/n,n]}(f^p)\|_\infty \leq n^{1/p}$. To see that each $f\chi_{(1/n,n]}(f^p)$ is in $L^1(\mathcal{M},\tau)$, recall that for $C_n = n^{(1-(1/p))}$, we have that $\lambda^{1/p} \leq C_n\lambda$ whenever $\lambda \geq \frac{1}{n}$. But then

$$f\chi_{(1/n,n]}(f^p) = \int_{1/n}^n \lambda^{1/p}\, de_\lambda(f^p) \leq C_n\int_{1/n}^n \lambda\, de_\lambda(f^p) = C_n f^p\chi_{(1/n,n]}(f^p)$$

whence $\tau(f\chi_{(1/n,n]}(f^p)) \leq C_n\tau(f^p\chi_{(1/n,n]}(f^p)) < \infty$ as required. $\square$

6.1.2 Convergence and completeness

Before proceeding with the proof of the completeness of these spaces, we pause to establish a dominated convergence theorem appropriate to the present context.

Theorem 6.13 *Let f_n ($n \in \mathbb{N}$) be a sequence of τ-measurable operators converging to $f \in \widetilde{\mathcal{M}}$ in the topology of convergence in measure on $\widetilde{\mathcal{M}}$. Suppose that there exist operators g_n ($n \in \mathbb{N}$) and g in $L^p(\mathcal{M},\tau)$ ($0 < p < \infty$), for which we have that:*

    (i) $\mathbf{m}_{f_n} \leq \mathbf{m}_{g_n}$ *for all n;*
    (ii) $\lim_{n\to\infty} \|g_n\|_p = \|g\|_p$*;*
    (iii) $\mathbf{m}_g(t) \leq \liminf_{n\to\infty} \mathbf{m}_{g_n}(t)$ *for each $t > 0$.*

Then, f_n $(n \in \mathbb{N})$ and f are all in $L^p(\mathcal{M}, \tau)$, with $f_n \to f$ in L^p-norm. In the case $p = 1$, we additionally have that $\tau(f) = \lim_{n \to \infty} \tau(f_n)$.

Proof.   It is palpably clear from (i) that for each n, we have that

$$\tau(|f_n|^p) = \int_0^\infty \mathbf{m}_{f_n}^p(s)\, ds \leq \int_0^\infty \mathbf{m}_{g_n}^p(s)\, ds = \tau(|g_n|^p)$$

and hence that each f_n belongs to $L^p(\mathcal{M}, \tau)$. We may next apply Lemma 5.28 and the usual Fatou lemma to see that

$$\tau(|f|^p) = \int_0^\infty \mathbf{m}_f^p(s)\, ds \leq \int_0^\infty \liminf_{n \to \infty} \mathbf{m}_{f_n}^p(s)\, ds$$

$$\leq \int_0^\infty \liminf_{n \to \infty} \mathbf{m}_{g_n}^p(s)\, ds \leq \liminf_{n \to \infty} \int_0^\infty \mathbf{m}_{g_n}^p(s)\, ds = \liminf_{n \to \infty} \|g_n\|_p^p = \|g\|_p^p.$$

Therefore, $f \in L^p(\mathcal{M}, \tau)$.

We proceed to prove that $\lim_{n \to \infty} \|f - f_n\|_p = 0$. First note that for any $t > 0$,

$$\mathbf{m}_{f-f_n}(t) \leq \mathbf{m}_f(t/2) + \mathbf{m}_{f_n}(t/2) \leq \mathbf{m}_f(t/2) + \mathbf{m}_{g_n}(t/2).$$

Hence, for $C_p = \max\{1, 2^{p-1}\}$, we have that

$$\mathbf{m}_{f-f_n}^p(t) \leq (\mathbf{m}_f(t/2) + \mathbf{m}_{g_n}(t/2))^p \leq C_p[\mathbf{m}_f^p(t/2) + \mathbf{m}_{g_n}^p(t/2)].$$

Thus, $t \mapsto C_p[\mathbf{m}_f^p(t/2) + \mathbf{m}_{g_n}^p(t/2)] - \mathbf{m}_{f-f_n}^p(t)$ is a non-negative function. Now observe that for any $t > 0$, we must have $\lim_{n \to \infty} \mathbf{m}_{f-f_n}(t) = 0$ by Proposition 5.23. If we combine this with assumption (iii), it now follows that

$$\liminf_{n \to \infty}\{C_p[\mathbf{m}_f^p(t/2) + \mathbf{m}_{g_n}^p(t/2)] - \mathbf{m}_{f-f_n}^p(t)\} \geq C_p[\mathbf{m}_f^p(t/2) + \mathbf{m}_g^p(t/2)]$$

for each $t > 0$. The standard Fatou lemma now ensures that

$$2C_p(\|f\|_p^p + \|g\|_p^p) = C_p \int_0^\infty [\mathbf{m}_f^p(t/2) + \mathbf{m}_g^p(t/2)]\, dt$$

$$\leq \int_0^\infty \liminf_{n \to \infty}\{C_p[\mathbf{m}_f^p(t/2) + \mathbf{m}_{g_n}^p(t/2)] - \mathbf{m}_{f-f_n}^p(t)\}\, dt$$

$$\leq \liminf_{n \to \infty} \int_0^\infty \{C_p[\mathbf{m}_f^p(t/2) + \mathbf{m}_{g_n}^p(t/2)] - \mathbf{m}_{f-f_n}^p(t)\}\, dt$$

$$= \liminf_{n \to \infty}[2C_p(\|f\|_p^p + \|g\|_p^p) - \|f - f_n\|_p^p]$$

or equivalently that $0 \leq -\limsup_{n\to\infty} \|f - f_n\|_p^p$. This clearly suffices to prove that $\lim_{n\to\infty} \|f - f_n\|_p = 0$ as required.

To see the claim regarding case $p = 1$, observe that we may then use Proposition 6.11 to see that $|\tau(f) - \tau(f_n)| \leq \tau(|f - f_n|)$. $\qquad\square$

Lemma 6.14 *Each of the spaces $L^p(\mathcal{M}, \tau)$ $(0 < p \leq \infty)$ injects continuously into $\widetilde{\mathcal{M}}$.*

Proof. In the case $p = \infty$, the claim is obvious. So, assume that $0 < p < \infty$, with (f_n) converging to f in $L^p(\mathcal{M}, \tau)$. For any $t > 0$, we may now use the fact that $\mathbf{m}_{f-f_n}$ is non-increasing to see that

$$t^{1/p}\mathbf{m}_{f-f_n}(t) \leq \left(\int_0^t \mathbf{m}_{f-f_n}^p(s)\, ds \right)^{1/p}$$

$$\leq \left(\int_0^\infty \mathbf{m}_{f-f_n}^p(s)\, ds \right)^{1/p}$$

$$= \|f - f_n\|_p.$$

The claim now follows from Proposition 5.23. $\qquad\square$

As a consequence of Theorem 6.13, we obtain the following important description of convergence in $L^p(\mathcal{M}, \tau)$.

Theorem 6.15 *Let f_n $(n \in \mathbb{N})$, f be elements of $L^p(\mathcal{M}, \tau)$, where $0 < p < \infty$. Then, the following are equivalent:*

(i) $\lim_{n\to\infty} \|f - f_n\|_p = 0$;

(ii) $\lim_{n\to\infty} \|f_n\|_p = \|f\|_p$, *and* $f_n \to f$ *in the topology of convergence in measure.*

Proof. The fact that (i)$\Rightarrow$(ii) easily follows from the lemma, whereas (ii)$\Rightarrow$(i) follows from Theorem 6.13 with $g_n = |f_n|$ and $g = |f|$. $\qquad\square$

We proceed with the proof of the completeness of the L^p-spaces.

Theorem 6.16 *Each of the spaces $L^p(\mathcal{M}, \tau)$ $(0 < p \leq \infty)$ is complete.*

Proof. In the case $p = \infty$, there is of course nothing to prove, so let $p < \infty$. We prove the case where $p \geq 1$. A similar proof, suitably modified, holds for the case $0 < p < 1$. Let (f_n) be a sequence in $L^p(\mathcal{M}, \tau)$ for which $\sum_{k=1}^{\infty} \|f_k\|_p < \infty$. We need to show that $\sum_{k=1}^{\infty} f_k$ converges in $L^p(\mathcal{M}, \tau)$. By separately considering the series $\sum_{k=1}^{\infty} \mathrm{Re}(f_k)$ and $\sum_{k=1}^{\infty} \mathrm{Im}(f_k)$, we can and do assume that each f_k is self-adjoint. Now, let $x_n = \sum_{k=1}^{n} f_k$ and $z_n = \sum_{k=1}^{n} |f_k|$ for each $n \in \mathbb{N}$. Given $n > m$, we have $\|z_n - z_m\|_p = \|\sum_{k=m+1}^{n} |f_n|\|_p \leq \sum_{k=m+1}^{\infty} \|f_k\|_p$. Since by assumption $\sum_{k=m}^{\infty} \|f_k\|_p \to 0$ as $m \to \infty$, the sequence (z_n) is Cauchy in $L^p(\mathcal{M}, \tau)$. By Lemma 6.14, (z_n) is then Cauchy in $\widetilde{\mathcal{M}}$. But $\widetilde{\mathcal{M}}$ is a complete linear metric space. So, there must exist some $z \in \widetilde{\mathcal{M}}$ which is the limit of (z_n) in the topology of convergence in measure. Notice that (z_n) is an increasing sequence of positive operators. So, by part (iii) of Proposition 5.12, we have $\mathbf{m}_z(t) = \lim_{n \to \infty} \mathbf{m}_{z_n}(t) = \sup_{n \in \mathbb{N}} \mathbf{m}_{z_n}(t)$. Equivalently, $\mathbf{m}_{z^p}(t) = \mathbf{m}_z^p(t) = \lim_{n \to \infty} \mathbf{m}_{z_n}^p(t) = \lim_{n \to \infty} \mathbf{m}_{z_n^p}(t)$. It therefore follows from Lemma 5.28 that $\|z\|_p = \tau(z^p)^{1/p} = \lim_{n \to \infty} \tau(z_n^p)^{1/p} = \lim_{n \to \infty} \|z_n\|$. But then Theorem 6.15 ensures that, in fact, $z_n \to z$ in $L^p(\mathcal{M}, \tau)$.

Next, observe that the self-adjointness assumption on the f_ks ensures that $0 \leq z_n + x_n \leq 2z_n$ for each n (and hence $\mathbf{m}_{z_n + x_n}(t) \leq \mathbf{m}_{2z_n}(t)$ for each n). Using the fact that for $n > m$, we have $\|x_n - x_m\|_p = \|\sum_{k=m+1}^{n} f_n\|_p \leq \sum_{k=m+1}^{\infty} \|f_k\|_p$, and so we may argue as before to conclude that the sequence (x_n) is Cauchy in $L^p(\mathcal{M}, \tau)$. Again, as before, the fact that $L^p(\mathcal{M}, \tau)$ continuously injects into $\widetilde{\mathcal{M}}$ ensures that (x_n) is Cauchy in $\widetilde{\mathcal{M}}$ and hence must have a limit x in $\widetilde{\mathcal{M}}$. The sequence $(x_n + z_n)$ therefore converges to $x + z$ in the topology of convergence in measure. Therefore, we may apply Theorem 6.13 to the pair of sequences $(x_n + z_n)$ and $(2z_n)$, to conclude that $(x_n + z_n)$ converges to $x + z$ in $L^p(\mathcal{M}, \tau)$ and hence that (x_n) converges to $x = (x + z) - z$ in $L^p(\mathcal{M}, \tau)$. This then proves the claim. $\qquad\square$

Remark 6.17 It now follows from Proposition 6.11 and the preceding theorem that $L^2(\mathcal{M}, \tau)$ is a Hilbert space.

6.1.3 The Schatten–von Neumann classes revisited

We briefly pause to indicate how the earlier construction of the Schatten–von Neumann classes harmonises with the construction of $L^p(\mathcal{M}, \tau)$-spaces.

Proposition 6.18 *For any $f \in B(H)$, we have $\mathbf{m}_f(t) = a_n(f)$ for all $n - 1 \leq t < n$. Therefore, $\mathrm{Tr}(|f|^p) = \int_0^\infty \mathbf{m}_f(t)^p \, dt = \sum_{n=1}^{\infty} a_n(f)^p$.*

Proof. This follows from Lemma 5.10 and the fact that for any projection e, $\mathrm{Tr}(e)$ is the rank of e. $\square$

Remark 6.19 Given $f \in B(H)$, the above fact shows that $\lim_{t\to\infty} \mathbf{m}_f(t) = 0$ if and only if $\lim_{n\to\infty} a_n(f) = 0$ if and only if f is compact. So, for the pair $(B(H), \mathrm{Tr})$, τ-compactness agrees with compactness. Since for any $f \in B(H)$ we have that $\mathrm{Tr}(|f|^p) = \int_0^\infty \mathbf{m}_f(t)^p\, dt = \sum_{n=1}^\infty a_n(f)^p$, it is now clear that $\mathrm{Tr}(|f|^p) < \infty$ if and only if $(a_n(f)) \in \ell_p(\mathbb{N})$ if and only if f is compact and a member of $\mathcal{S}_p(H)$. So, as vector spaces $\mathcal{S}_p(H) = L^p(B(H), \mathrm{Tr})$. The above trace formula also shows that $\sigma_p(f) = \|f\|_p$. Thus, the pth Schatten class is nothing but the space $L^p(B(H), \mathrm{Tr})$. From a structural point of view, these spaces may be regarded as noncommutative versions of the ℓ_p-spaces.

The following fact follows immediately from the properties of the trace on $B(H)$.

Remark 6.20 For any $f \in \widetilde{B(H)}$, we must have $\mathrm{Tr}(\chi_{(\epsilon,\infty)}(|f|)) \to 0$ as $\epsilon \to \infty$. So, there must exist some $\epsilon > 0$ for which $\mathrm{Tr}(\chi_{(\epsilon,\infty)}(|f|)) < 1$, whence $\mathrm{Tr}(\chi_{(\epsilon,\infty)}(|f|)) = 0$. Thus, $f \in B(H)$, showing that $B(H) = \widetilde{B(H)}$.

Remark 6.21 It is now clear from the above discussion that the space of Hilbert–Schmidt operators $\mathcal{S}_2(H)$ is nothing but the Hilbert space $L^2(B(H), \mathrm{Tr})$. Given an orthonormal basis (v_i), the fact that $\mathrm{Tr}(b) = \sum_i \langle bv_i, v_i \rangle$ ensures that the well-known inner product $\mathcal{S}_2(H) \times \mathcal{S}_2(H) \to \mathbb{C} : (f, g) \mapsto \sum_i \langle fv_i, gv_i \rangle$ on $\mathcal{S}_2(H)$ yielding σ_2 corresponds to $\langle f, g \rangle = \mathrm{Tr}(g^*f)$.

6.2 L^p-duality

We now come to the final ingredient in the development of the rudimentary theory of $L^p(\mathcal{M}, \tau)$ spaces, namely duality theory. A more refined understanding of the action of the trace on $L^1(\mathcal{M}, \tau)$ is crucial to this endeavour.

Proposition 6.22 *For any $x \in \mathcal{M}$ and $y \in L^1(\mathcal{M}, \tau)$, we have $\tau(xy) = \tau(yx)$. Similarly, for any $a, b \in L^2(\mathcal{M}, \tau)$, we have that $\tau(ab) = \tau(ba)$.*

Proof.   First, let $a, b \in L^2(\mathcal{M}, \tau)$ be given. By the polarisation identity

$$4ab = \sum_{k=0}^{3} i^k (a^* + i^k b)^* (a^* + i^k b) \qquad 4ba = \sum_{k=0}^{3} i^k (a^* + i^k b)(a^* + i^k b)^*$$

with each term in each of the sums an element of $L^1(\mathcal{M}, \tau)$. Hence, by the linearity of τ on $L^1(\mathcal{M}, \tau)$ and the known action of τ on $\widetilde{\mathcal{M}}_+$, we have that

$$\tau(ab) = \frac{1}{4} \sum_{k=0}^{3} i^k \tau((a^* + i^k b)^* (a^* + i^k b))$$

$$= \frac{1}{4} \sum_{k=0}^{3} i^k \tau((a^* + i^k b)(a^* + i^k b)^*) = \tau(ba).$$

Now, let $x \in \mathcal{M}$ and $y \in L^1(\mathcal{M}, \tau)$ be given. With $y = u|y|$ being the polar form of y, it is now an exercise to see that each of $u|y|^{1/2}$ and $|y|^{1/2}$ belong to $L^2(\mathcal{M}, \tau)$. Observe that for any $a \in L^2(\mathcal{M}, \tau)$, Hölder's inequality ensures that $ax, xa \in L^2(\mathcal{M}, \tau)$. If we combine this fact with what we have already proven regarding $L^2(\mathcal{M}, \tau)$, it follows that

$$\tau(xy) = \tau((xu|y|^{1/2})|y|^{1/2}) = \tau(|y|^{1/2}(xu|y|^{1/2}))$$

$$= \tau(((|y|^{1/2}x)u|y|^{1/2}) = \tau(u|y|^{1/2}(|y|^{1/2}x)) = \tau(yx)$$

as required. $\qquad\square$

The following lemma is now an easy consequence of the above result considered alongside the Hölder inequality.

Lemma 6.23   *For any $x \in L^1(\mathcal{M}, \tau)$, we have that*

$$\|x\|_1 = \sup\{|\tau(yx)| : y \in \mathcal{M}, \|y\| \le 1\} = \sup\{|\tau(xy)| : y \in \mathcal{M}, \|y\| \le 1\}.$$

Proof.   We prove the first equality. Hölder's inequality combined with the fact that $|\tau(yx)| \le \tau(|yx|)$ for each $x \in L^1(\mathcal{M}, \tau)$, $y \in \mathcal{M}$, ensures that $\sup\{|\tau(yx)| : y \in \mathcal{M}, \|y\| \le 1\} \le \|x\|_1$. Since for $y = u^*$, where $x = u|x|$ is the polar decomposition of x, we have that $\tau(yx) = \tau(|x|) = \|x\|_1$, it is clear that equality must hold. $\qquad\square$

Theorem 6.24 *The bilinear form*

$$L^1(\mathcal{M}, \tau) \times \mathcal{M} \to \mathbb{C} : (x, y) \mapsto \tau(yx)$$

defines a dual action of $\mathcal{M}$ on $L^1(\mathcal{M}, \tau)$ with respect to which $\mathcal{M}$ is identified with $(L^1(\mathcal{M}, \tau))^*$. Specifically, for each $x \in L^1(\mathcal{M}, \tau)$, the prescription $y \mapsto \tau(yx)$ defines a σ-weakly continuous linear functional ω_x on $\mathcal{M}$. Moreover, the mapping $x \mapsto \omega_x$ is a surjective isometry from $L^1(\mathcal{M}, \tau)$ onto $\mathcal{M}_*$.

Proof. The lemma ensures that the mapping $\iota : L^1(\mathcal{M}, \tau) \to \mathcal{M}_* :$ $x \mapsto \omega_x$ is a linear isometry. We therefore merely need to verify the surjectivity of this map, and that each ω_x is σ-weakly continuous. We saw in Remark 6.10 that $L^1(\mathcal{M}, \tau)$ is spanned by its positive elements. So, to prove the claim regarding the σ-weak continuity of the ω_xs, it is enough to do this for the case where $x \geq 0$. Hence, let this be the case, and let (y_i) be a net in $\mathcal{M}_+$ increasing to some $y \in \mathcal{M}_+$. Then $(x^{1/2} y_i x^{1/2})$ will of course increase to $x^{1/2} y x^{1/2}$. So, by the normality of the trace on $\widetilde{\mathcal{M}}$, we get that

$$\sup_i \omega_x(y_i) = \sup_i \tau(y_i x) = \sup_i \tau(x^{1/2} y_i x^{1/2})$$

$$= \tau(x^{1/2} y x^{1/2}) = \tau(yx) = \omega_x(y).$$

So, ω_x is, as required, a positive normal functional.

Since $L^1(\mathcal{M}, \tau)$ is complete and $\iota : L^1(\mathcal{M}, \tau) \to \mathcal{M}_*$ an isometric embedding, $\iota(L^1(\mathcal{M}, \tau)) = \{\omega_x : x \in L^1(\mathcal{M}, \tau)\}$ is a closed subspace of $\mathcal{M}_*$. If therefore we can show that $\iota(L^1(\mathcal{M}, \tau))$ is dense in $\mathcal{M}_*$, we will have that $\iota(L^1(\mathcal{M}, \tau)) = \mathcal{M}_*$, which would conclude the proof. Since $\iota(L^1(\mathcal{M}, \tau))$ is a linear subspace, the norm closure will agree with the $\sigma(\mathcal{M}_*, \mathcal{M})$-closure. By the bipolar theorem, $[(\iota(L^1(\mathcal{M}, \tau)))^\circ]_\circ$ in turn corresponds to the $\sigma(\mathcal{M}_*, \mathcal{M})$-closure. (For any $A \subseteq \mathcal{M}_*$ and $B \subseteq \mathcal{M}$, A° denotes the polar of A in $\mathcal{M}$, and $B_\circ$ the polar of B in $\mathcal{M}_*$.) We show that $(\iota(L^1(\mathcal{M}, \tau)))^\circ = \{0\}$ from which the theorem will then follow. For the sake of contradiction, suppose that $(\iota(L^1(\mathcal{M}, \tau)))^\circ$ contains a non-zero element a. Let $a = u|a|$ be the polar decomposition of a. By the semifiniteness of the trace, there exists a non-zero subprojection e of $\chi_{(0,\infty)}(|a|)$ with $\tau(e) < \infty$. Then, $e|a|e$ must of course be non-zero and hence $\tau(e|a|e) \neq 0$. Next observe that for $x = eu^*$, we have that $\tau(|x^*|) = \tau(e) < \infty$. Hence, x^*, and therefore x, belongs to

$L^1(\mathcal{M}, \tau)$. But by the assumption on a, we must in that case have that $\omega_x(a) = 0$. Since $\omega_x(a) = \tau(xa) = \tau(e|a|) = \tau(e|a|e)$, this is a clear contradiction. The space $(\iota(L^1(\mathcal{M}, \tau)))^\circ$ can therefore contain no non-zero elements. $\qquad\square$

We present two important corollaries to the above theorem. The first is an existence result for tracial algebras of what may be termed non-commutative conditional expectations. Once the requisite technology has been developed, we will revisit this issue and extend this result to general von Neumann algebras. (See Theorem 10.54.)

Corollary 6.25. [Tak02, V.2.36] *Let $\mathcal{M}$ be a von Neumann algebra equipped with a faithful normal semifinite trace τ, and $\mathcal{N}$ a von Neumann subalgebra. Then, $\tau{\restriction}\mathcal{N}$ is a semifinite trace on $\mathcal{N}$ if and only if there exists a contractive positive normal operator E from $\mathcal{M}$ onto $\mathcal{N}$ for which we have that $\mathrm{E}(abc) = a\mathrm{E}(b)c$ for all $a, c \in \mathcal{N}$ and $b \in \mathcal{M}$ and also that $\tau \circ \mathrm{E} = \tau$. We shall refer to this operator as the* conditional expectation from $\mathcal{M}$ onto $\mathcal{N}$ with respect to τ.

Proof. We leave the 'if' part as an exercise. It is not difficult to conclude from the first part of Remark 5.8 that $\widetilde{\mathcal{N}} \subset \widetilde{\mathcal{M}}$, and then from Proposition 5.13 that $L^1(\mathcal{N}, \tau{\restriction}\mathcal{N}) \subset L^1(\mathcal{M}, \tau)$. We define E as the dual of the canonical injection $\iota\colon L^1(\mathcal{N}, \tau{\restriction}\mathcal{N}) \hookrightarrow L^1(\mathcal{M}, \tau)$. The semifiniteness of $\tau{\restriction}\mathcal{N}$ on $\mathcal{N}$ ensures that we can select a net (e_i) in $\mathfrak{p}_\tau \subset L^1(\mathcal{N}, \tau{\restriction}\mathcal{N})$ increasing to $\mathbb{1}$ (see, for example, Proposition 3.31). Since ι preserves order, so does $\iota^* = \mathrm{E}$. For any $a \in \mathcal{M}_+$ we will, moreover, have that

$$\tau(a) = \sup_i \tau(a\iota(e_i)) = \sup_i \tau(\mathrm{E}(a)e_i) = \tau(\mathrm{E}(a)). \qquad\square$$

Corollary 6.26 *The space $L^1(\mathcal{M}, \tau) \cap \mathcal{M}$ is σ-weakly dense in $\mathcal{M}$.*

Proof. It is enough to show that any $a \in \mathcal{M}_+$ is in the σ-weak closure of $L^1(\mathcal{M}, \tau) \cap \mathcal{M}$. Given $a \in \mathcal{M}_+$, the semifiniteness of the trace ensures that we may select a net of projections (e_α) each with finite trace, increasing to $\chi_{(0,\infty)}(a)$. The existence of such a net is fairly well known, but for the sake of the reader, we pause to indicate how its existence may be verified. First, recall that the semifiniteness of the canonical trace τ on $\mathcal{M}$ ensures that each projection e in $\mathcal{M}$ admits a subprojection with finite trace. (This fact was verified in part(c) of the proof of Proposition 5.9.) One may then use Zorn's lemma to select a maximal family $\{f_\alpha\}$ of mutually orthogonal subprojections of $\chi_{(0,\infty)}(a)$, each with finite trace. The property just noted ensures that $\chi_{(0,\infty)}(a) = \sum_\alpha f_\alpha$. The net we seek consists of finite sums of elements of $\{f_\alpha\}$.

But then (e_α) converges to $\chi_{(0,\infty)}(a)$ in the σ-strong* topology and hence also in the σ-weak topology. For any $x \in L^1(\mathcal{M},\tau)$, the duality established in the preceding theorem therefore ensures that

$$\lim_\alpha \tau((ae_\alpha)x) = \lim_\alpha \tau(e_\alpha(xa)) = \tau(\chi_{(0,\infty)}(a)xa)$$

$$= \tau(a\chi_{(0,\infty)}(a)x) = \tau(ax).$$

Since for any α we have that $|ae_\alpha| \leq \|a\|_\infty e_\alpha$, it is clear that $\tau(ae_\alpha) = \tau(e_\alpha ae_\alpha) \leq \|a\|_\infty \tau(e_\alpha) < \infty$ for each α and hence $(ae_\alpha) \subseteq (L^1(\mathcal{M},\tau) \cap \mathcal{M})$. This proves the claim. $\qquad\square$

The following theorem significantly sharpens Proposition 6.22 and greatly improves our understanding of the action of the trace on $L^1(\mathcal{M},\tau)$. The proof presented here of the first statement is due to Brown and Kosaki [**BK90**, Theorem 17], with the proof of the second due to Dodds, Dodds and de Pagter [**DDdP93**, Proposition 3.4].

Theorem 6.27 *Let $x,y \in \widetilde{\mathcal{M}}$ be given with $xy, yx \in L^1(\mathcal{M},\tau)$. Then $\tau(xy) = \tau(yx)$. If in addition $x \geq 0$ and $y \geq 0$ with x satisfying the requirement $\lim_{t\to\infty} \mathbf{m}_x(t) = 0$, then $x^{1/2}yx^{1/2}, y^{1/2}xy^{1/2} \in L^1(\mathcal{M},\tau)$ with*

$$\tau(xy) = \tau(x^{1/2}yx^{1/2}) = \tau(y^{1/2}xy^{1/2}).$$

Proof. Let $x,y \in \widetilde{\mathcal{M}}$ be given with $xy, yx \in L^1(\mathcal{M},\tau)$. For $p := \chi_{(0,\infty)}(|x|)$ and $q := \chi_{(0,\infty)}(|x^*|)$, we define $p_n = \chi_{(1/n,n]}(|x|)$ and $q_n = \chi_{(1/n,n]}(|x^*|)$ for each $n \in \mathbb{N}$. By construction, $px = xq$ and $p_n x = x q_n$, with (p_n) and (q_n) increasing, respectively, to p and q in the σ-strong* topology. Since $p_n \leq n|x|$ for any n, we have $|p_n y| = (y^* p_n y)^{1/2} \leq n(y^* x^* x y)^{1/2} = |xy|$. In addition, since by assumption $xy \in L^1(\mathcal{M},\tau)$, we have $(p_n y) \subseteq L^1(\mathcal{M},\tau)$. The fact that $L^1(\mathcal{M},\tau)$ is an $\mathcal{M}$-bimodule by Theorem 6.2 additionally ensures that $(p_n yx) \subseteq L^1(\mathcal{M},\tau)$. Clearly, $q_n x, p_n \in \mathcal{M}$ for each n. Repeated use of Proposition 6.22 now shows that

$$\tau(q_n xy) = \tau(q_n x p_n y) = \tau(p_n y q_n x) = \tau(p_n y x p_n) = \tau(p_n y x).$$

The fact that both xy and yx are in $L^1(\mathcal{M}, \tau)$ now enables us to conclude from Theorem 6.24 that $\tau(xy) = \lim_{n\to\infty} \tau(q_n xy) = \lim_{n\to\infty} \tau(p_n y x) = \tau(yx)$.

Now move to the case where $x, y \geq 0$ with in addition $\lim_{t\to\infty} \mathbf{m}_x(t) = 0$. We remind the reader that in this case $d_x(s) < \infty$ for any $s > 0$ (Proposition 5.24). We clearly have that $p = q$ and $p_n = q_n$. Since $\frac{1}{n}p_n \leq x$, we then have that $|p_n x^{1/2} y|^2 = y x^{1/2} p_n x^{1/2} y \leq n|xy|^2$. But then $\tau(|p_n x^{1/2} y|) \leq \sqrt{n}\tau(|xy|) < \infty$, whence $(p_n x^{1/2} y) \in L^1(\mathcal{M}, \tau)$. Since each $p_n x^{1/2}$ is clearly bounded, we may now use Proposition 6.22 to see that

$$\tau(p_n xy) = \tau(p_n x^{1/2} p_n x^{1/2} y) = \tau(p_n x^{1/2} y p_n x^{1/2}) = \tau(p_n x^{1/2} y x^{1/2} p_n).$$

If we are able to show that $x^{1/2} y x^{1/2} \in L^1(\mathcal{M}, \tau)$, we would be able to use Theorem 6.24, to conclude from the above that $\tau(xy) = \tau(pxy) = \lim_{n\to\infty} \tau(p_n xy) = \lim_{n\to\infty} \tau(p_n x^{1/2} y x^{1/2} p_n) = \lim_{n\to\infty} \tau (p_n x^{1/2} y x^{1/2}) = \tau(p x^{1/2} y x^{1/2}) = \tau(x^{1/2} y x^{1/2})$. Since in this case

$$\tau(|y^{1/2} x^{1/2}|^2) = \tau(x^{1/2} y x^{1/2}) < \infty,$$

we would then clearly have that $y^{1/2} x^{1/2} \in L^2(\mathcal{M}, \tau)$. We could then use Proposition 6.22 to see that $\tau(x^{1/2} y x^{1/2}) = \tau(y^{1/2} x y^{1/2})$. It therefore remains to prove that $x^{1/2} y x^{1/2} \in L^1(\mathcal{M}, \tau)$. To see that this holds, notice that an easy modification of the argument used to prove the implication (i)$\Rightarrow$(iii) in Proposition 5.24 shows that $(p_n x^{1/2})$ converges to $x^{1/2}$ in measure. Therefore, $(p_n x^{1/2} y x^{1/2} p_n)$ converges to $x^{1/2} y x^{1/2}$ in measure. From what we have already proven, it is clear that $\tau(p_n x^{1/2} y x^{1/2} p_n) = \tau(p_n xy) \leq \tau(|p_n xy|) \leq \tau(|xy|)$. (Here, we used Proposition 6.11 and the fact that $xy \in L^1(\mathcal{M}, \tau)$.) This fact combined with Theorem 5.29 would then ensure that

$$\tau(x^{1/2} y x^{1/2}) \leq \liminf_{n\to\infty} \tau(p_n x^{1/2} y x^{1/2} p_n) \leq \tau(|xy|) < \infty.$$

Thus, as required, $x^{1/2} y x^{1/2} \in L^1(\mathcal{M}, \tau)$. $\qquad\square$

We proceed to developing the duality theory for the case $1 < p < \infty$.

Lemma 6.28 *Suppose $1 < p < \infty$. For any $x \in L^p(\mathcal{M}, \tau)$, we have that*

$$\|x\|_p = \sup\{|\tau(yx)| : y \in L^q(\mathcal{M}, \tau), \|y\|_q \le 1\}$$
$$= \sup\{|\tau(xy)| : y \in L^q(\mathcal{M}, \tau), \|y\|_q \le 1\}$$

where $1 = \frac{1}{p} + \frac{1}{q}$.

Proof. We prove the first equality. Hölder's inequality combined with the fact that $|\tau(yx)| \le \tau(|yx|)$ for each $x \in L^p(\mathcal{M}, \tau)$, $y \in L^q(\mathcal{M}, \tau)$ ensures that $\sup\{|\tau(yx)| : y \in L^q(\mathcal{M}, \tau), \|y\|_q \le 1\} \le \|x\|_p$.

For the converse, let $0 \ne x \in L^p(\mathcal{M}, \tau)$ be given. With $x = u|x|$ the polar decomposition of x, set $y = \|x\|_p^{-p/q}|x|^{p-1}u^*$. By Propositions 5.12 and 5.13, we have $\tau(|y|^q) = \tau(|y^*|^q) = \|x\|^{-p}\tau(|x|^{qp-q}) = \|x\|_p^{-p}\tau(|x|^p) = 1$ and hence $y \in L^q(\mathcal{M}, \tau)$ with $\|y\| = 1$. By construction

$$\tau(yx) = \|x\|_p^{-p/q}\tau(|x|^p) = \|x\|_p^{p-pq} = \|x\|_p.$$

Hence, equality must hold. $\qquad\qquad\qquad\qquad\qquad\qquad\qquad\square$

Lemma 6.29 *Let $1 \le p < \infty$, and let $a, b \in L^p_+(\mathcal{M}, \tau)$ be given. Then*

$$2^{1-p}\|a + b\|_p^p \le \|a\|_p^p + \|b\|_p^p \le \|a + b\|_p^p.$$

Proof. We know that $\|a+b\|_p \le \|a\|_p + \|b\|_p$. Since $1 \le p < \infty$, $t \mapsto t^p$ is a convex function on $[0, \infty)$, which then ensures that

$$\|a + b\|_p^p = (\|a\|_p + \|b\|_p)^p = 2^p\left(\frac{\|a\|_p}{2} + \frac{\|b\|_p}{2}\right)^p \le 2^p\left(\frac{1}{2}\left[\|a\|_p^p + \|b\|_p^p\right]\right).$$

This proves the first inequality. The second is an immediate consequence of part (ii) of Proposition 6.8. $\qquad\qquad\qquad\qquad\square$

Using the above lemma, we are now able to prove part of the famous Clarkson–McCarthy inequalities.

Proposition 6.30   *Let $2 \leq p < \infty$, and let $a, b \in L^p(\mathcal{M}, \tau)$ be given. Then*

$$\|a + b\|_p^q p + \|a - b\|_p^p \leq 2^{p-1}(\|a\|_p^p + \|b\|_p^p).$$

Proof.   On setting $r = p/2$, two applications of the lemma show that

$$
\begin{aligned}
\|a + b\|_p^p + \|a - b\|_p^p &= \| |a + b|^2 \|_r^r + \| |a - b|^2 \|_r^r \\
&\leq \| |a + b|^2 + |a - b|^2 \|_r^r \\
&= 2^r \| |a|^2 + |b|^2 \|_r^r \\
&\leq 2^r 2^{r-1} \left(\| |a|^2 \|_r^r + \| |b|^2 \|_r^r \right) \\
&\leq 2^{p-1} \left(\|a\|_p^p + \|b\|_p^p \right).
\end{aligned}
$$
$\qquad\square$

Remark 6.31   The 'other part' of the Clarkson–McCarthy inequalities is of course the claim that, for $1 < p \leq 2$, we will, for $q > 0$ such that $\frac{1}{p} + \frac{1}{q} = 1$ and any $a, b \in L^p(\mathcal{M}, \tau)$, have that

$$\|a + b\|_p^q + \|a - b\|_p^q \leq 2(\|a\|_p^p + \|b\|_p^p)^{q/p}.$$

These inequalities can also be shown to hold for even the most general von Neumann algebras. However, we shall have no need of this fact and hence will not prove it. Most proofs of these inequalities rely on the complex interpolation method. For a complex analytic proof using only the Hadamard three-line theorem, the reader should refer to [**FK86**, Theorem 5.3].

We now introduce the concept of uniformly convex Banach spaces.

Definition 6.32   A Banach space X is said to be uniformly convex if for every $0 < \epsilon \leq 2$, we can find a $\delta > 0$, so that for any two norm 1 vectors $x, y \in X$, the situation $\|x - y\| \geq \epsilon$ ensures that $\frac{\|x+y\|}{2} \leq 1 - \delta$.

It is an easy exercise to see that the Clarkson–McCarthy inequalities verified above ensure that for each $2 \leq p < \infty$, the space $L^p(\mathcal{M}, \tau)$ is uniformly convex. A crucial step in the development of a duality theory for L^p-spaces is showing that $L^p(\mathcal{M}, \tau)$ is reflexive (first for $2 \leq p < \infty$, then later for $1 \leq p < 2$ as well). For this step, the Milman–Pettis theorem, which asserts that all uniformly convex Banach spaces are reflexive, comes to the rescue. For the sake of a deeper understanding of the underlying principles, we will give a more self-contained proof of L^p-duality by embedding much of the proof of the Milman–Pettis theorem in that proof.

Theorem 6.33 *Let $1 < p \leq \infty$ and $1 \leq q < \infty$ be given with $1 = \frac{1}{p} + \frac{1}{q}$. The bilinear form*

$$L^p(\mathcal{M}, \tau) \times L^q(\mathcal{M}, \tau) \to \mathbb{C} : (x, y) \mapsto \tau(yx)$$

defines a dual action of $L^p(\mathcal{M}, \tau)$ on $L^q(\mathcal{M}, \tau)$ with respect to which $L^p(\mathcal{M}, \tau)$ is identified with $(L^q(\mathcal{M}, \tau))^*$. Specifically, for $x \in L^p(\mathcal{M}, \tau)$, the prescription $y \mapsto \tau(yx)$ defines a bounded linear functional ω_x on $L^q(\mathcal{M}, \tau)$. Moreover, the mapping $x \mapsto \omega_x$ is a surjective isometry from $L^p(\mathcal{M}, \tau)$ onto $(L^q(\mathcal{M}, \tau))^*$. In addition, $\omega_x \geq 0$ if and only if $x \geq 0$.

Proof. Apart from the final claim, the case where $q = 1$ corresponds to Theorem 6.24. In considering the first two claims, we may therefore assume that $q > 1$. It is clear from Lemma 6.28 that the mapping $\iota : L^p(\mathcal{M}, \tau) \to (L^q(\mathcal{M}, \tau))^* : x \mapsto \omega_x$ is a linear isometry from $L^p(\mathcal{M}, \tau)$ onto a subspace of $(L^q(\mathcal{M}, \tau))^*$. Since $L^p(\mathcal{M}, \tau)$ is complete, $\iota(L^p(\mathcal{M}, \tau))$ must be a closed subspace.

We will next show that $\iota(L^p(\mathcal{M}, \tau))$ is a weak*-dense subspace of $(L^q(\mathcal{M}, \tau))^*$. By the bipolar theorem, the $\sigma((L^q)^*, L^q)$-closure $\iota(L^p(\mathcal{M}, \tau))$, is given by $[(\iota(L^1(\mathcal{M}, \tau)))_\circ]^\circ$. (For an $A \subseteq L^q(\mathcal{M}, \tau)$ and $B \subseteq (L^q(\mathcal{M}, \tau))^*$, A° denotes the polar of A in $(L^q(\mathcal{M}, \tau))^*$, and $B_\circ$ the polar of B in $L^q(\mathcal{M}, \tau)$.) We show that $(\iota(L^p(\mathcal{M}, \tau)))_\circ = \{0\}$, from which the claim will then follow. For the sake of contradiction, suppose that $(\iota(L^p(\mathcal{M}, \tau)))^\circ$ contains a non-zero element a. Let $a = u|a|$ be the polar decomposition of a. From the proof of Lemma 6.28, we know that then $x = \|a\|_q^{-q/p} |a|^{q-1} u^*$ is a norm 1 element of $L^p(\mathcal{M}, \tau)$. Since $a \in (\iota(L^p(\mathcal{M}, \tau)))_\circ$, we must have $\omega_x(a) = 0$. But arguing as in the proof of Lemma 6.28, it follows that $\omega_x(a) = \tau(xa) = \|a\|_q \neq 0$. This is a clear contradiction. Therefore, $(\iota(L^p(\mathcal{M}, \tau)))^\circ = \{0\}$ as claimed.

Case 1 ($q \geq 2$): First consider the case where $q \geq 2$. We start by proving that $L^q(\mathcal{M}, \tau)$ is then reflexive. To see this, identify $L^q(\mathcal{M}, \tau)$ with the image of the natural embedding of $L^q(\mathcal{M}, \tau)$ into $(L^q(\mathcal{M}, \tau))^{**}$, and select an arbitrary norm 1 element $\widetilde{x}$ of $(L^q(\mathcal{M},\tau))^{**}$. By Goldstine's theorem, there must exist a net (x_α) in the closed unit ball of $L^q(\mathcal{M},\tau)$ converging to $\widetilde{x}$ in the weak*-topology on $(L^q(\mathcal{M},\tau))^{**}$. Given $\delta > 0$, we may select a norm 1 element x^* of $(L^q(\mathcal{M},\tau))^*$ so that $1 - \delta = \|\widetilde{x}\| - \delta < |\widetilde{x}(x^*)| \leq 1$. By the weak* convergence, we will then have that $1 - \delta < \lim_\alpha |x^*(x_\alpha)| \leq \liminf_\alpha \|x_\alpha\| \leq \limsup_\alpha \|x_\alpha\|_q \leq 1$. Since $\delta > 0$ was arbitrary, we must have that $\lim_\alpha \|x_\alpha\| = 1$.

Now, suppose that (x_α) is not Cauchy in norm, and let x^* be as before. Given $\epsilon > 0$, we may then inductively select an increasing sequence (α_n) of indices such that

$$\|x_{\alpha_{n+1}} - x_{\alpha_n}\| \geq \epsilon, \text{ and } |\widetilde{x}(x^*) - x^*(x_{\alpha_n})| \leq \frac{1}{n} \text{ for all } n \in \mathbb{N}.$$

Now use the Clarkson–McCarthy inequalities to conclude that

$$\frac{\|x_{\alpha_{n+1}} + x_{\alpha_n}\|^q}{2^q} + \frac{\|x_{\alpha_{n+1}} - x_{\alpha_n}\|^q}{2^q} \leq \frac{\|x_{\alpha_{n+1}}\|^q + \|x_{\alpha_n}\|^q}{2}$$

for all $n \in \mathbb{N}$. By the previously centred inequality, this in turn leads to

$$\frac{\|x_{\alpha_{n+1}} + x_{\alpha_n}\|^q}{2^q} + \frac{\epsilon^q}{2^q} \leq \frac{\|x_{\alpha_{n+1}} + x_{\alpha_n}\|^q}{2^q} + \frac{\|x_{\alpha_{n+1}} - x_{\alpha_n}\|^q}{2^q}$$

$$\leq \frac{\|x_{\alpha_{n+1}}\|^q + \|x_{\alpha_n}\|^q}{2} \leq 1.$$

Note that, by construction, $(x^*(x_{\alpha_{n+1}} + x_{\alpha_n}))$ will converge to $2\widetilde{x}(x^*)$ in the weak* topology. So, on arguing as before, we have $2 - 2\delta \leq |\widetilde{x}(x^*)| \leq \liminf_n \|x_{\alpha_{n+1}} + x_{\alpha_n}\| \leq 2$. But the previously centred inequality then leads to $(1-\delta)^q + \frac{\epsilon^q}{2^q} \leq 1$, which is impossible for appropriate choices of ϵ and δ. Thus, the assumption that (x_α) is not Cauchy in norm must be false. Being Cauchy, (x_α) must now, by completeness, converge to some element x of $L^q(\mathcal{M},\tau)$. The net (x_α) will then also converge to x in the weak* topology on $(L^q(\mathcal{M},\tau))^{**}$. By uniqueness of limits, we must have that $\widetilde{x} = x \in L^q(\mathcal{M},\tau)$. This shows that $L^q(\mathcal{M},\tau)$ is reflexive.

Since $L^q(\mathcal{M},\tau)$ is reflexive, so is $(L^q(\mathcal{M},\tau))^*$. But in that case the weak* closure of any subspace of $(L^q(\mathcal{M},\tau))^*$ will agree with its norm closure. Since $\iota(L^p(\mathcal{M},\tau))$ is then both norm-dense and closed, we have $\iota(L^p(\mathcal{M},\tau)) = (L^q(\mathcal{M},\tau))^*$ as required.

Case 2 ($1 < q < 2$): If $1 < q < 2$, then $p > 2$. The first part of the proof then shows that $L^q(\mathcal{M},\tau) = (L^p(\mathcal{M},\tau))^*$ via the tracial bilinear form. But then $L^q(\mathcal{M},\tau)$, is also reflexive, which means that the same argument as before shows that $\iota(L^p(\mathcal{M},\tau)) = (L^q(\mathcal{M},\tau))^*$.

For the final claim, the 'if' part follows from the observation that $\tau(xy) = \tau(y^{1/2}xy^{1/2}) \geq 0$ for any positive element y of $L^q(\mathcal{M},\tau)$. Conversely, assume that ω_x is positive. For any positive element y of $L^q(\mathcal{M},\tau)$, the fact that $\omega_x(y) \geq 0$ then enables us to conclude that $\omega_x(y) = \tau(xy) = \overline{\tau(xy)} = \tau(yx^*) = \omega_{x^*}(y)$. In view of the fact that any element of $L^q(\mathcal{M},\tau)$ may be written as a linear combination of four positive elements of $L^q(\mathcal{M},\tau)$, we therefore have that $\omega_x = \omega_{x^*}$ and hence that $x = x^*$ since $x \mapsto \omega_x$ is a bijection. Now consider the operator $|x|^{p-1}$. (In the case $p = 1$, we simply take $\mathbb{1}$ here.) Given that $\tau((|x|^{p-1})^q) = \tau(|x|^p) < \infty$, it is clear that $|x|^{p-1}$ belongs to $L^q(\mathcal{M},\tau)$. Hence, $|x|^{p-1}|x|^{p-1}\chi_{(-\infty,0)}(x)$ is a positive element of $L^q(\mathcal{M},\tau)$. But for this positive element, we have by the Borel functional calculus that $|x|^{p-1}x\chi_{(-\infty,0)}(x) = -|x|^p\chi_{(-\infty,0)}(x)$. So, if $\chi_{(-\infty,0)}(x)$, and hence $x\chi_{(-\infty,0)}(x)$ were non-zero, $|-x\chi_{(-\infty,0)}(x)|^p = |x|^p\chi_{(-\infty,0)}(x)$ would be non-zero, in which case we would have that $\omega_x(|x|^{p-1}\chi_{(-\infty,0)}(x)) = \tau(x|x|^{p-1}\chi_{(-\infty,0)}(x)) = -\tau(|x|^p\chi_{(-\infty,0)}(x)) < 0$, contradicting the positivity of ω_x. Hence, we must have $x \geq 0$. $\qquad\square$

6.3 Introduction to Orlicz spaces

The starting point of the theory of Orlicz spaces is the concept of a Young function (often also called an Orlicz function in the literature).

Definition 6.34 We say that a function $\Phi : [0,\infty) \to [0,\infty]$ is a Young function if:

- Φ is convex and increasing with $\Phi(0) = 0$;
- Φ is continuous on $[0, b_\Phi]$ where $b_\Phi = \sup\{t \in [0,\infty) : \Phi(t) < \infty\}$;
- Φ is neither identically zero nor infinite-valued on all of $(0,\infty)$. With a_Φ denoting the constant $\inf\{t \in [0,\infty) : \Phi(t) > 0\}$, neither of the situations $a_\Phi = \infty$, nor $b_\Phi = 0$ may therefore occur.

For any such Young function, we define the conjugate Young function to be Φ^* where for each $s > 0$, we set $\Phi^*(s) = \sup_{t>0}(st - \Phi(t))$.

Exercise 6.35 *Show that Φ^* is a Young function and that $\Phi^{**} = \Phi$.*

Remark 6.36

(a) If $b_\Phi < \infty$, then $\Phi^*(s) \leq b_\Phi s$ for any $s \geq 0$. To see this let $s > 0$ be given, and observe that if $t > b_\Phi$, then $st - \Phi(t) = -\infty$. Hence, $\Phi^*(s) = \sup_{b_\Phi \geq t > 0}(st - \Phi(t)) \leq b_\Phi s$ as claimed.

(b) There is an interesting alternative for computing Φ^*. Recall that in the proof of Theorem 5.21, we saw that a Young function Φ is of the form $\Phi(t) = \int_0^t \phi(s)\,ds$ for some non-negative left-continuous non-decreasing function ϕ on $[0,\infty)$, which is infinite-valued on (b_Φ, ∞), but neither identically 0 nor infinite-valued on all of $(0, \infty)$. Now, consider the function $\psi(t) = \inf\{s : \phi(s) \geq t\}$. The function ψ is then also a non-negative, non-decreasing, left-continuous function on $[0,\infty)$, which is finite on some portion $[0, r]$ of $[0, \infty)$, and neither identically 0 nor infinite-valued on all of $(0, \infty)$. Then, Ψ defined by $\Psi(t) = \int_0^\infty \psi(s)\,ds$ turns out to be nothing but the Young function Φ^*. The fact that Ψ as defined above agrees with Φ^* follows from [**BS88**, Theorem IV.8.12] (see also [**NP06**, §1.7].)

(c) Note that for any Young function Φ, the pair (Φ, Φ^*) will by the definition of Φ^*, satisfy the *Hausdorff–Young inequality*

$$st \leq \Phi(t) + \Phi^*(s), \quad \text{for all } s, t \geq 0.$$

The description of Φ and Φ^*, given above, forms the basis for the equality criteria for this inequality, namely that $st = \Phi(t) + \Phi^*(s)$ if and only if either $s = \phi(t)$, or $t = \psi(s)$. For a proof of this fact, the reader is referred to [**BS88**, Theorem IV.8.12].

(d) Given a Young function, we shall have occasion to use the 'right-continuous inverse' Φ^{-1} of Φ on $[0, \infty)$, given by

$$\Phi^{-1}(t) = \sup\{s : \Phi(s) \leq t\}.$$

It is only in the case where $a_\Phi = 0$ and $b_\Phi = \infty$ that this is an inverse in the true sense of the word. To see this, note that it is an exercise to see that $\Phi^{-1}(0) = a_\Phi$. Now, observe that the fact that Φ is strictly increasing on (a_Φ, b_Φ) ensures that for every $s \in (a_\Phi, b_\Phi)$, $\Phi(s) = t$ if and only if $s = \Phi^{-1}(t)$. If, in fact, $\Phi(b_\Phi) < \infty$, it is similarly an exercise to see that $\Phi^{-1}(t) = b_\Phi$ for every $t \geq \Phi(b_\Phi)$. So Φ^{-1} is a continuous function for which we have $\Phi \circ \Phi^{-1}(t) \leq t \leq \Phi^{-1} \circ \Phi(t)$ for all $t \geq 0$.

Exercise 6.37 *Prove that the conjugate Young function of $\Phi_1(t) = t$ is*

$$\Phi_\infty(t) = \begin{cases} 0, & \text{if } 0 \leq t \leq 1 \\ \infty, & \text{if } 1 < t \end{cases}$$

and that the conjugate function of $\cosh - 1$ *is* $\int_0^t \sinh^{-1}(s)\,ds = t \log(t + \sqrt{t^2 + 1}) - \sqrt{t^2 + 1} + 1$.

Definition 6.38 For a given Young function Φ, we define the noncommutative Orlicz space $L^\Phi(\mathcal{M}, \tau)$ to be the collection of all $f \in \widetilde{\mathcal{M}}$ for which there exists some $\alpha > 0$ such that $\tau(\Phi(\alpha|f|)) < \infty$. Note that even when Φ is infinite-valued on some part of the half-line, we can give meaning to $\Phi(\alpha|f|)$ as an element of the extended positive part of $\mathcal{M}$. Since the action of τ extends to the extended positive part, the requirement that $\tau(\Phi(\alpha|f|)) < \infty$ always makes sense.

Remark 6.39 We refine Remark 5.15 in the context of Young functions. Let $a \in {}^\eta\mathcal{M}$ be given, and let Φ be a general Young function. Then, $\Phi(|a|)$ will, in general, not be a member of $\widetilde{\mathcal{M}}$ (unless, of course, $a \in \widetilde{\mathcal{M}}$ and $\mathrm{sp}(|a|) \subseteq [0, b_\Phi)$). However, we are able to give meaning to $\Phi(|a|)$ as an element of $\widehat{\mathcal{M}}_+$ (and hence also give meaning to $\tau(\Phi(|a|))$). We pause to give some details of how this works: Suppose that $\mathcal{M}$ acts on the Hilbert space H. If $b_\Phi = \infty$, we have that $\mathrm{sp}(|a|) \subseteq [0, b_\Phi) = [0, \infty)$, in which case we can then use the continuous functional calculus to see that $\Phi(|a|) \in \widetilde{\mathcal{M}}$. Now, suppose that $b_\Phi < \infty$. There are two cases to consider here, namely $\Phi(b_\Phi) = \infty$, and $\Phi(b_\Phi) < \infty$. Suppose $\Phi(b_\Phi) = \infty$. If we attempt to use the spectral resolution $|a| = \int_0^\infty \lambda\, de_\lambda(|a|)$ to define $\Phi(|a|)$ by means of the prescription $\Phi(|a|) = \int_0^\infty \Phi(\lambda)\, de_\lambda(|a|)$, we find that $\Phi(|a|)$ exists as a densely-defined closed operator on $\chi_{[0,b_\Phi)}(|a|)(H)$ (which commutes with all unitaries in the commutant of $\mathcal{M}$), but that $\int_0^\infty \Phi(\lambda)\,d\langle e_\lambda(|a|)\xi,\xi\rangle = \infty$ for all $\xi \in \chi_{[b_\Phi,\infty)}(H)$. By Theorem 0.144, such objects are all part of $\widehat{\mathcal{M}}_+$. The only difference in the case $\Phi(b_\Phi) < \infty$ is that here $\Phi(|a|)$ makes sense as a densely defined operator on $\chi_{[0,b_\Phi]}(|a|)(H)$ not $\chi_{[0,b_\Phi)}(|a|)(H)$. As noted in Remark 5.15, we have $\tau(\Phi(|a|)) < \infty$ if and only if $\Phi(|a|)$ corresponds to an element of $L^1(\mathcal{M},\tau)$. However, more is true in this case. Recall that the right inverse Φ^{-1} is continuous on $[0,\infty)$ with $\Phi^{-1}(\Phi(t)) \geq t$ for all $t \geq 0$. Thus, if indeed $\Phi(|a|) \in L^1(\mathcal{M},\tau)$, then by the functional calculus for positive operators, $\Phi^{-1}(\Phi(|a|))$ will be a τ-measurable element of $\widehat{\mathcal{M}}$ such that $\Phi^{-1}(\Phi(|a|)) \geq |a|$. This ensures that $|a|$, and hence also a, is then a τ-measurable element of $\widehat{\mathcal{M}}$.

To sum up, in terms of the action of τ on $\widehat{\mathcal{M}}_+$, we have that a given $a \in {}^\eta\mathcal{M}$ will belong to $L^\Phi(\mathcal{M},\tau)$ if and only if $\tau_{\mathcal{M}}(\Phi(\alpha|a|)) < \infty$ for some $\alpha > 0$.

We now use the ideas described in the preceding remark to prove a deep fact regarding noncommutative Orlicz spaces. This fact is an extremely useful tool for lifting classical theory to the noncommutative context.

Theorem 6.40 *Let Φ be a Young function and let $f \in \widetilde{\mathcal{M}}$ be given. Then $\tau(\Phi(|f|)) = \int_0^\infty \Phi(\mathbf{m}_f(s))\,ds$. (Here, $\Phi(|f|)$ may not be in $\widetilde{\mathcal{M}}$, but is given meaning as an object in the extended positive part of $\mathcal{M}$.) Moreover, whenever $\tau(\Phi(|f|))$ is finite, we have that $\Phi(|f|) \in \widetilde{\mathcal{M}}$*

Proof. If indeed $\tau(\Phi(|f|))$ is finite, then by Remark 5.15, $\Phi(|f|) \in L^1(\mathcal{M},\tau)$. The fact that in this case $\tau(\Phi(|f|)) = \int_0^\infty \Phi(\mathbf{m}_f(s))\,ds$, then follows from Corollary 5.14. To conclude the proof, we need to show that $\tau(\Phi(|f|))$ will be finite, whenever $\int_0^\infty \Phi(\mathbf{m}_f(s))\,ds$ is finite. So, with this in mind, suppose we are given that $\int_0^\infty \Phi(\mathbf{m}_f(s))\,ds < \infty$. We consider two cases.

Case 1, $(b_\Phi = \infty)$: In this case, Φ is just a convex non-decreasing continuous function on $[0,\infty)$. It then follows from the continuous functional calculus that $\Phi(|f|)$ is again τ-measurable. But that means that we can apply Corollary 5.14 to conclude that $\tau(\Phi(|f|)) = \int_0^\infty \Phi(\mathbf{m}_f(s))\,ds < \infty$ as required.

Case 2, $(b_\Phi < \infty)$: Recall that Φ is infinite-valued on (b_Φ,∞). So, here the only way we could have $\int_0^\infty \Phi(\mathbf{m}_f(s))\,ds < \infty$, is if $\mathbf{m}_f(s) \leq b_\Phi$ for all $s > 0$. By the right-continuity of $s \mapsto \mathbf{m}_f(s)$, we will then have that $\|f\|_\infty = \mathbf{m}_f(0) \leq b_\Phi < \infty$. If in fact $\Phi(b_\Phi) < \infty$, it would then follow from the continuous functional calculus that $\Phi(|f|) \in \mathcal{M}$, in which case we could then apply Corollary 5.14 to see that $\tau(\Phi(|f|)) = \int_0^\infty \Phi(\mathbf{m}_f(s))\,ds < \infty$. Hence, assume that $\Phi(b_\Phi) = \infty$. For any $0 < \epsilon < 1$, we will then have that $\mathrm{sp}(\epsilon|f|) \subseteq [0, \epsilon\|f\|_\infty] \subseteq [0, b_\Phi)$. It will then follow from the continuous functional calculus that $\Phi(\epsilon|f|)$ even belongs to $\mathcal{M}$. Next, recall that $\Phi(|f|)$ may be realised as a member of the extended positive part of $\mathcal{M}$. Suppose that $\mathcal{M}$ acts on the Hilbert space H. Recall from Remark 6.39 that in this case, we could make sense of $\Phi(|f|)$ as a densely defined closed operator on $\chi_{[0,b_\Phi)}(|a|)(H)$, but that (formally) $\langle \Phi(|f|)^{1/2}\xi, \xi \rangle = \infty$ for all $\xi \in \chi_{[b_\Phi,\infty)}(|a|)(H)$. Now, select a sequence $(\epsilon_n) \subseteq (0,1)$ increasing to 1. Using the very specific structure of the function Φ in this case, it is then a somewhat non-trivial exercise to see that as members of the extended positive part $\widehat{\mathcal{M}}_+$, the operators $(\Phi(\epsilon_n|f|))$ increase to $\Phi(|f|)$. (Although infinite-valued on $[b_\Phi, \infty)$, Φ is here continuous on 'all' of $[0, \infty)$ in the sense that it is continuous on $[0, b_\Phi)$ with $\Phi(t)$ increasing to ∞ as t increases to b_Φ.) But the extension of the trace to $\widehat{\mathcal{M}}_+$, respects such suprema (Theorem 4.23). Therefore, $\tau(\Phi(|f|)) = \sup \tau(\Phi(\epsilon_n|f|))$. Since each $\Phi(\epsilon_n|f|)$ belongs to $\mathcal{M}$ and since $\mathbf{m}_{\epsilon_n f} \leq \mathbf{m}_f$, we may use Corollary 5.14 to see that $\tau(\Phi(\epsilon_n|f|)) = \int_0^\infty \Phi(\mathbf{m}_{\epsilon_n f}(s))\,ds \leq \int_0^\infty \Phi(\mathbf{m}_f(s))\,ds < \infty$. Hence, $\tau(\Phi(|f|)) \leq \int_0^\infty \Phi(\mathbf{m}_f(s))\,ds$, as required. $\qquad\square$

Definition 6.41 Let $L^\Phi(\mathcal{M}, \tau)$ be as before. Define the Luxemburg–Nakano norm on $L^\Phi(\mathcal{M}, \tau)$ to be

$$\|f\|_\Phi = \inf\{\epsilon > 0 \colon \tau(\Phi(\epsilon^{-1}|f|)) \leq 1\}.$$

The Orlicz norm is defined to be the quantity

$$\|f\|_\Phi^O = \inf\{\tau(|fg|) \colon g \in L^{\Phi^*}(\mathcal{M}, \tau), \tau(\Phi^*(|g|)) \leq 1\}.$$

The first task that now befalls us is to prove that $L^\Phi(\mathcal{M}, \tau)$ is a linear space and that these quantities are, in fact, norms. After that, we will compare these norms and investigate questions of completeness and duality. Our first result strengthens the link between the classical and noncommutative theory noted in Theorem 6.40 above.

Corollary 6.42 *Let Φ be a Young function and let $f \in \widetilde{\mathcal{M}}$ be given. Then $\mathbf{m}_f \in L^\Phi(0, \infty)$ if and only if $f \in L^\Phi(\mathcal{M}, \tau)$. Moreover, if indeed $f \in L^\Phi(\mathcal{M}, \tau)$, then $\|f\|_\Phi = \|\mathbf{m}_f\|_\Phi$.*

Theorem 6.43 *Let Φ be a Young function. Then, $L^\Phi(\mathcal{M},\tau)$ is a linear space, and $\|\cdot\|_\Phi$ a norm for $L^\Phi(\mathcal{M},\tau)$.*

Proof. Given $f \in L^\Phi(\mathcal{M},\tau)$, it is easy to see that for any $\alpha \in \mathbb{C}$, αf is again in $L^\Phi(\mathcal{M},\tau)$. Next, let $f, g \in L^\Phi(\mathcal{M},\tau)$ be given. By Corollary 6.42, we may select $\alpha_f, \alpha_g > 0$ so that $\int_0^\infty \Phi(\alpha_f \mathbf{m}_f(s))\, ds < \infty$ and $\int_0^\infty \Phi(\alpha_g \mathbf{m}_g(s))\, ds < \infty$. For $\alpha = \frac{1}{2}\min(\alpha_f, \alpha_g)$, it will then follow from Theorem 5.22 that

$$
\begin{aligned}
\int_0^\infty \Phi(\alpha \mathbf{m}_{f+g}(s))\, ds \;&=\; \int_0^\infty \Phi(\mathbf{m}_{\alpha(f+g)}(s))\, ds \\[4pt]
&\leq\; \int_0^\infty \Phi(\mathbf{m}_{\alpha f}(s) + \mathbf{m}_{\alpha g}(s))\, ds \\[4pt]
&\leq\; \int_0^\infty \Phi(\alpha(\mathbf{m}_f(s) + \mathbf{m}_g(s)))\, ds \\[4pt]
&\leq\; \int_0^\infty \Phi\Big(\tfrac{\alpha_f \mathbf{m}_f(s)}{2} + \tfrac{\alpha_g \mathbf{m}_g(s)}{2}\Big)\, ds.
\end{aligned}
$$

We may now use the convexity of Φ to see that we then further have that

$$
\begin{aligned}
\int_0^\infty \Phi(\alpha \mathbf{m}_{f+g}(s))\, ds \;&\leq\; \int_0^\infty \Phi\Big(\tfrac{\alpha_f \mathbf{m}_f(s)}{2} + \tfrac{\alpha_g \mathbf{m}_g(s)}{2}\Big)\, ds \\[4pt]
&\leq\; \int_0^\infty \tfrac{1}{2}[\Phi(\alpha_f \mathbf{m}_f(s)) + \Phi(\alpha_g \mathbf{m}_g(s)]\, ds \\[4pt]
&<\; \infty.
\end{aligned}
$$

But then by Corollary 6.42, $f + g \in L^\Phi(\mathcal{M},\tau)$.

We now show that $\|\cdot\|_\Phi$ is a norm. Let $f \in L^\Phi(\mathcal{M},\tau)$ be given with $\|f\|_\Phi = 0$, or equivalently $\|\mathbf{m}_f\|_\Phi = 0$. For any $\epsilon > 0$, we will then by definition have that $\int_0^\infty \Phi(\epsilon^{-1}\mathbf{m}_f(s))\, ds \leq 1$. By convexity and the fact that $\Phi(0) = 0$, we have that $\Phi(rt) = \Phi((1-r)0 + rt) \leq r\Phi(t)$ for any $0 < r \leq 1$, $t \geq 0$. Equivalently, $\Phi(\gamma t) \geq \gamma \Phi(t)$ for any $\gamma \geq 1$. Given $0 < \epsilon$, we will then for any $\gamma \geq 1$ have that

$$
\gamma \int_0^\infty \Phi(\epsilon^{-1}\mathbf{m}_f(s))\, ds \leq \int_0^\infty \Phi(\gamma \epsilon^{-1}\mathbf{m}_f(s))\, ds \leq 1.
$$

This can in turn only be true if $\int_0^\infty \Phi(\epsilon^{-1}\mathbf{m}_f(s))\, ds = 0$. That of course means that for any $\epsilon > 0$, $\Phi(\epsilon^{-1}\mathbf{m}_f(s))$ is 0 almost everywhere. Since $\mathbf{m}_f$ is non-increasing and Φ non-zero on some connected portion of

$(0, \infty)$, the only way this can be is if $\mathbf{m}_f$ is 0 on $(0, \infty)$. The right-continuity of $\mathbf{m}_f$ then ensures that $\|f\|_\infty = \mathbf{m}_f(0) = 0$, in other words, that $f = 0$. It is a simple exercise to see that for any $f \in L^\Phi(\mathcal{M}, \tau)$ and any $\gamma \in \mathbb{C}$, $\|\gamma f\|_\Phi = |\gamma| \, \|f\|_\Phi$. We proceed to prove the triangle inequality.

Let $f, g \in L^\Phi(\mathcal{M}, \tau)$ be given. For any $\epsilon > 0$, we will then, by the definition of the Luxemburg–Nakano norm, have $\int_0^\infty \Phi((\|f\|_\Phi + \epsilon)^{-1} \mathbf{m}_f(s)) \, ds \le 1$ and $\int_0^\infty \Phi((\|g\|_\Phi + \epsilon)^{-1} \mathbf{m}_g(s)) \, ds \le 1$. (Here, we silently used Corollary 6.42.) We may then use Theorem 5.22 and the convexity of Φ to see that

$$
\int_0^\infty \Phi((\|f\| + \|g\| + 2\epsilon)^{-1} \mathbf{m}_{f+g}(s)) \, ds
$$

$$
\le \int_0^\infty \Phi((\|f\| + \|g\| + 2\epsilon)^{-1}(\mathbf{m}_f(s) + \mathbf{m}_g(s))) \, ds
$$

$$
\le \int_0^\infty \Phi\left(\frac{\|f\| + \epsilon}{\|f\|_\Phi + \|g\| + 2\epsilon}(\|f\| + \epsilon)^{-1} \mathbf{m}_f(s) \right.
$$

$$
\left. + \frac{\|g\| + \epsilon}{\|f\| + \|g\| + 2\epsilon}(\|g\| + \epsilon)^{-1} \mathbf{m}_g(s) \right) ds
$$

$$
\le \frac{\|f\| + \epsilon}{\|f\|_\Phi + \|g\| + 2\epsilon} \int_0^\infty \Phi((\|f\| + \epsilon)^{-1} \mathbf{m}_f(s)) \, ds
$$

$$
+ \frac{\|g\| + \epsilon}{\|f\| + \|g\| + 2\epsilon} \int_0^\infty \Phi((\|g\| + \epsilon)^{-1} \mathbf{m}_g(s)) \, ds
$$

$$
\le \frac{\|f\| + \epsilon}{\|f\|_\Phi + \|g\| + 2\epsilon} + \frac{\|g\| + \epsilon}{\|f\| + \|g\| + 2\epsilon}
$$

$$
= 1.
$$

(Here, we dropped the subscripts of the norms for the sake of clarity.) The above clearly shows that $\|f + g\|_\Phi = \|\mathbf{m}_{f+g}\|_\Phi \le \|\mathbf{m}_f\|_\Phi + \|\mathbf{m}_g\|_\Phi + 2\epsilon = \|f\|_\Phi + \|g\|_\Phi + 2\epsilon$. Since $\epsilon > 0$ was arbitrary, we have $\|f + g\|_\Phi \le \|f\|_\Phi + \|g\|_\Phi$ as required. $\qquad\square$

Exercise 6.44 *Show that $L^p(\mathcal{M}, \tau)$ $(1 \le p \le \infty)$ are Orlicz spaces. Also show that the Orlicz space corresponding to the Young function*

$$
\Phi_\infty(t) = \begin{cases} 0, & \text{if } 0 \le t \le 1 \\ \infty, & \text{if } 1 < t \end{cases}
$$

is $L^\infty(\mathcal{M}, \tau)$.

Remark 6.45 In the theory of Orlicz spaces, two Young functions Φ and Ψ are said to be equivalent if there exists a constant $K > 0$ so that $K^{-1}\Phi \leq \Psi \leq K\Phi$. It is an interesting exercise to show that the norms $\|\cdot\|_\Phi$ and $\|\cdot\|_\Psi$ are equivalent whenever Φ and Ψ are.

We now show that $\|\cdot\|_\Phi^O$ is a seminorm and then use this fact to prove that $L^\Phi(\mathcal{M},\tau)$ injects continuously into $\widetilde{\mathcal{M}}$ and that it is in fact complete. After that, we will show that $\|\cdot\|_\Phi^O$ is in fact a norm, which is equivalent to $\|\cdot\|_\Phi$. To show that $\|\cdot\|_\Phi^O$ is a seminorm, all we need to do is to show that for each $f \in L^\Phi(\mathcal{M},\tau)$, $\|f\|_\Phi^O$ is finite. We need the following lemma to prove this fact. Apart from other considerations, this lemma shows that in the definition of the Orlicz norm, the requirement $\tau(\Phi^*(|g|)) \leq 1$ can be replaced with the requirement that $\|g\|_{\Phi^*} \leq 1$.

Lemma 6.46 *Let $f \in L^\Phi(\mathcal{M},\tau)$ be given. If $f \neq 0$, then $\tau(\Phi((\|f\|_\Phi)^{-1}|f|)) \leq 1$. If $\|f\|_\Phi \leq 1$, we will have that $\tau(\Phi(|f|)) \leq \|f\|_\Phi$, whilst if $\|f\|_\Phi > 1$, we will have that $\tau(\Phi(|f|)) \geq \|f\|_\Phi$. Therefore $\tau(\Phi(|f|)) \leq 1$ if and only if $\|f\|_\Phi \leq 1$.*

Proof. To prove the first claim, select a sequence $(\epsilon_n) \subseteq (\|f\|_\Phi, \infty)$ decreasing to $\|f\|_\Phi = \|\mathbf{m}_f\|_\Phi$. Then, by the monotone convergence theorem, $\int_0^\infty \Phi(\epsilon_n^{-1}\mathbf{m}_f(s))\,ds$ will increase to $\int_0^\infty \Phi(\|\mathbf{m}_f\|_\Phi^{-1}\mathbf{m}_f(s))\,ds$. Since by the definition of $\|\mathbf{m}_f\|_\Phi$ we have that $\int_0^\infty \Phi(\epsilon_n^{-1}\mathbf{m}_f(s))\,ds \leq 1$ for each n, it is clear that

$$\tau(\Phi(\|f\|^{-1}|f|)) = \int_0^\infty \Phi(\|\mathbf{m}_f\|_\Phi^{-1}\mathbf{m}_f(s))ds \leq 1.$$

For the second claim, suppose that $\|f\|_\Phi \leq 1$. If $\|f\|_\Phi = 0$, then $f = 0$, whence $\tau(\Phi(|f|)) = 0 \leq 1$. So, assume that $0 < \|f\|_\Phi \leq 1$. It then follows from the first part of the proof that $\tau(\Phi(\|f\|_\Phi^{-1}|f|)) \leq 1$ and hence that $\Phi(\|f\|_\Phi^{-1}|f|) \in \widetilde{\mathcal{M}}$. If we combine the convexity of Φ with the fact that $\Phi(0) = 0$, it is then clear that $\Phi(rt) \leq r\Phi(t)$ for any $t \geq 0$ and any $0 \leq r \leq 1$. Equivalently, $\Phi(\gamma t) \geq \gamma\Phi(t)$ for any $t \geq 0$ and any $\gamma \leq 1$. Since by assumption $0 < \|f\|_\Phi \leq 1$, we have $\Phi(\|f\|_\Phi^{-1}t) \geq \|f\|_\Phi^{-1}\Phi(t)$ for any $t \geq 0$, and hence $\Phi(\|f\|_\Phi^{-1}|f|) \geq \|f\|_\Phi^{-1}\Phi(|f|)$. Taking into account that $\Phi(\|f\|_\Phi^{-1}|f|) \in \widetilde{\mathcal{M}}$, we must therefore also have $\|f\|_\Phi^{-1}\Phi(|f|) \in \widetilde{\mathcal{M}}$, with $\|f\|_\Phi^{-1}\tau(\Phi(|f|)) \leq \tau(\Phi(\|f\|_\Phi^{-1}|f|)) \leq 1$, as required.

Now, suppose that $\|f\|_\Phi > 1$. The claimed inequality will clearly follow if $\tau(\Phi(|f|)) = \infty$. Therefore, assume that $\tau(\Phi(|f|)) < \infty$. Recall that this ensures that $\Phi(|f|) \in \widetilde{\mathcal{M}}$. Since $\|f\|_\Phi > 1$, we may select

$\epsilon > 0$ such that $(\|f\|_\Phi - \epsilon) > 1$. Since $(\|f\|_\Phi - \epsilon) < \|f\|_\Phi$, we must by the definition of the Luxemburg–Nakano norm have that $\tau(\Phi((\|f\|_\Phi - \epsilon)^{-1}|f|)) > 1$. We once again note that the convexity of Φ ensures that $\Phi(rt) \le r\Phi(t)$ for any $t \ge 0$ and any $r \le 1$. Given that $(\|f\|_\Phi - \epsilon) > 1$, we therefore have that $\Phi((\|f\|_\Phi - \epsilon)^{-1}t) \le (\|f\|_\Phi - \epsilon)^{-1})\Phi(t)$ for any $t \ge 0$ and hence that $\Phi((\|f\|_\Phi-\epsilon)^{-1}|f|) \le (\|f\|_\Phi-\epsilon)^{-1}\Phi(|f|)$. But then $\Phi((\|f\|_\Phi-\epsilon)^{-1}|f|)$ must belong to $\widetilde{\mathcal{M}}$ since $(\|f\|_\Phi-\epsilon)^{-1}\Phi(|f|)$ does. On applying the trace, it follows that $1 < \tau(\Phi((\|f\|_\Phi - \epsilon)^{-1}f)) \le (\|f\|_\Phi - \epsilon)^{-1}\tau(\Phi(f))$. Since $\epsilon > 0$ was arbitrary, we have that $\tau(\Phi(|f|)) \ge \|f\|_\Phi$ as required.

The one direction of the final claim clearly follows from the first claim, and the other from the definition of the Luxemburg–Nakano norm. $\square$

Proposition 6.47 *For any Young function Φ and any $f \in L^\Phi(\mathcal{M},\tau)$, we have that $\|f\|_\Phi^O \le 2\|f\|_\Phi$.*

Proof. Let f be a non-zero element of $L^\Phi(\mathcal{M},\tau)$. Recall that the pair (Φ, Φ^*) satisfies the Hausdorff–Young inequality $uv \le \Phi(u) + \Phi^*(v)$ $(u, v \ge 0)$. Given $g \in L^{\Phi^*}(\mathcal{M},\tau)$ and $\alpha \ge 0$, we may then combine this fact with Theorem 6.2 to see that

$$
\begin{aligned}
\tau(|\alpha f g|) &= \int_0^\infty (\mathbf{m}_{\alpha f g}(s))\, ds \\
&\le \int_0^\infty (\alpha \mathbf{m}_f(s)\mathbf{m}_g(s))\, ds \\
&\le \int_0^\infty \Phi(\alpha \mathbf{m}_f(s))\, ds + \int_0^\infty \Phi^*(\mathbf{m}_g(s))\, ds \\
&= \tau(\Phi(\alpha|f|)) + \tau(\Phi^*(|g|)).
\end{aligned}
$$

Recall that in the proof of the lemma, we showed that $\tau(\Phi(\|f\|_\Phi^{-1}|f|)) = \int_0^\infty \Phi(\|\mathbf{m}_f\|_\Phi^{-1}\mathbf{m}_f(s))\, ds \le 1$. So, if in addition $\tau(\Phi^*(|g|)) \le 1$, we will have $\tau(\|f\|_\Phi^{-1}fg|) \le 2$ and hence $\| \|f\|_\Phi^{-1}f\|_\Phi^O \le 2$. $\square$

The above proposition clearly shows that the Orlicz norm is finite on all of $L^\Phi(\mathcal{M},\tau)$. Having noted this fact, we are now ready for the following Hölder inequality for Orlicz spaces.

Corollary 6.48 *For any $f \in L^\Phi(\mathcal{M},\tau)$ and any $g \in L^{\Phi^*}(\mathcal{M},\tau)$ we have $fg, gf \in L^1(\mathcal{M},\tau)$, with $\tau(|fg|) \le \|f\|_\Phi^O\|g\|_{\Phi^*}$. In particular, $\|f\|_\Phi^O = \sup\{\tau(|fg|)\colon \tau(\Phi^*(|g|)) \le 1\} = \sup\{|\tau(fg)|\colon \tau(\Phi^*(|g|)) \le 1\} = \sup\{|\tau(gf)|\colon \tau(\Phi^*(|g|)) \le 1\} = \sup\{\tau(|gf|)\colon \tau(\Phi^*(|g|)) \le 1\}.$*

Proof. Let $f \in L^\Phi(\mathcal{M}, \tau)$ and $g \in L^{\Phi^*}(\mathcal{M}, \tau)$ be given. The fact that $\tau(|fg|) \leq \|f\|_\Phi^O \|g\|_{\Phi^*}$ follows fairly directly from Lemma 6.46 and the definition of the Orlicz norm. This clearly ensures that $fg \in L^1(\mathcal{M}, \tau)$. Since $\tau(|gf|) \leq \|f\|_\Phi \|g\|_{\Phi^*}^O$, we also have $gf \in L^1(\mathcal{M}, \tau)$. It then follows from Theorem 6.27 that

$$\sup\{|\tau(fg)| : \tau(\Phi^*(|g|)) \leq 1\} = \sup\{|\tau(gf)| : \tau(\Phi^*(|g|)) \leq 1\}.$$

We prove that

$$\sup\{|\tau(fg)| : \tau(\Phi^*(|g|)) \leq 1\} = \sup\{\tau(|fg|) : \tau(\Phi^*(|g|)) \leq 1\}.$$

(The proofs of the remaining equalities are similar.) Since $|\tau(fg)| \leq \tau(|fg|)$ we clearly have that

$$\sup\{|\tau(fg)| : \tau(\Phi^*(|g|)) \leq 1\} \leq \sup\{\tau(|fg|) : \tau(\Phi^*(|g|)) \leq 1\}.$$

Let $g_0 \in L^{\Phi^*}(\mathcal{M}, \tau)$ be given with $\tau(\Phi^*(|g_0|)) \leq 1$, and let u be the partial isometry in the polar form $fg_0 = u|fg_0|$ of fg_0. Since $\mathbf{m}_{g_0 u^*} \leq \|u^*\| \mathbf{m}_{g_0} \leq \mathbf{m}_{g_0}$, we clearly have that $\int_0^\infty \Phi^*(\alpha \mathbf{m}_{g_0 u^*}(s))\,ds \leq \int_0^\infty \Phi^*(\alpha \mathbf{m}_{g_0}(s))\,ds$ for any $\alpha > 0$ and hence that $g_0 u^* \in L^{\Phi^*}(\mathcal{M}, \tau)$. By construction, we then have that $\tau(|fg_0|) = \tau(u^* fg_0) = \tau(fg_0 u^*) \leq \sup\{|\tau(fg)| : \tau(\Phi^*(|g|) \leq 1\}$. In view of the fact that $g_0 \in L^{\Phi^*}(\mathcal{M}, \tau)$ was arbitrary, we are done. $\square$

Corollary 6.49 *For any $f \in \widetilde{\mathcal{M}}$, all the operators f, f^* and $|f|$ will belong to $L^\Phi(\mathcal{M}, \tau)$ whenever one of them does. In that case, $\|f\|_\Phi = \|f^*\|_\Phi = \| |f| \|_\Phi$ and $\|f\|_\Phi^O = \|f^*\|_\Phi^O = \| |f| \|_\Phi^O$.*

Proof. The first claim as well as the equality of the Luxemburg–Nakano norm is an immediate consequence of Proposition 5.12 considered alongside Corollary 6.42. On using what we have just verified, it now follows from Corollary 6.48 that

$$\|f\|_\Phi^O = \sup\{\tau(|fg|) : \tau(\Phi^*(|g|) \leq 1\}$$
$$= \sup\{\tau(|g^* f^*|) : \tau(\Phi^*(|g|) \leq 1\}$$
$$= \sup\{\tau(|g f^*|) : \tau(\Phi^*(|g|) \leq 1\} = \|f^*\|_\Phi^O.$$

Since for any $f, g \in \widetilde{\mathcal{M}}$ we have that $|fg| = ||f|g|$, it follows from the definition of the Orlicz norm that $\|f\|_\Phi^O = \| |f| \|_\Phi^O$. $\square$

As was the case with L^p-spaces, $\widetilde{\mathcal{M}}$ turns out to also be a natural superspace for Orlicz spaces.

Proposition 6.50 *For any Young function Φ, the space $L^\Phi(\mathcal{M}, \tau)$ continuously injects into $\widetilde{\mathcal{M}}$.*

Proof. Let (a_n) be a sequence in $L^\Phi(\mathcal{M}, \tau)$ converging to some $a \in L^\Phi(\mathcal{M}, \tau)$ in the norm $\|\cdot\|_\Phi$. Since $\mathbf{m}_{a-a_n}$ is non-increasing, we will for any $t > 0$ then have that

$$
\begin{aligned}
\mathbf{m}_{a-a_n}(t) \;&\leq\; \frac{1}{t} \int_0^t \mathbf{m}_{a-a_n}(s)\, ds \\
&=\; \frac{1}{t} \int_0^\infty \chi_{[0,t]}(s)\mathbf{m}_{a-a_n}(s)\, ds \\
&\leq\; \frac{1}{t} \|\chi_{[0,t]}\|_{\Phi*}^O \cdot \|\mathbf{m}_{a-a_n}\|_\Phi \\
&=\; \frac{1}{t} \|\chi_{[0,t]}\|_{\Phi*}^O \cdot \|a - a_n\|_\Phi.
\end{aligned}
$$

The claim now follows from Proposition 5.23. $\qquad\square$

With the above result at our disposal, we proceed with the proof of the completeness of $L^\Phi(\mathcal{M}, \tau)$.

Theorem 6.51 *Let Φ be a Young function. Then, $(L^\Phi(\mathcal{M}, \tau), \|\cdot\|_\Phi)$ is complete.*

Proof. Let (f_n) be a Cauchy sequence in $L^\Phi(\mathcal{M}, \tau)$. By Proposition 6.50, the sequence is Cauchy in the topology of convergence in measure. This topology is known to be complete, and hence, there exists $f \in \widetilde{\mathcal{M}}$ so that $f_n \to f$ in measure. Let $\epsilon > 0$ be given. For any fixed m, $(f_n - f_m)$ will trivially converge in measure to $f - f_m$. Recall that $\mathbf{m}_{f-f_m}$ is finite-valued and monotone on $(0, \infty)$. Since by the Lebesgue–Young theorem such functions are known to be differentiable almost everywhere (and hence continuous almost everywhere), we will by Lemma 5.28 then have that $\mathbf{m}_{f-f_m}(s) = \lim_{n\to\infty} \mathbf{m}_{f_n-f_m}(s)$ for almost every $s \geq 0$. Next select $N \in \mathbb{N}$ so that $\|f_n - f_m\|_\Phi = \|\mathbf{m}_{f_n-f_m}\|_\Phi < \epsilon$ for any $n, m \geq N$. We henceforth fix m as a natural number for which $m \geq N$. By the definition of the Luxemburg–Nakano norm (for $\mathbf{m}_{f_n-f_m}$), the fact that $\|\mathbf{m}_{f_n-f_m}\|_\Phi < \epsilon$ for any $n \geq N$ then means that

$$\int_0^\infty \Phi(\mathbf{m}_{\epsilon^{-1}(f_n-f_m)}(s))\,ds = \int_0^\infty \Phi(\epsilon^{-1}\mathbf{m}_{f_n-f_m}(s))\,ds \le 1 \text{ for all } n \ge N.$$

In the case where $b_\Phi = \infty$, Φ is continuous and finite-valued on $[0,\infty)$, and hence, in this case, we will have that $\lim_{n\to\infty}\Phi((2\epsilon)^{-1}\mathbf{m}_{f_n-f_m}(s)) = \Phi((2\epsilon)^{-1}\mathbf{m}_{f-f_m}(s))$ for almost every s. Now suppose that $b_\Phi < \infty$. Then, as in case 2 of the proof of Theorem 6.40, the fact that

$$\int_0^\infty \Phi(\mathbf{m}_{\epsilon^{-1}(f_n-f_m)}(s))\,ds \le 1 < \infty \text{ for all } n \ge N$$

means that $\mathbf{m}_{\epsilon^{-1}(f_n-f_m)}(0) = \epsilon^{-1}\|f_n - f_m\|_\infty \le b_\Phi$ for all $n \ge N$. Equivalently, $(2\epsilon)^{-1}\mathbf{m}_{f_n-f_m}(s) \le \frac{b_\Phi}{2}$ for all $s \ge 0$ and all $n \ge N$. Since Φ is continuous and finite-valued on $[0, \frac{b_\Phi}{2}]$, we will in this case also have that $\lim_{n\to\infty}\Phi((2\epsilon)^{-1}\mathbf{m}_{f_n-f_m}(s)) = \Phi((2\epsilon)^{-1}\mathbf{m}_{f-f_m}(s))$ for almost every s. We may, therefore, apply the standard Fatou lemma to see that

$$\int_0^\infty \Phi((2\epsilon)^{-1}\mathbf{m}_{f-f_m}(s))\,ds \le \liminf_{n\to\infty}\int_0^\infty \Phi((2\epsilon)^{-1}\mathbf{m}_{f_n-f_m}(s))\,ds \le 1$$

for all $n \ge N$. But by definition, this means that $\mathbf{m}_{f-f_m} \in L^\Phi(0,\infty)$ with $\|\mathbf{m}_{f-f_m}\|_\Phi \le 2\epsilon$. Corollary 6.42 informs us that this is equivalent to the statement that $f - f_m$ (and hence f) belongs to $L^\Phi(\mathcal{M},\tau)$, with $\|f - f_m\|_\Phi \le 2\epsilon$. Since $\epsilon > 0$ and $m \ge N$ were arbitrary, it follows by definition that (f_m) converges to f in the $\|\cdot\|_\Phi$ norm. $\square$

6.3.1 The Orlicz norm and Köthe duality for Orlicz spaces

We start by introducing the concept of Köthe duality. We first briefly review the concept of a Banach function space of measurable functions on a measure space (X,Σ,ν). Readers who wish to have a fuller account may consult one of [BS88] or[KPS82]. Though there are subtly different ways in which one can approach the theory, at its most basic level, one starts by defining a so-called Banach function norm ρ on $M_0(X,\Sigma,\nu)$ (the almost everywhere finite measurable functions) to be a mapping $\rho : M_0^+ \to [0,\infty]$ on the positive cone satisfying

[F1] $\rho(f) = 0$ if and only if $f = 0$ a.e.
[F2] $\rho(\lambda f) = \lambda\rho(f)$ for all $f \in M_0^+, \lambda > 0$.

[F3] $\rho(f + g) \leq \rho(f) + \rho(g)$ for all $f, g \in M_0^+$.

[F4] $f \leq g$ implies $\rho(f) \leq \rho(g)$ for all $f, g \in M_0^+$.

Such a ρ may be extended to all of M_0 by setting $\rho(f) = \rho(|f|)$, in which case we may then define $L^\rho(X, \Sigma, \nu) = \{f \in M_0(X, \Sigma, \nu): \rho(f) < \infty\}$. If indeed $L^\rho(X, \Sigma, \nu)$ turns out to be a Banach space when equipped with the norm $\| \cdot \|_\rho = \rho(\cdot)$, we refer to it as a Banach function space. If we add to the above list the so-called Fatou property, namely

[F5] for any sequence $(f_n) \subseteq M_0(X, \Sigma, \nu)$, we have that $0 \leq f_n \nearrow f$ implies $\rho(f_n) \nearrow \rho(f)$,

then $L^\rho(X, \Sigma, \nu)$ will automatically be complete. If further the situation $\mathbf{m}_f = \mathbf{m}_g$, $f \in L^\rho(X, \Sigma, \nu)$ and $g \in M_0$ ensures that $g \in L^\rho(X, \Sigma, \nu)$, we call $L^\rho(X, \Sigma, \nu)$ *rearrangement invariant*.

If we wish to ensure regular behaviour of the Banach function norm with respect to characteristic functions, we may additionally add the requirements that

[F6] for any measurable set E, $\nu(E) < \infty$ implies $\rho(\chi_E) < \infty$;

[F7] given any measurable set E with $\nu(E) < \infty$, there exists a constant $C_E > 0$ so that $\int_e f \, d\nu \leq C_E \rho(f)$ for any $f \in M_0(X, \Sigma, \nu)$.

For such a Banach function space, the Köthe dual is defined to be the space $L^{\rho'}(X, \Sigma, \nu) = \{f \in M_0(X, \Sigma, \nu): fg \in L^1(X, \Sigma, \nu) \text{ for all } g \in L^\rho(X, \Sigma, \nu)\}$, with the canonical norm being given by

$$\|f\|_{\rho'} = \sup\{\int |fg| \, d\nu: g \in L^\rho(X, \Sigma, \nu), \|g\|_\rho \leq 1\}.$$

The additional regularity criteria ensure that ρ' is in fact a Banach function norm, and $L^{\rho'}(X, \Sigma, \nu)$ the corresponding Banach function space. Whenever referring to Banach function spaces in the ensuing text, we shall generally assume that each of [F1] – [F7] holds.

However, our objective here is not to do a detailed study of Banach function spaces. Instead, we will show that for any Young function Φ, $L_\Phi(\mathcal{M}, \tau)$ is the noncommutative Köthe dual of $L^{\Phi^*}(\mathcal{M}, \tau)$, in the sense that as linear spaces $L^\Phi(\mathcal{M}, \tau) = \{f \in \widetilde{\mathcal{M}}: fg \in L^1(\mathcal{M}, \tau) \text{ for all } g \in$

$L^{\Phi^*}(\mathcal{M},\tau)\}$, with $\|f\|_{\Phi}^{O} = \sup\{\tau(|fg|)\colon g \in L^{\Phi^*}(\mathcal{M},\tau), \|g\|_{\Phi} \leq 1\}$. While proving this, we will also show that the Luxemburg–Nakano and Orlicz norms are equivalent.

Proposition 6.52 *Let Φ be a Young function.*

(a) *For any $g \in \widetilde{\mathcal{M}}$, the following are equivalent:*
   (i) *$gf \in L^1(\mathcal{M},\tau)$ for every $f \in L^{\Phi}(\mathcal{M},\tau)$;*
   (ii) *$fg \in L^1(\mathcal{M},\tau)$ for every $f \in L^{\Phi}(\mathcal{M},\tau)$;*
   (iii) *$\sup\{\tau(|gf|)\colon f \in L^{\Phi}(\mathcal{M},\tau), \tau(\Phi(|f|)) \leq 1\} < \infty$.*
(b) *Given some $g \in \widetilde{\mathcal{M}}$ satisfying the condition that $fg \in L^1(\mathcal{M},\tau)$ for every $f \in L^{\Phi}(\mathcal{M},\tau)$, we have that*

$$\sup\{\tau(|fg|)\colon f \in L^{\Phi}(\mathcal{M},\tau),\, \tau(\Phi(|f|)) \leq 1\}$$
$$= \sup\{|\tau(fg)|\colon f \in L^{\Phi}(\mathcal{M},\tau), \tau(\Phi(|f|)) \leq 1\}$$
$$= \sup\{|\tau(gf)|\colon f \in L^{\Phi}(\mathcal{M},\tau), \tau(\Phi(|f|)) \leq 1\}$$
$$= \sup\{\tau(|gf|)\colon f \in L^{\Phi}(\mathcal{M},\tau), \tau(\Phi(|f|)) \leq 1\}.$$

Moreover if g is as before and additionally $g \geq 0$, we then also have that

$$\sup\{|\tau(fg)|\colon f \in L^{\Phi}(\mathcal{M},\tau), \tau(\Phi(|f|)) \leq 1\}$$
$$= \sup\{\tau(fg)\colon f \in L^{\Phi}(\mathcal{M},\tau), f \geq 0, \tau(\Phi(f)) \leq 1\}.$$

Proof. We first prove the equivalence of (i) and (ii) in part (a). The proofs being similar, we only prove that (i)$\Rightarrow$(ii). Suppose that (i) holds and let $g^* = u|g^*|$ be the polar decomposition of g^*. For any $f \in L^{\Phi}(\mathcal{M},\tau)$, $|f|$ will, of course, also belong to $L^{\Phi}(\mathcal{M},\tau)$. Moreover, since $\mathbf{m}_{u|f|} \leq \mathbf{m}_{|f|}$, it is clear from Corollary 6.42 that in fact $u|f| \in L^{\Phi}(\mathcal{M},\tau)$. But then we must by hypothesis have $g(u|f|) = (|g^*|u^*)(u|f|) = |g^*|\,|f| \in L^1(\mathcal{M},\tau)$. On taking the adjoint, it follows that $|f|\,|g^*| \in L^1(\mathcal{M},\tau)$. Let v be the partial isometry in the polar decomposition $f = v|f|$ of f. Since $L^1(\mathcal{M},\tau)$ is an $L^{\infty}(\mathcal{M},\tau)$-bimodule, it follows that $fg = v|f|\,|g^*|u^* \in L^1(\mathcal{M},\tau)$ as required.

Having established the equivalence of (i) and (ii), the first half of part (b) now follows by the same argument used in Corollary 6.48. Next, we prove the second part of (b). Let $f \in L^{\Phi}(\mathcal{M},\tau)$ be given with $\tau(\Phi(|f|)) \leq 1$. We remind the reader that this condition is equivalent to requiring $\|f\|_{\Phi} \leq 1$ (see Lemma 6.46).

It is clear that

$$\sup\{\tau(fg)\colon f \in L^{\Phi}(\mathcal{M},\tau), f = f^*, \tau(\Phi(f)) \le 1\}$$
$$\le \sup\{|\tau(fg)|\colon f \in L^{\Phi}(\mathcal{M},\tau), \tau(\Phi(|f|)) \le 1\}.$$

We prove that equality holds. Let $f_0 \in L^{\Phi}(\mathcal{M},\tau)$ satisfy $\tau(\Phi(|f_0|)) \le 1$. For some $\alpha \in \mathbb{R}$, we have $\tau(e^{i\alpha}f_0 g) = |\tau(f_0 g)|$. Since for $\mathrm{Im}(e^{i\alpha}f_0)$ we have that $\tau(\mathrm{Im}(e^{i\alpha}f_0)g) = \tau(g^{1/2}\mathrm{Im}(e^{i\alpha}f_0)g^{1/2}) \in \mathbb{R}$ (and similarly $\tau(\mathrm{Im}(e^{i\alpha}f_0)g) \in \mathbb{R}$), the equality $\tau(e^{i\alpha}f_0 g) = |\tau(f_0 g)|$ ensures that

$$\tau(\mathrm{Im}(e^{i\alpha}f_0)g) = \mathrm{Im}(\tau(e^{i\alpha}f_0 g)) = 0.$$

Moreover, $\|\mathrm{Re}(e^{i\alpha}f_0)\|_{\Phi} \le \frac{1}{2}(\|e^{i\alpha}f_0\|_{\Phi} + \|(e^{i\alpha}f_0)^*\|_{\Phi}) = \|f_0\|_{\Phi}$. This inequality combined with Lemma 6.46 and the fact that $\tau(\Phi(|f_0|)) \le 1$, ensures that $\tau(\Phi(|\mathrm{Re}(e^{i\alpha}f_0)|)) \le 1$. Therefore, $|\tau(f_0 g)| = \tau(\mathrm{Re}(e^{i\alpha}f_0)g) \le \sup\{\tau(fg)\colon f \in L^{\Phi}(\mathcal{M},\tau), f = f^*, \tau(\Phi(f)) \le 1\}$. It is now clear that

$$\sup\{|\tau(fg)|\colon f \in L^{\Phi}(\mathcal{M},\tau), \tau(\Phi(|f|)) \le 1\}$$
$$= \sup\{\tau(fg)\colon f \in L^{\Phi}(\mathcal{M},\tau), f = f^*, \tau(\Phi(|f|)) \le 1\}.$$

As in Remark 6.10, we now set $f_+ = f\chi_{[0,\infty)}(f)$ and $f_- = -f\chi_{(-\infty,0)}(f)$. Recall that for f_+ and f_-, we have $f_{\pm} \ge 0$, $f = f_+ - f_-$ and $|f| = f_+ + f_-$, with $\mathbf{m}_{f_{\pm}} \le \mathbf{m}_f$. It is now an exercise to use Corollary 6.42 to show that this last fact ensures that $f_{\pm} \in L^{\Phi}(\mathcal{M},\tau)$ with $\|f_{\pm}\|_{\Phi} = \|\mathbf{m}_{f_{\pm}}\|_{\Phi} \le \|\mathbf{m}_f\|_{\Phi} = \|f\|_{\Phi} \le 1$. It now follows from Theorem 6.27 that $\tau((f_+)g) = \tau(g^{1/2}(f_+)g^{1/2}) \ge 0$ and similarly that $\tau((f_-)g) \ge 0$. Suppose that $\tau((f_+)g) \ge \tau((f_-)g)$. Then $|\tau(fg)| = |\tau((f_+)g) - \tau((f_-)g)| \le \tau((f_+)g)$. This then shows that

$$\sup\{\tau(fg)\colon f \in L^{\Phi}(\mathcal{M},\tau), f = f^*, \tau(\Phi(|f|)) \le 1\}$$
$$= \sup\{\tau(fg)\colon f \in L^{\Phi}(\mathcal{M},\tau), f \ge 0, \tau(\Phi(f)) \le 1\}$$

which together with the previous centred equality, proves the second part of (b).

It remains to prove the equivalence of (i) and (iii). The implication (iii)$\Rightarrow$(i) is obvious. Hence, we proceed to showing that (i)$\Rightarrow$(iii). Suppose by way of contradiction that for some fixed $g \in \widetilde{\mathcal{M}}$ (i) holds, but that (iii) fails. Let $g = u|g|$ be the polar decomposition of g. We clearly have that $|g|f = u^*gf \in L^1(\mathcal{M},\tau)$ for any $f \in L^{\Phi}(\mathcal{M},\tau)$.

Since, in addition, $|gf| = ||g|f|$ for any $f \in L^\Phi(\mathcal{M}, \tau)$, it follows that we may assume that $g \geq 0$. By the second part of (b) and the assumption regarding (iii), we then have that $\sup\{\tau(fg)\colon f \in L^\Phi(\mathcal{M}, \tau), f \geq 0, \tau(\Phi(f)) \leq 1\} = \infty$.

Taking note of Lemma 6.46, we may then select a sequence (f_n) of positive elements in the unit ball of $L^\Phi(\mathcal{M}, \tau)$, such that $\tau(gf_n) > n^3$ for each $n \in \mathbb{N}$. The formal sum $f_0 = \sum_{n=1}^\infty n^{-2} f_n$ converges absolutely in $L^\Phi(\mathcal{M}, \tau)$, and since this space is known to be a Banach space, f_0 must correspond to a well-defined element of $L^\Phi(\mathcal{M}, \tau)$. So, we must have $gf_0 \in L^1(\mathcal{M}, \tau)$ and hence that $\tau(gf_0) < \infty$. Since $f_0 = \sup_N \sum_{n=1}^N n^{-2} f_n$, we clearly have that $f_0 \geq n^{-2} f_n$ for any $n \in \mathbb{N}$. But by Theorem 6.27, this results in the situation that $\tau(gf_0) = \tau(g^{1/2} f_0 g^{1/2} f_0) \geq n^{-2} \tau(g^{1/2} f_n g^{1/2}) = n^{-2} \tau(gf_n) > n$ for any n. This is a clear contradiction. So, (iii) must hold if one of (ii) or (i) holds. $\qquad\square$

We need one more technical fact—important in its own right—before we are ready to prove the promised Köthe duality for the Orlicz spaces $L^\Phi(\mathcal{M}, \tau)$.

Lemma 6.53 *Let Φ be a Young function. We may formally extend the norms $\| \cdot \|_\Phi$ and $\| \cdot \|_\Phi^O$ to possibly infinite-valued quantities on $\widetilde{\mathcal{M}}$, by applying exactly the same prescriptions as those given in Definition 6.41. Denote these extensions by ρ_Φ and ρ_Φ^O, respectively. Given $f_0, f_1 \in \{f \in \widetilde{\mathcal{M}}\colon fg \in L^1(\mathcal{M}, \tau), g \in L^{\Phi^*}(\mathcal{M}, \tau)\}$ with $0 \leq f_0 \leq f_1$, we have that $\rho_\Phi(f_0) \leq \rho_\Phi(f_1)$ and $\rho_\Phi^O(f_0) \leq \rho_\Phi^O(f_1)$. More generally, if $(f_\alpha) \subseteq \{f \in \widetilde{\mathcal{M}}\colon fg \in L^1(\mathcal{M}, \tau), g \in L^{\Phi^*}(\mathcal{M}, \tau)\}$ is a net of positive elements increasing to $f_0 \in \{f \in \widetilde{\mathcal{M}}\colon fg \in L^1(\mathcal{M}, \tau), g \in L^{\Phi^*}(\mathcal{M}, \tau)\}$, then $(\rho_\Phi(f_\alpha))$ and $(\rho_\Phi^O(f_\alpha))$, respectively, increase to $\rho_\Phi(f_0)$ and $\rho_\Phi^O(f_0)$.*

Proof. We use the same notation ρ_Φ for the analogue of ρ_Φ on $M_0^+[0, \infty)$—the cone of non-negative finite almost everywhere Borel-measurable functions on $[0, \infty)$. The same argument used to prove Corollary 6.42 then suffices to prove that $\rho_\Phi(f) = \rho_\Phi(\mathbf{m}_f)$. If therefore we consider part (iii) of Proposition 5.12 alongside this fact, it is clear that in the case of the Luxemburg–Nakano norm, the claim follows from the corresponding fact for classical Orlicz spaces. Therefore, we need only prove the claim regarding the quantity ρ_Φ^O.

Let $f_0, f_1 \in \{f \in \widetilde{\mathcal{M}} : fg \in L^1(\mathcal{M}, \tau), g \in L^{\Phi^*}(\mathcal{M}, \tau)\}$ and $g_0 \in L^{\Psi^*}(\mathcal{M}, \tau)$ be given with $g_0 \geq 0$ and $0 \leq f_0 \leq f_1$. By Proposition 6.52, we may apply Theorem 6.27 to the products $f_0 g_0$ and $f_1 g_0$ to see that $\tau(f_0 g_0) = \tau(g_0^{1/2} f_0 g_0^{1/2}) \leq \tau(g_0^{1/2} f_1 g_0^{1/2}) = \tau(f_1 g_0)$. The fact that $\rho_\Phi^O(f_0) \leq \rho_\Phi^O(f_1)$ then follows from the final claim of part (b) of Proposition 6.52. Now, suppose that we are given a net $(f_\alpha) \subseteq \{f \in \widetilde{\mathcal{M}} : fg \in L^1(\mathcal{M}, \tau), g \in L^{\Phi^*}(\mathcal{M}, \tau)\}$ of positive elements increasing to $f_0 \in \{f \in \widetilde{\mathcal{M}} : fg \in L^1(\mathcal{M}, \tau), g \in L^{\Phi^*}(\mathcal{M}, \tau)\}$. It is clear from what we just proved that $\sup_\alpha \rho_\Phi^O(f_\alpha) \leq \rho_\Phi^O(f_0)$. It remains to prove the converse inequality.

Let $N \in (0, \rho_\Phi^O(f_0))$ be given. By part (b) of Proposition 6.52, $\rho_\Phi^O(f_0) = \sup\{\tau(f_0 g) : g \in L^{\Phi^*}(\mathcal{M}, \tau), g \geq 0, \tau(\Phi(g)) \leq 1\}$. We may therefore select $g_0 \in L^{\Phi^*}(\mathcal{M}, \tau)$ with $g_0 \geq 0$ and $\tau(\Phi(g_0)) \leq 1$, so that $\tau(f_0 g_0) > N$. Now notice that $g_0^{1/2} f_\alpha g_0^{1/2}$ increases to $g_0^{1/2} f_0 g_0^{1/2}$. By Proposition 5.17, we will then have $\sup_\alpha \tau(f_\alpha g_0) = \sup_\alpha \tau(g_0^{1/2} f_\alpha g_0^{1/2}) = \tau(g_0^{1/2} f_0 g_0^{1/2}) = \tau(f_0 g_0)$. So, there must exist an α such that $\tau(f_\alpha g_0) > N$, whence $\rho_\Phi^O(f_\alpha) > N$. Since $N \in (0, \rho_\Phi^O(f_0))$ was arbitrary, we have $\sup_\alpha \rho_\Phi^O(f_\alpha) \geq \rho_\Phi^O(f_0)$ as required. $\qquad\square$

Theorem 6.54 *Let Φ be a Young function. For any $f \in \widetilde{\mathcal{M}}$, we have that $\sup\{\tau(|fg|) : g \in L^{\Phi^*}(\mathcal{M}, \tau), \|g\|_\Phi^* \leq 1\} < \infty$ if and only if $f \in L^\Phi(\mathcal{M}, \tau)$, in which case $\|f\|_\Phi \leq \|f\|_\Phi^O \leq 2\|f\|_\Phi$. This, in particular, ensures that $L_\Phi(\mathcal{M}, \tau)$ is the Köthe dual of $L^{\Phi^*}(\mathcal{M}, \tau)$.*

Proof. We saw in Proposition 6.47 that $\|f\|_\Phi^O \leq 2\|f\|_\Phi$ for $f \in L^\Phi(\mathcal{M}, \tau)$. Notice that by Lemma 6.46, we then have $\sup\{\tau(|fg|) : g \in L^\Phi(\mathcal{M}, \tau), \|g\|_\Phi^* \leq 1\} = \|f\|_\Phi^O < \infty$. If therefore we are able to show that condition $\sup\{\tau(|fg|) : g \in L^\Phi(\mathcal{M}, \tau), \|g\|_\Phi^* \leq 1\} < \infty$ ensures $f \in L^\Phi(\mathcal{M}, \tau)$ and that in this case $\|f\|_\Phi \leq \|f\|_\Phi^O$, we will be done. We may clearly assume that $f \neq 0$.

Hence, let $f \in \widetilde{\mathcal{M}}$ be given with $\sup\{\tau(|fg|) : g \in L^\Phi(\mathcal{M}, \tau), \|g\|_\Phi^* \leq 1\} < \infty$. Recall that this is equivalent to requiring that $f \in \{a \in \widetilde{\mathcal{M}} : ag \in L^1(\mathcal{M}, \tau), g \in L^{\Phi^*}(\mathcal{M}, \tau)\}$. Note that $|fg| = ||f|g|$ for any $g \in L^\Phi(\mathcal{M}, \tau)$. When considering this fact alongside Corollary 6.49, it is clear that we may assume that $f = |f|$. Having made this assumption, we will actually prove that $\rho_\Phi(f) \leq \rho_\Phi^O(f)$. Since by Proposition 6.52 ρ_Φ^O is finite on $\{a \in \widetilde{\mathcal{M}} : ag \in L^1(\mathcal{M}, \tau), g \in L^{\Phi^*}(\mathcal{M}, \tau)\}$, this will force $\rho_\Phi(f) < \infty$, which ensures that $f \in L^\Phi(\mathcal{M}, \tau)$. Since then $\rho_\Phi(f) = \|f\|_\Phi$ and $\rho_\Phi^O(f) = \|f\|_\Phi^O$, this will prove the theorem.

Recall that any positive measurable function may be written as the increasing limit of a sequence of positive simple functions. If we combine this fact with the Borel functional calculus, it is clear that f can be written as an increasing limit of a sequence of operators (f_N)–all commuting with f–of the form $\sum_{k=1}^{n} \alpha_k e_k$ where the α_ks are positive real numbers and the e_ks mutually orthogonal projections. This ensures that in addition $f_N^2 \leq f^2$ for every N, and hence that $|f_N g|^2 \leq |fg|^2$ for every $g \in L^{\Phi^*}(\mathcal{M}, \tau)$. Taking square roots preserves the order, and hence for any $g \in L^{\Phi^*}(\mathcal{M}, \tau)$, $\tau(|f_N g|) \leq \tau(|fg|) < \infty$. So $(f_N) \subseteq \{a \in \widetilde{\mathcal{M}} : ag \in L^1(\mathcal{M}, \tau), g \in L^{\Phi^*}(\mathcal{M}, \tau)\}$. By Lemma 6.53, we may therefore pass to the case where $f = \sum_{k=1}^{n} \alpha_k e_k$, with the e_ks non-zero mutually orthogonal projections and the α_ks positive. The projection $e_1 \in \mathcal{M}$ may in turn be written as the supremum of an increasing net (e_β) of subprojections of e_1 with finite trace. (See the proof of Corollary 6.26 for the justification of this claim.) Therefore, the operators $f_\beta = e_\beta f = \alpha_1 e_\beta + \sum_{k=2}^{n} \alpha_k e_k$ increase to f. Since $\tau(|f_\beta|) = \tau(|e_\beta fg|) \leq \tau(|fg|)$, we may argue as before to see that we may assume that $\tau(e_1) < \infty$. When inductively applying the same argument to each of the e_ks, it is now clear that we may, in fact, assume that each of the e_ks has a finite trace.

For some $\gamma > 0$, we will have $\tau(\Phi^*(\gamma e_1)) = \Phi^*(\gamma)\tau(e_1) \leq 1$. This ensures that $0 < \alpha_1 \gamma \tau(e_1) = \tau(|f(\gamma e_1)|) \leq \rho_\Phi^O(f)$. We may now rescale f to pass to the case where $\rho_\Phi^O(f) = 1$. It is then incumbent on us to show that $\rho_\Phi(f) \leq 1$. By the definition of ρ_Φ, this will follow once we prove that $\tau(\Phi(\gamma f)) \leq 1$ for any $\gamma \in (0, 1)$.

We first claim that $f \leq b_\Phi \sum_{k=1}^{n} e_k$. This is of course trivial if $b_\Phi = \infty$, so assume that $b_\Phi < \infty$. The claimed operator inequality will hold if for each k, $\alpha_k \leq b_\Phi$. So if $f \leq b_\Phi \mathbb{1}$ were not true, there must then exist some k_0 and some $\epsilon > 0$, such that $\alpha_{k_0} \geq b_\Phi + \epsilon$. Now, consider the operator $h = (b_\Phi \tau(e_{k_0}))^{-1} e_{k_0}$. We may then use part (a) of Remark 6.36 to see that $\tau(\Phi^*(h)) \leq b_\Phi \tau((b_\Phi \tau(e_{k_0}))^{-1} e_{k_0}) \leq 1$. But this would force

$$\rho_\Phi^O(f) \geq \tau(|fh|) = (b_\Phi \tau(e_{k_0}))^{-1} \tau(\alpha_{k_0} e_{k_0}) = \frac{\alpha_{k_0}}{b_\Phi} \geq \frac{b_\Phi + \epsilon}{b_\Phi} > 1$$

which contradicts our assumption that $\rho_\Phi^O(f) = 1$. Hence, the claimed operator inequality holds. Since for any $0 < \gamma < 1$, we have $\Phi(\gamma b_\Phi) < \infty$, this in turn ensures that $\tau(\Phi(\gamma f)) \leq \Phi(\gamma b_\Phi)\tau(\sum_{k=1}^{n} e_k) < \infty$. (Recall that by Remark 5.15, this forces $\Phi(\gamma f) \in \widetilde{\mathcal{M}}$.)

Now, let $\gamma \in (0,1)$ be given. Recall that Φ is of the form $\Phi(t) = \int_0^t \phi(s)\,ds$ for some non-negative left-continuous non-decreasing function ϕ on $[0,\infty)$, which is infinite on (b_Φ,∞), but neither identically 0 nor infinite on all of $(0,\infty)$. Since ϕ is bounded and increasing on $[0,\gamma b_\Phi]$, the Borel functional calculus ensures that $g = \phi(\gamma f)$ is a well-defined element of $\mathcal{M}$, supported on $\sum_{k=1}^n e_k$. However, more is true. Suppose we are given $0 \le s, t < \infty$ with $s = \phi(t)$. It then follows from the equality criteria for the Hausdorff–Young inequality (see part (b) of Remark 6.36), that $\Phi^*(\phi(t)) = t\phi(t) - \Phi(t) \le t\phi(t)$. Thus, if ϕ is bounded on some interval $[0,r]$, then so is $\Phi^* \circ \phi$. Now, recall that ϕ is non-decreasing and finite on $[0,b_\Phi)$, and that by construction, $\gamma\|f\|_\infty < b_\Phi$. This ensures that ϕ, and therefore also $\Phi^* \circ \phi$, is bounded on $[0,\gamma\|f\|_\infty]$. It therefore follows that $\Phi^* \circ \phi(\gamma f) = \Phi^*(g) \in \mathcal{M}$.

Since the operators $\Phi(\gamma f)$ and $\Phi^*(g)$ are commuting operators affiliated to the von Neumann algebra generated by the spectral projections of f, we may apply the Borel functional calculus for f to the equality criteria for the Hausdorff–Young inequality (see part (b) of Remark 6.36), to see that $\gamma f g = \Phi(f) + \Phi^*(g)$. All operators in this expression belong to $\widetilde{\mathcal{M}}_+$ and hence we may apply the trace to see that

$$\tau(\gamma f g) = \tau(\Phi(\gamma f)) + \tau(\Phi^*(g)). \tag{6.1}$$

We have already seen that $\tau(\Phi(\gamma f)) < \infty$. Since by construction $\phi(\gamma f) \le \phi(\gamma\|f\|_\infty)\sum_{k=1}^n e_k$, we also have $0 \le \gamma f g \le \phi(\gamma\|f\|_\infty)\gamma\|f\|_\infty \sum_{k=1}^n e_k$, and hence $\tau(\gamma f g) < \infty$. Thus, by equation (6.1), we must have that $\tau(\Phi^*(g))$ is also finite. But then $g \in L^{\Phi^*}(\mathcal{M},\tau)$. By the definition of ρ_Φ^O, we then have $\tau(\gamma f g) \le \rho_\Phi^O(\gamma f)\|g\|_{\Phi^*} \le \|g\|_{\Phi^*}$. We now use Lemma 6.46 to see that

$$\tau(\gamma f g) \le \|g\|_{\Phi^*} \le \max(1, \tau(\Phi^*(g))) \le 1 + \tau(\Phi^*(g)). \tag{6.2}$$

Considering equations (6.1) and (6.2) alongside each other, the fact that all terms are finite ensures that $\tau(\Phi(\gamma f)) \le 1$. This proves that $f \in L^\Phi(\mathcal{M},\tau)$. However, since $\gamma \in (0,1)$ was arbitrary, we also have $\|f\|_\Phi \le 1$, as required. $\qquad\square$

Exercise 6.55 *Show that the space $L\log(L+1)(\mathcal{M},\tau)$ is isomorphic to the Köthe dual of $L^{\cosh-1}(\mathcal{M},\tau)$. (Here, $L\log(L+1)(\mathcal{M},\tau)$ is the space produced by the Young function $t \mapsto t\log(t+1)$.)*

6.3.2 The Orlicz spaces $L^1 \cap L^\infty$ and $L^1 + L^\infty$

We close this very brief introduction to Orlicz spaces with a description of spaces $L^1 \cap L^\infty$ and $L^1 + L^\infty$. Both of these will be shown to be Orlicz spaces. We will in particular also see that these spaces in a very real sense represent the smallest and largest of all Orlicz spaces. For this, we will need the concept of a *fundamental function* of a rearrangement invariant Banach function space. For our purposes, it is enough to at this point restrict attention to the measure space $((0,\infty), \mathcal{B}(0,\infty))$ equipped with Lebesgue measure.

Definition 6.56 Given a rearrangement invariant Banach function space $L^\rho(0,\infty)$, define the associated fundamental function $\mathbf{f}_\rho$: $[0,\infty) \to [0,\infty)$ by the prescription that $\mathbf{f}_\rho(t) = \|\chi_E\|_\rho$, where E is a Borel set with measure t.

Any two Borel sets E_1 and E_2 with the same finite measure will have the same decreasing rearrangement. Therefore, it is the rearrangement invariance of the space $L^\rho(0,\infty)$ that ensures that $\mathbf{f}_\rho$ is well defined. We proceed to list the basic properties of $\mathbf{f}_\rho$. We will merely sketch the proof. Readers who wish to have full details may refer to section II.5 of [**BS88**].

Proposition 6.57 *Let $L^\rho(0,\infty)$ be a rearrangement invariant Banach function space, and $L^{\rho'}(0,\infty)$ its Köthe dual. Both $\mathbf{f}_\rho$ and $\mathbf{f}_{\rho'}$ are so-called* quasi-concave functions, *meaning that they are non-decreasing, continuous on $(0,\infty)$, zero-valued at precisely 0 and with both $t \mapsto \frac{\mathbf{f}_\rho(t)}{t}$ and $t \mapsto \frac{\mathbf{f}_{\rho'}(t)}{t}$ non-increasing on $(0,\infty)$. In addition, $\mathbf{f}_\rho(t)\,\mathbf{f}_{\rho'}(t) = t$ for any $t \geq 0$.*

Sketch of proof. We shall not prove the final statement, but will merely indicate how that statement may be used to prove the rest of the claims. We first note that the non-degeneracy of the norm $\|\cdot\|_\rho$ ensures that $\mathbf{f}_\rho$ is zero-valued at precisely 0. If $0 \leq t_0 \leq t_1$, then $\chi_{[0,t_0]} \leq \chi_{[0,t_1]}$. It then follows from property [F4] in Subsection 6.3.1 that $\mathbf{f}_\rho(t_0) = \|\chi_{[0,t_0]}\|_\rho \leq \|\chi_{[0,t_1]}\|_\rho = \mathbf{f}_\rho(t_1)$. The fact that $t \mapsto \frac{\mathbf{f}_\rho(t)}{t}$ is non-increasing follows from the final claim combined with the fact that

$\mathbf{f}_{\rho'}$ is non-decreasing. To see that $\mathbf{f}_\rho$ is continuous, observe that it is a non-decreasing function that cannot have any jump discontinuities on $(0, \infty)$, since a jump discontinuity at a point $t_0 > 0$ would mean that $t \mapsto \frac{\mathbf{f}_\rho(t)}{t}$ fails to be non-increasing at that point. $\qquad\square$

Remark 6.58 We pause to note that the 'converse' of Proposition 6.57 is also true in that every quasi-concave function $\mathbf{f}$ appears as the fundamental function of some rearrangement invariant Banach function space. (See [**BS88**, Proposition II.5.8].)

Let us now apply the above ideas to the classical Orlicz space $L^\Phi(0, \infty)$. Both the Luxemburg–Nakano and Orlicz norms turn out to be rearrangement invariant Banach function norms. (See [**BS88**, section IV.8] for details.) For the Luxemburg–Nakano norm, the rearrangement invariance follows from Corollary 6.42. The fundamental function corresponding to the Luxemburg–Nakano and Orlicz norms will, respectively, be denoted by $\mathbf{f}^\Phi$ and $\mathbf{f}_\Phi$. We have the following very elegant formulae for these two fundamental functions.

Proposition 6.59 *Let Φ be a Young function. The fundamental function corresponding to the Luxemburg—Nakano norm for the space $L^\Phi(0, \infty)$ is given by the formula $\mathbf{f}^\Phi(t) = \frac{1}{\Phi^{-1}(1/t)}$, and the one corresponding to the Orlicz norm of the same space by the formula $\mathbf{f}_\Phi(t) = t(\Phi^*)^{-1}(1/t)$. (Here, Φ^{-1} and $(\Phi^*)^{-1}$ are as in part (d) of Remark 6.36.)*

Proof. It is clear from the definition that $\mathbf{f}^\Phi(0) = 0$. Now, let E be a Borel set of non-zero finite measure. First note that for any $0 < \alpha < b_\Phi$, the continuous functional calculus (applied to $L^\infty(0, \infty)$) ensures that $\Phi(\alpha\chi_E) = \Phi(\alpha)\chi_E$ and hence that

$$\int \Phi(\alpha\chi_E)\, d\mathbf{m}(t) = \Phi(\alpha) \int \chi_E \, d\mathbf{m}(t) = \Phi(\alpha)\mathbf{m}(E).$$

It is therefore clear that $\chi_E \in L^\Phi(0, \infty)$. To see the claim regarding $\mathbf{f}^\Phi$, we may use the above equality to see that

$$
\begin{aligned}
\|\chi_E\|_\Phi &= \inf\{\epsilon > 0 : \Phi\left(\frac{1}{\epsilon}\right) \int \chi_E \, d\mathrm{m}(t) \leq 1\} \\
&= \inf\left\{\epsilon > 0 : \Phi\left(\frac{1}{\epsilon}\right) \leq \left(\int \chi_E \, d\mathrm{m}(t)\right)^{-1}\right\} \\
&= \left[\sup\left\{\gamma > 0 : \Phi(\gamma) \leq \frac{1}{\int \chi_E \, d\mathrm{m}(t)}\right\}\right]^{-1} \\
&= \frac{1}{\Phi^{-1}(1/\int \chi_E \, d\mathrm{m}(t))}.
\end{aligned}
$$

Therefore, we have $\mathbf{f}^\Phi(t) = \frac{1}{\Phi^{-1}(1/t)}$ for all $t > 0$ as required.

The claim regarding $\mathbf{f}_\Phi$ follows from the fact that the fundamental function of the Köthe dual of $L^{\Phi^*}(0, \infty)$ must satisfy the relation $\mathbf{f}_\Phi(t)\,\mathbf{f}^{\Phi^*}(t) = t$. $\qquad\square$

Proposition 6.60 *Let Φ_1 and Φ_2 be Young functions. Algebraically, the Orlicz space $L^{\Phi_1 \vee \Phi_2}(\mathcal{M}, \tau)$ agrees with $L^{\Phi_1}(\mathcal{M}, \tau) \cap L^{\Phi_2}(\mathcal{M}, \tau)$. (Here, $\Phi_1 \vee \Phi_2$ is the Young function $(\Phi_1 \vee \Phi_2)(t) = \max(\Phi_1(t), \Phi_2(t))$.)*

Proof. If $f \in L^{\Phi_1 \vee \Phi_2}(\mathcal{M}, \tau)$, then by definition there exists $\alpha > 0$ so that $\tau(\Phi_1 \vee \Phi_2(\alpha|a|)) < \infty$. Since $\Phi_1 \vee \Phi_2$ majorises both Φ_1 and Φ_2, it is clear that we then also have that $\tau(\Phi_1(\alpha|a|)) < \infty$ and $\tau(\Phi_2(\alpha|a|)) < \infty$.

Conversely, if $f \in L^{\Phi_1}(\mathcal{M}, \tau) \cap L^{\Phi_2}(\mathcal{M}, \tau)$, there must exist $\alpha_1, \alpha_2 > 0$ such that $\tau(\Phi_1(\alpha_1|a|)) < \infty$ and $\tau(\Phi_2(\alpha_2|a|)) < \infty$. For $\alpha = \min(\alpha_1, \alpha_2)$, we will then clearly have $\tau(\Phi_1(\alpha|a|) + \Phi_2(\alpha|a|)) < \infty$. Since the function $\Phi_1 + \Phi_2$ majorises $\Phi_1 \vee \Phi_2$, we will then also have that $\tau((\Phi_1 \vee \Phi_2)(\alpha|a|)) < \infty$, as required. $\qquad\square$

We now apply the above to show that the spaces $(L^1 \cap L^\infty)(\mathcal{M}, \tau)$ and $(L^1 + L^\infty)(\mathcal{M}, \tau)$ may be realised as Orlicz spaces.

Proposition 6.61 *$L^1(\mathcal{M}, \tau) \cap L^\infty(\mathcal{M}, \tau)$ is an Orlicz space corresponding to the Young function*

$$
\Phi_{1 \cap \infty}(t) = \begin{cases} t, & \text{if } 0 \leq t \leq 1 \\ \infty, & \text{otherwise} \end{cases} .
$$

$L^1(\mathcal{M},\tau) + L^\infty(\mathcal{M},\tau)$ *is an Orlicz space corresponding to the conjugate Young function* $\Phi_{1+\infty} = \Phi^*_{1\cap\infty}$, *which is given by*

$$\Phi_{1+\infty}(t) = \begin{cases} 0, & \text{if } 0 \le t \le 1 \\ t-1, & \text{otherwise} \end{cases}.$$

Convention: *When considered as Orlicz spaces, we write* $L^{1\cap\infty}$ $(\mathcal{M},\tau)$ *for* $L^1(\mathcal{M},\tau) \cap L^\infty(\mathcal{M},\tau)$, *and* $L^{1+\infty}(\mathcal{M},\tau)$ *for* $L^1(\mathcal{M},\tau) + L^\infty(\mathcal{M},\tau)$.

Proof. The Young function generating $L^1(\mathcal{M},\tau)$ is clearly $\Phi(t) = t$. Hence, the fact that $L^1(\mathcal{M},\tau) \cap L^\infty(\mathcal{M},\tau)$ is an Orlicz space corresponding to the given Young function is a consequence of Exercise 6.44 and the preceding proposition.

We proceed with the claim regarding $L^1(\mathcal{M},\tau) + L^\infty(\mathcal{M},\tau)$. The fact that $\Phi_{1+\infty} = \Phi^*_{1\cap\infty}$ is as stated is left as an exercise. If $f \in L^1(\mathcal{M},\tau)$, we will clearly have $\tau(\Phi_{1+\infty}(|f|)) \le \tau(|f|) < \infty$. If on the other hand $f \in L^\infty(\mathcal{M},\tau)$, then by the definition of $\Phi_{1+\infty}$, we will have that $\Phi_{1+\infty}(\|f\|_\infty^{-1}|f|) = 0$. This then in turn ensures that $L^{\Phi_{1+\infty}}(\mathcal{M},\tau)$ contains $L^1(\mathcal{M},\tau)+L^\infty(\mathcal{M},\tau)$. Conversely, suppose that $f \in L^{\Phi_{1+\infty}}(\mathcal{M},\tau)$, and let $\alpha > 0$ be given such that $\tau(\Phi_{1+\infty}(\alpha|f|)) < \infty$. Next, observe that applying the Borel functional calculus to the definition of $\Phi_{1+\infty}$, we have that $\Phi_{1+\infty}(\alpha|f|) = (\alpha|f|-\mathbb{1})\chi_{(1,\infty)}(\alpha|f|)$. Therefore, requiring $\tau(\Phi_{1+\infty}(\alpha|f|)) < \infty$ is the same as requiring $(\alpha|f| - \mathbb{1})\chi_{(1,\infty)}(\alpha|f|) \in L^1(\mathcal{M},\tau)$. Since $(\alpha|f| - \mathbb{1})\chi_{[0,1]}(\alpha|f|) \in L^\infty(\mathcal{M},\tau)$, we therefore clearly have that $\alpha|f| - \mathbb{1} \in L^1(\mathcal{M},\tau) + L^\infty(\mathcal{M},\tau)$ and hence that $f \in L^1(\mathcal{M},\tau) + L^\infty(\mathcal{M},\tau)$. $\square$

Remark 6.62 If we apply Proposition 6.59 to Proposition 6.61, it follows that the fundamental function of the space $L^{1\cap\infty}(0,\infty)$ corresponding to the Luxemburg–Nakano norm is given by $\mathbf{f}^{1\cap\infty}(t) = \max(1,t)$. By the final claim in Proposition 6.57, the fundamental function of the space $L^{1+\infty}(0,\infty)$ corresponding to the Orlicz norm is given by $\mathbf{f}_{1+\infty}(t) = \min(1,t)$.

To compute the norms of $L^{1\cap\infty}(\mathcal{M},\tau)$ and $L^{1+\infty}(\mathcal{M},\tau)$, we need the following result.

Theorem 6.63   *For any $f \in (L^1 + L^\infty)(\mathcal{M}, \tau)$ and any $t > 0$, we have that*

$$\inf\{\|f_1\|_1 + t\|f_\infty\|_\infty : f = f_1 + f_\infty, f_1 \in L^1(\mathcal{M}, \tau), f_\infty \in L^\infty(\mathcal{M}, \tau)\}$$

$$= \int_0^t \mathbf{m}_f(s)\, ds.$$

In particular, for each $f \in (L^1 + L^\infty)(\mathcal{M}, \tau)$, $\int_0^t \mathbf{m}_f(s)\, ds$ is then finite for each $t > 0$

Proof.   Let $f \in (L^1 + L^\infty)(\mathcal{M}, \tau)$ be given and let $f = f_1 + f_\infty$ be an arbitrary decomposition of f with $f_1 \in L^1(\mathcal{M}, \tau)$ and $f_\infty \in L^\infty(\mathcal{M}, \tau)$. For $0 < \alpha < 1$ and $s > 0$, it follows from Proposition 5.12 that $\mathbf{m}_f(s) \leq \mathbf{m}_{f_1}(\alpha s) + \mathbf{m}_{f_\infty}((1 - \alpha)s) \leq \mathbf{m}_{f_1}(\alpha s) + \|f_\infty\|_\infty$. We may then apply Proposition 5.13 to see that

$$\int_0^t \mathbf{m}_f(s)\, ds \;=\; \int_0^t \mathbf{m}_{f_1}(\alpha s)\, ds + t\|f_\infty\|_\infty$$

$$\leq\; \int_0^\infty \mathbf{m}_{f_1}(\alpha s)\, ds + t\|f_\infty\|_\infty$$

$$=\; \alpha^{-1} \int_0^\infty \mathbf{m}_{f_1}(r)\, dr + t\|f_\infty\|_\infty$$

$$=\; \alpha^{-1}\|f_1\|_1 + t\|f_\infty\|_\infty.$$

On letting α increase to 1, it will follow that $\int_0^t \mathbf{m}_f(s)\, ds \leq \|f_1\|_1 + t\|f_\infty\|_\infty$. We may then take the infimum over all decompositions of f as a sum of elements from $L^1(\mathcal{M}, \tau)$ and $L^\infty(\mathcal{M}, \tau)$ to see that

$$\int_0^t \mathbf{m}_f(s)\, ds \leq \inf\{\|f_1\|_1 + t\|f_\infty\|_\infty : f = f_1 + f_\infty,$$

$$f_1 \in L^1(\mathcal{M}, \tau), f_\infty \in L^\infty(\mathcal{M}, \tau)\}.$$

This clearly also proves the final claim.

To prove the converse inequality, let $f = u|f|$ be the polar form of f, and let $|f| = \int_0^\infty \lambda\, de_\lambda$ be the spectral decomposition of $|f|$. For a fixed $t > 0$, we set $\alpha = \mathbf{m}_f(t)$ and then define f_1 and f_∞ by

$$f_1 = u \int_\alpha^\infty (\lambda - \alpha)\, de_\lambda \text{ and } f_\infty = f - f_1.$$

Since

$$g(\lambda) = \begin{cases} 0, & \text{if } 0 \leq \lambda \leq \alpha \\ \lambda - \alpha, & \text{if } \lambda \geq \alpha \end{cases}$$

is a continuous increasing function with $|f_1| = \int_0^\infty (\lambda - \alpha)\, de_\lambda = g(|f|)$, it follows from Proposition 5.12 that

$$\mathbf{m}_{f_1}(s) = g(\mathbf{m}_f(s)) = \begin{cases} \mathbf{m}_f(s) - \alpha, & \text{if } 0 < s < t \\ 0, & \text{if } s \geq t. \end{cases}$$

For f_∞, we clearly have that

$$f_\infty = f - f_1 = u \int_0^\infty \lambda\, de_\lambda - u \int_\alpha^\infty \alpha\, de_\lambda = u \int_0^\alpha \lambda\, de_\lambda$$

and hence that $f_\infty \in L^\infty(\mathcal{M}, \tau)$ with $\|f_\infty\|_\infty \leq \alpha$. It therefore follows that

$$\begin{aligned} \|f_1\|_1 + t\|f_\infty\|_\infty &\leq \int_0^\infty \mathbf{m}_{f_1}(s)\, ds + t\alpha \\ &= \int_0^t (\mathbf{m}_f(s) - \alpha)\, ds + t\alpha \\ &= \int_0^t \mathbf{m}_f(s)\, ds. \end{aligned}$$

Since $\int_0^t \mathbf{m}_f(s)\, ds$ must be finite, this also proves that $f_1 \in L^1(\mathcal{M}, \tau)$, which then proves the theorem. $\qquad\square$

It is a classical fact that if two Banach spaces X_0 and X_1 canonically embed into a Hausdorff topological vector space in such a way that one can give meaning to $X_0 \cap X_1$ and $X_0 + X_1$, then these spaces become Banach spaces when equipped, respectively, with the norms $\max(\|x\|_0, \|x\|_1)$ and $\inf\{\|x_0\|_0 + \|x_1\|_1 : x = x_0 + x_1, x_0 \in X_0, x_1 \in X_1\}$. In the following theorem, we show how for the pair $(L^{1\cap\infty}(\mathcal{M}, \tau), L^{1+\infty}(\mathcal{M}, \tau))$, these natural norms on $L^{1\cap\infty}(\mathcal{M}, \tau)$ and $L^{1+\infty}(\mathcal{M}, \tau)$ may be realised as Luxemburg–Nakano and Orlicz norms. We will denote these natural norms by $\|\cdot\|_{1\cap\infty}$ and $\|\cdot\|_{1+\infty}$, respectively.

Theorem 6.64 *For the spaces $L^{1\cap\infty}(\mathcal{M}, \tau)$ and $L^{1+\infty}(\mathcal{M}, \tau)$, the natural norms on these spaces agree, respectively, with the*

Luxemburg–Nakano and Orlicz norms. The spaces $L^{1 \cap \infty}(\mathcal{M}, \tau)$ and $L^{1 + \infty}(\mathcal{M}, \tau)$ are therefore Köthe duals of each other. The norm for $L^{1 \cap \infty}(\mathcal{M}, \tau)$ is given by $\|f\|_{1 \cap \infty} = \max(\|f\|_1, \|f\|_\infty)$, and for $L^{1 + \infty}(\mathcal{M}, \tau)$ by $\|f\|_{1 + \infty} = \int_0^1 \mathbf{m}_f(s)\, ds$.

Proof.   We start by proving that the Luxemburg–Nakano norm on $L^1(\mathcal{M}, \tau) \cap L^\infty(\mathcal{M}, \tau)$ is as stated. Let $f \in L^1(\mathcal{M}, \tau) \cap L^\infty(\mathcal{M}, \tau)$ be given and consider $\epsilon = \max(\|f\|_1, \|f\|_\infty)$. We then clearly have $\|\epsilon^{-1}|f|\,\|_\infty \leq 1$, which ensures that $\Phi_{1 \cap \infty}(\epsilon^{-1}|f|) = \epsilon^{-1}|f|$. Hence

$$\tau(\Phi_{1 \cap \infty}(\epsilon^{-1}|f|)) = \tau(\epsilon^{-1}|f|) = \epsilon^{-1}\|f\|_1 \leq 1.$$

This ensures that $\|f\|_{1 \cap \infty} \leq \max(\|f\|_1, \|f\|_\infty)$.

Now, suppose that $\epsilon < \max(\|f\|_1, \|f\|_\infty)$. Then, one of $\epsilon < \|f\|_1$ or $\epsilon < \|f\|_\infty$ must hold. If $\epsilon < \|f\|_\infty$, then for $\gamma = \frac{\|f\|_\infty}{\epsilon}$, the spectral projection $\chi_{(1,\gamma]}(\epsilon^{-1}|f|)$ must be non-zero. Since $\epsilon^{-1}|f| \geq \epsilon^{-1}|f| \chi_{(1,\gamma]}(\epsilon^{-1}|f|)$ with $\Phi_{1 \cap \infty}(\epsilon^{-1}|f|\chi_{(1,\gamma]}(\epsilon^{-1}|f|)) = \infty \cdot \chi_{(1,\gamma]}(\epsilon^{-1}|f|))$, we must then have that $\tau(\Phi_{1 \cap \infty}(\epsilon^{-1}|f|)) = \infty$. Now, suppose that $\|f\|_\infty \leq \epsilon$, but that $\|f\|_1 > \epsilon$. Since in this case $\|\epsilon^{-1}|f|\,\|_\infty \leq 1$, we will have that $\Phi_{1 \cap \infty}(\epsilon^{-1}|f|) = \epsilon^{-1}|f|$, which then ensures that $\tau(\Phi_{1 \cap \infty}(\epsilon^{-1}|f|)) = \tau(\epsilon^{-1}|f|) = \epsilon^{-1}\|f\|_1 > 1$. We must therefore have that $\|f\|_{1 \cap \infty} \geq \epsilon$, and hence that $\|f\|_{1 \cap \infty} \geq \max(\|f\|_1, \|f\|_\infty)$ as required.

We now use the above and Remark 6.62 to prove the claims regarding $L^{1 + \infty}(\mathcal{M}, \tau)$. Let $f \in L^{1 + \infty}(\mathcal{M}, \tau)$ be given and suppose that $f = f_1 + f_\infty$, where $f_1 \in L^1(\mathcal{M}, \tau)$ and $f_\infty \in L^\infty(\mathcal{M}, \tau)$. Given any $g \in L^{1 \cap \infty}(\mathcal{M}, \tau)$ with $\|g\|_{1 \cap \infty} \leq 1$, we may then select partial isometries u and v so that $|fg| \leq u|f_1 g|u^* + v|f_\infty g|v^*$ (Lemma 5.2). Therefore

$$\tau(|fg|) \; \leq \tau(u|f_1 g|u^* + v|f_\infty g|v^*) \leq \tau(|f_1 g|) + \tau(|f_\infty g|)$$
$$\leq \|f_1\|_1\|g\|_\infty + \|f_\infty\|_\infty\|g\|_1 \leq \|f_1\|_1 + \|f_\infty\|_\infty.$$

Taking the infimum over all decompositions of the form $f = f_1 + f_\infty$, we see that $\tau(|fg|) \leq \|f\|_{1 + \infty}$. Now take the supremum over all gs with $\|g\|_{1 \cap \infty} \leq 1$ to see that $\|f\|^O_{\Phi_{1 + \infty}} \leq \|f\|_{1 + \infty}$.

For the converse, notice that by Theorem 6.54, the space $L^{1 \cap \infty}(0, \infty)$ equipped with the norm $\|\cdot\|_{\Phi_{1 \cap \infty}}$ is the Köthe dual of $L^{1 + \infty}(0, \infty)$ equipped with the Orlicz norm. It now follows from Theorem 6.63 that

$$\begin{aligned}
\|f\|_{1+\infty} &= \int_0^1 \mathbf{m}_f(s)\, ds \\
&\leq \|\chi_{[0,1]}\|_{1\cap\infty}\|\mathbf{m}_f\|^O_{\Phi_{1+\infty}} \\
&= \mathbf{f}^{1\cap\infty}(1)\|\mathbf{m}_f\|^O_{\Phi_{1+\infty}} \\
&= \|\mathbf{m}_f\|^O_{\Phi_{1+\infty}} \\
&= \|f\|^O_{\Phi_{1+\infty}}.
\end{aligned}$$

Recall that we already know that $\|f\|^O_{\Phi_{1+\infty}} \leq \|f\|_{1+\infty}$. Hence, the norms are equal as claimed. $\qquad\square$

We close this chapter by justifying our earlier claim that $L^{1\cap\infty}(\mathcal{M},\tau)$ and $L^{1+\infty}(\mathcal{M},\tau)$ represent, respectively, the smallest and largest of all Orlicz spaces.

Theorem 6.65 *Let Φ be a Young function. Then, $L^{1\cap\infty}(\mathcal{M},\tau) \hookrightarrow L^{\Phi}(\mathcal{M},\tau) \hookrightarrow L^{1+\infty}(\mathcal{M},\tau)$ makes sense in the sense that $L^{1\cap\infty}(\mathcal{M},\tau)$ continuously injects into $L^{\Phi}(\mathcal{M},\tau)$, and $L^{\Phi}(\mathcal{M},\tau)$ continuously injects into $L^{1+\infty}(\mathcal{M},\tau)$.*

Proof. By definition, Φ is finite on some interval $[0,\delta]$. So, by convexity, there must exist some $K > 0$ so that $\Phi(t) \leq Kt$ for all $t \in [0,\delta]$. Equivalently $\Phi(\delta t) \leq K\delta t$ for all $t \in [0,1]$. It then clearly follows that $\Phi(\delta t) \leq (K\delta)\Phi_{1\cap\infty}(t)$ for all $t \geq 0$. We may of course assume that $K\delta \geq 1$. The fact that $\Phi_{1\cap\infty}$ is both convex and 0 at 0 then ensures that $\Phi(\delta t) \leq (K\delta)\Phi_{1\cap\infty}(t) \leq \Phi_{1\cap\infty}(K\delta t)$ for all $t \geq 0$. Equivalently, $\Phi(s) \leq \Phi_{1\cap\infty}(Ks)$ for all $s \geq 0$. So, if for some $f \in L^{1\cap\infty}(\mathcal{M},\tau)$ we are given an $\epsilon > 0$ for which $\tau(\Phi_{1\cap\infty}(\epsilon^{-1}|f|)) \leq 1$, we must then have that $\tau(\Phi((K\epsilon)^{-1}|f|)) \leq \tau(\Phi_{1\cap\infty}(\epsilon^{-1}|f|)) \leq 1$. This ensures that $\|f\|_{\Phi} \leq K\|f\|_{1\cap\infty}$, proving the first claim.

For the remaining injection, we show that if the above situation pertains, then $L_{\Phi^*}(\mathcal{M},\tau)$ continuously injects into $L^{1+\infty}(\mathcal{M},\tau)$. So choose $g \in L_{\Phi^*}(\mathcal{M},\tau)$ and assume that f is a non-zero element of $L^{1\cap\infty}(\mathcal{M},\tau)$. Using what we have just proven, it then follows that

$$\frac{\tau(|gf|)}{\|f\|_{1\cap\infty}} \leq K\frac{\tau(|gf|)}{\|f\|_{\Phi}} \leq \|g\|^O_{\Phi^*}.$$

By the preceding theorem, taking the supremum over all non-zero elements f of $L^{1\cap\infty}(\mathcal{M},\tau)$ yields the fact that $\|g\|_{1+\infty} \leq K\|g\|^O_{\Phi^*}$. Since $g \in L_{\Phi^*}(\mathcal{M},\tau)$ was arbitrary, we are done. $\qquad\square$

6.4 A brief history of noncommutative Orlicz spaces

Noncommutative Orlicz spaces were first mentioned by Muratov in 1978 [**Mur78**, **Mur79**]. Using the technology of Segal measurable operators, he introduced versions of these spaces for finite von Neumann algebras equipped with a tracial state for both the Orlicz and Luxemburg norm. The theory for tracial semifinite algebras really started taking off with the appearance of the papers, [**DDdP89**] and [**Kun90**], which appeared almost simultaneously. The paper [**DDdP89**] of Dodds, Dodds and de Pagter provided the framework for focused study of non-commutative rearrangement invariant Banach function spaces, which we will eventually meet in Section 7.2. All Orlicz spaces are of course examples of noncommutative rearrangement invariant Banach function spaces and hence are by default included in this study as special cases. As we shall see in chapter 7, the key construct on which this by now burgeoning theory is built is the noncommutative decreasing rearrangement. By contrast, the paper of Kunze [**Kun90**] focused specifically on Orlicz spaces and made no use whatsoever of decreasing rearrangements. We briefly outline his approach.

Let $\Phi\colon [0,\infty) \to [0,\infty]$ be a given Young function. This function may be extended to a function defined on all of $\mathbb{R}$ by simply defining the values of $(-\infty, 0)$ to be $\Phi(x) = \Phi(-x)$ for all $x \in (-\infty, 0)$. Having done this Kunze [**Kun90**] then defined the noncommutative Orlicz space associated with Φ to be

$$L^{\varphi}(\mathcal{M},\tau) = \cup_{n=1}^{\infty} n\{f \in \widetilde{\mathcal{M}} : \tau(\varphi(|f|)) \leq 1\}$$

and went on to show that this a linear space which becomes a Banach space when equipped with the Luxemburg–Nakano norm

$$\|f\|_{\varphi} = \inf\{\lambda > 0 : \tau(\varphi(|f|/\lambda)) \leq 1\}.$$

Using the linearity, it is not hard to see that Kunze's spaces may also be realised as

$$L^{\varphi}(\mathcal{M},\tau) = \{f \in \widetilde{\mathcal{M}} : \tau(\varphi(\lambda|f|)) < \infty \quad \text{for some} \quad \lambda = \lambda(f) > 0\}.$$

It now clearly follows from Theorem 6.40 and Corollary 6.42 that the two approaches lead to exactly the same spaces.

Remarkably, it was not until the appearance of [**LM11**] more than a decade later that this fact was explicitly noted. In the intervening period, these papers gave rise to related but somewhat distinct traditions. On the one hand, the paper [**DDdP89**] was the start of a systematic study of noncommutative rearrangement invariant Banach function spaces. On the other hand, the early 2000s saw the emergence of a specific focus on the utility of Orlicz space geometry for clarifying entropy strongly driven in its early years by Streater and Zegarlinski. It seems to have been exactly the paper of Kunze [**Kun90**] that in its early years formed the foundation of this effort. The difference in starting points and in focus between these fields meant that in this initial phase of 'separate development' somewhat distinct proof techniques were developed in the two fields which may yet lead to future mutual enrichment.

We briefly pause to recount the concerns that drove this programme. Following Streater [**Str04**, **Str08**], the argument in favour of Orlicz space geometry is based on the fact that for any given norm 1 element ϱ_0 of $\mathscr{S}_1(H)_+$ for which $\mathrm{Tr}(\varrho_0 \log(\varrho_0))$, one can find plenty of states with infinite entropy in any neighbourhood of ϱ_0. When this behaviour is considered alongside the thermodynamical rule that as a state function entropy should be increasing in time, this then leads to serious difficulties in explaining the phenomenon of return to equilibrium. Additional more sophisticated arguments supporting a focused study of Orlicz space geometry can be deduced from the hypercontractivity of quantum maps and log Sobolev techniques (see [**OZ99**] and Zegarlinski's lecture in [**Zeg02**]). The field has since grown in both scope of enquiry and footprint. The footprint now boasts contributions from as far afield as Iran [**Sad12**] with a more accurate description of the current scope of enquiry being something along the lines of 'Applications of Young function based convex analysis and Orlicz-like spaces'.

For a definitely non-exhaustive sampling of the development of the theory and the current issues being engaged, we refer the reader to [**Jen06**, **Zeg08**, **ARZ11**, **ARZ14**, **Zeg15**, **RZ22**, **WZ22**]. It is beyond the scope of the present work to give a full account of this field of enquiry, and we apologise if researchers worthy of mention have been unintentionally slighted.

Chapter 7
Real interpolation and monotone spaces

The dual objective of this chapter is to first show that the classical Marcinkiewicz interpolation theorem is also valid for the spaces $L^p(\mathcal{M}, \tau)$ and second to characterise the exact interpolation spaces of the pair $(L^1(\mathcal{M}, \tau), L^\infty(\mathcal{M}, \tau))$. These exact interpolation spaces turn out to be what one may call noncommutative monotone Riesz–Fischer spaces. The very first instance of the application of the real method to noncommutative L^p-spaces dates back to the 1975 paper of Peetre and Sparr [**PS75**] who showed that the Segal–Dixmier style L^p-spaces [**Dix53**, **Seg53**] form an interpolation scale. With the material presented in this chapter, the theory has in some sense come the full circle in that in Section 7.2, we here apply the K-method of interpolation to the so-called monotone interpolation spaces of the pair $(L^1(\mathcal{M}, \tau), L^\infty(\mathcal{M}, \tau))$ to actually define a large class of noncommutative Banach function spaces.

7.1 A noncommutative Marcinkiewicz interpolation theorem

We start our brief foray into the analysis of the real method of interpolation by presenting a noncommutative version of the Marcinkiewicz interpolation theorem. Throughout this section, $\mathcal{M}$ will be a semifinite von Neumann algebra equipped with a faithful normal semifinite trace τ. We start with a lemma showing how to compute the p-norm of elements of $L^p(\mathcal{M}, \tau)$ using distributions.

Lemma 7.1 *For any $f \in L^p(\mathcal{M}, \tau)$ where $1 \leq p < \infty$, we have that*
$$\|f\|_p^p = \tau(|f|^p) = \int_0^\infty p t^{p-1} d_f(t)\, dt$$

Noncommutative measures and L^p and Orlicz Spaces, with Applications to Quantum Physics. Stanisław Goldstein and Louis Labuschagne, Oxford University Press. © Stanisław Goldstein and Louis Labuschagne (2025).
DOI: 10.1093/oso/9780198950202.003.0009

Proof.   Let $\lambda \mapsto e_\lambda$ be the spectral resolution of $|f|$. The quantity $\tau(|f|^p)$ may then be written as

$$\tau(|f|^p) = \int_0^\infty \lambda^p \, d\tau(e_\lambda).$$

(This equality is clearly true for positive simple functions. The equality for general elements of $f \in L^p(\mathcal{M}, \tau)$, therefore, follows by using the Borel functional calculus to select a sequence of such functions increasing to $|f|$ and then appealing to the normality of both the trace and the integral.) We may now use Fubini's theorem to see that

$$\begin{aligned}
\int_0^\infty \lambda^p \, d\tau(e_\lambda) &= \int_0^\infty \left(\int_0^\lambda pt^{p-1} \, dt \right) d\tau(e_\lambda) \\
&= \int_0^\infty \left(\int_0^\infty \chi_{[0,\lambda]}(t) pt^{p-1} \, dt \right) d\tau(e_\lambda) \\
&= \int_0^\infty pt^{p-1} \left(\int_0^\infty \chi_{[0,\lambda]}(t) \, d\tau(e_\lambda) \right) dt \\
&= \int_0^\infty pt^{p-1} d_f(t) \, dt.
\end{aligned}$$

$\square$

Definition 7.2 We say that a mapping $T : L^p(\mathcal{M}, \tau) \supseteq \mathrm{dom}(T) \to L^p(\mathcal{M}, \tau)$ is sublinear if:

- $|T(\alpha f)| = |\alpha| \, |T(f)|$ for all $\alpha \in \mathbb{C}$, and all $f \in \mathrm{dom}(T)$;
- for any $f, g \in \mathrm{dom}(T)$, there exist partial isometries $u, v \in \mathcal{M}$ such that $|T(f + g)| \leq u|T(f)|u^* + v|T(g)|v^*$.

We say that such an operator is of *weak type* p where $1 \leq p < \infty$, if there exists a constant $C_p > 0$ such that

$$\lambda^p d_{T(f)}(\lambda) \leq C_p^p \|f\|_p^p \text{ for all } f \in \mathrm{dom}(T), \lambda \in [0, \infty).$$

The operator is of *strong type* p if there exists a constant $C_p > 0$ such that

$$\|T(f)\|_p \leq C_p \|f\|_p \text{ for all } f \in \mathrm{dom}(T).$$

Note that by Lemma 5.2, any linear operator on $L^p(\mathcal{M}, \tau)$ is sublinear.

We briefly pause to show that the concept of weak type may equivalently be described by rearrangements.

Proposition 7.3   *Let* Ψ *be a Young function, and* $\mathbf{f} = \mathbf{f}^{\Psi}(t) = 1/\Psi^{-1}(1/t)$ *the associated fundamental function corresponding to the Luxemburg-Nakano norm on* $L^{\Psi}(0,\infty)$. *For any* $f \in \widetilde{\mathcal{M}}$, *we then have that* $\sup\limits_{t>0} \mathbf{m}_f(t)\mathbf{f}(t) = \sup\limits_{s>0} s\mathbf{f}(d_f(s))$.

Proof.   Given $s > 0$, pick $\epsilon \in (0, s)$. Since $t \mapsto d_f(s)$ is non-decreasing, we have that $s \leq t \Rightarrow d_f(s) \geq d_f(t)$. So, by Proposition 5.7, $\mathbf{m}_f(d_f(s) - \epsilon) > s$. Hence

$$\sup_{t>0} \mathbf{m}_f(t)\mathbf{f}(t) \geq \mathbf{m}_f(d_f(s) - \epsilon)\mathbf{f}(d_f(s) - \epsilon) > s\mathbf{f}(d_f(s) - \epsilon).$$

Given that $\mathbf{f}$ is continuous on $(0, \infty)$, letting $\epsilon \to 0$ will yield the conclusion that

$$\sup_{t>0} \mathbf{m}_f(t)\mathbf{f}(t) \geq s\mathbf{f}(d_f(s))$$

for all $s > 0$.

Conversely, given $t > 0$ such that $\mathbf{m}_f(t) > 0$, we may select $\epsilon \in (0, \mathbf{m}_f(t))$. It is a simple exercise to conclude from Proposition 5.7 that $d_f(s) > t$ whenever $s < \mathbf{m}_f(t) = \inf\{r \geq 0 : d_f(r) \leq t\}$. Hence $d_f(\mathbf{m}_f(t) - \epsilon) > t > 0$, implying that

$$\sup_{s>0} s\mathbf{f}(d_f(s)) \geq (\mathbf{m}_f(t) - \epsilon)\mathbf{f}(d_f(\mathbf{m}_t(f) - \epsilon)) \geq (\mathbf{m}_f(t) - \epsilon)\mathbf{f}(t)$$

since $\mathbf{f}$ is nondecreasing. Letting $\epsilon \to 0$, we obtain

$$\sup_{s>0} s\mathbf{f}(d_f(s)) \geq \mathbf{m}_f(t)\mathbf{f}(t)$$

for all $t > 0$. If, on the other hand, $\mathbf{m}_f(t) = 0$, this inequality is trivially satisfied. So, in general, we have $\sup_{t>0} \mathbf{m}_f(t)\mathbf{f}(t) \leq \sup_{s>0} s\mathbf{f}(d_f(s))$. $\square$

Theorem 7.4   *Let* $T : (L^1 \cap L^{\infty})(\mathcal{M}, \tau) \supseteq \mathrm{dom}(T) \to \widetilde{\mathcal{M}}$ *be a sublinear operator of weak types* p *and* q, *where* $1 \leq p < q < \infty$. *Then, T is of strong type* r *for every* $r \in (p, q)$

Proof.   Let $f \in \mathrm{dom}(T)$ and $r \in (p, q)$ be given, and let $\lambda \mapsto e_\lambda$ be the spectral resolution of $|f|$. By hypothesis, there exist constants C_p and C_q so that

$$\lambda^p d_{T(f)}(\lambda) \le C_p^p \|f\|_p^p \text{ and } \lambda^q d_{T(f)}(\lambda) \le C_q^q \|f\|_q^q \text{ for all } f \in \mathrm{dom}(T),$$
$$\lambda \in [0, \infty).$$

We now set $C = C_p^{\frac{p}{q-p}} C_q^{\frac{-q}{q-p}}$. Given $f \in \mathrm{dom}(T)$, the sublinearity of T ensures that there exist partial isometries u and v so that

$$|T(f)| \le u|T(f\chi_{[0,Ct]}(|f|))|u^* + v|T(f\chi_{(Ct,\,\infty)}(|f|))|v^*.$$

By Proposition 5.3, we then have that

$$d_{T(f)}(2t) \le d_{T(f\chi_{[0,Ct]}(|f|))}(t) + d_{T(f\chi_{(Ct,\infty)}(|f|))}(t) \text{ for all } t > 0.$$

We may now combine the first and third centred inequalities of the proof to see that

$$d_{T(f)}(2t) \le t^{-p} C_p^p \|f\chi_{[0,Ct]}(|f|)\|_p^p + t^{-q} C_q^q \|f\chi_{(Ct,\infty)}(|f|)\|_q^q$$
$$= t^{-p} C_p^p \tau((|f|\chi_{[0,Ct]}(|f|))^p) + t^{-q} C_q^q \tau((|f|\chi_{(Ct,\infty)}(|f|))^q)$$
$$= t^{-p} C_p^p \int_0^\infty \lambda^p \chi_{[0,Ct]}(\lambda)\, d\tau(e_\lambda)$$
$$+ t^{-q} C_q^q \int_0^\infty \lambda^q \chi_{(Ct,\infty)}(\lambda)\, d\tau(e_\lambda).$$

Next apply the lemma to the above inequality to see that

$$\|T(f)\|_r^r = \int_0^\infty r t^{r-1} d_{T(f)}(t)\, dt$$
$$= r2^r \int_0^\infty t^{r-1} d_{T(f)}(2t)\, dt$$
$$\le r2^r \int_0^\infty t^{r-1} t^{-p} C_p^p \left(\int_0^\infty \lambda^p \chi_{(Ct,\infty)}(\lambda)\, d\tau(e_\lambda) \right) dt$$
$$+ r2^r \int_0^\infty t^{r-1} t^{-q} C_q^q \left(\int_0^\infty \lambda^q \chi_{[0,Ct]}(\lambda)\, d\tau(e_\lambda) \right) dt.$$

On using Fubini's theorem to change the order of integration, it then follows that

$$\|T(f)\|_r^r \le r2^r C_p^p \int_0^\infty \lambda^p \left(\int_0^\infty t^{r-p-1} \chi_{(0,\lambda/C)}(t)\, dt \right) d\tau(e_\lambda)$$

$$+ r2^r C_q^q \int_0^\infty \lambda^q \left(\int_0^\infty t^{r-q-1} \chi_{[\lambda/C,\infty)}(t)\, dt \right) d\tau(e_\lambda)$$

$$= r2^r \left(C_p^p C^{r-p} \cdot \frac{1}{r-p} + C_q^q C^{q-r} \cdot \frac{1}{q-r} \right) \int_0^\infty \lambda^r\, d\tau(e_\lambda)$$

$$= r2^r (C_p^p)^{\frac{q-r}{q-p}} (C_q^q)^{\frac{r-q}{q-p}} \left(\frac{1}{r-p} + \frac{1}{q-r} \right) \|f\|_r^r.$$

This then proves the theorem. $\qquad\square$

7.2 Noncommutative Banach function spaces and exact interpolation spaces of $(L^1(\mathcal{M},\tau), L^\infty(\mathcal{M},\tau))$

Our objective in this part of the chapter is to present that part of the theory of real interpolation that specifically relates to the theory of (noncommutative) Banach function spaces. The first description of the framework that would later become the foundation for the theory of non commutative Banach function spaces was given by von Neumann himself in [**VN37**]. The genesis of the modern theory of noncommutative symmetric and Banach function spaces may, however, be traced back to [**Med87, DDdP89**] with [**DDdP89**] positing a really robust framework for a noncommutative theory of Banach function spaces.

We will make absolutely no attempt to give a comprehensive introduction to real interpolation. For that, the interested reader may refer to [**BS88**] for the commutative theory and [**DDdP92**] for the noncommutative theory. We will rather focus on the K-method, more specifically, on monotone interpolation spaces and the K-method, since it is that theory that ties in very closely with the theory of Banach function spaces. Results that form part of the 'standard' theory of monotone interpolation spaces will be stated without proof. However, those results that show how this theory specialises to the setting of noncommutative Banach function spaces will be proven in full.

7.2.1 Basic concepts

Definition 7.5 Two Banach spaces X_0 and X_1 are called a compatible couple or Banach couple, denoted by (X_0, X_1), if there exists a Hausdorff topological vector space $\mathfrak{X}$ in which each of X_0 and X_1 are continuously embedded. For the sake of simplicity, we will consistently identify X_0 and X_1 with their images inside $\mathfrak{X}$.

For such a compatible couple (X_0, X_1) with associated Hausdorff topological vector superspace $\mathfrak{X}$, we may define the spaces $X_0 + X_1$ and $X_0 \cap X_1$ as follows:

- The space $X_0 + X_1$ is defined to be the formal sum of X_0 and X_1 inside $\mathfrak{X}$, that is all elements x of $\mathfrak{X}$, which are representable in the form $x = x_0 + x_1$ where $x_0 \in X_0$, $x_1 \in X_1$. For each $x \in X_0 + X_1$, we may set

$$\|x\|_{X_0+X_1} = \inf\{\|x_0\|_{X_0} + \|x_1\|_{X_1} : x = x_0 + x_1\}$$

  where the infimum is take over decompositions of x in the form $x = x_0 + x_1$ where $x_0 \in X_0$, $x_1 \in X_1$.
- For each element x in the intersection $X_0 \cap X_1$ of X_0 and X_1, we may set

$$\|x\|_{X_0\cap X_1} = \max\{\|x\|_{X_0}, \|x\|_{X_1}\}.$$

It is an easy exercise to see that the quantities $x \mapsto \|x\|_{X_0+X_1}$ and $x \mapsto \|x\|_{X_0\cap X_1}$ respectively define norms on $X_0 + X_1$ and $X_0 \cap X_1$. But in fact more is true. See, for example, the following theorem.

Theorem 7.6 ([BS88, Theorem III.1.3]) *Let (X_0, X_1) be a compatible couple of Banach spaces. Then, the spaces $X_0 + X_1$ and $X_0 \cap X_1$ are Banach spaces under the respective norms $x \mapsto \|x\|_{X_0+X_1}$ and $x \mapsto \|x\|_{X_0\cap X_1}$.*

We are now ready to introduce the key notions of admissible operators, intermediate spaces, and interpolation spaces for a compatible couple (X_0, X_1).

Definition 7.7 Let (X_0, X_1) and (Y_0, Y_1) be two compatible couples of Banach spaces, and let T be a linear operator from $X_0 + X_1$ into $Y_0 + Y_1$. Then, T is called an *admissible* operator if the restrictions of T to each of X_0 and X_1, respectively, induce bounded operators from

X_0 to Y_0, and from X_1 to Y_1. The space of all admissible operators for the two couples (X_0, X_1) and (Y_0, Y_1) is denoted by $\mathcal{A}(X_0, X_1; Y_0, Y_1)$ and may be normed by

$$\|T\|_{\mathcal{A}} = \max\{\|T\|_{B(X_0, Y_0)}, \|T\|_{B(X_1, Y_1)}\}.$$

In the case where $X_0 = Y_0$ and $X_1 = Y_1$, we simply write $\mathcal{A}(X_0, X_1)$. The unit ball of $\mathcal{A}(X_0, X_1)$ will be denoted by $\Sigma(X_0, X_1)$, or simply Σ if the pair (X_0, X_1) is understood.

Definition 7.8 Let (X_0, X_1) be a compatible couple of Banach spaces. A Banach space X is said to be an *intermediate space* of the pair (X_0, X_1) if $X_0 \cap X_1$ embeds continuously into X, and X embeds continuously into $X_0 + X_1$. We may pictorially write this requirement as

$$X_0 \cap X_1 \hookrightarrow X \hookrightarrow X_0 + X_1.$$

Intermediate spaces X which satisfy the additional property that every admissible operator $T \in \mathcal{A}(X_0, X_1; X_0, X_1)$ maps X back into itself are called *interpolation spaces*.

A very important property of admissible operators is that they have a continuous action on each interpolation space.

Proposition 7.9 ([BS88, Proposition III.1.7]) *Let (X_0, X_1) be a compatible couple of Banach spaces. For every interpolation space X, there exists a constant C depending only on the spaces X, X_0 and X_1, such that $\|T\|_{B(X,X)} \leq C\|T\|_{\mathcal{A}}$ for every admissible operator T.*

Definition 7.10 Let (X_0, X_1) be a compatible couple of Banach spaces. An interpolation space X is called an exact interpolation space for the couple (X_0, X_1), if the constant C in Proposition 7.9 can be taken as $C = 1$.

Remark 7.11 It is a fairly straightforward consequence of the definition that the norm on any exact interpolation space X of a compatible couple (X_0, X_1) of Banach spaces is unique. To see this observe that if, say, $\|\cdot\|_a$ and $\|\cdot\|_b$ are two norms with respect to which X is exact, the fact that the identity operator belongs to Σ then ensures that for each $x \in X$, we have that $\|x\|_a \leq \|x\|_b$ and $\|x\|_b \leq \|x\|_a$.

7.2.2 The K- and k-functionals

In this subsection, we introduce the reader to the basic concepts surrounding the K-method.

Definition 7.12 Let (X_0, X_1) be a compatible couple of Banach spaces. The K-functional $K : (X_0 + X_1) \times (0, \infty) \to [0, \infty)$ is for each $f \in X_0 + X_1$ and each $t > 0$ defined by

$$K(f, t; X_0, X_1) = \inf\{\|f_0\|_{X_0} + t\|f_1\|_{X_1} : f = f_0 + f_1\}$$

where for each $t > 0$, the infimum is taken over decompositions of f in the form $f = f_0 + f_1$ where $f_0 \in X_0$, $f_1 \in X_1$. Where the context is obvious, we will simply write $K(f, t)$ instead of $K(f, t; X_0, X_1)$

Proposition 7.13 ([BS88, Proposition V.1.2]) *Let (X_0, X_1) be a compatible couple of Banach spaces. For each $f \in X_0 + X_1$, the K-functional $K(f, t; X_0, X_1)$ is a non-negative concave function on $(0, \infty)$ satisfying*

$$t^{-1}K(f, t; X_0, X_1) = K(f, t^{-1}; X_1, X_0).$$

Moreover, $t \mapsto K(f, t; X_0, X_1)$ is increasing on $(0, \infty)$, and $t \mapsto t^{-1}K(f, t; X_0, X_1)$ decreasing.

The property regarding the K-functional noted in the preceding theorem now enables us to introduce the so-called k-functional.

Definition 7.14 Let (X_0, X_1) be a compatible couple of Banach spaces. The concavity of $t \mapsto K(f, t; X_0, X_1)$ on $[0, \infty)$ for each fixed $f \in X_0 + X_1$ ensures that $K(f, t; X_0, X_1)$ may be written in the form

$$K(f, t; X_0, X_1) = K(f, 0+; X_0, X_1) + \int_0^t k(f, s; X_0, X_1)\, ds$$

for some unique non-negative, right-continuous, decreasing function $k(f, s; X_0, X_1)$ of $s > 0$. We call the function $k(f, s; X_0, X_1)$ the k-functional.

We pause to characterise those K-functionals which may be reconstructed from their k-functionals.

Theorem 7.15 ([BS88, Theorem V.1.15]) *Let (X_0, X_1) be a compatible couple of Banach spaces, and let $f \in X_0 + X_1$ be given. Then, the following are equivalent:*

(1) $K(f, 0+; X_0, X_1) = 0$;
(2) $K(f, t; X_0, X_1) = \int_0^t k(f, s; X_0, X_1)\, ds$;
(3) $f \in \overline{(X_0 \cap X_1)}^{X_0} + X_1$;
(4) $f \in \overline{X_1}^{X_0 + X_1}$.

In particular, any one (and therefore all) of these conditions holds for each $f \in X_0 + X_1$, if and only if $(X_0 \cap X_1)$ is dense in X_0.

Note that we may conclude from the final claim above that $K(f, 0+; L^1, L^\infty) = 0$ for each $f \in L^1 + L^\infty$.

7.2.3 Monotone interpolation spaces

In this subsection, we will review the basic ingredients of the theory of monotone interpolation spaces, before in the next seeing how this theory unfolds in the noncommutative setting. As it turns out, monotone interpolation spaces are closely linked to Riesz–Fischer spaces. We therefore also need clarity on what exactly we mean by the term (monotone) Riesz–Fischer space.

Definition 7.16 Let $M_0^+([0, \infty), \mathrm{m})$ be the cone of a.e.-finite positive Borel measurable functions on $[0, \infty)$. A functional $\rho : M_0^+([0, \infty), \mathrm{m}) \to [0, \infty]$ is said to be a *Riesz–Fischer* norm on $([0, \infty), \mathscr{B}, \mathrm{m})$, if each of the following conditions hold for all $f, g, f_n \in M_0^+([0, \infty), \mathrm{m})$ $(n \in \mathbb{N})$ and all $\alpha > 0$.

(1) $\rho(f) = \rho(g)$ whenever f and g are equimeasurable. (Recall that f and g are equimeasurable precisely when $\mathrm{m}(\{x : |f(x)| > s\}) = \mathrm{m}(\{x : |g(x)| > s\})$ for all $s > 0$.)
(2) ρ behaves in a norm-like way in the sense that

- $\rho(f) = 0$ iff $f = 0$ a.e.;
- $\rho(f + g) \leq \rho(f) + \rho(g)$;
- $\rho(\alpha f) = \alpha \rho(f)$.

(3) $\rho(f) \leq \rho(g)$ whenever $f \leq g$ a.e.

(4) $\rho(\chi_E) < \infty$ whenever $\mathrm{m}(E) < \infty$.

(5) Whenever $\mathrm{m}(E) < \infty$, there exists a constant C_E that depends only on E and ρ, such that $\int_E f \, d\mathrm{m} \leq C_E \rho(f)$ for all f.

(6) $\rho(\sum_n f_n) \leq \sum_n \rho(f_n)$ for each sequence $(f_n) \subseteq M_0^+([0, \infty), \mathrm{m})$.

If, in addition, ρ satisfies the requirement that $\rho(f) \leq \rho(g)$ whenever $\int_0^t f^*(s) \, ds \leq \int_0^t g^*(s) \, ds$ for all $t > 0$, we say that ρ is a *monotone Riesz–Fischer norm*.

Definition 7.17 Let (X_0, X_1) be a compatible couple of Banach spaces, and let ρ be a monotone Riesz–Fischer norm on $([0, \infty), \mathscr{B}, \mathrm{m})$. The space $(X_0, X_1)_\rho$ is defined to be the space of all $f \in \overline{(X_0 \cap X_1)}^{X_0} + X_1$ for which $\rho(k(f, \cdot; X_0, X_1)) < \infty$.

The quantity $\|f\|_\rho = \rho(k(f, \cdot; X_0, X_1))$ in fact turns out to be a norm on $(X_0, X_1)_\rho$ under which $(X_0, X_1)_\rho$ is a Banach space. (The proof of this fact is contained in the proof of [**BS88**, Theorem V.1.19].) This space may be called the *Riesz–Fischer space associated with ρ* for the pair (X_0, X_1). As we shall shortly see, much more may be said about these spaces.

Definition 7.18 Let (X_0, X_1) be a compatible couple of Banach spaces. An intermediate space X of (X_0, X_1) is said to be *monotone* if and only if the condition

$$K(g, t) \leq K(f, t) \text{ for each } t > 0 \text{ where } f \in X, g \in X_0 + X_1$$

ensures that $g \in X$ with $\|g\|_X \leq \|f\|_X$.

Our first theorem shows how to construct monotone interpolation spaces.

Theorem 7.19 ([BS88, Theorem V.3.1]) *Let (X_0, X_1) be a compatible couple of Banach spaces, and let ρ be a monotone Riesz–Fischer norm on $([0, \infty), \mathscr{B}, \mathrm{m})$. Then, $X = (X_0, X_1)_\rho$ is a monotone interpolation space for the pair (X_0, X_1).*

We are now ready to state the pièce de résistance of this subsection. Under fairly mild restrictions, all monotone interpolation spaces are (up to an equivalent renorming) of the above form! The 'fairly mild restrictions' referred to essentially ensure that for a Banach couple (X_0, X_1), the scale of interpolation spaces includes the spaces X_0 and X_1 themselves.

Definition 7.20 Let (X_0, X_1) be a compatible couple of Banach spaces. This couple is defined to be a *Gagliardo couple* if:

- for any $f \in X_0 + X_1$ the quantity $\sup_{t>0} K(f,t) = \lim_{t\to\infty} K(f,t)$ is finite if and only if $f \in X_0$, in which case $\|f\|_0 = \sup_{t>0} K(f,t) = \lim_{t\to\infty} K(f,t)$;
- and similarly for any $f \in X_0 + X_1$, the quantity $\sup_{t>0} t^{-1}K(f,t) = \lim_{t\to 0} t^{-1}K(f,t)$ is finite if and only if $f \in X_1$, in which case $\|f\|_1 = \sup_{t>0} t^{-1}K(f,t) = \lim_{t\to 0} t^{-1}K(f,t)$.

Theorem 7.21 ([BS88, Theorem V.3.7])   *Let (X_0, X_1) be a Gagliardo couple. Then, X is a monotone interpolation space for the couple (X_0, X_1) if it is of one of the following forms.*

(1) *If $X \subseteq \overline{(X_0 \cap X_1)}^{X_0} + X_1$, then there is a monotone Riesz–Fischer norm ρ with respect to which X appears as an equivalent renorming of $(X_0, X_1)_\rho$.*

(2) *If $X \subseteq \overline{(X_0 \cap X_1)}^{X_1} + X_0$, then there is a monotone Riesz–Fischer norm ρ with respect to which X appears as an equivalent renorming of $(X_1, X_0)_\rho$.*

(3) *$X = X_0 + X_1$ otherwise.*

We close this subsection by establishing a link between the above results and the exactness of interpolation spaces. The proof of this result is essentially an adaptation of the proof of [**BS88**, Theorem III.2.1]. For the sake of the reader, we provide full details of the proof.

Theorem 7.22   *Let (X_0, X_1) be a compatible couple of Banach spaces. For any $T \in \mathcal{A}(X_0, X_1)$ and any $x \in X_0 + X_1$, we will then have $K(T(x), t; X_0, X_1) \leq \|T\|_{\mathcal{A}} K(x, t; X_0, X_1)$ for all $t > 0$. If the pair (X_0, X_1) is in fact a Gagliardo couple, then any linear operator T on $X_0 + X_1$ that satisfies the condition that there exists some $C > 0$ such that $K(T(x), t; X_0, X_1) \leq CK(T(x), t; X_0, X_1)$ for all $t > 0$ and for all $x \in X_0 + X_1$, will belong to $\mathcal{A}(X_0, X_1)$ with $\|T\|_{\mathcal{A}} \leq C$.*

Proof.   Suppose that we are given $T \in \mathcal{A}(X_0, X_1)$ and $x \in X_0 + X_1$. Select $x_0 \in X_0$ and $x_1 \in X_1$ such that $x = x_0 + x_1$, and let $t > 0$ be given.

It then follows that

$$
\begin{aligned}
K(T(x), t; X_0, X_1)) &= \inf\{\|y_0\|_{X_0} + t\|y_1\|_{X_1} : T(x) = y_0 + y_1\} \\
&\leq \|T(x_0)\|_{X_0} + t\|T(x_1)\|_{X_1} \\
&\leq \|T\|_{B(X_0)}\|x_1\|_{X_0} + t\|T\|_{B(X_1)}\|x_\infty\|_{X_1} \\
&\leq \max(\|T\|_{B(X_0)}, \|T\|_{B(X_1)})(\|x_1\|_{X_0} + t\|x_\infty\|_{X_1}) \\
&\leq \|T\|_{\mathcal{A}}(\|x_1\|_{X_0} + t\|x_\infty\|_{X_1}).
\end{aligned}
$$

If we now take the infimum over all decompositions of the form $x = x_0 + x_1$, we obtain the fact that $K(T(x), t; X_0, X_1) \leq \|T\|_{\mathcal{A}} K(x, t; X_0, X_1)$.

Conversely, suppose that (X_0, X_1) is a Gagliardo couple and that T is a linear operator on $X_0 + X_1$ for which we have that there exists some $C > 0$ such that $K(T(x), t; X_0, X_1) \leq CK(T(x), t; X_0, X_1)$ for all $t > 0$ and all $x \in X_0 + X_1$. First, observe that setting $t = 1$, we have that

$$
\|T(x)\|_{X_0 + X_1} = K(T(x), 1; X_0, X_1) \leq CK(x, 1; X_0, X_1) = C\|x\|_{X_0 + X_1}
$$

for all $x \in X_0 + X_1$. So, T is, in fact, a bounded operator on $X_0 + X_1$. Now suppose that $x \in X_0$. Since (X_0, X_1) is a Gagliardo couple, we must have $\|x\|_{X_0} = \sup_{t>0} K(x, t; X_0, X_1)$. But then

$$
\sup_{t>0} K(T(x), t; X_0, X_1) \leq C \sup_{t>0} K(x, t; X_0, X_1) = C\|x\|_{X_0} < \infty.
$$

By definition, this ensures that $T(x) \in X_0$ with $\|T(x)\|_{X_0} \leq C\|x\|_{X_0}$. Thus, T acts boundedly on X_0 with $\|T\|_{B(X_0)} \leq C$. Using the fact that $\|x\|_{X_1} = \sup_{t>0} t^{-1} K(x, t; X_0, X_1)$ whenever $x \in X_1$, we may similarly prove that T acts boundedly on X_1 with $\|T\|_{B(X_1)} \leq C$. This then clearly proves the claim. $\square$

Corollary 7.23   *Let (X_0, X_1) be a compatible couple of Banach spaces. Every space which is of the form $(X_0, X_1)_\rho$ for some monotone Riesz– Fischer norm is an exact interpolation space. If (X_0, X_1) is in fact a Gagliardo couple, then every exact monotone interpolation space will be of one of the following forms.*

(1) *If $X \subseteq \overline{(X_0 \cap X_1)}^{X_0} + X_1$, then there is a monotone Riesz–Fischer norm ρ such that $X = (X_0, X_1)_\rho$.*

(2) *If $X \subseteq \overline{(X_0 \cap X_1)}^{X_1} + X_0$, then there is a monotone Riesz–Fischer norm ρ such that $X = (X_1, X_0)_\rho$.*

(3) $X = X_0 + X_1$ *otherwise.*

7.2.4 The exact interpolation spaces of $(L^1(\mathcal{M}, \tau), L^\infty(\mathcal{M}, \tau))$

Thus far, we have worked in a very general context. Having laid the necessary groundwork, we are now finally ready to address the primary objective of this chapter, which is to describe the exact interpolation spaces of the pair $(L^1(\mathcal{M}, \tau), L^\infty(\mathcal{M}, \tau))$. Through this description, the reader will be introduced to a large class of noncommutative Banach function spaces. Readers wishing to have a fuller account of the theory of noncommutative Banach function spaces will find the monograph of Dodds, de Pagter and Sukochev [**DdPS23**] helpful. Returning to the matter at hand, since each of $L^1(\mathcal{M}, \tau)$ and $L^\infty(\mathcal{M}, \tau)$ trivially embeds into the space $\widetilde{\mathcal{M}}$ of τ-measurable operators, this pair is easily seen to be a compatible couple of Banach spaces. In our analysis of this pair, we now re-interpret Theorem 6.63. In terms of the current focus, we may reformulate that theorem as follows.

Theorem 7.24 *For the pair $(L^1(\mathcal{M}, \tau), L^\infty(\mathcal{M}, \tau))$, we have $K(f, t) = \int_0^t \mathbf{m}_f(s)\, ds$. In other words, for every $f \in (L^1 + L^\infty)(\mathcal{M}, \tau)$, we have $k(f, t) = \mathbf{m}_f(t)$ and $K(f, 0+) = 0$.*

Corollary 7.25 *The space $L^1(\mathcal{M}, \tau) \cap L^\infty(\mathcal{M}, \tau)$ is dense in $L^1(\mathcal{M}, \tau)$, and the pair $(L^1(\mathcal{M}, \tau), L^\infty(\mathcal{M}, \tau))$ is a Gagliardo couple.*

Proof. The first claim follows by applying the preceding theorem to Theorem 7.15. The second claim is left as an exercise. $\square$

As shown by the next very important theorem, the pair $(L^1(\mathcal{M}, \tau), L^\infty(\mathcal{M}, \tau))$ behaves very well with respect to monotone interpolation.

Theorem 7.26 *The exact interpolation spaces of the pair $(L^1(\mathcal{M}, \tau), L^\infty(\mathcal{M}, \tau))$ are all automatically monotone.*

The proof of the above theorem requires a fairly lengthy systematic development of suitable technology. Before expounding the proof of this theorem, we first indicate some of its important consequences.

Corollary 7.27 (Existence of noncommutative Riesz–Fischer spaces)   *A Banach space X is an exact interpolation space of the pair $(L^1(\mathcal{M},\tau), L^\infty(\mathcal{M},\tau))$ if and only if it is of the form $X = (L^1(\mathcal{M},\tau), L^\infty(\mathcal{M},\tau))_\rho$ for some monotone Riesz–Fischer norm ρ where*

$$(L^1(\mathcal{M},\tau), L^\infty(\mathcal{M},\tau))_\rho = \{f \in (L^\infty + L^1)(\mathcal{M},\tau) \colon \rho(\mathbf{m}_f) < \infty\}$$

with the norm given by $\|f\|_\rho = \rho(\mathbf{m}_f)$.

Proof.   The corollary is an obvious consequence of the preceding theorem considered alongside Corollary 7.23. The density of $L^1 \cap L^2$ in L^1 of course greatly simplifies the application of Corollary 7.23 in the present context. $\square$

Given a monotone Riesz–Fischer norm ρ, in the following, we write $L^\rho(\mathcal{M},\tau)$ for $(L^1(\mathcal{M},\tau), L^\infty(\mathcal{M},\tau))_\rho$.

Corollary 7.28   *Each monotone Riesz–Fischer space $L^\rho(\mathcal{M},\tau)$ embeds continuously into $(L^1 + L^\infty)(\mathcal{M},\tau)$ and hence into the τ-measurable operators.*

We proceed with the task of developing the technology required to prove Theorem 7.26. The first step is to study the closed unit ball Σ of the space $\mathcal{A}(L^1(\mathcal{M},\tau), L^\infty(\mathcal{M},\tau))$ of admissible operators for the pair $(L^1(\mathcal{M},\tau), L^\infty(\mathcal{M},\tau))$. So, Σ consists of all those operators T on $L^1(\mathcal{M},\tau) + L^\infty(\mathcal{M},\tau)$ that have L^1 and L^∞ as invariant subspaces and that satisfy the condition $\max(\|T\|_1, \|T\|_\infty) \leq 1$. We may, in a very natural way, equip this set with a topology under which it then appears as a compactum.

Definition 7.29   Given $T_0 \in \Sigma$, sets of the form

$$V(T_0; x_1, x_2, \ldots, x_n; y_1, y_2, \ldots, y_n; \epsilon)$$

$$= \{T \in \Sigma \colon |\tau((T(x_i) - T_0(x_i))y_i)| < \epsilon, 1 \leq i \leq n\}$$

where $\{x_i\}_{i=1}^n \in L^\infty(\mathcal{M}, \tau)$ and $\{y_i\}_{i=1}^n \in (L^1 \cap L^\infty)(\mathcal{M}, \tau)$, form a neighbourhood base at T_0. The topology generated by these neighbourhood bases will be denoted by $\mathscr{T}_\Sigma$.

Theorem 7.30 *The set Σ is a compactum under the topology $\mathscr{T}_\Sigma$.*

Proof.   The proof makes use of Tychonoff's theorem and follows well-worn paths. We start by observing that under the topology $\mathscr{T}_\Sigma$, the space Σ homeomorphically embeds into the compactum $\Pi(I_{xy})$ consisting of the product of the cells $I_{xy} = \{s \in \mathbb{C} \colon |s| \leq \|x\|_\infty \cdot \|y\|_{1 \cap \infty}\}$, by means of the prescription $T \mapsto \rho_T$, where $\rho_T(x, y) = \tau(T(x)y)$. If therefore we can show that as a subspace of this product Σ is closed, the claim will follow from Tychonoff's theorem. To this end, suppose that (T_λ) is a net in Σ for which (ρ_{T_λ}) converges to some $\widetilde{\rho}$ in $\Pi(I_{xy})$. We need to find some $\widetilde{T} \in \Sigma$ for which $\widetilde{\rho} = \rho_{\widetilde{T}}$.

Step 1 - Defining $\widetilde{T}$ on L^∞: Fix $x \in L^\infty$. The fact that $|\rho_{T_\lambda}(x, y)| \leq \|x\|_\infty \cdot \|y\|_1$ for each λ, also ensures that $|\widetilde{\rho}(x, y)| \leq \|x\|_\infty \cdot \|y\|_1$. Since $L^1 \cap L^\infty$ is dense in L^1 (by Proposition 6.12), the functional $\widetilde{\rho}(x, \cdot)$ has a unique norm-preserving extension to a linear functional on L^1, which we still denote by $\widetilde{\rho}(x, \cdot)$. Therefore, there must exist $z_x \in L^\infty$ such that $\widetilde{\rho}(x, y) = \tau(z_x y)$ for each $y \in L^1$. It is an exercise to see that the map $\widetilde{T}$ defined by $\widetilde{T}(x) = z_x$ is linear. Since by construction we have $|\tau(\widetilde{T}(x)y)| = |\widetilde{\rho}(x, y)| \leq \|x\|_\infty \cdot \|y\|_1$ for each $y \in L^1$, it follows that $\|\widetilde{T}(x)\|_\infty \leq \|x\|_\infty$.

Step 2 - Extending $\widetilde{T}$ to a map on L^1: Let $x \in L^1 \cap L^\infty$ be given. Since $|\rho_{T_\lambda}(x, y)| \leq \|x\|_1 \cdot \|y\|_\infty$ for each λ, we must then also have $|\widetilde{\rho}(x, y)| \leq \|x\|_1 \cdot \|y\|_\infty$. If in fact $\widetilde{T}(x) \notin L^1$, we will have $\tau(|\widetilde{T}(x)|) = \infty$. Let v be the partial isometry in the polar decomposition $\widetilde{T}(x) = v|\widetilde{T}(x)|$, and select a net (e_α) of projections with finite trace increasing to $\mathbb{1}$. (See the proof of Corollary 6.26 for the existence of such a net.) Setting $y_\alpha = e_\alpha v^*$, we then have $|\widetilde{\rho}(x, y_\alpha)| = \tau(\widetilde{T}(x)y_\alpha) = \tau(y_\alpha \widetilde{T}(x)) = \tau(e_\alpha |\widetilde{T}(x)|) \to \infty$ as $e_\alpha \to \mathbb{1}$. But this clearly contradicts the fact that $|\widetilde{\rho}(x, y_\alpha)| \leq \|x\|_1 \cdot \|y_\alpha\|_\infty \leq \|x\|_1$ for every α. Therefore, we must have that $\widetilde{T}(x) \in L^1 \cap L^\infty$, that is $\widetilde{T}$ maps $L^1 \cap L^\infty$ back into $L^1 \cap L^\infty$. This ensures that for each $x \in L^1 \cap L^\infty$, the functional $y \mapsto \widetilde{\rho}(x, y) = \tau(\widetilde{T}(x)y)$ is σ-weakly continuous on the subspace $L^1 \cap L^\infty$ of L^∞. Since $L^1 \cap L^\infty$ is σ-weakly dense in L^∞ (by Corollary 6.26), $y \mapsto \widetilde{\rho}(x, y)$ will for each $x \in L^1 \cap L^\infty$ admit a norm-preserving σ-weakly continuous extension to L^∞. This ensures that as a functional on L^1, $\tau(\widetilde{T}(x)\cdot)$ satisfies the equality $\|\tau(\widetilde{T}(x)\cdot)\| = \|\widetilde{T}(x)\|_1$. Since also $\|\tau(\widetilde{T}(x)\cdot)\| = \|\widetilde{\rho}(x,\cdot)\| \le \|x\|_1$, it is clear that $\widetilde{T}$ acts contractively on $L^1 \cap L^\infty$ with respect to the L^1-norm. But then the norm density of $L^1 \cap L^\infty$ in L^1 ensures that $\widetilde{T}$ extends to a contractive map on L^1.

The two facts demonstrated in Steps 1 and 2 are enough to ensure that $\widetilde{T}$ also extends to a contractive map on $L^1 + L^\infty$ and hence that $\widetilde{T} \in \Sigma$ as required. $\qquad\square$

Theorem 7.31 *For any* $x \in (L^1 + L^\infty)(\mathcal{M}, \tau)$, *the orbit* $\mathcal{O}_x = \{Tx : T \in \Sigma\}$ *of* x *with respect to* Σ *is* $\sigma((L^1 + L^\infty)(\mathcal{M}, \tau), (L^1 \cap L^\infty)(\mathcal{M}, \tau))$ *closed.*

Proof. We need to show that $(\Sigma, \mathscr{T}_\Sigma) \mapsto (L^1 + L^\infty, \sigma(L^1 + L^\infty, L^1 \cap L^\infty)) : T \mapsto Tx$ is continuous. The image will then be compact and hence closed. Hence, let $x \in L^1 + L^\infty$ be given and suppose that we have a net (T_λ) in Σ converging to some $T_0 \in \Sigma$ with respect to $\mathscr{T}_\Sigma$. That means that for any $z \in L^\infty$ and $y \in (L^1 \cap L^\infty)$, we have $\rho_{T_\lambda}(z, y) \to \rho_{T_0}(z, y)$. The norm density of $L^1 \cap L^\infty$ in L^1 ensures that L^∞ is norm dense in $L^1 + L^\infty$. So, we may select $(z_n) \subseteq L^\infty$ so that $z_n \to x$ in the $L^1 + L^\infty$-norm. Given any $\epsilon > 0$, we may therefore select $N \in \mathbb{N}$ so that $\|x - z_N\| < \epsilon/(4\|y\|_{1\cap\infty} + 1)$ and then further select λ_ϵ so that $|\rho_{T_\lambda}(z_N, y) - \rho_{T_0}(z_N, y)| < \epsilon/2$ for all $\lambda \ge \lambda_\epsilon$. For any $\lambda \ge \lambda_\epsilon$, we will then have that

$$|\rho_{T_\lambda}(x, y) - \rho_{T_0}(x, y)| \le |\rho_{T_\lambda}(x - z_N, y)| + |\rho_{T_\lambda}(z_N, y) - \rho_{T_0}(z_N, y)|$$
$$+ |\rho_{T_0}(x - z_N, y)|$$
$$\le 2\|x - z_N\|_{1+\infty}\cdot\|y\|_{1\cap\infty}$$
$$+ |\rho_{T_\lambda}(z_N, y) - \rho_{T_0}(z_N, y)|$$
$$< \epsilon.$$

(Here, we make use of the fact that $|\rho_S(x, y)| = |\tau(S(x)y)| \le \|x\|_{1+\infty}\cdot\|y\|_{1\cap\infty}$ for every $S \in \Sigma$.) This proves the continuity of the embedding and hence the theorem. $\qquad\square$

We need two more theorems to complete the grand preparation for the proof of Theorem 7.26. The first of these is taken from the paper **[DDdP93]**.

Lemma 7.32 ([DDdP93, Lemma 4.4]) *Let* $a = \sum_{i=1}^{n} \alpha_i e_i$ *and* $b = \sum_{j=1}^{m} \beta_j f_j$ *be given with* $\alpha_1 \geq \alpha_2 \geq \cdots \geq \alpha_n \geq 0$, $\beta_1 \geq \beta_2 \geq \cdots \geq \beta_m \geq 0$, *and* $\{e_1, e_2, \ldots, e_n\}$ *and* $\{f_1, f_2, \ldots, f_m\}$ *sets of mutually orthogonal projections in* $(L^1 \cap L^\infty)(\mathcal{M}, \tau)$. *Then, there exists* $T \in \Sigma_+$ *with* $\tau(bT(a)) = \int_0^\infty \mathbf{m}_a(t)\mathbf{m}_b(t)\,dt$.

Proof. On setting $s_k = \sum_{i=1}^{k} \tau(e_i)$ $(1 \leq k \leq n)$ and $r_k = \sum_{i=1}^{k} \tau(f_i)$ $(1 \leq k \leq m)$, and selecting s_0 and r_0 to be 0, it is an exercise to see that then $\mathbf{m}_a = \sum_{i=1}^{n} \alpha_i \chi_{[r_{i-1}, r_i)}$ and $\mathbf{m}_b = \sum_{j=1}^{m} \beta_j \chi_{[s_{j-1}, s_j)}$. We next define the action of the linear operator T on $(L^1 + L^\infty)(\mathcal{M}, \tau)$ by setting

$$T(x) = \sum_{j=1}^{m} \left(\sum_{i=1}^{n} \frac{\lambda_L([r_{i-1}, r_i) \cap [s_{j-1}, s_j))}{\tau(e_i)\tau(f_j)} \tau(xe_i) \right) f_j \text{ for all}$$

$$x \in (L^1 + L^\infty)(\mathcal{M}, \tau).$$

It is now an exercise to see that T does indeed belong to Σ_+ and that it satisfies the equality $\tau(bT(a)) = \int_0^\infty \mathbf{m}_a(t)\mathbf{m}_b(t)\,dt$. $\square$

Lemma 7.33 (Hardy's lemma) *Let* f_1 *and* f_2 *be two non-negative Borel-measurable functions on* $(0, \infty)$ *satisfying*

$$\int_0^t f_1(s)\,ds \leq \int_0^t f_2(s)\,ds \text{ for all } t > 0.$$

For any non-negative decreasing function g *on* $(0, \infty)$, *we then have that*

$$\int_0^\infty f_1(s)g(s)\,ds \leq \int_0^\infty f_2(s)g(s)\,ds.$$

Proof. Sketch of the proof The first step is to use the monotone convergence theorem to see that the result will hold, if it holds for the case where g is a step function. The next step is to note that non-negative decreasing step functions g may be written in the form $g = \sum_{k=1}^{n} a_k \chi_{(0, t_k)}$ where the a_ks are non-negative, and $t_1 \leq t_2 \leq \cdots \leq t_n$. By means of direct computation, we may then conclude from the hypothesis that

$$\int_0^\infty f_1(s)g(s)\,ds = \sum_{k=1}^{n} a_k \int_0^{t_k} f_1(s)\,ds \leq \sum_{k=1}^{n} a_k \int_0^{t_k} f_2(s)\,ds$$

$$= \int_0^\infty f_2(s)g(s)\,ds. \quad \square$$

Theorem 7.34 ([DDdP93, Theorem 4.5]) *Let $x \in (L^1 + L^\infty)$ $(\mathcal{M}, \tau)$ and $y \in (L^1 \cap L^\infty)(\mathcal{M}, \tau)$ be given. Then $\int_0^\infty \mathbf{m}_x(t)\mathbf{m}_y(t)\,dt = \sup_{T \in \Sigma} |\tau(yT(x))|$.*

Proof. Let x and y be as in the hypothesis. First, we show that $\int_0^\infty \mathbf{m}_x(t)\mathbf{m}_y(t)\,dt \geq \sup_{T \in \Sigma} |\tau(yT(x))|$. Let $T \in \Sigma$ and $t > 0$ be given, and select $x_1 \in L^1$ and $x_\infty \in L^\infty$ so that $x = x_1 + x_\infty$. We may then use Theorem 7.24 to see that

$$\int_0^t \mathbf{m}_{T(x)}(s)\,ds \leq \|T(x_1)\|_{L^1} + t\|T(x_\infty)\|_{L^\infty}$$

$$\leq \|x_1\|_{L^1} + t\|x_\infty\|_{L^\infty}.$$

If we now take the infimum over all decompositions of the form $x = x_1 + x_\infty$, and once again apply Theorem 7.24, it follows that $\int_0^t \mathbf{m}_{T(x)}(s)\,ds \leq \int_0^t \mathbf{m}_x(s)\,ds$. Since this holds for all $t > 0$, we may now use Hardy's lemma to conclude that $\int_0^\infty \mathbf{m}_x(t)\mathbf{m}_y(t)\,dt \geq \int_0^\infty \mathbf{m}_{T(x)}(t)\mathbf{m}_y(t)\,dt$. By next applying Theorem 6.2, we have $\int_0^\infty \mathbf{m}_x(t)\mathbf{m}_y(t)\,dt \geq \int_0^\infty \mathbf{m}_{yT(x)}(t)\,dt = \tau(|yT(x)|) \geq |\tau(yT(x))|$ as required.

Our first step in proving that equality holds when we take the supremum over all $T \in \Sigma$ is to show that we may assume that $x \geq 0$ and $y \geq 0$, and that $T \in \Sigma_+$. Let $x = u|x|$ and $y = v|y|$ be the polar decompositions of x and y. Now, it is an easy exercise to check that the operator $T_{xy} : a \mapsto T(u^*a)v^*$ belongs to Σ whenever $T \in \Sigma_+$. Given the fact that $\tau(|y|T(|x|)) = \tau(v^*yT(u^*x)) = \tau(yT_{xy}(x))$, it is now clear that the equality will hold for the pair (x, y) whenever it holds for the pair $(|x|, |y|)$. So we may assume that $x \geq 0$ and $y \geq 0$.

Next, let $0 \leq x_\alpha \leq x$ and $0 \leq y_\beta \leq y$ be nets increasing to x and y, respectively. We claim that by taking the supremum over all $T \in \Sigma_+$, equality holds for the pair (x, y) whenever it holds for each of the pairs (x_α, y_β). We prove this by means of the contrapositive. So, suppose that we can find some $C > 0$ so that in fact

$$\int_0^\infty \mathbf{m}_x(t)\mathbf{m}_y(t)\,dt > C \geq \sup_{T \in \Sigma_+} |\tau(yT(x))|.$$

For any of the pairs (x_α, y_β) and any $T \in \Sigma_+$, we will have that $y^{1/2}T(x)y^{1/2} \geq y^{1/2}T(x_\alpha)y^{1/2}$ and $T(x_\alpha)^{1/2}yT(x_\alpha)^{1/2} \geq T(x_\alpha)^{1/2} y_\beta T(x_\alpha)^{1/2}$ and hence that

$$\tau(yT(x)) = \tau(y^{1/2}T(x)y^{1/2})$$
$$\geq \tau(y^{1/2}T(x_\alpha)y^{1/2}) = \tau(T(x_\alpha)^{1/2}\,yT(x_\alpha)^{1/2})$$
$$\geq \tau(T(x_\alpha)^{1/2}y_\beta T(x_\alpha)^{1/2}) = \tau(y_\beta T(x_\alpha)).$$

Now, observe that the nets $(\mathbf{m}_{x_\alpha})$ and $(\mathbf{m}_{y_\beta})$, respectively, increase to $\mathbf{m}_x$ and $\mathbf{m}_y$. Therefore, we may use the monotone convergence theorem to first select x_α so that $\int_0^\infty \mathbf{m}_{x_\alpha}(t)\mathbf{m}_y(t)\,dt > C$ and then, with α fixed, select β so that $\int_0^\infty \mathbf{m}_{x_\alpha}(t)\mathbf{m}_{y_\beta}(t)\,dt > C$. For this specific pair (α, β), we then have that

$$\int_0^\infty \mathbf{m}_{x_\alpha}(t)\mathbf{m}_{y_\beta}(t)\,dt > C \geq \sup_{T\in\Sigma_+} |\tau(yT(x))| \geq \sup_{T\in\Sigma_+} |\tau(y_\beta T(x_\alpha))|$$

as required.

Since $x_n = x\chi_{[0,n]}(x)$ increases to x, it follows from the above that it suffices to prove the claimed equality for each of the pairs (x_n, y). That is, we may assume that $x \in L^\infty(\mathcal{M},\tau)$. We may then further select a net (x_γ) in $(L^1 \cap L^\infty)_+$ increasing to x. This can be done by selecting a net of projections (e_γ), all with finite trace, increasing to $\mathbb{1}$, and then setting $x_\gamma = x^{1/2}e_\gamma x^{1/2}$. (See the proof of Corollary 6.26 for the existence of such a net.) Since for each $t \geq 0$ we have $\mathbf{m}_{x_\gamma}(t) = \mathbf{m}_{x^{1/2}e_\gamma x^{1/2}}(t) \leq \|x\|_\infty \mathbf{m}_{e_\gamma}(t)$, we then clearly have that $\tau(x_\gamma) = \int_0^\infty \mathbf{m}_{x_\gamma}(t)\,dt \leq \|x\| \int_0^\infty \mathbf{m}_{e_\gamma}(t)\,dt = \|x\|\tau(e_\gamma) < \infty$ for each γ. So, once again applying what we have just proved, it follows that we may, in fact, assume that $x \in (L^1 \cap L^\infty)_+$. We now make this assumption and conclude the proof by constructing nets $0 \leq x_\alpha \leq x$ and $0 \leq y_\beta \leq y$ increasing to x and y, respectively, for which we do indeed have that

$$\int_0^\infty \mathbf{m}_{x_\alpha}(t)\mathbf{m}_{y_\beta}(t)\,dt = \sup_{T\in\Sigma_+} |\tau(y_\beta T(x_\alpha))|.$$

Let $x = \int_0^\infty \lambda\,de_\lambda = \int_0^{\|x\|_\infty} \lambda\,de_\lambda$ be the spectral decomposition of x. Since $x \in L^\infty(\mathcal{M},\tau)$, this decomposition is known to be the norm limit of 'Riemann sums'. Therefore, we may select progressively finer partitions $\mathcal{P}_n = \{t_{k(n)}: 0 \leq k(n) \leq m(n)\}$ of $[0, \|x\|_\infty]$ (i.e. $0 = t_0 < t_{1(n)} < t_{2(n)} < \cdots < t_{m(n)} = \|x\|$), so that the sequence of 'lower sums' $x_n = \sum_{0\leq k(n)\leq m(n)} t_{k(n-1)}e_{(t_{k(n-1)},t_{k(n)}]}$ not only increases to x, but in fact converges uniformly to x. Observe that for each fixed n, the projections $e_{(t_{k(n-1)},t_{k(n)}]}$ are mutually orthogonal, and also that for

$k(n) \geq 2$, $\tau(e_{(t_{k(n-1)}, t_{k(n)}]}) \leq \tau(e_{(t_{k(n-1)}, \infty)}) \leq t_{k(n-1)} \tau(x) < \infty$. Thus, each x_n is of the form described in Lemma 7.32. We may similarly construct a sequence (y_m) increasing to y in $L^1 \cap L^\infty$, which is also of the form described in Lemma 7.32. The fact that the required equality holds for each of the pairs (x_n, y_m) now follows from Lemma 7.32. This proves the theorem. $\qquad\square$

Theorem 7.35 *For any $x \in (L^1 + L^\infty)(\mathcal{M}, \tau)$, we have $y \in \mathcal{O}_x = \{Tx \colon T \in \Sigma\}$ if and only if $\int_0^t \mathbf{m}_y(s)\, ds \leq \int_0^t \mathbf{m}_x(s)\, ds$ for all $t > 0$.*

Proof. First suppose that we can find some $T \in \Sigma$ such that $y = Tx$. Let $t > 0$ be given. A similar argument to the one employed in the first part of the proof of the previous theorem now shows that $\int_0^t \mathbf{m}_y(s)\, ds \leq \int_0^t \mathbf{m}_x(s)\, ds$, as required.

Next, assume that we can find some $y \notin \mathcal{O}_x$ such that $\int_0^t \mathbf{m}_y(s)\, ds \leq \int_0^t \mathbf{m}_x(s)\, ds$ for all $t > 0$. The converse implication will follow if we can show that this leads to a contradiction. Since as proved in Theorem 7.31, $\mathcal{O}_x$ is a $\sigma((L^1 + L^\infty)(\mathcal{M}, \tau), (L^1 \cap L^\infty)(\mathcal{M}, \tau))$ closed subset of $(L^1 + L^\infty)(\mathcal{M}, \tau)$, we may use the Hahn–Banach separation theorem to select some $w \in (L^1 \cap L^\infty)(\mathcal{M}, \tau)$ such that

$$\tau(|wy|) = \tau(wy) > \sup_{T \in \Sigma} |\tau(wT(x))|.$$

On first applying Theorem 6.2 and then Theorem 7.34, it follows that

$$\int_0^\infty \mathbf{m}_w(s)\mathbf{m}_y(s)\, ds \; \geq \; \int_0^\infty \mathbf{m}_{wy}(s)\, ds$$
$$= \; \tau(|wy|)$$
$$> \; \int_0^\infty \mathbf{m}_w(s)\mathbf{m}_x(s)\, ds.$$

But if indeed we do have $\int_0^t \mathbf{m}_y(s)\, ds \leq \int_0^t \mathbf{m}_x(s)\, ds$ for all $t > 0$, then by Hardy's inequality, we should have $\int_0^\infty \mathbf{m}_w(s)\mathbf{m}_y(s)\, ds \leq \int_0^\infty \mathbf{m}_w(s)\mathbf{m}_x(s)\, ds$. This is a clear contradiction, which then establishes the theorem. $\qquad\square$

We are now finally ready to prove Theorem 7.26.

Proof of Theorem 7.26 Let X be an exact interpolation space for the pair $(L^1(\mathcal{M}, \tau), L^\infty(\mathcal{M}, \tau))$, and let x, y be given with $x \in X$,

$y \in (L^1 + L^\infty)(\mathcal{M}, \tau)$ satisfying the condition $K(y, t) \leq K(x, t)$ for all $t > 0$. On considering the preceding theorem alongside Theorem 7.24, it follows that there exists $T \in \Sigma$ such that $T(x) = y$. The fact that X is an interpolation space now ensures that $y = T(x) \in X$, while the exactness of X ensures that T acts contractively on X and hence that $\|y\| = \|T(x)\| \leq \|x\|$. This proves the theorem. $\qquad\square$

PART 3

GENERAL CASE

In this part of the book, we pass from the 'mildly' noncommutative setting of tracial von Neumann algebras to the 'wildly' noncommutative setting of general, possibly type III, von Neumann algebras. No theory of L^p-spaces would have been possible for these algebras were it not for modular theory. So, in Chapters 8 and 9, we review the foundations on which this general theory is built, before presenting the actual basic theory of Haagerup L^p-spaces in Chapter 10.

Of the earlier chapters, all of Chapters 5 and 6 need to be covered for this theory to be comprehensible. Readers wishing to read no further than Chapter 10 and only L^p-spaces could get away with skipping Section 6.3 (p. 276), but as soon as one progresses to Chapter 11, elements of this section become necessary.

A clear grasp of modular theory is absolutely necessary for anyone who aspires to ultimately being able to handle Haagerup L^p-spaces with a reasonable level of comfort. We therefore advise readers interested in the general theory to survey all of Chapter 8. Section 8.6 (p. 352) is not strictly necessary. However, a clear conception of how modular theory works in the setting of group algebras will help to build valuable intuition for the more tricky setting of general von Neumann algebras. Group algebras are also the gateway to many valuable more concrete examples of Haagerup L^p-spaces.

Crossed products are the construct within which the theory of Haagerup L^p-spaces is developed. A clear grasp of the dual weight construction and of Section 9.5 (p. 398) is therefore essential for the study of Haagerup L^p-spaces. However, Section 9.5 (p. 398) presupposes some of the earlier material in Chapter 9. If one reduces the prerequisites for Chapter 10 to the bare essentials, then, as mentioned in the Preface, the readers will at the very least need to be familiar with Theorems 9.17 (p. 379), 9.28 (p. 399), 9.31 (p. 401), 9.38 (p. 408), and 9.40 (p. 412) and Propositions 9.27 (p. 399), 9.33 (p. 404), and

9.36 (p. 406). However, anyone wishing to ultimately prove advanced theorems regarding Haagerup L^p-spaces is advised to at some stage wade through the technicalities of Chapter 9 since such investigations at times require delicate manipulations of the crossed product inside of which the relevant L^p-space is realised.

In Chapter 10, the primary objective of Section 10.1 (p. 413) is to clearly show how Haagerup L^p-spaces may be viewed as natural generalisations of their tracial cousins, as presented in Chapter 6. This is achieved by here computing the L^p and Orlicz spaces from Chapter 6 in a 'Haagerup way'. Basic facts about completeness, normability and duality are presented in Sections 10.2 (p. 418) and 10.3 (p. 434), with Section 10.4 (p. 441) devoted to describing the 'noncommutative simple functions' living inside L^p. Section 10.5 (p. 458) discusses conditional expectations and compressions and includes a new proof of the existence of conditional expectations from a given von Neumann algebra onto a von Neumann subalgebra with respect to some given weight (Theorem 10.54 (p. 458)). Finally, in Section 10.6 (p. 463), we show how, for any von Neumann algebra, the resultant L^p technology may be used to construct a standard form for that algebra.

Readers whose primary interest is in L^p-spaces can avoid the Orlicz content by identifying $L^\infty(\mathcal{M})$ with $\pi_\varphi(\mathcal{M})$ and taking Proposition 10.10 (p. 420), Theorem 10.11 (p. 421) and Example 10.15 (p. 423) for granted. This done, they may then pick up the exposition with Lemma 10.23 (p. 430) taking care to skip Theorem 10.55 (p. 461) along the way. In fact, a version of the second part of Theorem 10.11 (p. 421) valid for L^p with $0 < p < \infty$ may be recovered with little additional effort. This can be done by using part (2) of Lemma 10.23 (p. 430) as a substitute for Proposition 10.10 (p. 420) and suitably specialising the proof of Theorem 10.11 (p. 421) to the L^p $(0 < p < \infty)$ context. However, this shortcut comes at a cost with the price needing to be paid the somewhat surprising fact that the second claim in Theorem 10.11 (p. 421) is also true for $L^\infty(\mathcal{M})$.

Chapter 8
Basic elements of modular theory

A clear understanding of the fundamentals of Tomita–Takesaki modular theory, of Connes cocycles and of conditional expectations and operator-valued weights is absolutely crucial for the theory that will follow. For that reason, we will in this chapter briefly lay a suitable foundation regarding these theories, before proceeding with the development of the theory of crossed products. Our presentation in Section 8.1 of the foundational material regarding modular automorphism groups borrows very heavily from the matching presentation in [**BR87**]. In Sections 8.2 and 8.3, we present the essentials of Pedersen–Takesaki Radon–Nikodym derivatives and Connes cocycles (as introduced by Connes in his famous paper on the classification of type III factors [**Con73**]), with the bulk of Section 8.4 based on the material introduced by Haagerup in [**Haa79b, Haa79c**]. Then, in Section 8.5, we briefly outline the 'Hilbert space approach' to modular theory before in Section 8.6 investigating the interface of this theory with von Neumann group algebras and quantum groups. Where no proofs are offered, interested readers will find them in the indicated references. Didactic expediency has led us to, for the most part, focus on σ-finite von Neumann algebras in our presentation of the theory of modular automorphism groups in Section 8.1. Readers eager to, for this section, see proofs that hold for general von Neumann algebras may wish to, for example, consult [**Tak03a**] and [**Str81**]. Readers familiar with this theory may, of course, skip these sections and proceed directly to Chapter 9.

8.1 Modular automorphism groups

The key ingredient to developing a theory of L^p-spaces valid for possibly non-semifinite von Neumann algebras is unquestionably the theory of modular automorphism groups created by Minoru Tomita and Masamichi Takesaki. In view of this fact, we pause to review the

Noncommutative measures and L^p and Orlicz Spaces, with Applications to Quantum Physics. Stanisław Goldstein and Louis Labuschagne, Oxford University Press. © Stanisław Goldstein and Louis Labuschagne (2025).
DOI: 10.1093/oso/9780198950202.003.0010

foundational theory regarding modular automorphism groups that we shall need in the subsequent development of the theory. Although this theory is an essential background for Haagerup L^p-spaces, it is not a critical part of the core of that theory. We shall therefore merely survey the theory rather than proving all claims from first principles. Our exposition is very strongly based on the discussion of this material in [**BR87**] supplemented by some material from [**Tak03a, Haa75c**] to flesh out the exposition. Readers who wish to see detailed proofs may consult these references, as well as the very comprehensive review of modular theory presented in the classic work of Strătilă [**Str81**].

Unless otherwise stated, we will in this section for the most part assume that we are working with a von Neumann algebra $\mathcal{M}$ equipped with a faithful normal state ω. The essence of the theory is easier to convey, and formulation of results simpler in this case. However, all results stated have counterparts which hold for von Neumann algebras equipped with a faithful normal semifinite weight, rather than a state. This assumption is therefore being made for purely didactic reasons. The ensuing exposition builds on the material regarding the GNS representation in the Preliminaries, namely Definition 0.69, Remark 0.70, Proposition 0.99, Theorem 0.100, and Proposition 0.101. We urge the reader to revise this material before engaging the material hereafter.

8.1.1 Basic concepts

Recall that when a von Neumann algebra $\mathcal{M}$ equipped with a faithful normal state ω is identified with the GNS representation thereof engendered by ω, the state ω then becomes a vector state corresponding to a cyclic and separating vector Ω. The vector Ω is then in fact cyclic and separating for both $\mathcal{M}$ and $\mathcal{M}'$ (see [**BR87**, Propositions 2.5.3 and 2.5.6]). We may now use this vector to define both an involution and a product on each of $\mathcal{M}\Omega$ and $\mathcal{M}'\Omega$ by the prescriptions

$$(a\Omega)^\sharp = a^*\Omega \qquad (a\Omega)(b\Omega) = (ab\Omega).$$

The involution then induces antilinear operators S_0 and F_0 on the dense subspaces $\mathcal{M}\Omega$ and $\mathcal{M}'\Omega$ of H by means of the prescriptions

$$S_0(a\Omega) = (a\Omega)^\sharp, \quad F_0(a'\Omega) = (a'\Omega)^\sharp$$

where $a \in \mathcal{M}$ and $a' \in \mathcal{M}'$. The operators S_0 and F_0 both turn out to be closable, as can be seen from the following proposition.

Proposition 8.1 ([BR87, 2.5.9]) *The operators S_0 and F_0 defined above are both closable. In fact, $S_0^* = [F_0]$ and $F_0^* = [S_0]$ (square brackets denote the minimal closure). Also, for any $\xi \in D([S_0])$, there exists a densely defined closed operator q affiliated to $\mathcal{M}$ such that $q\Omega = \xi$ and $q^*\Omega = [S_0]\xi$, with a similar claim holding for $[F_0]$.*

We are now ready to formally introduce the definition of the modular operator.

Definition 8.3 We define the antilinear operators S and F to be $S = [S_0]$ and $F = [F_0]$. We let $\Delta_\omega = \Delta$ be the unique positive self-adjoint operator and $J_\omega = J$ the unique antiunitary operator occurring in the polar decomposition $S = J\Delta^{1/2}$. We refer to Δ as the *modular operator* and J as the *modular conjugation* for the pair $(\mathcal{M}, \Omega)$.

Remark 8.2 In the case where we have a normal semifinite weight ψ rather than a state, the Hilbert space H_ψ in the GNS construction for the pair $(\mathcal{M}, \psi)$ is constructed from the quotient space $\mathfrak{n}_\psi / N_\psi$, where $N_\psi \subseteq \mathfrak{n}_\psi$ is the left-ideal $N_\psi = \{x \in \mathcal{M} : \psi(x^*x) = 0\}$. This quotient space becomes a pre-Hilbert space when equipped with the inner product $\langle x + N_\psi, y + N_\psi \rangle = \psi(y^*x)$ $(x, y \in \mathfrak{n}_\psi)$. The Hilbert space H_ψ is then just the completion of this pre-Hilbert space with respect to the inner-product topology, with the prescription $\eta_\psi : x \mapsto x + N_\psi$ $(x \in \mathfrak{n}_\psi)$ defining a dense embedding of $\mathfrak{n}_\psi$ into H_ψ. As in the state case, there is a representation of $\mathcal{M}$ as a subalgebra of $B(H_\psi)$ realised by a *-homomorphism $\pi_\psi : \mathcal{M} \to B(H_\psi)$ satisfying $\langle \pi_\psi(a)\eta(b), \eta(c) \rangle = \psi(c^*ab)$ and $\pi_\psi(a)\eta(b) = \eta(ab)$ for all $a \in \mathcal{M}$ and $b, c \in \mathfrak{n}_\psi$. The triple $(\psi, H_\psi, \eta_\psi)$ is referred to as a semi-cyclic representation. Since in this case $\mathbb{1} \notin \mathfrak{n}_\psi$, it is clear that in this case the GNS construction corresponding to ψ cannot yield a cyclic and separating vector realising ψ as a state. In the case where we are dealing with a faithful normal semifinite weight φ, this construction is somewhat simpler, as the faithfulness of φ then ensures that $N_\varphi = \{0\}$ and that π_φ is a *-isomorphism.

Despite the absence of a cyclic and separating vector, one may nevertheless still develop a modular theory that closely rivals that of the σ-finite setting. The primary ingredient that is needed is a subspace of H_φ, which admits an involutive structure that we can use to define an analogue of the operators S and F. The subspace $\eta(\mathfrak{n}_\varphi \cap \mathfrak{n}_\varphi^*)$ turns out to be just such a subspace. We may specifically equip this subspace with product and involution operations defined by the prescriptions

$$\eta_\varphi(x)\eta_\varphi(y) = \eta_\varphi(xy)$$
$$\eta_\varphi(x)^\sharp = \eta_\varphi(x^*)$$

for all $x, y \in (\mathfrak{n}_\varphi \cap \mathfrak{n}_\varphi^*)$. Equipped with this structure $\eta_\varphi(\mathfrak{n}_\varphi \cap \mathfrak{n}_\varphi^*)$ then becomes a so-called full left Hilbert algebra [**Tak03a**, Theorem VII.2.6]. The completion of this full left Hilbert algebra then yields all of H_φ [**Tak03a**, Theorems VII.2.5 and VII.2.6]. In direct analogy with the state case, we may now densely define the operator S_0 on this subspace by means of the prescription $S_0 : \eta(a) \mapsto \eta(a)^\sharp = \eta(a^*)$. As before, this operator extends to a closed densely defined antilinear operator S with a densely defined adjoint. The modular operator Δ is then $\Delta = |S|^2$ with the modular conjugation J the antilinear isometry in the polar decomposition $S = J\Delta^{1/2}$ (consider the discussion preceding [**Tak03a**, Lemma VI.1.4] alongside [**Tak03a**, Lemma VI.1.5]). As we shall see shortly, the domain of S is in this case a little more difficult to describe. (Compare Propositions 8.1 and 8.4.)

We start our analysis by describing the domain of $\Delta^{1/2}$, before reviewing the basic properties and inter-relation of the operators S, F, Δ and J.

Proposition 8.4 ([Tak03a, Lemmata VI.1.4(i) and VI.1.5(iv)])
The domain of $\Delta^{1/2} = |S|$ consists of all vectors $\xi \in H$ for which there exists a sequence $(a_n) \subset \mathfrak{n}_\varphi^* \cap \mathfrak{n}_\varphi$ such that $(\eta(a_n^*)) \subset H$ is Cauchy and $\|\eta(a_n) - \xi\| \to 0$. In this case, $S(\xi) = \lim_n \eta(a_n^*)$.

Proposition 8.5 ([BR87, 2.5.11])   *The following relations between S, F, Δ and J are valid:*

$$\Delta = FS \text{ and } \Delta^{-1} = SF,$$
$$S = J\Delta^{1/2} \text{ and } F = J\Delta^{-1/2},$$
$$J = J^* \text{ and } J^2 = \mathbb{1},$$
$$\Delta^{-1/2} = J\Delta^{1/2}J.$$

An easy consequence of the above, which is nevertheless worth noting, is the fact that the vector Ω is an eigenvector of Δ corresponding to the eigenvalue 1. This follows from the formula $\Delta = FS$ and the fact that, by definition, $S\Omega = \Omega = F\Omega$. One of the great achievements of modular theory is the following theorem describing the action of the operators J and Δ on $\mathcal{M}$.

Theorem 8.6 ([Tomita–Takesaki theorem, cf. BR87, 2.5.14])
Let $\mathcal{M}$ be a von Neumann algebra equipped with a cyclic and separating vector Ω, and let Δ and J, respectively, be the modular operator and modular conjugation corresponding to Ω. Then, $J\mathcal{M}J = \mathcal{M}'$ with in addition $\Delta^{it}\mathcal{M}\Delta^{-it} = \mathcal{M}$ for all $t \in \mathbb{R}$.

The final fact noted in the above theorem now enables us to introduce the following definition. Due to its importance, we formulate this definition for the general case.

Definition 8.7 Let $\mathcal{M}$ be a von Neumann algebra equipped with a faithful normal semifinite weight φ. Let Δ_φ and J_φ be the modular operator and modular conjugation associated with the pair $(\pi_\varphi(\mathcal{M}), H_\varphi)$. The preceding theorem then ensures that the prescription $\sigma_t^\varphi(a) = \pi_\varphi^{-1}(\Delta_\varphi^{it}\pi_\varphi(a)\Delta_\varphi^{-it})$, where $a \in \mathcal{M}$ and $t \in \mathbb{R}$, yields a one-parameter group of $t \mapsto \sigma_t^\varphi$ of σ-weakly continuous *-automorphisms on $\mathcal{M}$, which we shall refer to as the *modular automorphism group associated with the pair $(\mathcal{M}, \varphi)$*.

To aid the reader in getting to grips with these concepts, we pause to present the following concrete example.

Example 8.8 Consider the case $\mathcal{M} = B(H)$ where H is a finite-dimensional Hilbert space. Any faithful normal state φ is then of the form $\varphi = \mathrm{Tr}(f\cdot)$ for some positive norm 1 element of $L^1(B(H), \mathrm{Tr})$. Since in this case $B(H)$ is basically just a copy of the algebra of $n \times n$ matrices, the faithfulness of $\varphi = \mathrm{Tr}(f\cdot)$ ensures that here f is, in fact, an invertible element of $B(H)$. The GNS Hilbert space H_φ is then just a copy of $L^2(B(H), \mathrm{Tr})$ (with norm given by $\|\xi\|_2 = \mathrm{Tr}(f\xi^*\xi f)^{1/2}$). Regarding $B(H)$ as an algebra of left multiplication operators on this space, we find that $\Omega = f^{1/2}$ with the action of Δ and J given by $\Delta(\xi) = f\xi f^{-1}$ and $J(af^{1/2}) = f^{1/2}a^*$, respectively. The action of the modular automorphism group σ_t^φ is given by $\sigma_t^\varphi(a) = f^{it}af^{-it}$. (See the introductions to [**LMM06**, §4] and [**Ara73**].) for details.)

Remark 8.9

(1) The σ-weak continuity noted above follows from the fact that the unitary group $t \mapsto \Delta^{it}$ is strongly continuous by Stone's theorem. The automorphism group $t \mapsto \Delta_\varphi^{it}\pi_\varphi(\cdot)\Delta_\varphi^{-it}$ is, therefore, strong operator continuous.

(2) It is worth noting that we will, in the general case, have $\Delta^{it}(\eta_\varphi(x)) = \eta_\varphi(\sigma_t^\varphi(x))$ for all $x \in \mathfrak{n}_\varphi(\mathcal{M}) \cap \mathfrak{n}_\varphi(\mathcal{M})^*$. (To see this, apply [**vD74**, Theorem 5.3(ii)] to [**Tak03a**, Theorem VII.2.6].)

(3) The faithful normal semifinite weight φ is invariant under the action of the modular group σ_t^φ. In the case where φ is a state, this can easily be seen to follow from the fact noted earlier that $\Delta\Omega = \Omega$ and hence that $\Delta^{it}\Omega = \Omega$ for all $t \in \mathbb{R}$.

(4) Let $\mathcal{M}$ be a von Neumann algebra equipped with a faithful normal state ω corresponding to some cyclic and separating vector Ω. The defining property of traces is that operators 'commute' under the trace. However, even if φ is not a trace, then for all $a, b \in \mathcal{M}$ for which $\Delta^{1/2}a\Delta^{-1/2}$ and $\Delta^{1/2}b\Delta^{-1/2}$ uniquely extend to elements of $\mathcal{M}$, there is a sense in which the modular automorphism group may be used to swop the order of multiplication under φ. Specifically, for such $a, b \in \mathcal{M}$, we have that

$$\langle(\Delta^{-1/2}b\Delta^{1/2})(\Delta^{1/2}a\Delta^{-1/2})\Omega, \Omega\rangle = \langle\Delta^{1/2}a\Omega, \Delta^{1/2}b^*\Omega\rangle$$
$$= \langle JSa\Omega, JSb^*\Omega\rangle$$
$$= \langle Ja^*\Omega, Jb\Omega\rangle$$
$$= \overline{\langle a^*\Omega, b\Omega\rangle}$$
$$= \langle b\Omega, a^*\Omega\rangle$$
$$= \langle ab\Omega, \Omega\rangle.$$

(Here, we used the fact that the adjoint formula for antilinear operators $T : H \to H$ is of the form $\langle T\xi, \eta\rangle = \overline{\langle \xi, T^*\eta\rangle}$.) This may formally be written as the claim that $\omega(\sigma_{i/2}^\omega(b)\, \sigma_{-i/2}^\omega(a)) = \omega(ab)$. (See [**Str81**, Proposition 2.17] for details.)

5. It is not difficult to see that the modular automorphism group of the pair $(\mathcal{M}, \tau)$, where τ is a faithful normal semifinite trace, is trivial. For the sake of lucidity of exposition, we proceed to substantiate this for the case where $\tau(a) = \langle a\Omega, \Omega \rangle$ for some cyclic and separating vector $\Omega \in H$. With τ being tracial, we will in this case for any $a \in \mathcal{M}$ have $\|a\Omega\|^2 = \langle a^*a\Omega, \Omega \rangle = \tau(a^*a) = \tau(aa^*) = \langle aa^*\Omega, \Omega \rangle = \|a^*\Omega\|^2$. Thus, here S is itself an antilinear isometry, ensuring that $\Delta = |S| = \mathbb{1}$.

8.1.2 The KMS condition and analyticity

One of the crowning achievements of modular theory is Takesaki's theorem. This important theorem shows that the modular automorphism group satisfies the so-called KMS condition.

Theorem 8.10 (cf. [Tak03a, Theorem VIII.1.2]) *Let φ be a faithful normal semifinite weight on a von Neumann algebra. Then the automorphism group σ_t^{φ} satisfies the KMS condition, (Kubo-Martin-Schwinger) namely that it is the unique *-automorphism group $\sigma_t^{\varphi} = \sigma_t$ which satisfies the condition that:*

(1) $\varphi \circ \sigma_t = \varphi$ for all $t \in \mathbb{R}$;
(2) for any $x, y \in (\mathfrak{n}_{\varphi} \cap \mathfrak{n}_{\varphi}^*)$, there exists a bounded continuous function $F_{x,y}$ on the closed strip $S = \{z \in \mathbb{C} \colon 0 \leq \mathrm{Im}(z) \leq 1\}$ which is analytic on the interior $S_o = \{z \in \mathbb{C} \colon 0 < \mathrm{Im}(z) < 1\}$ such that

$$F_{x,y}(t) = \varphi(\sigma_t(x)y) \text{ and } F_{x,y}(t + i) = \varphi(y\sigma_t(x)) \text{ for all } t \in \mathbb{R}.$$

On a similar note, one may now introduce the following notion.

Definition 8.11 Let $t \mapsto \sigma_t$ be a one-parameter *-automorphism group on $\mathcal{M}$ for which $t \mapsto \sigma_t(a)$ is σ-weakly continuous for every $a \in \mathcal{M}$. An element $a \in \mathcal{M}$ is said to be σ_t-analytic if there exists a strip $S_\gamma = \{z \in \mathbb{C} \colon |\Im(z)| < \gamma\}$ in $\mathbb{C}$, and a function $F : S_\gamma \to \mathcal{M}$ such that:

- $F(t) = \sigma_t(a)$ for each $t \in \mathbb{R}$;
- with $z \mapsto \rho(F(z))$ analytic for every $\rho \in \mathcal{M}_*$.

In such a case, we write $\sigma_z(a)$ for $F(z)$. If F even extends to an entire-analytic function, we say that $a \in \mathcal{M}$ is entire-analytic.

The above definition raises the question of just how many analytic elements there are in $\mathcal{M}$. This has a very elegant answer.

Proposition 8.12 ([BR87, Proposition 2.5.22]) *For each $a \in \mathcal{M}$, the elements of the sequence (a_n) defined by*

$$a_n = \sqrt{\frac{n}{\pi}} \int \sigma_t(a) e^{-nt^2} \, dt$$

are entire-analytic with $\|a_n\| \le \|a\|$ for each $n \in \mathbb{N}$, and with $a_n \to a$ in the σ-weak topology.

The set of entire analytic elements of $\mathcal{M}$ actually forms a σ-weakly dense subalgebra. We specifically have the following.

Theorem 8.13 ([Tak03a, Theorem VIII.2.3]) *Let $t \mapsto \sigma_t$ be a σ-weakly continuous one-parameter $*$-automorphism group on $\mathcal{M}$, and let $\mathcal{M}_\sigma^a$ be the set of all entire analytic elements. Then, $\mathcal{M}_\sigma^a$ is a σ-weakly dense $*$-subalgebra of $\mathcal{M}$. Moreover, for any $a, b \in \mathcal{M}_\sigma^a$ and $z, w \in \mathbb{C}$, we have that:*

- $\sigma_z(ab) = \sigma_z(a)\sigma_z(b)$;
- $\sigma_{z+w}(a) = \sigma_z(a)\sigma_w(a)$;
- $\sigma_{\bar{z}}(a) = \sigma_z(a^*)^*$.

Any point to σ-weakly continuous one-parameter $*$-automorphism group $t \mapsto \sigma_t$ on $\mathcal{M}$ will, by general operator semigroup theory, admit an infinitesimal generator $\delta : \mathcal{M} \supseteq \mathrm{dom}(\delta) \to \mathcal{M}$ for which $\mathrm{dom}(\delta) = \{a \in \mathcal{M} : \frac{d\sigma_t(a)}{dt}\big|_{t=0} \text{exists}\}$. For each $a \in \mathrm{dom}(\delta)$, δ has the action of mapping a to $\frac{d\sigma_t(a)}{dt}\big|_{t=0}$. This operator turns out to be a σ-weakly closed operator with $\mathrm{dom}(\delta)$ a σ-weakly dense $*$-subalgebra of $\mathcal{M}$. The operator δ is a so-called $*$-derivation in that $\delta(ab) = a\delta(b) + \delta(a)b$ and $\delta(a^*) = \delta(a)^*$ for any $a, b \in \mathrm{dom}(\delta)$. It is a useful observation to note that for any $t \in \mathbb{R}$ and any $a \in \mathrm{dom}(\delta)$, we have that $\sigma_t(\mathrm{dom}(\delta)) = \mathrm{dom}(\delta)$ with $\delta(\sigma_t(a)) = \sigma_t(\delta(a))$. (See chapter 3 of

[**BR87**] for details.) One may alternatively define analyticity in terms of this infinitesimal generator, as described below.

Definition 8.14 Let $t \mapsto \sigma_t$ be a σ-weakly continuous one-parameter *-automorphism group on $\mathcal{M}$ and let δ be the infinitesimal generator. We define an element $a \in \mathcal{M}$ to be an analytic element for δ if $a \in \mathrm{dom}(\delta^n)$ for each $n \in \mathbb{N}$ and if for some $t > 0$, we have that $\sum_{n=0}^{\infty} \frac{t}{n!} \|\delta^n(a)\| < \infty$. If in fact $\sum_{n=0}^{\infty} \frac{t}{n!} \|\delta^n(a)\| < \infty$ for all $t > 0$, we say that a is entire-analytic (for δ).

It is a beautiful and elegant fact that this definition of analyticity is entirely equivalent to the earlier one. If, for example, an element a is analytic in the above sense, then the function F defined by $F(t + z) = \sum_{n=0}^{\infty} \frac{z^n}{n!} \sigma_t(\delta^n(a))$ will be analytic for z in the radius of analyticity of $\sum_{n=0}^{\infty} \frac{z^n}{n!} \|\delta^n(a)\|$. It can now be verified that a then satisfies the criteria for σ_t-analyticity. If conversely a is σ_t-analytic in the strip $S_\gamma = \{z \in \mathbb{C} : |\mathrm{Im}(z)| < \gamma\}$, then the well-known Cauchy inequalities yield

$$\|\delta^n(a)\| = \|\sigma_t(\delta^n(a))\| = \|\frac{d^n}{dt^n}\sigma_t(a)\| \leq \frac{n!M}{\gamma^n}$$

for some $M > 0$. But then $\sum_{n=0}^{\infty} \frac{|z|^n}{n!} \|\delta^n(a)\| \leq M \sum_{n=0}^{\infty} (\frac{|z|}{\gamma}) < \infty$ for all $|z| < \gamma$. (See the discussion following [**BR87**, Definition 3.1.17] for fuller details.)

In closing this subsection, we describe uniqueness criteria in more general contexts than for modular automorphism groups, namely for groups of isometries on von Neumann algebras. Here, the general approach borrows heavily from the ideas initiated in the preceding theory. The fastidious reader may find details and proofs of the claims made below in [**CZ76**]. (In particular, note [**CZ76**, Theorems 2.4 and 4.4].) Let $t \mapsto \sigma_t$ be a σ-weakly continuous group of isometries on some von Neumann algebra $\mathcal{R}$. For any $w \in \mathbb{C}$ with $\mathrm{Im}(w) > 0$, we may then define $D(\sigma_w)$ to be the set of all $a \in \mathcal{R}$ for which the map $t \mapsto \sigma_t(a)$ $(t \in \mathbb{R})$ may be extended to a σ-weakly continuous function f_w on the strip $\{z \in \mathbb{C} : 0 \leq \mathrm{Im}(z) \leq \mathrm{Im}(w)\}$ which is analytic on the interior of that strip. We then define σ_w on $D(\sigma_w)$ by setting $\sigma_w(a) = f_w(a)$ for any $a \in D(\sigma_w)$. In the case where $\mathrm{Im}(w) < 0$, the map σ_w is defined similarly using the strip $\{z \in \mathbb{C} : 0 \geq \mathrm{Im}(z) \geq \mathrm{Im}(w)\}$. The maps σ_w

can then be shown to be σ-weakly closed and σ-weakly densely defined linear operators on $\mathcal{R}$, for which we have $\sigma_{w_1}\sigma_{w_2} \subseteq \sigma_{w_1+w_2}$ for all $w_1, w_2 \in \mathbb{C}$, with equality holding whenever $\mathrm{Im}(w_1).\mathrm{Im}(w_2) \geq 0$. In particular $\sigma_{-w} = \sigma_w^{-1}$. Among these maps, σ_{-i} plays a particularly crucial role and is referred to as the analytic generator of the group. As we see below, two such groups coincide if their analytic generators coincide.

Proposition 8.15 ([Tak03a, VIII.3.24]/[Haa79b, Lemma 4.4])
Let $\mathcal{M}$ and $\mathcal{M}_0$ be von Neumann algebras, with $\mathcal{M}_0$ a von Neumann subalgebra of $\mathcal{M}$. Let σ_t and $\sigma_t^{(0)}$ (where $t \in \mathbb{R}$) be σ-weakly continuous one-parameter groups of isometries on $\mathcal{M}$ and $\mathcal{M}_0$, respectively. If $\sigma_{-i} \subseteq \sigma_{-i}^{(0)}$, then $\sigma_t^{(0)}(a) = \sigma_t(a)$ for all $a \in \mathcal{M}_0$ and all $t \in \mathbb{R}$.

If these ideas are to be applied to modular automorphism groups, we shall need a criterion that describes membership of $D(\sigma^{\varphi}_{-i/2})$ where (σ^{φ}) is the modular automorphism group of φ. This is provided by the following result.

Lemma 8.16 ([Haa79b, Lemma 3.3]/[Tak03a, VIII.3.18(1)])
Let φ be a faithful normal semifinite weight on a von Neumann algebra $\mathcal{M}$. For any $a \in \mathcal{M}$ and $k \geq 0$, the following are equivalent:

- $\varphi(a \cdot a^*) \leq k^2 \varphi$;
- $a \in D(\sigma^{\varphi}_{-i/2})$ *and* $\|\sigma^{\varphi}_{-i/2}(a)\| \leq k$.

In the same setting, membership of a pair (a, b) to the graph of σ^{φ}_{-i} is very elegantly described by the following result.

Theorem 8.17 ([Tak03a, VIII.3.25])   *Let φ be a faithful normal semifinite weight on a von Neumann algebra $\mathcal{M}$. For any $a, b \in \mathcal{M}$, the following are equivalent:*

- *(a, b) belongs to the graph $\mathcal{G}(\sigma^{\varphi}_{-i})$, that is we have $a \in D(\sigma^{\varphi}_{-i})$ with $\sigma^{\varphi}_{-i}(a) = b$;*
- *$a\mathfrak{n}^*_{\varphi} \subseteq \mathfrak{n}^*_{\varphi}$, $\mathfrak{n}_{\varphi}b \subseteq \mathfrak{n}_{\varphi}$, and $\varphi(ax) = \varphi(xb)$ for all $x \in \mathfrak{m}_{\varphi}$.*

Together, the preceding two results then provide us with alternative criteria for identifying modular automorphism groups with each other.

8.2 Centralisers and Radon–Nikodym derivatives

The observations made in Remark 8.9 regarding the remnants of 'trace-like' behaviour for weights really come into their own on the portion of $\mathcal{M}$ on which (as is the case for a trace) the modular automorphism group is trivial. It is with respect to this portion that we see stronger evidence of trace-like behaviour. This is captured in the following definition and the theorem that follows it (which, in view of its importance, we once again formulate for weights).

Definition 8.18 Let $\mathcal{M}$ be a von Neumann algebra equipped with a faithful normal semifinite weight φ. We define the *centraliser* of the pair $(\mathcal{M}, \varphi)$ to be the subalgebra $\mathcal{M}_\varphi = \{a \in \mathcal{M} \colon \sigma_t^\varphi(a) = a \text{ for all } t \in \mathbb{R}\}$.

Theorem 8.19 ([Tak03a, VIII.2.6]) *Let φ be a faithful normal semifinite weight on a von Neumann algebra $\mathcal{M}$. A necessary and sufficient condition for $a \in \mathcal{M}$ to belong to the centraliser $\mathcal{M}_\varphi$ is that*

- $a\mathfrak{m}_\varphi \subseteq \mathfrak{m}_\varphi$ *and* $\mathfrak{m}_\varphi a \subseteq \mathfrak{m}_\varphi$,
- *and that* $\varphi(ax) = \varphi(xa)$ *for all* $x \in \mathfrak{m}_\varphi$.

With some basic technology regarding the centraliser of a weight at our disposal in the form of the above theorem, we are now able to extend the theory regarding Radon–Nikodym derivatives for traces presented in Theorem 4.26 to the case of weights. We shall only sketch the theory, including samplings of proofs from [**PT73**] here and there, for the sake of building intuition. Readers eager for a fuller account of the theory are referred to the iconic paper of Pedersen and Takesaki on this topic [**PT73**]. This paper is the basis for the subsequent discussion. (See also [**Tak03a**, §VIII.2].) We start by showing how one may use positive elements of the centraliser $\mathcal{M}_\varphi$ to generate normal semifinite weights.

Lemma 8.20 *Let φ be a faithful normal semifinite weight on $\mathcal{M}$. Given any $h \in \mathcal{M}_\varphi^+$, the prescription $\varphi_h(a) = \varphi(h^{1/2}ah^{1/2})$ $(a \in \mathcal{M}^+)$ will yield a normal semifinite weight. Moreover, the map*

$h \mapsto \varphi_h$ *is a monotone affine map from* $\mathcal{M}_\varphi^+$ *into the normal semifinite weights on* $\mathcal{M}_+$.

Proof. Let $h \in \mathcal{M}_\varphi^+$ be given. The normality of φ_h is easily seen. Specifically, given a net (a_γ) increasing to a, we clearly have that

$$\varphi_h(\sup_\gamma a_\gamma) = \varphi(h^{1/2}(\sup_\gamma a_\gamma)h^{1/2}) = \varphi(\sup_\gamma(h^{1/2}a_\gamma h^{1/2}))$$

$$= \sup_\gamma \varphi(h^{1/2}a_\gamma h^{1/2}) = \sup_\gamma \varphi_h(a_\gamma).$$

To see the semifiniteness, notice that Theorem 8.19 ensures $h^{1/2}(\mathfrak{p}_\varphi) h^{1/2} \subseteq \mathfrak{p}_\varphi$, and hence that $\mathfrak{p}_\varphi \subseteq \mathfrak{p}_{\varphi_h}$. This clearly ensures that $\mathfrak{m}_{\varphi_h}$ is σ-weakly dense. Now, let $h, k \in \mathcal{M}_\varphi^+$ be given. We clearly have that $h^{1/2}, k^{1/2} \le (h+k)^{1/2}$. So, by Proposition 3.90, there exist contractions $c_h, c_k \in \mathcal{M}_\varphi^+$ supported on $\mathbf{s}(h+k)$ such that $(h+k)^{1/2}c_h(h+k)^{1/2} = h$ and $(h+k)^{1/2}c_k(h+k)^{1/2} = k$. It is clear that we must also have $c_h + c_k = \mathbf{s}(h+k)$. We now select partial isometries v_h and v_k so that $v_h c_h^{1/2}(h+k)^{1/2} = h^{1/2}$ and $v_k c_k^{1/2}(h+k)^{1/2} = k^{1/2}$. So, for $u_h = v_h u_h^{1/2}$ and $u_k = v_k u_k^{1/2}$, we have that $h^{1/2} = u_h(h+k)^{1/2} = (h+k)^{1/2}u_h^*$, $k^{1/2} = u_k(h+k)^{1/2} = (h+k)^{1/2}u_k^*$ and $u_h^*u_h + u_k^*u_k = \mathbf{s}(h+k)$.

Let $a \in \mathcal{M}_+$ be given and suppose that $\varphi_{(h+k)}(a) = \varphi((h+k)^{1/2}a(h+k)^{1/2}) < \infty$. But then $(h+k)^{1/2}a(h+k)^{1/2} \in \mathfrak{m}_\varphi$. Since u_h and u_k belong to $\mathcal{M}_\varphi$, we may now use Theorem 8.19 to conclude that both $u_h(h+k)^{1/2}a(h+k)^{1/2}u_h^*$ and $u_k(h+k)^{1/2}a(h+k)^{1/2}u_k^*$ will then belong to $\mathfrak{m}_\varphi$. So, in this case

$$\varphi_h(a) + \varphi_k(a) = \varphi(h^{1/2}ah^{1/2} + k^{1/2}ak^{1/2})$$

$$= \varphi(u_h(h+k)^{1/2}a(h+k)^{1/2}u_h^* + u_k(h+k)^{1/2}$$

$$a(h+k)^{1/2}u_h^*)$$

$$= \varphi((u_h^*u_h + u^k u_k)(h+k)^{1/2}a(h+k)^{1/2})$$

$$= \varphi(\mathbf{s}(h+k)(h+k)^{1/2}a(h+k)^{1/2})$$

$$= \varphi((h+k)^{1/2}a(h+k)^{1/2}) = \varphi_{(h+k)}(a).$$

Now, suppose that $\varphi_h(a) < \infty$ and $\varphi_k(a) < \infty$. That of course means that $h^{1/2}ah^{1/2}$ and $k^{1/2}ak^{1/2}$ both belong to $\mathfrak{m}_\varphi$. Using the fact that $kah + hak \le hah + kak$, it follows that

$$(h+k)^{1/2}a(h+k)^{1/2} = \lim_{\epsilon\searrow0}(h+k+\epsilon\mathbb{1})^{-1/2}(h+k)a(h+k)$$

$$(h+k+\epsilon\mathbb{1})^{-1/2}$$

$$\leq 2\lim_{\epsilon\searrow0}(h+k+\epsilon\mathbb{1})^{-1/2}(hah+kak)$$

$$(h+k+\epsilon\mathbb{1})^{-1/2}.$$

Now, observe that $\lim_{\epsilon\searrow0}(h+k+\epsilon\mathbb{1})^{-1/2}(hah)(h+k+\epsilon\mathbb{1})^{-1/2}$ equals

$$\lim_{\epsilon\searrow0}(h+k+\epsilon\mathbb{1})^{-1/2}((h+k)^{1/2}u_h^*h^{1/2}ah^{1/2}u_h(h+k)^{1/2})$$

$$(h+k+\epsilon\mathbb{1})^{-1/2}$$

which in turn equals $u_h^*h^{1/2}ah^{1/2}u_h$ and similarly that $\lim_{\epsilon\searrow0}(h+k+\epsilon\mathbb{1})^{-1/2}(kak)(h+k+\epsilon\mathbb{1})^{-1/2} = u_k^*k^{1/2}ak^{1/2}u_k$. We, therefore, obtain that

$$(h+k)^{1/2}a(h+k)^{1/2} \leq 2(u_h^*h^{1/2}ah^{1/2}u_h + u_k^*k^{1/2}ak^{1/2}u_k).$$

But, by Theorem 8.19, $(u_h^*h^{1/2}ah^{1/2}u_h + u_k^*k^{1/2}ak^{1/2}u_k) \in \mathfrak{m}_\varphi$. Therefore, the above inequality ensures that also $(h+k)^{1/2}a(h+k)^{1/2} \in \mathfrak{m}_\varphi$ and hence that $\varphi_{(h+k)}(a) < \infty$. Thus, in general, $\varphi_{(h+k)}(a) = \varphi_h(a) + \varphi_k(a)$.

Given $h, k \in \mathcal{M}_\varphi$ with $0 \leq k \leq h$, the monotonicity now follows upon noting that $\varphi_h = \varphi_k + \varphi_{(h-k)} \geq \varphi_k$. $\qquad\square$

The preceding lemma now enables us to prove the following partial analogue of Theorem 4.26.

Theorem 8.21 ([PT73, Proposition 4.2]) *Let φ be a faithful normal semifinite weight on $\mathcal{M}$. For any $h\,\eta\,\mathcal{M}_\varphi^+$, the prescription $\varphi_h = \lim_{\epsilon\searrow0}\varphi_{h_\epsilon}$ where for each $\epsilon > 0$ we set $h_\epsilon = h(\mathbb{1}+\epsilon h)^{-1}$ is a normal semifinite weight on $\mathcal{M}$ satisfying $\varphi_h \circ \sigma_t^\varphi = \varphi_h$ for each $t \in \mathbb{R}$. The mapping $h \mapsto \varphi_h$ (where $h\,\eta\,\mathcal{M}_\varphi^+$) is in fact order preserving. Moreover, $\mathbf{s}(h) = \mathrm{supp}\,(\varphi_h)$.*

Sketch of proof. Let $h\,\eta\,\mathcal{M}_\varphi^+$ be given. For any $\epsilon > 0$, we set $h_\epsilon = h(\mathbb{1}+\epsilon h)^{-1}$. The net (h_ϵ) turns out to be a net in $\mathcal{M}_\varphi$ which increases to h as $\epsilon \searrow 0$. By the preceding lemma, φ_{h_ϵ} is then an increasing net of normal weights. On setting $\varphi_h = \lim_{\epsilon\searrow0}\varphi_{h_\epsilon}$, the weight φ_h is then clearly normal.

We show that it is also semifinite. For any $n \in \mathbb{N}$, $h^{1/2}\chi_{[0,n]}(h)$ is a bounded positive operator in $\mathcal{M}_\varphi$. Moreover, $h_\epsilon^{1/2}\chi_{[0,n]}(h) = h^{1/2}(\mathbb{1} + \epsilon h)^{-1/2}\chi_{[0,n]}(h) \to h^{1/2}\chi_{[0,n]}(h)$ as $\epsilon \searrow 0$. Since by Theorem 8.19

$$h^{1/2}\chi_{[0,n]}(h)(\mathfrak{p}_\varphi)h^{1/2}\chi_{[0,n]}(h) \subseteq \mathfrak{p}_\varphi$$

it is therefore clear that

$$\varphi_h(\chi_{[0,n]}(h)a\chi_{[0,n]}(h)) = \varphi((h^{1/2}\chi_{[0,n]}(h))a(h^{1/2}\chi_{[0,n]}(h))) < \infty$$

for every $a \in \mathfrak{p}_\varphi$. In other words, $\chi_{[0,n]}(h)(\mathfrak{p}_\varphi)\chi_{[0,n]}(h) \subseteq \mathfrak{p}_{\varphi_h}$ holds for every $n \in \mathbb{N}$. But $\chi_{[0,n]}(h)(\mathfrak{m}_\varphi)\chi_{[0,n]}(h)$ can be shown to be σ-weakly dense in $\chi_{[0,n]}(h)\mathcal{M}\chi_{[0,n]}(h)$, and $\cup_{n \geq 1}(\chi_{[0,n]}(h)\mathcal{M}\chi_{[0,n]}(h))$ σ-weakly dense in $\mathcal{M}$. So $\mathfrak{m}_{\varphi_h}$ σ-weakly dense in $\mathcal{M}$, and φ_h therefore semifinite.

Since each h_ϵ belongs to $\mathcal{M}_\varphi$, it is clear that $\sigma_t^\varphi(h_\epsilon^{1/2}) = h_\epsilon^{1/2}$ for each $\epsilon > 0$. So, for any $a \in \mathcal{M}^+$ and any $\epsilon > 0$, we have that

$$\begin{aligned}
\varphi_{h_\epsilon}(\sigma_t^\varphi(a)) &= \varphi(h_\epsilon^{1/2}\sigma_t^\varphi(a)h_\epsilon^{1/2}) \\
&= \varphi(\sigma_t^\varphi(h_\epsilon^{1/2}ah_\epsilon^{1/2})) \\
&= \varphi(h_\epsilon^{1/2}ah_\epsilon^{1/2}) \\
&= \varphi_{h_\epsilon}(a).
\end{aligned}$$

Letting $\epsilon \searrow 0$ now yields the fact that $\varphi_h \circ \sigma_t^\varphi = \varphi_h$ for each $t \in \mathbb{R}$.

To show that $h \mapsto \varphi_h$ is order-preserving, one only needs to show that if $h \leq k$, then $h_\epsilon \leq k_\epsilon$ for each ϵ and appeal to the lemma.

To see the final claim, let $a \in \mathcal{M}_+ \backslash \{0\}$ be given. Since the net (φ_{h_ϵ}) is increasing, we have that $\varphi_h(a) = 0$ if and only if $\varphi_{h_\epsilon}(a) = 0$ for each $\epsilon > 0$. But the faithfulness of φ now ensures that for a given $\epsilon > 0$, we have that $\varphi_{h_\epsilon}(a) = 0$ if and only if $h_\epsilon^{1/2}ah_\epsilon^{1/2} = 0$ if and only if $\mathbf{s}(a) \perp \mathbf{s}(h_\epsilon^{1/2})$. Taking into account that $\mathbf{s}(h_\epsilon^{1/2}) = \mathbf{s}(h)$, it now follows that $\mathbf{s}(h) = \mathrm{supp}(\varphi_h)$. $\qquad\square$

Given a normal semifinite weight ψ on $\mathcal{M}^+$, the requirement that $\psi \circ \sigma_t^\varphi = \psi$ for each $t \in \mathbb{R}$ is, therefore, clearly a necessary requirement for the existence of an operator $h \,\eta\, \mathcal{M}_\varphi^+$ such that $\psi = \varphi_h$. The remarkable achievement of Pedersen and Takesaki (see [**PT73**]) was showing that this requirement is also sufficient. Building on the above, they were able to establish the following converse to the preceding theorem:

Theorem 8.22 ([PT73, Theorem 5.11]) *Let φ be a faithful normal semifinite weight on $\mathcal{M}$. For any normal semifinite weight ψ on $\mathcal{M}^+$,* *satisfying $\psi \circ \sigma_t^\varphi = \psi$ for all $t \in \mathbb{R}$, there exists a unique $h \, \eta \, \mathcal{M}_\varphi^+$ such that $\psi = \varphi_h$. We call this operator the Radon–Nikodym derivative of ψ with respect to φ and will denote it by $h = \frac{d\psi}{d\varphi}$. This derivative satisfies $\mathbf{s}(\frac{d\psi}{d\varphi}) = \mathrm{supp}(\psi)$.*

Given normal semifinite weights φ and ψ on $\mathcal{M}$ with φ also faithful, the requirement that $\psi \circ \sigma_t^\varphi = \psi$ for each $t \in \mathbb{R}$ is undoubtedly of cardinal importance in the theory of Radon–Nikodym derivatives of weights. This property has by now therefore earned its own terminology, with ψ commonly said to *commute* with φ when this requirement is satisfied.

We close this discussion of Radon–Nikodym derivatives by noting that one may use this theory to establish a very elegant relationship between the modular automorphism groups of two commuting faithful normal semifinite weights. We specifically have the following.

Theorem 8.23 ([PT73, Theorem 4.6])   *Let φ be a faithful normal semifinite weight on $\mathcal{M}$, and h a non-singular positive operator affiliated to $\mathcal{M}_\varphi$ (where non-singular means $\mathbf{s}(h) = \{0\}$). Then, the modular automorphism group of the weight $\psi = \varphi_h$ is given by $\sigma_t^\psi(a) = h^{it}\sigma_t^\varphi(a)h^{-it}$ for each $a \in \mathcal{M}$, $t \in \mathbb{R}$.*

8.3 Connes cocycle derivatives

For any serious study of modular automorphism groups, the theory of Connes cocycles is an essential companion, as it is par excellence the theory which provides us with the technology to compare the automorphism groups of two distinct faithful normal semifinite weights on a fixed von Neumann algebra. This technology will prove to be a vital ingredient in our development of the theory of crossed products. We therefore pause to briefly review the essentials of Connes cocycles as they relate to modular automorphism groups. In his paper on the classification of type III factors, Connes proved the following very important theorem:

Theorem 8.24 ([Tak03a, VIII.3.3])   *Let $\mathcal{M}$ be a von Neumann algebra and ψ_1 and ψ_2 faithful semifinite normal weights on $\mathcal{M}$.*

Then, there exists a unique σ-strongly continuous one-parameter family $\{u_t\}$ of unitaries in $\mathcal{M}$ with the following properties.

- *$u_{s+t} = u_s \sigma_s^{\psi_2}(u_t)$ for all $s, t \in \mathbb{R}$.*
- *$u_s \sigma_s^{\psi_2}(\mathfrak{n}_{\psi_1}^* \cap \mathfrak{n}_{\psi_2}) = \mathfrak{n}_{\psi_1}^* \cap \mathfrak{n}_{\psi_2}$ for all $s \in \mathbb{R}$.*
- *For each $a \in \mathfrak{n}_{\psi_1} \cap \mathfrak{n}_{\psi_2}^*$ and $b \in \mathfrak{n}_{\psi_1}^* \cap \mathfrak{n}_{\psi_2}$, there exists a function F which is bounded and continuous on the closed strip $S = \{z \in \mathbb{C} : 0 \leq \Im(z) \leq 1\}$ and analytic on the open strip $S^o = \{z \in \mathbb{C} : 0 < \Im(z) < 1\}$, such that $F(t) = \psi_1(u_t \sigma_t^{\psi_1}(b)a)$ and $F(t+i) = \psi_2(au_t \sigma_t^{\psi_1}(b))$ for all $s \in \mathbb{R}$.*
- *$\sigma_t^{\psi_1}(a) = u_t \sigma_t^{\psi_2}(a)u_t^*$ for all $a \in \mathcal{M}, t \in \mathbb{R}$.*

In view of the uniqueness, we make the following definition.

Definition 8.25 The family $\{u_t\}$ described in the preceding theorem is called the cocycle derivative of ψ_1 with respect to ψ_2 and is denoted by $(D\psi_1 : D\psi_2)_t = u_t$ $(t \in \mathbb{R})$.

Remark 8.26 Let φ and ψ be faithful normal semifinite weights on $\mathcal{M}$ such that $\psi \circ \sigma_t^{\varphi} = \psi$ for each $t \in \mathbb{R}$. We then know from Theorem 8.22 that $\frac{d\psi}{d\varphi} = h$ exists as a positive operator affiliated to $\mathcal{M}_\varphi$ which is non-singular since $\operatorname{supp}(\frac{d\psi}{d\varphi}) = \operatorname{supp}(\psi) = \mathbb{1}$. By Theorem 8.23, the modular automorphism group of ψ is then of the form $\sigma_t^\psi(a) = (\frac{d\psi}{d\varphi})^{it} \sigma_t^\varphi(a)(\frac{d\psi}{d\varphi})^{-it}$ for each $a \in \mathcal{M}$, $t \in \mathbb{R}$. The claims of Theorem 8.23 may in fact be further refined to show that in this case, $(\frac{d\psi}{d\varphi})^{it} = (D\psi : D\varphi)_t$ for all $t \in \mathbb{R}$. Cocycles may, therefore, justifiably be regarded as a theory of derivatives of faithful normal semifinite weights with respect to each other, which generalises the Pedersen–Takesaki theory.

One may now use the uniqueness clause in Theorem 8.24 to verify the following fact. This was first observed by Digernes [**Dig75**, Corollary 2.3].

Corollary 8.27 *If α is a $*$-automorphism of $\mathcal{M}$ and ψ_1 and ψ_2 faithful semifinite normal weights, then $(D\psi_1 \circ \alpha : D\psi_2 \circ \alpha)_t = \alpha^{-1}((D\psi_1 : D\psi_2)_t)$ for all $t \in \mathbb{R}$.*

The following technical lemma is often useful.

Lemma 8.28 ([Con73, Lemma 1.2.3(c)]) *Let ψ be a faithful semifinite normal weight on $\mathcal{M}$, and u a unitary in $\mathcal{M}$. For the weight ψ_u defined by $\psi_u(a) = \psi(uau^*)$ for all $a \in \mathcal{M}_+$, we have $\sigma_t^\psi(u) = u(D(\psi_u) : D\psi)_t$ for all $t \in \mathbb{R}$.*

The cocycle derivatives satisfy the following very elegant chain rule.

Theorem 8.29 ([Tak03a, VIII.3.7]) *Let ψ_1, ψ_2 and ψ_3 be faithful semifinite normal weights on $\mathcal{M}$. Then*

$$(D\psi_1 : D\psi_2)_t = (D\psi_1 : D\psi_3)_t(D\psi_3 : D\psi_2)_t$$

for all $t \in \mathbb{R}$.

There is a kind of converse to Theorem 8.24 in the form of the following theorem.

Theorem 8.30 ([Tak03a, VIII.3.8]) *Let ψ be a faithful semifinite normal weight on $\mathcal{M}$, and $\{u_t\}$ $(t \in \mathbb{R})$ a σ-strongly continuous family of unitaries in $\mathcal{M}$ that satisfies the cocycle identity $u_{s+t} = u_s\sigma_s^\psi(u_t)$ for all $s, t \in \mathbb{R}$. Then, there exists a faithful semifinite normal weight ψ_0 on $\mathcal{M}$ for which $u_t = (D\psi : D\psi_0)_t$ for all $t \in \mathbb{R}$.*

The theory of cocycle derivatives also yields the following very deep and useful characterisation of semifinite algebras. The beauty of this result is that it not only characterises semifiniteness, but also gives a prescription for constructing a faithful normal semifinite trace on the given algebra. (The final claim is not formulated in [**Tak03a**], but can easily be seen to hold from a consideration of the proof of that theorem.)

Theorem 8.31 ([Tak03a, VIII.3.14]) *For a von Neumann algebra $\mathcal{M}$, the following are equivalent.*

- *$\mathcal{M}$ is semifinite.*
- *For every faithful semifinite normal weight ψ on $\mathcal{M}$, the modular automorphism group σ_t^ψ $(t \in \mathbb{R})$ is inner in the sense that there exists a strongly continuous unitary group u_t $(t \in \mathbb{R})$ in $\mathcal{M}$ such that $\sigma_t^\psi(a) = u_t a u_t^*$ for all $a \in \mathcal{M}$ and all $t \in \mathbb{R}$.*

- *There exists a faithful semifinite normal weight φ on $\mathcal{M}$ for which the modular automorphism group σ_t^φ $(t \in \mathbb{R})$ is inner in the above sense.*

If the above conditions hold, then by Stone's theorem, the unitary group implementing σ_t^φ $(t \in \mathbb{R})$ is of the form $u_t = h^{it}$ for some positive non-singular operator affiliated to the centraliser $\mathcal{M}_\varphi$, with the prescription $\tau = \lim_{\epsilon \searrow 0} \varphi((h_\epsilon^{-1})^{1/2} \cdot (h_\epsilon^{-1})^{1/2})$ where $h_\epsilon^{-1} = h^{-1}(1 + \epsilon h^{-1})^{-1}$, then yielding a faithful semifinite normal trace on $\mathcal{M}$.

When using the cocycle derivative to describe the domination of one weight by another, the following theorem is very useful.

Theorem 8.32 ([Tak03a, VIII.3.17])   *For a pair ψ_1, ψ_2 of faithful semifinite normal weights on $\mathcal{M}$, the following conditions are equivalent.*

(1) *There exists $K > 0$ such that*

$$\psi_2(x) \leq K\psi_1(x), \quad x \in \mathcal{M}_+.$$

(2) *The cocycle derivative $(D\psi_1 : D\psi_2)_t \equiv u_t$ can be extended to an $\mathcal{M}$-valued σ-weakly continuous bounded function on the horizontal strip $\overline{D}_{\frac{1}{2}} = \{z \in \mathbb{C} \colon -\frac{1}{2} \leq \mathrm{Im}(z) \leq 0\}$ for which $\|u_{-\frac{i}{2}}\| \leq K^{1/2}$, and which is analytic on the interior of the strip.*

If these conditions hold, then

$$\psi_2(x) = \psi_1(u^*_{-\frac{i}{2}} x u_{-\frac{i}{2}}), \quad x \in \mathfrak{m}_{\psi_1}.$$

8.4 Conditional expectations and operator-valued weights

We briefly encountered the notion of a noncommutative conditional expectation in Corollary 6.25, albeit only for semifinite algebras. For the further development of the theory, we need a basic working knowledge of these objects in the context of general von Neumann algebras and in particular of their relationship to operator-valued weights, introduced below. First, we present a more general definition of the notion introduced in Corollary 6.25.

Definition 8.33 Let φ be a faithful semifinite normal weight on a von Neumann algebra $\mathcal{M}$, and $\mathcal{N}$ a von Neumann subalgebra for which the restriction $\varphi\lceil\mathcal{N}$ of φ to $\mathcal{N}$ is still semifinite. A linear map E from $\mathcal{M}$ onto $\mathcal{N}$ for which we have that

- $\|\mathrm{E}(a)\| \leq \|a\|$ for all $a \in \mathcal{M}$,
- $\mathrm{E}(a) = a$ for all $a \in \mathcal{N}$,
- $\varphi \circ \mathrm{E} = \varphi$

is called the *conditional expectation of $\mathcal{M}$ onto $\mathcal{N}$ with respect to φ.*

Remark 8.34 Let $\mathcal{C}$ be a unital C^*-algebra, and $\mathcal{B}$ a unital subalgebra. It is a well-known result of Tomiyama [**Tom70**] that any unital contractive projection P from $\mathcal{C}$ onto $\mathcal{B}$ is completely contractive and satisfies $P(abc) = aP(b)c$ for all $a, c \in \mathcal{B}$ and all $b \in \mathcal{C}$. A linear map $T : \mathcal{C} \to \mathcal{B}$ is said to be completely contractive (resp. completely positive) if for any n the induced map $T_n \colon [a_{ij}] \mapsto [T(a_{ij})]$ from $M_n(\mathcal{C})$ to $M_n(\mathcal{B})$ is again contractive (resp. positive). (We will explore these issues in more detail in § 12.1.) Applying this to a conditional expectation E of $\mathcal{M}$ onto $\mathcal{N}$ with respect to φ, it trivially follows that any such conditional expectation is completely contractive and satisfies the condition that $\mathrm{E}(abc) = a\mathrm{E}(b)c$ for all $a, c \in \mathcal{N}$ and all $b \in \mathcal{M}$. Since by assumption $\mathrm{E}(\mathbb{1}) = \mathbb{1}$ with $\|\mathrm{E}\| = \|\mathrm{E}(\mathbb{1})\| = 1$, E must in fact be positivity preserving by Proposition 0.63. Applying this same proposition to the induced maps from $M_n(\mathcal{C})$ to $M_n(\mathcal{B})$ one easily sees that E is in fact completely positive (not just completely contractive). We also have $\mathrm{E}(a^*)\mathrm{E}(a) \leq \mathrm{E}(a^*a)$ for all $a \in \mathcal{M}$. This can be seen by applying E to the inequality $(a^* - \mathrm{E}(a^*))(a - \mathrm{E}(a)) \geq 0$.

The question of existence of such maps now arises. This will ultimately be answered by Theorem 10.54, which will only be proved once the requisite technology is in place. We will not actually need this result until chapter 12, but record it here for the sake of completeness.

Theorem 10.54 ([Tak03a, Theorem IX.4.2]) *Let φ be a faithful semifinite normal weight on a von Neumann algebra $\mathcal{M}$, and $\mathcal{N}$ a von Neumann subalgebra for which the restriction $\varphi\lceil\mathcal{N}$ of φ to $\mathcal{N}$ is still semifinite. We then have that $\sigma_t^\varphi(\mathcal{N}) = \mathcal{N}$ for each $t \in \mathbb{R}$ if and only if there exists a unique normal (σ-weakly continuous) conditional expectation E of $\mathcal{M}$ onto $\mathcal{N}$ with respect to φ.*

Remark 8.35

(i) Let E be a conditional expectation of $\mathcal{M}$ onto $\mathcal{N}$ with respect to φ. We then have $\mathrm{E} \circ \sigma_t^\varphi = \sigma_t^\varphi \circ \mathrm{E}$. To see this, note that for any $a \in \mathfrak{m}_\varphi$, we may conclude from the facts $\varphi \circ \mathrm{E} = \varphi$ and $\varphi \circ \sigma_t^\varphi = \varphi$, that $\varphi(|\mathrm{E}(\sigma_t^\varphi(a)) - \sigma_t^\varphi(\mathrm{E}(a))|^2) = 0$ for all t, and hence that $\mathrm{E} \circ \sigma_t^\varphi = \sigma_t^\varphi \circ \mathrm{E}$ on $\mathfrak{m}_\varphi$. The normality and σ-weak density of $\mathfrak{m}_\varphi$ in $\mathcal{M}$ now yields the claim.

(ii) As we saw in Corollary 6.25, the criteria for the existence of a conditional expectation are much simpler in the case where $\mathcal{M}$ is equipped with a faithful semifinite normal trace τ. This stems from the fact that the modular automorphism group of a trace is trivial.

Of course, a conditional expectation of the above type may not always exist. In such cases, a so-called operator-valued weight is often a good substitute.

Definition 8.36 Let φ be a faithful semifinite normal weight on a von Neumann algebra $\mathcal{M}$, and $\mathcal{N}$ a von Neumann subalgebra of $\mathcal{M}$. An *operator-valued weight* from $\mathcal{M}$ to $\mathcal{N}$ is a mapping $\mathscr{W} : \mathcal{M}_+ \to \widehat{\mathcal{N}}_+$ such that:

(1) $\mathscr{W}(\gamma x) = \gamma \mathscr{W}(x) \quad \gamma \geq 0, x \in \mathcal{M}_+$;
(2) $\mathscr{W}(x + y) = \mathscr{W}(x) + \mathscr{W}(y) \quad x, y \in \mathcal{M}_+$;
(3) $\mathscr{W}(a^* x a) = a^* \mathscr{W}(x) a \quad x \in \mathcal{M}_+, a \in \mathcal{N}$.

We say that $\mathscr{W}$ is *normal* if $x_i \nearrow x \Rightarrow \mathscr{W}(x_i) \nearrow \mathscr{W}(x) \quad x_i, x \in \mathcal{M}_+$.

By analogy with ordinary weights, we set

$$\mathfrak{n}_{\mathscr{W}} = \{x \in \mathcal{M} \colon \|\mathscr{W}(x^* x)\| < \infty\}$$

$$\mathfrak{m}_{\mathscr{W}} = \mathfrak{n}_{\mathscr{W}}^* \mathfrak{n}_{\mathscr{W}} = \operatorname{span}\{x^* y \colon x, y \in \mathfrak{n}_{\mathscr{W}}\}.$$

The weight $\mathscr{W}$ is called *faithful* if $\mathscr{W}(x^* x) = 0$ implies $x = 0$, and *semifinite* if $\mathfrak{n}_{\mathscr{W}}$ is σ-weakly dense in $\mathcal{M}$.

Remark 8.37 In the case where $\mathscr{W}$ is normal, it allows for a natural extension to an affine normal map from $\widehat{\mathcal{M}}_+$ onto $\widehat{\mathcal{N}}_+$ [**Haa79b**, Proposition 2.5]. Given any $m \in \widehat{\mathcal{M}}_+$, the construction is done by picking $(x_i) \subseteq \mathcal{M}_+$ such that $x_i \nearrow m$ and then defining $\mathscr{W}(m)$ as $\sup_i \mathscr{W}(x_i)$ once the uniqueness of this supremum has been verified.

When working with operator-valued weights, the following technical facts often come in handy.

Proposition 8.38 ([Tak03a, IX.4.13]) *Let $\mathscr{W}$ be as in the preceding definition. Then, the following holds.*

- *$\mathfrak{m}_{\mathscr{W}}$ is spanned by its positive part $\mathfrak{p}_{\mathscr{W}} = \{x \in \mathcal{M}_+ : \|\mathscr{W}(x)\| < \infty\}$.*
- *$\mathfrak{n}_{\mathscr{W}}$ and $\mathfrak{m}_{\mathscr{W}}$ are two-sided modules over $\mathcal{N}$.*
- *The restriction of $\mathscr{W}$ to $\mathfrak{p}_{\mathscr{W}}$ extends to a linear map $\dot{\mathscr{W}} : \mathfrak{m}_{\mathscr{W}} \to \mathcal{N}$ which satisfies the 'expectation-like' property that*

$$\dot{\mathscr{W}}(axb) = a\dot{\mathscr{W}}(x)b \ \text{for all } a, b \in \mathcal{N} \text{ and all } x \in \mathfrak{m}_{\mathscr{W}}.$$

In particular, if $\mathscr{W}(\mathbb{1}) = \mathbb{1}$, then $\dot{\mathscr{W}}$ is a contractive projection from $\mathcal{M}$ onto $\mathcal{N}$ (in which case Tomiyama's result applies—see Remark 8.34).

We close this survey by noting the behaviour of operator-valued weights with respect to tensor products. We first remind the reader that the tensor product $\psi_1 \otimes \psi_2$ of two semifinite normal weights ψ_1 and ψ_2 on von Neumann algebras $\mathcal{M}_1$ and $\mathcal{M}_2$ is the unique semifinite normal weight on $\mathcal{M}_1 \overline{\otimes} \mathcal{M}_2$ with $\mathrm{supp}(\psi_1 \otimes \psi_2) = \mathrm{supp}(\psi_1) \otimes \mathrm{supp}(\psi_2)$ and with $(\psi_1 \otimes \psi_2)(a \otimes b) = \psi_1(a)\psi_2(b)$ for all simple tensors $a \otimes b$ with $a \in \mathcal{M}_1^+$ and $b \in \mathcal{M}_2^+$. (For a fairly complete introduction to tensor products of weights see [**Cuc93**, §3]. Section VIII.4 of [**Tak03a**] is also helpful.)

Theorem 8.39 ([Haa79c, Theorem 5.5]) *Let $\mathcal{M}_1$ and $\mathcal{M}_2$ be von Neumann algebras, and $\mathcal{N}_1$ and $\mathcal{N}_2$, respectively, be von Neumann subalgebras of $\mathcal{M}_1$ and $\mathcal{M}_2$. For each $i \in \{1, 2\}$, let $\mathscr{W}_i$ be an operator-valued weight from $\mathcal{M}_i^+$ to $\widehat{\mathcal{N}_i}^+$. Then, there is a unique operator-valued weight $\mathscr{W}$ from $(\mathcal{M}_1 \overline{\otimes} \mathcal{M}_2)_+$ to $(\widehat{\mathcal{N}_1 \overline{\otimes} \mathcal{N}_2})_+$ such that*

$$(\psi_1 \otimes \psi_2) \circ \mathscr{W} = (\psi_1 \circ \mathscr{W}_1) \otimes (\psi_2 \circ \mathscr{W}_2)$$

for any pair (ψ_1, ψ_2) of f.n.s. weights on the pair $(\mathcal{N}_1, \mathcal{N}_2)$.

8.5 A Hilbert space approach

Modular theory may also be studied at Hilbert space level. For this part of the theory, we shall not go into any measure of detail, but rather content ourselves with the very rudiments.

Definition 8.40 Let $\mathcal{M}$ be a von Neumann algebra equipped with a cyclic and separating vector Ω. We define the *natural positive cone* $\mathscr{P}^\natural$ associated with the pair $(\mathcal{M}, \Omega)$ as the closure of the set $\{aj(a)\Omega \colon a \in \mathcal{M}\}$, where $j \colon \mathcal{M} \to \mathcal{M}'$ is the antilinear *-isomorphism given by $j(a) = JaJ$ (where $a \in \mathcal{M}$).

For the space H, $\mathscr{P}^\natural$ plays the same role that the cone of positive elements in $L^2(X, \Sigma, \mu)$ plays in this space. We present some basic technical facts regarding $\mathscr{P}^\natural$ before presenting a result substantiating this claim.

Proposition 8.41 ([BR87, 2.5.26]) *The closed subset $\mathscr{P}^\natural$ of H has the following properties.*

(1) $\mathscr{P}^\natural \;=\; \overline{\Delta^{1/4}\mathcal{M}_+\Omega} \;=\; \overline{\Delta^{-1/4}\mathcal{M}'_+\Omega} \;=\; \overline{\Delta^{1/4}\overline{\mathcal{M}_+\Omega}} \;=\; \overline{\Delta^{-1/4}\overline{\mathcal{M}'_+\Omega}}$, *and hence $\mathscr{P}^\natural$ is convex.*

(2) $\Delta^{it}\mathscr{P}^\natural = \mathscr{P}^\natural$ *for all $t \in \mathbb{R}$.*

(3) *For any positive-definite function f, we have that $f(\log \Delta)\,\mathscr{P}^\natural \subseteq \mathscr{P}^\natural$.*

(4) *For any $\xi \in \mathscr{P}^\natural$, we have that $J\xi = \xi$.*

(5) *For any $a \in \mathcal{M}$, we have that $aj(a)\mathscr{P}^\natural \subseteq \mathscr{P}^\natural$.*

We end this discussion of $\mathscr{P}^\natural$ with the promised presentation of the geometric properties of $\mathscr{P}^\natural$.

Proposition 8.42 ([BR87, 2.5.28])

(1) *$\mathscr{P}^\natural$ is a self-dual cone in the sense that $\xi \in \mathscr{P}^\natural$ if and only if $\langle \xi, \eta \rangle \geq 0$ for all $\eta \in \mathscr{P}^\natural$.*

(2) *$\mathscr{P}^\natural$ is a pointed cone in the sense that $\mathscr{P}^\natural \cap (-\mathscr{P}^\natural) = \{0\}$.*

(3) *Any $\xi \in H$ for which we have that $J\xi = \xi$ admits a unique decomposition $\xi = \xi_1 - \xi_2$, where $\xi_1, \xi_2 \in \mathscr{P}^\natural$ with $\xi_1 \perp \xi_2$.*

(4) *The Hilbert space H is linearly spanned by $\mathscr{P}^\natural$.*

We close this section with a discussion regarding the uniqueness of the GNS representation of the pair $(\mathcal{M}, \varphi)$ where φ is a faithful normal semifinite weight on the von Neumann algebra $\mathcal{M}$. Haagerup proved a very deep theorem essentially showing that any representation of $\mathcal{M}$ which admits objects that mimic the action of J_φ and $\mathscr{P}_\varphi^\natural$ is a faithful copy of the GNS representation of the pair $(\mathcal{M}, \varphi)$. This claim may be made exact with the following definition.

Definition 8.43 Given a von Neumann algebra $\mathcal{M}$ equipped with a faithful normal semifinite weight φ, a quadruple $(\pi_0(\mathcal{M}), H_0, J, \mathscr{P})$ where π_0 is a faithful representation of $\mathcal{M}$ on the Hilbert space H_0, $J : H_0 \to H_0$ anti-linear isometric involution, and $\mathscr{P}$ a self-dual cone of H_0 is said to be a *standard form* of $\mathcal{M}$ if the following conditions hold:

- $J\mathcal{M}J = \mathcal{M}'$ (the commutant of $\mathcal{M}$);
- $JzJ = z^*$ for all z in the centre of $\mathcal{M}$;
- $J\xi = \xi$ for all $\xi \in \mathscr{P}$;
- $a(JaJ)\mathscr{P} \subseteq \mathscr{P}$ for all $a \in \mathcal{M}$.

(Recall that when we say that $\mathscr{P}$ is a self-dual cone, we mean that $\xi \in \mathscr{P}$ if and only if $\langle \xi, \zeta \rangle \geq 0$ for all $\zeta \in \mathscr{P}$.)

The value of the above concept is derived from the following very deep and useful theorem.

Theorem 8.44 ([Haa75c]) *The standard form of a von Neumann algebra $\mathcal{M}$ is unique in the sense that if*

$$(\pi_0(\mathcal{M}), H_0, J, \mathscr{P}) \text{ and } (\widetilde{\pi}_0(\widetilde{\mathcal{M}}), \widetilde{H}_0, \widetilde{J}, \widetilde{\mathscr{P}})$$

are two standard forms, and $\alpha : \pi_0(\mathcal{M}) \to \widetilde{\pi}_0(\widetilde{\mathcal{M}})$ is a $*$-isomorphism, then there exists a unique unitary operator $u : H_0 \to \widetilde{H}_0$ such that

- $\alpha(x) = uxu^*$ *for $x \in \pi_0(\mathcal{M})$;*
- $\widetilde{J} = uJu^*$;
- $\widetilde{\mathscr{P}} = u\mathscr{P}$.

Readers wishing to understand tensor products of standard forms will find very useful information in [**Cuc93**].

8.6 Group algebras—a rich store of examples

In this section, we give a brief overview of von Neumann group algebras culminating in a presentation of von Neumann algebraic quantum groups. A basic knowledge of the rudiments of locally compact groups will here be helpful. The reasons for this are manifold. Not only is there a long history of mutual enrichment through exchange of ideas between the theories of von Neumann algebras and abstract harmonic analysis, but in addition the fields of enquiry that lie at the interface of these fields (quantum harmonic analysis and quantum groups) are not only currently very active, but also make widespread use of noncommutative L^p-spaces. A detailed exposition of this field requires some knowledge of not only locally compact groups, but also left Hilbert algebras. We shall, however, strive to present the material in a manner which is palatable to the non-specialist, and yet also concise and accurate. The overview of group von Neumann algebras is a compilation of material taken from [**Ped79**], [**Sut78**] and [**Tak03a**]. The material on quantum groups is based on expositions by van Daele [**vD07, vD14**] and Vaes [**Vae01a**]. For a general introduction to abstract harmonic analysis, the book of, for example, Folland [**Fol16**] is helpful. In Takesaki's presentation, much of the detailed development of the theory is done in the language of left Hilbert algebras. The close relation between left Hilbert algebras and group algebras (see [**Tak03a**, Example VI.1.2 and Proposition VII.3.1]) nevertheless makes this a valuable reference work for material regarding even the most general locally compact groups.

8.6.1 A brief look at abstract Harmonic analysis

Let G be a not necessarily abelian locally compact group. Each such group, of course, admits a left (respectively, right) Haar measure μ_H (respectively, $_H\mu$) (the subscript H serves to honour the inventor of this measure, Alfréd Haar), where 'left' means that for any $s \in G$ and any measurable subset E, we have that $\mu_H(sE) = \mu_H(E)$. Such measures are unique up to a positive scalar multiple. Moreover, each left Haar measure μ_H canonically corresponds to a right Haar measure given by $_H\mu(E) = \mu_H(E^{-1})$ where $E^{-1} = \{s \colon s^{-1} \in E\}$. If G is compact, we also require that $\mu_H(G) = 1$. In the case where G is discrete, we similarly require that $\mu_H(\{s\}) = 1$ for each $s \in G$. These restrictions ensure that, in the case of compact or discrete groups, the Haar measure as described above is unique. For finite groups, this convention is, of course, self-contradictory since such groups are both compact and discrete. When dealing with such groups, we will, therefore, a priori, have to decide whether we wish to classify the group as compact or discrete and then stick to that convention.

For non-abelian groups, the left and right Haar measures are linked by means of the so-called *modular function* δ_G, which is a continuous homomorphism from G into the positive reals $(0, \infty)$, such that $\mu_H(Es) = \delta_G(s)\mu_H(E)$ for any $s \in G$ and any measurable subset E of G. Writing ds for $d\mu_H(s)$, left invariance of Haar measure ensures that $d(ts) = ds$. By now bringing the modular function into play, we further have that

$$d\mu_H(st) = \delta_G(t)d\mu_H(s), \qquad d\mu_H(s^{-1}) = \delta_G(s)^{-1}d\mu_H(s) = d_H\mu(s).$$

Since therefore integration against right Haar measure can be done in terms of $\delta_G(s)^{-1}d\mu_H(s)$, we shall henceforth not make any explicit use of right Haar measure and will simply write ds for $d\mu_H(s)$. A locally compact group is said to be *unimodular* if δ_G is the constant function 1. So, unimodular groups are those groups for which the left Haar measure is also a right Haar measure.

Given $f \in L^1(G)$ and $g \in L^p(G)$, the prescription $(f * g)(s) = \int_G f(t)g(t^{-1}s)\,dt$ defines an element $f * g$ of $L^p(G)$ for which we have that $\|f * g\|_p \leq \|f\|_1\|g\|_p$. The element $f * g$ is referred to as the *convolution* of f and g. With respect to convolution, $L^1(G)$ then becomes a Banach algebra. We point out that for f and g as above, the integral representing $(f * g)(s)$ can be expressed in several different forms. We specifically have $\int_G f(t)g(t^{-1}s)\,dt = \int_G f(ts)g(t^{-1})\,dt = \int_G f(t^{-1})g(ts)\,\delta_G(t^{-1})dt = \int_G f(st^{-1})g(t)\,\delta_G(t^{-1})dt$. This information is particularly useful for a deeper analysis of the properties of convolution.

Now, suppose that the group G is abelian. In this setting, the group $\widehat{G}$ is defined to be the group of all characters of G (continuous group-homomorphisms from G into $\mathbb{T}$), with $\widehat{G}$ itself again proving to be a locally compact abelian group. Using this fact, one may introduce the following definition of the abstract Fourier transform.

Definition 8.45 We may define the Fourier transform $\mathcal{F}$ on $L^1(G)$ by the prescription

$$\mathcal{F}(f)(\gamma) = \int_G \overline{\gamma(g)}f(g)\,dg, \qquad \gamma \in \widehat{G}.$$

We list some of the properties of this Fourier transform. For proofs of these facts, refer to, for example, [**Tak03a**, Theorem VII.3.14].

- $\mathcal{F}$ is just the so-called Gelfand transform on $L^1(G)$ and maps $L^1(G)$ into $C_0(\widehat{G})$ (the functions vanishing at infinity).
- **Abstract Plancherel formula:** $\mathcal{F}$ preserves the L^2-norm on $L^1(G) \cap L^2(G)$ and extends to a unitary from $L^2(G)$ to $L^2(\widehat{G})$. For every $f \in L^1(\widehat{G}) \cap L^2(\widehat{G})$, the map $\mathcal{F}^* = \mathcal{F}^{-1}$ agrees with the inverse Fourier transform defined by $\widehat{\mathcal{F}}(f)(g) = \int_{\widehat{G}} \gamma(g) f(\gamma)\, d\gamma$, where $f \in L^1(\widehat{G})$. (Here, integration is with respect to the so-called *dual measure* on $\widehat{G}$, which is the unique left Haar measure for which the stated claim is true.)

8.6.2 Basic concepts: algebra, commutant and predual

We proceed to the construction of group von Neumann algebras in $B(L^2(G))$ for general locally compact groups. For each $s \in G$, one may define a corresponding unitary operator $\lambda(s)$ by means of the prescription $(\lambda(s)\xi)(t) = \xi(s^{-1}t)$ for all $\xi \in L^2(G)$ and all $s \in G$. It is an exercise to see that $\lambda(s)^* = \lambda(s^{-1})$. Right multiplication may similarly be used to define unitaries $\rho(s)$ where $s \in G$. These unitaries are defined by the prescription $(\rho(s)\xi)(t) = \delta_G(s)^{1/2}\xi(ts)$ for all $\xi \in L^2(G)$ and all $s \in G$. The *left* group von Neumann algebra $\mathrm{VN}_l(G)$ is then defined to be the von Neumann subalgebra of $B(L^2(G))$ generated by the unitaries $\lambda(s)$ $(s \in G)$, with the *right* group von Neumann algebra $\mathrm{VN}_r(G)$ being the von Neumann subalgebra generated by the unitaries $\rho(s)$. The left and right group algebras are related by the fact that $\mathrm{VN}_l(G)' = \mathrm{VN}_r(G)$ and $\mathrm{VN}_r(G)' = \mathrm{VN}_l(G)$ [**Tak03a**, Proposition VII.3.1].

If G is abelian (and therefore also unimodular), it is clear that then $\rho(s) = \lambda(s^{-1}) = \lambda(s)^*$ for each $s \in G$. So, in this case, $\mathrm{VN}_l(G) = \mathrm{VN}_r(G)$. Therefore, we shall simply write $\mathrm{VN}(G)$ if G is abelian. For this group algebra, we have that $\mathrm{VN}(G)$ is a copy of $L^\infty(\widehat{G})$. More properly,

if G is abelian, then $\mathcal{F}(\mathrm{VN}(G))\mathcal{F}^{-1}$ agrees with the von Neumann algebra of multiplication operators on $L^2(\widehat{G})$ with symbols in $L^\infty(\widehat{G})$ [**Tak03a**, Theorem VII.3.14].

Although we shall not really have need of this fact, we briefly pause to describe the predual of $\mathrm{VN}_l(G)$. As a bookkeeping device, we first define the operation $f^\vee(s) = f(s^{-1})$ $(s \in G)$ on functions on G. Next note that for any $\xi, \zeta \in L^2(G)$, the function $\bar{\xi} * \zeta^\vee$ can be shown to be a continuous function on G vanishing at infinity for which we also have that

$$\langle \lambda(s)\zeta, \xi \rangle = \bar{\xi} * \zeta^\vee(s) \text{ for all } s \in G.$$

The set of functions $A(G) = \{\bar{\xi} * \zeta^\vee : \xi, \zeta \in L^2(G)\}$ then turns out to be a dense *-subalgebra of $C_0(G)$—the continuous functions on G vanishing at infinity [**Tak03a**, Lemma VII.3.7]. This algebra may be regarded as a subset of the vector functionals on $\mathrm{VN}_l(G)$ by means of the correspondence $\bar{\xi} * \zeta^\vee \leftrightarrow \langle \cdot \zeta, \xi \rangle$ described in the displayed equation above. When equipped with the topology inherited from the vector states, $A(G)$ can then be seen to be a commutative Banach algebra corresponding to the predual of $\mathrm{VN}_l(G)$. This realisation of $A(G)$ as a Banach algebra is referred to as the *Fourier algebra* of G.

For any $f \in L^1(G)$, the prescription $\xi \to f * \xi$ yields a bounded operator on $L^2(G)$, which we shall denote by $\lambda(f)$. One finds that the von Neumann algebra generated by these $\lambda(f)$s again yields $\mathrm{VN}_l(G)$ (see Remark 7.2.1 in [**Ped79**]). Given $f \in L^1(G)$, one may define $f^* \in L^1(G)$ by $f^*(s) = \delta_G(s)^{-1}\overline{f(s^{-1})}$ for all $s \in G$. It is then an instructive exercise to show that $\lambda(f)^* = \lambda(f^*)$ for all $f \in L^1(G)$. Thus, the adjoint operation on $\mathrm{VN}_l(G)$ may be encoded at the group level.

8.6.3 The Plancherel weight

Loosely speaking, an element $f \in L^2(G)$ is called *left-bounded* if as for elements of $L^1(G)$, the formal prescription $\xi \mapsto f * \xi$ yields a bounded operator on $L^2(G)$. More precisely, $f \in L^2(G)$ is said to be left-bounded if there exists a constant $C > 0$ such that $\|f * g\|_2 \leq C\|g\|_2$ for all $g \in C_c(G)$. The unique bounded extension of this densely defined operator to all of $L^2(G)$ will, as for the case where symbols are in $L^1(G)$, be denoted by $\lambda(f)$. On considering terms of the form $g * f$, one may similarly define a notion of right-bounded operators. If then indeed $f \in L^2(G)$ induces a right-bounded operator, we will denote that operator by $\rho(f)$. The main result of this subsection is the fact that the group algebra $\mathrm{VN}_l(G)$ admits a canonical weight—the so-called *Plancherel weight* (alt. left Haar weight)—which in a sense encodes left Haar integration at the algebra level. (More on this later.) We specifically have the following.

Theorem 8.46 *The prescription*

$$\psi_G(x^*x) = \begin{cases} \|f\|_2^2, & \text{if } x = \lambda(f) \text{ for some left-bounded } f \in L^2(G) \\ \infty, & \text{otherwise} \end{cases}$$

defines a faithful normal semifinite weight on $\mathrm{VN}_l(G)$.

The existence of the Plancherel weight may, of course, be deduced from the theory of left Hilbert algebras. To do this, one first needs to check that the notion of left-bounded as defined above will for the present context (see [**Tak03a**, Example VI.1.2]) correspond exactly to the definition presented in [**Tak03a**, Definition VI.1.7′]. The Plancherel weight ψ_G may then be defined as in [**Tak03a**, Definition VII.3.2]. When this definition is considered alongside [**Tak03a**, Theorem VII.2.5], it is then clear from [**Tak03a**, Theorem VII.3.4] that for any left-bounded element $f \in L^2(G)$, we have that

$$\psi_G(\lambda(f)^*\lambda(f)) = \|f\|_2^2.$$

An application of a suitable polarisation identity then ensures that in fact

$$\psi_G(\lambda(f)^*\lambda(g)) = \langle g, f \rangle \text{ for all left-bounded } f, g \text{ in } L^2(G). \qquad (8.1)$$

We will, however, show how the existence of this weight may be directly verified using only group theoretic (rather than left Hilbert algebra) machinery. For the case of separable locally compact groups, this was done by Pedersen (see [**Ped79**, Theorem 7.2.7]). Using a very different proof, we now provide a construction valid for general locally compact groups. To effect the proof, we need some additional tools. We specifically introduce the operator $j : L^2(G) \to L^2(G)$ given by $(jh)(t) = \delta_G(t)^{-1/2}\overline{h(t^{-1})}$. (See also [**Ped79**, 7.2.3] for details.) It is an exercise to see that $h \mapsto jh$ is an antiunitary operator for which we have $j(h * g) = (jg) * (jh)$. For functions h on G satisfying $jh \in L^1(G)$, the formal operator $\rho(h)(g) = g * h$ for all $g \in L^2(G)$ is well defined and bounded. The well-definiteness and boundedness of these operators follows from the fact that $g * h = j((jh) * (jg))$ for all $g \in L^2(G)$. To see this, notice that then $\|g * h\|_2 = \|(jh) * (jg)\|_2 \leq \|jh\|_1\|jg\|_2 = \|jh\|_1\|g\|_2$. These operators all belong to $\mathrm{VN}_r(G)(= \mathrm{VN}_l(G)')$, with duality given by the formula $\rho(h)^* = \rho(\widetilde{h})$, where $\widetilde{h}$ is defined by $\widetilde{h}(t) = \overline{h(t^{-1})}$. For the operation $h \mapsto \widetilde{h}$, we similarly have $\widetilde{h * g} = \widetilde{g} * \widetilde{h}$.

Proof.   Once we are able to show that such a weight exists and is normal, the faithfulness and semifiniteness will be self-evident. (The semifiniteness follows upon noting that $\{\lambda(f) \colon f \in L^2(G)$ left-bounded$\}$ must be weak* dense since all continuous functions of compact support are left-bounded.) Our task is therefore to prove existence and normality. To start with, we select a neighbourhood base $\mathcal{U}$ of compact neighbourhoods of the group unit $e \in G$ whose elements are symmetric in the sense that $U = U^{-1}$ for each $U \in \mathcal{U}$ and for which the collection $\{U.U \colon U \in \mathcal{U}\}$ is again a neighbourhood base of e. (This is possible by [**Fol16**, Proposition 2.1(b)].) On writing ξ_U for $\frac{1}{|U|}\chi_U$, we know that for any $f \in L^2(G)$, the nets $(f * \xi_U)_{U \in \mathcal{U}}$ and $(\xi_U * f)_{U \in \mathcal{U}}$ will both converge in norm to f where the limit is taken by letting U decrease to $\{e\}$. See [**Fol16**, Proposition 2.44] for these facts. In fact, the same is true for the net $(\xi_U * \xi_U)$. We proceed to verify this fact. These functions can easily be seen to still be symmetric in the sense that $(\xi_U * \xi_U)(t) = (\xi_U * \xi_U)(t^{-1})$ for all $t \in G$ with in addition

$$\int_G (\xi_U * \xi_U)(s)\, ds = \frac{1}{|U|^2} \int_G \int_G \chi_U(t)\chi_U(t^{-1}s)\, dt\, ds$$

$$= \frac{1}{|U|^2} \int_G \int_G \chi_U(t)\chi_{tU}(s)\, dt\, ds$$

$$= \frac{1}{|U|^2} \int_G \int_G \chi_U(t)\chi_{tU}(s)\, ds\, dt$$

$$= 1.$$

We finally also have that

$$(\xi_U * \xi_U)(s) = \frac{1}{|U|^2} \int_G \chi_U(t)\chi_U(t^{-1}s)\, dt$$

$$= \frac{1}{|U|^2} \int_G \chi_U(t)\chi_U(s^{-1}t)\, dt$$

$$= \frac{1}{|U|^2} \int_G \chi_U(t)\chi_{sU}(t)\, dt$$

$$= \frac{1}{|U|^2} \int_G \chi_{U \cap sU}(t)\, dt$$

$$= \frac{|U \cap sU|}{|U|^2}.$$

We leave it as an exercise to verify that $U \cap sU$ is non-empty if and only if $s \in U.U$ and hence that $\xi_U * \xi_U$ is supported on $U.U$. The claim regarding the net $(\xi_U * \xi_U)$ now follows by applying [**Fol16**, Proposition 2.44] to the fact that the collection of sets $\{U.U : U \in \mathcal{U}\}$ is again a neighbourhood base of e.

Now, notice that when combined with the compactness of the Us, the continuity of $t \mapsto \delta_G(t)$ ensures that $\gamma(U) = \sup_{t \in U} \delta_G(t)$ decreases to $\delta_G(e) = 1$ as U decreases to $\{e\}$. On term-for-term replacing the ξ_Us with $\zeta_U = \frac{1}{|U|\gamma(U)^{1/2}}\chi_U$, we will therefore still have a net of symmetric functions for which the nets $(f * \zeta_U)_{U \in \mathcal{U}}$ and $(\zeta_U * f)_{U \in \mathcal{U}}$ will for any $f \in L^2(G)$ both converge in norm to f as U decreases to $\{e\}$. Again, as before, the net $(\zeta_U * \zeta_U)$ is still a net of symmetric functions for which the nets $(f * \zeta_U * \zeta_U)$ and $(\zeta_U * \zeta_U * f)$ will for any $f \in L^2(G)$ converge in norm to f.

For any left-bounded element of $L^2(G)$, it now clearly follows from the above that

$$\|\lambda(f)(\zeta_U)\|_2 = \|f * \zeta_U\|_2 \to \|f\|_2 \text{ as } U \searrow \{e\}$$

and second that

$$\begin{aligned}
\|\lambda(f)(\zeta_U)\|_2 &= \|f * \zeta_U\|_2 \\
&= \|(j\zeta_U) * (jf)\|_2 \\
&\leq \|j\zeta_U\|_1 . \|f\|_2 \\
&= \left(\frac{1}{|U|\gamma(U)^{1/2}} \int \delta_G(t^{-1})^{1/2}\chi_U \, dt\right) . \|f\|_2 \\
&\leq \left(\frac{1}{|U|} \int \chi_U \, dt\right) . \|f\|_2 \\
&= \|f\|_2.
\end{aligned} \tag{8.2}$$

So, on setting

$$\psi_G(x^*x) = \sup_{U \in \mathcal{U}} \langle x^*x\zeta_U, \zeta_U \rangle = \sup_{U \in \mathcal{U}} \|x\zeta_U\|_2^2$$

the result will by Theorem 4.4 clearly be a normal weight for which we have that

$$\psi_G(\lambda(f)^*\lambda(f)) = \|f\|_2^2 \text{ whenever } f \in L^2(G) \text{ is left-bounded.} \tag{8.3}$$

To conclude the proof, we therefore need to show that whenever $\psi_G(x^*x) < \infty$, we necessarily have that $x = \lambda(f)$ for some left-bounded $f \in L^2(G)$.

The first piece of technology we need for this part of the proof is the fact that the set $\{\lambda(f) \colon f \in L^2(G) \text{ left-bounded}\}$ is a left ideal. To see this, let some left-bounded f and $a \in \mathrm{VN}_l(G)$ be given. For any continuous function h of compact support, we may use the fact that $\mathrm{VN}_l(G)' = \mathrm{VN}_r(G)$ to see that

$$a(f) * h = \rho(h)(a(f)) = a(\rho(h)(f)) = a(f * h) = a \circ \lambda(f)(h). \quad (8.4)$$

Since in addition

$$\|a(f) * h\|_2 = \|a \circ \lambda(f)(h)\|_2 \leq \|a\| \cdot \|\lambda(f)\| \cdot \|h\|_2,$$

it is clear that $a(f)$ is in fact left-bounded, with $\lambda(a(f)) = a \circ \lambda(f)$. This then proves the ideal claim.

We are now able to use the above fact to provide an elegant formula for computing the value $\psi_G(x^*x)$ for all $x \in \mathfrak{n}_{\psi_G}$. For any $U, V \in \mathcal{U}$, we will by inequality (8.2) have that

$$\|x(\zeta_V) * \zeta_U\|_2^2 \leq \|x(\zeta_V)\|_2^2.$$

It then follows from equation (8.4), inequality (8.3) and the L^2 convergence of $(\zeta_V * \zeta_U)_V$ to ζ_U that

$$\|x(\zeta_U)\|_2^2 = \lim_V \|x(\zeta_V * \zeta_U)\|_2^2 \leq \liminf_V \|x(\zeta_V)\|_2^2$$

which then leads to the conclusion that

$$\psi_G(x^*x) = \sup_U \|x(\zeta_U)\|_2^2 \leq \liminf_V \|x(\zeta_V)\|_2^2.$$

If $\psi_G(x^*x) = \infty$, we clearly have equality, while in the case where $x \in \mathfrak{n}_{\psi_G}$, the fact that

$$\psi_G(x^*x) \leq \liminf_V \|x(\zeta_V)\|_2^2 \leq \limsup_V \|x(\zeta_V)\|_2^2 \leq \sup_V \|x(\zeta_V)\|_2^2$$

$$= \psi_G(x^*x) < \infty$$

ensures that we then have that

$$\psi_G(x^*x) = \lim_V \|x(\zeta_V)\|_2^2.$$

It then follows from a suitable polarisation identity that we, in fact, have that

$$\psi_G(y^*x) = \lim_V \langle x(\zeta_V), y(\zeta_V)\rangle \text{ for all } x, y \in \mathfrak{n}_{\psi_G}. \tag{8.5}$$

This formula is the second piece of technology that we need to complete the proof.

Let $x \in \mathfrak{n}_{\psi_G}$ be given. For any $g \in C_c(G)$, we have that

$$|\psi_G(x^*\lambda(g))|^2 \leq \psi_G(x^*x)\psi(\lambda(g)^*\lambda(g)) = \psi_G(x^*x)\|g\|_2^2.$$

The prescription $g \mapsto \psi_G(x^*\lambda(g))$ therefore extends to a bounded linear functional on $L^2(G)$ with norm majorised by $\psi_G(x^*x)^{1/2}$. So, by the Riesz representation theorem, the density of $C_c(G)$ in $L^2(G)$ ensures that there exists a unique $f \in L^2(G)$ such that

$$\psi_G(x^*\lambda(g)) = \langle g, f\rangle \text{ for all } g \in C_c(G).$$

Given $g, h \in C_c(G)$, we therefore have by equation (8.5) that

$$\begin{aligned}
\langle h, f * g\rangle &= \langle h * \widetilde{g}, f\rangle \\
&= \psi_G(x^*\lambda(h * \widetilde{g})) \\
&= \lim_V \langle x^*\lambda(h * \widetilde{g})(\zeta_V), \zeta_V\rangle \\
&= \lim_V \langle x^*(h * \widetilde{g} * \zeta_V), \zeta_V\rangle \\
&= \lim_V \langle \rho(\widetilde{g} * \zeta_V)x^*(h), \zeta_V\rangle \\
&= \lim_V \langle x^*(h), \rho(\widetilde{\zeta_V} * g)(\zeta_V)\rangle \\
&= \lim_V \langle x^*(h), \rho(\zeta_V * g)(\zeta_V)\rangle \\
&= \lim_V \langle x^*(h), \zeta_V * \zeta_V * g\rangle \\
&= \langle x^*(h), g\rangle \\
&= \langle h, x(g)\rangle.
\end{aligned}$$

It clearly follows that the operator x is an extension of the prescription $g \mapsto f * g$ $(g \in C_c(G))$. Hence, f is left-bounded with $\lambda(f) = x$. Therefore, the result follows. $\qquad\square$

Notice that it clearly follows from the definition of the Plancherel weight that the prescription of mapping left-bounded elements f of $L^2(G)$ to the element $\eta(\lambda(f))$ of the GNS Hilbert space H_{ψ_G} corresponding to the pair $(\mathrm{VN}_l(G), \psi_G)$ extends to a unitary from $L^2(G)$ to H_{ψ_G}. (See also [**Ped79**, Theorem 7.2.7], where G is assumed to be separable.)

If G is unimodular, we will for any $f, g \in L^1(G)$ have $(f * g)(e) = \int_G f(t)g(t^{-1})\,dt = \int_G f(t)g(t^{-1})\,d(t^{-1}) = \int_G f(s^{-1})g(s)\,ds = (g * f)(e)$. For functions $f, g \in C_c(G)$, we will therefore, in particular, have $\psi_G(\lambda(f)^*\lambda(g)) = ((f^*) * g))(e) = (g * (f^*))(e) = \psi_G(\lambda(g^*)^*\lambda(f^*)) = \psi_G(\lambda(g)\lambda(f)^*)$. This ensures that ψ_G will be a trace in the unimodular case. The converse is also true, that is, if ψ_G is a trace, G is unimodular. To see this, we shall follow the argument of [**Ped79**, Proposition 7.2.8]. Given $f \in C_c(G)$, define $g \in C_c(G)$ by $g(s) = \delta_G(t)f(st)$. It is then an exercise to see that $\|g\|_2^2 = \delta_G(t)\|f\|_2^2$, and that $\lambda(g) = \lambda(f)\lambda_t^*$. When these facts are in place, it then follows from equation (8.1) that $\psi_G(\lambda_t|\lambda(f)|^2\lambda_t^*) = \delta_g(t)\psi_G(|\lambda(f)|^2)$ for all $f \in C_c(G)$. Thus, G must clearly be unimodular if ψ_G is a trace. Therefore, we arrive at the conclusion that

G is unimodular if and only if ψ_G is a trace.

8.6.4 The modular group of a group algebra

In view of the fact noted at the end of the previous subsection, the truly interesting modular theory for $\mathrm{VN}_l(G)$ may, therefore, be found in the non-unimodular case. We show how in this case each of modular conjugation J and the modular operator Δ may be realised at the group level.

It is clear from Theorem 8.46 that the simple prescription $f \mapsto \lambda(f)$ $(f \in C_c(G))$ extends to a unitary that identifies $L^2(G)$ with the Hilbert space H_{ψ_G} constructed from the pair $(\mathrm{VN}_l(G), \psi_G)$ by means of the GNS process. At the von Neumann algebra level, the operator S is of course the antilinear operator, which appears as the closed extension of the prescription $S(\eta(\lambda(f))) = \eta(\lambda(f)^*) = \eta(\lambda(f^*))$. With respect to the above-mentioned unitary, this operator will at the $L^2(G)$ level correspond to the closed extension of the prescription $\widetilde{S}(f) = f^*$, with $\widetilde{\Delta}_G = |\widetilde{S}|^2$ then corresponding to the modular operator $\Delta_G = |S|^2$ of

$\mathrm{VN}_l(G)$. Given $f, g \in C_c(G)$, we will for $\widetilde{\Delta}_G$ have

$$\langle |\widetilde{S}|^2 f, g \rangle = \langle g^*, f^* \rangle = \int_G \delta_G(s^{-1})^2 \, \overline{g(s^{-1})} f(s^{-1}) \, ds$$

$$= \int_G \delta_G(t) f(t) \overline{g(t)} \, dt = \langle \delta_G f, g \rangle.$$

So, on $L^2(G)$, the operator $\widetilde{\Delta}_G$ is nothing but the operator with natural domain $\{f \in L^2(G) \colon \int_G \delta_G(s)^2 |f(s)|^2 \, ds < \infty\}$, and action described by $\widetilde{\Delta}_G(f)(s) = \delta_G(s) f(s)$. The modular operator Δ_G, therefore, corresponds very directly to the modular function δ_G.

It remains to describe modular conjugation J at the level of $L^2(G)$. For this, we turn to the operator $j \colon L^2(G) \to L^2(G)$ introduced immediately before the proof of Theorem 8.46. We have already noted that $f \mapsto jf$ is an antiunitary operator. Recall that at the level of $L^2(G)$, the action of the operator S is given by $\widetilde{S}(f)(t) = f^*(t) = \delta_G(t)^{-1}\overline{f(t^{-1})}$, which can be rewritten as $\widetilde{S}(f) = j\widetilde{\Delta}_G^{1/2}(f)$. If now we compare this to the fact that $S = J\Delta_G^{1/2}$ at the algebra level, then by the uniqueness of the polar decomposition, we must have that J corresponds to j under the natural unitary identification of H_{ψ_G} with $L^2(G)$.

The facts observed above now yield the following serendipitous observation regarding the Plancherel weight. The proof is based on an idea of a Mathematics Stack Exchange user by the name of Mogget [**Mog**].

Proposition 8.47 *The Plancherel weight ψ_G is finite if and only if the group G is discrete.*

Proof. Notice that the unit $\mathbb{1}$ corresponds to λ_e in $\mathrm{VN}_l(G)$. It then follows from Theorem 8.46 that when G is discrete, we have that $\psi_G(\mathbb{1}) = \|\chi_{\{e\}}\|_2^2 = \mu_H(\{e\})^2 = 1$.

If, conversely, $\psi_G(\mathbb{1}) < \infty$, then we can rescale it so that it is a normal state on $\mathrm{VN}_l(G)$. The facts noted above ensure that in its action on $L^2(G)$, the algebra $\mathrm{VN}_l(G)$ is, in fact, in standard form with respect to ψ_G. Therefore, there must be a unit vector $\xi \in L^2(G)$ such that $\psi_G(\cdot) = \langle (\cdot)\xi, \xi \rangle$. Next, let $\mathcal{U}$ be the neighbourhood base used in the proof of Theorem 8.46. If $g \in G$ is distinct from the group unit e, then there must exist some $U_0 \in \mathcal{U}$ such that for all $U \in \mathcal{U}$ with $U \subseteq U_0$, we have $gU \cap U = \emptyset$. Recalling that λ_g maps χ_U onto χ_{gU}, it therefore follows from equation (8.5) that then $\psi_G(\lambda_g) = 0$. So, for arbitrary points $g \in G$, we have $\langle \lambda_g \xi, \xi \rangle = \psi_G(\lambda_g) = \delta_{g,e}$. Since the function $G \to \mathbb{C} : g \mapsto \langle \lambda_g \xi, \xi \rangle$ is continuous, the set $A = \{g \in G : |1 - \langle \lambda_g \xi, \xi \rangle| < 1/2\} = \{e\}$ is open, which then suffices to show that G is discrete. $\qquad\square$

8.6.5 Encoding the group structure at the algebra level

We now turn to the matter of encoding the structure of the underlying group at the algebra. For this, we will first take stock of what we have already discovered regarding $\mathrm{VN}_l(G)$ and then see what still needs to be added.

- We have already noted how the modular operator Δ on $\mathrm{VN}_l(G)$ encodes the action of the modular function of G at the algebra level.
- The Plancherel weight ψ_G on $\mathrm{VN}_l(G)$ encodes integration against left Haar measure in the sense that the value on any a^*a where $a \in \mathrm{VN}_l(G)$ is given by $\psi_G(a^*a) = \int_G |f(s)|^2 \, ds$ whenever $a = \lambda(f)$ for some left-bounded element of $L^2(G)$, and $\psi_G(a^*a) = \infty$ otherwise. In particular, we then also have that $\mathfrak{n}_{\psi_G} = \{\lambda(f) : f \in L^2(G)\,\text{left-bounded}\}$ and that the formal prescription $f \mapsto \lambda(f)$ extends to a unitary from $L^2(G)$ to H_{ψ_G}, with the action of ψ_G on $\mathfrak{m}_{\psi_G}^+$ in a very real sense corresponding to integration against the left Haar measure. As we shall shortly see, this enables us to encode left-invariance of the Haar measure at the algebra level. Therefore, we may also justifiably refer to this weight as the *left Haar weight* of $\mathrm{VN}_l(G)$.
- The algebra $\mathrm{VN}_l(G)$ also admits a *right Haar weight*. To see this, we may once again use the left-bounded elements of $L^2(G)$ to define an alternative weight φ_G on $\mathrm{VN}_l(G)$ by defining the value on any a^*a to be $\varphi_G(a^*a) = \int_G |f^\vee(s)|^2 \, ds$ whenever $a = \lambda(f)$ for some left-bounded element of $L^2(G)$, and $\psi_G(a^*a) = \infty$ otherwise. As above, the manner in which this weight is constructed enables us to encode right-invariance of the Haar measure at the algebra level. We may justifiably refer to this weight as the right Haar weight.

The challenges we are left with are to describe group multiplication and inversion, and the left (right) invariance of left (right) Haar measure at the algebra level. As far as group multiplication and inversion is concerned, we define an operator W on $L^2(G \times G) = L^2(G) \otimes L^2(G)$ by means of the prescription $Wf(s,t) = f(s, s^{-1}t)$. This operator is basically a variation of the *structure operator* considered by Takesaki in [**Tak03a**, VII.3.5-VII.3.10] and is nowadays referred to as the Kac–Takesaki operator. Minor modifications of some of the arguments in [**Tak03a**] will reveal the properties of this operator. In particular, W turns out to be a unitary, for which the prescription $a \mapsto W(a \otimes \mathbb{1})W^*$ defines a normal unital *-homomorphism Δ from $\mathrm{VN}_l(G)$ to $\mathrm{VN}_l(G)\overline{\otimes}\mathrm{VN}_l(G)$ such that $\Delta(\lambda_t \otimes \mathbb{1}) = (\lambda_t \otimes \lambda_t)$ for all $t \in G$ and also that $(\Delta \otimes \iota)\Delta = (\iota \otimes \Delta)\Delta$ where ι is just the identity operator on $\mathrm{VN}_l(G)$. The formula encoded by means of the operator Δ is known as *comultiplication* since it in principle encodes the group product at the algebra level. It also satisfies the pentagonal equation: $W_{12}W_{13}W_{23} = W_{23}W_{12}$, which in some sense encodes coassociativity of the coproduct. However, to fully encode structures like the group unit and group inversion at the algebra level, we need a little more than what is offered by the above. That 'little more' is offered by the way the left(right) Haar weight interacts with Δ. It is these invariance properties that, in essence, encode the left (right) invariance of left (right) Haar measure at the algebra level. We specifically need

$$\psi_G((\omega \otimes \iota)\Delta(a)) = \omega(\mathbb{1})\psi_G(a), \text{ for any } a \in \mathrm{VN}_l(G)_+, \quad \omega \in \mathrm{VN}_l(G)_*^+;$$

$$\varphi_G((\iota \otimes \omega)\Delta(a)) = \omega(\mathbb{1})\varphi_G(a), \text{ for any } a \in \mathrm{VN}_l(G)_+, \quad \omega \in \mathrm{VN}_l(G)_*^+.$$

As is explained in, for example, [**Vae01a**], [**vD07**, Example 1.8], this is enough to fully encode the structure of G at the algebra level. In Hopf algebra language, one would rather postulate the existence of an anti-homomorphism $S : \mathrm{VN}_l(G) \to \mathrm{VN}_l(G)$ encoding group inversion (the so-called antipode), a co-unit $\varepsilon \colon \mathrm{VN}_l(G) \to \mathbb{C}$, and a multiplication map $\mathrm{m}\colon \mathrm{VN}_l(G) \otimes \mathrm{VN}_l(G) \to \mathrm{VN}_l(G)$, which are related by the expression $\mathrm{m}(S \otimes \iota)\Delta(a) = \varepsilon(a)\mathbb{1} = \mathrm{m}(\iota \otimes S)\Delta(a)$. However, as Vaes points out in the discussion preceding [**Vae01a**, Definition 2.1], on the one hand, a realisation of such a process in a von Neumann algebra context will necessarily be discontinuous, and on the other that although at first sight the formulae displayed above seem to weaker, they are nevertheless sufficient to complete the encoding of the group structure at the algebra level [**KV00**, **KV03**].

8.6.6 Locally compact quantum groups

The objects used in the preceding subsection to encode the structure of the group at the algebra level represent a reliable way of 'quantising' the notion of a group. The formalisation of these structures is, in fact, precisely the modern von Neumann algebraic conception of a locally compact quantum group as defined by Kustermans and Vaes [**KV03**]. Since its origins in the work of Drinfeld and Woronowicz, the field of quantum groups has burgeoned. We will in no way attempt to survey the field. We shall rather content ourselves with raising the reader's awareness of this field and indicating the parallel of the theory with classical locally compact groups. Readers who wish to have a fairly quick yet detailed introduction to the modern von Neumann algebraic approach can refer to, for example, [**vD14**]. The definition of quantum groups that emerges from the preceding subsection and the above comments is the following.

Definition 8.48 A locally compact quantum group $\mathbf{G}$ is a quadruple $\mathbf{G} = (\mathcal{M}; \Delta; \psi; \varphi)$ consisting of:

- a von Neumann algebra $\mathcal{M}$;
- a comultiplication in the form of a normal unital *-homomorphism $\Delta : \mathcal{M} \to \mathcal{M}\overline{\otimes}\mathcal{M}$ satisfying the relations $(\Delta \otimes \iota)\Delta = (\iota \otimes \Delta)\Delta$ where ι is the identity operator on $\mathcal{M}$;
- two faithful normal semifinite weights $\psi : \mathcal{M} \to [0,\infty]$ and $\varphi : \mathcal{M} \to [0,\infty]$ (respectively referred to as the left and right Haar weight) satisfying

$$\psi((\omega \otimes \iota)\Delta(a)) = \psi(a), \text{ for any } a \in \mathcal{M}_+, \omega \in \mathcal{M}_*^+ \qquad (8.16)$$

$$\varphi((\iota \otimes \omega)\Delta(a)) = \varphi(a), \text{ for any } a \in \mathcal{M}_+, \omega \in \mathcal{M}_*^+. \qquad (8.17)$$

For any locally compact group G, the quadruple $(\mathrm{VN}_l(G); \Delta; \psi_G; \varphi_G)$ constructed in the previous subsection is, of course, an example of what we may call a 'concrete' quantum group. The quadruple $(L^\infty(G); \Delta; \psi; \varphi)$ where now ψ and φ are just integration against left and right Haar measure, respectively, is similarly an example of a quantum group. Those familiar with the Hopf algebra definition of quantum groups will have noticed that no mention is made of a co-unit and antipode in the above definition (see the discussion at the end of the previous subsection). However, as van Daele shows in [**vD14**, section 2], one is able to construct a well-defined, albeit unbounded, antipode from the information encoded in the above definition.

We note that, as with the theory of locally compact groups, the left and right Haar weights must be unique up to a scalar multiple [**vD14**, Theorem 3.5]. Moreover, each locally compact group $\mathbb{G}$ admits a well-defined dual quantum group $\widehat{\mathbb{G}}$ with the duality satisfying Pontrjagin duality in the sense that $\mathbb{G} = \widehat{\widehat{\mathbb{G}}}$. (See [**vD14**, Theorem 4.18] and the analysis preceding it for details.) Given a locally compact group G, the dual quantum group of $(\mathrm{VN}_l(G); \Delta; \psi_G; \varphi_G)$ will just be $(L^\infty(G); \Delta; \psi; \varphi)$ and vice versa [**KV00**]. Upon noting that in the case of abelian groups $\mathrm{VN}_l(G)$ is unitarily equivalent to $L^\infty(\widehat{G})$ (see Subsection 8.6.2), it is then clear that this notion of duality canonically extends the notion of Pontrjagin duality for locally compact abelian groups. To date, the category of locally compact quantum groups, as defined above, has been the largest known category canonically containing the locally compact groups, which satisfies the requirement of Pontrjagin duality.

Chapter 9
Crossed products

In Chapter 6, we were introduced to the very elegant theory of L^p and Orlicz spaces for semifinite algebras. What is clear from that chapter is the central role that the algebra of τ-measurable operators played in that development. In trying to extend that theory to general algebras, a major difficulty we need to overcome is the fact that many von Neumann algebras do not admit a faithful normal semifinite trace. Hence, for these von Neumann algebras, no direct construction of an algebra of τ-measurable operators is possible. To overcome this challenge, we appeal to the theory of crossed products. Using the theory of crossed products, any von Neumann algebra may, in a canonical way, be enlarged to an algebra which does admit a faithful normal semifinite trace. Via this enlarged algebra, one may then gain access to the technology of τ-measurable operators. It is this specific aspect that is the focus of our interest in crossed products. We will therefore in no way attempt to give a comprehensive introduction to crossed products, but will content ourselves with familiarising the reader with those aspects essential to the theory of Haagerup L^p-spaces. Throughout this chapter, $\mathcal{M}$ will be a von Neumann algebra acting on a Hilbert space H, equipped with a faithful normal semifinite weight φ. Readers who wish to get to the nuts and bolts of Haagerup L^p-spaces as quickly as possible may at a first reading merely familiarise themselves with the content of Theorems 9.17, 9.28, 9.31, 9.38, and 9.40 and Propositions 9.27, 9.33, and 9.36, and then move on to Chapter 10.

The material in Section 9.1 is, for the most part, based on similar material in [**vD78**], with Section 9.3 borrowing heavily from [**Haa78a**]. We do, however, note that the dual weight construction, as presented in Theorem 9.24, extends the dual weight construction as presented by Haagerup [**Haa78a**, **Haa78b**] and, in the context of crossed products with modular groups, by Terp [**Ter81**]. Haagerup demonstrated the validity of the dual weight construction for possibly non-abelian groups, but only considered faithful and semifinite normal weights. In a more restricted context, Terp managed to demonstrate the validity of

Noncommutative measures and L^p and Orlicz Spaces, with Applications to Quantum Physics. Stanisław Goldstein and Louis Labuschagne, Oxford University Press. © Stanisław Goldstein and Louis Labuschagne (2025).
DOI: 10.1093/oso/9780198950202.003.0011

Theorem 9.24 for normal semifinite but not necessarily faithful weights. Theorem 9.24 shows that only normality is required. The final section is an extension and modification of similar material to that in [**Ter81**].

9.1 Crossed products with general group actions

Given a locally compact group G, we shall consistently assume that the associated Haar measure is localisable. The reader may refer to Theorem 1.45 and the discussion surrounding it, to see the choices we have in this regard. We start by introducing the concept of a group action.

Definition 9.1 Let G be a locally compact group. We define an action of G on $\mathcal{M}$, to be a point to σ-weakly continuous mapping α from G into the group of $*$-automorphisms on $\mathcal{M}$, which respects the group action in the sense that $\alpha_s \circ \alpha_t = \alpha_{st}$ for all $s, t \in G$.

Throughout this section, G will denote a locally compact group admitting an action on the von Neumann algebra $\mathcal{M}$. The following proposition is a vital ingredient in the construction of the crossed product and seems to be part of mathematical folklore by now. The interested reader may find a proof in the book of van Daele [**vD78**].

Proposition 9.2 *Let $L^2(G)$ be the Hilbert space of square Haar-integrable functions on G. The Hilbert space tensor product $H \otimes L^2(G)$ is a copy of $L^2(G, H)$, the space of square Bochner-integrable functions from G to H. Hence, the simple tensors $x \otimes f$ ($x \in H$, $f \in L^2(G)$) may be thought of as functions of the form $G \to H : s \mapsto f(s)x$.*

Theorem 9.3 *For every $a \in \mathcal{M}$, the prescription*

$$\pi_\alpha(a)(\xi)(s) = \alpha_{s^{-1}}(a)(\xi(s)) \qquad \xi \in L^2(G, H)$$

is a well-defined bounded map on $L^2(G, H)$. Moreover, the map

$$\pi_\alpha : \mathcal{M} \to B(L^2(G, H)) : a \mapsto \pi_\alpha(a) \qquad a \in \mathcal{M}$$

is a normal $*$-isomorphism from $\mathcal{M}$ into $B(L^2(G, H))$. (Hence, $\pi_\alpha(\mathcal{M})$ is a von Neumann algebra.) Convention: *When the group action α is understood, we shall simply write π for π_α.*

Proof. For every element $\xi \in L^2(G, H)$, we have that

$$\|\pi(a)(\xi)\|^2 = \int_G \|\alpha_{s^{-1}}(a)(\xi(s))\|^2\, ds \leq \|a\|^2 \int_G \|\xi(s)\|^2\, ds = \|a\|^2 \|\xi\|^2.$$

This shows that $\pi(a)$ is bounded map on $L^2(G, H)$. However, this computation also shows that $\|\pi(a)\| \leq \|a\|$. Hence, the map $\pi : \mathcal{M} \to B(L^2(G, H)) : a \mapsto \pi(a)$ is contractive. It is now an exercise to see that π is a $*$-homomorphism. To conclude the proof, we show that it is injective and hence a $*$-isomorphism. So, let $0 \neq a \in \mathcal{M}$ be given, and select $x, z \in H$ such that $\langle ax, z \rangle \neq 0$. By continuity, we will then have that $0 \neq \langle \alpha_{s^{-1}}(a)x, z \rangle$ for all s in some compact neighbourhood K of the group unit e. That in turn ensures that $0 \neq \alpha_{s^{-1}}(a)x$ for all $s \in K$. We may in fact arrange matters so that for some $\epsilon > 0$, we have that $\|\alpha_{s^{-1}}(a)x\| \geq \epsilon$ on K. Now consider $\pi(a)(x \otimes \chi_K)$. Based on what we have noted thus far, direct computation now shows that $\|\pi(a)(x \otimes \chi_K)\|^2 = \int_G \|\alpha_{s^{-1}}(a)(x)\|^2 \chi_K(s)\, ds = \int_K \|\alpha_{s^{-1}}(a)x\|^2\, ds \geq \epsilon \mu_H(K) > 0$. Thus, $a \mapsto \pi(a)$ is injective, and the claim therefore follows.

It remains to prove the normality of π_α. To this end, let (a_γ) be a net in $\mathcal{M}_+$ increasing to $a \in \mathcal{M}$. The normality of the α_ss ensures that for any $s \in G$ and any $\xi \in L^2(G, H)$, we will then clearly have $\sup_\gamma \alpha_{s^{-1}}(a_\gamma)\xi(s) = \alpha_{s^{-1}}(a)\xi(s)$. Using the fact that integration against μ_H is a normal weight (see Proposition 1.55), it now follows that we will for any $\xi \in L^2(G, H)$ have that

$$\sup_\gamma \langle \pi_\alpha(a_\gamma)\xi, \xi \rangle = \sup_\gamma \int_G \langle \alpha_{s^{-1}}(a_\gamma)\xi(s), \xi(s) \rangle\, ds$$

$$= \int_G \langle \alpha_{s^{-1}}(a)\xi(s), \xi(s) \rangle\, ds = \langle \pi_\alpha(a)\xi, \xi \rangle.$$

We therefore clearly have that $(\pi_\alpha(a_\gamma))$ increases to $\pi_\alpha(a)$, proving that $a \mapsto \pi_\alpha(a)$ is normal. $\qquad \square$

Definition 9.4 For every $g \in G$, define $\lambda_g \in B(L^2(G, H))$ to be the map $\lambda_g(\xi)(s) = \xi(g^{-1}s)$, where $\xi \in L^2(G, H)$.

Proposition 9.5 *The prescription $g \mapsto \lambda_g$ is a strongly continuous unitary representation of G on $L^2(G, H)$, with $\lambda_{g^{-1}} = \lambda_g^*$. Furthermore, $\lambda_g \pi(a) \lambda_g^* = \pi(\sigma_g(a))$ for any $g \in G$ and any $a \in \mathcal{M}$.*

Proof. It is a straightforward exercise to conclude from the left-translation invariance of Haar measure that each λ_g is a unitary operator with $\lambda_{g^{-1}} = \lambda_g^*$. We prove the claim regarding strong continuity. For any simple tensor $(x \otimes f)$ where $x \in H$ and $f \in L^2(G)$, $\pi(a)(x \otimes f)$, we have that $\lambda_g(x \otimes f)(s) = f(g^{-1}s)x$. Denoting left translation by elements of G on $L^2(G)$ by ℓ_g, we therefore have that $\|\lambda_g(x \otimes f) - (x \otimes f))\| = \|x\|\|f - \ell_g(f)\|_2$. From the basic theory of Haar measure, we know that $\|f - \ell_g(f)\|_2 \to 0$ as g tends to the group unit e. This proves strong continuity on the simple tensors. We may now use this fact to prove the claim for general elements of $L^2(G, H)$, by suitably approximating such elements with linear combinations of simple tensors.

Finally, let $a \in \mathcal{M}$ and $\xi \in L^2(G, H)$ be given. On fixing $g \in G$, direct checking now shows that $\pi(a)(\lambda_g^*\xi)(s) = \sigma_{s^{-1}}(a)(\lambda_g^*\xi)(s) = \sigma_{s^{-1}}(a)\xi(g^{-1}s)$ and hence that

$$\lambda_g(\pi(a)\lambda_g^*\xi)(s) = \sigma_{gs^{-1}}(a)\xi(s) = \sigma_{s^{-1}}(\sigma_g(a))\xi(s) = \pi(\sigma_g(a))\xi(s). \quad \square$$

Definition 9.6 We define the crossed product of $\mathcal{M}$ with the group action of G, to be the von Neumann algebra on $L^2(G, H)$ generated by $\pi(\mathcal{M})$ and the translation operators λ_g where $g \in G$. We will denote this von Neumann algebra by $\mathcal{M} \rtimes_\alpha G$.

Remark 9.7 Let G and $\mathcal{M}$ be as above. It is clear from Proposition 9.5 that $a \mapsto \lambda_g a \lambda_g^*$ defines an implemented action of the group G on $\mathcal{M} \rtimes_\alpha G$. This same proposition then also shows that in a very real sense, this action may be thought of as an extension of the a priori given action of G in $\mathcal{M}$.. Using the fact that $\lambda_g\pi(a)\lambda_g^* = \pi(\sigma_g(a))$ for any $g \in G$ and any $a \in \mathcal{M}$, it is an exercise to see that any algebraic combination of the λ_gs and elements from $\pi(\mathcal{M})$ may be written as a linear combination of terms of the form $\pi(a)\lambda_g$. Hence, the crossed product corresponds to the σ-weak closure of $\mathrm{span}\{\pi(a)\lambda_g : a \in \mathcal{M}, g \in G\}$ in $B(L^2(G, H))$.

In view of the fact that $L^2(G, H) = H \otimes L^2(G)$, the question of how $\mathcal{M} \rtimes_\alpha G$ compares to $B(H)\overline{\otimes}B(L^2(G))$ now arises. This is answered by the following proposition.

Proposition 9.8 ([vD78, Part I: Proposition 2.12 and Lemma 3.1]) *The crossed product $\mathcal{M} \rtimes_\alpha G$ is a subspace of the von Neumann algebra tensor product $\mathcal{M}\overline{\otimes}B(L^2(G))$. In particular, each λ_g $(g \in G)$*

is of the form $\mathbb{1} \otimes \ell_g$, where ℓ_g is the left-translation operator defined on $L^2(G)$ by $\ell_g(f)(s) = f(g^{-1}s)$. Furthermore, if the action α of G on $\mathcal{M}$ is implemented by some strongly continuous unitary group $\{u_g\} \subseteq B(H)$ (in the sense that $\alpha_g(a) = u_g a u_g^*$ for each $a \in \mathcal{M}$ and each $g \in G$), then for any $a \in \mathcal{M}$, we have that $U^*(a \otimes 1)U = \pi_\alpha(a)$, where U is the unitary defined by $U(\xi)(s) = u_s \xi(s)$ for each $\xi \in L^2(G)$

The requirement in the last part of the above proposition that α be implemented is not too onerous. If the identity of the specific Hilbert space underlying $\mathcal{M}$ is not important, we may always arrange matters in such a way that this holds. We may, for example, replace $\mathcal{M}$ by $\pi(\mathcal{M})$ to pass to a context where α is implemented (see Proposition 9.5).

Proof.   We start by verifying the second claim. Specifically, for any simple tensor $(x \otimes f) \in H \otimes L^2(G)$, we have $\lambda_g(x \otimes f)(s) = f(g^{-1}s)x = \ell_g(f)(s)x = \ell_g((x \otimes f)(s))$. By continuity, this equality extends to the closure of the span of the simple tensors, namely all of $H \otimes L^2(G)$. Therefore, it will follow that $\mathcal{M} \rtimes_\alpha G \subset \mathcal{M} \overline{\otimes} B(L^2(G))$, if we are able to show that $\pi(\mathcal{M}) \subset \mathcal{M} \otimes B(L^2(G))$. The commutant of $\mathcal{M} \otimes B(L^2(G))$ is $\mathcal{M}' \otimes \mathbb{1}$. If therefore we are able to show that $\pi(\mathcal{M}) \subset (\mathcal{M}' \otimes \mathbb{1})'$, the claim will follow from the von Neumann double commutant theorem. Let $c \in \mathcal{M}'$ be given. For any simple tensor $(x \otimes f) \in H \otimes L^2(G)$ and any $s \in G$, we then have $(c \otimes \mathbb{1})\pi(a)(x \otimes f)(s) = f(s)c\sigma_{s^{-1}}(a)x = f(s)\sigma_{s^{-1}}(a)cx = \pi(a)(cx \otimes f)(s)$ for each $a \in \mathcal{M}$. By continuity, we then have $(c \otimes \mathbb{1})\pi(a) = \pi(a)(c \otimes \mathbb{1})$ on all of $H \otimes L^2(G)$ as required.

Finally, suppose that α is implemented by a unitary group $\{u_g\}$ on H, and let U be defined as in the hypothesis. By continuity, it is once again sufficient to check the final set of claims on simple tensors. For any $(x \otimes f) \in H \otimes L^2(G)$ and any $s \in G$, we have that

$$U^*(a \otimes \mathbb{1})U((x \otimes f)(s) = f(s)u_s^* a u_s x = f(s)u_{s^{-1}} a u_{s^{-1}}^* x$$

$$= f(s)\sigma_{s^{-1}}(a)x = \pi(a)(x \otimes f)(s)$$

as required. $\square$

With the previous proposition as background, we are now able to prove that *-isomorphic copies of a von Neumann algebra will yield *-isomorphic copies of the crossed product $\mathcal{M} \rtimes_\alpha G$. This then shows that up to *-isomorphic equivalence the crossed product is independent of the particular copy of a von Neumann algebra being used. We will revisit this issue of uniqueness when we analyse the structure of crossed products with modular automorphism groups of a given canonical weight.

Theorem 9.9 *Let $\mathcal{M}$ and $\mathcal{N}$ be two von Neumann algebras and $\mathcal{I}$ a ∗-isomorphism from $\mathcal{M}$ onto $\mathcal{N}$. Let G be a locally compact group that admits actions α and β on $\mathcal{M}$ and $\mathcal{N}$, respectively. If for all $g \in G$ and $a \in \mathcal{M}$ we have that $\mathcal{I}(\alpha_g(a)) = \beta_g(\mathcal{I}(a))$, then $\widetilde{\mathcal{I}} = \mathcal{I} \otimes \mathrm{Id}$ is a ∗-isomorphism $\widetilde{\mathcal{I}}$ from $\mathcal{M} \rtimes_\alpha G$ onto $\mathcal{N} \rtimes_\beta G$, for which we have that $\widetilde{\mathcal{I}}(\pi_\alpha(a)) = \pi_\beta(\mathcal{I}(a))$ for all $a \in \mathcal{M}$, and also that $\widetilde{\mathcal{I}}(\lambda_g) = \widetilde{\lambda}_g$ for all $g \in G$ where λ_g and $\widetilde{\lambda}_g$, respectively, denote the left-shift operators corresponding to $\mathcal{M} \rtimes_\alpha G$ and $\mathcal{N} \rtimes_\beta G$.*

Proof. We know, from the standard theory of von Neumann algebra tensor products, that $\widetilde{\mathcal{I}} = \mathcal{I} \otimes \mathrm{Id}$ is a *-isomorphism from $\mathcal{M} \overline{\otimes} B(L^2(G))$ to $\mathcal{N} \overline{\otimes} B(L^2(G))$. We need to show that $\widetilde{\mathcal{I}}$ maps the subspace $\mathcal{M} \rtimes_\alpha G$ onto the subspace $\mathcal{N} \rtimes_\beta G$ in the manner described in the hypothesis. Let λ_g denote the left-shift operators in $\mathcal{M} \rtimes_\alpha G$, and $\widetilde{\lambda}_g$ those in $\mathcal{N} \rtimes_\beta G$. Now recall that by Proposition 9.8, $\lambda_g = \mathbb{1}_\mathcal{M} \otimes \ell_g$ and $\widetilde{\lambda}_g = \mathbb{1}_\mathcal{N} \otimes \ell_g$. Thus, we clearly have that $\widetilde{\mathcal{I}}(\lambda_g) = \widetilde{\lambda}_g$ for all $g \in G$. To complete the proof, we need to show that $\widetilde{\mathcal{I}}(\pi_\alpha(a)) = \pi_\beta(\mathcal{I}(a))$ for all $a \in \mathcal{M}$, since then $\widetilde{\mathcal{I}}$ will clearly map the von Neumann algebra generated by $\{\pi_\alpha(a), \lambda_g : a \in \mathcal{M}, g \in G\}$ (namely $\mathcal{M} \rtimes_\alpha G$) onto the von Neumann algebra generated by $\{\pi_\beta(b), \widetilde{\lambda}_g : b \in \mathcal{N}, g \in G\}$ (that is $\mathcal{N} \rtimes_\alpha G$).

For this part of the proof, we need the fact that any element of $\mathcal{M} \overline{\otimes} B(L^2(G))$ may be represented as some sort of matrix. Select an orthonormal basis $\{f_i\}$ of $L^2(G)$. Any element $\widetilde{a} \in \mathcal{M} \overline{\otimes} B(L^2(G))$ may be written as the sum $\sum_{i,j \in I}(\mathbb{1} \otimes e_i)\widetilde{a}(\mathbb{1} \otimes e_j)$ where the e_is are the projections $\langle \cdot, f_i \rangle f_i$, and convergence is in the σ-strong topology. In the specific case where $\widetilde{a} = \pi_\alpha(a)$ for some $a \in \mathcal{M}$, it is not altogether trivial to see that here $(\mathbb{1} \otimes e_i)\pi_\alpha(a)(\mathbb{1} \otimes e_j) = (\pi_\alpha(a)_{ij} \otimes u_{i,j})$ with $\pi_\alpha(a)_{ij} = \int_G \alpha_{s^{-1}}(x)\overline{f_i}(s)f_j(s)\,ds$ and $u_{i,j} = \langle \cdot, f_j \rangle f_i$. The $u_{i,j}$s are a set of 'matrix units' in that $u_{i,j}^* = u_{j,i}$ and $u_{i,j}u_{j,k} = u_{i,k}$ with $\sum_{i \in I} u_{i,i} = \sum_{i \in I} e_i = \mathbb{1}_{B(L^2(G))}$ (where the sum converges in the σ-strong* topology). So, for such an a, we must have $\pi_\alpha(a) = \sum_{i,j} \pi_\alpha(a)_{ij} \otimes u_{i,j}$.

Given $a \in \mathcal{M}$, the normality of $\widetilde{\mathscr{I}}$ then ensures that $\widetilde{\mathscr{I}}(\pi_\alpha(a)) = \sum_{i,j} \widetilde{\mathscr{I}}(\pi_\alpha(a)_{ij} \otimes u_{i,j}) = \sum_{i,j} \mathscr{I}(\pi_\alpha(a)_{ij}) \otimes u_{i,j}$. Now, observe that for each $i,j \in I$, we have that

$$\begin{aligned}
\mathscr{I}\big((\pi_\alpha(a))_{ij}\big) &= \int_G \mathscr{I}(\alpha_{s^{-1}}(a))\overline{f_i}(s)f_j(s)\,ds \\
&= \int_G \beta_{s^{-1}}(\mathscr{I}(a))\overline{f_i}(s)f_j(s)\,ds \\
&= \big(\pi_\beta(\mathscr{I}(a))\big)_{ij}.
\end{aligned}$$

We, therefore, have that $\widetilde{\mathscr{I}}(\pi_\alpha(a)) = \pi_\beta(\mathscr{I}(a))$ as was required. $\qquad\square$

9.2 Applications—a brief interlude on the utility of crossed products

One of the most profound applications of crossed products is the structural classification of type III factors. For the sake of enabling the reader to appreciate this fact, we pause to briefly outline the relevant theory. The theory outlined here is not essential to the theory developed in the rest of the book, and hence, many details will be deferred to the literature.

Prior to the creation of modular theory, the structure of type III algebras appeared to be almost impenetrable. However, shortly after the appearance of modular theory, Connes in his by now famous work [**Con73**] succeeded in classifying type III factors. A key tool in that classification was the identification of the following von Neumann algebra invariant:

$$S(\mathcal{M}) = \cap\{\mathrm{sp}(\Delta_\psi) \colon \psi \text{ an f.n.s. weight on } \mathcal{M}\}. \tag{9.1}$$

Within the class of type III factors, one may identify the following subclasses.

Definition 9.10 We say that a type III von Neumann algebra factor:

- is type III_0 if $S(\mathcal{M}) = \{0,1\}$;
- type III_λ (where $0 < \lambda < 1$) if $S(\mathcal{M}) = \{\lambda^n \colon n \in \mathbb{Z}\}$;
- and type III_1 if $S(\mathcal{M}) = [0,\infty)$.

Connes' triumph was in showing that any von Neumann algebra factor of type III must belong to one of these classes. The subclasses defined above is therefore a complete classification of type III algebras. The subsequent work by both Connes and Takesaki then yielded the structural description of the various classes described by Theorem 9.11. This theorem is a summary of the contents of [**Tak03a**, Theorems XII.1.1, XII.2.1 and XII.3.7]. (The validity of part (3) of the theorem follows upon first observing that [**Tak03a**, Theorem XII.1.1] records the analysis of section 8 of [**Tak73**] and then applying [**Tak73**, Corollary 9.7].)

Theorem 9.11

(1) *Let $\mathcal{N}$ be a type II_∞ factor, and let θ be an automorphism of $\mathcal{N}$ strictly ergodic on the centre of $\mathcal{N}$. Suppose also that $\mathcal{N}$ admits an f.n.s. trace τ such that $\tau \circ \theta \leq \delta\tau$ for some $0 < \delta < 1$. The crossed product $\mathcal{N} \rtimes_\theta \mathbb{Z}$ of $\mathcal{N}$ with the action $z \mapsto \theta^z$ of $\mathbb{Z}$ induced by θ is a factor of type III_0. Moreover, all factors of type III_0 arise in this way. This decomposition is unique in the sense that $\mathcal{N}_1 \rtimes_{\theta_1} \mathbb{Z}$ and $\mathcal{N}_2 \rtimes_{\theta_2} \mathbb{Z}$ are *-isomorphic if and only if there exist nonzero projections e_1 and e_2, respectively, in the centres of $\mathcal{N}_1$ and $\mathcal{N}_2$ such that automorphisms $\theta_1^{e_1}$ and $\theta_2^{e_2}$ which are in the sense of Kakutani, respectively, induced by θ_1 and θ_2 on e_1 and e_2, are the same (see part (iii) of [**Tak03a**, Theorem XII.3.7] for details).*

(2) *Let $0 < \lambda < 1$ be given, let $\mathcal{N}$ be a type II_∞ factor equipped with a faithful normal semifinite trace τ, and θ an automorphism on $\mathcal{N}$ for which $\tau \circ \theta = \lambda\tau$. Then, $\mathcal{M} = \mathcal{N} \rtimes_\theta \mathbb{Z}$ is a factor of type III_λ. Moreover, any type of III_λ factor is up to *-isomorphism of this form. In this decomposition, the covariant system $\{\mathcal{N}, \theta, \mathbb{Z}\}$ is unique.*

(3) *Let $\mathcal{N}$ be a type II_∞ factor and $(\theta_t)_{t \in \mathbb{R}}$ a one-parameter automorphism group of $\mathcal{N}$ such that $\tau \circ \theta_t = e^{-t}\tau$ for some f.n.s. trace on $\mathcal{N}$. Then, the crossed product of $\mathcal{N}$ by the action $t \mapsto \theta_t$ is a factor of type III_1. Any factor of type III_1 arises in this way, and the decomposition is unique in the same sense as described above.*

The above result is also useful in constructing group von Neumann algebras of various types. In order to fully appreciate this statement, some background is necessary. We noted in Subsection 8.6.3 that G is unimodular if and only if ψ_G is trace. Unimodularity therefore ensures that $\mathrm{VN}_l(G)$ is semifinite. However, when it comes to identifying the type of a group von Neumann algebra, much more can be said. In the cases where either $\delta_G(G) = (0, \infty)$ or $\delta_G(G)$ is a singly generated subgroup, the central type decomposition of $\mathrm{VN}_l(G)$ may be fully described in terms of the subgroup $G_0 = \delta_G^{-1}(1)$. For the type III portion, one may even use group structures to effect a decomposition into the different III_λ types. Details of this process may be found in [**Sut78**, Theorems 3.4 and 4.4]. The interested reader will also find some very helpful examples and discussions on this topic in chapters 13 and 14 of the recent book of Bekka and de la Harpe [**BdlH20**].

As far as examples are concerned, any finite group will, of course, produce a type I group algebra. Less trivially, the group consisting of complex matrices of the form $\left[\begin{smallmatrix} a & b \\ 0 & 1 \end{smallmatrix}\right]$ is known to produce a factor of type I_∞. Murray and von Neumann gave examples of group algebras of type II_1 as early as 1943. See [**KR86**, section 6.7] for a more recent exposition of some of these examples. Examples of type III group algebras took a bit longer to produce. Godement [**God51**] seems to have been the first to provide such an example. The group algebra in Godement's example has proven to be a factor of type III_1. Using the theory developed in [**Sut78**], Sutherland was able to modify Godement's construction to produce families $\{G_\lambda\}$, $\{G_{\lambda,0}\}$ with $\lambda \in (0, 1]$ of groups, yielding, respectively, factors of type III_λ and of type III_0. None of these algebras are, however, hyperfinite. Parallel to Sutherland's analysis, Connes gave an example of a family of groups which for each $0 \le \lambda \le 1$ yields a hyperfinite group algebra of type III_λ. Details of these examples may also be found in [**Sut78**].

We finally come to the identification of a process by means of which we may construct groups that generate group algebras of type III_λ. A primary ingredient in this process is the construction of semidirect products of locally compact groups. For the rest of this discussion, we shall for the sake of simplicity restrict ourselves to separable groups. Suppose that we are given two multiplicative locally compact groups G and Γ with G admitting a continuous action α of Γ (meaning that there exist continuous automorphisms α_γ ($\gamma \in \Gamma$) on G respecting the group action of Γ, for which the map $G \times \Gamma : (s, \gamma) \mapsto \alpha_\gamma(s)$ is continuous with respect to the product topology on $G \times \Gamma$). We then construct a new locally compact group $G \times_\alpha \Gamma$ from the given groups by equipping the Cartesian product $G \times \Gamma$ with the product topology and a product defined by $(s_1, \gamma_1)(s_2, \gamma_2) = (s_1 \alpha_{\gamma_1}(s_2), \gamma_1 \gamma_2)$. The result is yet another locally compact group—the so-called semidirect product $G \times_\alpha \Gamma$. (For multiplication defined as above, the inverse of (s, γ) will, of course, be $(\alpha_\gamma^{-1}(s^{-1}), \gamma^{-1})$.)

Parallel to the construction of the semidirect product, one may also use the fact that if G admits an action α of Γ, then that action can be lifted to an action $\widetilde{\alpha}_\gamma$ $(\gamma \in \Gamma)$ of Γ on $\mathrm{VN}_l(G)$. (On the elements $\lambda(s) \in \mathrm{VN}_l(G)$ $(s \in G)$, this action is defined by $\widetilde{\alpha}_\gamma(\lambda(s)) = \lambda(\alpha_\gamma(s))$. The following remarkably elegant fact then pertains:

$$\mathrm{VN}_l(G) \rtimes_{\widetilde{\alpha}} \Gamma = \mathrm{VN}_l(G \times_\alpha \Gamma).$$

(See, for example, [**Sut78**, Proposition 2.2].)

The above fact then gives us access to Theorem 9.11. Specifically, if we can first find a unimodular locally compact group G for which $\mathrm{VN}_l(G)$ is a factor of type II_∞ and then construct an action α of $\mathbb{Z}$ on G for which the induced action $\widetilde{\alpha}$ on $\mathrm{VN}_l(G)$ satisfies $\psi_G \circ \widetilde{\alpha} = \lambda \psi_G$ for some $0 < \lambda < 1$, then by Theorem 9.11 and the correspondence noted above, the group algebra $\mathrm{VN}_l(G \times_\alpha \mathbb{Z})$ will be a factor of type III_λ. While the idea is simple, getting it to work is no walk in the park. Nevertheless using this approach, with sufficient care and skill, one is able to effectively 'construct' locally compact groups for which the associated group von Neumann algebra is a factor of type III_λ. Several examples produced using this method may be found in [**Sut78**].

9.3 Crossed products with abelian locally compact groups

For the rest of this chapter, we shall assume that G is abelian. Our ultimate interest here is to first introduce the notion of a dual action, and then to use that to construct an operator-valued weight from $\mathcal{M} \rtimes_\alpha G$ to $\mathcal{M}$. Although most of these constructs remain valid for more general groups when given suitable interpretations, the theory is more easily accessible in the setting of abelian groups. As a starting point, we shall in this context introduce the notion of a dual action of G on $\mathcal{M} \rtimes_\alpha G$, or more properly an action $\widehat{\alpha}$ of the dual group $\widehat{G}$ on $\mathcal{M} \rtimes_\alpha G$, and then show how this action may be used to describe $\pi(\mathcal{M})$ as a subspace of $\mathcal{M} \rtimes_\alpha G$.

Let $v_\gamma : L^2(G) \to L^2(G)$ $(\gamma \in \widehat{G})$ be the operator defined by $v_\gamma(f)(s) = \overline{\gamma(s)}f(s)$ for each $f \in L^2(G)$ and each $s \in G$. It is an exercise to see that these maps are actually unitaries. The maps $w_\gamma = \mathbb{1} \otimes v_\gamma$ on $H \otimes L^2(G) = L^2(G, H)$ are then also unitaries. It is these maps that we use to define an action $\widehat{\alpha}$ of the dual group $\widehat{G}$ on $\mathcal{M} \rtimes_\alpha G$. In their action on $L^2(G, H)$, they fulfil the prescription

$$w_\gamma(\xi)(s) = \overline{\gamma(s)}\xi(s) \qquad \xi \in L^2(G, H), s \in G, \gamma \in \widehat{G}.$$

(It is the unimodularity of the numbers $\overline{\gamma(s)}$, which ensures that each w_γ is a unitary.) For each simple tensor $x \otimes f$ $(x \in H,\, f \in L^2(G))$, we have that

$$\int_G \|(\mathbb{1} - w_\gamma)(x \otimes f)(s)\|^2 \, ds = \|x\|^2 \int_G |1 - \overline{\gamma(s)}|^2 |f(s)|^2 \, ds.$$

Standard estimates show that $\int_G |1 - \overline{\gamma(s)}|^2 |f(s)|^2 \, ds \to 0$ as γ tends to the group unit of $\widehat{G}$. Hence, for each simple tensor $x \otimes f$, $\gamma \mapsto w_\gamma(x \otimes f)$ is continuous in $L^2(G, H)$-norm. By suitably approximating general elements of $L^2(G, H)$ with linear combinations of simple tensors, we may then show that $\gamma \mapsto w_\gamma$ is strongly continuous on all of $L^2(G, H)$. It is now easy to check that in addition $w_{\gamma_1} w_{\gamma_2} = w_{\gamma_1 \gamma_2}$. If now for each $b \in B(L^2(G, H))$ we define $\widehat{\alpha}_\gamma(b)$ as $w_\gamma b w_\gamma^*$, the maps $\widehat{\alpha}_\gamma$ can easily be shown to be a group action of $\widehat{G}$ on $B(L^2(G, H))$. However, more is true. Further checking reveals that

$$\widehat{\alpha}_\gamma(\pi(a)) = \pi(a) \text{ and } \widehat{\alpha}_\gamma(\lambda_g) = \overline{\gamma(g)}\lambda_g \text{ for each } a \in \mathcal{M} \text{ and } g \in G. \tag{9.2}$$

Since each of the maps $\widehat{\alpha}_\gamma$ map the generators of $\mathcal{M} \rtimes_\alpha G$ back into $\mathcal{M} \rtimes_\alpha G$, they must each preserve $\mathcal{M} \rtimes_\alpha G$. Thus, the action $\widehat{\alpha}$ restricts to an action on $\mathcal{M} \rtimes_\alpha G$.

Definition 9.12 The restriction of $\widehat{\alpha}$ to $\mathcal{M} \rtimes_\alpha G$ is defined to be the dual action of $\widehat{G}$ on $\mathcal{M} \rtimes_\alpha G$.

Our first order of business is to note that in some sense the dual action is also independent of the particular representation of $\mathcal{M}$ (as was the case with the computation of $M \rtimes_\alpha G$). This fact follows from a modification of the proof of Theorem 9.9.

Corollary 9.13 ([vD78, Part I: Proposition 4.5]) *Let $\mathcal{M}$ and $\mathcal{N}$ and $\mathscr{I}$ be as in Theorem 9.9, with G abelian. If as before we for all*

$g \in G$ and $a \in \mathcal{M}$ have that $\mathscr{I}(\alpha_g(a)) = \beta_g(a)$, then in addition to what was noted in Theorem 9.9, we will for all $\widetilde{a} \in \mathcal{M} \rtimes_\alpha G$ also have that $\widetilde{\mathscr{I}}(\widehat{\alpha}_\gamma(\widetilde{a})) = \widehat{\beta}_\gamma(\widetilde{\mathscr{I}}(\widetilde{a}))$ where as before $\widetilde{\mathscr{I}}$ is a restriction of $\mathscr{I} \otimes \mathrm{Id}$ to $\mathcal{M} \rtimes_\alpha G$ onto $\mathcal{N} \rtimes_\beta G$.

Proof. Let $\widetilde{a} \in \mathcal{M} \rtimes_\alpha G$ and $\gamma \in \widehat{G}$ be given. Taking into account that $\widetilde{\mathscr{I}}$ is a restriction of the *-isomorphism $\mathscr{I} \otimes \mathrm{Id}$ on $\mathcal{M} \overline{\otimes} B(L^2(G))$, we may write $\widetilde{\mathscr{I}}(\widehat{\alpha}_\gamma(\widetilde{a})) = \widetilde{\mathscr{I}}(w_\gamma \widetilde{a} w_\gamma^*) = \widetilde{\mathscr{I}}(w_\gamma)\, \widetilde{\mathscr{I}}(\widetilde{a})\, \widetilde{\mathscr{I}}(w_\gamma)^*$. But for any $\gamma \in \widehat{G}$ we have that $\widetilde{\mathscr{I}}(w_\gamma) = (\mathscr{I} \otimes \mathrm{Id})(\mathbb{1} \otimes v_\gamma) = (\mathbb{1} \otimes v_\gamma) = w_\gamma$. Hence, as claimed, we have that $\widetilde{\mathscr{I}}(\widehat{\alpha}_\gamma(\widetilde{a})) = w_\gamma \widetilde{\mathscr{I}}(\widetilde{a}) w_\gamma^* = \widehat{\beta}_\gamma(\widetilde{\mathscr{I}}(\widetilde{a}))$. $\qquad\square$

We have seen that the action $\widehat{\alpha}$ leaves $\pi(\mathcal{M})$ invariant. One of the primary results in this section asserts that the elements of $\pi(\mathcal{M})$ are the *only* fixed points of the dual action. Proving this result is not difficult as such, but it does rely on a few rather non-trivial facts regarding Fourier analysis on locally compact groups. Readers not familiar with the finer points of that theory may take the result at face value, and proceed with the rest of the analysis. The result first appeared in print in a paper of Haagerup [**Haa78a**, Lemma 3.6], but seems to be due to Landstad. Some preparation is needed before we can present the proof of the result.

Theorem 9.14 *Let X be a locally compact space equipped with a Radon measure μ_R, and $\mathcal{A}$ the von Neumann algebra consisting of all multiplication operators on the Hilbert space $L^2(X, \mu_R)$, with symbols in $L^\infty(X, \mu_R)$. Then $\mathcal{A}$ is maximal abelian, that is $\mathcal{A} = \mathcal{A}'$. (This is proved in Takesaki [**Tak02**], Theorem III.1.2, for the case where μ_R is finite. That proof readily adapts to the present setting.)*

Lemma 9.15 *Put $(X, d\mu_R) = (G, ds)$ in Theorem 9.14. Then $\mathcal{A} = \{v_\gamma : \gamma \in \widehat{G}\}''$.*

Proof. Let ρ be a σ-weakly continuous functional on A such that $\rho(v_\gamma) = 0$ for all $\gamma \in \widehat{G}$. Let $h \in L^1(G)$ be the density $h = \frac{d\rho}{ds}$ of ρ with respect to ds. Then $\rho(v_\gamma) = \int \overline{\gamma(s)} h(s)\, ds = 0$, that is $\widehat{h}(\gamma) = 0$ for all $\gamma \in \widehat{G}$ where $\widehat{h}$ is the Fourier transform of h. Since the Fourier transform is injective on $L^1(G, ds)$, we have $h = 0$, whence $\rho = 0$. The span of $\{v_\gamma \colon \gamma \in \widehat{G}\}$ is therefore σ-weakly dense in $\mathcal{A}$, with the result then following from the von Neumann double commutant theorem $\square$

Lemma 9.16 $\{\ell_s, v_\gamma \colon s \in G, \gamma \in \widehat{G}\}'' = B(L^2(G))$.

Proof.   Let $x \in \{\ell_s, v_\gamma \colon s \in G, p \in \widehat{G}\}'$. By Theorem 9.14, x must be a multiplication operator by some function $f \in L^\infty(G, ds)$. If now $x\lambda_s = \lambda_s x$ for all $s \in G$, then, as

$$((x\lambda_s)(g))(t) = f(t)g(ts^{-1})$$

and

$$((\lambda_s x)(g))(t) = x(g)(ts^{-1}) = f(ts^{-1})g(ts^{-1})$$

we have $f(t)g(ts^{-1}) = f(ts^{-1})g(ts^{-1})$ for all $g \in L^2(G)$. It follows that f is constant almost everywhere, so that x is a multiple of the identity. $\square$

Theorem 9.17   *As a subspace of $\mathcal{M} \rtimes_\alpha G$, the algebra $\pi(\mathcal{M})$ corresponds to the fixed points of the dual action $\widehat{\alpha}$.*

Proof.   By the comment following Proposition 9.8, we may assume that the action α of G on $\mathcal{M}$ is implemented. Having made this assumption, let U be as in the hypothesis of Proposition 9.8.

We have already noted that $\pi(\mathcal{M}) \subseteq \mathcal{N} = \{\widetilde{a} \in \mathcal{M} \rtimes_\alpha G \colon \widehat{\alpha}_\gamma(\widetilde{a}) = \widetilde{a}, \gamma \in \widehat{G}\}$. Conversely, let $\widetilde{a} \in \mathcal{N} \subseteq \mathcal{M} \rtimes_\alpha G$.

Let the v_γs and w_γs ($\gamma \in \widehat{G}$) be as in the discussion preceding Definition 9.12. The fact that $\widehat{\alpha}_\gamma(\widetilde{a}) = \widetilde{a}$ for each $\gamma \in \widehat{G}$ then corresponds to the claim that $\widetilde{a}$ commutes with each $w_\gamma = \mathbb{1} \otimes v_\gamma$. Direct checking now reveals $U^*(\mathbb{1} \otimes v_\gamma)U = (\mathbb{1} \otimes v_\gamma)$ for each $\gamma \in \widehat{G}$. We therefore have $\widetilde{a} \in \{U^*(\mathbb{1} \otimes v_\gamma)U \colon \gamma \in \widehat{G}\}'$.

Since the group is abelian, the operators λ_g and their adjoints all commute with each other. Note, for example, that $\lambda_s \lambda_t^* = \lambda_{st^{-1}} = \lambda_{t^{-1}s} = \lambda_t^* \lambda_s$. Now, let the operators ℓ_g ($g \in G$) be as in Proposition 9.8. Each $\mathbb{1} \otimes \ell_g$ clearly commutes with each $u_g \otimes \mathbb{1}$, and hence, we must have $U^*(\mathbb{1} \otimes \ell_g)U = (\mathbb{1} \otimes \ell_g)$ for each $g \in G$, or in different notation, that $U^*\lambda_g U = \lambda_g$. Since $\pi(\mathcal{M}) = U^*(\mathcal{M} \times \mathbb{1})U$, each element of $\pi(\mathcal{M})$ must therefore clearly commute with each $U^*(\mathbb{1} \otimes \ell_g)U$. With all the generators of $\mathcal{M} \rtimes_\alpha G$ now commuting with each $U^*(\mathbb{1} \otimes \ell_g)U$, it follows that all of $\mathcal{M} \rtimes_\alpha G$ must commute with each $U^*(\mathbb{1} \otimes \ell_g)U$.

If we combine the conclusions of the previous two paragraphs, we have $U\widetilde{a}U^* \in \{1\otimes\ell_g, 1\otimes v_\gamma\colon g\in G, \gamma\in\widehat{G}\}'$. However, by Lemma 9.16, $\{1\otimes\ell_g, 1\otimes v_\gamma\colon g\in G, \gamma\in\widehat{G}\}' = B(H)\otimes 1$. In other words, $\widetilde{a}$ is of the form $U^*(b\otimes 1)U$ for some $b\in B(H)$.

Finally, observe that each $\lambda_g = 1\otimes\ell_g$ commutes with $\mathcal{M}'\otimes 1$. Recall that in the proof of Proposition 9.8, we saw that in addition $\pi(\mathcal{M})\subseteq(\mathcal{M}'\otimes 1)'$. Therefore, $\mathcal{M}\rtimes_\alpha G\subseteq(\mathcal{M}'\otimes 1)'$. Let $c\in\mathcal{M}'$ be given. As an element of $\mathcal{M}\rtimes_\alpha G$, $\widetilde{a}=U^*(b\otimes 1)U$ must commute with $c\otimes 1$. Therefore, for each simple tensor $x\otimes f\in H\otimes L^2(G)$ and each $s\in G$, we must have $(cu_s^*bu_sx)f(s) = (c\otimes 1)U^*(b\otimes 1)U(x\otimes f)(s) = U^*(b\otimes 1)U(c\otimes 1)(x\otimes f)(s) = (u_s^*bu_scx)f(s)$. If we take s to be the group unit e, we may conclude from this that b commutes with c, and hence that $b\in\mathcal{M}$. Therefore, $\widetilde{a}=U^*(b\otimes 1)U\in U^*(\mathcal{M}\otimes 1)U=\pi(\mathcal{M})$ as required. $\qquad\square$

Although not directly relevant for our purposes, we pause to note one additional remarkable fact regarding the dual action, namely the so-called duality theorem. Here, we will closely follow the argument presented in [**Tak03a**, Theorem X.2.3(iii)]. The proof requires some technical information about Bochner L^1-spaces. We merely formulate what we need. Readers may find details in [**Tak02**, Proposition IV.7.16 and Theorem IV.7.17].

Lemma 9.18 *Let $(X,\mathscr{B},\mu)$ be a Radon measure space, and E a Banach space.*

(1) *For any bounded function $A\colon X\to E^*$ that satisfies the requirement that $X\ni x\mapsto\langle b, A(x)\rangle$ is measurable for all $b\in L^1(X,\mathscr{B},\mu)$, the prescription $B\mapsto\int_X\langle A(x), B(x)\rangle\,d\mu(x)$ yields a bounded linear functional on $L^1((X,\mathscr{B},\mu);E)$.*

(2) *On interpreting $L^\infty(X,\mathscr{B},\mu)$ as a von Neumann algebra, we will, for any von Neumann algebra $\mathcal{M}$, have that $(\mathcal{M}\overline{\otimes}L^\infty(X,\mathscr{B},\mu))_* = L^1((X,\mathscr{B},\mu);\mathcal{M}_*)$.*

Theorem 9.19 *The crossed product $(\mathcal{M}\rtimes_\alpha G)\rtimes_{\widehat{\alpha}}\widehat{G}$, is unitarily equivalent to $\pi_\alpha(\mathcal{M})\overline{\otimes}B(L^2(G))$. The unitary $U\colon L^2(G\times\widehat{G};H)\to L^2(G\times G;H)$ that realises this equivalence transforms the dual action*

$\widehat{\widehat{\alpha}}$ *of* $G = \widehat{\widehat{G}}$ *on* $(\mathcal{M} \rtimes_\alpha G) \rtimes_{\widehat{\alpha}} \widehat{G}$ *to the action*

$$U\widehat{\widehat{\alpha}}_s U^*(\pi_\alpha(a) \otimes b) = \pi_\alpha(\alpha_s(a)) \otimes \ell_s^* b \ell_s, \quad s \in G$$

where ℓ_s *is the unitary defined by* $\ell_s(f)(t) = f(s^{-1}t)$ *(*$s, t \in G$*).*

Proof. We will merely sketch the proof, leaving some of the more 'rote' details as exercises. We remind the reader that the dual action $\widehat{\alpha}$ is implemented by the unitaries $w_\gamma = \mathbb{1} \otimes v_\gamma$ $(\gamma \in \widehat{G})$ where v_γ is the unitary on $L^2(G)$ given by $v_\gamma(f)(s) = \overline{\gamma(s)}f(s)$ for each $s \in G$ and $f \in L^2(G)$.

By definition, $\mathcal{M} \rtimes_\alpha G \subset \mathcal{M}\overline{\otimes}B(L^2(G))$ is just $\{\pi_\alpha(a), (\mathbb{1} \otimes \ell_g): a \in \mathcal{M}, g \in G\}''$ with $(\mathcal{M} \rtimes_\alpha G) \rtimes_{\widehat{\alpha}} \widehat{G}$ realised as $\{\pi_{\widehat{\alpha}}(\widetilde{a}), (\mathbb{1} \otimes \mathbb{1} \otimes \ell_\gamma): \widetilde{a} \in (\mathcal{M} \rtimes_\alpha G), \gamma \in \widehat{G}\}''$ (acting on $(H \otimes L^2(G) \otimes L^2(\widehat{G}))$). We thus see that $(\mathcal{M} \rtimes_\alpha G) \rtimes_{\widehat{\alpha}} \widehat{G}$ may be realised as $\{\pi_{\widehat{\alpha}}(\pi_\alpha(a)), \pi_{\widehat{\alpha}}(\mathbb{1} \otimes \ell_g), (\mathbb{1} \otimes \mathbb{1} \otimes \ell_\gamma): a \in \mathcal{M}, g \in G, \gamma \in \widehat{G}\}''$. As far as the dual action $\widehat{\widehat{\alpha}}$ on the second crossed product is concerned, we know from Theorem 9.17 that this action will fix the generators of $\pi_{\widehat{\alpha}}(\mathcal{M} \rtimes_\alpha G)$ (namely the $\pi_{\widehat{\alpha}}(\pi_\alpha(a))$s and $\pi_{\widehat{\alpha}}(\mathbb{1} \otimes \ell_g)$s) with $\widehat{\widehat{\alpha}}_g(\mathbb{1} \otimes \mathbb{1} \otimes \ell_\gamma) = \overline{\gamma(g)}(\mathbb{1} \otimes \mathbb{1} \otimes \ell_\gamma)$ for each $g \in G$ and $\gamma \in \widehat{G}$.

The fact that $\pi_\alpha(M)$ corresponds to the fixed points of the dual action $\widehat{\alpha}$, ensures that

$$\pi_{\widehat{\alpha}}(\pi_\alpha(a)) = \pi_\alpha(a) \otimes \mathbb{1} \text{ for each } a \in \mathcal{M}.$$

On writing $(H \otimes L^2(G) \otimes L^2(\widehat{G}))$ as $L^2(G \times \widehat{G}; H)$, one may also check to see that

$$\pi_{\widehat{\alpha}}(\mathbb{1} \otimes \ell_g)\xi(s, \gamma) = \gamma(g)(\lambda_g \otimes 1)\xi(s, \gamma) = (\mathbb{1} \otimes \ell_g \otimes M_{\langle g, \cdot\rangle})\xi(s, \gamma)$$

for all $s, g \in G$, $\gamma \in \widehat{G}$, where $M_{\langle g, \cdot\rangle}$ is the multiplication operator with symbol $\langle g, \cdot\rangle$ on $L^2(\widehat{G})$. Writing $\mathcal{F}$ for the Fourier transform on $L^2(G)$, we note that $\mathcal{F}^{-1}M_{\langle g, \cdot\rangle}\mathcal{F} = \ell_g$ and that $\mathcal{F}^{-1}\ell_\gamma\mathcal{F} = v_\gamma$ for all $g \in G$ and $\gamma \in \widehat{G}$. (The first claim is essentially proved in [**Fol16**, page 116], with the second following by a similar (slightly easier) argument.) Now let V be the unitary $\mathbb{1} \otimes \mathbb{1} \otimes \mathcal{F}^{-1}$ mapping $H \otimes L^2(G) \otimes L^2(\widehat{G})$ onto $H \otimes L^2(G) \otimes L^2(G)$. The prescription $\mathfrak{a} \mapsto V\mathfrak{a}V^*$ then defines a *-isomorphism on $(\mathcal{M} \rtimes_\alpha G) \rtimes_{\widehat{\alpha}} \widehat{G}$ which transforms this algebra onto an algebra acting on $H \otimes L^2(G) \otimes L^2(G)$. The generators of $(\mathcal{M} \rtimes_\alpha G) \rtimes_{\widehat{\alpha}} \widehat{G}$

are then for all $a \in \mathcal{M}$, $g \in G$ and $\gamma \in \widehat{G}$, transformed as shown below:

$$V(\pi_\alpha(a) \otimes \mathbb{1})V^* = (\pi_\alpha(a) \otimes \mathbb{1})$$

$$V(\pi_{\widehat{\alpha}}(\mathbb{1} \otimes \ell_g))V^* = (\mathbb{1} \otimes \ell_g \otimes \ell_g)$$

$$V(\mathbb{1} \otimes \mathbb{1} \otimes \ell_\gamma)V^* = (\mathbb{1} \otimes \mathbb{1} \otimes v_\gamma).$$

Thus, $(\mathcal{M} \rtimes_\alpha G) \rtimes_{\widehat{\alpha}} \widehat{G}$ is then identified with the algebra $\{(\pi_\alpha(a) \otimes \mathbb{1}), (\mathbb{1} \otimes \ell_g \otimes \ell_g), (\mathbb{1} \otimes \mathbb{1} \otimes v_\gamma) \colon a \in \mathcal{M}, g \in G, \gamma \in \widehat{G}\}''$. Our task is therefore to show that this algebra is of the required form. The first two sets of generators described above will, of course, be fixed by the transformed dual action, with the transformed action on the third set given by $V\widehat{\widehat{\alpha}}_g V^*(\mathbb{1} \otimes \mathbb{1} \otimes v_\gamma) = \overline{\gamma(g)}(\mathbb{1} \otimes \mathbb{1} \otimes v_\gamma)$ for each $g \in G$ and $\gamma \in \widehat{G}$. Given that $\overline{\gamma(g)}(\mathbb{1} \otimes \mathbb{1} \otimes v_\gamma) = (\mathbb{1} \otimes \mathbb{1} \otimes \ell_g^*)(\mathbb{1} \otimes \mathbb{1} \otimes v_\gamma)(\mathbb{1} \otimes \mathbb{1} \otimes \ell_g)$ for each $g \in G$ and $\gamma \in \widehat{G}$, the transformed dual action $V\widehat{\widehat{\alpha}}V^*$ can here be seen to be of the form

$$V\widehat{\widehat{\alpha}}_g V^*(\cdot) = (\mathbb{1} \otimes \mathbb{1} \otimes \ell_g^*)(\cdot)(\mathbb{1} \otimes \mathbb{1} \otimes \ell_g) \tag{9.3}$$

for each $g \in G$.

To achieve the objective of identifying the above algebra with $\pi_\alpha(\mathcal{M})\overline{\otimes}B(L^2(G))$, we write $H \otimes L^2(G) \otimes L^2(G)$ as $L^2(G \times G; H)$ and make use of the unitary W on $L^2(G \times G; H)$ defined by $W(\xi)(t,s) = \xi(ts,s)$ for every $\xi \in L^2(G \times G; H)$ and $t, s \in G$. Further checking now reveals that we will for all $a \in \mathcal{M}$, $\xi \in L^2(G \times G; H)$ and $g, t, s \in G$ have that

$$W(\pi_\alpha(a) \otimes \mathbb{1})W^*(\xi)(t,s) = (\alpha_{(ts)^{-1}}(a) \otimes \mathbb{1})\xi(t,s)$$

$$= (\pi_\alpha(\alpha_{s^{-1}}(a)) \otimes \mathbb{1})\xi(t,s)$$

$$W(\mathbb{1} \otimes \ell_g \otimes \ell_g)W^* = (\mathbb{1} \otimes \mathbb{1} \otimes \ell_g)$$

$$W(\mathbb{1} \otimes \mathbb{1} \otimes v_\gamma)W^* = (\mathbb{1} \otimes \mathbb{1} \otimes v_\gamma).$$

The unitary U mentioned in the hypothesis of the theorem is then nothing but $U = WV$. We momentarily focus our attention on $\pi_\alpha(\mathcal{M})\overline{\otimes}L^\infty(G)$. It is easily verified that $W(\mathbb{1} \otimes \mathbb{1} \otimes L^\infty(G))W^* = (\mathbb{1} \otimes \mathbb{1} \otimes L^\infty(G))$. Moreover, from the centred formulae above, it is clear that $W(\pi_\alpha(a) \otimes \mathbb{1})W^*$ may be realised as the σ-weakly continuous function $G \ni s \mapsto (\pi_\alpha(\alpha_{s^{-1}}(a)) \otimes \mathbb{1}) \in (\pi_\alpha(\mathcal{M}) \otimes \mathbb{1})$. Lemma 9.18 then ensures that $W(\pi_\alpha(a) \otimes \mathbb{1})W^* \in \pi_\alpha(\mathcal{M})\overline{\otimes}L^\infty(G)$. Combining these facts then yields the conclusion that $W(\pi_\alpha(\mathcal{M})\overline{\otimes}L^\infty(G))W^* \subset (\pi_\alpha(\mathcal{M})\overline{\otimes}L^\infty(G))$. However, a similar computation to that used in the previously displayed formulae show that $W^*(\pi_\alpha(a)\otimes\mathbb{1})W$ may be realised as the σ-weakly continuous function $G \ni s \mapsto (\pi_\alpha(\alpha_s(a))\otimes\mathbb{1}) \in (\pi_\alpha(\mathcal{M})\otimes\mathbb{1})$. An entirely analogous argument to the one above shows that we also have $W^*(\pi_\alpha(\mathcal{M})\overline{\otimes}L^\infty(G))W \subset (\pi_\alpha(\mathcal{M})\overline{\otimes}L^\infty(G))$ and hence that $W(\pi_\alpha(\mathcal{M})\overline{\otimes}L^\infty(G))W^* = (\pi_\alpha(\mathcal{M})\overline{\otimes}L^\infty(G))$.

On appealing to Lemma 9.15, it is clear that $\{(\pi_\alpha(a)\otimes\mathbb{1}), (\mathbb{1}\otimes\ell_g\otimes\ell_g), (\mathbb{1}\otimes\mathbb{1}\otimes v_\gamma)\colon a \in \mathcal{M},\, g \in G,\, \gamma \in \widehat{G}\}''$ may also be written as $\{\widetilde{a}, (\mathbb{1}\otimes\ell_g\otimes\ell_g)\colon \widetilde{a} \in (\pi_\alpha(\mathcal{M})\overline{\otimes}L^\infty(G)),\, g \in G\}''$, which by the facts verified above will be mapped onto $\{\widetilde{a}, (\mathbb{1}\otimes\mathbb{1}\otimes\ell_g)\colon \widetilde{a} \in (\pi_\alpha(\mathcal{M})\overline{\otimes}L^\infty(G)),\, g \in G\}''$ by the prescription $\mathfrak{a} \mapsto W\mathfrak{a}W^*$. Applying a combination of Lemmas 9.15 and 9.16, it now follows that $\{\widetilde{a}, (\mathbb{1}\otimes\mathbb{1}\otimes\ell_g)\colon \widetilde{a} \in (\pi_\alpha(\mathcal{M})\overline{\otimes}L^\infty(G)),\, g \in G\}''$ is precisely $\pi_\alpha(\mathcal{M})\overline{\otimes}B(L^2(G))$, as required.

With regard to the transformed action $U\widehat{\alpha}U^*$, we note that $W(\mathbb{1}\otimes\mathbb{1}\otimes\ell_g)W^* = (\mathbb{1}\otimes\ell_g^*\otimes\ell_g) = (\lambda_g^*\otimes\ell_g)$ for each $g \in G$. Thus, by this fact and equation (9.3), the action of $U\widehat{\alpha}U^*$ on $\pi_\alpha(\mathcal{M})\overline{\otimes}B(L^2(G))$ is as claimed for all $g \in G$ described by

$$U\widehat{\alpha}_g U^*(\pi_\alpha(a)\otimes b) = (\lambda_g\otimes\ell_g^*)(\pi_\alpha(a)\otimes b)(\lambda_g^*\otimes\ell_g)$$

$$= \pi_\alpha(\alpha_g(a))\otimes\ell_g^* b\ell_g. \qquad \square$$

9.4 The dual weight construction

Let $\mathcal{M}$ be a von Neumann algebra admitting an action α of some abelian locally compact group G. In this section, we show that every normal weight on $\mathcal{M}$ admits a canonical dual weight on the crossed product $\mathcal{M} \rtimes_\alpha G$. The key to proving this is the construction of a canonical operator-valued weight from $\mathcal{M} \rtimes_\alpha G$ to $\mathcal{M}$. As we shall see, the foundation for this construction is Theorem 9.17.

Definition 9.20 We formally define the operator-valued weight $\mathscr{W}_G$ from $(\mathcal{M} \rtimes_\alpha G)_+$ onto the extended positive part of $\pi(M)$ by the prescription $\mathscr{W}_G(a) = \int_{\widehat{G}} \widehat{\alpha}_\gamma(a)\, d\gamma$ where $a \in (\mathcal{M} \rtimes_\alpha G)_+$.

Haagerup's result confirms that the above definition serves the purpose for which it was formulated.

Proposition 9.21 ([Haa78b, Theorem 1.1]) *The prescription $a \mapsto \mathscr{W}_G(a)$ defined above yields a faithful normal semifinite operator-valued weight from $(\mathcal{M} \rtimes_\alpha G)_+$ onto $\widehat{\pi(\mathcal{M})}$ for which we have $\mathscr{W}_G \circ \widehat{\alpha}_\gamma = \mathscr{W}_G$ for each $\gamma \in \widehat{G}$.*

Proof. Any compactum $\widehat{K} \subseteq \widehat{G}$ has finite measure. Hence, given any $\widetilde{a} \in (\mathcal{M} \rtimes_\alpha G)_+$, it is clear that $\mathscr{W}_{\widehat{K}}(\widetilde{a}) = \int_{\widehat{K}} \widehat{\alpha}_\gamma(a)\, d\gamma$ is a positive operator affiliated to $(\mathcal{M} \rtimes_\alpha G)_+$. If now we order these compacta by inclusion, we obtain a net $\{\mathscr{W}_{\widehat{K}}(\widetilde{a})\}$ that increases to $\mathscr{W}_G(\widetilde{a})$. Hence, $\mathscr{W}_G(\widetilde{a})$ is in the extended positive part of $(\mathcal{M} \rtimes_\alpha G)_+$.

Next, recall that any member m of the extended positive part of $(\mathcal{M} \rtimes_\alpha G)_+$ may be represented by a projection $p \in (\mathcal{M} \rtimes_\alpha G)_+$ (corresponding to the value ∞ in the spectral resolution of m), and a positive operator f densely defined on $(\mathbb{1} - p)L^2(G, H)$, and affiliated to $(\mathbb{1} - p)(\mathcal{M} \rtimes_\alpha G)(\mathbb{1} - p)$. Using this fact, it is clear that the action of $\widehat{\alpha}$ may be extended to the extended positive part of $(\mathcal{M} \rtimes_\alpha G)$. By the same token, the validity of Theorem 9.17 then also extends to the extended positive part of $(\mathcal{M} \rtimes_\alpha G)_+$. In fact, with care, even aspects of the Borel functional calculus may be applied to such objects.

Since Haar measure is translation invariant, it is clear that for any $a \in (\mathcal{M} \rtimes_\alpha G)_+$, and any $\delta \in \widehat{G}$, we have that $\widehat{\alpha}_\delta(\mathscr{W}_G(a)) = \int_{\widehat{G}} \widehat{\alpha}_{\delta\gamma}(a)\, d\gamma = \int_{\widehat{G}} \widehat{\alpha}_\gamma(a)\, d\gamma = \mathscr{W}_G(a))$ as required. So, $\mathscr{W}_G(a)$ in fact belongs to the extended positive part of $\pi(\mathcal{M})$.

It is easy to check from the definition that $\mathscr{W}_G$ is additive and positive homogeneous on $(\mathcal{M} \rtimes_\alpha G)_+$, and also that for any $a \in (\mathcal{M} \rtimes_\alpha G)_+$ and $b \in \pi(\mathcal{M})$, we have $b^*\mathscr{W}_G(a)b = \mathscr{W}_G(b^*ab)$. Thus, $\mathscr{W}_G$ is indeed an operator-valued weight.

Now, let $\{a_\alpha\} \subseteq (\mathcal{M} \rtimes_\alpha G)$ be a net of positive operators increasing monotonically to $a \in (\mathcal{M} \rtimes_\alpha G)_+$. For any $\gamma \in \widehat{G}$ and any $\xi \in L^2(G, H)$, we then have that $\sup_\alpha \langle \widehat{\alpha}_\gamma(a_\alpha)\xi, \xi \rangle = \langle \widehat{\alpha}_\gamma(a)\xi, \xi \rangle$. Given $\xi \in L^2(G, H)$, let ω_ξ denote the functional $b \mapsto \langle b\xi, \xi \rangle$ on $B(L^2(G, H))$. We may then use this fact to see that for any $\xi \in L^2(G, H)$, we have that

$$
\begin{aligned}
\sup_\alpha \mathscr{W}_G(a_\alpha)(\omega_\xi) &= \sup_\alpha \int_{\widehat{G}} \langle \widehat{\alpha}_\gamma(a_\alpha)\xi, \xi \rangle\, d\gamma \\
&= \int_{\widehat{G}} \sup_\alpha \langle \widehat{\alpha}_\gamma(a_\alpha)\xi, \xi \rangle\, d\gamma \\
&= \int_{\widehat{G}} \langle \widehat{\alpha}(a)\xi, \xi \rangle\, d\gamma \\
&= \mathscr{W}_G(a)(\omega_\xi).
\end{aligned}
$$

This suffices to show that $\mathscr{W}_G$ is normal.

Next, suppose that for some $a \in (\mathcal{M} \rtimes_\alpha G)_+$, we have $\mathscr{W}_G(a) = 0$. For any $\xi \in L^2(G, H)$, we will then have that $0 = \langle \mathscr{W}_G(a)\xi, \xi \rangle = \int_{\widehat{G}} \langle \widehat{\alpha}(a_\alpha)\xi, \xi \rangle \, d\gamma$. But by assumption, $\gamma \mapsto \langle \widehat{\alpha}(a_\alpha)\xi, \xi \rangle$ is a (non-negative) continuous function for each ξ. We must therefore have that this function is the zero function. It is clear that we then have $\langle a\xi, \xi \rangle = 0$ for each ξ and hence that $a = 0$. So $\mathscr{W}_G$ is faithful.

We prove that $\mathscr{W}_G$ is semifinite. Let $f : G \to \mathbb{C}$ be a function of compact support, and define λ_f by $\lambda_f = \int_G f(s)\lambda_s \, ds$. It is clear that the σ-weak closure of $\mathrm{span}\{a\lambda_f : a \in (\mathcal{M} \rtimes_\alpha G), f \text{ of compact support}\}$ must include all terms of the form $a\lambda_g$, where $a \in (\mathcal{M} \rtimes_\alpha G)$ and $g \in G$, and hence that it must be all of $\mathcal{M} \rtimes_\alpha G$. If therefore we can show that $\mathscr{W}_G(\lambda_f^* a^* a \lambda_f) \in \pi(\mathcal{M})_+$ for f and a as above, we will have achieved our goal. Since $a^* a \leq \|a\|^2 \mathbb{1}$, this will in turn follow if we can show that $\mathscr{W}_G(\lambda_f^* \lambda_f) \in \pi(\mathcal{M})_+$ for each f as above.

Let $\rho \in \mathfrak{M}_*$ be a norm 1 normal functional. Since $\widehat{\alpha}_\gamma(\lambda_g) = \overline{\gamma(g)}\lambda_g$, we then have $\rho(\widehat{\alpha}_\gamma(\lambda_f)) = \rho(\int_G f(s)\widehat{\alpha}_\gamma(\lambda_s) \, ds) = \int_G f(s)\overline{\gamma(s)}\rho(\lambda_s) \, ds$. If we write f_ρ for the function $G \to \mathbb{C} : s \mapsto f(s)\rho(\lambda_s)$, then $\rho(\widehat{\alpha}_\gamma(\lambda_f))$ is therefore just $\mathcal{F}(f_\rho)(\gamma)$—the value of the abstract Fourier transform of f_ρ at γ. Now notice that

$$\int_G |f_\rho(s)|^2 \, ds = \int_G |f(s)\rho(\lambda_s)|^2 \, ds \leq \|f\|_\infty \mu_H(\mathrm{supp}(f)) < \infty$$

where μ_H denotes Haar measure on G. Thus, $f_\rho \in L^2(G)$, and hence, we may use the abstract Plancherel formula to see that

$$\int_{\widehat{G}} |\rho(\widehat{\alpha}_\gamma(\lambda_f))|^2 \, d\gamma = \int_{\widehat{G}} |\mathcal{F}(f_\rho)(\gamma)|^2 \, d\gamma = \int_G |f(s)\rho(\lambda_s)|^2 \, ds$$

$$\leq \|f\|_\infty^2 \mu_H(\mathrm{supp}(f)).$$

For any compactum $\widehat{K}$, $\mathscr{W}_{\widehat{K}}(\lambda_f^* \lambda_f) = \int_{\widehat{K}} |\widehat{\alpha}_\gamma(\lambda_f)|^2 \, d\gamma$ is a well-defined element of $\mathcal{C}_+$ for which we then have that

$$\rho(\mathscr{W}_{\widehat{K}}(\lambda_f^* \lambda_f)) = \int_{\widehat{K}} |\rho(\widehat{\alpha}_\gamma(\lambda_f))|^2 \, d\gamma \leq \int_{\widehat{G}} |\rho(\widehat{\alpha}_\gamma(\lambda_f))|^2 \, d\gamma$$

$$\leq \|f\|_\infty^2 \mu_H(\mathrm{supp}(f)).$$

Since ρ was arbitrary, this can only be the case if $\|\mathscr{W}_{\widehat{K}}(\lambda_f^* \lambda_f)\|_\infty \leq \|f\|_\infty^2 \mu_H(\mathrm{supp}(f))$. Since $\mathscr{W}_G(\lambda_f^* \lambda_f)$ is the limit of the net $(\mathscr{W}_{\widehat{K}}(\lambda_f^* \lambda_f))$, it now follows that $\mathscr{W}_G(\lambda_f^* \lambda_f) \leq \|f\|_\infty^2 \mu_H(\mathrm{supp}(f))\mathbb{1}$. $\qquad\square$

Corollary 9.22 *If the group G is discrete, the operator-valued weight $\mathscr{W}_G$ defined above is a positive scalar multiple of a faithful normal* *conditional expectation from $\mathcal{M} \rtimes_\alpha G$ onto $\pi(\mathcal{M})$. The action of this conditional expectation is uniquely determined by the formula*

$$\mathscr{W}_G(\lambda_g \pi(a)) = \begin{cases} \pi(a), & \text{if } g = 0 \\ 0, & \text{otherwise} \end{cases} \qquad g \in G, a \in \mathcal{M}.$$

Proof. We have already observed that the group G is discrete if and only if the dual group $\widehat{G}$ is compact. Therefore, Haar measure on $\widehat{G}$ will be finite. It is clear that in this case, $\mathscr{W}_G(a) = \int_{\widehat{G}} \widehat{\alpha}_\gamma(a) \, d\gamma$ will be an element of $\mathcal{M}$ for each $a \in \mathcal{M} \rtimes_\alpha G$. In fact, upon rescaling, we may assume Haar measure on $\widehat{G}$ to be a probability measure, in which case $\mathscr{W}_G(\mathbb{1}) = \mathbb{1}$. The fact that the action of the conditional expectation on terms of the form $\lambda_g \pi(a)$ (where $g \in G, a \in \mathcal{M}$) uniquely determines the expectation follows from Remark 9.7 and the noted normality of this expectation. Given such an element, we may apply Theorem 9.17 to see that $\mathscr{W}_G(\lambda_g \pi(a)) = \int_{\widehat{G}} \widehat{\alpha}_\gamma(\lambda_g \pi(a)) \, d\gamma = \lambda_g \pi(a) \int_{\widehat{G}} \overline{\gamma(g)} \, d\gamma$. Assuming G to be additive, the claim now follows from the known fact that

$$\int_{\widehat{G}} \overline{\gamma(g)} \, d\gamma = \begin{cases} 1, & \text{if } g = 0 \\ 0, & \text{otherwise} \end{cases}.$$

(See Exercise VII.5.6 of [**Kat04**].) $\qquad\qquad\qquad\qquad\qquad\qquad\qquad\square$

We are now ready to introduce the notion of a dual weight and study its properties.

Definition 9.23 Given any normal weight ψ on $\mathcal{M}$, we define $\widetilde{\psi} = \widehat{\psi} \circ \widehat{\pi^{-1}} \circ \mathscr{W}_G$ to be the corresponding dual weight on $(\mathcal{M} \rtimes_\alpha G)$. Here, $\widehat{\psi}$ is the extension of ψ to $\widehat{\mathcal{M}}$ and $\widehat{\pi^{-1}}$ the extension of π^{-1}, to $\widehat{\pi(\mathcal{M})}$.

We hasten to point out that Haagerup has demonstrated the existence of an operator-valued weight from $(\mathcal{M} \rtimes_\alpha G)_+$ to $\widehat{\mathcal{M}}_+$ and of dual weights for general possibly non-abelian locally compact groups ([**Haa78b**], [**Haa78a**, Definition 3.1]).

We note that the fact proved below that $\psi \mapsto \widetilde{\psi}$ is a bijection on the full set of normal weights seems to be new. (See the remark at the start of this chapter.)

Theorem 9.24 *The mapping $\psi \mapsto \tilde{\psi}$ is a bijection between the set of all normal weights on $\mathcal{M}$ and the set of normal weights on $\mathcal{M} \rtimes_\alpha G$ which are $\hat{\alpha}$-invariant in the sense that $\tilde{\psi} \circ \hat{\alpha}_\gamma = \tilde{\psi}$ for all $\gamma \in \hat{G}$.*

For any two normal weights ψ_1 and ψ_2 on $\mathcal{M}$, and any $a \in \mathcal{M}$, we moreover have that:

(i) *$\psi_1 \leq \psi_2$ if and only if $\tilde{\psi}_1 \leq \tilde{\psi}_2$, and in addition $\psi_\alpha \nearrow \psi$ if and only if $\widetilde{\psi}_\alpha \nearrow \tilde{\psi}$;*

(ii) *$\widetilde{(\psi_1 + \psi_2)} = \tilde{\psi}_1 + \tilde{\psi}_2$;*

(iii) *$\pi(a^*).\tilde{\psi}.\pi(a) = \widetilde{a^*.\psi.a}$;*

(iv) *$e_0(\tilde{\psi}) = \pi(e_0(\psi))$;*

(v) *$e_\infty(\tilde{\psi}) = \pi(e_\infty(\psi))$.*

It follows that ψ is faithful (respectively, semifinite) if and only if $\tilde{\psi}$ is. Hence, the map $\psi \mapsto \tilde{\psi}$ restricts to a bijection between the set of all normal semifinite weights on $\mathcal{M}$, and the set of normal semifinite $\hat{\alpha}$-invariant weights on $\mathcal{M} \rtimes_\alpha G$.

The following proof is based on the argument presented in [**Ter81**, Lemma II.1].

Proof. For the sake of simplicity, we will in the proof identify $\mathcal{M}$ with $\pi(\mathcal{M})$, and write $\mathscr{W}$ for $\mathscr{W}_G$ where convenient.

First, note that if ψ is normal, then so is its extension $\hat{\psi}$ to the extended positive part of $\mathcal{M}$. The normality of $\mathscr{W}_G$ then ensures that $\tilde{\psi} = \hat{\psi} \circ \mathscr{W}_G$ is indeed normal. In addition, the fact that $\mathscr{W}_G$ is $\hat{\alpha}$-invariant clearly ensures that the same is true of $\tilde{\psi}$.

Since $\mathscr{W}_G$ maps onto $\widehat{\mathcal{M}}_+$, it is clear that $\psi_1 \leq \psi_2 \Leftrightarrow \hat{\psi}_1 \leq \hat{\psi}_2 \Leftrightarrow \tilde{\psi}_1 \leq \tilde{\psi}_2$ and similarly that $\psi_\alpha \nearrow \psi \Leftrightarrow \hat{\psi}_\alpha \nearrow \hat{\psi} \Leftrightarrow \tilde{\psi}_\alpha \nearrow \tilde{\psi}$. Hence, (i) follows.

Given two normal semifinite weights ψ_1 and ψ_2 on $\mathcal{M}$, it easily follows from the basic properties of the extension of a weight to the extended positive part that $\widehat{\psi_1 + \psi_2} = \hat{\psi}_1 + \hat{\psi}_2$ (see [**Haa79b**, Proposition 1.10]). The validity of (ii) then follows by definition.

Let $a \in \mathcal{M}$ be given. Recall that for any m in the extended positive part $\widehat{M}_+$ of $\mathcal{M}$, a^*ma is defined to be the member of $\widehat{M}_+$ which maps a positive linear functional ρ on $\mathcal{M}$, onto $m(a^*\rho a)$, where $a^*\rho a$ is the positive linear functional given by $\rho(a \cdot a^*)$. Given a normal semifinite weight ψ on $\mathcal{M}$, we may then define $a^*\hat{\psi}a$ to be the map on $\widehat{M}_+$ which maps each $m \in \widehat{M}_+$ onto $\widehat{\psi}(a^*ma)$. On once again considering the basic properties of the extension of a weight to the extended positive part, careful checking now reveals that $a^*\widehat{\psi}a = \widehat{a^*\psi a}$. If we combine this fact with the observation that $\mathscr{W}_G(a^*fa) = a^*\mathscr{W}_G(f)a$ for any $f \in (M \rtimes_\alpha G)_+$, the validity of (iii) then follows by definition.

Let $p_0 \in M$ and $q_0 \in M \rtimes_\alpha G$ be the projections for which $Mp_0 = N_\psi = \{a \in M : \psi(a^*a) = 0\}$ and $(M \rtimes_\alpha G)q_0 = N_{\widetilde{\psi}} = \{f \in (M \rtimes_\alpha G) : \widetilde{\psi}(f^*f) = 0\}$. We first show that in fact $q_0 \in M$. Since $\widetilde{\psi}$ is $\widehat{\alpha}$-invariant, it easily follows that $f \in N_{\widetilde{\psi}}$ if and only if for any $\gamma \in \widehat{G}$, we have that $\widehat{\alpha}_\gamma(f) \in N_{\widetilde{\psi}}$. That means that $(M \rtimes_\alpha G)q_0 = N_{\widetilde{\psi}}$ is $\widehat{\alpha}$-invariant, which can only be the case if q_0 itself is $\widehat{\alpha}$-invariant. But by Theorem 9.17, we then have $q_0 \in M$.

We next claim that $\mathfrak{n}_{\mathscr{W}}N_\psi \subseteq N_{\widetilde{\psi}}$. To see this, observe that for any $a \in N_\psi$ and any $f \in \mathfrak{n}_{\mathscr{W}}$, we have that $\widetilde{\psi}(a^*f^*fa^*) = \widehat{\psi}(\mathscr{W}_G(a^*f^*fa)) = \widehat{\psi}(a^*\mathscr{W}_G(f^*f)a) \leq \|\mathscr{W}_G(f^*f)\|\widehat{\psi}(a^*a) = 0$. Since $\mathfrak{n}_{\mathscr{W}}$ is σ-weakly dense in $M \rtimes_\alpha G$ and $N_\psi = Mp_0$, we therefore have that $\overline{\mathfrak{n}_{\mathscr{W}}N_\psi}^{w^*} = (M \rtimes_\alpha G)p_0 \subseteq (M \rtimes_\alpha G)q_0$. This can of course only be the case if $p_0 \leq q_0$.

Recall that, by definition, $\psi(\dot{\mathscr{W}}_G(f)) = \widetilde{\psi}(f)$ for all $f \in \mathfrak{m}_{\mathscr{W}}$. Let $a, b \in \mathfrak{n}_{\mathscr{W}}$ be given. For any $k \in \{0, 1, 2, 3\}$, we will therefore have that

$$0 = \widetilde{\psi}(q_0(a + i^k b)^*(a + i^k b)q_0) = \psi(\dot{\mathscr{W}}_G(q_0(a + i^k b)^*(a + i^k b)q_0))$$
$$= \psi(q_0\dot{\mathscr{W}}_G((a + i^k b)^*(a + i^k b))q_0).$$

Therefore, $q_0\dot{\mathscr{W}}_G((a+i^k b)^*(a+i^k b))q_0 \in N_\psi^* N_\psi$ for each $k \in \{0, 1, 2, 3\}$. We may now use the identity $q_0 b^* a q_0 = \frac{1}{4}\sum_{k=0}^3 q_0(a + i^k b)^*(a + i^k b)q_0$ to conclude that in fact $q_0\dot{\mathscr{W}}_G(b^*a)q_0 \in N_\psi^* N_\psi$ and hence that $q_0\dot{\mathscr{W}}_G(\mathfrak{m}_{\mathscr{W}})q_0 \subseteq N_\psi^* N_\psi$. Since $\dot{\mathscr{W}}_G(\mathfrak{m}_{\mathscr{W}})$ is σ-weakly dense in M, it follows from the fact that $N_\psi = Mp_0$ that the above inclusion can only hold if $q_0 Mq_0 \subseteq p_0 Mp_0$. Therefore, $q_0 \leq p_0$, and hence, equality holds. This then proves (iv).

With regard to (v), we will first show that $e_\infty(\psi) \leq e_\infty(\widetilde{\psi})$ and then prove that equality holds. To be able to do this, we first need to show that if ψ is semifinite, then so is $\widetilde{\psi}$. So, let ψ be a given normal semifinite weight on M. The semifiniteness of $\mathscr{W}_G$ ensures that $\mathfrak{n}_{\mathscr{W}} = \{a \in M \rtimes_\alpha G : \mathscr{W}_G(a^*a) \in M\}$ is σ-weakly dense in $M \rtimes_\alpha G$. By assumption, $\mathfrak{n}_\psi = \{f \in M : \psi(f^*f) < \infty\}$ is σ-weakly dense in M. Hence, we may select a net $(f_i) \subseteq \mathfrak{n}_\psi$ which is σ-weakly convergent to $\mathbb{1}$. For any $a \in \mathfrak{n}_{\mathscr{W}}$, the net (af_i) will then be σ-weakly convergent to a. In addition

$$\begin{aligned}
\widetilde{\psi}((af_i)^*(af_i)) &= \psi(\mathscr{W}_G(f_i^* a^* a f_i)) \\
&= \psi(f_i^* \mathscr{W}_G(a^* a) f_i) \\
&\leq \|\mathscr{W}_G(a^* a)\| \psi(f_i^* f_i) < \infty.
\end{aligned}$$

In other words, $(af_i) \subseteq \mathfrak{n}_{\widetilde{\psi}}$. But then $\mathfrak{n}_{\widetilde{\psi}}$ must be σ-weakly dense in $\mathfrak{n}_{\mathscr{W}}$, which in turn we know to be σ-weakly dense in $\mathcal{M} \rtimes_\alpha G$. Hence, $\mathfrak{n}_{\widetilde{\psi}}$ is σ-weakly dense in $\mathcal{M} \rtimes_\alpha G$. Thus, if ψ is semifinite, then so is $\widetilde{\psi}$.

Returning to the general case, recall that for any normal weight ψ, the weight $e_\infty(\psi).\psi.e_\infty(\psi)$ is semifinite. It now follows from part (iii) and what we have just shown that $e_\infty(\psi).\widetilde{\psi}.e_\infty(\psi)$ will then be semifinite. Writing e_∞ for $e_\infty(\psi)$, this means that $\mathfrak{n}_{e_\infty.\widetilde{\psi}.e_\infty}$ is σ-weakly dense in $\mathcal{M} \rtimes_\alpha G$. It is now an easy exercise to see that $\mathfrak{n}_{e_\infty.\widetilde{\psi}.e_\infty} e_\infty \subseteq \mathfrak{n}_{\widetilde{\psi}}$. On taking the σ-weak closure of both sides of this inclusion, we obtain $(\mathcal{M} \rtimes_\alpha G)e_\infty \subseteq (\mathcal{M} \rtimes_\alpha G)e_\infty(\widetilde{\psi})$, which can only be the case if $e_\infty(\psi) \leq e_\infty(\widetilde{\psi})$.

We go on to show that equality holds. Let $f \in \mathfrak{n}_\psi$ be given. For ease of notation, write e_0 for $e_0(\psi)$. Observe that as a member of $\widehat{\mathcal{M}}_+$, $(\mathbb{1} - e_0)\mathscr{W}_G(f^* f)(\mathbb{1} - e_0)$ has a spectral resolution of the form

$$(\mathbb{1} - e_0)\mathscr{W}_G(f^* f)(\mathbb{1} - e_0) = \int_0^\infty \lambda \, de_\lambda + \infty.p$$

where each e_λ is orthogonal to p. Since $(\mathbb{1} - e_0)\mathscr{W}_G(f^* f)(\mathbb{1} - e_0)$ is 'supported' on $\mathbb{1} - e_0$, we in particular also have that $p \leq (\mathbb{1} - e_0)$.

Since ψ is faithful on $(\mathbb{1} - e_0)\mathcal{M}(\mathbb{1} - e_0)$, and since

$$\psi((\mathbb{1} - e_0)\mathscr{W}_G(f^* f)(\mathbb{1} - e_0)) = \psi(\mathscr{W}_G(f^* f)) = \widetilde{\psi}(f^* f) < \infty$$

we must have that $\psi(p) = 0$. Given that ψ is faithful on $(\mathbb{1} - e_0)\mathcal{M}(\mathbb{1} - e_0)$, this ensures that $p = 0$. But then $(\mathbb{1} - e_0)\mathscr{W}_G(f^* f)(\mathbb{1} - e_0)$ is a densely defined operator affiliated to $\mathcal{M}$ with spectral resolution $(\mathbb{1} - e_0)\mathscr{W}_G(f^* f)(\mathbb{1} - e_0) = \int_0^\infty \lambda \, de_\lambda$. Setting $e_n = \chi_{[0,n]}((\mathbb{1} - e_0)\mathscr{W}_G(f^* f)(\mathbb{1} - e_0))$, we will for each $n \in \mathbb{N}$ then have that $e_n(\mathbb{1} - e_0)\mathscr{W}_G(f^* f)(\mathbb{1} - e_0)e_n = \int_0^n \lambda \, de_\lambda \in \mathcal{M}$ with in addition

$$\psi(e_n(\mathbb{1}-q_0)\mathscr{W}_G(f^*f)(\mathbb{1}-q_0)e_n) \ \leq \ \psi((\mathbb{1}-q_0)\mathscr{W}_G(f^*f)(\mathbb{1}-q_0))$$
$$= \ \widehat{\psi}(\mathscr{W}_G(f^*f)) = \widetilde{\psi}(f^*f) < \infty$$

where we used the fact that $e_n(\mathbb{1}-e_0)\mathscr{W}_G(f^*f)(\mathbb{1}-e_0)e_n = \int_0^n \lambda\,de_\lambda \leq \int_0^\infty \lambda\,de_\lambda = (\mathbb{1}-e_0)\mathscr{W}_G(f^*f)(\mathbb{1}-e_0)$. Hence, $e_n(\mathbb{1}-e_0)\mathscr{W}_G(f^*f)(\mathbb{1}-e_0)e_n \in \mathfrak{m}_\psi \subseteq e_\infty \mathcal{M} e_\infty$ for each $n \in \mathbb{N}$. If now we let $n \to \infty$, it will follow that $(\mathbb{1}-e_0)\mathscr{W}_G(f^*f)(\mathbb{1}-e_0)\,\eta\,e_\infty \mathcal{M} e_\infty$. Recall that $e_0 \leq e_\infty$, and that e_∞ and e_0 must therefore commute. So, for any $f \in \mathfrak{n}_\psi$, we will then have that

$$\widetilde{\psi}(f^*f) = \widehat{\psi}(\mathscr{W}_G(f^*f)) \ = \ \widehat{\psi}(e_\infty(\mathbb{1}-e_0)\mathscr{W}_G(f^*f)(\mathbb{1}-e_0)e_\infty)$$
$$= \ \widehat{\psi}(\mathscr{W}_G(e_\infty(\mathbb{1}-e_0)f^*f(\mathbb{1}-e_0)e_\infty))$$
$$= \ \widetilde{\psi}(e_\infty(\mathbb{1}-e_0)f^*f(\mathbb{1}-e_0)e_\infty)$$
$$= \ \widetilde{\psi}((\mathbb{1}-e_0)e_\infty f^*f e_\infty(\mathbb{1}-e_0))$$
$$= \ \widetilde{\psi}(e_\infty f^*f e_\infty).$$

Recall that $\mathfrak{m}_{\widetilde{\psi}} = \mathrm{span}\{g^*f : f,g \in \mathfrak{n}_{\widetilde{\psi}}\}$ and that $g^*f = \frac{1}{4}\sum_{k=0}^3 (f+i^k g)^*(f+i^k g)$ for any $f,g \in \mathfrak{n}_{\widetilde{\psi}}$. The equalities displayed above therefore show that $\widetilde{\psi}$ and $e_\infty \widetilde{\psi} e_\infty$ agree on $\mathfrak{m}_{\widetilde{\psi}}$, and hence also on $\overline{\mathfrak{m}_{\widetilde{\psi}}}^{w^*} = e_\infty(\widetilde{\psi})(\mathcal{M}\rtimes_\alpha G)e_\infty(\widetilde{\psi})$. But since $e_\infty = e_\infty(\psi) \leq e_\infty(\widetilde{\psi})$, we must then have that

$$0 = (e_\infty.\widetilde{\psi}.e_\infty)(e_\infty(\widetilde{\psi}) - e_\infty) = \widetilde{\psi}(e_\infty(\widetilde{\psi}) - e_\infty)$$
$$= \widetilde{\psi}((\mathbb{1}-e_0)(e_\infty(\widetilde{\psi}) - e_\infty)(\mathbb{1}-e_0)).$$

(Here, we used the fact that $\widetilde{\psi}(x) = \widetilde{\psi}((\mathbb{1}-e_0)x(\mathbb{1}-e_0))$ for any $x \in (\mathcal{M}\rtimes_\alpha G)_+$.) Since $\widetilde{\psi}$ is faithful on $(\mathbb{1}-e_0)(\mathcal{M}\rtimes_\alpha G)(\mathbb{1}-e_0)$, it follows that $0 = (\mathbb{1}-e_0)(e_\infty(\widetilde{\psi}) - e_\infty)(\mathbb{1}-e_0)$. Given that $e_0 = e_0(\widetilde{\psi}) \leq e_\infty = e_\infty(\psi) \leq e_\infty(\widetilde{\psi})$, it now easily follows that $(\mathbb{1}-e_0)(e_\infty(\widetilde{\psi}) - e_\infty)(\mathbb{1}-e_0) = e_\infty(\widetilde{\psi}) - e_\infty$ and hence that $e_\infty(\widetilde{\psi}) = e_\infty = e_\infty(\psi)$. Thus, the validity of (v) is established.

Since a normal weight ψ is semifinite if and only if $e_\infty(\psi) = \mathbb{1}$ and faithful if and only if $e_0(\psi) = 0$ (and similarly for $\widetilde{\psi}$), it is now clear that ψ is faithful (respectively, semifinite) if and only if $\widetilde{\psi}$ is.

We show that the map $\psi \mapsto \widetilde{\psi}$ is injective. Hence, suppose that for normal weights ψ_1 and ψ_2 on $\mathcal{M}$, we have that $\widetilde{\psi}_1 = \widetilde{\psi}_2$. These weights extend uniquely to weights $\widehat{\widetilde{\psi}}_1 = \widehat{\widetilde{\psi}}_2$ on $(\widehat{M\rtimes_\alpha G})_+$. This uniqueness, of course, ensures that these extensions must be of the form

$$\widehat{\widetilde{\psi}}_1 = \widehat{\psi}_1 \circ \widehat{\pi^{-1}} \circ \widehat{\mathscr{W}_G}, \quad \widehat{\widetilde{\psi}}_2 = \widehat{\psi}_2 \circ \widehat{\pi^{-1}} \circ \widehat{\mathscr{W}_G}$$

where $\widehat{\mathscr{W}_G}$ is the unique extension of $\mathscr{W}_G$ to all of $(\widehat{M \rtimes_\alpha G})_+$. Recall that the normality of $\mathscr{W}_G$ ensures that $\widehat{\mathscr{W}_G}$ actually maps onto $\widehat{\mathcal{M}}_+$ (see Remark 8.37). So, given $x \in \mathcal{M}_+$, there must exist some $m_x \in (\widehat{M \rtimes_\alpha G})_+$ such that $x = \widehat{\mathscr{W}_G}(m_x)$. By the formulas displayed previously, we will then have that $\psi_1(x) = \widehat{\widetilde{\psi}}_1(m_x) = \widehat{\widetilde{\psi}}_2(m_x) = \psi_2(x)$. Therefore, we have $\psi_1 = \psi_2$ as was required.

We need some additional technology before we are able to prove the claim regarding the surjectivity of the map $\psi \mapsto \widetilde{\psi}$. We, therefore, defer the proof of surjectivity until after the requisite technology has been developed. $\qquad\square$

The proof of the following theorem is due to Haagerup. The proofs of the two bullets are taken, respectively, from [**Haa79b**] and [**Haa78a**].

Theorem 9.25 *For any faithful normal semifinite weight ψ on $\mathcal{M}$, the action of $\sigma_t^{\widetilde{\psi}}$ on $\mathcal{M} \rtimes_\alpha G$ is uniquely determined by the prescriptions:*

- $\sigma_t^{\widetilde{\psi}}(\pi(a)) = \pi(\sigma_t^{\psi}(a))$ *for all $a \in \mathcal{M}$ and all $t \in \mathbb{R}$;*
- $\sigma_t^{\widetilde{\psi}}(\lambda_g) = \lambda_g(D\widetilde{\psi} \circ \alpha_g : D\widetilde{\psi})_t$ *for all $g \in G$, with $\sigma_t^{\widetilde{\psi}}(\lambda_g) = \lambda_g$ if ψ is α_g invariant.*

Proof. For the sake of simplicity, we identify $\mathcal{M}$ and $\pi(\mathcal{M})$. We first prove the second bullet. Given any $f \in \mathcal{M} \rtimes_\alpha G$ and any $\xi \in L^2(G, H)$, it follows from the definition of the dual action $\widehat{\alpha}$ that $\widehat{\alpha}_\gamma(\lambda_g f \lambda_g^*)\xi(s) = \lambda_g \widehat{\alpha}_\gamma(f)\lambda_g^* \xi(s)$ for every $\gamma \in \widehat{G}$, and all $s, g \in G$ (see equation (9.2)).

For any $f \in (\mathcal{M} \rtimes_\alpha G)_+$, we may now apply Proposition 9.5 to see that

$$\mathscr{W}_G(\lambda_g f \lambda_g^*) = \int_{\widehat{G}} \widehat{\alpha}_\gamma(\lambda_g f \lambda_g^*)\, d\gamma = \lambda_g \int_{\widehat{G}} \widehat{\alpha}_\gamma(f)\, d\gamma\, \lambda_g^*$$

$$= \lambda_g \mathscr{W}_G(f)\lambda_g^* = \sigma_g(\mathscr{W}_G(f)).$$

But then $\widetilde{\psi}(\lambda_g f \lambda_g^*) = \widehat{\psi}(\mathscr{W}_G(\lambda_g f \lambda_g^*)) = \widehat{\psi}(\alpha_g(\mathscr{W}_G(f))) = \widetilde{\psi \circ \alpha_g}(f)$. The claim now follows from Lemma 8.28.

We proceed to prove the first bullet. By Proposition 8.15, it will be enough to show that in their action on $\mathcal{M}$, we have that $\sigma_{-i}^{\psi} \subseteq$

$\sigma^{\widetilde{\psi}}_{-i}$. In principle, we therefore need to show that if $(a,b) \in \mathcal{G}(\sigma^{\psi}_{-i})$ (the graph of σ^{ψ}_{-i}), then $(a,b) \in \mathcal{G}(\sigma^{\widetilde{\psi}}_{-i})$. So, let $(a,b) \in \mathcal{G}(\sigma^{\psi}_{-i})$ be given. By the discussion preceding the theorem, we then have that $a \in D(\sigma^{\psi}_{-i}) \subseteq D(\sigma^{\psi}_{-i/2})$, and $b \in D(\sigma^{\psi}_{i}) \subseteq D(\sigma^{\psi}_{i/2})$. (The claim about b follows since σ^{ψ}_{i} is the inverse of σ^{ψ}_{-i}.) Since for any $t \in \mathbb{R}$ we have that $\sigma^{\psi}_{t}(b^*) = \sigma^{\psi}_{t}(b)^*$, careful checking shows that this ensures that then $b^* \in D(\sigma^{\psi}_{-i/2})$. By Lemma 8.16, we then have that there exists some $k > 0$ such that

$$\psi(axa^*) \le k^2\psi(x) \text{ and } \psi(b^*xb) \le k^2\psi(x) \text{ for every } x \in \mathcal{M}_+.$$

Any element of the extended positive part of $\mathcal{M}$ may be written as the limit of an increasing net in $\mathcal{M}_+$. Since the extension $\widehat{\psi}$ of ψ to $\widehat{\mathcal{M}}_+$ is normal, it therefore follows that

$$\widehat{\psi}(ama^*) \le k^2\widehat{\psi}(m) \text{ and } \widehat{\psi}(b^*mb) \le k^2\widehat{\psi}(m) \text{ for every } m \in \widehat{\mathcal{M}}_+.$$

On using the fact that for any $x \in (\mathcal{M}\rtimes_\alpha G)_+$ we have that $\mathscr{W}_G(axa^*) = a\mathscr{W}_G(x)a^*$ and $\mathscr{W}_G(b^*xb) = b^*\mathscr{W}_G(x)b$, it is then clear that

$$\widetilde{\psi}(axa^*) \le k^2\widetilde{\psi}(x) \text{ and } \widetilde{\psi}(b^*xb) \le k^2\widetilde{\psi}(x) \text{ for every } x \in (\mathcal{M}\rtimes_\alpha G)_+. \tag{9.4}$$

(Note, for example, that

$$\widetilde{\psi}(axa^*) = \widehat{\psi}(\mathscr{W}_G(axa^*)) = \widehat{\psi}(a\mathscr{W}_G(x)a^*) \le k^2\widehat{\psi}(\mathscr{W}_G(x)) = k^2\widetilde{\psi}(x)$$

when $x \in (\mathcal{M}\rtimes_\alpha G)_+$.) It now trivially follows from the above inequalities that $\mathfrak{n}_{\widetilde{\psi}}a^* \subseteq \mathfrak{n}_{\widetilde{\psi}}$, and $\mathfrak{n}_{\widetilde{\psi}}b \subseteq \mathfrak{n}_{\widetilde{\psi}}$. So, if we can show that

$$\widetilde{\psi}(ax) = \widetilde{\psi}(xb) \text{ for all } x \in \mathfrak{m}_{\widetilde{\psi}} \tag{9.5}$$

we would, by Theorem 8.17, then have that $(a,b) \in \mathcal{G}(\sigma^{\widetilde{\psi}}_{-i})$, which would prove the theorem. So, it remains to show that equation (9.5) holds. We first show that equation (9.5) holds for all x in the subspace $(\mathfrak{n}_{\widetilde{\psi}} \cap \mathfrak{n}_{\mathscr{W}})^*(\mathfrak{n}_{\widetilde{\psi}} \cap \mathfrak{n}_{\mathscr{W}}) = \text{span}\{g^*f : f,g \in (\mathfrak{n}_{\widetilde{\psi}} \cap \mathfrak{n}_{\mathscr{W}})\}$. The first step in doing this is showing that $\mathscr{W}_G$ maps $(\mathfrak{n}_{\widetilde{\psi}}\cap\mathfrak{n}_{\mathscr{W}})^*(\mathfrak{n}_{\widetilde{\psi}}\cap\mathfrak{n}_{\mathscr{W}})$ into $\mathfrak{n}_{\psi}^*\mathfrak{n}_{\psi} = \mathfrak{m}_{\psi}$. To see this, let $f,g \in (\mathfrak{n}_{\widetilde{\psi}} \cap \mathfrak{n}_{\mathscr{W}})$ be given. For any $k \in \{0,1,2,3\}$, we then have that

$$\psi(\dot{\mathscr{W}}_G((f+i^k g)^*(f+i^k g))) = \widetilde{\psi}((f+i^k g)^*(f+i^k g)) < \infty$$

and hence that $\dot{\mathscr{W}}_G((f+i^k g)^*(f+i^k g)) \in \mathfrak{m}_\psi$. Since $g^* f = \frac{1}{4}\sum_{k=0}^3 (f+i^k g)^*(f+i^k g)$, the same is then true of $\dot{\mathscr{W}}_G(g^* f)$, which proves the claim. Using what we know about ψ, we, therefore, have that

$$\begin{aligned} \widetilde{\psi}(a x_0) &= \psi(\dot{\mathscr{W}}_G(a x_0)) = \psi(a\dot{\mathscr{W}}_G(x_0)) \\ &= \psi(\dot{\mathscr{W}}_G(x_0)b) = \psi(\dot{\mathscr{W}}_G(x_0 b)) \\ &= \widetilde{\psi}(x_0 b) \end{aligned} \tag{9.6}$$

for all $x_0 \in (\mathfrak{n}_{\widetilde{\psi}} \cap \mathfrak{n}_{\mathscr{W}})^*(\mathfrak{n}_{\widetilde{\psi}} \cap \mathfrak{n}_{\mathscr{W}})$. We will use this equality to prove equation (9.5). Observe that since $\mathfrak{m}_{\widetilde{\psi}} = \mathrm{span}\{g^* f \colon f, g \in \mathfrak{n}_{\widetilde{\psi}}\}$, and since $g^* f = \frac{1}{4}\sum_{k=0}^3 (f+i^k g)^*(f+i^k g)$ for any $f, g \in \mathfrak{n}_{\widetilde{\psi}}$, it is enough to prove that equation (9.5) holds for terms of the form $x = f^* f$ where $f \in \mathfrak{n}_{\widetilde{\psi}}$.

Let such an f be given. Recall that as a member of $\widehat{\mathcal{M}}_+$, $\mathscr{W}_G(f^* f)$ has a spectral resolution of the form $\mathscr{W}_G(f^* f) = \int_0^\infty \lambda\, de_\lambda + \infty.p$ where each e_λ is orthogonal to p. Since $\widehat{\psi}(\mathscr{W}_G(f^* f)) = \widetilde{\psi}(f^* f) < \infty$, we must have that $\widehat{\psi}(p) = 0$ and hence that $p = 0$. But then $\mathscr{W}_G(f^* f)$ is a densely defined operator affiliated to $\mathcal{M}$. For each $n \in \mathbb{N}$, we then have that

$$\mathscr{W}_G(e_n f^* f e_n) = e_n \mathscr{W}_G(f^* f) e_n = \int_0^n \lambda\, de_\lambda \in \mathcal{M}$$

with in addition

$$\begin{aligned} \widetilde{\psi}(e_n f^* f e_n) &= \widehat{\psi}(\mathscr{W}_G(e_n f^* f e_n)) = \widehat{\psi}(e_n \mathscr{W}_G(f^* f)e_n) \\ &\leq \widehat{\psi}(\mathscr{W}_G(f^* f)) = \widetilde{\psi}(f^* f) < \infty \end{aligned}$$

since $e_n \mathscr{W}_G(f^* f) e_n \leq \mathscr{W}_G(f^* f)$. Hence, $f e_n \in (\mathfrak{n}_{\widetilde{\psi}} \cap \mathfrak{n}_{\mathscr{W}})$ for each $n \in \mathbb{N}$. Also notice that by the normality of ψ, $\widetilde{\psi}(e_n f^* f e_n) = \psi(e_n \mathscr{W}_G(f^* f)e_n) = \psi(\int_0^n \lambda\, de_\lambda)$ increases to $\widetilde{\psi}(f^* f) = \psi(\mathscr{W}_G(f^* f)) = \psi(\int_0^\infty \lambda\, de_\lambda)$ as $n \nearrow \infty$. Since $\psi(\int_0^\infty \lambda\, de_\lambda) < \infty$, this ensures that

$$\begin{aligned} \widetilde{\psi}((\mathbb{1} - e_n)f^* f(\mathbb{1} - e_n)) &= \psi(\mathscr{W}_G((\mathbb{1} - e_n)f^* f(\mathbb{1} - e_n))) \\ &= \psi((\mathbb{1} - e_n)\mathscr{W}_G(f^* f)(\mathbb{1} - e_n)) \\ &= \psi\left(\int_n^\infty \lambda\, de_\lambda\right) \to 0 \text{ as } n \nearrow \infty. \end{aligned}$$

We now use this fact to show that $\lim_{n\to\infty} \widetilde{\psi}(ae_n f^* f e_n) = \widetilde{\psi}(af^* f)$, and $\lim_{n\to\infty} \widetilde{\psi}(e_n f^* f e_n b) = \widetilde{\psi}(f^* f b)$.

For any $n \in \mathbb{N}$, we may write

$$\widetilde{\psi}(af^* f - ae_n f^* f e_n) = \widetilde{\psi}(a(\mathbb{1} - e_n)f^* f) + \widetilde{\psi}(ae_n f^* f(\mathbb{1} - e_n)).$$

By the Cauchy–Schwarz inequality for $\widetilde{\psi}$ on $\mathfrak{m}_{\widetilde{\psi}}$ and equation (9.4), we have that

$$\begin{aligned}
|\widetilde{\psi}(a(\mathbb{1} - e_n)f^* f)| &\leq \widetilde{\psi}(a(\mathbb{1} - e_n)f^* f(\mathbb{1} - e_n)a^*)^{1/2}\widetilde{\psi}(f^* f)^{1/2} \\
&\leq k\widetilde{\psi}((\mathbb{1} - e_n)f^* f(\mathbb{1} - e_n))^{1/2}\widetilde{\psi}(f^* f)^{1/2}.
\end{aligned}$$

This clearly ensures that $\widetilde{\psi}(a(\mathbb{1} - e_n)f^* f) \to 0$ as $n \to \infty$. An entirely similar proof shows that $\widetilde{\psi}(ae_n f^* f(\mathbb{1} - e_n))$ also tends to 0 as $n \to \infty$. This then ensures that $\lim_{n\to\infty} \widetilde{\psi}(ae_n f^* f e_n) = \widetilde{\psi}(af^* f)$, as required. To prove the second limit formula, we write $\widetilde{\psi}(f^* f b - e_n f^* f e_n b) = \widetilde{\psi}((\mathbb{1} - e_n)f^* f b) + \widetilde{\psi}(be_n f^* f(\mathbb{1} - e_n)b)$ and argue along similar lines.

By equation (9.6), we have that $\widetilde{\psi}(ae_n f^* f e_n) = \widetilde{\psi}(e_n f^* f e_n b)$ for every $n \in \mathbb{N}$. If we consider this fact alongside the limit formulae we have just proven, we have that $\widetilde{\psi}(af^* f) = \widetilde{\psi}(f^* f b)$ as required. $\square$

Given two faithful normal semifinite weights ψ_1 and ψ_2 on $\mathcal{M}$, our next result compares the Connes cocycle derivatives of the pair (ψ_1, ψ_2) to that of the pair $(\widetilde{\psi}, \widetilde{\psi}_2)$.

Theorem 9.26 *For any two faithful normal semifinite weights ψ_1 and ψ_2 on $\mathcal{M}$, we have that $(D\widetilde{\psi}_1 : D\widetilde{\psi}_2)_t = \pi((D\psi_1 : D\psi_2)_t)$ for every $t \in \mathbb{R}$*

Haagerup provided two proofs of this fact, one in [**Haa78a**], and the other in [**Haa79b**]. We outline a modified version of the first proof.

Proof. The proof relies on tricks that are fairly standard in the theory of Connes cocycle derivatives. However, although standard, this theory is somewhat outside the scope of this book. Therefore, we provide only an outline of the proof. The interested reader may find details of these tricks in §VIII.3 of [**Tak03a**], and also in [**Haa78a**].

Let M_2 be the 2×2 matrices over $\mathbb{C}$, and let $\{e_{i,j} : 1 \leq i,j \leq 2\}$ be the standard basis for M_2. Now, consider $\mathcal{M}\overline{\otimes}M_2$. The action α of G

on $\mathcal{M}$, then lifts to an action $\beta = \alpha \otimes \mathrm{id}$ of G on $\mathcal{M}\overline{\otimes}M_2$ (where id is the identity operator on M_2). One may then further show that

$$(\mathcal{M}\overline{\otimes}M_2) \rtimes_\beta G = (\mathcal{M} \rtimes_\alpha G)\overline{\otimes}M_2.$$

The next step is to define a weight ψ on $(\mathcal{M}\overline{\otimes}M_2)_+$ by setting

$$\psi\left(\begin{bmatrix} x_{11} & x_{1,2} \\ x_{21} & x_{22} \end{bmatrix}\right) = \psi_1(x_{11}) + \psi_2(x_{22}) \text{ for any } [x_{ij}] \in (\mathcal{M}\overline{\otimes}M_2)_+.$$

Next note that by Theorem 8.39, the canonical operator-valued weight on $((\mathcal{M}\overline{\otimes}M_2) \rtimes_\beta G)_+$ is just $\mathscr{W}_G \otimes \mathrm{id}$. One may then use this fact to show that the dual weight $\widetilde{\psi}$ on $((\mathcal{M}\overline{\otimes}M_2) \rtimes_\beta G)_+$ is of the form

$$\widetilde{\psi}\left(\begin{bmatrix} y_{11} & y_{1,2} \\ y_{21} & y_{22} \end{bmatrix}\right) = \widetilde{\psi}_1(y_{11}) + \widetilde{\psi}_2(y_{22})$$

$$\text{for any } [y_{ij}] \in ((\mathcal{M}\overline{\otimes}M_2) \rtimes_\beta G)_+.$$

As we have done many times before, we will in the rest of the proof again identify $\mathcal{M}$ with $\pi(\mathcal{M})$ to simplify notation. Having done this, we next observe that the one-parameter modular automorphism group induced by $\widetilde{\psi}$ on $(\mathcal{M}\overline{\otimes}M_2) \rtimes_\beta G$ is of the form

$$\sigma_t^{\widetilde{\psi}}\left(\begin{bmatrix} y_{11} & y_{1,2} \\ y_{21} & y_{22} \end{bmatrix}\right) = \left(\begin{bmatrix} \sigma_t^{\widetilde{\psi}_1}(y_{11}) & \sigma_t^{\widetilde{\psi}_1,\widetilde{\psi}_2}(y_{1,2}) \\ \sigma_t^{\widetilde{\psi}_2,\widetilde{\psi}_1}(y_{21}) & \sigma_t^{\widetilde{\psi}_1}(y_{22}) \end{bmatrix}\right)$$

for any $[y_{ij}] \in (\mathcal{M}\overline{\otimes}M_2) \rtimes_\beta G$, where $\sigma_t^{\widetilde{\psi}_1,\widetilde{\psi}_2}$ and $\sigma_t^{\widetilde{\psi}_2,\widetilde{\psi}_1}$ are groups of isometries on $(\mathcal{M}\overline{\otimes}M_2) \rtimes_\beta G$. (See the discussion preceding [**Tak03a**, Lemma VIII.3.5].) Similarly, for σ_t^{ψ}, we have that

$$\sigma_t^{\psi}\left(\begin{bmatrix} x_{11} & x_{1,2} \\ x_{21} & x_{22} \end{bmatrix}\right) = \left(\begin{bmatrix} \sigma_t^{\psi_1}(x_{11}) & \sigma_t^{\psi_1,\psi_2}(x_{1,2}) \\ \sigma_t^{\psi_2,\psi_1}(x_{21}) & \sigma_t^{\psi_1}(x_{22}) \end{bmatrix}\right)$$

for any $[x_{ij}] \in \mathcal{M}\overline{\otimes}M_2$. In fact the cocycle derivatives $(D\widetilde{\psi}_1 : D\widetilde{\psi}_2)_t$ and $(D\psi_1 : D\psi_2)_t$ may, respectively, be *defined* by the prescriptions $\sigma_t^{\widetilde{\psi}_1,\widetilde{\psi}_2}(\mathbb{1}) = (D\widetilde{\psi}_1 : D\widetilde{\psi}_2)t$ and $\sigma_t^{\widetilde{\psi}_1,\widetilde{\psi}_2}(\mathbb{1}) = (D\psi_1 : D\psi_2)_t$. (Once again see the discussion preceding [**Tak03a**, Lemma VIII.3.5].) Having noted these facts, we may now use Theorem 9.25 to see that for any $t \in \mathbb{R}$

$$\begin{bmatrix} 0 & (D\widetilde{\psi_1} : D\widetilde{\psi_2})t \\ 0 & 0 \end{bmatrix} = \begin{bmatrix} 0 & \sigma_t^{\widetilde{\psi_1},\widetilde{\psi_2}}(\mathbb{1}) \\ 0 & 0 \end{bmatrix} = \sigma_t^{\widetilde{\psi}}\left(\begin{bmatrix} 0 & \mathbb{1} \\ 0 & 0 \end{bmatrix}\right)$$

$$= \sigma_t^{\psi}\left(\begin{bmatrix} 0 & \mathbb{1} \\ 0 & 0 \end{bmatrix}\right) = \begin{bmatrix} 0 & \sigma_t^{\psi_1,\psi_2}(\mathbb{1}) \\ 0 & 0 \end{bmatrix} = \begin{bmatrix} 0 & (D\psi_1 : D\psi_2)t \\ 0 & 0 \end{bmatrix}.$$

The result now clearly follows. $\qquad\qquad\square$

We are now finally ready to complete the proof of Theorem 9.24.

Proof of the surjectivity claim in Theorem 9.24. The proof uses the technology of Connes cocycle derivatives. The reader unfamiliar with this theory should review the relevant material in Section 8.3. Let ϑ be a normal semifinite $\widehat{\alpha}$-invariant weight on $M \rtimes_\alpha G$.

We first consider the case where ϑ is faithful. Let ψ be any faithful normal semifinite weight on M. By what we have already proven, $\widetilde{\psi}$ is then an $\widehat{\alpha}$-invariant faithful normal semifinite weight on $M \rtimes_\alpha G$. It then directly follows from Corollary 8.27 that $\widehat{\alpha}_\gamma((D\vartheta : D\widetilde{\psi})_t) = (D\vartheta : D\widetilde{\psi})_t$ for any $t \in \mathbb{R}$ and any $\gamma \in \widehat{G}$. By Theorem 9.17, this ensures that each $(D\vartheta : D\widetilde{\psi})_t$ belongs to $\pi(M)$. Now let $u_t = \pi^{-1}((D\vartheta : D\widetilde{\psi})_t)$ for each $t \in \mathbb{R}$. We may now use Theorems 8.24 and 9.25 to see that

$$\pi(u_{s+t}) = (D\vartheta : D\widetilde{\psi})_{s+t} = (D\vartheta : D\widetilde{\psi})_s \sigma_s^{\widetilde{\psi}}((D\vartheta : D\widetilde{\psi})_t)$$

$$= (D\vartheta : D\widetilde{\psi})_s \pi(\sigma_s^{\psi}(u_t)) = \pi(u_s \sigma_s^{\psi}(u_t))$$

or equivalently that $u_{s+t} = u_s \sigma_s^{\psi}(u_t)$ for all $s, t \in \mathbb{R}$. By Theorem 8.30, there must then exist a faithful normal semifinite weight ψ_0 on M, such that $u_t = (D\psi_0 : D\psi)_t$ for all $t \in \mathbb{R}$. An application of Theorem 9.26 now allows us to conclude that

$$(D\vartheta : D\widetilde{\psi})_t = \pi(u_t) = \pi((D\psi_0 : D\psi)_t) = (D\widetilde{\psi_0} : D\widetilde{\psi})_t.$$

One may now use Theorem 8.29 to conclude that $(D\vartheta : D\widetilde{\psi_0})_t = \mathbb{1}$ for all $t \in \mathbb{R}$. But then Theorem 8.32 ensures that we must have that $\widetilde{\psi_0} \leq \vartheta \leq \widetilde{\psi_0}$. In other words, $\widetilde{\psi_0} = \vartheta$.

Now, let ϑ be a normal $\widehat{\alpha}$-invariant weight. Recall that, by definition, we then have that $(M \rtimes_\alpha G)e_0(\vartheta) = \{x \in M \rtimes_\alpha G : \vartheta(x^*x) = 0\}$. Since ϑ is $\widehat{\alpha}$-invariant, we clearly have that $\widehat{\alpha}_\gamma(\{x \in M \rtimes_\alpha G : \vartheta(x^*x)$

$= 0\}) = \{x \in \mathcal{M} \rtimes_\alpha G : \vartheta(x^*x) = 0\}$ for each $\gamma \in \widehat{G}$ and hence that $\widehat{\alpha}_\gamma((\mathcal{M} \rtimes_\alpha G)e_0(\vartheta)) = (\mathcal{M} \rtimes_\alpha G)e_0(\vartheta)$ for each $\gamma \in \widehat{G}$. This can only be the case if $\widehat{\alpha}_\gamma(e_0(\vartheta)) = e_0(\vartheta)$ for each $\gamma \in \widehat{G}$, which ensures that in fact $e_0(\vartheta) \in \pi(\mathcal{M})$. The $\widehat{\alpha}$-invariance of ϑ similarly ensures that $\widehat{\alpha}_\gamma(\mathfrak{n}_\vartheta) = \mathfrak{n}_\vartheta$ for each $\gamma \in \widehat{G}$, which on taking the σ-weak closure, yields the fact that $\widehat{\alpha}_\gamma((\mathcal{M} \rtimes_\alpha G)e_\infty(\vartheta)) = (\mathcal{M} \rtimes_\alpha G)e_\infty(\vartheta)$ for each $\gamma \in \widehat{G}$. As before, we may conclude from this that in fact $e_\infty(\vartheta) \in \mathcal{M}$. Having noted this fact, we will in the subsequent analysis simply write e_0 and e_∞, for $e_0(\vartheta)$ and $e_\infty(\vartheta)$.

Now, suppose that ϑ is a normal semifinite weight (so $e_\infty = \mathbb{1}$) for which $e_0(\vartheta) \neq 0$. Let ψ_1 be any normal semifinite weight on $\mathcal{M}$ with $e_0(\psi_1) = \mathbb{1} - e_0$. It then follows from what we have already shown that $\widetilde{\psi}_1$ is a normal semifinite weight on $(\mathcal{M} \rtimes_\alpha G)_+$ with $e_0(\widetilde{\psi}_1) = \mathbb{1} - e_0$. It is now an exercise to see that $\nu = \vartheta + \widetilde{\psi}_1$ is then normal and faithful. Since $\vartheta = (\mathbb{1} - e_0).\vartheta.(\mathbb{1} - e_0)$ and $\widetilde{\psi}_1 = e_0\widetilde{\psi}_1 e_0$, it is clear that $\mathfrak{n}_\vartheta(\mathbb{1} - e_0) \subseteq \mathfrak{n}_\vartheta$ and $\mathfrak{n}_{\widetilde{\psi}_1} e_0 \subseteq \mathfrak{n}_{\widetilde{\psi}_1}$ and that $\mathfrak{n}_\vartheta(\mathbb{1} - e_0) + \mathfrak{n}_{\widetilde{\psi}_1} e_0 \subseteq \mathfrak{n}_\nu$. By the semifiniteness of ϑ and $\widetilde{\psi}_1$, each of $\mathfrak{n}_\vartheta$ and $\mathfrak{n}_{\widetilde{\psi}_1}$ is σ-weakly dense in $\mathcal{M} \rtimes_\alpha G$, and hence so is $\mathfrak{n}_\vartheta(\mathbb{1} - e_0) + \mathfrak{n}_{\widetilde{\psi}_1} e_0$. The weight ν is, therefore, also semifinite. By construction ν is $\widehat{\alpha}$-invariant. Hence, there must exist a faithful normal semifinite weight ς on $\mathcal{M}$ such that $\nu = \widetilde{\varsigma}$. By construction and part (iii) of the present theorem, we must then have that

$$\vartheta = (\mathbb{1} - e_0)\nu(\mathbb{1} - e_0) = (\mathbb{1} - e_0) \cdot \widetilde{\varsigma} \cdot (\mathbb{1} - e_0) = \widetilde{(\mathbb{1} - e_0) \cdot \varsigma \cdot (\mathbb{1} - e_0)}.$$

This shows that for the normal weight $\varrho = (\mathbb{1} - e_0) \cdot \varsigma \cdot (\mathbb{1} - e_0)$ on $\mathcal{M}$, we have $\vartheta = \widetilde{\varrho}$. (Since $\vartheta = \widetilde{\varrho}$ is semifinite, we in fact have that ϱ is semifinite.)

Finally, suppose that ϑ is merely normal and $\widehat{\alpha}$-invariant. Since $e_\infty \cdot \vartheta \cdot e_\infty$ is semifinite and still $\widehat{\alpha}$-invariant (by the fact that $e_\infty \in \mathcal{M}$), there exists a normal semifinite weight ς on $\mathcal{M}$ such that $\widetilde{\varsigma} = e_\infty \cdot \vartheta \cdot e_\infty$. We will construct a normal weight ν from ς for which $\widetilde{\nu} = \vartheta$. We first note that $\{x : (e_\infty \cdot \vartheta \cdot e_\infty)(x^*x) = 0\} = N_{e_\infty \cdot \vartheta \cdot e_\infty} = (\mathcal{M} \rtimes_\alpha G)(e_0 + (\mathbb{1} - e_\infty))$. To see this, observe that $x \in N_{e_\infty \cdot \vartheta \cdot e_\infty}$ if and only if $xe_\infty \in N_\vartheta = (\mathcal{M} \rtimes_\alpha G)e_0$. Since we trivially have that $(\mathcal{M} \rtimes_\alpha G)(\mathbb{1} - e_\infty) \subseteq N_{e_\infty \cdot \vartheta \cdot e_\infty}$, the claim follows. By part (iv), we therefore have $e_0(\varsigma) = e_0(e_\infty \cdot \vartheta \cdot e_\infty) = e_0 + (\mathbb{1} - e_\infty)$.

It clearly follows that $\mathbb{1} - e_\infty \leq e_0(\varsigma)$, or equivalently, that $\mathbb{1} - e_0(\varsigma) \leq e_\infty$. Since $(\mathbb{1} - e_0(\varsigma)) \cdot \varsigma \cdot (\mathbb{1} - e_0(\varsigma)) = \varsigma$, this ensures that $e_\infty \cdot \varsigma \cdot e_\infty = \varsigma$.

Thus, ς is in particular 0-valued on $(1 - e_\infty)\mathcal{M}_+(1 - e_\infty)$. We now define a new weight ϱ on $\mathcal{M}$ with the prescription that it must be infinite valued on all the non-zero elements of $(1 - e_\infty)\mathcal{M}_+(1 - e_\infty)$ and equal ς on $e_\infty\mathcal{M}_+e_\infty$, with $\varrho(x) = \varsigma(e_\infty x e_\infty) + \varrho((1 - e_\infty)x(1 - e_\infty))$ for $x \in \mathcal{M}_+$. It is not difficult to verify that ϱ is normal. In addition (using the fact that $e_\infty.\varsigma.e_\infty = \varsigma$), it is also clear that $e_\infty \cdot \varrho \cdot e_\infty = \varsigma$.

The fact we have just noted clearly ensures that for any $x \in \mathcal{M}$, $\varsigma(x^*x) < \infty$ if and only if $\varrho(e_\infty x^* x e_\infty) < \infty$. So $\mathfrak{n}_\varsigma e_\infty \subseteq \mathfrak{n}_\varrho$. Also if $\varrho(x^*x) < \infty$, then we must have that $(1 - e_\infty)x^*x(1 - e_\infty) = 0$, in which case $x \in \mathcal{M}e_\infty$. Hence, $\mathfrak{n}_\varrho \subseteq \mathcal{M}e_\infty$. Taken together, these facts ensure that the σ-weak closure of $\mathfrak{n}_\varrho$ is precisely $\mathcal{M}e_\infty$ and hence that $e_\infty(\varrho) = e_\infty$. Since $N_\varrho = \{x \colon \varrho(x^*x) = 0\} \subseteq \mathfrak{n}_\varrho$, it is clear from the above that $N_\varrho \subseteq \mathcal{M}e_\infty$. Using the facts that ϱ and ς agree on $e_\infty\mathcal{M}e_\infty$, we may then conclude that $N_\varrho = N_\varsigma e_\infty$. By part (iv), we have that $e_0(\varsigma) = e_0(\widetilde{\varsigma}) = e_0(e_\infty \cdot \vartheta \cdot e_\infty) = e_0 + (1 - e_\infty)$. Hence, $N_\varrho = N_\varsigma e_\infty = \mathcal{M}(e_0 + (1 - e_\infty))e_\infty = \mathcal{M}e_0$; that is $e_0(\varrho) = e_0$. For the weight ϱ, we therefore have by parts (iv) and (v) that $e_0(\widetilde{\varrho}) = e_0 = e_0(\vartheta)$ and $e_\infty(\widetilde{\varrho}) = e_\infty = e_\infty(\vartheta)$. In addition

$$e_\infty \cdot \widetilde{\varrho} \cdot e_\infty = e_\infty \cdot \widetilde{\varrho \cdot e_\infty} = \widetilde{\varsigma} = e_\infty \cdot \vartheta \cdot e_\infty.$$

These facts are enough to ensure that $\widetilde{\varrho} = \vartheta$ as required. $\qquad\square$

9.5 Crossed products with modular automorphism groups

Let the von Neumann algebra be equipped with a faithful normal semifinite weight φ. Tomita–Takesaki theory then informs us that this weight induces a canonical one-parameter group of point to σ-weak continuous $*$-automorphisms σ_t^φ $(t \in \mathbb{R})$ on $\mathcal{M}$—the so-called modular automorphism group—for which we have $\varphi \circ \sigma_t^\varphi = \varphi$ for all $t \in \mathbb{R}$. This group is of course an action of $\mathbb{R}$ on $\mathcal{M}$, and hence, we may construct the crossed product with respect to this action. In this case, we will denote this crossed product by $\mathcal{M} \rtimes_\varphi \mathbb{R}$. This crossed product is absolutely central to everything that follows and will be repeatedly used. It is now commonly referred to as the core of $\mathcal{M}$ in the literature. We will for the sake of brevity often simply write $\mathfrak{M}$ for $\mathcal{M} \rtimes_\varphi \mathbb{R}$. The dual group of $\mathbb{R}$ is of course again a copy of $\mathbb{R}$, with the characters in the 'dual group' of the form $\gamma_t \colon \mathbb{R} \to \mathbb{T} \colon s \mapsto e^{ist}$ (here $\mathbb{T}$ is the circle group

$\{z \in \mathbb{C}: |z| = 1\}$). So, in this case, the unitary group w_t $(t \in \mathbb{R})$ acting on $L^2(\mathbb{R}, H)$ that induces the dual action on $\mathfrak{M}$ is of the form

$$w_t(\xi)(s) = e^{-its}\xi(s) \qquad \xi \in L^2(\mathbb{R}, H), s, t \in \mathbb{R}.$$

Following convention, we will in this case write θ_t $(t \in \mathbb{R})$ for the dual action, with the action on the generators of $\mathfrak{M}$ being given by

$$\theta_t(\pi(a)) = a, \quad \theta_t(\lambda_s) = e^{-ist}\lambda_s, \qquad a \in \mathcal{M}, \quad s, t \in \mathbb{R}.$$

In this particular case, the crossed product has some special features not shared by crossed products with more general groups. Much of this follows from the fact that in this case the modular automorphism group of the dual weight $\widetilde{\varphi}$ on $\mathfrak{M}$ is implemented. (Note that by Theorem 9.24, $\widetilde{\varphi}$ is indeed a faithful normal semifinite weight on $\mathfrak{M}$.) By combining Proposition 9.5 and Theorem 9.25, it is clear that the modular automorphism group $\sigma_t^{\widetilde{\varphi}}$ is implemented by $\{\lambda_t\} \subseteq \mathfrak{M}$. Due to its significance, we state this as a proposition.

Proposition 9.27 *The modular automorphism group $\sigma_t^{\widetilde{\varphi}}$ $(t \in \mathbb{R})$ on $\mathfrak{M}$ corresponding to the dual weight $\widetilde{\varphi}$ is implemented by the unitary group $\{\lambda_t\} \subseteq \mathfrak{M}$ in the sense that $\sigma_t^{\widetilde{\varphi}}(a) = \lambda_t a \lambda_t^*$ for all $t \in \mathbb{R}$ and all $a \in \mathfrak{M}$.*

The above fact has two very far-reaching consequences, which we summarise in the following theorem.

Theorem 9.28 *Let $\mathcal{M}$ and $\mathfrak{M}$ be as above.*

(1) *The centre of $\pi(\mathcal{M})$ is contained in the centre of $\mathfrak{M}$.*
(2) *There exists a positive non-singular operator h affiliated with $\mathfrak{M}_{\widetilde{\varphi}}$ (the centraliser of $\widetilde{\varphi}$), such that $\lambda_t = h^{it}$ for all $t \in \mathbb{R}$. For this operator, the derived weight $\tau = \widetilde{\varphi}(h^{-1}\cdot)$ is a faithful normal semifinite trace which satisfies the identity $\tau \circ \theta_s = e^{-s}\tau$ for all $s \in \mathbb{R}$. When equipped with this trace, h is just the Radon–Nikodym derivative $\frac{d\widetilde{\varphi}}{d\tau}$.*

Proof. To see that (1) holds, notice that on the centre of $\mathcal{M}$, φ behaves like a trace. Since traces are known to induce trivial modular automorphism groups, it is no surprise to find that the elements of the centre are fixed points of the automorphism group σ_t^φ ($t \in \mathbb{R}$) (see Theorem 8.19). When passing to $\pi(\mathcal{M})$, we see from Proposition 9.5 that this means that for every element a of the centre, we have that $\lambda_t \pi(a) \lambda_t* = \pi(a)$ for all t, or equivalently that $\lambda_t \pi(a) = \pi(a)\lambda_t$ for all t. Since $\pi(a)$ commutes with each λ_t and also each element of $\pi(\mathcal{M})$, it must in fact commute with each element of the algebra generated by these objects, namely $\mathfrak{M}$. We proceed with proving the second claim.

The fact that h is a positive non-singular operator affiliated with $\mathfrak{M}_{\widetilde{\varphi}}$, and τ a faithful normal semifinite trace, follows directly from Theorem 8.31. Since τ is a trace, its modular automorphism group is trivial. In addition by the choice of τ, $\widetilde{\varphi}$ is then just $\widetilde{\varphi} = \tau(h\cdot)$. So, by definition, h is the Radon–Nikodym derivative $\frac{d\widetilde{\varphi}}{d\tau}$. It remains to check the claim that $\tau \circ \theta_s = e^{-s}\tau$ for all $s \in \mathbb{R}$. In this regard, note that for a fixed s, we have that $\theta_s(\lambda_t) = \theta_s(h^{it}) = (\theta_s(h))^{it}$ for each $t \in \mathbb{R}$. However, we also have that $\theta_s(\lambda_t) = e^{-ist}h^{it} = (e^{-s}h)^{it}$. Hence, both $\theta_s(h)$ and $e^{-s}h$ are generators of the unitary group $\{\theta_s(\lambda_t)\}$ ($t \in \mathbb{R}$). By the uniqueness of such generators, we must have that $\theta_s(h) = e^{-s}h$. Since s was arbitrary, this holds for all s. But then for any s, $\tau \circ \theta_s = \widetilde{\varphi}(h^{-1}\theta_s(\cdot)) = e^{-s}\widetilde{\varphi}(\theta_s(h^{-1}\cdot))$. Since $\widetilde{\varphi}$ is invariant with respect to the dual action, we have that $\tau \circ \theta_s = e^{-s}\widetilde{\varphi}(\theta_s(h^{-1}\cdot)) = e^{-s}\widetilde{\varphi}(h^{-1}\cdot) = e^{-s}\tau$ as required. $\qquad\square$

Corollary 9.29 *The dual action $\{\theta_t \colon t \in \mathbb{R}\}$ extends to a continuous action on $\widetilde{\mathfrak{M}}$.*

Proof. The equality $\tau \circ \theta_s = e^{-s}\tau$ ensures that in their action on the projection lattice of $\mathfrak{M}$, we have that $\tau \circ \theta_s$ and τ are mutually absolutely continuous with respect to each other in an ϵ-δ sense. Hence, the claim follows from Proposition 3.98. $\qquad\square$

We close this discussion of the trace on $\mathfrak{M}$ with the following observation which will prove to be an important tool in investigating the uniqueness of the crossed product.

Proposition 9.30 *Let τ be the canonical trace on $\mathfrak{M}$ as described above. Then $\lambda_t = (D\widetilde{\varphi} : D\tau)_t$ for all $t \in \mathbb{R}$.*

Proof. Recall that the modular automorphism group of $\widetilde{\varphi}$ is implemented by the λ_ts, and that τ is a trace, so that its modular automorphism group is trivial. We leave it as an exercise to verify that the λ_ts meet all the requirements stipulated for the $(D\widetilde{\varphi} : D\tau)_t$s in Theorem 8.24. So, by the uniqueness criterion in that theorem, we must have $\lambda_t = (D\widetilde{\varphi} : D\tau)_t$ as required. $\qquad\square$

We pass to proving that in a very concrete sense, the algebra $\mathfrak{M} = \mathcal{M} \rtimes_\varphi \mathbb{R}$ is, almost surprisingly, independent of the particular faithful normal semifinite weight used to construct it!

Theorem 9.31 *Let φ_1 and φ_2 be two f.n.s. weights on $\mathcal{M}$. Both crossed products $\mathcal{M} \rtimes_{\varphi_1} \mathbb{R} = \mathfrak{M}_1$ and $\mathcal{M} \rtimes_{\varphi_2} \mathbb{R} = \mathfrak{M}_2$, are realised on the same Hilbert space $L^2(\mathbb{R}, H)$ and share the same shift operators λ_t $(t \in \mathbb{R})$. The action θ_s of the dual group therefore shares the same implementation for each of the crossed products. In addition, there is a *-isomorphism $\mathscr{I}$ from $\mathcal{M} \rtimes_{\varphi_1} \mathbb{R} = \mathfrak{M}_1$ onto $\mathcal{M} \rtimes_{\varphi_2} \mathbb{R} = \mathfrak{M}_2$ implemented by a unitary element u of $B(L^2(\mathbb{R}, H))$ for which we have that $\tau_1 = \tau_2 \circ \mathscr{I}$. The *-isomorphism $\mathscr{I}$ extends to a *-isomorphism which homeomorphically maps $\widetilde{\mathfrak{M}}_1$ onto $\widetilde{\mathfrak{M}}_2$, and which leaves the dual action invariant in the sense that $\mathscr{I} \circ \theta_t = \theta_t \circ \mathscr{I}$ for every $t \in \mathbb{R}$.*

Proof. A consideration of Definitions 9.6 and 9.12 reveals that the first claim is by construction. We now define the unitary u on $L^2(\mathbb{R}, H)$ by

$$(u\xi)(t) = (D\varphi_2 : D\varphi_1)_{-t}\xi(t) \qquad \text{for all } t \in \mathbb{R} \text{ and all } \xi \in L^2(\mathbb{R}, H).$$

For each $a \in \mathcal{M}$ and each $\xi \in L^2(\mathbb{R}, H)$, it now follows from Theorem 8.24 that

$$u\pi_1(a)u^*\xi(t) = (D\varphi_2 : D\varphi_1)_{-t}\sigma_{-t}^{\varphi_1}(a)(D\varphi_2 : D\varphi_1)^*_{-t}\xi(t)$$
$$= \sigma_{-t}^{\varphi_2}(a)\xi(t) = \pi_2(a)\xi(t).$$

We proceed to compute $u\lambda_t u^*$ for all $t \in \mathbb{R}$. For this, we need the fact that the chain rule for cocycle derivatives (Theorem 8.29) ensures that $\mathbb{1} = (D\varphi_1 : D\varphi_1)_t = (D\varphi_1 : D\varphi_2)_t(D\varphi_2 : D\varphi_1)_t$ for each t and hence that $(D\varphi_1 : D\varphi_2)^*_t = (D\varphi_2 : D\varphi_1)_t$ Given $s \in \mathbb{R}$, we have that

$$\begin{aligned}
(u\lambda_s u^*)\xi(t) &= (D\varphi_2 : D\varphi_1)_{-t}(\lambda_s u^*)\xi(t) \\
&= (D\varphi_2 : D\varphi_1)_{-t}(u^*\xi)(t-s) \\
&= (D\varphi_2 : D\varphi_1)_{-t}(D\varphi_2 : D\varphi_1)^*_{s-t}\xi(t-s)
\end{aligned}$$

for all $t \in \mathbb{R}$ and all $\xi \in L^2(\mathbb{R}, H)$. Since by Theorem 8.24

$$(D\varphi_2 : D\varphi_1)_{-t}\sigma^{\varphi_1}_{-t}((D\varphi_2 : D\varphi_1)_s) = (D\varphi_2 : D\varphi_1)_{s-t} \text{ for all } s, t \in \mathbb{R},$$

the above may be rewritten as

$$\begin{aligned}
&(u\lambda_s u^*)\xi(t) \\
&= (D\varphi_2 : D\varphi_1)_{-t}[(D\varphi_2 : D\varphi_1)_{-t}\sigma^{\varphi_1}_{-t}((D\varphi_2 : D\varphi_1)_s)]^*\xi(t-s) \\
&= (D\varphi_2 : D\varphi_1)_{-t}\sigma^{\varphi_1}_{-t}((D\varphi_2 : D\varphi_1)^*_s)(D\varphi_2 : D\varphi_1)^*_{-t}\xi(t-s) \\
&= (D\varphi_2 : D\varphi_1)_{-t}\sigma^{\varphi_1}_{-t}((D\varphi_1 : D\varphi_2)_s)(D\varphi_2 : D\varphi_1)^*_{-t}\xi(t-s) \\
&= \sigma^{\varphi_2}_{-t}((D\varphi_1 : D\varphi_2)_s)\xi(t-s) \\
&= \sigma^{\varphi_2}_{-t}((D\varphi_1 : D\varphi_2)_s)\lambda_s\xi(t) \\
&= \pi_2((D\varphi_1 : D\varphi_2)_s)\lambda_s\xi(t).
\end{aligned}$$

(In the fourth equality, we silently applied Theorem 8.24 once again.) We therefore have $u\lambda_s u^* = \pi_2((D\varphi_1 : D\varphi_2)_s)\lambda_s$ for each s. This is clearly an element of $\mathcal{M} \rtimes_{\varphi_2} \mathbb{R}$, and hence, the prescription $a \mapsto uau^*$ maps $\mathfrak{M}_1$—the algebra generated by $\pi_1(\mathcal{M})$ and the λ_ts—into $\mathfrak{M}_2$. By now swopping the roles of φ_1 and φ_2, we can similarly show that the prescription $a \mapsto u^*au$ maps $\mathfrak{M}_2$ into $\mathfrak{M}_1$. Hence, we must have that $u\mathfrak{M}_1 u^* = \mathfrak{M}_2$. The prescription $\mathscr{I}(a) = uau^*$ therefore clearly defines a *-isomorphism from $\mathfrak{M}_1$ onto $\mathfrak{M}_2$.

It is now a simple matter to check that on $\mathfrak{M}_1$, $\theta_s \circ \mathscr{I} = \mathscr{I} \circ \theta_s$ for each $s \in \mathbb{R}$. Specifically, Theorem 9.17 ensures that for any $a \in \pi_1(\mathcal{M})$ we have that $\theta_s \circ \mathscr{I}(a) = \mathscr{I}(a) = \mathscr{I} \circ \theta_s(a)$. For any $t \in \mathbb{R}$, we additionally have that $\theta_s \circ \mathscr{I}(\lambda_t) = \theta_s(\pi_2((D\varphi_1 : D\varphi_2)_t)\lambda_t) = e^{-ist}\pi_2((D\varphi_1 : D\varphi_2)_t)\lambda_t = \mathscr{I} \circ \theta_s(\lambda_t)$. Hence, the claim follows. If we can show that $\tau_2 \circ \mathscr{I} = \tau_1$, then the fact that the *-isomorphism $\mathscr{I}$ extends to a bicontinuous *-isomorphism from $\widetilde{\mathfrak{M}}_1$ onto $\widetilde{\mathfrak{M}}_2$ will follow from Proposition 3.98. By continuity, the extension will still satisfy $\theta_s \circ \mathscr{I} = \mathscr{I} \circ \theta_s$ for each $s \in \mathbb{R}$.

We proceed to show that indeed $\tau_2 \circ \mathscr{I} = \tau_1$. The first technical fact we need is the observation that for each $a \in \mathfrak{M}_1^+$

$$\mathscr{W}(\mathscr{I}(a)) = \int_{\mathbb{R}} \theta_s(\mathscr{I}(a))\, ds = \int_{\mathbb{R}} \mathscr{I}(\theta_s(a))\, ds = u\mathscr{W}(a)u^*.$$

A subtlety we need to contend with here is that with two weights, we now have two ways in which to define the corresponding dual weight. Let $\mathscr{W}$ be the operator-valued weight from $\mathfrak{M}_1$ to $\pi_1(\mathcal{M})_+$. This weight is defined by $\mathscr{W}(a) = \int_{\mathbb{R}} \theta_s(a)\, ds$. Observe that exactly the same prescription is used to define the operator-valued weight from $\mathfrak{M}_2$ to $\pi_2(\mathcal{M})_+$. We will, therefore, use the same notation for both versions. The standard way of defining $\widetilde{\varphi}_1$ is in terms of $\mathfrak{M}_1$ by means of the prescription $\widehat{\varphi}_1 \circ \pi_1^{-1} \circ \mathscr{W}$. However, we can also define a dual version of φ_1 in terms of $\mathfrak{M}_2$. We shall write $\widetilde{\varphi}_1^{(2)}$ for this alternative version. In this case, the prescription for $\widetilde{\varphi}_1^{(2)}$ is $\widehat{\varphi}_1 \circ \pi_2^{-1} \circ \mathscr{W}$. Using the technical facts we verified above, we can now see that these two versions are related by the formula

$$\begin{aligned}
\widetilde{\varphi}_1(a) &= \widehat{\varphi}_1 \circ \pi_1^{-1} \circ \mathscr{W}(a) \\
&= \widehat{\varphi}_1 \circ \pi_2^{-1}(\pi_2(\pi_1^{-1}(\mathscr{W}(a)))) \\
&= \widehat{\varphi}_1 \circ \pi_2^{-1}(u\pi_1(\pi_1^{-1}(\mathscr{W}(a)))u^*) \\
&= \widehat{\varphi}_1 \circ \pi_2^{-1}(u\mathscr{W}(a)u^*) \\
&= \widehat{\varphi}_1 \circ \pi_2^{-1}(\mathscr{W}(\mathscr{I}(a))) \\
&= \widetilde{\varphi}_1^{(2)}(\mathscr{I}(a))
\end{aligned}$$

for each $a \in \mathfrak{M}_1^+$. We saw earlier that $\mathscr{I}(\lambda_t) = \pi_2((D\varphi_1 : D\varphi_2)_t)\lambda_t$. By reversing the roles of φ_1 and φ_2 in the proof of that fact, we may show that $\mathscr{I}^{-1}(\lambda_t) = u^*\lambda_t u = \pi_1((D\varphi_2 : D\varphi_1)_t)\lambda_t$ for each t. We may now use these two facts alongside Corollary 8.27 to see that for each t

$$(D\widetilde{\varphi}_1 : D(\tau_2 \circ \mathscr{I}))_t = (D(\widetilde{\varphi}_1^{(2)} \circ \mathscr{I}) : D(\tau_2 \circ \mathscr{I}))_t = \mathscr{I}^{-1}((D\widetilde{\varphi}_1^{(2)} : D\tau_2)_t).$$

On bringing Theorems 8.29 and 9.26 and Proposition 9.30 into play, it then follows that

$$
\begin{aligned}
(D\widetilde{\varphi}_1 : D(\tau_2 \circ \mathscr{I}))_t &= \mathscr{I}^{-1}((D\widetilde{\varphi}_1^{(2)} : D\tau_2)_t) \\
&= \mathscr{I}^{-1}((D\widetilde{\varphi}_1^{(2)} : D\widetilde{\varphi}_2)_t).\mathscr{I}^{-1}((D\widetilde{\varphi}_2 : D\tau_2)_t) \\
&= \mathscr{I}^{-1}(\pi_2((D\varphi_1 : D\varphi_2)_t)).\mathscr{I}^{-1}(\lambda_t) \\
&= \pi_1((D\varphi_1 : D\varphi_2)_t).\pi_1((D\varphi_2 : D\varphi_1)_t)\lambda_t \\
&= \lambda_t \\
&= (D\widetilde{\varphi}_1 : D\tau_1)_t.
\end{aligned}
$$

This ensures that $(D(\tau_2 \circ \mathscr{I}) : D\tau_1)_t = (D(\tau_2 \circ \mathscr{I}) : D\widetilde{\varphi}_1)_t (D\widetilde{\varphi}_1 : D\tau_1)_t = (D\widetilde{\varphi}_1 : D\tau_1)_t^*(D\widetilde{\varphi}_1 : D\tau_1)_t = \mathbb{1}$ for all t. Thus, by Theorem 8.32, we have $\tau_2 \circ \mathscr{I} = \tau_1$ as required. $\qquad\square$

The presence of a trace on $\mathfrak{M}$, now enables us to reinterpret the dual weight map $\psi \mapsto \widetilde{\psi}$. Specifically with Theorem 4.26 as background, we are now able to prove the promised reformulation of Theorem 9.24.

Definition 9.32 For each normal weight ψ on $\mathcal{M}$, we define m_ψ to be the unique element of $\widehat{\mathfrak{M}}_+$ corresponding to the dual weight $\widetilde{\psi}$ by means of the bijection described in Theorem 4.26.

Proposition 9.33 *The mapping $\psi \mapsto m_\psi$ defined above is a bijection from the set of all normal weights on $\mathcal{M}$ onto the set of all elements m of $\widehat{\mathfrak{M}}_+$ satisfying $\theta_s(m) = e^{-s}m$. For normal weights ψ, ψ_1 and ψ_2 on $\mathcal{M}$, and any $a \in \mathcal{M}$, we moreover have that:*

> (1) *$m_{\psi_1} \le m_{\psi_2}$ if and only if $\psi_1 \le \psi_2$, and in addition $m_{\psi_\alpha} \nearrow m_\psi$ if and only if $\psi_\alpha \nearrow \psi$;*
> (2) *$m_{\psi_1 + \psi_2} = m_{\psi_1} + m_{\psi_2}$;*
> (3) *$a^* \cdot m_\psi \cdot a = m_{a^* \cdot \psi \cdot a}$;*
> (4) *with $\int_0^\infty \lambda\, de_\lambda + \infty \cdot p$ denoting the spectral resolution of m_ψ, we have that $\mathbb{1} - p = \pi(e_\infty(\psi))$ and $e_0 = \pi(e_0(\psi))$.*

Proof. As we have done before, we will in this proof also identify $\pi(\mathcal{M})$ with $\mathcal{M}$. Properties (1)–(4) are fairly immediate consequences of Theorems 9.24 and 4.26. It therefore remains to prove the first claim. It is clear from Theorems 9.24 and 4.26 that we may prove that claim by proving that any normal weight ψ on $\mathfrak{M}$ is θ_s-invariant if and only if for each s we have that $\theta_s(m_\psi) = e^{-s}m_\psi$. By Theorem 4.26, a typical normal weight on $\mathfrak{M}$ is of the form ψ_m for some $m \in \widehat{\mathfrak{M}}_+$. For such a weight, for any $s \in \mathbb{R}$ and any $a \in \mathfrak{M}$, we will have that

$$\psi_{e^s\theta_s(m)}(a^*a) = e^s(\tau \circ \theta_s)(\theta_{-s}(a)m\theta_{-s}(a^*))$$
$$= \tau(\theta_{-s}(a)m\theta_{-s}(a^*))$$
$$= (\psi_m \circ \theta_{-s})(a^*a).$$

This clearly ensures that

$$e^{-s}m = \theta_s(m) \Leftrightarrow \psi_{e^s\theta_s(m)} = \psi_m \Leftrightarrow \psi_m = \psi_m \circ \theta_{-s}.$$

$\square$

Remark 9.34 It follows from part (3) of the preceding proposition that the mapping $\psi \mapsto m_\psi$ restricts to a bijection from the normal semifinite weights on $\mathcal{M}$, to the positive operators h affiliated with $\mathfrak{M}$ for which we have that $\theta_s(h) = e^{-s}h$ for all $s \in \mathbb{R}$. (This is the case where $\mathbb{1} - p = \pi(e_\infty(\psi)) = 0$.) *To distinguish these two settings, we shall when working with the restriction of this bijection to the set of normal semifinite weights on $\mathcal{M}$, denote the bijection by $\psi \mapsto h_\psi$.* This bijection then restricts further to a bijection from the faithful normal semifinite weights on $\mathcal{M}$ to the positive non-singular operators affiliated with $\mathfrak{M}$ for which we have that $\theta_s(h) = e^{-s}h$ for all $s \in \mathbb{R}$. (This is the case where in addition $e_0 = \pi(e_0(\psi)) = 0$.)

We proceed to the final topic of this section, which is to show that the dual action $\{\theta_s\}$ may be used to identify isometric copies of both $\mathcal{M}$ and $\mathcal{M}_*$ inside $\widetilde{\mathfrak{M}}$. Our first result in this regard is a refinement of Theorem 9.17.

Proposition 9.35 *For any $x \in \widetilde{\mathfrak{M}}$, we have that $\theta_s(x) = x$ for each $s \in \mathbb{R}$ if and only if $x \in \pi(\mathcal{M})$*

Proof. The result will clearly follow from Theorem 9.17 if we are able to prove that if for some $x \in \widetilde{\mathfrak{M}}$, we have that $\theta_s(x) = x$ for each $s \in \mathbb{R}$, then x must belong to $\mathfrak{M}$. So, let $x \in \widetilde{\mathfrak{M}}$ be given such that $\theta_s(x) = x$ for each $s \in \mathbb{R}$. Since x is τ-measurable, there must exist some $\lambda > 0$ such that $\tau(\chi_{(\lambda,\infty)}(|x|)) < \infty$. Given some $s \neq 0$, we may then apply Theorem 9.28 to see that

$$\tau(\chi_{(\lambda,\infty)}(|x|)) = \tau(\chi_{(\lambda,\infty)}(|\theta_s(x)|))$$
$$= \tau(\theta_s(\chi_{(\lambda,\infty)}(|x|)))$$
$$= e^{-s}\tau(\chi_{(\lambda,\infty)}(|x|)).$$

The only way this can be is if $\tau(\chi_{(\lambda,\infty)}(|x|))=0$ and hence $\chi_{(\lambda,\infty)}(|x|) = 0$. Thus, $|x|$, and hence x, is bounded as required. $\quad\square$

We will now refine the flow of ideas in Remark 9.34 and work toward establishing technology which will allow us to obtain a bijection on the positive normal functionals on $\mathcal{M}$ by further restricting the map in Proposition 9.33. This will then equip us with the necessary tools for the completion of the task of finding an isometric copy of $\mathcal{M}_*$ inside $\widetilde{\mathfrak{M}}$. The key result in this quest is an important result of Haagerup.

Proposition 9.36 *Let ψ be a normal semifinite weight on $\mathcal{M}$. For any $\gamma > 0$, we will then have $\tau_{\mathfrak{M}}(\chi_{(\gamma,\infty)}(h_\psi)) = \gamma^{-1}\psi(\mathbb{1})$.*

Proof. Since for any $\gamma > 0$, $\tau_{\mathfrak{M}}(\chi_{(\gamma,\infty)}(h_\psi)) = \tau_{\mathfrak{M}}(\chi_{(1,\infty)}(\gamma^{-1}h_\psi))$ where $\gamma^{-1}\psi$ corresponds to $\gamma^{-1}h_\psi$ by means of the bijection defined earlier, it suffices to prove the proposition for the case $\gamma = 1$. Let g_ψ be the positive operator affiliated to $\mathfrak{M}$ and commuting with h_ψ, for which we have that $g_\psi h_\psi = \mathbf{s}(h_\psi)$. It is clear from Theorem 9.17 and Proposition 9.33(4) that $\mathbf{s}(h_\psi)$ is invariant under the action of the θ_t's. So, since $\theta_t(h_\psi) = e^{-t}h_\psi$ for each t, we must have $\theta_t(g_\psi) = e^t g_\psi$ for each t.

Let $h_\psi = \int_0^\infty \lambda\, de_\lambda$ be the spectral resolution of h_ψ. Then, g_ψ is of course just $\int_0^\infty \lambda^{-1}\, de_\lambda$. Let $\xi \in L^2(\mathbb{R}, H)$ be a unit vector, and write $\rho_{\xi,\xi}$ for the functional $a \mapsto \langle a\xi, \xi\rangle$. We then have that

$$
\begin{aligned}
\mathscr{W}_{\mathbb{R}}(g_\psi \chi_{(1,\infty)}(h_\psi))(\rho_{\xi,\xi}) &= \int_0^\infty \theta_s(g_\psi \chi_{(1,\infty)}(h_\psi))(\rho_{\xi,\xi})\, ds \\
&= \int_0^\infty (\theta_s(g_\psi)\chi_{(1,\infty)}(\theta_s(h_\psi))(\rho_{\xi,\xi})\, ds \\
&= \int_0^\infty (e^s g_\psi \chi_{(1,\infty)}(e^{-s}h_\psi)(\rho_{\xi,\xi})\, ds \\
&= \int_0^\infty \int_0^\infty e^s \lambda^{-1} \chi_{(1,\infty)}(e^{-s}\lambda)\, d\langle e_\lambda \xi, \xi\rangle\, ds \\
&= \int_0^\infty \lambda^{-1} \left(\int_{(-\infty,\log(\lambda))} e^s\, ds \right) d\langle e_\lambda \xi, \xi\rangle \\
&= \int_0^\infty \lambda^{-1}\lambda\, d\langle e_\lambda \xi, \xi\rangle \\
&= \| \mathbf{s}(h_\psi)(\xi)\|^2 .
\end{aligned}
$$

Since ξ was an arbitrary unit vector, we therefore have that

$$
\mathscr{W}_{\mathbb{R}}(g_\psi \chi_{(1,\infty)}(h_\psi)) = \mathbf{s}(h_\psi) = e_0(\psi).
$$

Recalling that $\widetilde{\psi} = \tau(h_\psi \cdot)$, it therefore follows that

$$
\begin{aligned}
\tau(\chi_{(1,\infty)}(h_\psi)) &= \tau(h_\psi^{1/2}(g_\psi \chi_{(1,\infty)}(h_\psi))h_\psi^{1/2}) \\
&= \widetilde{\psi}(g_\psi \chi_{(1,\infty)}(h_\psi)) \\
&= \psi(\mathscr{W}_{\mathbb{R}}(g_\psi \chi_{(1,\infty)}(h_\psi))) \\
&= \psi(e_0(\psi)) \\
&= \psi(\mathbb{1})
\end{aligned}
$$

as claimed. $\qquad\square$

The corollary stated below follows on noticing that both conditions correspond to the finiteness (for some γ) of the quantity discussed in the above proposition.

Corollary 9.37 *Let ψ be a normal semifinite weight on $\mathcal{M}$. Then, ψ belongs to $\mathcal{M}_*$ if and only if $h_\psi \in \widetilde{\mathfrak{M}}$.*

Theorem 9.38   *The mapping $\mathcal{M}_*^+ \to \widetilde{\mathfrak{M}}_+ : \omega \mapsto h_\omega$ extends to a linear bijection from $\mathcal{M}_*$ onto $\{h \in \widetilde{\mathfrak{M}}: \theta_s(h) = e^{-s}h \text{ for all } s \in \mathbb{R}\}$. For all $\omega \in \mathcal{M}_*$ and all $a, b \in \mathcal{M}$, this bijection satisfies the following properties:*

> (1) $h_{x \cdot \omega \cdot y} = \pi(x)h_\omega \pi(y)$;
> (2) $h_{\omega^*} = h_\omega^*$, *where ω^* denotes the normal functional defined by* $\omega^*(a) = \overline{\omega(a^*)}$;
> (3) *If $\omega = u \cdot |\omega|$ is the polar decomposition of ω, then $h_\omega = u|h_\omega|$ is the polar decomposition of h_ω.*

Proof.   As is our wont, we will also in this proof identify $\mathcal{M}$ with $\pi(\mathcal{M})$. Let $\omega_1, \omega_2 \in \mathcal{M}_*^+$ be given, and let $h_{\omega_1} = h_1$ and $h_{\omega_2} = h_2$ be the image of these functionals under the bijection described in Proposition 9.33. By Corollary 9.37, both these operators are τ-measurable. This in turn ensures that their strong sum is again a positive self-adjoint operator. So, by Proposition 4.28, that strong sum $h_1 \bar{+} h_2$ must be $h_1 \widehat{+} h_2$. Hence, we have that $h_{\omega_1 + \omega_2} = h_{\omega_1} + h_{\omega_2}$, where the right-hand side represents the sum in $\widetilde{\mathfrak{M}}$. Similar considerations show that for any $\omega \in \mathcal{M}_*$ and any $a \in \mathcal{M}$, we will have that $h_{a \cdot \omega \cdot a^*} = a \bar{\cdot} h_\omega \bar{\cdot} a^* = ah_\omega a^*$ where the right-hand product is in $\widetilde{\mathfrak{M}}$. If in this final formula we take a to be $\gamma^{1/2}\mathbb{1}$, it is now clear that the prescription $\omega \mapsto h_\omega$ is an affine map from the positive cone $\mathcal{M}_*^+$, onto the positive cone $\{h \in \widetilde{\mathfrak{M}}_+ : \theta_s(h) = e^{-s}h \text{ for all } s \in \mathbb{R}\}$. Hence, this map will extend to a linear map from $\mathcal{M}_*$ onto $\mathrm{span}\{h \in \widetilde{\mathfrak{M}}_+ : \theta_s(h) = e^{-s}h \text{ for all } s \in \mathbb{R}\} = \{h \in \widetilde{\mathfrak{M}}: \theta_s(h) = e^{-s}h \text{ for all } s \in \mathbb{R}\}$. (It is an exercise to see that $\{h \in \widetilde{\mathfrak{M}}: \theta_s(h) = e^{-s}h \text{ for all } s \in \mathbb{R}\}$ is spanned by its positive elements.) To see that this map is well defined, let $\omega \in \mathcal{M}_*$ be a Hermitian functional with $\omega = \omega_1 - \omega_2 = \rho_1 - \rho_2$, where $\omega_1, \omega_2, \rho_1$ and ρ_2 are positive normal functionals. Then, of course, $\omega_1 + \rho_2 = \rho_1 + \omega_2$ by Proposition 9.33, in which case we then have $h_{\omega_1} + h_{\rho_2} = h_{\rho_1} + h_{\omega_2}$. Thus $h_{\omega_1} - h_{\omega_2} = h_{\rho_1} - h_{\rho_2}$, which ensures that h_ω is well defined. A similar argument shows that if we are given $\omega \in \mathcal{M}_*$ such that $\omega = \omega_1 + i\omega_2 = \rho_1 + i\rho_2$ where $\omega_1, \omega_2, \rho_1$ and ρ_2 are now Hermitian normal functionals, then $h_{\omega_1} + ih_{\omega_2} = h_{\rho_1} + ih_{\rho_2}$.

Given $\omega \in \mathcal{M}_*$, we may then write ω as the linear combination $\omega = \omega_1 - \omega_2 + i\omega_3 - i\omega_4$ of four positive normal functionals ω_i $(i = 1, 2, 3, 4)$. It is not difficult to see that then $\omega^* = \omega_1 - \omega_2 - i\omega_3 + i\omega_4$. So writing

h_i for h_{ω_i}, ω will by linearity map onto $h_1 - h_2 + ih_3 - ih_4$, and ω^* onto $h_1 - h_2 - ih_3 + ih_4 = (h_1 - h_2 + ih_3 - ih_4)^*$, thereby verifying (2). We know that for any $x \in \mathcal{M}$, we have $xh_ix^* = h_{x\cdot\omega_i\cdot x^*}$. So also by linearity the same must then be true of $h_\omega = h_1 - h_2 + ih_3 - ih_4$. Given $a, b \in \mathcal{M}$, we may then use this fact and the polarisation identities $ah_\omega b = \frac{1}{4}\sum_{k=0}^{3}(b + i^k a^*)^* h(b + i^k a^*)$ and $a \cdot \omega \cdot b = \frac{1}{4}\sum_{k=0}^{3}(b + i^k a^*)^* \cdot \omega \cdot (b + i^k a^*)$ to see that (1) holds.

Now, let $\omega = u \cdot |\omega|$ be the polar decomposition of ω. It then follows from (1) that $h_\omega = h_{u\cdot|\omega|} = uh_{|\omega|}$. Notice that the initial projection for the partial isometry u is $e_0(\omega)$. By Proposition 9.33, this is precisely $\mathsf{s}(h_{|\omega|})$. Therefore, it follows that $h_\omega^* h_\omega = h_{|\omega|}^2$ and hence that $|h_\omega| = h_{|\omega|}$. Thus, $h_\omega = h_{u\cdot|\omega|} = uh_{|\omega|}$ is indeed the polar decomposition of h_ω.

It remains to prove the injectivity of this map. To see this, note that if $h_\omega = 0$, then $0 = |h_\omega| = h_{|\omega|}$. But then $|\omega| = 0$ and hence $\omega = 0$ by the injectivity of the affine map from $\mathcal{M}_*^+$ to $\{h \in \widetilde{\mathfrak{M}}_+ : \theta_s(h) = e^{-s}h$ for all $s \in \mathbb{R}\}$. $\qquad\square$

Remark 9.39 Proposition 9.35 and Theorem 9.38 show how both $\mathcal{M}$ and $\mathcal{M}_*$ may be realised as concrete spaces of operators within the same algebra of τ-measurable operators and provide strong circumstantial evidence that this algebra may be a natural home for a theory of L^p-spaces for type III von Neumann algebras. To gain some intuition of how such a type III theory might look, we will first see how the well-understood L^p and Orlicz spaces of semifinite algebras, may be described using the crossed product technology developed in this chapter.

9.5.1 Crossed products of semifinite algebras

For the sake of facilitating the objective outlined in the above remark, we close this chapter by investigating the structure of $\mathcal{M} \rtimes_\tau \mathbb{R}$ in the case where $\mathcal{M}$ is semifinite, and $\tau_\mathcal{M}$ a faithful normal semifinite trace on $\mathcal{M}$. We shall write $\tau_{\mathfrak{M}}$ for the canonical trace on $\mathcal{M} \rtimes_\tau \mathbb{R}$.

For a trace, the modular automorphism group is of the form $\sigma_t^\tau = \mathrm{Id}$ for every t. So U as defined in Proposition 9.8 is just $\mathbb{1}$. Thus, by Proposition 9.8, $\pi(\mathcal{M})$ is just $\mathcal{M} \otimes 1$. The von Neumann algebra generated by the left shift operators $\lambda_t = \mathbb{1} \otimes \ell_t$ is of course just $\mathbb{1} \otimes \mathrm{VN}(\mathbb{R})$, where $\mathrm{VN}(\mathbb{R})$ is the group von Neumann algebra generated by the operators ℓ_t on $L^2(\mathbb{R})$. So, $\mathcal{M} \rtimes_\tau \mathbb{R}$ is just $\mathcal{M} \overline{\otimes} \mathrm{VN}(\mathbb{R})$. The unitary $\mathbb{1} \otimes \mathcal{F}$ clearly commutes with $\mathcal{M} \otimes \mathbb{1}$. Using the fact that $\mathcal{F}\mathrm{VN}(\mathbb{R})\mathcal{F}^{-1} \equiv L^\infty(\mathbb{R})$ (see the discussion in Subsection 8.6.2), it therefore follows that $(\mathbb{1} \otimes \mathcal{F})(\mathcal{M} \rtimes_\tau \mathbb{R})(\mathbb{1} \otimes \mathcal{F}^{-1}) = (\mathbb{1} \otimes \mathcal{F})(\mathcal{M} \overline{\otimes} \mathrm{VN}(\mathbb{R})(\mathbb{1} \otimes \mathcal{F}^{-1}) \equiv \mathcal{M} \overline{\otimes} L^\infty(\mathbb{R})$. (Recall that properly $\mathcal{F}\mathrm{VN}(\mathbb{R})\mathcal{F}^{-1}$ is the algebra of multiplication operators with symbols in $L^\infty(\mathbb{R})$.) It is a refreshing exercise to show that for any $f \in L^2(\mathbb{R})$, we have that $v_t\mathcal{F}(f) = \mathcal{F}(\ell_t f)$ and hence that $\ell_t f = (\mathcal{F}^{-1}v_t\mathcal{F})(f)$. We claim that this same unitary transformation transforms the dual action to the action given by $\tilde{a} \mapsto \lambda_t \tilde{a}$ for all $\tilde{a} \in \mathcal{M} \overline{\otimes} L^\infty(\mathbb{R})$ $(t \in \mathbb{R})$. We pause to justify this fact.

Given some $f \in L^\infty(\mathbb{R})$, let M_f be the associated multiplication operator on $L^2(\mathbb{R})$. It can then be shown that in their action on $L^2(\mathbb{R})$, $\ell_t M_f \ell_t^*$ agrees with $M_{\ell_t f}$. For simplicity of notation, we write $U_\mathcal{F}$ for $\mathbb{1} \otimes \mathcal{F}$, and $\alpha_\mathcal{F}$ for the map $\mathcal{M} \overline{\otimes} L^\infty(\mathbb{R}) \to U_\mathcal{F}^* \mathcal{M} \overline{\otimes} L^\infty(\mathbb{R})U_\mathcal{F} = \mathcal{M} \rtimes_\tau \mathbb{R}$. Together, the above observations ensure that when passing from $\mathcal{M} \rtimes_\tau \mathbb{R} = \mathcal{M} \overline{\otimes} \mathit{VN}(\mathbb{R})$ to $\mathcal{M} \overline{\otimes} L^\infty(\mathbb{R})$ by means of the unitary transformation described above, the transformed dual action $\tilde{\theta}_t = U_\mathcal{F}\theta_t(U_\mathcal{F}^*(\cdot)U_\mathcal{F})U_\mathcal{F}^*$ will have the form $\tilde{\theta}_t(a \otimes f) = (a \otimes \ell_t f) = \lambda_t(a \otimes f)$ on the simple tensors of $\mathcal{M} \overline{\otimes} L^\infty(\mathbb{R})$. Since the simple tensors are σ-weakly dense in $\mathcal{M} \overline{\otimes} L^\infty(\mathbb{R})$, this formula will by continuity hold on all of $\mathcal{M} \overline{\otimes} L^\infty(\mathbb{R})$.

It remains to compute the precise forms of $\tilde{\tau}_\mathcal{M} \circ \alpha_\mathcal{F}$ and $\tau_\mathfrak{M} \circ \alpha_\mathcal{F}$.

It follows from Proposition 9.30 that $(D\tilde{\tau}_\mathcal{M} : D\tau_\mathfrak{M})_t = \lambda_t = \mathbb{1} \otimes \ell_t$ for each $t \in \mathbb{R}$. In the context of $\mathcal{M} \rtimes_\tau \mathbb{R}$, the density $h = \frac{d\tilde{\tau}_\mathcal{M}}{d\tau_\mathfrak{M}}$ is the unique positive non-singular operator for which $h^{it} = \lambda_t$ for each $t \in \mathbb{R}$. (See Theorem 9.28.)

Recall that $\mathcal{F}\ell_t\mathcal{F}^{-1} = v_t$, or equivalently that $\alpha_\mathcal{F}^{-1}(\mathbb{1} \otimes \ell_t) = \mathbb{1} \otimes v_t$ for each $t \in \mathbb{R}$. By Corollary 8.27, this then ensures that $(D(\tilde{\tau}_\mathcal{M} \circ \alpha_\mathcal{F}) : D(\tau_\mathfrak{M} \circ \alpha_\mathcal{F}))_t = \alpha_\mathcal{F}^{-1}(\mathbb{1} \otimes \ell_t) = \mathbb{1} \otimes v_t$. It is a not altogether trivial exercise to now use this equality to conclude that the density $h_\mathcal{F} = \frac{d(\tilde{\tau}_\mathcal{M} \circ \alpha_\mathcal{F})}{d(\tau_\mathfrak{M} \circ \alpha_\mathcal{F})}$ is similarly the unique positive non-singular operator for which $h_\mathcal{F}^{it} = \mathbb{1} \otimes v_t$. (This involves a technical modification of the proof of Theorem 9.28.) Now observe that in this particular case, the operators v_t are of the form $v_t(g)(s) = e^{-its}g(s)$ for each $g \in L^2(G)$ and each $s, t \in \mathbb{R}$. We require a clear understanding of the manner in which the operators v_t act on the multiplication operators M_f where $f \in L^\infty(\mathbb{R})$. For each $f \in L^\infty(\mathbb{R})$ and $g \in L^2(\mathbb{R})$, and each $s, t \in \mathbb{R}$, we have that $v_t(M_f g)(s) = v_t(fg)(s) = e^{-its}f(s)g(s) = (e^{-s})^{it}f(s)g(s)$.

The positive non-singular operator $h_{\mathcal{F}}$ for which $h_{\mathcal{F}}^{it} = \mathbb{1} \otimes v_t$ is therefore nothing but

$$h_{\mathcal{F}} = \mathbb{1} \otimes \frac{1}{\exp}. \tag{9.7}$$

Using the 'transformed' version of the dual action described above, the prescription for the dual weight will on simple tensors $a \otimes f$ of $(\mathcal{M} \overline{\otimes} L^\infty(\mathbb{R}))_+$ translate to the prescription $\mathscr{W}_{\mathcal{F}}(a \otimes f) = \int_{\mathbb{R}} \widetilde{\theta}_t(a \otimes f)\, dt = \int_{\mathbb{R}}(a \otimes \ell_t f)\, d\mathrm{m}_L(t) = a \otimes (\int_{\mathbb{R}}(\ell_t f\, d\mathrm{m}(t))$. (Here, m denotes Lebesgue measure.) Recall that the translation invariance of Lebesgue measure ensures that for each fixed $s \in \mathbb{R}$ we have $\int_{\mathbb{R}} f(s - t)\, d\mathrm{m}(t) = \int_{\mathbb{R}} f(-t)\, d\mathrm{m}(t)$. So, $\mathscr{W}_{\mathcal{F}}(a \otimes f)$ is just $a \otimes (\int_{\mathbb{R}} f(-t)\, d\mathrm{m}(t))$ (or more properly $a \otimes (\int_{\mathbb{R}} f(-t)\, d\mathrm{m}(t))\mathbb{1})$. (In the case where $\int_{\mathbb{R}} f(-t)\, d\mathrm{m}(t) = \infty$, we can give meaning to this object in the extended positive part of $\mathcal{M} \overline{\otimes} L^\infty(\mathbb{R})$.) Therefore

$$(\widetilde{\tau_{\mathcal{M}}} \circ \alpha_{\mathcal{F}})(a \otimes f) = \tau_{\mathcal{M}}(\mathscr{W}_{\mathcal{F}}(a \otimes f)) = \tau_{\mathcal{M}}(a) \cdot \int_{\mathbb{R}} f(-t)\, d\mathrm{m}(t).$$

However, we know that $\widetilde{\tau_{\mathcal{M}}} \circ \alpha_{\mathcal{F}} = \tau_{\mathfrak{M}} \circ \alpha_{\mathcal{F}}(h_{\mathcal{F}}(\cdot))$ where $h_{\mathcal{F}}$ is as in equation (9.7) above. So, on simple tensors $(a \otimes f) \in (\mathcal{M} \overline{\otimes} L^\infty(\mathbb{R}))_+$, $\tau_{\mathfrak{M}} \circ \alpha_{\mathcal{F}}$ must then have the form

$$(\tau_{\mathfrak{M}} \circ \alpha_{\mathcal{F}})(a \otimes f) = (\widetilde{\tau_{\mathcal{M}}} \circ \alpha_{\mathcal{F}})(h_{\mathcal{F}}^{-1}(a \otimes f))$$
$$= \tau_{\mathcal{M}}(a) \cdot \int_{\mathbb{R}} e^t f(-t)\, d\mathrm{m}(t).$$

Since for any Borel set $E \subseteq \mathbb{R}$ we have that $\mathrm{m}(E) = \mathrm{m}(-E)$, we will for any positive Borel function g on $\mathbb{R}$ have $\int_{\mathbb{R}} g(-t)\, d\mathrm{m}(t) = \int_{\mathbb{R}} g(t)\, d\mathrm{m}(t)$. Hence, we have that

$$(\widetilde{\tau_{\mathcal{M}}} \circ \alpha_{\mathcal{F}})(a \otimes f) = \tau_{\mathcal{M}}(a) \cdot \int_{\mathbb{R}} f(t)\, d\mathrm{m}(t)$$

and

$$(\tau_{\mathfrak{M}} \circ \alpha_{\mathcal{F}})(a \otimes f) = \tau_{\mathcal{M}}(a) \cdot \int_{\mathbb{R}} f(t)\, e^{-t} d\mathrm{m}(t).$$

With $\widetilde{\tau_{\mathcal{M}}} \circ \alpha_{\mathcal{F}}$ and $\tau_{\mathfrak{M}} \circ \alpha_{\mathcal{F}}$ represented in this form, the sign change effected in the variable of the second coordinate, ensures that in this context the density $\frac{d(\widetilde{\tau_{\mathcal{M}}} \circ \alpha_{\mathcal{F}})}{d(\tau_{\mathfrak{M}} \circ \alpha_{\mathcal{F}})}$ is (by slight abuse of notation) just the simple tensor $\mathbb{1} \otimes e^t$ $(t \in \mathbb{R})$. We collate the observations made above in the following result:

Theorem 9.40 *Let $\mathcal{M}$ be a semifinite von Neumann algebra equipped with a faithful normal semifinite trace $\tau_{\mathcal{M}}$. Write $\tau_{\mathfrak{M}}$ for the canonical trace on $\mathcal{M} \rtimes_\tau \mathbb{R}$. Up to Fourier transform, we may then identify $\mathcal{M} \rtimes_\tau \mathbb{R}$ with the von Neumann algebra tensor product $\mathcal{M}\overline{\otimes}L^\infty(\mathbb{R})$. Under this identification, the dual action takes the form $\widetilde{\theta}_t(a \otimes f) = (a \otimes \ell_t f)$ on the simple tensors of $\mathcal{M}\overline{\otimes}L^\infty(\mathbb{R})$. In their action on $\mathcal{M} \rtimes_\tau \mathbb{R}$, the dual trace $\widetilde{\tau}_{\mathcal{M}}$ and $\tau_{\mathfrak{M}}$ will under this identification, respectively, be of the form*

$$\widetilde{\tau}_{\mathcal{M}} \equiv \tau_{\mathcal{M}} \otimes \int_{\mathbb{R}} (\cdot)\, d\mathfrak{m}_L(t) \ \text{ and } \ \tau_{\mathfrak{M}} \equiv \tau_{\mathcal{M}} \otimes \int_{\mathbb{R}} (\cdot)\, e^{-t} d\mathfrak{m}(t)$$

where $\mathfrak{m}$ denotes Lebesgue measure. In this representation, the derivative $\frac{d\widetilde{\tau}_{\mathcal{M}}}{d\tau_{\mathfrak{M}}}$ may (by slight abuse of notation) be identified with $\mathbb{1}\otimes e^t$, which clearly commutes with each element of $\mathcal{M}\overline{\otimes}L^\infty(\mathbb{R})$ (with respect to the strong product).

Chapter 10
L^p and Orlicz spaces for general von Neumann algebras

We are now finally ready to develop the theory of Haagerup L^p and Orlicz spaces. We hasten to once again point out that there is no significant reduction in effort to be had by proving the foundational results in this chapter for L^p-spaces only. On the other hand, access to the construction of Orlicz spaces for general von Neumann algebras will prove to be invaluable tool in the next chapter, as it is this very technology which will be used to construct the superspace within which the interpolation described there takes place.

10.1 The semifinite setting revisited

In this section, we will indicate how the theory of L^p and Orlicz spaces for semifinite algebras may be realised using crossed product techniques. So, throughout this section, we will assume that $\mathcal{M}$ is a semifinite von Neumann algebra equipped with a faithful normal semifinite trace $\tau_{\mathcal{M}}$. Theorem 9.40 will play a crucial role in our analysis. In fact, throughout this section, we will freely identify $\mathcal{M} \rtimes_\tau \mathbb{R}$ with $\mathcal{M} \overline{\otimes} L^\infty(\mathbb{R})$. Therefore, readers are advised to keep an eye on Theorem 9.40 as they read this section. The trace on $\mathcal{M} \overline{\otimes} L^\infty(\mathbb{R})$ will be denoted by $\tau_{\mathfrak{M}}$.

The following adaptation of Proposition 9.36 is the fountainhead of the theory developed here.

Lemma 10.1 *Let f and g be commuting positive operators affiliated to an arbitrary von Neumann algebra $\mathcal{M}$. Also let Ψ be an Orlicz function and $\mathbf{f}^\Psi$ be the fundamental function of $L^\Psi(\mathbb{R})$ equipped with the Luxemburg norm. Then $\chi_{(1,\infty)}(\Psi(g)f) = \chi_{(1,\infty)}(\mathbf{f}^\Psi(f)g)$.*

Noncommutative measures and L^p and Orlicz Spaces, with Applications to Quantum Physics. Stanisław Goldstein and Louis Labuschagne, Oxford University Press. © Stanisław Goldstein and Louis Labuschagne (2025).
DOI: 10.1093/oso/9780198950202.003.0012

Proof. Let $\alpha, \beta > 0$ be given. We proceed to show that $\alpha\Psi(\beta) \leq 1 \Leftrightarrow \beta \leq \Psi^{-1}(\frac{1}{\alpha})$. Since $\Psi^{-1}(\Psi(t)) \geq t$ (see part(d) of Remark 6.36), it is clear that

$$\alpha\Psi(\beta) \leq 1 \Leftrightarrow \Psi(\beta) \leq \frac{1}{\alpha} \Rightarrow \beta \leq \Psi^{-1}(1/\alpha),$$

and hence it only remains to show that $\beta \leq \Psi^{-1}(\frac{1}{\alpha}) \Rightarrow \alpha\Psi(\beta) \leq 1$. Clearly

$$\beta \leq \Psi^{-1}(1/\alpha) \Rightarrow \Psi(\beta) \leq \Psi(\Psi^{-1}(1/\alpha)).$$

Since $\Psi(\Psi^{-1}(1/\alpha)) \leq 1/\alpha$ (by part(d) of Remark 6.36), we have that $\beta \leq \Psi^{-1}(\frac{1}{\alpha}) \Rightarrow \alpha\Psi(\beta) \leq 1$ as required. It therefore follows that

$$\alpha\Psi(\beta) \leq 1 \Leftrightarrow \beta\frac{1}{\Psi^{-1}(1/\alpha)} = \beta\mathbf{f}^{\Psi}(\alpha) \leq 1$$

or equivalently

$$\alpha\Psi(\beta) > 1 \Leftrightarrow \beta\mathbf{f}_{\Psi}(\alpha) > 1.$$

Since f and g are commuting positive operators affiliated to $\mathcal{M}$, the von Neumann algebra generated by their spectral projections is abelian and hence may be represented as some $L^{\infty}(\Omega, \Sigma, \nu)$-space. By the Borel functional calculus, both f and g then appear as almost everywhere finite Borel functions on the measure space (Ω, Σ, ν). However, given two positive almost everywhere finite Borel functions f and g on $\mathbb{R}$, the above equivalence ensures that the sets $\{t \in \mathbb{R} : f(t)\Psi(g(t)) > 1\}$ and $\{t \in \mathbb{R} : \mathbf{f}_{\Psi}(f(t))g(t) > 1\}$ differ by a set of measure 0 and hence that the characteristic functions corresponding to these sets agree almost everywhere. Thus, as elements of $L^{\infty}(\Omega, \Sigma, \nu)$, they must agree. But in the context of the von Neumann algebra $\mathcal{R}$, these characteristic functions are just the spectral projections $\chi_{(1,\infty)}(\Psi(g)f)$ and $\chi_{(1,\infty)}(\mathbf{f}^{\Psi}(f)g)$. This suffices to prove the lemma. $\qquad\square$

Theorem 10.2 *Let $a\,\eta\,\mathcal{M}$ be given. Let Ψ be an Orlicz function and let $\mathbf{f}^{\Psi}$ be the fundamental function of $L^{\Psi}(\mathbb{R})$ equipped with the Luxemburg norm. For any $\epsilon > 0$, we then have that*

$$d_{(a\otimes\mathbf{f}^{\Psi}(\exp))}(\epsilon) = \tau_{\mathcal{M}}(\Psi(|a|/\epsilon)).$$

Proof. Observe that the operators $(\mathbb{1} \otimes \exp)$ and $|a \otimes 1| = |a| \otimes 1$ are commuting positive operators affiliated to $\mathcal{M}\overline{\otimes}L^{\infty}(\mathbb{R})$. We may now apply the Lemma to this pair, with $(\mathbb{1} \otimes \exp)$ playing the role of f, and $|a| \otimes 1$ the role of g, to see that

$$
\begin{aligned}
\chi_{(1,\infty)}(|a \otimes \mathbf{f}^{\Psi}(\exp)|) &= \chi_{(1,\infty)}((|a| \otimes 1)(\mathbb{1} \otimes \mathbf{f}^{\Psi}(\exp))) \\
&= \chi_{(1,\infty)}((|a| \otimes 1)\mathbf{f}^{\Psi}(\mathbb{1} \otimes \exp)) \\
&= \chi_{(1,\infty)}(\Psi(|a| \otimes 1)(\mathbb{1} \otimes \exp)) \\
&= \chi_{(1,\infty)}((\Psi(|a|) \otimes 1)(\mathbb{1} \otimes \exp)) \\
&= \chi_{(1,\infty)}(\Psi(|a|) \otimes \exp)
\end{aligned}
$$

By Proposition 9.36, we have that

$$
\tau_{\mathfrak{M}}(\chi_{(1,\infty)}(|a \otimes \mathbf{f}^{\Psi}(\exp)|)) = \psi(\mathbb{1})
$$

where ψ is the weight $f \mapsto \tau_{\mathcal{M}}(\Psi(|a|)^{1/2} f \Psi(|a|)^{1/2})$. (Note that Proposition 9.33 ensures that Proposition 9.36 is applicable to $\Psi(|a|) \otimes \exp$.) In other words

$$
\tau_{\mathfrak{M}}(\chi_{(1,\infty)}(|a \otimes \mathbf{f}^{\Psi}(\exp)|)) = \tau_{\mathcal{M}}(\Psi(|a|)).
$$

Given $\epsilon > 0$, we therefore have that

$$
\begin{aligned}
d_{a \otimes \mathbf{f}^{\Psi}(\exp)}(\epsilon) &= \tau_{\mathfrak{M}}(\chi_{(\epsilon,\infty)}(|a \otimes \mathbf{f}^{\Psi}(\exp)|)) \\
&= \tau_{\mathfrak{M}}(\chi_{(\epsilon,\infty)}(|a| \otimes \mathbf{f}^{\Psi}(\exp))) \\
&= \tau_{\mathfrak{M}}(\chi_{(1,\infty)}((|a|/\epsilon) \otimes \mathbf{f}^{\Psi}(\exp))) \\
&= \tau_{\mathcal{M}}(\Psi(|a|/\epsilon))
\end{aligned}
$$

as required. $\qquad\qquad\qquad\qquad\qquad\qquad\qquad\qquad\qquad\qquad\qquad\qquad\square$

The first consequence of this theorem gives us some clue as to how $L^{\Psi}(\mathcal{M}, \tau_{\mathcal{M}})$ may be realised inside $\widetilde{\mathfrak{M}}$.

Corollary 10.3 (Luxemburg norm) *Let Ψ be an Orlicz function. Given $a \,\eta\, \mathcal{M}$, we have $a \in L^{\Psi}(\mathcal{M}, \tau_{\mathcal{M}})$ if and only if $a \otimes \mathbf{f}^{\Psi}(\exp)$ belongs to $\widetilde{\mathfrak{M}}$. Moreover, for any $a \in L^{\Psi}(\mathcal{M}, \tau_{\mathcal{M}})$, we have the following formula for the Luxemburg norm:*

$$
\|a\|_{\Psi} = \mathbf{m}_{(a \otimes \mathbf{f}_{\Psi}(\exp))}(1).
$$

Proof. Recall that by Remark 6.39, we have $a \in L^{\Psi}(\mathcal{M}, \tau_{\mathcal{M}})$ if and only if $\tau_{\mathcal{M}}(\Psi(\alpha|a|)) < \infty$ for some $\alpha > 0$. But by the theorem this is the same as saying that $a \in L^\Psi(\mathcal{M}, \tau_\mathcal{M})$ if and only if $d_{(a \otimes \mathbf{f}^\Psi(\exp))}(1/\alpha) < \infty$ for some $\alpha > 0$. From the basic theory of τ-measurable operators, we further have $d_{(a \otimes \mathbf{f}^\Psi(\exp))}(\epsilon) < \infty$ for some $\epsilon > 0$ if and only if $(a \otimes \mathbf{f}_\Psi(\exp))$ is $\tau_\mathfrak{M}$-measurable. Combining these facts leads to the conclusion that $a \in L^\Psi(\mathcal{M}, \tau_\mathcal{M})$ if and only if $a \otimes \mathbf{f}^\Psi(\exp)$ is $\tau_\mathfrak{M}$-measurable.

Let $a \in L^\Psi(\mathcal{M}, \tau_\mathcal{M})$ be given. To see the second claim, we use the theorem to conclude that

$$
\begin{aligned}
\|a\|_\Psi &= \inf\{\epsilon > 0 \colon \tau_\mathcal{M}(\Psi(|a|/\epsilon) \le 1\} \\
&= \inf\{\epsilon > 0 \colon d_{(a \otimes \mathbf{f}^\Psi(\exp))}(\epsilon) \le 1\} \\
&= \inf\{\epsilon \ge 0 \colon d_{(a \otimes \mathbf{f}^\Psi(\exp))}(\epsilon) \le 1\} \\
&= \mathbf{m}_{(a \otimes \mathbf{f}^\Psi(\exp))}(1).
\end{aligned}
$$

(Here, the second to last equality follows from the fact that the function $s \mapsto d_{(a \otimes \mathbf{f}^\Psi(\exp))}(s)$ is right-continuous.) $\qquad\square$

The preceding theorem and corollary describe how one may use the simple tensors in $\widetilde{\mathfrak{M}}$, to realise an isometric copy of $L^\Psi(\mathcal{M}, \tau_\mathcal{M})$ inside $\widetilde{\mathfrak{M}}$. What remains to be done is to find a reliable test to identify those elements of $\widetilde{\mathfrak{M}}$ that are of the form $a \otimes \mathbf{f}^\Psi(\exp)$ for some $a \in L^\Psi(\mathcal{M}, \tau_\mathcal{M})$.

Theorem 10.4 *Let a Young function* Ψ*, and* $\widetilde{a} \in \widetilde{\mathfrak{M}}$ *be given. Then,* $\widetilde{a}$ *is of the form* $\widetilde{a} = a \otimes \mathbf{f}^\Psi(\exp)$ *for some* $a \in L^\Psi(\mathcal{M}, \tau_\mathcal{M})$ *if and only if* $\widetilde{\theta}_s(\widetilde{a}) = v_s \widetilde{a}$ *for all* $s \in \mathbb{R}$*, where* $v_s = \mathbf{f}^\Psi(e^{-s}h)\mathbf{f}^\Psi(h)^{-1}$*. (Here,* h *is the density* $\frac{d\widetilde{\tau}_\mathcal{M}}{d\tau_\mathfrak{M}} = \mathbb{1} \otimes \exp$*.)*

We pause to point out that the operators v_s are all bounded. This fact will be verified in a more general context in Lemma 10.7.

Proof.   The 'if' part is easy to see. If indeed $a \otimes \mathbf{f}^\Psi(\exp) \in \widetilde{\mathfrak{M}}$, then for any $s \in \mathbb{R}$,

$$
\begin{aligned}
\widetilde{\theta}_s(a \otimes \mathbf{f}^\Psi(\exp)) &= a \otimes \ell_s \mathbf{f}^\Psi(\exp) \\
&= a \otimes \mathbf{f}^\Psi(e^{-s} \exp) \\
&= v_s(a \otimes \mathbf{f}^\Psi(\exp)).
\end{aligned}
$$

Now, suppose that we are given some $\widetilde{a} \in \widetilde{\mathfrak{M}}$ such that $\widetilde{\theta}_s(\widetilde{a}) = v_s\widetilde{a}$ for all $s \in \mathbb{R}$. Recall that $h = \mathbb{1} \otimes \exp$ commutes with every element of $\widetilde{\mathfrak{M}}$, and therefore also with each operator v_s. (See Theorem 9.40.) On the basis of this fact, it is now fairly easy to see that then $\widetilde{\theta}_s(|\widetilde{a}|^2) = \widetilde{\theta}_s(\widetilde{a}^*)\widetilde{\theta}_s(\widetilde{a}) = v_s^2|\widetilde{a}|^2$ and hence $\widetilde{\theta}_s(|\widetilde{a}|) = v_s|\widetilde{a}|$ for all $s \in \mathbb{R}$. Let $\widetilde{a} = \widetilde{u}|\widetilde{a}|$ be the polar decomposition of $\widetilde{a}$. For a given $s \in \mathbb{R}$, it is clear that $\widetilde{\theta}_s(\widetilde{u})\widetilde{\theta}_s(|\widetilde{a}|)$ is a polar decomposition of $\widetilde{\theta}_s(\widetilde{a})$. But since $\widetilde{\theta}_s(|\widetilde{a}|) = v_s|\widetilde{a}|$, we have $\widetilde{\theta}_s(\widetilde{a}) = v_s\widetilde{a} = \widetilde{u}v_s|\widetilde{a}| = \widetilde{u}\widetilde{\theta}_s(|\widetilde{a}|)$. Thus, $\widetilde{\theta}_s(\widetilde{a}) = \widetilde{u}\widetilde{\theta}_s(|\widetilde{a}|)$ is then also a polar decomposition of $\widetilde{a}$. The uniqueness of the polar decomposition then ensures that $\widetilde{\theta}_s(\widetilde{u}) = \widetilde{u}$. This holds for every $s \in \mathbb{R}$, which, by Proposition 9.35, ensures that $\widetilde{u} \in \mathcal{M} \otimes 1$. That is $\widetilde{u} = u \otimes 1$ for some partial isometry $u \in \mathcal{M}$. Thus, if we can prove that the claim holds for $|\widetilde{a}|$, it will also hold for $\widetilde{a}$. We therefore may and do assume that $\widetilde{a} \geq 0$.

By Theorem 9.40, $h = \mathbb{1} \otimes \exp$ and $\widetilde{a}$ are commuting affiliated operators, and hence so are $\mathbf{f}^{\Psi}(h)^{-1}$ and $\widetilde{a}$. By the Borel functional calculus, the strong product $\widetilde{b} = \mathbf{f}^{\Psi}(h)^{-1}\widetilde{a}$ is a densely defined positive operator affiliated to $\mathfrak{M}$. For this operator, we have that

$$
\begin{aligned}
\widetilde{\theta}_s(\widetilde{b}) &= \mathbf{f}^{\Psi}(\widetilde{\theta}_s h)^{-1}v_s\widetilde{a} \\
&= \mathbf{f}^{\Psi}(e^{-s}h)^{-1}[\mathbf{f}^{\Psi}(e^{-s}h)\mathbf{f}^{\Psi}(h)^{-1}]\widetilde{a} \\
&= \mathbf{f}^{\Psi}(h)^{-1}\widetilde{a} \\
&= \widetilde{b}
\end{aligned}
$$

for each $s \in \mathbb{R}$. Since $\widetilde{b} = \mathbf{f}^{\Psi}(h)^{-1}\widetilde{a}$ is positive, the operator $e^{-\widetilde{b}}$ is bounded. For this operator, we still have $\widetilde{\theta}_s(e^{-\widetilde{b}}) = e^{-\widetilde{\theta}_s(\widetilde{b})} = e^{-\widetilde{b}}$ for all $s \in \mathbb{R}$. So, by Proposition 9.35, there must then exist some $b \in \mathcal{M}_+$ such that $e^{-\widetilde{b}} = b \otimes 1$. But then $\mathbf{f}^{\Psi}(h)^{-1}\widetilde{a} = \widetilde{b} = -\log(b \otimes 1) = -(\log(b) \otimes 1)$. Given that $\mathbf{f}^{\Psi}(h)^{-1} = \mathbf{f}^{\Psi}(\mathbb{1} \otimes \exp)^{-1} = (\mathbb{1} \otimes \mathbf{f}^{\Psi}(\exp))^{-1}$, we therefore have that $\widetilde{a} = -(\log(b) \otimes \mathbf{f}^{\Psi}(\exp))$. By Theorem 10.2, the $\tau_{\mathfrak{M}}$-measurability of $\widetilde{a}$ now ensures that $\log(b) \in L^{\Psi}(\mathcal{M}, \tau_{\mathcal{M}})$, and that $\widetilde{a}$ is of the required form. $\qquad\square$

We now gather the preceding analysis in the following theorem:

Theorem 10.5 *Let Ψ be a Young function and let $h = \frac{d\widetilde{\tau_{\mathcal{M}}}}{d\tau_{\mathfrak{M}}}$. Then, the quantity $\widetilde{a} \mapsto \mathbf{m}_{\widetilde{a}}(1)$ is a norm on the space*

$$
L^{\Psi}(\mathcal{M}) = \{\widetilde{a} \in \widetilde{\mathfrak{M}} : \widetilde{\theta}_s(\widetilde{a}) = v_s\widetilde{a} \text{ for all } s \in \mathbb{R}\}
$$

where the operators v_s are defined by $v_s = \mathbf{f}^\Psi(e^{-s}h)\mathbf{f}^\Psi(h)^{-1}$ for each s. The mapping $(L^\Psi(\mathcal{M}, \tau_\mathcal{M}), \|\cdot\|_\Psi) \to (L^\Psi(\mathcal{M}), \mathbf{m}_{(\cdot)}(1)) : a \mapsto a \otimes \mathbf{f}^\Psi(\exp)$ is then a surjective isometric isomorphism.

In this section, we have one final task to perform, namely to describe the topology on $(L^\Psi(\mathcal{M}), \mathbf{m}_{(\cdot)}(1))$ more fully.

Theorem 10.6 *For a Young function Ψ, the norm topology on the space $L^\Psi(\mathcal{M})$ is homeomorphic to the relative topology induced by the topology of convergence in measure on $\widetilde{\mathfrak{M}}$.*

Note that this result also holds for L^∞!

Proof. We remind the reader that the basic neighbourhoods of zero for the topology of convergence in measure on $\widetilde{\mathfrak{M}}$ are of the form

$$\mathcal{N}(\epsilon, \delta) = \{\widetilde{a} : d_{\widetilde{a}}(\epsilon) \leq \delta\}.$$

Let $1 > \epsilon > 0$ be given and suppose that $\mathbf{m}_{\widetilde{a}}(1) < \epsilon$ for some $\widetilde{a} = a \otimes \mathbf{f}^\Psi(\exp) \in L^\Psi(\mathcal{M})$. Then, there must exist an $0 < \alpha < \epsilon$ such that $d_{\widetilde{a}}(\alpha) \leq 1$. By Theorem 10.2, this ensures that $\tau_\mathcal{M}(\Psi(|a|/\alpha)) \leq 1$. Next notice that $\frac{1}{\sqrt{\alpha}} \geq 1$, since by assumption $0 < \alpha < \epsilon < 1$. We may therefore use the convexity of Ψ to conclude that $\frac{1}{\sqrt{\alpha}}\tau_\mathcal{M}(\Psi(|a|/\sqrt{\alpha})) \leq \tau_\mathcal{M}(\Psi(|a|/\alpha)) \leq 1$; in other words $\tau_\mathcal{M}(\Psi(|a|/\sqrt{\alpha})) \leq \sqrt{\alpha}$. Once again using Theorem 10.2, this can be reformulated as the claim that $d_{(a \otimes \mathbf{f}^\Psi(\exp))}(\sqrt{\alpha}) \leq \sqrt{\alpha}$. This ensures that $\widetilde{a} = a \otimes \mathbf{f}^\Psi(\exp) \in \mathcal{N}(\sqrt{\alpha}, \sqrt{\alpha})$.

Conversely, suppose that for some $\epsilon, \delta > 0$ with $\delta \leq 1$, we have $\widetilde{a} = a \otimes \mathbf{f}^\Psi(\exp) \in \mathcal{N}(\epsilon, \delta)$. Then $d_{(a \otimes \mathbf{f}^\Psi(\exp))}(\epsilon) \leq \delta \leq 1$. It then follows that $\mathbf{m}_{(a \otimes \mathbf{f}^\Psi(\exp))}(1) \leq \epsilon$. Thus, (a_n) converges to 0 in the norm topology on $L^\Psi(\mathcal{M}, \tau_\mathcal{M})$ if and only if $((a_n \otimes \mathbf{f}^\Psi(\exp)))$ converges to 0 in the topology of convergence in measure on $\widetilde{\mathfrak{M}}$.

The norm topology on $L^\Psi(\mathcal{M})$ must, therefore, be homeomorphic to the relative topology induced by the topology of convergence in measure on $\widetilde{\mathfrak{M}}$. $\qquad\square$

10.2 Definition and normability of general L^p and Orlicz spaces

Throughout the rest of this chapter, $\mathcal{M}$ will be a von Neumann algebra and φ the 'reference' faithful normal semifinite weight on $\mathcal{M}$.

The crossed product $\mathfrak{M} = \mathcal{M} \rtimes_\varphi \mathbb{R}$ will play a crucial role in the development of the theory of Orlicz and L^p-spaces for general von Neumann algebras. Therefore, readers are advised to make sure that they are familiar with the basic structural theory of this algebra as presented in Chapter 9 before attempting to make sense of the theory presented here. In view of the repeated use of $\mathfrak{M}$ in developing this theory, to simplify notation, we will identify $\mathcal{M}$ with the copy $\pi(\mathcal{M})$ living inside $\mathfrak{M}$.

Based on the previous section, we are now ready to rigorously define L^p and Orlicz spaces for general von Neumann algebras in a manner which is a natural extension of the definition of these spaces for semifinite algebras equipped with a faithful normal semifinite trace. Before doing that, we need the following lemma.

Lemma 10.7 *Let φ be a faithful normal semifinite weight on a von Neumann algebra $\mathcal{M}$, and let $h = \frac{d\widetilde{\varphi}}{d\tau}$ be the density of the dual weight $\widetilde{\varphi}$ on the crossed product $\mathfrak{M}$. Let $\mathbf{f} : [0, \infty) \to [0, \infty)$ be any quasi-concave function. Then, the operators $v_s = \mathbf{f}(e^{-s}h)\mathbf{f}(h)^{-1}$ ($s \in \mathbb{R}$) are always bounded, with norm between 1 and e^{-s}.*

Proof. This can be proven by noting that the facts that $t \mapsto \mathbf{f}(t)$ is increasing and $t \mapsto \frac{\mathbf{f}(t)}{t}$ decreasing ensure that for any $t > 0$, the number $\frac{\mathbf{f}(e^{-s}t)}{\mathbf{f}(t)}$ lies between 1 and e^{-s}. The Borel functional calculus does the rest. $\qquad\square$

Definition 10.8 Let φ be a fixed (canonical) faithful normal semifinite weight on the von Neumann algebra $\mathcal{M}$, and let $h = \frac{d\widetilde{\varphi}}{d\tau}$ be the density of the dual weight $\widetilde{\varphi}$ on the crossed product $\mathfrak{M}$. Given a Young function Ψ, the Orlicz space $L^\Psi(\mathcal{M})$ associated with $\mathcal{M}$ (corresponding to the Luxemburg–Nakano norm) is defined to be

$$L^\Psi(\mathcal{M}) = \{a \in \widetilde{\mathfrak{M}} : \theta_s(a) = v_s^{1/2} a v_s^{1/2} \text{ for all } s \in \mathbb{R}\}$$

where $v_s = \mathbf{f}^\Psi(e^{-s}h)\mathbf{f}^\Psi(h)^{-1}$. We formally define the Luxemburg–Nakano 'norm' on $L^\Psi(\mathcal{M})$ to be the quantity $\|a\|_\Psi = \mathbf{m}_a(1)$. In the case where $\Psi(t) = t^p$ ($1 \leq p$), we will denote this quantity by $\|\cdot\|_p$. In this case, we have that $\mathbf{f}^\Psi(e^{-s}h)\mathbf{f}^\Psi(h)^{-1} = e^{-s/p}\mathbb{1}$ for all $s \in \mathbb{R}$.

So, by analogy with the above, we may for all $0 < p < \infty$ define $L^p(\mathcal{M})$ to be the space

$$L^p(\mathcal{M}) = \{a \in \widetilde{\mathfrak{M}} : \theta_s(a) = e^{-s/p}a \text{ for all } s \in \mathbb{R}\}.$$

We note that the definition of Orlicz spaces for general von Neumann algebras given in [**Lab13**] differs from that given above. However, the two definitions were shown to be equivalent in [**LM20**, Lemma 4.10].

Remark 10.9 Having defined Orlicz spaces for general von Neumann algebras, our task now is to further justify the choice of $\mathbf{m}_f(1)$ as the natural Luxemburg–Nakano norm for these spaces. We saw in Theorem 5.6 that $\rho(f) = d_f(1)$ is the natural modular on $\widetilde{\mathfrak{M}}$. So, the obvious candidate for a Luxemburg–Nakano 'norm' on an Orlicz space $L^\Psi(\mathcal{M})$, would be

$$\inf\{\epsilon > 0 : \rho(\epsilon^{-1}|f|) \le 1\} = \inf\{\epsilon > 0 : d_f(\epsilon) \le 1\} = \mathbf{m}_f(1).$$

A fact which strongly supports this proposal is the fact that the analysis in the first section of this chapter shows that in the case where φ is a faithful normal semifinite trace, the quantity $\mathbf{m}_{(\cdot)}(1)$ is indeed a norm on $L^\Psi(\mathcal{M})$, and that when equipped with this norm, the space $(L^\Psi(\mathcal{M}), \mathbf{m}_{(\cdot)}(1))$ is an isometric copy of the space $(L^\Psi(\mathcal{M}, \varphi), \|\cdot\|_\Psi)$.

We now show that the quantity $\mathbf{m}_{(\cdot)}(1)$ is indeed a quasi-norm on each of the spaces introduced above and describe the quasi-norm topology on those spaces. For this, the following result is crucial.

Proposition 10.10   *Let Ψ be a Young function. For any $a \in L^\Psi(\mathcal{M})$, we have that*

$$t\mathbf{m}_a(t) \le \mathbf{m}_a(1) \qquad for\ all \qquad 0 < t \le 1.$$

Proof. Let $a \in L^\Psi(\mathcal{M})$ and $0 < t$ be given. We may, of course, write such a t as $t = e^s$ where $s \in \mathbb{R}$. Observe that for any $r > 0$, we then have that

$$\begin{aligned}
\frac{1}{t}d_a(r) &= e^{-s}\tau(\chi_{(r,\infty)}(|a|)) \\
&= \tau(\theta_s(\chi_{(r,\infty)}(|a|))) \\
&= \tau(\chi_{(r,\infty)}(\theta_s|a|)) \\
&= \tau(\chi_{(r,\infty)}(|\theta_s(a)|)) \\
&= d_{\theta_s(a)}(r).
\end{aligned}$$

With v_s as in Definition 10.8, it now follows that

$$\begin{aligned}
\mathbf{m}_a(t) &= \inf\{r > 0 : d_a(r) \le t\} \qquad\qquad (10.1) \\
&= \inf\{r > 0 : d_{\theta_s(a)}(r) \le 1\} \\
&= \mathbf{m}_{\theta_s(a)}(1) \\
&= \mathbf{m}_{v_s^{1/2}av_s^{1/2}}(1)
\end{aligned}$$

for all $t = e^s$.

For any $c \in \widetilde{\mathfrak{M}}$ and $b \in \mathfrak{M}$, it is a simple matter to see that

$$\mathbf{m}_{bc}(1) = \mathbf{m}_{c^*|b|^2c}(1)^{1/2} \le \|b\| \cdot \mathbf{m}_{c^*c}(1)^{1/2} = \|b\| \cdot \mathbf{m}_c(1)$$

and similarly that $\mathbf{m}_{cb}(1) \le \|b\| \cdot \mathbf{m}_c(1)$. With s and t as before, we now pass to the case where $s \le 0$, or equivalently where $0 < t \le 1$. In this case, $\|v_s\| \le e^{-s}$ by Lemma 10.7. As required, it therefore follows from equation (10.1) and the above computations that $\mathbf{m}_a(t) \le \|v_s\| \cdot \mathbf{m}_a(1) \le \frac{1}{t}\mathbf{m}_a(1)$ for all such t. $\qquad\square$

The above proposition now yields the following important theorem.

Theorem 10.11 *The quantity $\mathbf{m}_a(1)$ ($a \in L^\Psi(\mathcal{M})$) is a quasi-norm for $L^\Psi(\mathcal{M})$. The topology induced on $L^\Psi(\mathcal{M})$ by this quasi-norm is complete and homeomorphic to the topology of convergence in measure inherited from $\widetilde{\mathfrak{M}}$.*

Proof. First, suppose that we are given $a \in L^\Psi(\mathcal{M})$ with $\mathbf{m}_a(1) = 0$. By the preceding proposition, we then clearly have that $\mathbf{m}_a(t) = 0$ for all $0 < t \le 1$. Since the decreasing rearrangement is right-continuous, this then yields the fact that $\|a\| = \lim_{t\searrow 0}\mathbf{m}_a(t) = 0$, which can only be the case if $a = 0$.

The properties of the decreasing rearrangement ensure that $\mathbf{m}_{\lambda a}(1) = |\lambda|\mathbf{m}_a(1)$ for any scalar λ and any $a \in L^\Psi(\mathcal{M})$. It remains to show that

$\mathbf{m}_{(\cdot)}(1)$ satisfies a generalised triangle inequality. Given $a, b \in L^\Psi(\mathcal{M})$, we may conclude from the properties of the decreasing rearrangement and the preceding theorem that

$$\mathbf{m}_{(a+b)}(1) \leq 2\left(\frac{1}{2}\mathbf{m}_a(1/2) + \frac{1}{2}\mathbf{m}_b(1/2)\right) \leq 2(\mathbf{m}_a(1) + \mathbf{m}_b(1)).$$

We now show that $L^\Psi(\mathcal{M})$ is a closed subspace of $\widetilde{\mathfrak{M}}$ with respect to the topology of convergence in measure. Then, once we have established that the topology on $L^\Psi(\mathcal{M})$ induced by $\mathbf{m}_{(\cdot)}(1)$ agrees with the topology of convergence in measure, this closedness will suffice to prove the completeness of $L^\Psi(\mathcal{M})$. It is clear from the definition of $L^\Psi(\mathcal{M})$ that the membership of an element a of $\widetilde{\mathfrak{M}}$ to $L^\Psi(\mathcal{M})$ can be rephrased as the claim that a belongs to the intersection of the kernels of the operators

$$T_s : \widetilde{\mathfrak{M}} \to \widetilde{\mathfrak{M}} : a \mapsto \theta_s(a) - v_s^{1/2} a v_s^{1/2} \quad s \in \mathbb{R}$$

where v_s is as in Definition 10.8. The operation $a \mapsto v_s^{1/2} a v_s^{1/2}$ is clearly continuous with respect to the topology of convergence in measure, whereas the operation $a \mapsto \theta_s(a)$ was shown to be similarly continuous in Corollary 9.29. Thus, by continuity, the kernels must be closed as claimed.

It remains to prove that the topology induced on $L^\Psi(\mathcal{M})$ by $\mathbf{m}_a(1)$ is precisely the topology of convergence in measure. The fact that convergence in measure implies convergence in the quasi-norm $a \mapsto \mathbf{m}_a(1)$ follows from Proposition 5.23. For the converse, fix $0 < \epsilon \leq 1$, and suppose that we are given some $a \in L^\Psi(\mathcal{M})$ with $\mathbf{m}_a(1) < \epsilon^2$. By the preceding theorem, we have $\mathbf{m}_a(\epsilon) \leq \frac{1}{\epsilon}\mathbf{m}_a(1) < \epsilon$. We may now once again use Proposition 5.23 to conclude that a then belongs to the basic neighbourhood of zero $\mathcal{N}(\epsilon, \epsilon)$ of $\widetilde{\mathfrak{M}}$. Thus, any sequence in $L^\Psi(\mathcal{M})$ that converges to zero in the quasi-norm $a \mapsto \mathbf{m}_a(1)$ converges to zero in measure. $\qquad\square$

Inspired by the above theorem, we now make the following definition.

Definition 10.12 For each $L^\Psi(\mathcal{M})$, we define $\mathbf{m}_{(\cdot)}(1)$ to be the Luxemburg–Nakano quasi-norm on $L^\Psi(\mathcal{M})$ and will henceforth denote this quasi-norm by $\|\cdot\|_\Psi$ where appropriate. In the case where $\Psi(t) = t^p$ for some $p \geq 1$, we will write $\|\cdot\|_p$ for this quasi-norm.

Remark 10.13 It is a simple matter to check that $a \in \widetilde{\mathfrak{M}}$ belongs to $L^\Psi(\mathcal{M})$ if and only if a^* does. In addition whenever a (and therefore also a^*) belongs to $L^\Psi(\mathcal{M})$, it is clear from (ii) of Proposition 5.12 that then $\|a\|_\Psi = \|a^*\|_\Psi$

A fact worth noting at this stage is the following uniqueness theorem.

Proposition 10.14 *Let φ_1 and φ_2 be two faithful normal semifinite weights on $\mathcal{M}$, and let $\mathfrak{M}_1 = \mathcal{M} \rtimes_{\varphi_1} \mathbb{R}$ and $\mathcal{M}_2 = \mathcal{M} \rtimes_{\varphi_2} \mathbb{R}$. Then, the $*$-isomorphism $\mathscr{I}$ from $\widetilde{\mathfrak{M}}_1$ onto $\widetilde{\mathfrak{M}}_2$ constructed in Theorem 9.31 will for each $1 \le p < \infty$ restrict to a linear isometry from $L^p(\mathcal{M}, \varphi_1)$ onto $L^p(\mathcal{M}, \varphi_1)$.*

Proof. Since $\mathscr{I}$ $*$-isomorphically identifies $\widetilde{\mathfrak{M}}_1$ with $\widetilde{\mathfrak{M}}_2$, it is clear that for any $a \in \widetilde{\mathfrak{M}}_1$, we have $\mathbf{m}_a(1) = \mathbf{m}_{\mathscr{I}(a)}(1)$. So, all that needs to be checked is that $\mathscr{I}(L^p(\mathcal{M}, \varphi_1)) = L^p(\mathcal{M}, \varphi_2)$. This in turn follows from Definition 10.8 and the fact that $\mathscr{I} \circ \theta_t = \theta_t \circ \mathscr{I}$. $\square$

Example 10.15 We remind the reader that $L^\infty(\mathcal{M}) = \mathcal{M}$ itself is an Orlicz space generated by the Young function

$$\Psi_\infty(t) = \begin{cases} 0, & \text{if } 0 \le t \le 1 \\ \infty, & \text{if } 1 < t \end{cases}.$$

This then raises the question of how the Luxemburg–Nakano quasi-norm computed in the preceding theorem compares to the operator norm on $\mathcal{M}$. Therefore, we proceed to compute $\|\cdot\|_{\Psi_\infty}$ for this space. First notice that for Ψ_∞ as above,

$$\mathbf{f}^{\Psi_\infty}(t) = \frac{1}{(\Psi_\infty)^{-1}(1/t)} = \begin{cases} 0, & \text{if } t = 0 \\ 1, & \text{if } 0 < t \end{cases}.$$

We leave the verification of this fact as an exercise. However, what is clear from this fact is that in this setting, $v_s = \mathbf{f}^{\Psi_\infty}(e^{-s}h).\mathbf{f}^{\Psi_\infty}(h)^{-1} = \mathbb{1}$. So, by equation (10.1), we will in this setting have $\mathbf{m}_a(t) = \mathbf{m}_a(1)$ for all $t \in (0, 1]$ and all $a \in \mathcal{M}$. We may now finally use the right-continuity of $t \mapsto \mathbf{m}_a(t)$ to conclude that $\|a\| = \lim_{t \searrow 0} \mathbf{m}_a(t) = \mathbf{m}_a(1) = \|a\|_{\Psi_\infty}$ for all $a \in \mathcal{M}$.

Remark 10.16 Despite the elegance of the preceding example, it is not clear at this stage that these Orlicz spaces are in general normable. For now, the strongest statement we may deduce from Theorem 10.11 is that in this generality, Orlicz spaces may only admit a quasi-norm instead of a norm. To understand why this might be the case, we look at the modular approach to Orlicz spaces (see Definition 5.4). Given some Young function Ψ, in its action on $L^\Psi(X, \Sigma, \nu)$, the quantity $\rho(|f|) = \int \Psi(|f|)\, d\nu$ can be shown to be a convex modular with $L^\Psi(X, \Sigma, \nu)$ a modular space. The prescription $\|f\| = \inf\{\epsilon > 0 \colon \rho(\epsilon^{-1} x) \leq 1\}$ then yields the Luxemburg–Nakano norm on $L^\Psi(X, \Sigma, \nu)$. In the theory of modular spaces, the so-called Amemiya norm is given by the formula $\|f\|_\Psi^A = \inf\{(1 + \rho(kf))/k \colon k > 0\}$. It was only fairly recently that Hudzik and Maligranda showed that in the measure space setting described above, this norm corresponds to the Orlicz norm whenever the measure space is σ-finite [**HM00**]. These facts suggest that if in the type III case we can identify the correct modular, we may be able to construct both a Luxemburg–Nakano and Orlicz norm. Since the quasi-norm topology on $L^\Psi(\mathcal{M})$ agrees with the topology of convergence in measure, it is clear from Theorem 5.6 that the appropriate modular to use in this setting is nothing but $\rho(f) = d_f(1)$—the semi-modular that determines the topology of $\widetilde{\mathfrak{M}}$. It is precisely here that the problem lies. With the structure currently at our disposal, there is no obvious way of proving that in this generality, $\rho(f) = d_f(1)$ is indeed a convex modular. The best we can do at this stage is to show that it is a semi-modular. Without such convexity, the theory regarding the Amemiya norm fails, and the Luxemburg–Nakano 'norm' cannot be expected to be a norm (see, for example, the prerequisites for [**Mus83**, Theorem I.10]). For algebras equipped with a faithful normal semifinite trace, the situation is more regular. We know from the previous chapter that in this setting, we will have for any $f \in L^\Psi(\mathcal{M}, \tau)$ that $d_{f \otimes \mathbf{f}^\Psi(f)}(1) = \tau(\Psi(|f|))$—a fact which ensures that in its action on $\{f \otimes \mathbf{f}^\Psi(f) \colon f \in L^\Psi(\mathcal{M}, \tau)\} \subseteq (\mathcal{M}\widetilde{\overline{\otimes}L^\infty}(\mathbb{R}))$, the quantity $d_{(\cdot)}(1)$ is in this case indeed a convex modular.

The negative tone of the previous remark aside, there is in fact a remnant of the equivalence of the Orlicz and Luxemburg–Nakano norms that survives the transition to the general setting, and also a class of more regular Orlicz spaces which do turn out to be normed. We discuss each of these issues in turn. To achieve this objective, we will need the following classical fact.

Proposition 10.17 ([BS88, Lemma IV.8.16]) *Let Ψ be a Young function. For any $t \geq 0$, we will then have $t \leq \Psi^{-1}(t)(\Psi^*)^{-1}(t) \leq 2t$.*

Corollary 10.18 *Given a Young function Ψ, let $L_\Psi(\mathcal{M})$ denote the space defined by*

$$L_\Psi(\mathcal{M}) = \{a \in \widetilde{\mathfrak{M}} \colon \theta_s(a) = \widetilde{v}_s^{1/2} a \widetilde{v}_s^{1/2} \text{ for all } s \leq 0\}$$

where $\widetilde{v}_s = \mathbf{f}_\Psi(e^{-s}h)\mathbf{f}_\Psi(h)^{-1}$. (Here, $\mathbf{f}_\Psi$ is the fundamental function corresponding to the Orlicz norm on $L^\Psi(0, \infty)$.) The quantity $\|a\|_\Psi^O = \mathbf{m}_a(1)$ is a quasi-norm for $L_\Psi(\mathcal{M})$, with the quasi-normed topology homeomorphic to the topology of convergence in measure. Moreover, there is a canonical bijection ι from $L^\Psi(\mathcal{M})$ onto $L_\Psi(\mathcal{M})$ for which we have $\|a\|_\Psi \leq \|\iota(a)\|_\Psi^O \leq 2\|a\|_\Psi$ for all $a \in L^\Psi(\mathcal{M})$.

Proof. The first claim may be proved using similar arguments as those used for $L^\Psi(\mathcal{M})$. To see the second, observe that on replacing t with $\frac{1}{u}$, the inequalities in the preceding proposition may be rephrased as the claim that

$$\frac{1}{\Psi^{-1}(1/u)} \leq u(\Psi^*)^{-1}(1/u) \leq 2\frac{1}{\Psi^{-1}(1/u)} \text{ for all } u > 0$$

or equivalently that

$$\mathbf{f}^\Psi(u) \leq \mathbf{f}_\Psi(u) \leq 2\mathbf{f}^\Psi(u) \text{ for all } u > 0.$$

Next, consider the mapping $\iota : a \mapsto w_\Psi^{1/2} a w_\Psi^{1/2}$ $(a \in \widetilde{\mathfrak{M}})$ where $w_\Psi = \mathbf{f}_\Psi(h)\mathbf{f}^\Psi(h)^{-1}$. The previously centred inequality ensures that $\mathbb{1} \leq w_\Psi \leq 2\mathbb{1}$. Now, observe that for v_s as in Definition 10.8 and $\widetilde{v}_s$ as in the hypothesis, we have that

$$
\begin{aligned}
\theta_s(w_\Psi)v_s &= \mathbf{f}_\Psi(\theta_s(h))^{1/2}\mathbf{f}^\Psi(\theta_s(h))^{-1}.\mathbf{f}^\Psi(e^{-s}h)\mathbf{f}^\Psi(h)^{-1} \\
&= \mathbf{f}_\Psi(e^{-s}h)^{1/2}\mathbf{f}^\Psi(e^{-s}h)^{-1}.\mathbf{f}^\Psi(e^{-s}h)\mathbf{f}^\Psi(h)^{-1} \\
&= \mathbf{f}_\Psi(e^{-s}h)^{1/2}\mathbf{f}^\Psi(h)^{-1} \\
&= \widetilde{v}_s w_\Psi.
\end{aligned}
$$

This ensures that for any $a \in L^\Psi(\mathcal{M})$, we have that

$$\theta_s(w_\Psi^{1/2} a w_\Psi^{1/2}) = \theta_s(w_\Psi^{1/2})v_s^{1/2} a v_s^{1/2}\theta_s(w_\Psi^{1/2}) = \widetilde{v}_s^{1/2}(w_\Psi^{1/2} a w_\Psi^{1/2})\widetilde{v}_s^{1/2}.$$

In other words, the prescription $a \mapsto w_\Psi^{1/2} a w_\Psi^{1/2}$ maps $L^\Psi(\mathcal{M})$ into $L_\Psi(\mathcal{M})$. A similar argument now shows that the prescription

$a \mapsto w_\Psi^{-1/2} a w_\Psi^{-1/2}$ maps $L_\Psi(\mathcal{M})$ into $L^\Psi(\mathcal{M})$, and hence that the original map ι is a bijection from $L^\Psi(\mathcal{M})$ to $L_\Psi(\mathcal{M})$. Given $f \in L^\Psi(\mathcal{M})$, we have that $\|\iota(f)\|_\Psi^O = \mathbf{m}_{w_\Psi^{1/2} a w_\Psi^{1/2}}(1) \leq \|w_\Psi\| \mathbf{m}_f(1) \leq 2\|f\|_\Psi$ and that

$$\|f\|_\Psi = \mathbf{m}_f(1) \leq \|w_s^{-1}\| \mathbf{m}_{w_\Psi^{1/2} a w_\Psi^{1/2}}(1) \leq \|\iota(f)\|_\Psi^O. \qquad \square$$

For the next result, we need the concept of a fundamental index of an Orlicz space. This definition makes sense for all rearrangement invariant Banach function spaces, but that is outside the scope of this book. For the spaces under consideration, these indices may be defined as follows.

Definition 10.19 Let Ψ be a Young function and let

$$M_\Psi(t) = \sup_{s>0} \frac{\mathbf{f}^\Psi(st)}{\mathbf{f}^\Psi(s)}.$$

Then, the lower and upper fundamental indices of $L^\Psi(\mathcal{M})$ are defined to be

$$\underline{\beta}_{L^\Psi} = \lim_{s \to 0+} \frac{\log M_\Psi(s)}{\log s} \quad \text{and} \quad \overline{\beta}_{L^\Psi} = \lim_{s \to \infty} \frac{\log M_\Psi(s)}{\log s}$$

respectively.

Proposition 10.20 *Let $L^\Psi(\mathcal{M})$ be an Orlicz space with upper fundamental index strictly less than 1. Then $L^\Psi(\mathcal{M}) \subseteq (L^\infty + L^1)(\mathfrak{M}, \tau_\mathfrak{M})$. Moreover, the canonical topology on $L^\Psi(\mathcal{M})$ then agrees with the subspace topology inherited from $(L^\infty + L^1)(\mathfrak{M}, \tau_\mathfrak{M})$.*

Proof. If indeed the upper fundamental index of $L^\Psi(\mathcal{M})$ is strictly less than 1, then there must exist some $t_0 > 1$ and some $0 < \delta < 1$ such that $\frac{\log M_\Psi(t)}{\log t} \leq \delta$ for all $t > t_0$. Since for $t \geq 1$ we have that $M_\Psi(t) \geq 1$, this can be shown to be equivalent to the claim that $M_\Psi(t) \leq t^\delta$ for all $t \geq t_0$ or equivalently that $M_\Psi(\frac{1}{t}) \leq (\frac{1}{t})^\delta$ for all $0 < t \leq \frac{1}{t_0}$.

Now, recall that by equation (10.1), we have that $\mathbf{m}_a(t) = \mathbf{m}_{v_s^{1/2} a v_s^{1/2}}(1)$ for all $t = e^{st}$ where $s_t < 0$, and $v_{s_t} = \mathbf{f}^\Psi(e^{-s_t} h) \mathbf{f}^\Psi(h)^{-1}$. If indeed $s_t \leq -\log(t_0)$ (equivalently $0 < t < \frac{1}{t_0}$), we will have that $\frac{\mathbf{f}^\Psi(t^{-1} r)}{\mathbf{f}^\Psi(r)} \leq M_\Psi(\frac{1}{t}) \leq (\frac{1}{t})^\delta$ for all $r > 0$, and hence that $\|v_{s_t}\| \leq (\frac{1}{t})^\delta$.

It therefore follows that $\mathbf{m}_a(t) = \mathbf{m}_{v_s^{1/2} a v_s^{1/2}}(1) \leq \|v_s\| \mathbf{m}_a(1) \leq (\frac{1}{t})^\delta \mathbf{m}_a(1)$ for all $0 < t \leq \frac{1}{t_0}$, with $t \mathbf{m}_a(t) \leq \mathbf{m}_a(1)$ on $[t_0^{-1}, 1]$.

Alongside this fact, we may now use the fact that $t \mapsto \mathbf{m}_t(a)$ is non-increasing to see that

$$\mathbf{m}_a(1) \leq \int_0^1 \mathbf{m}_a(t)\, dt \leq \left[\int_0^{1/t_0} (\tfrac{1}{t})^\delta\, dt + \int_{1/t_0}^1 \frac{1}{t}\, dt \right] \mathbf{m}_a(1).$$

The result now follows from Theorem 6.64. $\qquad\qquad\qquad\qquad\square$

Example 10.21 We show that the upper fundamental index of the space $L^{\cosh\,-1}(0,\infty)$ is $\frac{1}{2}$. This space is therefore one of the Orlicz spaces for which the canonical topology is a norm topology. The first step in proving this claim is to note that isomorphic Orlicz spaces share the same indices. We leave the verification of this fact as an exercise. So, it is sufficient to prove this for a space isomorphic to $L^{\cosh\,-1}(0,\infty)$. We show how to construct such a space before proving the claim. It is easy to see that the graphs of e^t and $\frac{e^2}{4}t^2$ are tangent at $t = 2$. This fact ensures that

$$\Psi_e(t) = \begin{cases} \dfrac{e^2}{4}t^2, & \text{if } \ 0 \leq t \leq 2 \\[2ex] e^t, & \text{if } \ 2 < t \end{cases}$$

is a Young function. Using Maclaurin series, it is easy to see that

$$\lim_{t \to 0+} \frac{\cosh(t) - 1}{\Psi_e(t)} = \frac{2}{e^2}.$$

Since we also have $\lim_{t \to \infty} \frac{\cosh(t)-1}{\Psi_e(t)} = \lim_{t \to \infty} \frac{e^t + e^{-t} - 2}{2e^t} = \frac{1}{2}$, it is clear that $\Psi_e \approx \cosh - 1$. (To see this, note that the limit formulae ensure that we may find $0 < \alpha < \beta < \infty$ so that $\frac{1}{e^2} < \frac{\cosh(t)-1}{\Psi_e(t)} < \frac{3}{e^2}$ on $[0,\alpha]$, and $\frac{1}{4} < \frac{\cosh(t)-1}{\Psi_e(t)} < \frac{3}{4}$ on $[\beta,\infty)$). Since the function $\frac{\cosh(t)-1}{\Psi_e(t)}$ has both a minimum and a maximum on the interval $[\alpha,\beta]$, a combination of these facts ensures that we can find positive constants $0 < m < M < \infty$ so that $m\Psi_e(t) < \cosh(t) - 1 < M\Psi_e(t)$ for all $t \in [0,\infty)$.) This is clearly enough to ensure that $L^{\Psi_e}(0,\infty) \equiv L^{\cosh\,-1}(0,\infty)$.

It remains to compute the fundamental indices of $L^{\Psi_e}(0,\infty)$, where we assume that $L^{\Psi_e}(0,\infty)$ is equipped with the Luxemburg–Nakano norm. Since

$$\Psi_e^{-1}(t) = \begin{cases} \frac{2}{e}t^{1/2}, & \text{if} \quad 0 \leq t \leq e^2 \\ \log(t), & \text{if} \quad e^2 < t \end{cases},$$

it now follows that the fundamental function $\frac{1}{\Psi_e^{-1}(1/t)}$ of $L^{\Psi_e}(0,\infty)$ is given by

$$\mathbf{f}_e(t) = \begin{cases} \frac{e}{2}t^{1/2}, & \text{if} \quad t \geq e^{-2} \\ \frac{1}{-\log(t)}, & \text{if} \quad t < e^{-2}. \end{cases}$$

We proceed to compute the function $M_{\Psi_e}(s) = \sup_{t>0} \frac{\mathbf{f}_e(st)}{\mathbf{f}_e(t)}$. In computing this function, we first consider the case where $0 < s \leq 1$. Since $\mathbf{f}_e$ is increasing, we then have that

$$\frac{\mathbf{f}_e(st)}{\mathbf{f}_e(t)} \leq \frac{\mathbf{f}_e(t)}{\mathbf{f}_e(t)} = 1$$

for any $t > 0$. Since we also have that

$$\lim_{t \to 0} \frac{\mathbf{f}_e(st)}{\mathbf{f}_e(t)} = \lim_{t \to 0} \frac{\log(t)}{\log(st)} = \lim_{t \to 0} \frac{\log(t)}{\log(s) + \log(t)} = 1,$$

it is clear that $M_{\Psi_e}(s) = 1$ in this case.

Now, let s be given with $s > 1$. We then have

$$\frac{\mathbf{f}_e(st)}{\mathbf{f}_e(t)} = \begin{cases} s^{1/2}, & \text{if} \quad t > \frac{1}{e^2} \\ -\frac{e}{2}s^{1/2}t^{1/2}\log(t), & \text{if} \quad \frac{1}{e^2} > t > \frac{1}{se^2} \\ \frac{\log(t)}{\log(st)}, & \text{if} \quad t < \frac{1}{se^2}. \end{cases}$$

It is not too difficult to see that on the interval $(0,1)$, the function $t \mapsto -\frac{e}{2}s^{1/2}t^{1/2}\log(t)$ has a maximum of $s^{1/2}$ at $t = e^{-2}$. So, for $t \in (\frac{1}{se^2}, \frac{1}{e^2})$, the supremum of the above quotient is $s^{1/2}$. Finally, consider the function

$$t \mapsto \frac{\log(t)}{\log(st)} = \frac{\log(t)}{\log(s)+\log(t)} = 1 - \frac{\log(s)}{\log(s)+\log(t)} = 1 - \frac{\log(s)}{\log(st)}.$$

It is easy to see that

$$\frac{d}{dt}\left(1 - \frac{\log(s)}{\log(s)+\log(t)}\right) = \frac{\log(s)}{t(\log(st))^2} > 0$$

on $t \in (0, \frac{1}{se^2})$. Hence, on $(0, \frac{1}{se^2}]$, $t \mapsto \frac{\log(t)}{\log(st)}$ attains a maximum of $1 + \frac{1}{2}\log(s)$ at $t = \frac{1}{se^2}$. Using the fact that $1 + \log(t) \leq t$, it is now easy to see that $1 + \frac{1}{2}\log(s) = 1 + \log(s^{1/2}) \leq s^{1/2}$. Putting all these facts together leads to the conclusion that $M_{\Psi_e}(s) = \sup_{t>0} \frac{\mathbf{f}_e(st)}{\mathbf{f}_e(t)} = s^{1/2}$ in this case. We, therefore, have that

$$\overline{\beta}_{\Psi_e} = \lim_{s\to\infty} \frac{\log M_{\Psi_e}(s)}{\log s} = \lim_{s\to\infty} \frac{\log s^{1/2}}{\log s} = \frac{1}{2}$$

as claimed. Similarly

$$\underline{\beta}_{\Psi_e} = \lim_{s\to 0+} \frac{\log M_{\Psi_e}(s)}{\log s} = \lim_{s\to 0+} \frac{\log 1}{\log s} = 0.$$

We conclude this discussion of Orlicz spaces for general von Neumann algebras by showing that as in the semifinite setting, the spaces $L^{1\cap\infty}(\mathcal{M})$ and $L^{1+\infty}(\mathcal{M})$ (constructed using the Young functions described in Proposition 6.61) are in a very concrete sense, respectively, the smallest and largest of all the Orlicz spaces. With $\Psi_{1\cap\infty}$ and $\Psi_{1+\infty}$ as in Proposition 6.61, we know from Remark 6.62 that the fundamental functions corresponding to these spaces are respectively

$$\mathbf{f}^{1\cap\infty}(t) = \max(1,t) \text{ and } \mathbf{f}_{1+\infty}(t) = \min(1,t).$$

Now, recall that for an arbitrary Young function Ψ, the fundamental function $\mathbf{f}^{\Psi}$ is quasi-concave; that is $\mathbf{f}^{\Psi}$ is increasing, continuous on $(0,\infty)$, zero valued at precisely zero, and with $t \mapsto \frac{\mathbf{f}^{\Psi}(t)}{t}$ decreasing. This ensures that $\frac{\mathbf{f}^{\Psi}(t)}{t} \geq \mathbf{f}^{\Psi}(1)$ whenever $0 \leq t \leq 1$, and that $\frac{\mathbf{f}^{\Psi}(t)}{t} \leq \mathbf{f}^{\Psi}(1)$ whenever $t \geq 1$. Put differently, this ensures that for any $t \geq 0$, we have that

$$\mathbf{f}^{\Psi}(1)\mathbf{f}_{1+\infty}(t) \leq \mathbf{f}^{\Psi}(t) \leq \mathbf{f}^{\Psi}(1)\mathbf{f}^{1\cap\infty}(t).$$

These inequalities are the cornerstone of the following result:

Proposition 10.22 *Let Ψ be a Young function, and let $\zeta^{\Psi}(t)$ and $\zeta_{\Psi}(t)$, respectively, be the functions defined by $\zeta^{\Psi}(t) = \frac{\mathbf{f}_{1+\infty}(t)}{\mathbf{f}^{\Psi}(t)}$ and $\zeta_{\Psi}(t) = \frac{\mathbf{f}^{\Psi}(t)}{\mathbf{f}^{1\cap\infty}(t)}$. These functions are, respectively, bounded by $\frac{1}{\mathbf{f}^{\Psi}(1)}$ and $\mathbf{f}^{\Psi}(1)$. The operators $\zeta^{\Psi}(h)$ and $\zeta_{\Psi}(h)$ are therefore bounded operators (with the same bounds). The prescriptions $\iota_{\Psi} : a \mapsto \zeta_{\Psi}(h)^{1/2}a\zeta_{\Psi}(h)^{1/2}$ and $\iota^{\Psi} : a \mapsto \zeta^{\Psi}(h)^{1/2}a\zeta^{\Psi}(h)^{1/2}$, where $a \in \widetilde{\mathfrak{M}}$, are then continuous maps on $\widetilde{\mathfrak{M}}$ which, respectively, restrict to continuous embeddings of $L^{1\cap\infty}(\mathcal{M})$ into $L^{\Psi}(\mathcal{M})$, and $L^{\Psi}(\mathcal{M})$ into $L^{1+\infty}(\mathcal{M})$.*

Proof. Except for the final claim, all statements follow fairly immediately from the discussion preceding the proposition. We prove the final claim. First assume that $a \in L^\Psi(\mathcal{M})$. For any $s \in \mathbb{R}$, we will then have

$$
\begin{aligned}
\theta_s&(\zeta^\Psi(h)^{1/2} a \zeta^\Psi(h)^{1/2}) \\
&= \theta_s(\zeta^\Psi(h)^{1/2}) \theta_s(a) \theta_s(\zeta^\Psi(h)^{1/2}) \\
&= \theta_s(\mathbf{f}_{1+\infty}(h)^{1/2} \mathbf{f}^\Psi(h)^{-1/2}) \cdot \theta_s(a) \cdot \theta_s(\mathbf{f}^\Psi(h)^{-1/2} \mathbf{f}_{1+\infty}(h)^{1/2}) \\
&= [\mathbf{f}_{1+\infty}(e^{-s}h)^{1/2} \mathbf{f}^\Psi(e^{-s}h)^{-1/2}] \\
&\qquad \cdot [\mathbf{f}^\Psi(e^{-s}h)^{1/2} \mathbf{f}^\Psi(h)^{-1/2} a \mathbf{f}^\Psi(h)^{-1/2} \mathbf{f}^\Psi(e^{-s}h)^{1/2}] \\
&\qquad \cdot [\mathbf{f}^\Psi(e^{-s}h)^{-1/2} \mathbf{f}_{1+\infty}(e^{-s}h)^{1/2}] \\
&= [\mathbf{f}_{1+\infty}(e^{-s}h)^{1/2} \mathbf{f}^\Psi(h)^{-1/2}] a [\mathbf{f}^\Psi(h)^{-1/2} \mathbf{f}_{1+\infty}(e^{-s}h)^{1/2}] \\
&= [\mathbf{f}_{1+\infty}(e^{-s}h)^{1/2} \mathbf{f}_{1+\infty}(h)^{-1/2}] \cdot [\mathbf{f}_{1+\infty}(h)^{1/2} \mathbf{f}^\Psi(h)^{-1/2}] \cdot a \\
&\qquad \cdot [\mathbf{f}^\Psi(h)^{-1/2} \mathbf{f}_{1+\infty}(h)^{1/2}] \cdot [\mathbf{f}_{1+\infty}(h)^{-1/2} \mathbf{f}_{1+\infty}(e^{-s}h)^{1/2}] \\
&= [\mathbf{f}_{1+\infty}(e^{-s}h)^{1/2} \mathbf{f}_{1+\infty}(h)^{-1/2}][\zeta^\Psi(h)^{1/2} a \zeta^\Psi(h)^{1/2}] \\
&\qquad [\mathbf{f}_{1+\infty}(h)^{-1/2} \mathbf{f}_{1+\infty}(e^{-s}h)^{1/2}].
\end{aligned}
$$

By definition, $\zeta^\Psi(h)^{1/2} a \zeta^\Psi(h)^{1/2}$ must then be an element of $L^{1+\infty}(\mathcal{M})$. The proof of the other case runs along similar lines. $\square$

Even in this generality, the L^p-spaces of $\mathcal{M}$ have a much more regular structure than the class of Orlicz spaces. In the rest of this chapter, we will therefore focus on refining the theory of L^p-spaces for general von Neumann algebras. We already noted in Example 10.15 that the Luxemburg–Nakano norm on $L^\infty(\mathcal{M})$ agrees with the operator norm on $\mathcal{M}$. We now pass to analysing the (p-)norm on $L^p(\mathcal{M})$ where $0 < p < \infty$. For $0 < p < 1$, t^p is of course not convex and hence $L^p(\mathcal{M})$ not an Orlicz space. However, by analogy with the theory already developed, we will here also write $\| \cdot \|_p$ for the action of $\mathbf{m}_{(\cdot)}(1)$ on $L^p(\mathcal{M})$. The following lemma provides useful technical information for the analysis of the (p-)norm $\| \cdot \|_p$.

Lemma 10.23 *Let $0 < p < \infty$ be given.*

(1) *For any $a \in \widetilde{\mathfrak{M}}$ with polar decomposition $a = u|a|$, the following are equivalent:*

- $a \in L^p(\mathcal{M})$;
- $a^* \in L^p(\mathcal{M})$;
- $|a| \in L^p(\mathcal{M})$ and $u \in \mathcal{M}$;
- $|a|^p \in L^1(\mathcal{M})$ and $u \in \mathcal{M}$.

  If any one (and therefore all) of these conditions hold, we moreover have that $\|a\|_p = \|a^*\|_p = \| \, |a| \, \|_p$.

(2) For any $0 \neq a \in L^p(\mathcal{M})$, we have that:
- $\frac{1}{t} d_a(r) = d_a(t^{1/p} r)$ for all $t, r > 0$;
- $\|a\|_p = \mathbf{m}_a(1) = d_a(1)^{1/p}$;
- $t^{-1/p} \mathbf{m}_a(1) = \mathbf{m}_a(t)$ for all $t > 0$.

Proof. First consider part (1). Recall that a being an element of $L^p(\mathcal{M})$ means that $\theta_s(a) = e^{-s/p} a$ for any $s \in \mathbb{R}$. But then we must clearly have that $\theta_s(|a|) = |\theta_s(a)| = |e^{-s/p} a| = e^{-s/p} |a|$ for any $s \in \mathbb{R}$. So, $|a| \in L^p(\mathcal{M})$ as claimed. Similarly, $a^* \in L^p(\mathcal{M})$. As far as u is concerned, it is clear that for any $s \in \mathbb{R}$, $\theta_s(u)\theta_s(|a|)$ is the polar decomposition of $\theta_s(a)$. But from what has already been verified, this means that $\theta_s(u)|a|$ is the polar decomposition of a. From the uniqueness of the polar decomposition, it follows that $\theta_s(u) = u$ for all $s \in \mathbb{R}$. But then $u \in \mathcal{M}$. This then proves the equivalence of the first three bullets. The proof that the fourth is equivalent to the third follows on noting that the definition ensures that $|a| \in L^p(\mathcal{M})$ if and only if $|a|^p \in L^1(\mathcal{M})$. The final claim is an easy consequence of (ii) of Proposition 5.12.

Now consider part (2). These claims may be verified by modifying the technique used in the proof of Proposition 10.10. For the sake of clarity, we repeat the essentials of that argument. Let $a \in L^p(\mathcal{M})$ and $0 < t$ be given. We may then select $s \in \mathbb{R}$ so that $t = e^s$. For any $r > 0$, we then have that

$$
\begin{aligned}
\frac{1}{t} d_a(r) &= e^{-s} \tau(\chi_{(r,\infty)}(|a|)) \\
&= \tau(\theta_s(\chi_{(r,\infty)}(|a|))) \\
&= \tau(\chi_{(r,\infty)}(\theta_s|a|)) \\
&= \tau(\chi_{(r,\infty)}(|\theta_s(a)|)) \\
&= d_{\theta_s(a)}(r) \\
&= d_{e^{-s/p} a}(r) \\
&= d_{t^{-1/p} a}(r) \\
&= d_a(t^{1/p} r).
\end{aligned}
$$

Note that since $a \neq 0$, we must have $d_a(s) \neq 0$ for some $s > 0$. If we combine this with the equality just verified, it follows that in fact $d_a(s) \neq 0$ for all $s > 0$. Now, observe that the above equality on the one hand gives $r^p d_a(r) = d_a(1)$, and on the other hand that $d_a(d_a(1)^{1/p}) = \frac{1}{d_a(1)} d_a(1) = 1$ with $d_a((d_a(1) - \epsilon)^{1/p}) = \frac{1}{d_a(1)-\epsilon} d_a(1) > 1$ for any $0 < \epsilon < d_a(1)$. Hence, $\mathbf{m}_a(1) = \inf\{r > 0 \colon d_a(r) \leq 1\} = d_a(1)^{1/p}$. We may now use the first bullet of part (2) to conclude that $t^{1/p}\mathbf{m}_a(t) = t^{1/p} \inf\{r > 0 \colon d_a(r) \leq t\} = t^{1/p} \inf\{r > 0 \colon d_a(t^{1/p}r) \leq 1\} = \inf\{s > 0 \colon d_a(s) \leq 1\} = \mathbf{m}_a(1)$. $\qquad\square$

Armed with the above lemma, we are now able to prove a Hölder and Minkowski inequality for the present context.

Proposition 10.24 *Let $p, q, r > 0$ be given such that $\frac{1}{r} = \frac{1}{p} + \frac{1}{q}$. For any $a \in L^p(\mathcal{M})$ and $b \in L^q(\mathcal{M})$, we will have $ab \in L^r(\mathcal{M})$ with $\|ab\|_r \leq \|a\|_p \|b\|_q$.*

Proof. Let $a \in L^p(\mathcal{M})$ and $b \in L^q(\mathcal{M})$ be given. For any $s \in \mathbb{R}$, we will have that $\theta_s(ab) = e^{-s/p}a \cdot e^{-s/q}b = e^{-s/r}ab$. So, by definition, $ab \in L^r(\mathcal{M})$. It now follows from Theorem 6.2 that

$$\exp \int_0^\infty \log(\mathbf{m}_{ab}(t))\, dt \leq \exp \int_0^\infty \log(\mathbf{m}_a(t))\, dt \cdot \exp \int_0^\infty \log(\mathbf{m}_b(t))\, dt.$$

By Lemma 10.23, this may be rewritten as

$$\exp \int_0^\infty \log(t^{-1/r}\mathbf{m}_{ab}(1))\, dt$$

$$\leq \exp \int_0^\infty \log(t^{-1/p}\mathbf{m}_a(1))\, dt \cdot \exp \int_0^\infty \log(t^{-1/q}\mathbf{m}_b(1))\, dt$$

which yields

$$e^{1/r}\|ab\|_r = e^{1/r}\mathbf{m}_{ab}(1) \leq e^{1/p}\mathbf{m}_a(1) \cdot e^{1/q}\mathbf{m}_b(1)$$

$$= e^{1/r}\|a\|_p\|b\|_q. \qquad\square$$

Proposition 10.25 *For any $a, b \in L^p(\mathcal{M})$, we have that:*

- $\|a + b\|_p \leq \|a\|_p + \|b\|_p$ *whenever* $1 < p \leq \infty$;
- *and* $\|a + b\|_p^p \leq \|a\|_p^p + \|b\|_p^p$ *whenever* $0 < p \leq 1$.

The quantity $\| \cdot \|_p = \mathbf{m}_{(\cdot)}(1)$ is therefore a norm when $1 \leq p \leq \infty$, and a p-norm when $0 < p < 1$.

Proof. The case $p = \infty$ follows from Example 10.15. Now, consider the case where $1 < p < \infty$. It then follows from Theorem 5.22 that $\int_0^1 \mathbf{m}_{a+b}(t)\, dt \leq \int_0^1 \mathbf{m}_a(t)\, dt + \int_0^1 \mathbf{m}_b(t)\, dt$. The final bullet in Lemma 10.23 now ensures that this inequality corresponds to the claim that

$$\frac{p}{p-1}\mathbf{m}_{a+b}(1) \leq \frac{p}{p-1}(\mathbf{m}_a(1) + \mathbf{m}_b(1)).$$

The claim follows.

Now, let $0 < p \leq 1$. For any $0 < r < p$, it then follows from Theorem 6.9 that

$$\int_0^1 \mathbf{m}_{|a+b|^r}(t)\, dt = \int_0^1 \mathbf{m}_{a+b}(t)^r\, dt$$

$$\leq \int_0^1 \mathbf{m}_a(t)^r\, dt + \int_0^1 \mathbf{m}_b(t)^r\, dt$$

$$= \int_0^1 \mathbf{m}_{|a|^r}(t)\, dt + \int_0^1 \mathbf{m}_{|b|^r}(t)\, dt.$$

Now, if say $f \in L^p(\mathcal{M})$, then $\theta_s(|f|^r) = |\theta_s(f)|^r = |e^{-s/p}f|^r = e^{-s/(p/r)}|f|$ for all $s \in \mathbb{R}$. In other words, we then have $|f| \in L^{p/r}(\mathcal{M})$. This ensures that $\mathbf{m}_{|f|^r}(t) = t^{-r/p}\mathbf{m}_{|f|^r}(t) = t^{-r/p}\mathbf{m}_f(t)^r$ for all $t > 0$. If we apply this fact to the inequality

$$\int_0^1 \mathbf{m}_{|a+b|^r}(t)\, dt \leq \int_0^1 \mathbf{m}_{|a|^r}(t)\, dt + \int_0^1 \mathbf{m}_{|b|^r}(t)\, dt$$

we get $\frac{p-r}{p}\mathbf{m}_{a+b}(1)^r \leq \frac{p-r}{p}(\mathbf{m}_a(1)^r + \mathbf{m}_b(1)^r)$. On dividing throughout by $\frac{p-r}{p}$ and letting $r \nearrow p$, we obtain $\mathbf{m}_{a+b}(1)^p \leq \mathbf{m}_a(1)^p + \mathbf{m}_b(1)^p$ as required.

Regarding the final claim, recall that in the case $1 \leq p < \infty$, we already know that $\| \cdot \|_p$ is a quasi-norm. The validity of the triangle inequality now ensures that it is a norm. In the case $0 < p < 1$, the one fact regarding a p-norm that is not immediately obvious is the fact that here too $\| \cdot \|_p$ is non-degenerate. To see this, note that given some $a \in L^p(\mathcal{M})$ for which $\mathbf{m}_a(1) = 0$, Lemma 10.23 informs us that then $\mathbf{m}_a(t) = 0$ for all $t > 0$. Therefore, $\|a\|_\infty = \lim_{t \searrow 0} \mathbf{m}_t(a) = 0$, or equivalently $a = 0$. $\qquad \square$

Remark 10.26 It is clear from the above result that the $L^p(\mathcal{M})$-spaces are in fact Banach spaces whenever $p \geq 1$.

10.3 The trace functional and tr-duality for L^p-spaces

Although we have defined L^p-spaces for general von Neumann algebras and have even proved a Hölder and Minkowski inequality for these spaces, at present they bear little resemblance to their classical counterparts. We now remedy this by introducing the so-called trace functional on $L^1(\mathcal{M})$. This functional is a crucial tool for the development of duality theory in this context.

Definition 10.27 We define tr to be the linear functional on $L^1(\mathcal{M})$ given by $\mathrm{tr}(a) = \omega_a(\mathbb{1})$, where $\omega_a \in \mathcal{M}_*$ is the normal functional on $\mathcal{M}$ corresponding to $a \in L^1(\mathcal{M})$ by means of the bijection described in Theorem 9.38.

Proposition 10.28 *Let $0 < p < \infty$ be given. For any $a \in L^p(\mathcal{M})$, the quantity $\|a\|_p = \mathbf{m}_a(1)$ agrees with $\mathrm{tr}(|a|^p)^{1/p}$.*

Proof. First let $a \in L^1(\mathcal{M})$. Let $\omega_{|a|}$ be the functional in $\mathcal{M}_*$ corresponding to $|a|$ by means of the bijection in Theorem 9.38. Since $\omega_{|a|}(\mathbb{1}) = d_{|a|}(1)$ by Theorem 9.38, it then follows from Lemma 10.23 that $\mathrm{tr}(|a|) = \mathbf{m}_a(1)$ in this setting. Now, suppose that $a \in L^p(\mathcal{M})$. It is easy to see that then $|a|^p \in L^1(\mathcal{M})$. We may then use what we have noted regarding $L^1(\mathcal{M})$ to conclude that $\mathrm{tr}(|a|^p)^{1/p} = \mathbf{m}_{|a|^p}(1)^{1/p} = \mathbf{m}_{|a|}(1) = \mathbf{m}_a(1)$. $\qquad\square$

The above proposition now enables us to show that $L^1(\mathcal{M})$ is an isometric copy of $\mathcal{M}_*$, and to show that $L^2(\mathcal{M})$ is in fact a Hilbert space.

Proposition 10.29 *For any $a \in L^1(\mathcal{M})$, we have that $\overline{\mathrm{tr}(a)} = \mathrm{tr}(a^*)$ and $|\mathrm{tr}(a)| \leq \mathrm{tr}(|a|) = \|\omega_a\|$, where ω_a is the normal functional corresponding to a (Theorem 9.38). The quantity $\mathrm{tr}(|\cdot|) = \|\cdot\|_1$ is a norm on $L^1(\mathcal{M})$, and when equipped with this norm, $L^1(\mathcal{M})$ is isometrically isomorphic to $\mathcal{M}_*$.*

Proof. Let a and ω_a be as in the hypothesis. All claims follow fairly immediately from Theorem 9.38. For the sake of the reader, we provide suitable details. For the first claim, note that $\overline{\mathrm{tr}(a)} = \overline{\omega_a(\mathbb{1})} = \omega_a^*(\mathbb{1}) = \omega_{a^*}(\mathbb{1}) = \mathrm{tr}(a^*)$. For the second claim, note that $|\mathrm{tr}(a)| = |\omega_a(\mathbb{1})| \leq \|\omega_a\| = \|\,|\omega_a|\,\| = \|\omega_{|a|}\| = |\omega_{|a|}(\mathbb{1})| = \mathrm{tr}(|a|)$. The equality $\mathrm{tr}(|\cdot|) = \|\cdot\|_1$ was proved in the preceding proposition. Given $a, b \in L^1(\mathcal{M})$, we may use this equality to see that $\|a + b\|_1 = \mathrm{tr}(|a + b|) = \|\omega_{a+b}\| = \|\omega_a + \omega_b\| \leq \|\omega_a\| + \|\omega_b\| = \mathrm{tr}(|a|) + \mathrm{tr}(|b|) = \|a\|_1 + \|b\|_1$. The final claim is now an immediate consequence of Theorem 9.38. $\qquad\square$

Proposition 10.30 *The prescription $\langle a, b \rangle = \mathrm{tr}(b^*a)$ defines an inner product on $L^2(\mathcal{M})$ for which $\langle a, a \rangle = \|a\|_2^2$. The space $L^2(\mathcal{M})$ is therefore a Hilbert space.*

Proof. For any $f \in L^1(\mathcal{M})$, we will again write ω_f for the normal functional corresponding to f by means of the bijection described in Theorem 9.38. For any $a_1, a_2, b \in L^2(\mathcal{M})$ and any $\gamma \in \mathbb{R}$, we have $\langle a_1 + a_2, b \rangle = \mathrm{tr}(b^*(a_1 + \gamma a_2)) = \omega_{b^*(a_1 + \gamma a_2)} = \omega_{b^*a_1} + \gamma\omega_{b^*a_2} = \mathrm{tr}(b^*a_1) + \gamma\mathrm{tr}(b^*a_2) = \langle a_1, b \rangle + \gamma\langle a_2, b \rangle$. The remaining properties of an inner product may be proved by similar techniques. Given $a \in L^2(\mathcal{M})$, the claim regarding the norm follows by applying Proposition 10.28 to the equality $\langle a, a \rangle = \mathrm{tr}(|a|^2)$. $\qquad\square$

We proceed to show that tr satisfies a trace-like property. Once that is done, L^p-duality will follow fairly quickly.

Lemma 10.31 *For any $f \in \widetilde{\mathfrak{M}}_+$, the mapping $z \mapsto f^z$ is a differentiable $\widetilde{\mathfrak{M}}$-valued map on the open right half-plane $\mathbb{C}_+^o = \{z \colon \mathrm{Re}(z) > 0\}$.*

Proof. Let $z_0 \in \mathbb{C}_+^o$ and $f \in \widetilde{\mathfrak{M}}_+$ be given. We will show that $f^{z_0} \log(f) \in \widetilde{\mathfrak{M}}$ with $\frac{d}{dz} f^z \big|_{z_0} = f^{z_0} \log(f)$.

First, consider the case where $f \in \mathfrak{M}_+$. Notice that $t \mapsto t^{z_0} \log(t)$ extends to a function which is continuous on $[0, \|f\|]$, and 0-valued at 0. Since $\mathrm{sp}(f) \subseteq [0, \|f\|]$, it therefore follows from the continuous functional calculus that we also have $f^{z_0} \log(f) \in \mathfrak{M}$. It is an exercise to see that as $z \to z_0$, the expression $\frac{t^z - t^{z_0}}{z - z_0} - t^{z_0} \log(t)$ will converge to 0 uniformly on $[0, \|f\|]$. We may then once again apply the continuous functional calculus to see that $\frac{1}{z-z_0}(f^z - f^{z_0}) - f^{z_0}\log(f)$ converges to 0 in norm. Thus, in this case, $\frac{d}{dz}f^z\big|_{z_0} = f^{z_0}\log(f)$ as required.

Now, suppose that $f \in \widetilde{\mathfrak{M}}_+$. It then follows from what we have just proven that $f^{z_0}\log(f)\chi_{[0,\gamma]}(f) \in \mathfrak{M}$ for any $\gamma > 0$. If therefore we can show that $\tau(\chi_{(\epsilon,\infty)}(f^{x_0}|\log(f)|)) \to 0$ as $\epsilon \to \infty$, it will follow that $f^{z_0}\log(f) \in \widetilde{\mathfrak{M}}$. Let x_0 denote $\mathrm{Re}(z_0)$, and notice that $|f^{z_0}\log(f)| = f^{x_0}|\log(f)|$. The function $t \mapsto t^{x_0}\log(t)$ has a minimum of $-\frac{1}{ex_0}$ on $(0,\infty)$. So, if we choose $\gamma > 0$ large enough so that $\gamma^{x_0}\log(\gamma) > \frac{1}{ex_0}$, this would ensure that all of the statements

$$t^{x_0}\log(t) > \gamma^{x_0}\log(\gamma), \quad t^{x_0}\log(t) > \gamma^{x_0}\log(\gamma), \quad t > \gamma$$

are equivalent. So, for such a γ, the Borel functional calculus ensures that $\chi_{(\gamma,\infty)}(f) = \chi_{(\gamma^{x_0}\log(\gamma),\infty)}(f^{x_0}|\log(f)|)$. We may now use this equality to conclude from the known fact that $\tau(\chi_{(\gamma,\infty)}(f)) \to 0$ as $\gamma \to \infty$, that also $\tau(\chi_{(\gamma,\infty)}(f^{x_0}|\log(f)|)) \to 0$ as $\gamma \to \infty$.

Now, let $\epsilon > 0$ be given and select γ so that $\tau(\chi_{(\gamma,\infty)}(f)) < \epsilon$. With e denoting $e = \chi_{[0,\gamma]}(f)$, the operator fe is of course bounded. Therefore, it follows from the first part of the proof that $\|\frac{1}{z-z_0}((fe)^z - (fe)^{z_0}) - (fe)^{z_0}\log(fe)\| \le \epsilon$ for z close enough to z_0. But this means that $\frac{1}{z-z_0}(f^z - f^{z_0}) - f^{z_0}\log(f) \in N(\epsilon,\epsilon)$ for z close enough to z_0. So, by definition, $\frac{1}{z-z_0}(f^z - f^{z_0}) - f^{z_0}\log(f)$ converges to 0 in measure as $z \to z_0$, which then proves the lemma. $\qquad\square$

Lemma 10.32 *Let S^o be the open strip $S^o = \{z \in \mathbb{C}: 0 < \mathrm{Re}(z) < 1\}$. Let $f,g \in L^1(\mathcal{M})$ be given. For any $z \in S^o$, we have $f^z g^{1-z} \in L^1(\mathcal{M})$. Moreover, the map $S^o \to L^1(\mathcal{M}): z \mapsto f^z g^{1-z}$ is analytic.*

Proof. Let $z \in S^o$ be given. It is not difficult to see that each of f^z and g^{1-z} belongs to $\widetilde{\mathfrak{M}}$. Now, observe that for each $s \in \mathbb{R}$, we have $\theta_s(f^z g^{1-z}) = (\theta_s(f))^z(\theta_s(g))^{1-z} = (e^{-sz}f^z)(e^{-s(1-z)}g^{1-z}) = e^{-s}f^z g^{1-z}$. Thus, by definition $f^z g^{1-z} \in L^1(\mathcal{M})$. We know from the previous lemma that as maps into $\widetilde{\mathfrak{M}}$, each of $z \mapsto f^z$ and $z \mapsto g^{1-z}$ is analytic on S^o. (Here, we used the fact that $1 - z \in \mathbb{C}^o_+$ if $z \in S^o$.) It is not difficult to show that the product rule holds in this context, from which we may then conclude that $z \mapsto f^z g^{1-z}$ is analytic as a map into $\widetilde{\mathfrak{M}}$. But we know that this map is $L^1(\mathcal{M})$-valued. So since by Theorem 10.11 the norm topology on $L^1(\mathcal{M})$ agrees with the topology of convergence in measure inherited from $\widetilde{\mathfrak{M}}$, we are done. $\qquad\square$

Lemma 10.33 *Let $t \in \mathbb{R}$ be given and let $L^{1/((1/2)+it)}(\mathcal{M})$ denote the vector space*

$$L^{1/((1/2)+it)}(\mathcal{M}) = \{a \in \widetilde{\mathfrak{M}} : \theta_s(a) = e^{-s((1/2)+it)}a \text{ for all } s \in \mathbb{R}\}.$$

For all $a, b \in L^{1/((1/2)+it)}(\mathcal{M})$, we then have that $b^*a, ab^* \in L^1(\mathcal{M})$ with $\mathrm{tr}(b^*a) = \mathrm{tr}(ab^*)$.

Proof. Given any $a, b \in L^{1/((1/2)+it)}(\mathcal{M})$ and any $s \in \mathbb{R}$, it is easy to check that

$$\theta_s(b^*a) = (\theta_s(b))^*\theta_s(a) = e^{-s((1/2)-it)}b^* \cdot e^{-s((1/2)+it)}a = e^{-s}b^*a.$$

Similarly, $\theta_s(ab^*) = e^{-s}ab^*$ for any $s \in \mathbb{R}$. So, by definition, $b^*a, ab^* \in L^1(\mathcal{M})$. We may now apply Lemmas 10.23 and 10.28 to see that $\mathrm{tr}(a^*a) = d_{a^*a}(1) = d_{aa^*}(1) = \mathrm{tr}(aa^*)$ for any $a \in L^{1/((1/2)+it)}(\mathcal{M})$. Given $a, b \in L^{1/((1/2)+it)}(\mathcal{M})$, we may then apply this fact to the polarisation identities

$$b^*a = (1/4)\sum_{k=0}^{3}(a + i^k b)^*(a + i^k b) \text{ and } ab^* = (1/4)\sum_{k=0}^{3}(a + i^k b)(a + i^k b)^*$$

to see that $\mathrm{tr}(b^*a) = \mathrm{tr}(ab^*)$. $\square$

Theorem 10.34 *Let $p, q \geq 1$ be given with $1 = \frac{1}{p} + \frac{1}{q}$. For any $a \in L^p(\mathcal{M})$ and $b \in L^q(\mathcal{M})$, we have $\mathrm{tr}(ab) = \mathrm{tr}(ba)$.*

Proof. Let p, q and a, b be as in the hypothesis. We already noted in Proposition 10.24 that then $ab, ba \in L^1(\mathcal{M})$. So, tr is well-defined on both these products. First, consider the case where say $p = \infty$. Let $\omega_b \in \mathcal{M}_*$ be the functional corresponding to $b \in L^1(\mathcal{M})$ by means of the bijection described in Theorem 9.38. It then follows from Theorem 9.38 and the definition of tr that $\mathrm{tr}(ab) = \omega_{ab}(\mathbb{1}) = a \cdot \omega_b(\mathbb{1}) = \omega_b(\mathbb{1}) \cdot a = \omega_{ba}(\mathbb{1}) = \mathrm{tr}(ba)$.

Next, suppose that $1 < p, q < \infty$. We know from Lemma 10.23 that if $a \in L^p(\mathcal{M})$, then each of a^* and $|a|$ also belong to $L^p(\mathcal{M})$. Using these facts, it is easy to see that a can then be written as a linear combination of positive elements of $L^p(\mathcal{M})$, specifically $a = (|\mathrm{Re}(a)| + \mathrm{Re}(a)) - (|\mathrm{Re}(a)| - \mathrm{Re}(a)) + i(|\mathrm{Im}(a)| + \mathrm{Im}(a)) - i(|\mathrm{Im}(a)| - \mathrm{Im}(a))$. If therefore we can show that in the case where a and b are positive we have that $\mathrm{tr}(ab) = \mathrm{tr}(ba)$, the same equality will then by linearity hold for general elements a and b. We may therefore without loss of generality assume that $a, b \geq 0$. Since then $a^p, b^q \in L^1(\mathcal{M})$, Lemma 10.32 ensures that the functions $F(z) = \mathrm{tr}(a^{pz} b^{q(1-z)})$ and $G(z) = \mathrm{tr}(b^{q(1-z)} a^{pz})$ are analytic on S^o. Now, notice that for any fixed $t \in \mathbb{R}$, we have that $\theta_s(a^{p((1/2)+it)}) = (\theta_s(a))^{p((1/2)+it)} = (e^{-s/p}a)^{p((1/2)+it)} = e^{-s((1/2)+it)} a^{p((1/2)+it)}$ for all $s \in \mathbb{R}$. Hence, $a^{p((1/2)+it)} \in L^{((1/2)+it)^{-1}}(\mathcal{M})$. We similarly have that $b^{q((1/2)+it)} \in L^{((1/2)+it)^{-1}}(\mathcal{M})$. Now, consider the functions F and G defined on the open strip $S^o = \{z \in \mathbb{C} : 0 < \mathrm{Re}(z) < 1\}$, by $F(z) = \mathrm{tr}(a^{pz} b^{q(1-z)})$, and $G(z) = \mathrm{tr}(b^{q(1-z)} a^{pz})$. It then follows from Lemma 10.32 that F and G are well-defined analytic functions on the domain S^o. Now, observe that by Lemma 10.33,

$$
\begin{aligned}
F((1/2) + it) &= \mathrm{tr}(a^{p((1/2)+it)} b^{q(1-[(1/2)+it])}) \\
&= \mathrm{tr}(a^{p((1/2)+it)} b^{q((1/2)-it)}) \\
&= \mathrm{tr}(a^{p((1/2)+it)} (b^{q((1/2)+it)})^*) \\
&= \mathrm{tr}((b^{q((1/2)+it)})^* a^{p((1/2)+it)}) \\
&= \mathrm{tr}(b^{q((1/2)-it)} a^{p((1/2)+it)}) \\
&= \mathrm{tr}(b^{q(1-[(1/2)+it])} a^{p((1/2)+it)}) \\
&= G((1/2) + it)
\end{aligned}
$$

for any $t \in \mathbb{R}$. Thus, F and G are analytic functions on the domain S^o, which agree on the line $(1/2) + it$ $(t \in \mathbb{R})$. That ensures that F and G agree on all of S^o, and in particular that $\mathrm{tr}(ab) = F(1/p) = G(1/p) = \mathrm{tr}(ba)$. $\qquad \square$

We are now ready to start the development of a duality theory for the general case. We follow essentially the same strategy as in Chapter 6. Most of the proofs are minor modifications of the earlier ones. For the sake of the reader, we provide occasional details.

Lemma 10.35 *Suppose $1 \leq p < \infty$. For any $a \in L^p(\mathcal{M})$, we have* $\|a\|_p = \sup\{|\mathrm{tr}(ab)| : b \in L^q(\mathcal{M}), \|b\|_q \leq 1\}$, *where* $1 = \frac{1}{p} + \frac{1}{q}$.

Proof. Hölder's inequality combined with the fact that $|\mathrm{tr}(ab)| \leq \mathrm{tr}(|ab|)$ for each $a \in L^p$, $b \in L^q$ ensures that

$$\sup\{|\mathrm{tr}(ab)| : b \in L^q(\mathcal{M}), \|b\|_q \leq 1\} \leq \|a\|_p.$$

For the converse inequality, let $0 \neq a \in L^p$ be given. In the case where $p = 1$, the inequality $\sup\{|\mathrm{tr}(ab)| : b \in L^\infty(\mathcal{M}), \|b\|_\infty \leq 1\} \leq \|a\|_1$ is obvious. We simply choose $b \in L^\infty$ to be $b = u^*$ where u is the partial isometry in the polar decomposition $a = u|a|$ of a, to see that $\mathrm{tr}(ab) = \mathrm{tr}(ba) = \mathrm{tr}(|a|) = \|a\|_1$. In the case where $1 < p < \infty$, we set $b = \|a\|_p^{-p/q}|a|^{p-1}u^*$, where $a = u|a|$ is the polar decomposition of a. For any $s \in \mathbb{R}$, we have that $\theta_s(b) = \|a\|_p^{-p/q}|\theta_s(a)|^{p-1}\theta_s(u^*) = \|a\|_p^{-p/q}|e^{-s/p}a|^{p-1}u^* = e^{-s/q}\|a\|_p^{-p/q}|a|^{p-1}u^* = e^{-s/q}b$, ensuring that $b \in L^q(\mathcal{M})$. Observe that $\|b\|_q^q = \mathbf{m}_b(1)^q = \mathbf{m}_{b^*}(1)^q = \mathrm{tr}(|b^*|^q) = \|a\|_p^{-p}\mathrm{tr}(|a|^{qp-q}) = \|a\|_p^{-p}\mathrm{tr}(|a|^p) = 1$, and hence that $b \in L^q$ with $\|b\|_q = 1$. By construction, $\mathrm{tr}(ba) = \|a\|_p^{-p/q}\tau(|a|^p) = \|a\|_p^{p-pq} = \|a\|_p$. Hence, equality must hold. $\qquad\square$

We will now prove that part of the Clarkson–McCarthy inequalities also hold in this context. For this, we need the following lemma.

Lemma 10.36 *Let $1 \leq p < \infty$, and let $f, g \in L^p_+(\mathcal{M})$ be given. Then*

$$2^{1-p}\|f + g\|_p^p \leq \|f\|_p^p + \|g\|_p^p \leq \|f + g\|_p^p.$$

Proof. The first inequality may be proven in exactly the same way as in Lemma 6.29. To prove the second, we need to modify the proof of part (ii) of Proposition 6.8, to ensure that it goes through for tr instead of τ. Recall that in the proof of Proposition 6.8, we showed that there exist contractions $a, b \in \mathfrak{M}_+$ so that $a(f + g)^{1/2} = f^{1/2}$, $b(f + g)^{1/2} = g^{1/2}$ and $a^*a + b^*b = \mathbf{s}(f + g)$. We claim that both these contractions are in $\mathcal{M}$. To see this, note that for any $s \in \mathbb{R}$, we have that $e^{-2s/p}a(f + g)^{1/2} = e^{-2s/p}f^{1/2} = \theta_s(f^{1/2}) = \theta_s(a(f + g)^{1/2}) = \theta_s(a)\theta_s((f + g)^{1/2}) = e^{-2s/p}\theta_s(a)(f + g)^{1/2}$. This can of course only be the case if $\theta_s(a) = a$ for all $s \in \mathbb{R}$, in which case $a \in \mathcal{M}$. The proof that $b \in \mathcal{M}$ is entirely analogous. It then easily follows from Lemma 6.7 that

$$\begin{aligned}
\|f\|_p + \|g\|_p^p &= \mathbf{m}_1((a(f+g)a^*))^p + \mathbf{m}_1((b(f+g)b^*))^p \\
&= \mathbf{m}_1((a(f+g)a^*)^p) + \mathbf{m}_1((b(f+g)b^*)^p) \\
&\leq \mathbf{m}_1(a(f+g)^p a^*) + \mathbf{m}_1(b(f+g)^p b^*).
\end{aligned}$$

Now, recall that for any $p \in \widetilde{\mathfrak{M}}$, we have that $\mathbf{m}_{p^*p}(1) = \mathbf{m}_p(1)^2 = \mathbf{m}_{p^*}(1)^2 = \mathbf{m}_{pp^*}(1)$. In addition, it can easily be shown that both $(f+g)^{p/2} a^* a (f+g)^{p/2}$ and $(f+g)^{p/2} b^* b (f+g)^{p/2}$ are in $L^1(\mathcal{M})$. We may now use these two facts to conclude from the above inequality that

$$\begin{aligned}
\|f\|_p^p + \|g\|_p^p &\leq \mathbf{m}_1(a(f+g)^p a^*) + \mathbf{m}_1(b(f+g)^p b^*) \\
&= \mathbf{m}_1((f+g)^{p/2} a^* a (f+g)^{p/2}) + \mathbf{m}_1((f+g)^{p/2} b^* b (f+g)^{p/2})) \\
&= \mathrm{tr}((f+g)^{p/2} a^* a (f+g)^{p/2}) + \mathrm{tr}((f+g)^{p/2} b^* b (f+g)^{p/2})) \\
&= \mathrm{tr}((f+g)^{p/2}(a^* a + b^* b)(f+g)^{p/2}) \\
&= \mathrm{tr}((f+g)^p) \\
&= \|f+g\|_p^p. \qquad \qquad \square
\end{aligned}$$

Armed with the above lemma, we are now able to prove a generalised version of Proposition 6.30. The proof is entirely analogous to the former proof, the only difference being that we use the above lemma, in place of the earlier semifinite version.

Proposition 10.37 *Let $2 \leq p < \infty$, and let $a, b \in L^p(\mathcal{M})$ be given. Then*

$$\|a+b\|_p^p + \|a-b\|_p^p \leq 2^{p-1}(\|a\|_p^p + \|b\|_p^p).$$

We are now finally ready to prove the tr-duality of $L^p(\mathcal{M})$-spaces.

Theorem 10.38 *Let $1 < p \leq \infty$ and $1 \leq q < \infty$ be given with $1 = \frac{1}{p} + \frac{1}{q}$. The bilinear form*

$$L^p(\mathcal{M}) \times L^q(\mathcal{M}) \to \mathbb{C} : (b, a) \mapsto \mathrm{tr}(ba)$$

defines a dual action of $L^p(\mathcal{M})$ on $L^q(\mathcal{M})$ with respect to which $L^p(\mathcal{M}) = (L^q(\mathcal{M}))^*$. Specifically, for each $a \in L^p(\mathcal{M})$, the prescription $b \mapsto \mathrm{tr}(ba)$ defines a bounded linear functional ω_a on $L^q(\mathcal{M})$. Moreover, the mapping $a \mapsto \omega_a$ is a surjective isometry from $L^p(\mathcal{M})$ onto $(L^q(\mathcal{M}))^*$. In addition, $\omega_a \geq 0$ if and only if $a \geq 0$.

Proof. By Proposition 10.29, the case where $p = \infty$ is just a restatement of the well-known duality of $\mathcal{M}$ and $\mathcal{M}_*$. Hence, we may assume that $1 < p < \infty$. For this case, the proof is almost identical to the proof of Theorem 6.33. The only changes needed are to replace τ with tr, $L^p(\mathcal{M}, \tau)$ and $L^p(\mathcal{M}, \tau)$ with $L^p(\mathcal{M})$ and $L^q(\mathcal{M})$, and references to Lemma 6.28, with references to Lemma 10.35. The one aspect that gets used in the last part of the proof which may be less obvious is the fact that given $a \in L^p(\mathcal{M})$ with $a = a^*$, we will find that $|a|^{p-1}\chi_{(0,-\infty)}(a)$ is a positive element of $L^q(\mathcal{M}, \tau)$. The positivity of this element is clear. The membership of $L^q(\mathcal{M})$ can be seen by noting that for every $s \in \mathbb{R}$, $\theta_s(|a|^{p-1}) = |\theta_s(a)|^{p-1} = e^{-s(p-1)/p}|a| = e^{-s/q}|a|$ and $\theta_s(\chi_{(0,-\infty)}(a)) = \chi_{(0,-\infty)}(\theta_s(a)) = \chi_{(0,-\infty)}(e^{-s/p}a) = \chi_{(0,-\infty)}(a)$. Hence, $|a|^{p-1} \in L^q$ and $\chi_{(0,-\infty)}(a) \in L^\infty(\mathcal{M})$, which ensures that $|a|^{p-1}\chi_{(0,-\infty)}(a) \in L^q(\mathcal{M})$. $\qquad\square$

10.4 Dense subspaces of L^p-spaces

One of our main objectives in this section is to show that the non-commutative analogue of simple functions is dense in each $L^p(\mathcal{M})$ $(1 \leq p < \infty)$. Formally, these simple functions are linear combinations of terms of the form $h^{1/2p}eh^{1/2p}$, where $h = \frac{d\widetilde{\varphi}}{d\tau}$ is the density of the dual weight, and $e \in \mathcal{M}$ a projection with $\varphi(e) < \infty$. The main challenge that we need to overcome here is the fact that, in general, h is not τ-measurable! We therefore need to be extremely careful when working with these operators, and for this reason will in this section depart from the notational conventions we have been using for τ-measurable operators. Given two affiliated operators a and b, we shall denote the operator product by ab. If the product is in fact closed, we shall where necessary indicate this by writing (ab). If the product is closable, the minimal closed extension will be denoted by $[ab]$.

Our first task is to describe those elements of $\mathcal{M}$ for which products of the form $h^{1/q}a$ are τ-measurable.

Lemma 10.39 *Let $a \in \mathcal{M}$ be given, and let $h = h_\varphi = \frac{d\widetilde{\varphi}}{d\tau}$ be the density of the dual weight $\widetilde{\varphi}$ of the canonical faithful normal semifinite weight φ on $\mathcal{M}$. Recall that we may then make sense of the form product $a \widehat{\cdot} h \widehat{\cdot} a^*$ as an element of the extended positive part $\widehat{\mathfrak{M}}_+$ of $\mathfrak{M}$ (see the discussion preceding Proposition 4.28). Then, the following holds.*

(a) *The partially defined operator $h_\varphi^{1/2} a^*$ is densely defined if and only if $a\,\widehat{\cdot}\,h_\varphi\,\widehat{\cdot}\,a^*$ is a (non-negative self-adjoint) operator. In this case, $a\,\widehat{\cdot}\,h_\varphi\,\widehat{\cdot}\,a^* = \left| h_\varphi^{1/2} a^* \right|^2$.*

(b) *$a \in \mathfrak{n}_\varphi$ if and only if $a\,\widehat{\cdot}\,h_\varphi\,\widehat{\cdot}\,a^* \in L^1(\mathcal{M})$, in which case $\varphi(a^* a) = \mathrm{tr}(|h^{1/2} a^*|^2)$.*

Proof. Let H be the Hilbert space for which $\mathfrak{M} \subseteq B(H)$. For any $\xi \in H$, let ω_ξ be the positive functional defined by $a \mapsto \langle a\xi, \xi \rangle$. For such a $\xi \in H$, we will then in the notation of Proposition 9.33 have that

$$(a\,\widehat{\cdot}\,h_\varphi\,\widehat{\cdot}\,a^*)(\omega_\xi) = h_\varphi(a\omega_\xi a^*) = h_\varphi(\omega_{a^*\xi})$$

$$= \begin{cases} \left\| h_\varphi^{1/2} a^* \xi \right\|^2, & \text{if } a^*\xi \in \mathrm{dom}(h_\varphi^{1/2}), \\ \infty, & \text{otherwise.} \end{cases}$$

Therefore, $h_\varphi^{1/2} a^*$ is clearly densely defined if and only if $a\,\widehat{\cdot}\,h_\varphi\,\widehat{\cdot}\,a^*$ corresponds to an operator in the sense that the projection p_∞ in the spectral resolution $a\,\widehat{\cdot}\,h_\varphi\,\widehat{\cdot}\,a^* = \int_0^\infty \lambda\, de_\lambda + \infty.p_\infty$ is 0. Therefore, the above equality also shows that $a\,\widehat{\cdot}\,h_\varphi\,\widehat{\cdot}\,a^* = |h_\varphi^{1/2} a^*|^2$.

Now, recall that the action of the automorphisms θ_s extend to the extended positive part $\widehat{\mathfrak{M}}_+$. In their action on $\widehat{\mathfrak{M}}_+$ we have that

$$\theta_s(a\,\widehat{\cdot}\,h_\varphi\,\widehat{\cdot}\,a^*) = \theta_s(a)\,\widehat{\cdot}\,\theta_s(h_\varphi)\,\widehat{\cdot}\,\theta_s(a)^* = e^{-s}a\,\widehat{\cdot}\,h_\varphi\,\widehat{\cdot}\,a^*.$$

Therefore, it is clear that $(a\,\widehat{\cdot}\,h_\varphi\,\widehat{\cdot}\,a^*) \in L^1(\mathcal{M})$ if and only if $a\,\widehat{\cdot}\,h_\varphi\,\widehat{\cdot}\,a^* = h_{a^*\varphi a}$ is a τ-measurable operator. By Corollary 9.37, this is in turn equivalent to the assertion that $a^*\varphi a \in \mathcal{M}_*$. We therefore have that

$$(a\,\widehat{\cdot}\,h_\varphi\,\widehat{\cdot}\,a^*) \in L^1(\mathcal{M}) \Leftrightarrow \varphi(a^* a) < \infty \Leftrightarrow a \in \mathfrak{n}_\varphi$$

as required. Notice that we then also have $|h_\varphi^{1/2} a^*| \in L^2(\mathcal{M})$. It then follows from the definition of tr that

$$\mathrm{tr}(|h_\varphi^{1/2} a^*|^2) = \mathrm{tr}(a\,\widehat{\cdot}\,h_\varphi\,\widehat{\cdot}\,a^*) = \mathrm{tr}(h_{a^*\varphi a}) = (a^*\varphi a)(\mathbb{1}) = \varphi(a^* a). \qquad \square$$

Proposition 10.40 *Let $q \in [2, \infty)$. If $a \in \mathfrak{n}_\varphi$, then $ah^{1/q}$ is closable with $[ah^{1/q}]$, $h^{1/q} a^* \in L^q(\mathcal{M})$, and $[ah^{1/q}] = (h^{1/q} a^*)^*$.*

Proof. Let $q \in [2, \infty)$ be given. By Lemma 10.39, the closed operator $h^{1/2}a^*$ is densely defined, with $|h^{1/2}a^*| \in L^2(\mathcal{M})$. Hence, $|h^{1/2}a^*|$ and therefore also $h^{1/2}a^*$ is τ-measurable. In the case where $q > 2$, we may write $h^{1/2}a^*$ as $h^{1/2}a^* = h^{1/r} \cdot h^{1/q}a^*$ where $r > 2$ is given such that $\frac{1}{2} = \frac{1}{q} + \frac{1}{r}$. This clearly shows that $\mathrm{dom}(h^{1/2}a^*) \subseteq \mathrm{dom}(h^{1/q}a^*)$. The τ-measurability of $h^{1/2}a^*$ ensures that $\mathrm{dom}(h^{1/2}a^*)$ is τ-dense. But then the same must be true for $\mathrm{dom}(h^{1/q}a^*)$. Therefore, the closed operator $h^{1/q}a^*$ is also τ-measurable. We therefore need only confirm that $\theta_s(h^{1/q}a^*) = e^{-s/q}$ for each $s \in \mathbb{R}$ to prove that $h^{1/q}a^* \in L^q(\mathcal{M})$.

Given $\gamma > 0$, we have $\chi_{[0,\gamma]}(h)(h^{1/q}a^*) \subseteq [\chi_{[0,\gamma]}(h)h^{1/q}]a^*$, and hence that $[\chi_{[0,\gamma]}(h)(h^{1/q}a^*)] = [\chi_{[0,\gamma]}(h)h^{1/q}]a^*$ since all the (bracketed) operators in the product are τ-measurable. So, for any $s \in \mathbb{R}$ and any $n \in \mathbb{N}$, we have

$$
\begin{aligned}
\chi_{[0,e^s n]}(h)\theta_s(h^{1/q}a^*) &= \chi_{[0,n]}(e^{-s}h)\theta_s(h^{1/q}a^*) \\
&= \theta_s(\chi_{[0,n]}(h))\theta_s(h^{1/q}a^*) \\
&= \theta_s(\chi_{[0,n]}(h)(h^{1/q}a^*)) \\
&\subseteq \theta_s([\chi_{[0,n]}(h)h^{1/q}]a^*) \\
&= \theta_s([\chi_{[0,n]}(h)h^{1/q}])\theta_s(a^*) \\
&= e^{-s/q}[\chi_{[0,n]}(e^{-s}h)h^{1/q}]a^* \\
&= e^{-s/q}[\chi_{[0,e^s n]}(h)h^{1/q}]a^* \\
&= e^{-s/q}[\chi_{[0,e^s n]}(h)(h^{1/q}a^*)].
\end{aligned}
$$

The τ-measurability of $h^{1/q}a^*$ ensures that the τ-measurable extension of $\chi_{[0,e^s n]}(h)\theta_s(h^{1/q}a^*)$ agrees with $e^{-s/q}[\chi_{[0,e^s n]}(h)(h^{1/q}a^*)]$ (see Proposition 3.71). We may therefore combine Proposition 5.12 with the fact that $\chi_{[0,e^s n]}(h) \nearrow \mathbb{1}$ as $n \nearrow \infty$, to see that

$$
\begin{aligned}
&\mathbf{m}_{\theta_s(h^{1/q}a^*) - e^{-s/q}(h^{1/q}a^*)}(t) \\
&= \mathbf{m}_{|\theta_s(h^{1/q}a^*) - e^{-s/q}(h^{1/q}a^*)|^2}(t)^{1/2} \\
&= \sup_{n \in \mathbb{N}} \mathbf{m}_{|\chi_{[0,e^s n]}(h)(\theta_s(h^{1/q}a^*) - e^{-s/q}(h^{1/q}a^*))|^2}(t)^{1/2} \\
&= 0
\end{aligned}
$$

for all $t > 0$, and hence that

$$
\|\theta_s(h^{1/q}a^*) - e^{-s/q}(h^{1/q}a^*)\|_\infty = \lim_{t \searrow 0} \mathbf{m}_{\theta_s(h^{1/q}a^*) - e^{-s/q}(h^{1/q}a^*)}(t) = 0.
$$

To see the claims regarding $[ah^{1/q}]$, we first note that $(ah^{1/q})^* = h^{1/q}a^*$. Hence, the second adjoint of $(ah^{1/q})^*$ exists and agrees with $(h^{1/q}a^*)^*$. This shows that $ah^{1/q}$ is closable with $[ah^{1/q}] = (h^{1/q}a^*)^*$. $\qquad\square$

Remark 10.41 If we apply a suitable polar formula to the equality in part (b) of Lemma 10.39, then the added technology provided by Proposition 10.40 enables us to conclude that $\varphi(b^*a) = \mathrm{tr}((h^{1/2}b^*)[ah^{1/2}])$ for all $a, b \in \mathfrak{n}_\varphi$.

Lemma 10.39 shows that any $a \in \mathfrak{n}_\varphi$ realises a density $\|[ah^{1/2}]\|^2$ majorised by h. The converse of this statement also holds, as is shown by the following proposition.

Proposition 10.42 (Radon–Nikodym type lemma) *Let $1 \leq p < \infty$ and let $x \in L^p(\mathcal{M})$ be such that $0 \leq x \leq h^{1/p}$. Then, there is a contractive element $b \in \mathfrak{n}^{(2p)}$ such that $bh^{1/2p}$ is closable and $[bh^{1/2p}] \in L^{2p}(\mathcal{M})$, with in addition*

$$x^{1/2} = h^{1/2p}b^* = [bh^{1/2p}]. \tag{10.2}$$

In the case $p = 1$, we have $b \in \mathfrak{n}_\varphi$.

Proof. Define an operator b_0 on the underlying Hilbert space H by

$$\mathrm{dom}(b_0) = \mathrm{ran}(h^{1/2p}) \, ; \; b_0(h^{1/2p}\xi) = x^{1/2}\xi.$$

The operator b_0 is clearly densely defined. To see that it is also contractive, observe that for any $\xi \in \mathrm{dom}(h^{1/2p})$ we have $\|b_0(h^{1/2p}\xi)\|^2 = \|x^{1/2}\xi\|^2 = \langle x\xi.\xi\rangle \leq \langle h^{1/p}\xi.\xi\rangle = \|h^{1/2p}\xi\|^2$. Let b denote its continuous extension to H. Since $b_0 = x^{1/2}\left(h^{1/2p}\right)^{-1}$ is affiliated to $\mathfrak{M} = \mathcal{M} \rtimes_\varphi \mathbb{R}$, we must therefore have that $b \in \mathfrak{M}$.

Observe that $bh^{1/2p} = b_0h^{1/2p} \subset x^{1/2}$ with $bh^{1/2p}$ a densely defined operator affiliated to $\mathfrak{M}$. Since $x^{1/2}$ is τ-measurable, we therefore have $[bh^{1/2p}] = x^{1/2}$ by Proposition 3.71. Since $x^{1/2}$ is self-adjoint, equation (10.2) holds, and hence, it only remains to prove that $b \in \mathcal{M}$. Let $e_n = \chi_{[0,n]}(h^{1/2p})$. Then $\theta_t\left(h^{1/2p}e_nb^*\right) = e^{-t/2p}h^{1/2p}e_{n(t)}\theta_t(b^*)$, where $n(t) = e^{t/2p}n$. Notice that $\theta_t(e_n) = \theta_t(\chi_{[0,n]}(h^{1/2p})) = \chi_{[0,n]}(\theta_t(h^{1/2p}))$

$$= \chi_{[0,n]}(e^{-t/2p}h^{1/2p}) = \chi_{[0,n(t)]}(h^{1/2p}) = e_{n(t)}.$$ We therefore have that

$$\theta_t\left(e_n\bar{\cdot}h^{1/2p}b^*\right) = \theta_t\left(e_n\bar{\cdot}x^{1/2}\right) = e_{n(t)}\bar{\cdot}e^{-t/2p}x^{1/2}$$

$$= e^{-t/2p}e_{n(t)}\bar{\cdot}h^{1/2p}b^* = e^{-t/2p}h^{1/2p}e_{n(t)}b^* = \theta_t(h^{1/2p}e_n)b^*.$$

Using the injectivity of the operator $h^{1/2p}$ and letting $n \to \infty$, we therefore have that $\theta_t(b^*) = b^*$ for each $t \in \mathbb{R}$, whence $b \in \mathcal{M}$. The final claim is an easy consequence of Lemma 10.39. $\qquad\square$

We need one more piece of mathematical technology before we are able to prove our first density result. Since $\mathfrak{n}_\varphi$ is a left-ideal, we know from Theorem 0.60 that it admits a right approximate identity. The semifiniteness of φ ensures that the projection to which such a right approximate identity converges must be the identity $\mathbb{1}$. However, elegant this fact may seem, we shall need a right approximate identity of $\mathfrak{n}_\varphi$ with more refined properties. In particular, we shall need a net inside $\mathfrak{n}_\varphi$ that increases to $\mathbb{1}$ and which consists of analytic elements. In the development of noncommutative integration theory, similar techniques have been used by many authors [**PT73**, **Ter82**, **Vae01b**]. We, however, require some very particular facts. With Proposition 8.12 as starting point, we pause to give some hints on how the approximate identity we need is constructed.

Proposition 10.43 ([Vae01b, Lemma 3.1], [Ter82, Lemma 9])
There exists a net (f_λ) of positive entire analytic elements in $\mathfrak{n}_\varphi$ converging strongly to $\mathbb{1}$, and for which

(a) $\sigma_z^\varphi(f_\lambda) \in \mathfrak{n}_\varphi \cap \mathfrak{n}_\varphi^*$ *for each $z \in \mathbb{C}$ and each λ;*
(b) $\|\sigma_z^\varphi(f_\lambda)\| \leq e^{\delta(\mathrm{Im}(z))^2}$ *for each $z \in \mathbb{C}$ and each λ;*
(c) $(\sigma_z^\varphi(f_\lambda))$ *is σ-weakly convergent to $\mathbb{1}$ for each $z \in \mathbb{C}$.*

Outline of proof. We will give details as appropriate, but merely sketch some parts of the proof.

One starts by selecting any right approximate identity (g_λ) that increases σ-strongly to $\mathbb{1}$ as λ increases. Fixing some $\delta > 0$, one now defines the net (f_λ) by means of the prescription $f_\lambda = \sqrt{\dfrac{\delta}{\pi}} \int \sigma_t^\varphi(g_\lambda)e^{-\delta t^2}\, dt.$

The next step is to show that the function

$$F : \mathbb{C} \to \mathcal{M} : z \mapsto \sqrt{\frac{\delta}{\pi}} \int \sigma_t^\varphi(g_\lambda) e^{-\delta(t-z)^2} \, dt$$

fulfils the criteria of Definition 8.11. (Details of this part may be found in the proof of [**BR87**, Proposition 2.5.22].) Having verified this fact, the values $\sigma_z^\varphi(f_\lambda)$ are then given by $\sigma_z^\varphi(f_\lambda) = \sqrt{\frac{\delta}{\pi}} \int \sigma_t^\varphi(g_\lambda) e^{-\delta(t-z)^2} \, dt$. This then enables us to conclude that

$$\|\sigma_z^\varphi(f_\lambda)\| \le \sqrt{\frac{\delta}{\pi}} \int \|\sigma_t^\varphi(g_\lambda)\| |e^{-\delta(t-z)^2}| \, dt$$

$$\le \sqrt{\frac{\delta}{\pi}} \int |e^{-\delta(t-z)^2}| \, dt$$

$$= e^{\delta(\mathrm{Im}(z))^2}.$$

The quick way to see that $\sigma_z^\varphi(f_\lambda) \in \mathfrak{n}_\varphi \cap \mathfrak{n}_\varphi^*$ for each $z \in \mathbb{C}$ is to appeal to the technology of left Hilbert algebras. The connection of $\mathfrak{n}_\varphi \cap \mathfrak{n}_\varphi^*$ to left Hilbert algebras may be found in, for example, Theorem VII.2.6 of [**Tak03a**]. The fact that $\sigma_z^\varphi(f_\lambda) \in \mathfrak{n}_\varphi \cap \mathfrak{n}_\varphi^*$ for each $z \in \mathbb{C}$, then follows from, for example, [**SZ79**, Corollary, p272]. The verification of this fact is also embedded in the proof of [**Tak03a**, Theorem VI.2.2(i)] (see p 25 of that reference). For the sake of the reader, we provide the skeleton of a direct proof of this fact. First, let $R > 0$ be given and let $S_N = \sqrt{\frac{\delta}{\pi}} \sum_{k=1}^N \int e^{-\delta(\widetilde{t}_k-z)^2} \sigma_{t_k}^\varphi(g_\lambda) \Delta t_k$ be a Riemann sum of $\sqrt{\frac{\delta}{\pi}} \int_{-R}^R \sigma_t^\varphi(g_\lambda) e^{-\delta(t-z)^2} \, dt$. Next, recall that in its action on $\mathfrak{n}_\varphi$, φ satisfies a Cauchy–Schwarz inequality. If we combine this fact with the fact that $\varphi \circ \sigma_t^\varphi = \varphi$ for all $t \in \mathbb{R}$, then for any $s, t \in \mathbb{R}$ and any λ, we will have that

$$|\varphi(\sigma_s^\varphi(g_\lambda^*)\sigma_t^\varphi(g_\lambda))| \le \varphi(|\sigma_s^\varphi(g_\lambda)|^2)^{1/2} \cdot \varphi(|\sigma_t^\varphi(g_\lambda)|^2)^{1/2} = \varphi(|g_\lambda|^2) < \infty.$$

One may then use this fact to see that

$$\varphi(S_N^* S_N) \le \frac{\delta}{\pi} \left(\sum_{k=1}^N |e^{-\delta(\widetilde{t}_k-z)^2}| \Delta t_k \right)^2 \varphi(|g_\lambda|^2).$$

Taking the limit yields

$$\frac{\delta}{\pi} \varphi\left(\left| \int_{-R}^R \sigma_t^\varphi(g_\lambda) e^{-\delta(t-z)^2} \, dt \right|^2 \right) \le \frac{\delta}{\pi} \left(\int_{-R}^R |e^{-\delta(t-z)^2}| \, dt \right)^2 \varphi(|g_\lambda|^2).$$

Now, let $R \to \infty$ to see that

$$\varphi(|\sigma_z^\varphi(f_\lambda)|^2) \le \frac{\delta}{\pi} \left(\int |e^{-\delta(t-z)^2}| \, dt \right)^2 \varphi(|g_\lambda|^2) = (e^{\delta(\mathrm{Im}(z))^2})^2 \varphi(|g_\lambda|^2) < \infty.$$

It remains to show that (f_λ) converges strongly to 1. Let H be the Hilbert space on which $\mathcal{M}$ acts. For any $\xi \in H$, we then have that

$$
\begin{aligned}
\lim_\lambda \langle f_\lambda \xi, \xi \rangle &= \lim_\lambda \left\langle \left(\sqrt{\frac{\delta}{\pi}} \int \sigma_t^\varphi(g_\lambda) e^{-\delta t^2} \, dt \right) \xi, \xi \right\rangle \\
&= \lim_\lambda \sqrt{\frac{\delta}{\pi}} \int \langle \sigma_t^\varphi(g_\lambda) \xi, \xi \rangle e^{-\delta t^2} \, dt \\
&= \sqrt{\frac{\delta}{\pi}} \int \langle \sigma_t^\varphi(1) \xi, \xi \rangle e^{-\delta t^2} \, dt \\
&= \sqrt{\frac{\delta}{\pi}} \int e^{-\delta t^2} \, dt . \|\xi\|^2 \\
&= \|\xi\|^2.
\end{aligned}
$$

If we combine the above formula with the fact that $\|f_\lambda\| \le 1$, that then enables us to conclude that

$$\limsup_\lambda \|f_\lambda \xi - \xi\|^2 = \limsup_\lambda \left(\|f_\lambda \xi\|^2 - \langle f_\lambda \xi, \xi \rangle - \langle \xi, f_\lambda \xi \rangle + \|\xi\|^2 \right) \le 0$$

which proves the claim regarding the strong convergence of (f_λ).

We proceed to proving (c). For any $\xi, \zeta \in H$, we have that

$$
\begin{aligned}
\lim_\lambda \langle \sigma_z^\varphi(f_\lambda) \xi, \zeta \rangle &= \lim_\lambda \left\langle \left(\sqrt{\frac{\delta}{\pi}} \int \sigma_t^\varphi(g_\lambda) e^{-\delta(t-z)^2} \, dt \right) \xi, \zeta \right\rangle \\
&= \lim_\lambda \sqrt{\frac{\delta}{\pi}} \int \langle \sigma_t^\varphi(g_\lambda) \xi, \zeta \rangle e^{-\delta(t-z)^2} \, dt \\
&= \sqrt{\frac{\delta}{\pi}} \int \langle \sigma_t^\varphi(1) \xi, \zeta \rangle e^{-\delta(t-z)^2} \, dt \\
&= \sqrt{\frac{\delta}{\pi}} \int e^{-\delta(t-z)^2} \, dt \cdot \langle \xi, \zeta \rangle \\
&= \langle \xi, \zeta \rangle.
\end{aligned}
$$

Thus, $(\sigma_z^\varphi(f_\lambda))$ converges to 1 in the weak topology. But the σ-weak topology and the weak topology agree on the unit ball of $\mathcal{M}$. Hence, by part (b), $(\sigma_z^\varphi(f_\lambda))$ is σ-weakly convergent to 1 as claimed. $\qquad\square$

With the existence of such a net verified, we now prove the technical lemma which will unlock the first density result.

Lemma 10.44   *Let $a \in \mathcal{M}$ be entire analytic with respect to the modular automorphism group σ_t^φ. Then, for any $z \in \mathbb{C}$ with $\mathrm{Re}(z) \geq 0$, we have that*

$$ah^z \subseteq h^z \sigma_{iz}^\varphi(a).$$

Proof.   Let $D = \bigcap_{n \in \mathbb{N}} \mathrm{dom}(h^n)$ and $\mathbb{C}_+ = \{z \in \mathbb{C} : \mathrm{Re}(z) \geq 0\}$. Then, for each $z \in \mathbb{C}_+$, D is a core for h^z. For $\xi, \zeta \in D$, define the following two functions $\mathbb{C}_+ \to \mathbb{C}$:

$$F \colon \alpha \mapsto \langle \xi, ah^z \zeta \rangle, \qquad G \colon \alpha \mapsto \langle h^{\bar{z}} \xi, \sigma_{i\alpha}(a) \zeta \rangle.$$

These are continuous on $\mathbb{C}_+$, analytic in the interior of $\mathbb{C}_+$, and satisfy

$$F(it) = \langle \xi, ah^{it} \zeta \rangle = \langle \xi, h^{it} \sigma_{-t}(a) \zeta \rangle = \langle h^{-it} \xi, \sigma_{-t}(a) \zeta \rangle = G(it)$$

for all $t \in \mathbb{R}$. (Note that $\sigma_{-t}^\varphi(a) = h^{-it} a h^{it}$ by Proposition 9.5 and Theorem 9.28). Therefore, F and G coincide. Now, let $\xi \in \mathrm{dom}(h^z)$. If (ζ_n) is a sequence in D converging to ζ in the graph norm of h^z, then

$$\langle \xi, ah^z \zeta \rangle = \lim \langle \xi, ah^z \zeta_n \rangle = \lim \langle h^{\bar{z}} \xi, \sigma_{iz}(a) \zeta_n \rangle = \langle h^{\bar{z}} \xi, \sigma_{iz}(a) \zeta \rangle$$

for each $\xi \in D$. Since D is a core for $h^{\bar{z}}$, the result follows. $\square$

Corollary 10.45   *Let $a \in \mathfrak{n}_\varphi \cap \mathfrak{n}_\varphi^*$ be an entire analytic element for which $\sigma_z^\varphi(a) \in \mathfrak{n}_\varphi \cap \mathfrak{n}_\varphi^*$ for each $z \in \mathbb{C}$. Given $z \in \mathbb{C}$ with $0 \leq \mathrm{Re}(z) \leq 1/2$, we have that*

$$[ah^z] = h^z \sigma_{iz}(a).$$

Proof.   Set $z = s + it$ where $s \in [0, 1/2]$ and $t \in \mathbb{R}$, and let $a \in \mathfrak{n}_\infty$. Lemma 10.44 then informs us that $ah^\alpha \subseteq h^\alpha \sigma_{i\alpha}(a)$. Moreover, since $a \in \mathfrak{n}$, we know from Proposition 10.40 that $ah^\alpha = (ah^s) h^{it}$ is a product of a unitary and a τ-premeasurable operator. Similarly, since $\sigma_{i\alpha}(a) \in \mathfrak{n}^*$, $h^\alpha \sigma_{i\alpha}(a) = h^{it}(h^s \sigma_{i\alpha}(a))$ will by the same proposition also be a product of a unitary and a τ-premeasurable operator, the result therefore follows from the uniqueness of the τ-measurable extension (Proposition 3.71). $\square$

Theorem 10.46

(a) *For any $q \in [2, \infty)$, $\{[ah^{1/q}] : a \in \mathfrak{n}_\varphi\}$ is dense in $L^q(\mathcal{M})$.*

(b) *For any $p \in [1, \infty)$, $\{(h^{1/2p}a^{1/2})[a^{1/2}h^{1/2p}] : a \in \mathfrak{p}_\varphi\}$ is dense in $L^p_+(\mathcal{M})$.*

Proof. *Part (a):* Let $r > 1$ be given so that $1 = \frac{1}{q} + \frac{1}{r}$. Suppose that $z \in L^r(\mathcal{M})$. If we can show that we must have $z = 0$ whenever $\mathrm{tr}(z[ah^{1/q}]) = 0$ for each $a \in \mathfrak{n}_\varphi$, then by the duality theory developed in the previous section, $\{[ah^{1/q}] : a \in \mathfrak{n}_\varphi\}$ will be weakly dense in $L^q(\mathcal{M})$, and hence norm dense. So, suppose that we do indeed have $\mathrm{tr}(z[ah^{1/q}]) = 0$ for each $a \in \mathfrak{n}_\varphi$. For each $b \in \mathcal{M}$, the fact that $\mathfrak{n}_\varphi$ is a left-ideal ensures that $ba \in \mathfrak{n}_\varphi$ for each $a \in \mathfrak{n}_\varphi$, and hence that $\mathrm{tr}(z[bah^{1/q}]) = 0$ for each $a \in \mathfrak{n}_\varphi$. Now, notice that for any $a \in \mathfrak{n}_\varphi$, we have that $b[ah^{1/q}] \supseteq bah^{1/q}$. By the uniqueness of the τ-measurable extension (Proposition 3.71), we have that the τ-measurable operator corresponding to the product $b \cdot [ah^{1/q}]$ is just $[bah^{1/q}]$. That means that for any $a \in \mathfrak{n}_\varphi$ and any $b \in \mathcal{M}$, we have that $0 = \mathrm{tr}(z[bah^{1/q}]) = \mathrm{tr}([bah^{1/q}]z) = \mathrm{tr}(b([ah^{1/q}]z))$. The duality between L^1 and L^∞ then ensures that this can only be the case if $[ah^{1/q}]z = 0$ for each $a \in \mathfrak{n}_\varphi$, or equivalently that $z^*(h^{1/q}a^*) = 0$ for each $a \in \mathfrak{n}_\varphi$. It therefore remains to show that $z = 0$ (equivalently $z^* = 0$) if $z^*(h^{1/q}a^*) = 0$ for each $a \in \mathfrak{n}_\varphi$.

It is easy to see that $z^*(h^{1/q}a^*) = 0$ if and only if $|z^*|(h^{1/q}a^*) = 0$. Since trivially $z^* = 0$ if and only if $|z^*| = 0$, it follows that we may assume that $z^* \geq 0$. Having made this assumption, one may then further note that $z^* = 0$ if and only if $(z^*\chi_{[0,\gamma]}(z^*)) = 0$ for every $\gamma > 0$. Since in this setting the equality $z^*(h^{1/q}a^*) = 0$ ensures that $0 = \chi_{[0,\gamma]}(z^*)z^*(h^{1/q}a^*)$ and hence that $0 = (z^*\chi_{[0,\gamma]}(z^*))(h^{1/q}a^*)$, it follows that we may further assume z^* to be bounded.

Now, apply Proposition 10.43 to select a net (f_λ) of positive entire analytic elements in $\mathfrak{n}_\varphi$ which increase to $\mathbb{1}$, and for which $\sigma^\varphi_{i/q}(f_\lambda) \in \mathfrak{n}_\varphi \cap \mathfrak{n}_\varphi^*$ for each λ. Assuming z^* to be bounded, we have $0 = z^*(h^{1/q}\sigma^\varphi_{i/q}(f_\lambda))$ for each λ. But by Corollary 10.45, $[f_\lambda h^{1/q}] = (h^{1/q}\sigma^\varphi_{i/q}(f_\lambda))$. Hence, given $\xi \in \mathrm{dom}(h^{1/q}) \subseteq \mathrm{dom}([f_\lambda h^{1/q}])$, we have $\xi \in \mathrm{dom}(h^{1/q}\sigma_{i/q}(f_\lambda))$ and $h^{1/q}\sigma_{i/q}(f_\lambda)\xi = f_\lambda h^{1/q}\xi \to h^{1/q}\xi$. The fact that $0 = z^*(h^{1/q}\sigma^\varphi_{i/q}(f_\lambda))$ for each λ, therefore, ensures that $z^* = 0$ on the range of $h^{1/q}$, which must be dense by the fact that h is non-singular and positive. Therefore, as required, $z^* = 0$.

Part (b): This claim easily follows from part (a). To see this, let $f \in L^p_+(\mathcal{M})$ be given. Then $f^{1/2} \in L^{2p}(\mathcal{M})$. So, by part (a), we may select a sequence $(a_n) \subseteq \mathfrak{n}_\varphi$ such that $[a_n h^{1/2p}] \to f^{1/2}$ (or equivalently that $(h^{1/2p} a_n^*) = [a_n h^{1/2p}]^* \to f^{1/2}$). Let u_n be the partial isometry of the polar decomposition of a_n. It is not difficult to see that then $(h^{1/2p} a_n^*)[a_n h^{1/2p}] = (h^{1/2p} a_n^*) u_n [|a_n| h^{1/2p}] = (h^{1/2p}|a_n|)[|a_n| h^{1/2p}]$. To see this, note that since $\mathrm{ran}([|a_n| h^{1/2p}]) \subseteq \ker(h^{1/2p}|a_n|)^\perp = \ker(|a_n|)^\perp = \overline{\mathrm{ran}}(|a_n|)$, the operator product $u_n[|a_n| h^{1/2p}]$ is closed and hence τ-measurable. (Here, we silently used the fact that $\mathfrak{n}_\varphi$, being a left ideal, is invariant under the absolute value map.) But then $u_n[|a_n| h^{1/2p}]$ and $[a_n h^{1/2p}]$ must agree, since both are τ-measurable extensions of $a_n h^{1/2p}$ (Proposition 3.71). We may now apply Hölder's inequality to see that

$$\|f - (h^{1/2p}|a_n|)[|a_n| h^{1/2p}]\|_p$$
$$= \|f - (h^{1/2p} a_n^*)[a_n h^{1/2p}]\|_p$$
$$= \|f - f^{1/2}[a_n h^{1/2p}] + f^{1/2p}[a_n h^{1/2p}] - (h^{1/2p} a_n^*)[a_n h^{1/2p}]\|_p$$
$$\leq \|f^{1/2}(f^{1/2} - [a_n h^{1/2p}])\|_p + \|(f^{1/2} - (h^{1/2p} a_n^*))[a_n h^{1/2p}]\|_p$$
$$\leq \|f^{1/2}\|_{2p}\|f^{1/2} - [a_n h^{1/2p}]\|_{2p} + \|f^{1/2} - (h^{1/2p} a_n^*)\|_{2p}$$
$$\|[a_n h^{1/2p}]\|_{2p}$$

from which it follows that $(h^{1/2p}|a_n|)[|a_n| h^{1/2p}] \to f$ as $n \to \infty$. $\square$

In addition to answering some questions regarding dense subspaces, the preceding theorem also raises questions.

For example, given $a_1, a_2 \in \mathfrak{p}_\varphi$, how do

$$(h^{1/2p} a_1^{1/2})[a_1^{1/2} h^{1/2p}] + (h^{1/2p} a_2^{1/2})[a_2^{1/2} h^{1/2p}]$$

and

$$(h^{1/2p}(a_1 + a_2)^{1/2})[(a_1 + a_2)^{1/2} h^{1/2p}]$$

compare? We now address these issues before concluding this section with an analysis of 'simple functions'.

Definition 10.47 For $q \in [2, \infty)$, define the map

$$\mathrm{j}^{(q)} : \mathfrak{n}_\varphi \ni a \mapsto \left[ah^{1/q} \right] \in L^q(\mathcal{M}).$$

For $p \in [1, \infty)$, define the map

$$\mathrm{i}^{(p)} : \mathfrak{p}_\varphi \ni a \mapsto \mathrm{j}^{(2p)}(a^{1/2})^* \mathrm{j}^{(2p)}(a^{1/2}) \in L^p(\mathcal{M}).$$

For the task we have set for ourselves, the following lemma is crucial.

Lemma 10.48 *Let $a, b \in \mathfrak{n}_\varphi$ and $r_i, s_i \in [2, \infty)$ be given with $r_1^{-1} + s_1^{-1} = r_2^{-1} + s_2^{-1}$. Then*

$$([ah^{1/r_1}](h^{1/s_1}b^*)) = ([ah^{1/r_2}](h^{1/s_2}b^*)).$$

Proof. Assume without loss of generality that $s_1 \leq s_2$, so that $r_1 \geq r_2$. Then

$$h^{1/r_2}a^* = h^{(1/r_2 - 1/r_1)}(h^{1/r_1}a^*) \subseteq ([ah^{1/r_1}]h^{(1/r_2 - 1/r_1)})^*,$$

so that $[ah^{1/r_2}] \supseteq [ah^{1/r_1}]h^{(1/s_1 - 1/s_2)}$, and

$$[ah^{1/r_2}]h^{1/s_2}b^* \supseteq [ah^{1/r_1}]h^{(1/s_1 - 1/s_2)}h^{1/s_2}b^* = [ah^{1/r_1}]h^{1/s_1}b^*.$$

Since each of $h^{1/s_2}b^*$ and $h^{1/s_1}b^*$ is τ-measurable by Proposition 10.40, both sides of the formula represent τ-premeasurable operators. The claim, therefore, follows from the uniqueness of the τ-measurable extension (Proposition 3.71). $\qquad\square$

Proposition 10.49 *For $q \in [2, \infty)$, each of the maps $\mathrm{j}^{(q)}$ is linear and injective. Moreover, they are related through the estimate*

$$\|\mathrm{j}^{(q)}(a)\|_q \leq \|\mathrm{j}^{(2)}(a)\|_2^{2/q}\|a\|_\infty^{1-2/q} \tag{10.3}$$

for $q \in \{2^k : k \in \mathbb{N}\}$ and $a \in \mathfrak{n}_\varphi$.

Proof. For $a, b \in \mathfrak{n}$ and $\lambda \in \mathbb{C}$, both $\mathrm{j}^{(q)}(a + \lambda b)$ and the strong sum $\mathrm{j}^{(q)}(a) + \lambda \mathrm{j}^{(q)}(b)$ are closed extensions of the τ-premeasurable operator $ah^{1/q} + \lambda bh^{1/q}$, and so linearity follows from the uniqueness of the τ-measurable extension (Proposition 3.71). Injectivity of the map $\mathrm{j}^{(q)}$ follows from the injectivity of the operator $h^{1/q}$. By Lemma 10.48 and Hölder's inequality, we have that

$$\left\| j^{(2q)}(a) \right\|_{2q}^{2q} = \mathrm{tr}\left(\left(h^{1/2q} a^* \left[a h^{1/2q} \right] \right)^q \right)$$

$$= \mathrm{tr}\left(\left(\left[a h^{1/2q} \right] h^{1/2q} a^* \right)^q \right)$$

$$= \mathrm{tr}\left(\left(\left[a h^{1/q} \right] a^* \right)^q \right)$$

$$= \left\| j^{(q)}(a) a^* \right\|_q^q \le \left\| j^{(q)}(a) \right\|_q^q \| a \|_\infty^q$$

for any $q \ge 2$, $a \in \mathfrak{n}_\varphi$. The final estimate follows by iterating this inequality. $\qquad\square$

Theorem 10.50   *For $p \in [1, \infty)$, the map*

$$i^{(p)} : \mathfrak{p}_\varphi \to L^p(\mathcal{M})$$

is additive and injective—in particular, it has a unique extension to a linear map (also called $i^{(p)}$) from $\mathfrak{m}_\varphi$ into $L^p(\mathcal{M})$. The extension is injective and positivity preserving, and any $a, b \in \mathfrak{n}_\varphi$ satisfies the formula

$$j^{(2p)}(a)^* \bar{j}^{(2p)}(b) = i^{(p)}(a^* b).$$

As a positive map, it is normal in the sense that if a net $(a_\lambda) \subseteq \mathfrak{p}_\varphi$ increases to $a \in \mathfrak{p}_\varphi$, then $i^{(p)}(a_\lambda)$ increases to $i^{(p)}(a)$.

Proof.   Suppose that $a, b \in \mathfrak{p}_\varphi$ and

$$h^{1/2p} a^{1/2} \left[a^{1/2} h^{1/2p} \right] = h^{1/2p} b^{1/2} \left[b^{1/2} h^{1/2p} \right].$$

By the injectivity of $h^{1/2p}$, we have $a^{1/2}\left[a^{1/2} h^{1/2p} \right] = b^{1/2}\left[b h^{1/2p} \right]$ and so $a h^{1/2p} = b h^{1/2p}$. But $h^{1/2p}$ has a dense range, and a and b are bounded, and so $a = b$. Hence, the map is injective. To prove additivity, we must show that for $a, b \in \mathfrak{p}_\varphi$,

$$h^{1/2p}(a+b)^{1/2}[(a+b)^{1/2} h^{1/2p}]$$

$$= h^{1/2p} a^{1/2}[a^{1/2} h^{1/2p}] + h^{1/2p} b^{1/2}[b^{1/2} h^{1/2p}] \qquad (*)$$

where the sum on the right is in the strong sense. Since $x_1 := (a+b)^{1/2}\left[(a+b)^{1/2} h^{1/2p} \right]$ and $x_2 := a^{1/2}\left[a^{1/2} h^{1/2p} \right] + b^{1/2}\left[b^{1/2} h^{1/2p} \right]$ are closable τ-premeasurable extensions of the densely defined operator $(a+b) h^{1/2p}$, their closures coincide by the uniqueness of the

τ-measurable extension (Proposition 3.71). Since then $h^{1/2p}[x_1] \supseteq h^{1/2p}(a+b)^{1/2}\left[(a+b)^{1/2}h^{1/2p}\right]$ and

$$h^{1/2p}[x_2] \;\supseteq\; h^{1/2p}\left[a^{1/2}\left[a^{1/2}h^{1/2p}\right] + b^{1/2}\left[b^{1/2}h^{1/2p}\right]\right]$$

$$\supseteq\; h^{1/2p}a^{1/2}\left[a^{1/2}h^{1/2p}\right] + h^{1/2p}b^{1/2}\left[b^{1/2}h^{1/2p}\right],$$

$(*)$ holds on the intersection of the domains of $i^{(p)}(a+b)$, $i^{(p)}(a)$ and $i^{(p)}(b)$. But this intersection is τ-dense, and hence, uniqueness of the τ-measurable extension (Proposition 3.71) demands that $(*)$ holds unreservedly. Therefore, $i^{(p)}$ is additive. Since $\mathfrak{m}_\varphi$ is linearly spanned by its non-negative elements, the prescription

$$(a_1-a_2)+i(a_3-a_4) \mapsto i^{(p)}(a_1)-i^{(p)}(a_2)+i\left(i^{(p)}(a_3) - i^{(p)}(a_4)\right), \quad a_k \in \mathfrak{p}_\varphi$$

then gives a well-defined extension of $i^{(p)}$ to a linear map $\mathfrak{m}_\varphi \to L^p(\mathcal{M})$. Clearly, this is the promised unique linear map extension of $i^{(p)}$ to all of $\mathfrak{m}_\varphi$, which is moreover injective and positivity preserving.

We now prove the stated formula for realising the extension. First consider the case where $b = a$. This case follows from the fact that $y_1 := ((h^{1/2p}a^*)[ah^{1/2p}])$ and $y_2 := ((h^{1/2p}|a|)[|a|h^{1/2p}]$ coincide. (This was verified at the beginning of the proof of part (b) of Theorem 10.46.) Since each of the maps $(a,b) \mapsto i^{(p)}(a^*b)$ and $(a,b) \mapsto j^{(2p)}(a)^*\overline{j}^{(2p)}(b)$ is sesquilinear (by the linearity of $i^{(p)}$ and $j^{(2p)}$), the full result follows from the polarisation identity.

Finally, let $a, a_\lambda \in \mathfrak{p}_\varphi$ $(\lambda \in \Lambda)$ be given with $a_\lambda \nearrow a$. The fact that the map $i^{(p)}$ is order preserving ensures that $0 \leq i^{(p)}(a_\lambda) \leq i^{(p)}(a)$ for each λ, and hence that $\sup_\lambda i^{(p)}(a_\lambda) = g$ exists as an element of $\widetilde{\mathfrak{M}}_+$, for which $g \leq i^{(p)}(a)$ (see Proposition 3.86). For any $\xi \in H$, we have

$$\langle a_\lambda^{1/2}h^{1/2p}\xi, a_\lambda^{1/2}h^{1/2p}\xi\rangle \nearrow \langle a^{1/2}h^{1/2p}\xi, a^{1/2}h^{1/2p}\xi\rangle.$$

We shall use this fact to prove that $g = i^{(p)}(a)$, but we need some technical information before we are able to do so.

Let f be a closable operator on the Hilbert space H, with minimal closure $\overline{f}$. Let $\mathscr{G}(\overline{f})$ be the graph of $\overline{f}$, and P the bounded mapping $P : \mathscr{G}(\overline{f}) \to H : (u,v) \mapsto u$. By [**KR83**, Remark 2.7.7] P^* has a dense range, which is contained in $\mathrm{dom}((\overline{f})^*\overline{f})$ and which is a core for $\overline{f}$. By the closability of f, $\mathscr{G}(f)$ is dense in $\mathscr{G}(\overline{f})$, and so by continuity, $P^*(\mathscr{G}(f))$ is dense in $P^*(\mathscr{G}(\overline{f}))$, and therefore also in H. The subspace

$P^*(\mathscr{G}(f))$ is clearly contained in $P^*(\mathscr{G}(\overline{f}))$ and hence in $\mathrm{dom}((\overline{f})^*\overline{f})$, and is by definition a core for $\overline{f}$.

By Proposition 10.40, $h^{1/2p}a^{1/2}$ is τ-measurable and $a^{1/2}h^{1/2p}$ densely defined and closable, with the τ-measurable closure given by $[a^{1/2}h^{1/2p}] = (h^{1/2p}a^{1/2})^*$. As can be seen from the preceding discussion, $\mathrm{dom}(a^{1/2}h^{1/2p}) = \mathrm{dom}(h^{1/2p})$ contains a core $\mathcal{C}$ of $[a^{1/2}h^{1/2p}]$, which is also contained in

$$\mathrm{dom}((h^{1/2p}a^{1/2})[a^{1/2}h^{1/2p}]) = \mathrm{dom}(\mathfrak{i}^{(p)}(a)).$$

Therefore, the previously centred equation may be reformulated as the claim that

$$\langle \mathfrak{i}^{(p)}(a_\lambda)\xi, \xi \rangle \nearrow \langle \mathfrak{i}^{(p)}(a)\xi, \xi \rangle \text{ for all } \xi \in \mathcal{C}.$$

This means that $\langle \mathfrak{i}^{(p)}(a)\xi, \xi \rangle = \langle g\xi, \xi \rangle$ for all $\xi \in \mathcal{C}$. But by the polarisation identity, this in turn ensures that $\langle \mathfrak{i}^{(p)}(a)\xi, \zeta \rangle = \langle g\xi, \zeta \rangle$ for all $\xi, \zeta \in \mathcal{C}$, and hence that $\mathfrak{i}^{(p)}(a)\xi = g\xi$ for all $\xi \in \mathcal{C}$. We therefore have two τ-measurable operators agreeing on a dense subspace of H. By Proposition 3.71, this suffices to ensure that $\mathfrak{i}^{(p)}(a) = g$. $\square$

Remark 10.51 With the technicalities regarding the map $\mathfrak{i}^{(p)}$ now taken care of, we will in the rest of these notes, for a given $f \in \mathfrak{m}_\varphi$, simply write $h^{1/2p}fh^{1/2p}$ for $\mathfrak{i}^{(p)}(f)$ where convenient.

Proposition 10.52 *Given $p \in [1, \infty)$, let*

$$\mathfrak{s}_\varphi = \mathrm{span}\{\mathfrak{i}^{(p)}(e) \colon e \in \mathbb{P}(\mathcal{M}), \varphi(e) < \infty\}.$$

The positive cone $\mathfrak{s}_\varphi^+$ is σ-weakly dense in $\mathcal{M}_+$ and $\mathfrak{i}^{(p)}(\mathfrak{s}_\varphi^+)$ norm dense in $L_+^p(\mathcal{M})$.

Proof. We shall prove that the closure of $\mathfrak{i}^{(p)}(\mathfrak{s}_\varphi^+)$ contains $\mathfrak{i}^{(p)}(\mathfrak{p}_\varphi)$. The claim will then follow from Theorem 10.46. So, let $a \in \mathfrak{p}_\varphi$ be given. With e_λ denoting the spectral resolution of a, we of course have $a = \int_0^{\|a\|} \lambda\, de_\lambda$. We may now select a sequence (g_n) of Riemann sums increasing to a. These Riemann sums are, of course, of the form $a_n = \sum_{k_n=1}^{N_n} \gamma_{k_n} e(k_n)$ where the $e(k_n)$s are mutually orthogonal spectral projections of a, and the γ_{k_n}s non-negative reals. Next, observe that for each $\epsilon > 0$, we have $\varphi(\chi_{(\epsilon, \infty)}(a)) \le \epsilon^{-1}\varphi(a) < \infty$. Using this fact, we now replace each g_n with $a_n = g_n\chi_{(n^{-1}, \infty)}(a)$. The result is a positive 'simple function' for which all the projections now have finite weight. Since $g_n \nearrow a$ and $\chi_{(n^{-1},\infty)}(a) \nearrow \mathbb{1}$, we may use the Borel functional calculus to conclude that $a_n \nearrow a$. For the case $p = \infty$, this shows that $\mathfrak{s}_\varphi^+$ is σ-strong* dense in $\mathfrak{p}_\varphi$ and hence σ-weakly dense in $\mathcal{M}_+$. For the case $1 \le p < \infty$, we will show that $\mathfrak{i}^{(p)}(a_n)$ converges to $\mathfrak{i}^{(p)}(a)$. The case $p = 1$ is somewhat simpler to describe. We therefore first deal with this case, before showing how that argument may be modified to prove the general case.

Consider the case $p = 1$, and let $b \in \mathcal{M}_+$ be given. It then follows from Proposition 3.88 that $b^{1/2}\mathfrak{i}^{(p)}(a_n)b^{1/2} \nearrow b^{1/2}\mathfrak{i}^{(p)}(a)b^{1/2}$. We may therefore use part (iii) of Proposition 5.12 to conclude that

$$
\begin{aligned}
\mathrm{tr}(b\mathfrak{i}^{(p)}(a_n)) &= \mathrm{tr}(b^{1/2}\mathfrak{i}^{(p)}(a_n)b^{1/2}) \\
&= \mathbf{m}_{b^{1/2}\mathfrak{i}^{(p)}(a_n)b^{1/2}}(1) \\
&\nearrow \mathbf{m}_{b^{1/2}\mathfrak{i}^{(p)}(a)b^{1/2}}(1) \\
&= \mathrm{tr}(b^{1/2}\mathfrak{i}^{(p)}(a)b^{1/2}) \\
&= \mathrm{tr}(b\mathfrak{i}^{(p)}(a)).
\end{aligned}
$$

Since each $b \in \mathcal{M}$ is a linear combination of four positive elements, it follows that $\mathrm{tr}(b\mathfrak{i}^{(p)}(a_n)) \to \mathrm{tr}(b\mathfrak{i}^{(p)}(a))$ for each $b \in \mathcal{M}$. Therefore, $\mathfrak{i}^{(p)}(a)$ is in the weak closure of $\mathfrak{i}^{(p)}(\mathfrak{s}_\varphi^+)$, which by convexity must agree with the norm closure.

Now, suppose that $1 < p < \infty$, and let $1 < q < \infty$ be given with $1 = p^{-1} + q^{-1}$. Given $b \in \mathfrak{p}_\varphi$, it will in this case follow that

$$
\mathfrak{j}^{(2q)}(b^{1/2})\mathfrak{i}^{(p)}(a_n)\mathfrak{j}^{(2q)}(b^{1/2}) \nearrow \mathfrak{j}^{(2q)}(b^{1/2})\mathfrak{i}^{(p)}(a)\mathfrak{j}^{(2q)}(b^{1/2}).
$$

As before we may then use part (iii) of Proposition 5.12 to conclude that

$$
\begin{aligned}
\mathrm{tr}(\mathfrak{i}^{(q)}(b)\mathfrak{i}^{(p)}(a_n)) &= \mathrm{tr}(\mathfrak{j}^{(2q)}(b^{1/2})\mathfrak{i}^{(p)}(a_n)\mathfrak{j}^{(2q)}(b^{1/2})^*) \\
&= \mathbf{m}_{\mathfrak{j}^{(2q)}(b^{1/2})\mathfrak{i}^{(p)}(a_n)\mathfrak{j}^{(2q)}(b^{1/2})^*}(1) \\
&\nearrow \mathbf{m}_{\mathfrak{j}^{(2q)}(b^{1/2})\mathfrak{i}^{(p)}(a)\mathfrak{j}^{(2q)}(b^{1/2})^*}(1) \\
&= \mathrm{tr}(\mathfrak{j}^{(2q)}(b^{1/2})\mathfrak{i}^{(p)}(a)\mathfrak{j}^{(2q)}(b^{1/2})^*) \\
&= \mathrm{tr}(\mathfrak{i}^{(q)}(b)\mathfrak{i}^{(p)}(a)).
\end{aligned}
$$

Again as before we use the fact $b \in \mathfrak{m}_\varphi$ is a linear combination of four positive elements of $\mathfrak{m}_\varphi$, to see that $\mathrm{tr}(\mathfrak{i}^{(q)}(b)\mathfrak{i}^{(p)}(a_n)) \to$

$\mathrm{tr}(\mathfrak{i}^{(q)}(b)\mathfrak{i}^{(p)}(a))$ for each $b \in \mathfrak{m}_\varphi$. For a general $f \in L^q(\mathcal{M})$, let $\epsilon > 0$ be given, and select $b \in \mathfrak{m}_\varphi$ so that $\|f - \mathfrak{i}^{(q)}(b)\|_q \leq \epsilon$. Notice that the fact that $0 \leq \mathfrak{i}^{(p)}(a_n) \leq \mathfrak{i}^{(p)}(a)$ ensures that $\|\mathfrak{i}^{(p)}(a_n)\|_p = \mathbf{m}_{\mathfrak{i}^{(p)}(a_n)}(1) \leq \mathbf{m}_{\mathfrak{i}^{(p)}(a)}(1) = \|\mathfrak{i}^{(p)}(a)\|_p$. It therefore follows that

$$\begin{aligned}
&|\mathrm{tr}(f(\mathfrak{i}^{(p)}(a) - \mathfrak{i}^{(p)}(a_n)))| \\
&\quad \leq |\mathrm{tr}((f - \mathfrak{i}^{(q)}(b))(\mathfrak{i}^{(p)}(a) - \mathfrak{i}^{(p)}(a_n)))| \\
&\qquad + |\mathrm{tr}(\mathfrak{i}^{(q)}(b)(\mathfrak{i}^{(p)}(a) - \mathfrak{i}^{(p)}(a_n)))| \\
&\quad \leq 2\epsilon\|\mathfrak{i}^{(p)}(a)\|_p + |\mathrm{tr}(\mathfrak{i}^{(q)}(b)(\mathfrak{i}^{(p)}(a) - \mathfrak{i}^{(p)}(a_n)))|.
\end{aligned}$$

By the first part of the proof, there must then exist $N \in \mathbb{N}$ so that $|\mathrm{tr}(f(\mathfrak{i}^{(p)}(a) - \mathfrak{i}^{(p)}(a_n)))| \leq \epsilon(2\|\mathfrak{i}^{(p)}(a)\|_p + 1)$ for all $n \geq N$. Thus, by definition

$$\lim_{n \to \infty} |\mathrm{tr}(f(\mathfrak{i}^{(p)}(a) - \mathfrak{i}^{(p)}(a_n)))| = 0.$$

So, as in the former case, we find that $\mathfrak{i}^{(p)}(a)$ is in the weak closure of $\mathfrak{i}^{(p)}(\mathfrak{s}_\varphi^+)$ and hence by convexity in the norm closure. $\qquad\square$

In the case where φ is a state rather than a weight, one can obtain a much stronger result with significantly less difficulty. We close this section with a consideration of this case.

Proposition 10.53 *Let φ be a faithful normal state on $\mathcal{M}$. For any $p \in (0, \infty)$, $h^{1/p} \in L^p(\mathcal{M})$ where $h = \frac{d\widetilde{\varphi}}{d\tau}$. In addition, for any $p \in (0, \infty)$ and $c \in [0, 1]$, $\mathrm{span}(h^{c/p}\mathbb{P}(\mathcal{M})h^{(1-c)/p})$ is dense in $L^p(\mathcal{M})$.*

We note that it is precisely the fact that in this case h is τ-measurable, which renders this case significantly simpler to deal with.

Proof. Notice that the first claim is a direct consequence of Theorem 9.38. We shall prove the second claim in several stages, the first of which is the claim that for any $p \in [1, \infty)$, $h^{1/p}\mathcal{M}$ is norm-dense in $L^p(\mathcal{M}, \tau)$. It is a simple matter to check that $h^{1/p}\mathcal{M} \subseteq L^p(\mathcal{M})$. Now, let $q \geq 1$ be given with $1 = p^{-1} + q^{-1}$, and let $g \in L^q(\mathcal{M})$ be given with $\mathrm{tr}(gh^{1/p}a) = 0$ for all $a \in \mathcal{M}$. It is easy to check that $gh^{1/p} \in L^1(\mathcal{M})$, and hence, the duality theory developed earlier shows that we must then have that $gh^{1/p} = 0$. But h is a positive non-singular element of $\widetilde{\mathfrak{M}}$, which ensures that $\mathrm{ran}(h^{1/p})$ is a dense subspace of the underlying Hilbert space. The two τ-measurable operators g and 0 agree on this dense subspace, and so by the uniqueness of the τ-measurable extension (Proposition 3.71), we must have that $g = 0$. By the theory of L^p-duality developed earlier, this means that $h^{1/p}\mathcal{M}$ is weakly dense in $L^p(\mathcal{M})$. But since $h^{1/p}\mathcal{M}$ is convex, the norm closure must agree with the weak closure, proving the claim in this case. Now, suppose that $p \in [\frac{1}{2}, 1)$. It is a simple matter to use Hölder's inequality to see that for any $a \in \mathcal{M}$, the embedding $L^{2p}(\mathcal{M}) \to L^p(\mathcal{M}) : f \mapsto h^{1/2p} f a$ is well-defined and continuous. It therefore follows from the former case that $h^{1/p}\mathcal{M} = h^{1/2p}(h^{1/2p}\mathcal{M})\mathcal{M}$ is dense in $h^{1/2p}L^{2p}(\mathcal{M})\mathcal{M}$, and hence that $\overline{h^{1/p}\mathcal{M}} \supseteq h^{1/2p}L^{2p}(\mathcal{M})\mathcal{M} \supseteq h^{1/2p}(\mathcal{M}h^{1/2p})\mathcal{M}$. We may now once again use the Hölder inequality to see that the bilinear map $L^{2p}(\mathcal{M}) \times L^{2p}(\mathcal{M}) \to L^p(\mathcal{M}) : (f, g) \mapsto fg$ is well-defined and continuous, from which it then follows that $\overline{h^{1/p}\mathcal{M}} \supseteq \overline{(h^{1/2p}\mathcal{M})(h^{1/2p}\mathcal{M})} \supseteq L^{2p}(\mathcal{M}).L^{2p}(\mathcal{M}) = L^p(\mathcal{M})$. Given that $h^{1/p}\mathcal{M} \subseteq L^p(\mathcal{M})$, this clearly suffices to prove the claim.

To prove the claim for $p \in (0, 1)$, one simply iterates the above procedure.

Since the process of taking adjoints is continuous with respect to the L^p-topology, it now trivially follows that for each $p \in (0, \infty)$, $\mathcal{M}h^{1/p} = (h^{1/p}\mathcal{M})^*$ is dense in $L^p(\mathcal{M})$.

Next, let $c \in (0, 1)$ be given. As before, we may use Hölder's inequality to see that the bilinear map

$$L^{p/c}(\mathcal{M}) \times L^{p/(1-c)}(\mathcal{M}) \to L^p(\mathcal{M}) : (f, g) \mapsto fg$$

is well-defined and continuous. Using this fact, the respective density of $h^{c/p}\mathcal{M}$ and $\mathcal{M}h^{(1-c)/p}$ in $L^{p/c}(\mathcal{M})$ and $L^{p/(1-c)}(\mathcal{M})$ ensures that $h^{c/p}\mathcal{M}h^{(1-c)/p} = (h^{c/p}\mathcal{M})(\mathcal{M}h^{(1-c)/p})$ is dense in $L^{p/c}(\mathcal{M}).L^{p/(1-c)}(\mathcal{M}) = L^p(\mathcal{M})$.

We have therefore proven that for each $p \in (0, \infty)$ and each $c \in [0, 1]$, $h^{c/p}\mathcal{M}h^{(1-c)/p}$ is dense in $L^p(\mathcal{M})$. To conclude the proof, we need to show that $h^{c/p}\mathcal{M}h^{(1-c)/p}$ is contained in the closure of $\mathrm{span}(h^{c/p}\mathbb{P}(\mathcal{M})h^{(1-c)/p})$. For this, it suffices to consider some $a \in \mathcal{M}_+$ and show that $h^{c/p}ah^{(1-c)/p}$ is in this closure. Given $a \in \mathcal{M}_+$, we may argue as in the proof of the previous proposition to select a sequence $(a_n) \subseteq \mathrm{span}(\mathbb{P}(\mathcal{M}))$ of (noncommutative) Riemann sums converging to a in the L^∞-norm. Hölder's inequality once again informs us that the embedding $\mathcal{M} \to L^p(\mathcal{M}) : b \mapsto h^{c/p}bh^{(1-c)/p}$ is well-defined and continuous. Hence, $(h^{c/p}a_n h^{(1-c)/p})$ must converge to $h^{c/p}ah^{(1-c)/p}$ in the L^p-topology. $\qquad\square$

10.5 Existence of conditional expectations and compressions

The focus of this section is to establish conditions for the existence of faithful normal conditional expectations on general von Neumann algebras, and also for the existence of well-behaved compressions of the Orlicz spaces associated with such algebras.

As far as conditional expectations are concerned, we present an alternative proof of Takesaki's existence result for conditional expectations using only Haagerup L^p-space technology. As in Takesaki's original proof, the 'if' part uses only basic modular theory. The big difference is with the 'only if' part of the proof. Takesaki by contrast here uses advanced left Hilbert algebra theory. Readers will notice that this part of the proof follows the same line of attack as the semifinite proof offered in Corollary 6.25

Theorem 10.54 ([Tak03a, IX.4.2]) *Let φ be a faithful semifinite normal weight on a von Neumann algebra $\mathcal{M}$, and $\mathcal{N}$ a von Neumann subalgebra for which the restriction $\varphi{\upharpoonright}\mathcal{N}$ of φ to $\mathcal{N}$ is still semifinite. We then have $\sigma_t^\varphi(\mathcal{N}) = \mathcal{N}$ for each $t \in \mathbb{R}$ if and only if there exists a unique normal (σ-weakly continuous) conditional expectation E of $\mathcal{M}$ onto $\mathcal{N}$ with respect to φ (meaning that $\varphi \circ \mathrm{E} = \varphi$.)*

Proof. We first prove the 'if' part. Hence, suppose that we are given a normal conditional expectation E from $\mathcal{M}$ onto $\mathcal{N}$ for which $\varphi \circ \mathrm{E} = \varphi$. By Tomiyama's result (Remark 8.34), E is completely positive. For any $x \in \mathcal{M}$, we will therefore have that $\varphi(\mathrm{E}(x)^*\mathrm{E}(x)) \leq \varphi(\mathrm{E}(x^*x)) = \varphi(x^*x)$. This inequality shows that

$$\mathrm{E}(\mathfrak{n}_\varphi(\mathcal{M})) = \mathfrak{n}_\varphi(\mathcal{N}).$$

We know that $\mathfrak{n}_\varphi(\mathcal{M})$ is σ-weakly dense in $\mathcal{M}$. The σ-weak continuity of E, therefore, ensures that $\mathrm{E}(\mathfrak{n}_\varphi(\mathcal{M})) = \mathfrak{n}_\varphi(\mathcal{N})$ is then σ-weakly dense in $\mathrm{E}(\mathcal{M}) = \mathcal{N}$, i.e. that $\varphi{\upharpoonright}\mathcal{N}$ is a semifinite weight on $\mathcal{N}$

Now, let P_{E} be the orthogonal projection from $\overline{\eta_\varphi(\mathfrak{n}_\varphi(\mathcal{M}))} = H_\varphi$ to $\overline{\eta_\varphi(\mathfrak{n}_\varphi(\mathcal{N}))}$. For any $x \in \mathfrak{n}_\varphi(\mathcal{M})$ and $y \in \mathfrak{n}_\varphi(\mathcal{N})$, we have that

$$\langle \eta_\varphi(x), \eta_\varphi(y)\rangle = \nu(y^*x) = \nu(\mathrm{E}(y^*x)) = \nu(y^*\mathrm{E}(x)) = \langle \eta_\varphi(\mathrm{E}(x)), \eta_\varphi(y)\rangle.$$

This equality shows that

$$P_{\mathrm{E}}(\eta_\varphi(x)) = \eta_\varphi(\mathrm{E}(x)) \text{ for every } x \in \mathfrak{n}_\varphi(\mathcal{N}).$$

Now, recall that the domain of the anti-linear operator S defined as the closed extension of $\eta_\varphi(x) \mapsto \eta_\varphi(x^*)$ ($x \in (\mathfrak{n}_\varphi \cap \mathfrak{n}_\varphi^*)(\mathcal{M})$), corresponds to $\{f \in H_\varphi : \exists(a_n) \subset (\mathfrak{n}_\varphi \cap \mathfrak{n}_\varphi^*)(\mathcal{M}), \|\eta_\varphi(a_n) - f\|_\varphi \to 0, (\eta(a_n^*)) \text{ Cauchy}\}$.

From the above displayed equations, it is now easily verifiable that at the Hilbert space level involution satisfies $P_{\mathrm{E}}(\eta_\varphi(x))^\sharp = P_{\mathrm{E}}(\eta_\varphi(x)^\sharp)$ for any $x \in (\mathfrak{n}_\varphi \cap \mathfrak{n}_\varphi^*)(\mathcal{M})$, and that $P_{\mathrm{E}}(\mathrm{Dom}(S)) \subset \mathrm{Dom}(S)$. When combined with the norm boundedness of P_{E}, these facts then ensure that $P_{\mathrm{E}}S \subseteq SP_{\mathrm{E}}$ and hence that $(\mathbb{1} - 2P_{\mathrm{E}})S \subseteq S(\mathbb{1} - 2P_{\mathrm{E}})$. But then also $(\mathbb{1} - 2P_{\mathrm{E}})S(\mathbb{1} - 2P_{\mathrm{E}}) \subseteq S(\mathbb{1} - 2P_{\mathrm{E}})^2 = S$, from which we may conclude that

$$S = (\mathbb{1} - 2P_{\mathrm{E}})^2 S(\mathbb{1} - 2P_{\mathrm{E}})^2 \subseteq (\mathbb{1} - 2P_{\mathrm{E}})S(\mathbb{1} - 2P_{\mathrm{E}}) \subseteq S$$

and hence that $S = (\mathbb{1} - 2P_{\mathrm{E}})S(\mathbb{1} - 2P_{\mathrm{E}})$. By duality, we then have that $F = S^* = (\mathbb{1} - 2P_{\mathrm{E}})S^*(\mathbb{1} - 2P_{\mathrm{E}}) = (\mathbb{1} - 2P_{\mathrm{E}})F(\mathbb{1} - 2P_{\mathrm{E}})$, and therefore that

$$\Delta = FS = (\mathbb{1} - 2P_{\mathrm{E}})F(\mathbb{1} - 2P_{\mathrm{E}})^2 S(\mathbb{1} - 2P_{\mathrm{E}})$$
$$= (\mathbb{1} - 2P_{\mathrm{E}})FS(\mathbb{1} - 2P_{\mathrm{E}}) = (\mathbb{1} - 2P_{\mathrm{E}})\Delta(\mathbb{1} - 2P_{\mathrm{E}}).$$

This proves that Δ and P_{E} (and hence also Δ^{it} and P_{E}) commute strongly. The formulae mentioned immediately prior to the proof now enable us to conclude that

$$
\begin{aligned}
\eta_\varphi(\sigma_\varphi^t((\mathfrak{n}_\varphi \cap \mathfrak{n}_\varphi^*)(\mathcal{N}))) &= \Delta^{it}\eta_\varphi((\mathfrak{n}_\varphi \cap \mathfrak{n}_\varphi^*)(\mathcal{N})) \\
&= \Delta^{it}P_{\mathrm{E}}\eta_\varphi((\mathfrak{n}_\varphi \cap \mathfrak{n}_\varphi^*)(\mathcal{M})) \\
&= P_{\mathrm{E}}\Delta^{it}\eta_\varphi((\mathfrak{n}_\varphi \cap \mathfrak{n}_\varphi^*)(\mathcal{M})) \\
&= P_{\mathrm{E}}\eta_\varphi(\sigma_t^\varphi(\mathfrak{n}_\varphi \cap \mathfrak{n}_\varphi^*)(\mathcal{M})) \\
&= P_{\mathrm{E}}\eta_\varphi((\mathfrak{n}_\varphi \cap \mathfrak{n}_\varphi^*)(\mathcal{M})) \\
&= \eta_\varphi((\mathfrak{n}_\varphi \cap \mathfrak{n}_\varphi^*)(\mathcal{N})).
\end{aligned}
$$

So σ_t^φ preserves $(\mathfrak{n}_\varphi \cap \mathfrak{n}_\varphi^*)(\mathcal{N})$. The semifiniteness of φ on $\mathcal{N}$ ensures that this subspace is σ-weakly dense in $\mathcal{N}$, from which we may then conclude that σ_t^φ also preserves $\mathcal{N}$.

We proceed to prove the 'only if' part. With that in mind, suppose that $\sigma_t^\varphi(\mathcal{N}) = \mathcal{N}$ for all $t \in \mathbb{R}$. When $\mathcal{N}$ is equipped with the restriction of the weight φ to $\mathcal{N}$, the associated modular group will therefore be the restriction of the modular group of $\mathcal{M}$. In comparing the crossed products $\mathcal{M} \rtimes_\varphi \mathbb{R}$ and $\mathcal{N} \rtimes_{\varphi \restriction \mathcal{N}} \mathbb{R}$, we first note that since both von Neumann algebras act on the same Hilbert space with the modular group of $\mathcal{N}$ being nothing more than the restriction of the modular group of $\mathcal{M}$, then when their modular groups are used to represent them as algebras acting on $L^2(\mathbb{R}, H)$,

- $\pi_\varphi(\mathcal{N})$ will appear as a subalgebra of $\pi_\varphi(\mathcal{M})$ (see Proposition 9.3);
- they will share the same left-shift operators (see Definition 9.4).

Therefore, $\mathfrak{N} = \mathcal{N} \rtimes_{\varphi \restriction \mathcal{N}} \mathbb{R}$ will be a subspace of $\mathfrak{M} = \mathcal{M} \rtimes_\varphi \mathbb{R}$ for which the modular automorphism group of the dual weight is a restriction of the modular automorphism group of $\mathfrak{M}$ (see Proposition 9.27). In addition, $\frac{d\widetilde{\varphi}}{d\tau_\mathfrak{M}} = \frac{d\widetilde{\varphi}_\mathcal{N}}{d\tau_\mathfrak{N}}$ by part (2) of Theorem 9.28. Therefore, by the same theorem, the canonical trace on $\mathfrak{N}$ will then be a restriction of the trace on $\mathfrak{M}$, which in turn ensures that $\widetilde{\mathfrak{N}}$ is a subspace $\widetilde{\mathfrak{M}}$. It is also clear from the discussion at the beginning of Section 9.5 that the dual action $(\theta_t^\mathcal{N})$ corresponding to $\mathcal{N}$ is just the restriction of the dual action of $\mathcal{M}$. By Definitions 9.20 and 9.23, this then further ensures that the dual weight on $\mathfrak{N}$ is just the restriction of the dual weight on $\mathfrak{M}$. Given any $a \in \mathfrak{p}(\mathcal{N})_\varphi$ we will then by Remark 10.41 have that

$$\mathrm{tr}_\mathcal{N}(\mathfrak{i}^{(1)}(a)) = \varphi(a) = \mathrm{tr}_\mathcal{M}(\mathfrak{i}^{(1)}(a)).$$

By Theorem 10.46, equality will then hold on all of $L^1_+(\mathcal{N})$, and hence by linearity on all of $L^1(\mathcal{N})$. So, $\mathrm{tr}_\mathcal{N}$ is clearly nothing more than a restriction of $\mathrm{tr}_\mathcal{M}$. Thus, one may safely just write tr without danger of confusion.

However, more is true. From the above, we clearly have $\mathfrak{p}(\mathcal{N})_\varphi \subset \mathfrak{p}(\mathcal{M})_\varphi$ and hence (on taking closures of $\mathfrak{i}^{(1)}(\mathfrak{m}(\mathcal{N}))$ and $\mathfrak{i}^{(1)}(\mathfrak{m}(\mathcal{M}))$ in the topology of convergence in measure on $\widetilde{\mathcal{M} \rtimes_\varphi \mathbb{R}}$) that $L^1(\mathcal{N})$ is a subspace of $L^1(\mathcal{M})$. With ι denoting the inclusion map from $L^1(\mathcal{N})$ into $L^1(\mathcal{M})$, one defines the normal conditional expectation $\mathbf{E}$ from $\mathcal{M}$ onto $\mathcal{N}$ to be the dual of this inclusion. The proof of this part is then concluded by using L^p-duality to show that the map $\mathbb{E}$ does everything it should. $\qquad\square$

We pass to the second objective of this section, which is to provide criteria for well-behaved compressions of Orlicz spaces.

Theorem 10.55 *Let $\mathcal{M}$ be a von Neumann algebra equipped with an f.n.s. weight φ, and let $e \in \mathcal{M}$ be a projection in the centraliser $\mathcal{M}_\varphi$ for which the restriction $\varphi{\upharpoonright}e\mathcal{M}e$ is semifinite. If $e\mathcal{M}e$ is equipped with the restriction $\varphi{\upharpoonright}e\mathcal{M}e$ of the reference weight, then for any Orlicz function Ψ, the space $eL^\Psi(\mathcal{M})e$ is a subspace of $L^\Psi(\mathcal{M})$ and agrees with $L^\Psi(e\mathcal{M}e)$.*

The proof is fairly straightforward, primarily involving lots of careful checking of the construction of the relevant spaces.

Proof. Our first task is to describe the modular automorphism group of $e\mathcal{M}e$. In the subsequent analysis, we will write φ_e for $\varphi{\upharpoonright}e\mathcal{M}e$. Using the fact that e is fixed point of σ_t^φ, it is not difficult to see that $a \mapsto e\sigma_t^\varphi(a)e$ is a *-automorphism group on $e\mathcal{M}e$. It is clear from Theorem 8.10 that the triple $(\mathcal{M}, \varphi, \sigma_t^\varphi)$ satisfies the criteria of that theorem. Using the facts that e is a fixed point of σ_t^φ and that $(\mathfrak{n}\cap\mathfrak{n}^*)(e\mathcal{M}e) \subset (\mathfrak{n}\cap\mathfrak{n}^*)(\mathcal{M})$, we may conclude from what we just noted that the triple $(e\mathcal{M}e, \varphi_e, e\sigma_t^\varphi(\cdot)e)$ also fulfils the criteria of Theorem 8.10. The uniqueness claim in Theorem 8.10 then ensures that the restriction of $(e\sigma_t^\varphi(\cdot)e)$ to $e\mathcal{M}e$ is the modular automorphism group of the pair $(e\mathcal{M}e, \varphi_e)$.

Next, we compare $e\mathcal{M}e \rtimes_{\varphi_e} \mathbb{R}$ and $\pi_\varphi(e)(\mathcal{M} \rtimes_\varphi \mathbb{R})\pi_\varphi(e)$. First, note that since e is a fixed point of the modular group, it follows from the definition of the embedding π_φ of e into $B(L^2(\mathbb{R}, H)) \equiv B(H \otimes L^2(\mathbb{R}))$, that $\pi_\varphi(e) = e\otimes\mathbb{1}$. Alongside the previously noted structure of the modular automorphism group of $e\mathcal{M}e$, a careful check of the definition of the embedding $\pi_{\varphi_e} : e\mathcal{M}e \to B(eHe \otimes L^2(\mathbb{R}))$ reveals that for each $a \in \mathcal{M}$, it is of the form $\pi_{\varphi_e}(eae) = e\pi_\varphi(a)e$. As far as left shift operators are concerned, it is not difficult to see that each $\lambda_t \in L^2(H, \mathbb{R}) \equiv H\otimes L^2(\mathbb{R})$ is of the form $\lambda_t = \mathbb{1}\otimes\ell_t$ where ℓ_t is the corresponding left shift operator on $L^2(\mathbb{R})$. So, $\pi_\varphi(e) = e\otimes\mathbb{1}$ clearly commutes with all the λ_ts. The shift operators $\widetilde{\lambda}_t$ on $L^2(eH, \mathbb{R}) \equiv eH \otimes L^2(\mathbb{R})$ are then easily seen to be of the form $\widetilde{\lambda}_t = (e \otimes \mathbb{1})\lambda_t$. The description of both $\pi_{\varphi_e}(e\mathcal{M}e)$ and the associated shift operators now reveals that $e\mathcal{M}e \rtimes_{\varphi_e} \mathbb{R}$ is effectively just $\pi_\varphi(e)(\mathcal{M} \rtimes_\varphi \mathbb{R})\pi_\varphi(e)$. (This can be seen by invoking the observation made in Remark 9.7). It is then also easy to check that the dual action (θ_s) of $\mathbb{R}$ on $e\mathcal{M}e \rtimes_{\varphi_e} \mathbb{R}$ (as described in the discussion preceding Proposition 9.27) is just the restriction of the matching dual action on $\mathcal{M} \rtimes_\varphi \mathbb{R}$. By Definition 9.20, that fact in turn has the consequence that the canonical operator-valued weight from $e\mathcal{M}e \rtimes_{\varphi_e} \mathbb{R}$ to $e\mathcal{M}e$ is just the restriction to $\pi_\varphi(e)(\mathcal{M} \rtimes_\varphi \mathbb{R})\pi_\varphi(e)$ of the matching operator-valued weight from $\mathcal{M} \rtimes_\varphi \mathbb{R}$ to $\mathcal{M}$. Therefore, by Definition 9.23, the dual weight $\widetilde{\varphi}_e$ on $e\mathcal{M}e \rtimes_{\varphi_e} \mathbb{R}$, is just the restriction of the dual weight $\widetilde{\varphi}$ to $\pi_\varphi(e)(\mathcal{M} \rtimes_\varphi \mathbb{R})\pi_\varphi(e)$.

As far as comparing the crossed products is concerned, it remains to compute the trace on $e\mathcal{M}e \rtimes_{\varphi_e} \mathbb{R}$. In this regard, note that by Proposition 9.27 and Theorem 9.28, the fact that e is a fixed point of the modular group of $(\mathcal{M}, \varphi)$ means that inside the crossed product, $\pi_\varphi(e)$ will for any t commute with h^{it} where h is the density of the dual weight $\widetilde{\varphi}$ with respect to the canonical trace on $\mathcal{M} \rtimes_\varphi \mathbb{R}$. The Borel functional calculus then ensures that h must commute strongly with $\pi_\varphi(e)$. Now, note that by Theorem 9.28 and the preceding analysis, the shift operators on $L^2(eH, \mathbb{R})$ are of the form $\widetilde{\lambda}_t = \pi_\varphi(e)\lambda_t = (\pi_\varphi(e)h)^{it}$. Therefore, again from Theorem 9.28, it follows that the canonical f.n.s. trace τ_e on $e\mathcal{M}e \rtimes_{\varphi_e} \mathbb{R} = \pi_\varphi(e)(\mathcal{M} \rtimes_\varphi \mathbb{R})\pi_\varphi(e)$ is for any $a \in \pi_\varphi(e)(\mathcal{M} \rtimes_\varphi \mathbb{R})\pi_\varphi(e)$ of the form $\widetilde{\varphi}((\pi_\varphi(e)h)^{-1/2}(\pi_\varphi(e)a\pi_\varphi(e))(\pi_\varphi(e)h)^{-1/2}) = \widetilde{\varphi}(h^{-1/2}(\pi_\varphi(e)a\pi_\varphi(e))h^{-1/2})$. So, τ_e is just the restriction of the canonical trace τ on $\mathcal{M} \rtimes_\varphi \mathbb{R}$ to the compression $\pi_\varphi(e)(\mathcal{M} \rtimes_\varphi \mathbb{R})\pi_\varphi(e)$, with the density of the dual weight $\widetilde{\varphi}_e$ with respect to τ_e corresponding to $[\pi_\varphi(e)h]$.

We finally compare the topologies of convergence in measure on the two crossed products. For simplicity of notation, we shall write $\mathfrak{M}$ for $\mathcal{M} \rtimes_\varphi \mathbb{R}$, and $\mathfrak{M}_e$ for $e\mathcal{M}e \rtimes_{\varphi_e} \mathbb{R}$. Given an element $\pi_\varphi(e)a\pi_\varphi(e)$ of $\mathfrak{M}_e \equiv \pi_\varphi(e)\mathfrak{M}\pi_\varphi(e)$, it is trivially clear that $(\mathbb{1} - \pi_\varphi(e)) \leq$ $\mathrm{supp}(|\pi_\varphi(e)a\pi_\varphi(e)|)$. So, for any $\epsilon > 0$, the spectral projection $\chi_{(\epsilon,\infty)}$ $(|\pi_\varphi(e)a\pi_\varphi(e)|)$ is automatically an element of $\mathfrak{M}_e$ and, hence as a spectral projection of $|\pi_\varphi(e)a\pi_\varphi(e)|$, remains unchanged in the passage from $\mathfrak{M}$ to $\mathfrak{M}_e$. Since τ_e is just a restriction of τ, the distributions $\tau_e(\chi_{(\epsilon,\infty)}(|\pi_\varphi(e)a\pi_\varphi(e)|))$ and $\tau(\chi_{(\epsilon,\infty)}(|\pi_\varphi(e)a\pi_\varphi(e)|))$ therefore agree. Therefore, Lemma 3.73 then ensures that the topology of convergence in measure generated by τ_e on $\mathfrak{M}_e$ is just the subspace topology inherited from the corresponding topology of convergence in measure on $\mathfrak{M}$. If we now pass to the completions of these spaces, it follows that equipped with its topology of convergence in measure, $\widetilde{\mathfrak{M}_e}$ is effectively just a topological subspace of $\widetilde{\mathfrak{M}}$. In fact, using the fact that compression with $\pi_\varphi(e)$ is continuous in the topology of convergence in measure, it is clear that $\widetilde{\mathfrak{M}}$ is nothing but $\pi_\varphi(e)\widetilde{\mathfrak{M}}\pi_\varphi(e)$. The observation made earlier that distributions of elements of the form $\pi_\varphi(e)a\pi_\varphi(e)$ remain unchanged in the passage from $\mathfrak{M}$ to its compression clearly also hold for elements of $\pi_\varphi(e)\widetilde{\mathfrak{M}}\pi_\varphi(e)$.

With all the preparation done, we are now finally ready to prove the claim of the theorem. Let Ψ be a given Young function and consider the function $v_s(t) = \mathbf{f}^\Psi(e^{-s}t)\mathbf{f}^\Psi(t)^{-1}$. The strong commutation of h with $\pi_\varphi(e)$ of course ensures that $\pi_\varphi(e)v_s(h) = \pi_\varphi(e)v_s(\pi_\varphi(e)h)$. Therefore, it is clear from Definition 10.8 that algebraically $L^\Psi(e\mathcal{M}e) \subset \widetilde{\mathfrak{M}_e}$ agrees with $\pi_\varphi(e)L^\Psi(\mathcal{M})\pi_\varphi(e)$ and moreover that the latter is a subspace of $L^\Psi(\mathcal{M})$. By Proposition 5.7, the observation made regarding the distributions in the preceding paragraph ensures that for any $a \in \widetilde{\mathfrak{M}}$, the value of $\mathbf{m}_{\pi_\varphi(e)a\pi_\varphi(e)}(1)$ remains unchanged in the passage from $\widetilde{\mathfrak{M}}$ to $\pi_\varphi(e)\widetilde{\mathfrak{M}}\pi_\varphi(e)$. But these quantities are, respectively, the quasi-norms of $\pi_\varphi(e)L^\Psi(\mathcal{M})\pi_\varphi(e)$ and $L^\Psi(e\mathcal{M}e)$. Hence, these spaces agree isometrically. $\qquad\qquad\square$

10.6 $L^2(\mathcal{M})$ and the standard form of a von Neumann algebra

Let $\mathcal{M}$ be a von Neumann algebra, and φ a faithful normal semifinite weight on $\mathcal{M}$. Our primary goal in this section is to prove that the theory of $L^p(\mathcal{M})$-spaces is rich enough to allow for the realisation of a standard form of $\mathcal{M}$ in the sense of Definition 8.43 within this framework.

Definition 10.56 For each $p \in [1, \infty]$, we define left λ_p and right ρ_p actions of $\mathcal{M}$ on $L^p(\mathcal{M})$, by the prescriptions

$$\lambda_p(a)f = af, \qquad f \in L^p(\mathcal{M}),$$
$$\rho_p(a)f = fa, \qquad f \in L^p(\mathcal{M}).$$

Given $a \in \mathcal{M}$, it follows from Proposition 10.24 that

$$\|\lambda_p(a)f\|_p \leq \|a\|_\infty \|f\|_p \text{ and } \|\rho_p(a)f\|_p \leq \|a\|_\infty \|f\|_p$$

for each $f \in L^p(\mathcal{M})$, and so each of $\lambda_p(a)$ and $\rho_p(a)$ continuously maps $L^p(\mathcal{M})$ into $L^p(\mathcal{M})$.

Proposition 10.57 *For each $p \in [1, \infty]$, the following holds.*

(a) *λ_p is faithful representation and ρ_p a faithful anti-representation of $\mathcal{M}$ on the Banach space $L^p(\mathcal{M})$.*

(b) *For all $a \in \mathcal{M}$, we have that $J_p \lambda_p(a) J_p = \rho_p(a^*)$ and $J_p \rho_p(a) J_p = \lambda_p(a^*)$, where J_p denotes the anti-linear isometric involution $f \mapsto f^*$ on $L^p(\mathcal{M})$.*

(c) *For any z in the centre of $\mathcal{M}$, we have that $\lambda_p(z) = \rho_p(z)$.*

Proof. *Part (a):* Suppose that for some $a \in \mathcal{M}$, we have that $\lambda_p(a) = 0$. Let $1 \leq q$ be given so that $1 = p^{-1} + q^{-1}$. For any $f \in L^p(\mathcal{M})$ and any $g \in L^q(\mathcal{M})$, we then have that $\mathrm{tr}(afg) = \mathrm{tr}((\lambda_p(a)f)g) = 0$. It is not difficult to show that $L^p(\mathcal{M}).L^q(\mathcal{M}) = L^1(\mathcal{M})$. Therefore, by duality we must then have that $a = 0$. To see the multiplicativity of λ_p, notice that for any $a, b \in \mathcal{M}$ and $f \in L^p(\mathcal{M})$ we have that $\lambda_p(ab)f = abf = a\lambda_p(b)(f) = (\lambda_p(a) \circ \lambda_p(b))(f)$. We leave the proof of the linearity of the map $a \mapsto \lambda_p(a)$ as an exercise. To see that this map is continuous, observe that it follows from the discussion preceding this proposition that $\|\lambda_p(a)\| \leq \|a\|_\infty$. The proof that $a \mapsto \rho_p(a)$ is an anti-representation runs along similar lines.

Part (b): For all $a \in \mathcal{M}$ and $f \in L^p(\mathcal{M})$, we have that $J_p \lambda_p(a) J_p(f) = (af^*)^* = fa^* = \rho_p(a^*)(f)$ and that $J_p \rho_p(a) J_p(f) = (f^*a)^* = a^*f = \lambda_p(a^*)(f)$.

Part (c): It follows from Theorem 9.28 that z is in the centre of $\mathfrak{M}$ and hence (by the density of $\mathfrak{M}$ in $\widetilde{\mathfrak{M}}$) in the centre of $\widetilde{\mathfrak{M}}$. Having noted this, it is now trivial to see that for any $f \in L^p(\mathcal{M})$, we have $\lambda_p(z)f = zf = fz = \rho_p(z)(f)$. $\qquad\square$

Theorem 10.58 *Let $p \in [1, \infty]$. For any subset $\mathcal{S} \subseteq B(L^p(\mathcal{M}))$, we will write $\mathcal{S}'$ for the set of all bounded linear operators on $L^p(\mathcal{M})$*

commuting with all the elements of $\mathcal{S}$. We have $\lambda_p(\mathcal{M}) = \rho_p(\mathcal{M})'$ and $\rho_p(\mathcal{M}) = \lambda_p(\mathcal{M})'$.

Proof. For all $a, b \in \mathcal{M}$ and $f \in L^p(\mathcal{M})$, we have that $\lambda_p(a)(\rho_p(b)(f)) = afb = \rho_p(b)(\lambda_p(a)(f))$. Hence, $\lambda_p(\mathcal{M}) \subseteq \rho_p(\mathcal{M})'$ and $\rho_p(\mathcal{M}) \subseteq \lambda_p(\mathcal{M})'$. We show that $\lambda_p(\mathcal{M})' \subseteq \rho_p(\mathcal{M})$. The inclusion $\rho_p(\mathcal{M})' \subseteq \lambda_p(\mathcal{M})$ follows by a similar argument. This will then suffice to prove the theorem.

Case 1 ($p = \infty$): Let $T \in \lambda_\infty(\mathcal{M})'$ be given. For any $f \in L^\infty(\mathcal{M})$, we then have $T(f) = T(f\mathbb{1}) = T(\lambda_\infty(f)(\mathbb{1})) = \lambda_\infty(f)(T(\mathbb{1})) = fT(\mathbb{1}) = \rho_\infty(T(\mathbb{1}))(f)$. Clearly $T = \rho_\infty(T(\mathbb{1})) \in \rho_\infty(\mathcal{M})$, which proves the claim in this case.

Case 2 ($p = 1$): Let $T \in \lambda_1(\mathcal{M})'$ be given. Let $S = T^\star \in B(L^\infty(\mathcal{M}))$ be the Banach adjoint of T. So, by tr-duality, we then have that $\mathrm{tr}(T(f)a) = \mathrm{tr}(fS(a))$ for every $f \in L^1(\mathcal{M})$ and $a \in L^\infty(\mathcal{M})$. For any $a, b \in L^\infty(\mathcal{M})$ and $f \in L^1(\mathcal{M})$, we have that $\mathrm{tr}(S(ab)f) = \mathrm{tr}(abT(f)) = \mathrm{tr}(a\lambda_1(b)(T(f))) = \mathrm{tr}(aT(\lambda_1(b)(f))) = \mathrm{tr}(aT(bf)) = \mathrm{tr}(S(a)bf)$. Therefore, it follows that $S(ab) = S(a)b$ for all $a, b \in L^\infty(\mathcal{M})$. This equality may be reformulated as the claim that $S \circ \rho_\infty(b) = \rho_\infty(b) \circ S$. It therefore follows from case 1 that $S = \lambda_\infty(x)$ for some $x \in L^\infty(\mathcal{M})$. But for every $a \in L^\infty(\mathcal{M})$ and $f \in L^1(\mathcal{M})$, we then have that $\mathrm{tr}(T(f)a) = \mathrm{tr}(fS(a)) = \mathrm{tr}(f\lambda_\infty(x)(a)) = \mathrm{tr}(fxa) = \mathrm{tr}(\rho_1(x)(f)a)$. This ensures that $T = \rho_1(x)$.

Case 3 ($1 < p < \infty$): Let $T \in \lambda_p(\mathcal{M})'$ be given and select $q > 1$ so that $1 = p^{-1} + q^{-1}$. Our strategy will be to reduce the problem to the setting of case 2. To achieve this objective, we formally define a linear map $S : L^1(\mathcal{M}) \to L^1(\mathcal{M})$ by the prescription $S(\sum_{k=1}^n b_k a_k) = \sum_{k=1}^n b_k T(a_k)$ for all n-tuples $a_1, a_2, \ldots, a_n \in L^p(\mathcal{M})$ and $b_1, b_2, \ldots, b_n \in L^q(\mathcal{M})$.

We first establish the well-definiteness of S. Assume $a_1, a_2, \ldots, a_n \in L^p(\mathcal{M})$ and $b_1, b_2, \ldots, b_n \in L^q(\mathcal{M})$ are given with $\sum_{k=1}^n b_k a_k = 0$. Consider the term $a = (\sum_{k=1}^n a_k^* a_k)^{1/2}$. Since $a_m^* a_m \leq a^2$ for each $1 \leq m \leq n$, it then follows from Proposition 3.90 that there exist contractions $c_1, c_2, \ldots, c_n \in \mathcal{M}_+$ supported on $\mathbf{s}(a)$, such that $ac_m a = a_m^* a_m$ for each $1 \leq m \leq n$. Since $a(\sum_{k=1}^n c_k)a = a^2$, we must have that $\sum_{k=1}^n c_k = \mathbf{s}(a)$. Now observe that $|c_m^{1/2} a| = |a_m|$ for each $1 \leq m \leq n$.

We may therefore select partial isometries u_k $(1 \leq k \leq n)$ so that for $v_k = u_k c_k^{1/2}$ $(1 \leq k \leq n)$, we have that $v_k a = a_k$ $(1 \leq k \leq n)$, and $\sum_{k=1}^{n} |v_k|^2 = \mathbf{s}(a)$. Now, consider the term $\sum_{k=1}^{n} b_k v_k$. By construction, we then have that $\mathbf{s}_r(\sum_{k=1}^{n} b_k v_k) \leq \mathbf{s}_r(\sum_{k=1}^{n} v_k) \leq \mathbf{s}(\sum_{k=1}^{n} |v_k|^2) = \mathbf{s}(a)$. But we also have that $(\sum_{k=1}^{n} b_k v_k)a = \sum_{k=1}^{n} b_k a_k = 0$. Together these two facts ensure that $\sum_{k=1}^{n} b_k v_k = 0$. It therefore follows that

$$
\begin{aligned}
\sum_{k=1}^{n} b_k T(a_k) &= \sum_{k=1}^{n} b_k T(v_k a) \\
&= \sum_{k=1}^{n} b_k T(\lambda_p(v_k)a) \\
&= \sum_{k=1}^{n} b_k \lambda_p(v_k)(T(a)) \\
&= \left(\sum_{k=1}^{n} b_k v_k \right) T(a) \\
&= 0,
\end{aligned}
$$

as required.

The linearity of S is clear. Hence, we next show that S is, in fact, bounded. Let $0 \neq c \in L^1(\mathcal{M})$ be given, with $c = u|c|$ being the polar form of c. We then set $a = u|c|^{1/p}$ and $b = |c|^{1/q}$. It is clear that $ab = c$. However, we also have $\|a\|_p \|b\|_q = \mathrm{tr}(|a|^p)^{1/p}\mathrm{tr}(|b|^q)^{1/q} = \mathrm{tr}(|c|)^{1/p}\mathrm{tr}(|c|)^{1/q} = \mathrm{tr}(|c|) = \|c\|_1$. We may therefore use Hölder's inequality to conclude that $\|S(c)\|_1 = \|bT(a)\|_1 \leq \|b\|_q\|T(a)\|_p \leq \|T\|\|b\|_q\|a\|_p = \|T\|.\|c\|_1$.

With the existence of S verified, we now use S to prove that $T \in \rho_p(\mathcal{M})$. First, note that for any $f \in \mathcal{M}$, $a \in L^p(\mathcal{M})$ and $b \in L^q(\mathcal{M})$, it follows from the definition of S that $S(fba) = fbT(a) = fS(ba)$. This translates to the claim that $S \circ \lambda_1(f) = \lambda_1(f) \circ S$ for each $f \in L^\infty(\mathcal{M})$. It therefore follows from case 2 that $S = \rho_1(x)$ for some $x \in L^\infty(\mathcal{M})$. But for all $a \in L^p(\mathcal{M})$ and $b \in L^q(\mathcal{M})$, we then have that

$$
\mathrm{tr}(bT(a)) = \mathrm{tr}(S(ba)) = \mathrm{tr}((ba)x) = \mathrm{tr}(b(ax)) = \mathrm{tr}(b\rho_p(x)(a))
$$

and hence that $T(a) = \rho_p(x)$ for all $a \in L^p(\mathcal{M})$.   □

Our primary interest is in the case $p = 2$. So, for this case, we shall respectively simply write λ, ρ and J, for λ_2, ρ_2 and J_2.

Theorem 10.59

(a) λ *is a faithful normal* $*$*-representation, and* ρ *a faithful normal* $*$*-anti-representation of* $\mathcal{M}$ *on the Hilbert space* $L^p(\mathcal{M})$.

(b) *The von Neumann algebras* $\lambda(\mathcal{M})$ *and* $\rho(\mathcal{M})$ *are commutants of each other, with* $\rho(\mathcal{M}) = J\lambda(\mathcal{M})J$.

(c) *The quadruple* $(\lambda(\mathcal{M}), L^2(\mathcal{M}), J, L^2_+(\mathcal{M}))$ *is a* standard form *of* $\mathcal{M}$ *in the sense of Definition 8.43.*

Proof.   *Part (a):* Due to the similarity of the proofs, we will prove only one of the claims. We know from Proposition 10.57 that $\lambda(\mathcal{M})$ is a representation of $\mathcal{M}$. We just need to check that it preserves adjoints and that it is normal. For any $a \in \mathcal{M}$ and $f, g \in L^2(\mathcal{M})$, we have that

$$\langle \lambda(a)f, g \rangle = \langle af, g \rangle = \mathrm{tr}(g^*af) = \mathrm{tr}((a^*g)^*f) = \langle f, a^*g \rangle = \langle f, \lambda(a^*)g \rangle.$$

Hence, $\lambda(a^*) = \lambda(a)^*$. Now, suppose that $(a_\alpha) \subseteq \mathcal{M}_+$ increases to $a \in \mathcal{M}_+$. For any $f \in L^2(\mathcal{M})$, $(f^*a_\alpha f)$ will then increase to f^*af by Proposition 3.88. But for any $f \in L^2(\mathcal{M})$, we will then have that

$$\sup_\alpha \langle \lambda(a_\alpha)f, f \rangle = \sup_\alpha \mathrm{tr}(f^*a_\alpha f) = \sup_\alpha \mathbf{m}_{f^*a_\alpha f}(1) = \mathbf{m}_{f^*af}(1)$$

$$= \mathrm{tr}(f^*af) = \langle \lambda(a)f, f \rangle.$$

In other words $\sup_\alpha \lambda(a_\alpha) = \lambda(a)$

Part (b): To see that these claims hold, notice that combining part (b) of Proposition 10.57 and Theorem 10.58, we have $J\lambda(\mathcal{M})J = \rho(\mathcal{M}) = \lambda(\mathcal{M})'$ and $J\rho(\mathcal{M})J = \lambda(\mathcal{M}) = \rho(\mathcal{M})'$.

Part (c): The fact that $L^2_+(\mathcal{M})$ is a self-dual cone follows from the final claim of Theorem 10.38. Therefore, it remains to show that the last three criteria mentioned in Definition 8.43 hold when this cone is used. We investigate these in turn.

- By Proposition 10.57, we have that $J\lambda(z)J = \rho(z^*) = \lambda(z^*)$ for all z in the centre of $\mathcal{M}$.
- For any $f \in L^2_+(\mathcal{M})$, we trivially have that $Jf = f^* = f$.

- For any $a \in \mathcal{M}$ and $f \in L^2_+(\mathcal{M})$, we have by part (b) of Proposition 10.57, that $\lambda(a)(J\lambda(a)J)f = \lambda(a)(\rho(a^*)f) = afa^* \in L^2_+(\mathcal{M})$.

Thus, the quadruple $(\lambda(\mathcal{M}), L^2(\mathcal{M}), J, L^2_+(\mathcal{M}))$ is indeed a standard form of $\mathcal{M}$. $\qquad\square$

The identification of $L^2_+(\mathcal{M})$ as a substitute for $\mathscr{P}^\natural$ bears further comment. In the case where φ is a faithful normal state with $\mathcal{M}$ in its GNS representation, $\mathscr{P}^\natural$ may be defined as the closure of $\{\Delta^{1/4}a\Omega\colon a \in \mathcal{M}\} = \{\Delta^{1/4}a\Delta^{-1/4}\Omega\colon a \in \mathcal{M}_+\}$, where Ω is a cyclic and separating vector representing φ. In this setting, $\mathfrak{M}_\varphi = \mathcal{M}$, with $h = \frac{d\widetilde{\varphi}}{d\tau}$ an element of $L^1(\mathcal{M})$ (see Corollary 9.37 and the definition of $L^1(\mathcal{M})$). On passing to the noncommutative L^p-picture, $h^{1/2}$ proves to be a suitable substitute for Ω, since it too is cyclic and separating. The cyclicity follows from Theorem 10.46. To see that it is separating, let $a \in \mathcal{M}_+$ be given. Since then $\langle ah^{1/2}, h^{1/2}\rangle = \operatorname{tr}(h^{1/2}ah^{1/2}) = \|h^{1/2}ah^{1/2}\|_1$, it is clear that $\langle ah^{1/2}, h^{1/2}\rangle = 0 \Rightarrow h^{1/2}ah^{1/2} = 0 \Rightarrow a = 0$. (The last implication follows from the injectivity of $\mathrm{i}^{(1)}$.) On the other hand, by Proposition 9.5 and Theorem 9.28, h may be regarded as a substitute for Δ_φ. So, the formula

$$\mathscr{P}^\natural = \overline{\{\Delta^{1/4}a\Omega\colon a \in \mathcal{M}_+\}} = \overline{\{\Delta^{1/4}a\Delta^{-1/4}\Omega\colon a \in \mathcal{M}_+\}}$$

translates to the formula $\mathscr{P} = \overline{h^{1/4}\mathcal{M}_+h^{1/4}}$ in the $L^2(\mathcal{M})$ picture, which by Theorem 10.46 is exactly $L^2_+(\mathcal{M})$.

10.7 A brief history of noncommutative L^p-spaces in the general case

In the Introduction, we briefly described the history of noncommutative L^p-spaces in the tracial case. The need for a good definition of general noncommutative L^p-spaces was evident early on, but it was not clear how to deal with the difficulties involved. For example, the 'triangle inequality', valid for a trace (cf. Theorem 6.6 with $p = 1$), does not hold, in general, for a state. Further progress would have been impossible without the modular theory of Minoru Tomita and Masamichi

Takesaki [**Tak70**]. In the modern approach, a predominant approach is to construct L^p-spaces consisting of operators, but there have also been attempts to construct L^p-spaces consisting of vectors or bilinear forms. It was the L^p-spaces consisting of bilinear forms that were constructed earliest, and we will begin our story by describing this trend. The reader should be warned that many of the publications described below appeared only in Russian.

Three of the L^p-spaces, namely L^1, L^2 and L^∞ have a special status. Whatever the construction, we expect $L^\infty(\mathcal{M})$ to be a von Neumann algebra isomorphic to the von Neumann algebra $\mathcal{M}$, $L^1(\mathcal{M})$ to be isometric to the predual $\mathcal{M}_*$, and $L^2(\mathcal{M})$ to be a Hilbert space in which $\mathcal{M}$ can be standardly represented, that is so that $\pi(\mathcal{M})$ is anti-isomorphic to its commutant $\pi(\mathcal{M}')$ (cf 8.43).

Between 1972 and 1978, a group of Russian mathematicians from Kazan State University, along with their leader, A. N. Sherstnev, described $L^1(\mathcal{M}, \varphi)$ and $L^2(\mathcal{M}, \varphi)$ spaces for general von Neumann algebras $\mathcal{M}$ acting on H with a faithful semifinite normal weight φ as spaces of bilinear forms defined on the so-called *lineal* of φ. When φ is a state, the lineal is the set of vectors $\xi \in H$ such that the vector state $\omega_\xi \leq c\varphi$ for some positive constant c. The $L^1(\varphi)$-spaces for states and weights φ were constructed by Sherstnev in [**She74, She77, She78**]. The theory was further developed by Trunov and Sherstnev [**TS78a, TS78b**], where it was proved that the spaces are isometric and order-isomorphic to the predual of the von Neumann algebra. It was also shown that if a bilinear form $a \in L^1(\varphi)$ is invariant under the modular group of φ, then it can be represented by an operator affiliated with the centraliser of $(\mathcal{M}, \varphi)$ (see Definition 8.18). L^2-spaces of bilinear forms were defined by Trunov [**Tru79**].

N. V. Trunov [**Tru81**] succeeded in constructing L^p-spaces with respect to a strongly semifinite weight φ on a semifinite von Neumann algebra $\mathcal{M}$ equipped with a f.n.s. trace τ. The pth norm was first defined on $\mathcal{M}$ as

$$\|a\|_p := \tau((h^{1/2p}ah^{1/2p})^p)^{1/p}$$

where h is the Radon–Nikodym derivative $\frac{d\varphi}{d\tau}$, and then extended to a proper space of bilinear forms. The assumption on the weight φ implies, by Theorem 4.32, that h is locally measurable, so the formula above makes perfect sense.

The most comprehensive study of the use of bilinear forms in noncommutative measure and integral theory is given in Sherstnev's book *Methods of Bilinear Forms in Noncommutative Measure and Integral Theory* [**She08**] (in Russian). For those who do not read Russian, we recommend the survey article [**BS21**], which cites a lot of other papers and alternative approaches (like the real L^p-spaces constructed by O. E. Tikhonov [**Tik82**]). One should point out that the construction of Trunov has, in fact, been generalised to general von Neumann algebras with a state by Carlo Cecchini [**Cec86**], a work which went virtually unnoticed.

A successful use of modular theory led to the first constructions of noncommutative L^p-spaces for general von Neumann algebra at the end of 1970s. They were due to Uffe Haagerup [**Haa79a**] and Alain Connes [**Con80**]. The Haagerup construction was beautifully presented in the notes of Marianne Terp [**Ter81**], and the Connes construction by Michel Hilsum [**Hil81**]. The Connes–Hilsum spatial spaces $L^p(\mathcal{M}, \psi)$ depend on a choice of a weight ψ on the commutant $\mathcal{M}'$ of $\mathcal{M}$ and consist of unbounded operators acting on the Hilbert space on which the von Neumann algebra is represented. Dealing with unbounded operators means that even showing that $L^p(\mathcal{M}, \psi)$ is a linear space is a nontrivial matter, and to date the only known way of proving this fact is to exploit the fact that these spaces are isometric to the Haagerup L^p-spaces which then enables one to extract the linearity from the Haagerup setting. Thorough discussion of these spaces can be found in Hiai's recent book [**Hia21**]. These are far from being the only known constructions. One may, for example, note the ingenious construction of Huzihiro Araki and Tetsuya Masuda [**AM82**] using only relative modular operators. However, the constructions of Connes and Haagerup are certainly the most successful ones. Especially, Haagerup's L^p-spaces are now 'standard'—if we speak about noncommutative L^p-spaces for general von Neumann algebras without giving any names, we certainly mean Haagerup's spaces. Their main advantage is that they are entirely autonomous—we never need to leave the realm of Haagerup spaces when dealing with questions originating there.

One more approach to noncommutative L^p-spaces needs to be mentioned here: they can all be viewed as interpolation spaces. For the spaces of Sherstnev and Trunov, this was established in a series of papers by A. A. Zolotarev [**Zol82**, **Zol85**, **Zol88**]. Marianne Terp [**Ter82**] first proved this fact in the setting of Connes–Hilsum L^p-spaces with Hideki Kosaki then shortly afterward publishing a 'Haagerup oriented' interpolative construction of L^p-spaces in the state case (σ-finite algebras) [**Kos84a**]. The most general case of an arbitrary von Neumann algebra with a faithful normal semifinite weight was elaborated by Hideaki Izumi [**Izu96**, **Izu97**, **Izu98b**, **Izu98a**]. It should be noted that all of these approaches produce spaces which are isometric copies of each other.

One tantalising difference between the general theory as it stands now and the tracial theory is the almost complete absence of a theory of Banach function spaces in the general setting. The difficulties involved here are first to find the 'correct' way of defining decreasing rearrangements in the general case, and second to develop a modified approach to real interpolation which returns the 'correct' spaces. (It is known that when the 'standard' K-method is applied to the general case, prescriptions which in tracial case will return copies of $L^p(\mathcal{M}, \tau)$ fail to do so in the general case.) On the positive side, as elucidated in this book, the Haagerup approach to noncommutative L^p-spaces was in 2013 successfully extended by Labuschagne [**Lab13**] to the point where it can produce Orlicz spaces for general von Neumann, which are a direct generalisation of those we met earlier in the tracial setting. Since then other markedly different approaches have been developed for positing Orlicz structures for general von Neumann algebras (see, for example, [**Jen06**]). In this paper, the notion of a Young function on a real Banach spaces is formally defined, with the resultant theory then applied to the real Banach space $\mathcal{M}_h$ for the purpose of constructing of a Banach manifold structure on the set of faithful normal states of $\mathcal{M}$. These two approaches follow different ideologies which we shall now elucidate. Starting with a Young function on a real Banach space Jencova uses convex analysis of Minkowski functionals to define analogues of the Orlicz and Luxemburg norms on a subspace of the given Banach space, from which the associated Orlicz spaces are constructed. This approach is therefore well suited to very general settings where no a priori 'quantum integral' is in view. The approach [**Lab13**] is to rather use a quantised version of Musielak's approach[**Mus83**]. Specifically, to rather use the existing noncommutative measure theoretic structures on von Neumann algebras to construct appropriate noncommutative modulars and to then define the associated Orlicz spaces in terms of these modulars (see Remark 10.16). Given this difference, it is at this stage not all clear if these two approaches are in any way related to each other. The emergence of this kind of analysis nevertheless raises hopes that in time a theory of Banach function spaces for general von Neumann algebras may yet be achieved.

PART 4

ADVANCED THEORY AND APPLICATIONS

In this part, we come to the advanced theory requiring a certain level of mastery of the earlier part. Chapter 11 has a singular focus which is to show that the Haagerup L^p-spaces form a complex interpolation scale. The proof is quite intricate, demanding a high level of perspicacity. There are a number of novel aspects to the proof, and readers interested in interpolation theory would be well advised to read this chapter carefully. For example, the theory of Orlicz spaces introduced in Chapter 10 is needed to construct the superspace within which we construct the analogues of $L^1 + L^\infty$ and $L^1 \cap L^\infty$. This is also the first time the fact that Haagerup L^p-spaces form a complex interpolation scale has been verified using only techniques internal to the theory of Haagerup L^p-spaces. Readers with no abiding interest in complex interpolation should, nevertheless, take note of the main theorem, as it is an essential tool for the theory presented in Section 12.2 (p. 528).

Chapter 12 contains a set of results that should form part of the toolkit of any reader interested in serious applications of the theory of Haagerup L^p-spaces to mathematical physics. All such readers are strongly encouraged to explore this chapter. Section 12.1 (p. 518) gets the ball rolling by describing the operator space structure of the Haagerup L^p-spaces while at the same time refining some of the embeddings introduced in Section 10.4 (p. 441). These structures are then used to describe how completely bounded maps from one von Neumann algebra $\mathcal{M}$ to another, say $\mathcal{N}$, which intertwines group actions of LCGs G and H on $\mathcal{M}$ and $\mathcal{N}$, respectively, may be extended to a completely bounded map from $\mathcal{M} \rtimes G$ to $\mathcal{N} \rtimes H$. In Section 12.2 (p. 528), we analyse the notion of integrable operators. By means of the embeddings introduced in Section 10.4 (p. 441), any operator on a von Neumann algebra $\mathcal{M}$ that preserves $\mathfrak{m}_\varphi$ may naively be lifted to a densely defined endomorphism on $L^p(\mathcal{M})$ ($1 \leq p < \infty$). The so-called p-integrable operators are those operators for which this lifted operator admits a continuous extension to all of L^p. Sufficient conditions are provided for a positive map to be integrable. Finally, in Section 12.3 (p. 543), we define what it means for an operator on $L^p(\mathcal{M})$ to be Markov and then proceed with a comprehensive analysis of this notion. One of the crowning achievements of this section is to show that the classes of Markov operators on, respectively, L^∞ and L^2, mirror each other. In view of the importance of Markov dynamics for mathematical physics, all readers with a serious interest in the applications to Quantum Physics should take the trouble to study this section.

As was the case with Chapter 11, Chapter 13 too has a singular goal, namely to prove Haagerup's reduction theorem – first for L^∞ and then for L^p. The basic goal of this theorem is to show that any von Neumann algebra $\mathcal{M}$ is an expected subalgebra of a von Neumann superalgebra $\mathcal{R}$ for which the associated L^p-spaces may be approximated by an increasing sequence of tracial L^p-spaces. Haagerup's intention with this theorem was to set up a formalism which could be used to lift certain somewhat technical results from the tracial case to the general case. To date, it has found widespread application in the theory of noncommutative H^p-spaces, a theory beyond the scope of this monograph.

Chapter 14 has been written in such a way that all readers who have at least mastered L^p-spaces for tracial algebras will be able to engage this chapter with benefit. However, the portion of the chapter readers will be able to engage will of course depend on how much they have read. Section 14.1 (p. 580) introduces the reader to the algebraic approach to quantum mechanics and can be read by all. This section starts with a brief review of the emergence of the Heisenberg and Schrödinger approaches to quantum mechanics and the inherent mathematical challenges. We in particular also discuss the emergence of von Neumann algebras as a formalism, the process of canonical quantisation as conceptualised by Born, Jordan and Dirac, and the relevance of the theory presented in this book as a further means for meeting the challenge of quantum mechanics.

Then in Section 14.2 (p. 590) we briefly review the work of Boltzmann, which was the inspiration behind the definition of von Neumann entropy for $B(H)$. We then describe how this concept can be canonically extended to first tracial algebras and then general algebras. The reader will see that the class of states that admit good entropy all live in the Orlicz space $L^1 \cap L\log(L+1)$. Familiarity with the content of Chapters 6 and 10 for respectively the tracial and general aspects of the theory presented here is therefore essential. We then continue with the theme of entropy by in Section 14.3 (p. 599) discussing the relative entropy of two states (representing the likelihood of a change of state from one to the other). We present a definition of this concept based on Connes cocycle derivatives which is a more "dynamic" definition of this concept than the original definition proposed by Araki. In Theorem 14.5 (p. 601) we then prove the equivalence of these two definitions before in Theorem 14.8 (p. 605) reaffirming the connection of this theory with its classical roots by establishing a limit formula for relative entropy with a very classical flavour. To be able to master this section, readers will need to be very comfortable with the content of Chapter 10 and the material on Connes cocycle derivatives in Chapter 8.

In Section 14.4 (p. 610) the focus shifts from states to observables, specifically regular observables namely the class of (possibly unbounded) observables for which all their moments are finite. This turns out to be a dual concept to the notion of states with "good" entropy encountered in Section 14.2 (p. 590) in the sense that while those states found their home in a subspace of the Orlicz space $L\log(L+1)$, regular observables find their home in $L^{\cosh - 1}$ - the Köthe dual of $L\log(L+1)$. As a formalism for quantum mechanics, the pairing $(L\log(L+1), L^{\cosh - 1})$ as a home for states and observables is, therefore, a valid alternative to the more classical pairing of (L^1, L^∞). The picture is then completed by the fact that complete Markov dynamics extends to $L^{\cosh - 1}$ (Proposition 14.16 (p. 614)). Readers with only Chapter 6 as background will probably be able to appreciate the general thrust of the theory, but mastery of Chapters 10 and 12 is needed to fully comprehend the latter details.

In Section 14.5 (p. 614) the focus shifts to quantum field theory, clearly demonstrating that the theory developed in this book is relevant not only for quantum statistical mechanics but also for field theory. This section starts by reviewing the two formalisms that emerged in the 50s and 60s, namely the Gårding-Wightman and Haag-Kastler axiomatic frameworks, and the subsequent research which focused on how these two approaches could be harmonised. On this same theme or harmonisation, we then show that for a large number of models produced by means of the so-called Osterwalder-Schrader reconstruction, the associated field operators actually canonically embed into $L^{\cosh-1}$ — the home for regular observables. Readers will need to be familiar with Chapter 10 to master this section.

Chapter 11

Complex interpolation of noncommutative L^p spaces

In [**Ter82**], Terp showed that for general von Neumann algebras, Connes–Hilsum L^p-spaces ($1 \leq p \leq \infty$) form a complex interpolation scale. Here, we modify Terp's method to obtain a direct proof of the validity of this fact for Haagerup L^p-spaces. This fact is a very important tool in the advanced theory. We start by carefully describing the structure of the spaces that will play the role of the intersection and sum of L^1 and L^∞. Next, we spend some time clearly describing the context of the complex interpolation we will perform, before going on to first show that appropriate copies of the L^p-spaces are intermediate spaces of copies of L^1 and L^∞, and then finally showing that these intermediate spaces actually form a complex interpolation scale.

11.1 A description of $\mathscr{L}^1 \cap \mathscr{L}^\infty$ and $\mathscr{L}^1 + \mathscr{L}^\infty$

Recall that by Proposition 10.22, each L^p may be canonically embedded into $L^{1+\infty}(\mathcal{M})$ by means of the map $\iota^{(p)}$. We write $\mathscr{L}^p$ for $\iota^{(p)}(L^p)$. The embedding of $\mathcal{M}$ into $L^{1+\infty}(\mathcal{M})$ takes the form $\iota^{(\infty)}(a) = \mathbf{f}_{1+\infty}(h)^{1/2}a\mathbf{f}_{1+\infty}(h)^{1/2}$ for all $a \in \mathcal{M}$, while for $L^1(\mathcal{M})$ the embedding takes the form

$$\iota^{(1)}(a) = [\mathbf{f}_{1+\infty}(h)^{1/2}h^{-1/2}]a[\mathbf{f}_{1+\infty}(h)^{1/2}h^{-1/2}]$$
$$= \mathbf{f}^{1\cap\infty}(h)^{-1/2}a\mathbf{f}^{1\cap\infty}(h)^{-1/2}$$

for all $a \in L^1(\mathcal{M})$. (Here, we used the facts that $\mathbf{f}^{1\cap\infty}(t) = \max(1,t)$ and $\mathbf{f}_{1+\infty}(t) = \min(1,t)$ (where $t \geq 0$), whence $\mathbf{f}^{1\cap\infty}(t)\mathbf{f}_{1+\infty}(t) = t$ for all $t \geq 0$.) We define the natural norm on $\mathscr{L}^p$, to be the norm inherited from L^p. The spaces $\mathscr{L}^1 \cap \mathscr{L}^\infty$ and $\mathscr{L}^1 + \mathscr{L}^\infty$ will play the role of the intersection and sum of L^1 and L^∞. Classical arguments then show that

Noncommutative measures and L^p and Orlicz Spaces, with Applications to Quantum Physics. Stanisław Goldstein and Louis Labuschagne, Oxford University Press. © Stanisław Goldstein and Louis Labuschagne (2025).
DOI: 10.1093/oso/9780198950202.003.0013

both $\mathscr{L}^1 \cap \mathscr{L}^\infty$ and $\mathscr{L}^1 + \mathscr{L}^\infty$ are Banach spaces when, respectively, equipped with the norms

$$\|g\|_{(\cap)} = \max(\|g\|_{\mathscr{L}^1}, \|g\|_{\mathscr{L}^\infty})$$

and

$$\|a\|_{(+)} = \inf\{\|a_1\|_{\mathscr{L}^1} + \|a_\infty\|_{\mathscr{L}^\infty} :$$
$$a = a_1 + a_\infty, a_1 \in \mathscr{L}^1 \text{ and } a_\infty \in \mathscr{L}^\infty\}.$$

A typical element g of $\mathscr{L}^1 \cap \mathscr{L}^\infty$ is of course given by a pair $g_1 \in \mathscr{L}^1$ and $g_\infty \in \mathscr{L}^\infty$ satisfying $\iota^{(1)}(g_1) = g = \iota^{(\infty)}(g_\infty)$. Given such a $g \in \mathscr{L}^1 \cap \mathscr{L}^\infty$ and $f \in \mathscr{L}^1 + \mathscr{L}^\infty$ with $a = \iota^{(1)}(b_a) + \iota^{(\infty)}(c_a)$ where $b_a \in L^1$ and $c_a \in L^\infty$, we may attempt to define a bilinear functional on the Cartesian product $(\mathscr{L}^1 \cap \mathscr{L}^\infty) \times (\mathscr{L}^1 + \mathscr{L}^\infty)$ by the prescription

$$\langle g, a \rangle_{\cap,+} = \mathrm{tr}(g_\infty b_a + g_1 c_a).$$

This does indeed turn out to be a well-defined well-behaved bilinear functional, which may be used to study the duality between these two spaces, as can be seen from the following theorem.

Theorem 11.1 *The formal prescription given above yields a well-defined bilinear form on $(\mathscr{L}^1 \cap \mathscr{L}^\infty) \times (\mathscr{L}^1 + \mathscr{L}^\infty)$ for which we have $|\langle g, a \rangle| \leq \|g\|_{(\cap)} \|a\|_{(+)}$.*

Proving this theorem requires some rather substantial preparation. It is clear that $\iota^{(\infty)}(\mathfrak{m}_\varphi)$ is a subspace of $\mathscr{L}^1 \cap \mathscr{L}^\infty$. We start our assault on the proof of Theorem 11.1, with an analysis of how $\mathscr{L}^1 \cap \mathscr{L}^\infty$ may be recovered from this subspace. To achieve this end, we need some technical information regarding the behaviour of $\mathfrak{m}_\varphi$ with respect to $\mathbf{f}^{1\cap\infty}(h)$ and $\mathbf{f}_{1+\infty}(h)$.

Remark 11.2 It is clear that $\mathbf{f}^{1\cap\infty}(t)\mathbf{f}_{1+\infty}(t) = t$ for all $t \geq 0$. It is not difficult to show that $\frac{1}{2}(\sqrt{t} + 1) \leq \mathbf{f}^{1\cap\infty}(t)^{1/2} \leq \sqrt{t} + 1$. At the operator level, this means that $[\mathbf{f}^{1\cap\infty}(h)^{1/2}(\sqrt{h} + \mathbb{1})^{-1})]$ (where the square brackets denote the minimal closure) is an invertible operator. Given any $a \in \mathfrak{n}_\varphi$, it is clear from Proposition 10.40 that $a(\sqrt{h} + \mathbb{1})$ will be closable, with both $[a(\sqrt{h} + \mathbb{1})]$ and $(\sqrt{h} + \mathbb{1})a^*$ τ-measurable. Upon, respectively, composing these operators on the right and left with the operator $[\mathbf{f}^{1\cap\infty}(h)^{1/2}$ $(\sqrt{h}+\mathbb{1})^{-1})]$, it follows that $a\mathbf{f}^{1\cap\infty}(h)^{1/2}$ is closable, with both $[a\mathbf{f}^{1\cap\infty}(h)^{1/2}]$ and $\mathbf{f}^{1\cap\infty}(h)^{1/2}a^*$ τ-measurable, and $[a\mathbf{f}^{1\cap\infty}(h)^{1/2}]^* = \mathbf{f}^{1\cap\infty}(h)^{1/2}a^*$. Given any $a_0, a_1 \in \mathfrak{n}_\varphi$, one may easily show that $(\mathbf{f}^{1\cap\infty}(h)^{1/2}a_0^*)[a_1\mathbf{f}^{1\cap\infty}(h)^{1/2}]$ satisfies the membership criterion for $L^{1\cap\infty}(\mathcal{M})$. We shall where convenient simply write $\mathbf{f}^{1\cap\infty}(h)^{1/2}(a_0^*a_1)\mathbf{f}^{1\cap\infty}(h)^{1/2}$ for $(\mathbf{f}^{1\cap\infty}(h)^{1/2}a_0^*)[a_1\mathbf{f}^{1\cap\infty}(h)^{1/2}]$.

Lemma 11.3 *Let $f \in L^1(\mathcal{M})$ be given, and let (e_α) be a net in $\mathcal{M}$ such that $\|e_\alpha\| \leq 1$ for all α, with $e_\alpha \to \mathbb{1}$ strongly. Then, $(e_\alpha f e_\alpha^*)$ converges to f in the L^1-norm.*

Proof. Let $f = u|f|$ be the polar form of f. Then, of course, $f^* = u^*|f^*|$ with each of $u|f|^{1/2}$ $u^*|f^*|^{1/2}$, $|f|^{1/2}$ and $|f^*|^{1/2}$ belonging to $L^2(\mathcal{M})$. For each α, we let u_α and v_α be the partial isometries in the polar forms $(e_\alpha f e_\alpha^* - f e_\alpha^*) = u_\alpha|(e_\alpha f e_\alpha^* - f e_\alpha^*)|$ and $(e_\alpha - \mathbb{1})f^* = v_\alpha|(e_\alpha - \mathbb{1})f^*|$. We then have that

$$\mathrm{tr}(|e_\alpha f e_\alpha^* - f|)$$
$$\leq \mathrm{tr}(|e_\alpha f e_\alpha^* - f e_\alpha^*|) + \mathrm{tr}(|f e_\alpha^* - f|)$$
$$= \mathrm{tr}(|e_\alpha f e_\alpha^* - f e_\alpha^*|) + \mathrm{tr}(|e_\alpha f^* - f^*|)$$
$$= \mathrm{tr}(u_\alpha^*(e_\alpha f e_\alpha^* - f e_\alpha^*)) + \mathrm{tr}(v_\alpha^*(e_\alpha - \mathbb{1})f^*)$$
$$= \mathrm{tr}([(e_\alpha - \mathbb{1})u|f|^{1/2}](|f|^{1/2}e_\alpha^* u_\alpha^*))$$
$$\quad + \mathrm{tr}([(e_\alpha - \mathbb{1})v|f^*|^{1/2}][|f^*|^{1/2}v_\alpha^*])$$
$$\leq \|(e_\alpha - \mathbb{1})u|f|^{1/2}\|_2 \cdot \||f|^{1/2}\|_2 + \|(e_\alpha - \mathbb{1})u|f|^{1/2}\|_2 \cdot \||f^*|^{1/2}\|_2.$$

The result now clearly follows from the strong convergence of (e_α) applied to the Haagerup standard form for $L^2(\mathcal{M})$. $\square$

The following lemma complements Proposition 10.40.

Lemma 11.4 *Let $h = h_\varphi = \frac{d\widetilde{\varphi}}{d\tau}$ be the density of the dual weight $\widetilde{\varphi}$ of the canonical faithful normal semifinite weight φ on $\mathcal{M}$.*

(1) *Let $\mathbf{f}^{1\cap\infty}(h)$ be as described in the discussion preceding Proposition 10.22. Then, for any $a \in \mathfrak{n}_\varphi$, we have that $a\mathbf{f}^{1\cap\infty}(h)^{1/2}$ is closable with $[a\mathbf{f}^{1\cap\infty}(h)^{1/2}] \in \widetilde{\mathfrak{M}}$. In addition, $\mathbf{f}^{1\cap\infty}(h)^{1/2}a^* = [a\mathbf{f}^{1\cap\infty}(h)^{1/2}]^*$.*

(2) *Let $\iota^{(1)}$ and $\iota^{(\infty)}$ be the embeddings of $\mathcal{M}$ and $L^1(\mathcal{M})$ into $L^{1+\infty}(\mathcal{M})$, as described in Proposition 10.22. For any $a, b \in \mathfrak{n}_\varphi$, $x \in \mathcal{M}$ and $y \in L^1(\mathcal{M})$, we then have that*

$$[a\mathbf{f}^{1\cap\infty}(h)^{1/2}]\iota^{(\infty)}(x)(\mathbf{f}^{1\cap\infty}(h)^{1/2}b^*) = [ah^{1/2}]x(h^{1/2}b^*)$$

and

$$[a\mathbf{f}^{1\cap\infty}(h)^{1/2}]\iota^{(1)}(y)(\mathbf{f}^{1\cap\infty}(h)^{1/2}b^*) = ayb^*.$$

Proof. Part (1) basically records the final part of Remark 11.4. With part (1) as the foundation, the proof of part (2) runs along much the same lines as the proof of Lemma 10.48. We leave this as an exercise. $\qquad\square$

Lemma 11.5 *There exists a net $(a_\alpha) \subset \mathfrak{n}_\varphi$ such that $(a_\alpha^* a_\alpha)$ increases to $\mathbb{1}$. For any such net, we have that $\mathbb{1} = \vee_\alpha \mathbf{s}_l(\mathbf{f}^{1\cap\infty}(h)^{1/2}a_\alpha^*) = \vee_\alpha \mathbf{s}_r([a\mathbf{f}^{1\cap\infty}(h)^{1/2}]).$*

Proof. First note that with φ represented as in part (6) of Theorem 4.4, the partial sums from that representation yield a net of functionals $(\omega_\alpha) \subset \mathcal{M}_*^+$ increasing to φ. It then follows from Proposition 9.33 that the densities $h_\alpha = \frac{d\widetilde{\omega}_\alpha}{d\tau} \in L^1(\mathcal{M})_+$ increase to $h = h_\varphi = \frac{d\widetilde{\varphi}}{d\tau}$ - the density of the dual weight $\widetilde{\varphi}$. By Proposition 10.42, each h_α is then for some $a_\alpha \in \mathfrak{n}_\varphi$ of the form $h_\alpha = \mathfrak{i}^{(1)}(a_\alpha^* a_\alpha)$.

Given $b \in \mathfrak{m}_\varphi$, we may then use Lemma 10.48 to see that $tr(bh_\alpha) = tr(b\mathfrak{i}^{(1)}(a_\alpha^* a_\alpha)) = tr(\mathfrak{i}^{(1)}(b)a_\alpha^* a_\alpha)$ for each α. In particular, if $\alpha \geq \beta$ and $b \geq 0$, we will then have $tr(\mathfrak{i}^{(1)}(b)(a_\alpha^* a_\alpha - a_\beta^* a_\beta)) = tr(b(h_\alpha - h_\beta)) \geq 0$. Since $\mathfrak{i}^{(1)}(\mathfrak{m}_\varphi^+)$ is norm dense in $L_+^1(\mathcal{M})$ (see Theorem 10.46), we may therefore conclude from L^p-duality that $a_\alpha^* a_\alpha \geq a_\beta^* a_\beta$.

Since for each $b \in \mathfrak{m}_\varphi^+$ we have $tr(\mathfrak{i}^{(1)}(b)a_\alpha^* a_\alpha) = tr(b^{1/2}h_\alpha b^{1/2}) \nearrow tr(b^{1/2} \cdot h \cdot b^{1/2}) = tr([b^{1/2}h^{1/2}](h^{1/2}b^{1/2})) = tr((h^{1/2}b^{1/2})[b^{1/2}h^{1/2}]) = tr((\mathfrak{i}^{(1)}(b)))$, a similar application of Theorem 10.46 and L^p-duality then shows that $(a_\alpha^* a_\alpha)$ increases to $\mathbb{1}$.

We proceed to proving the second claim. So, let $(a_\alpha) \subset \mathfrak{n}_\varphi$ be a net for which $(a_\alpha^* a_\alpha)$ increases monotonically to $\mathbb{1}$. We show that the left supports $\mathbf{s}_l(\mathbf{f}^{1\cap\infty}(h)^{1/2}a_\alpha^*)$ must then increase monotonically to $\mathbb{1}$; that is, $\mathbb{1} = \vee_\alpha \mathbf{s}_l(\mathbf{f}^{1\cap\infty}(h)^{1/2}a_\alpha^*)$. To see this, notice that $\mathbf{s}_l(\mathbf{f}^{1\cap\infty}(h)^{1/2}a_\alpha^*) = \mathbf{s}((\mathbf{f}^{1\cap\infty}(h)^{1/2}a_\alpha^*)[a_\alpha\mathbf{f}^{1\cap\infty}(h)^{1/2}])$. Now, let ρ_α and ρ be the normal weights on $\mathfrak{M}$ associated with $(\mathbf{f}^{1\cap\infty}(h)^{1/2}a_\alpha^*)[a_\alpha\mathbf{f}^{1\cap\infty}(h)^{1/2}]$ and $\mathbf{f}^{1\cap\infty}(h)$ respectively, as described in Theorem 4.26. Since by construction $(\mathbf{f}^{1\cap\infty}(h)^{1/2}a_\alpha^*)[a_\alpha\mathbf{f}^{1\cap\infty}(h)^{1/2}]$ increases monotonically to $\mathbf{f}^{1\cap\infty}(h)$, ρ_α must by the same theorem increase to ρ. Now, notice that since h is injective with dense range, so is $\mathbf{f}^{1\cap\infty}(h)$. Therefore, again by Theorem 4.26, the weight ρ is faithful. Let e be the projection $\mathbb{1} - \vee_\alpha \mathbf{s}((\mathbf{f}^{1\cap\infty}(h)^{1/2}a_\alpha^*)[a_\alpha\mathbf{f}^{1\cap\infty}(h)^{1/2}])$. Since by Theorem 4.26 $\mathrm{supp}(\rho_\alpha) = \mathbf{s}((\mathbf{f}^{1\cap\infty}(h)^{1/2}a_\alpha^*)[a_\alpha\mathbf{f}^{1\cap\infty}(h)^{1/2}])$, we must then have that $\rho_\alpha(e) = 0$ for each α, which in turn ensures that $\rho(e) = \sup_\alpha \rho_\alpha(e) = 0$. The faithfulness of ρ then ensures that $e = 0$, which proves the claim. $\qquad\square$

Lemma 11.6

(a) *For any $x, y \in \mathfrak{m}_\varphi$, we will in the notation of Section 10.4 have $\mathrm{tr}(\mathfrak{i}^{(1)}(x)y) = \mathrm{tr}(x\mathfrak{i}^{(1)}(y))$. (When required, we shall for the sake of convenience write this as $\mathrm{tr}([h^{1/2}xh^{1/2}]y) = \mathrm{tr}(x[h^{1/2}yh^{1/2}])$.)*

(b) *Let $g_\infty \in M$ and $g_1 \in L^1(\mathcal{M})$ be given. Then $\iota^{(\infty)}(g_\infty) = \iota^{(1)}(g_1)$ (whence $\iota^{(\infty)}(g_\infty) \in \mathscr{L}^1 \cap \mathscr{L}^\infty$) if and only if $\mathrm{tr}(\mathfrak{i}^{(1)}(y)g_\infty) = \mathrm{tr}(yg_1)$ for all $y \in \mathfrak{m}_\varphi$.*

Proof. Part (a) follows from a combination of Lemma 10.48 and Theorem 10.50.

Next, we prove the 'if' part of part (b). Let $g_\infty \in M$ and $g_1 \in L^1(\mathcal{M})$ be given with $\mathrm{tr}(\mathfrak{i}^{(1)}(y)g_\infty) = \mathrm{tr}(yg_1)$ for all $y \in \mathfrak{m}_\varphi$. For (a_α) as in Lemma 11.5, we will for any $b \in M$ and any α and β then have that

$$\mathrm{tr}(b[a_\beta\mathbf{f}^{1\cap\infty}(h)^{1/2}]\iota^{(\infty)}(g_\infty)(\mathbf{f}^{1\cap\infty}(h)^{1/2}a_\alpha))$$

$$= \mathrm{tr}(b[a_\beta\mathbf{f}^{1\cap\infty}(h)^{1/2}](\mathbf{f}_{1+\infty}(h)^{1/2}g_\infty\mathbf{f}_{1+\infty}(h)^{1/2})(\mathbf{f}^{1\cap\infty}(h)^{1/2}a_\alpha))$$

$$= \mathrm{tr}(b[a_\beta h^{1/2}]g_\infty(h^{1/2}a_\alpha))$$

$$= \mathrm{tr}((h^{1/2}a_\alpha)b[a_\beta h^{1/2}]g_\infty)$$

$$= \mathrm{tr}(\mathfrak{i}^{(1)}(a_\alpha b a_\beta)g_\infty)$$

$$= \mathrm{tr}(a_\alpha b a_\beta g_1)$$

$$= \mathrm{tr}(ba_\beta g_1 a_\alpha)$$

$$= \mathrm{tr}(b[a_\beta\mathbf{f}^{1\cap\infty}(h)^{1/2}](\mathbf{f}^{1\cap\infty}(h)^{-1/2}g_1\mathbf{f}^{1\cap\infty}(h)^{-1/2})(\mathbf{f}^{1\cap\infty}(h)^{1/2}a_\alpha))$$

$$= \mathrm{tr}(b[a_\beta\mathbf{f}^{1\cap\infty}(h)^{1/2}]\iota^{(1)}(g_1)(\mathbf{f}^{1\cap\infty}(h)^{1/2}a_\alpha)).$$

In the second and last equality, we used the following fact: Since $e_1 \mathbf{f}^{1 \cap \infty}(h)^{1/2}$ is closable and $\mathbf{f}_{1+\infty}(h)^{1/2}$ bounded, we must have that $[e_1 \mathbf{f}^{1 \cap \infty}(h)^{1/2}] \mathbf{f}_{1+\infty}(h)^{1/2}$ is closed, and hence an extension of the minimal closure of $e_1 \mathbf{f}^{1 \cap \infty}(h)^{1/2} \mathbf{f}_{1+\infty}(h)^{1/2} = e_1 h^{1/2}$. Since the equality verified above holds for all $b \in \mathcal{M}$, we must have that

$$[a_\beta \mathbf{f}^{1 \cap \infty}(h)^{1/2}] \iota^{(\infty)}(g_\infty)(\mathbf{f}^{1 \cap \infty}(h)^{1/2} a_\alpha)$$
$$= [a_\beta \mathbf{f}^{1 \cap \infty}(h)^{1/2}] \iota^{(1)}(g_1)(\mathbf{f}^{1 \cap \infty}(h)^{1/2} a_\alpha)$$

for all α, β. It then follows from the second claim in Lemma 11.5 that $\iota^{(\infty)}(g_\infty) = \iota^{(1)}(g_1)$.

Conversely, suppose that we are given $g_\infty \in \mathcal{M}$ and $g_1 \in L^1(\mathcal{M})$ with $\iota^{(\infty)}(g_\infty) = \iota^{(1)}(g_1)$. For any $y_1, y_0 \in \mathfrak{n}_\varphi$, we then have that

$$\begin{aligned}
& \operatorname{tr}((y_0^* y_1) g_1) \\
= \; & \operatorname{tr}(y_1 g_1 y_0^*) \\
= \; & \operatorname{tr}([y_1 \mathbf{f}^{1 \cap \infty}(h)^{1/2}](\mathbf{f}^{1 \cap \infty}(h)^{-1/2} g_1 \mathbf{f}^{1 \cap \infty}(h)^{-1/2})(\mathbf{f}^{1 \cap \infty}(h)^{1/2} y_0^*)) \\
= \; & \operatorname{tr}([y_1 \mathbf{f}^{1 \cap \infty}(h)^{1/2}] \iota^{(1)}(g_1)(\mathbf{f}^{1 \cap \infty}(h)^{1/2} y_0^*)) \\
= \; & \operatorname{tr}([y_1 \mathbf{f}^{1 \cap \infty}(h)^{1/2}] \iota^{(\infty)}(g_\infty)(\mathbf{f}^{1 \cap \infty}(h)^{1/2} y_0^*)) \\
= \; & \operatorname{tr}([y_1 \mathbf{f}^{1 \cap \infty}(h)^{1/2}](\mathbf{f}_{1+\infty}(h)^{1/2} g_\infty \mathbf{f}_{1+\infty}(h)^{1/2})(\mathbf{f}^{1 \cap \infty}(h)^{1/2} y_0^*)) \\
= \; & \operatorname{tr}([y_1 h^{1/2}] g_\infty (h^{1/2} y_0^*)) \\
= \; & \operatorname{tr}((h^{1/2} y_0^*) a [y_1 h^{1/2}] g_\infty) \\
= \; & \operatorname{tr}(\mathbf{i}^{(1)}(y_0^* y_1) g_\infty),
\end{aligned}$$

which proves the claim. $\qquad\square$

Lemma 11.7 *Let $(x_\alpha) \subseteq \mathcal{M}$ be given such that $(\iota^{(\infty)}(x_\alpha))$ is a $\| \cdot \|_{(\cap)}$-bounded net in $\mathscr{L}^1 \cap \mathscr{L}^\infty$. Let $x \in \mathcal{M}$ be given such that $\iota^{(\infty)}(x) \in \mathscr{L}^1 \cap \mathscr{L}^\infty$. If $\lim_\alpha \|x_1 - x_{\alpha,1}\|_1 = 0$ where $(x_{\alpha,1})$ and x_1 are those unique elements of $L^1(\mathcal{M})$ for which $\iota^{(1)}(x_1) = \iota^{(\infty)}(x)$ and $\iota^{(1)}(x_{\alpha,1}) = \iota^{(\infty)}(x_\alpha)$ for each α, then (x_α) is σ-weakly convergent to x.*

Proof. Let (x_α), x, x_1 and $x_{\alpha,1}$ be as in the hypothesis. For any $y \in \mathfrak{m}_\varphi$, we may then use part (b) of the preceding lemma, and the convergence assumption on $(x_{\alpha,1})$, to see that then

$$\mathrm{tr}(\mathrm{i}^{(1)}(y)x_\alpha) = \mathrm{tr}(yx_{\alpha,1}) \to \mathrm{tr}(yx_1) = \mathrm{tr}(\mathrm{i}^{(1)}(y)x_\alpha).$$

On using the facts that $\mathrm{i}^{(1)}(\mathfrak{m}_\varphi)$ is norm dense in $L^1(\mathcal{M})$ and (x_α) $\|\cdot\|_{(\cap)}$-bounded (and hence also $\|\cdot\|_{\mathscr{L}^1}$-bounded), we may conclude from this fact that $\mathrm{tr}(bx_\alpha) \to \mathrm{tr}(bx)$ for all $b \in L^1(\mathcal{M})$. Equivalently, (x_α) is σ-weakly convergent to x. $\qquad\square$

Theorem 11.8 *Let $g \in \mathscr{L}^1 \cap \mathscr{L}^\infty$ be given. Then, there exists a net $(g_\alpha) \subseteq \mathfrak{m}_\varphi$ such that the following holds:*

- (i) $\sup_\alpha \|\iota^{(\infty)}(g_\alpha)\|_{(\cap)} < \infty$;
- (ii) *with $g_\infty \in \mathcal{M}$ chosen so that $\iota^{(\infty)}(g_\infty) = g$, the net (g_α) is σ-weakly convergent to g_∞;*
- (iii) *with $g_1 \in L^1(\mathcal{M})$ chosen so that $\iota^{(1)}(g_1) = g$, we have that $\lim_\alpha \|\mathrm{i}^{(1)}(g_\alpha) - g_1\|_1 = 0$.*

Proof. Given $\delta > 0$, select (e_α) as in Proposition 10.43. For each α, set $g_\alpha = \sigma^\varphi_{i/2}(e_\alpha) g \sigma^\varphi_{i/2}(e_\alpha)^*$. Moreover, since $(\sigma^\varphi_{i/2}(e_\alpha)) \subseteq \mathfrak{n}_\varphi \cap \mathfrak{n}_\varphi^*$, it is clear that $(g_\alpha) \subseteq \mathfrak{m}_\varphi$. By part (b) of Proposition 10.43, we have that $\sup_\alpha \|g_\alpha\|_\infty \le e^{\delta/2}\|g\|_\infty$. Next, let $a \in \mathfrak{m}_\varphi$ be given. Then, by Corollary 10.45 (and mild abuse of notation for clarity)

$$\mathrm{tr}(a\mathrm{i}^{(1)}(g_\alpha)) = \mathrm{tr}(\mathrm{i}^{(1)}(a)\sigma^\varphi_{i/2}(e_\alpha)g\sigma^\varphi_{i/2}(e_\alpha)^*)$$

$$= \mathrm{tr}(\sigma^\varphi_{i/2}(e_\alpha)^*(h^{1/2}ah^{1/2})\sigma^\varphi_{i/2}(e_\alpha)g)$$

$$= \mathrm{tr}((h^{1/2}e_\alpha^* a e_\alpha h^{1/2})g)$$

$$= \mathrm{tr}(\mathrm{i}^{(1)}(e_\alpha^* a e_\alpha)g)$$

$$= \mathrm{tr}(e_\alpha^* a e_\alpha g_1).$$

Thus, the functionals $a \mapsto \mathrm{tr}(a\mathrm{i}^{(1)}(g_\alpha))$ and $a \mapsto \mathrm{tr}(e_\alpha^* a e_\alpha g_1)$ agree on a σ-weakly dense subspace of $\mathcal{M}$. Since both functionals are σ-weakly continuous, they must therefore agree on all of $\mathcal{M}$. Having noted this, it now follows from Lemma 11.3 that $\lim_\alpha \|\mathrm{i}^{(1)}(g_\alpha) - g_1\|_1 = 0$. Since for any α

$$\|\mathrm{i}^{(1)}(g_\alpha)\|_1 = \|\mathrm{tr}(e_\alpha^*(\cdot)e_\alpha g_1)\| \le \|e_\alpha\|_\infty^2 \|g_1\|_1 \le \|g_1\|_1$$

the net $(\iota^{(\infty)}(g_\alpha)) \subseteq$ is $\|\cdot\|_{(\cap)}$-bounded. Thus, part (ii) follows from Lemma 11.7. $\qquad\square$

Corollary 11.9 *For any $f, g \in \mathscr{L}^1 \cap \mathscr{L}^\infty$, we have that* $\operatorname{tr}(g_1 f_\infty) = \operatorname{tr}(g_\infty f_1)$.

Proof.   Select (g_α) as in the preceding theorem. By Lemma 11.6, we have that $\operatorname{tr}((i)^{(1)}(g_\alpha)f_\infty) = \operatorname{tr}(g_\alpha f_1)$. The claim now follows by passing to the limit. $\qquad\square$

Proof. Proof of Theorem 11.1   Let $g \in \mathscr{L}^1 \cap \mathscr{L}^\infty$ and $a \in \mathscr{L}^1 + \mathscr{L}^\infty$ be given. Then, a is of the form $a = \iota^{(\infty)}(b) + \iota^{(1)}(c)$ for some $b \in \mathcal{M}$ and $c \in L^1(\mathcal{M})$, while for g we may find (unique) elements $g_\infty \in \mathcal{M}$ and $g_1 \in L^1(\mathcal{M})$ such that $g = \iota^{(\infty)}(g_\infty) = \iota^{(1)}(g_1)$. Now, let $b_0 \in \mathcal{M}$ and $c_0 \in L^1(\mathcal{M})$ be another pair of elements for which $a = \iota^{(\infty)}(b_0) + \iota^{(1)}(c_0)$. But if that is the case, then surely $\iota^{(\infty)}(b_0 - b) = \iota^{(1)}(c - c_0)$, in which case $\iota^{(\infty)}(b_0 - b) = \iota^{(1)}(c - c_0)$ corresponds to an element of $\mathscr{L}^1 \cap \mathscr{L}^\infty$. We may, therefore, use Corollary 11.9 to see that

$$\operatorname{tr}(g_\infty b + g_1 c) = \operatorname{tr}(g_\infty b + g_1 c_0 + g_1(c - c_0))$$
$$= \operatorname{tr}(g_\infty b + g_1 c_0 + g_\infty(b_0 - b))$$
$$= \operatorname{tr}(g_\infty b_0 + g_1 c_0).$$

The value $\langle g, a \rangle_{\cap,+} = \operatorname{tr}(g_\infty b + g_1 c)$ is therefore uniquely defined, and independent of the particular representation of a. Next, observe that for b and c as above, it is an easy exercise to see that

$$|\langle g, a \rangle_{\cap,+}| = |\operatorname{tr}(g_\infty b + g_1 c)| \leq \|g\|_{(\cap)}(\|b\|_1 + \|c\|_\infty).$$

On taking the infimum over all possible decompositions of a, it follows that $|\langle g, a \rangle_{\cap,+}| \leq \|g\|_{(\cap)}\|a\|_{(+)}$ as required. $\qquad\square$

With the existence of the promised bilinear form established, we proceed with an analysis of the duality between $\mathscr{L}^1 \cap \mathscr{L}^\infty$ and $\mathscr{L}^1 + \mathscr{L}^\infty$. Consider the space $\mathcal{M} \times L^1(\mathcal{M})$ equipped with the norm $\|(a, b)\|_{max} = \max(\|a\|_\infty, \|b\|_1)$. It is then clear that the subspace $\{(a, i^{(1)}(a)) : a \in \mathfrak{m}_\varphi\}$ of $\mathcal{M} \times L^1(\mathcal{M})$ is an isometric copy of the subspace $(\iota^{(\infty)}(\mathfrak{m}_\varphi), \|\cdot\|_{(\cap)})$ of $\mathscr{L}^1 \cap \mathscr{L}^\infty$. We may equip $\mathcal{M} \times L^1(\mathcal{M})$ with the product topology generated by the weak* topology $\sigma(\mathcal{M}, \mathcal{M}_*)$ (σ-weak topology) on $\mathcal{M}$, and the norm topology on $L^1(\mathcal{M})$. It then follows from Theorem 11.8 that the closure of $\{(a, i^{(1)}(a)) : a \in \mathfrak{m}_\varphi\}$ in this topology is just $\{(g_\infty, g_1) : g_\infty \in \mathcal{M}, g_1 \in L^1(\mathcal{M}) \text{ with } \iota^{(\infty)}(g_\infty) = \iota^{(1)}(g_1)\}$, which is clearly an isometric copy of $\mathscr{L}^1 \cap \mathscr{L}^\infty$. Thus, as a subspace of $\mathcal{M} \times L^1(\mathcal{M})$, this copy of $\mathscr{L}^1 \cap \mathscr{L}^\infty$ represents the 'diagonal' of $\mathcal{M} \times L^1(\mathcal{M})$. We now use this structure to introduce the following concept:

Definition 11.10 The relative topology on

$$\{(g_\infty, g_1) : g_\infty \in \mathfrak{m}_\varphi, g_1 \in L^1(\mathcal{M}) \text{ with } \iota^{(\infty)}(g_\infty) = \iota^{(1)}(g_1)\}$$

inherited from the $(\sigma(\mathcal{M}, \mathcal{M}_*) \times \|\cdot\|_1)$-product topology canonically determines a topology on $\mathscr{L}^1 \cap \mathscr{L}^\infty$ by means of the isometry noted above. We call this topology the $(\sigma - \|\cdot\|)$-topology.

Using this topology, we may now realise $\mathscr{L}^1 + \mathscr{L}^\infty$ as a space of functionals on $\mathscr{L}^1 \cap \mathscr{L}^\infty$.

Theorem 11.11 *Let ρ be a linear functional on $\mathscr{L}^1 \cap \mathscr{L}^\infty$. Then, the following are equivalent.*

> (i) *The functional ρ is of the form $\rho = \langle \cdot, a \rangle_{\cap,+}$ for some $a \in \mathscr{L}^1 + \mathscr{L}^\infty$.*
> (ii) *The functional ρ is $(\sigma - \|\cdot\|)$-continuous on $\mathscr{L}^1 \cap \mathscr{L}^\infty$.*
> (iii) *The functional ρ is $(\sigma - \|\cdot\|)$-continuous on $\|\cdot\|_{(\cap)}$-bounded subsets of $\mathscr{L}^1 \cap \mathscr{L}^\infty$.*

If one (and, therefore, all) of the above conditions hold, then $\|\rho\| = \|a\|_{(+)}$

We require some technical lemmata before we are able to prove this result.

Lemma 11.12 *Let ρ be a linear functional on $\mathscr{L}^1 \cap \mathscr{L}^\infty$ which is $(\sigma - \|\cdot\|)$-continuous on $\|\cdot\|_{(\cap)}$-bounded subsets of $\mathscr{L}^1 \cap \mathscr{L}^\infty$, and let $y \in \mathfrak{n}_\varphi$ be given. For any $f \in \mathcal{M}$, we have $\iota^{(\infty)}(y^* f y) \in \mathscr{L}^1 \cap \mathscr{L}^\infty$. The functional ϑ defined on $\mathcal{M}$ by $\vartheta(f) = \rho(\iota^{(\infty)}(y^* f y))$ $(f \in \mathcal{M})$ belongs to $\mathcal{M}_*$.*

Proof.   The first claim is an obvious consequence of the fact that $y \in \mathfrak{n}_\varphi$ ensures that $y^* \mathcal{M} y \subseteq \mathfrak{m}_\varphi$. To see that $\vartheta \in \mathcal{M}_*$, it suffices to show that ϑ is σ-strongly continuous on the unit ball of $\mathcal{M}$. So, let (f_α) be a net in the unit ball of $\mathcal{M}$ converging σ-strongly to $f \in \mathcal{M}$. If we are able to show that $(y^* f_\alpha y)$ is then $(\sigma - \|\cdot\|)$-convergent to $y^* f y$, the claim will follow from the assumption on ρ. Let $a \in \mathcal{M}$ be given. For any α, we then have that $\langle \iota^{(\infty)}(y^* f_\alpha y), \iota^{(\infty)}(a) \rangle_{\cap,+} = \mathrm{tr}((h^{1/2} y^* f_\alpha y h^{1/2})a)$, and similarly that

$$\langle \iota^{(\infty)}(y^* f y), \iota^{(\infty)}(a) \rangle_{\cap,+} = \mathrm{tr}((h^{1/2} y^* f y h^{1/2})a).$$

We may now use Hölder's inequality to see that

$$\mathrm{tr}(|(h^{1/2} y^* (f_\alpha - f) y h^{1/2})|) \leq \|h^{1/2} y^*\|_2 \|(f_\alpha - f)(y h^{1/2})\|_2$$

where $(y h^{1/2}) \in L^2(\mathcal{M})$. When applied to the Haagerup standard form for $L^2(\mathcal{M})$, the σ-strong convergence to 0 of $(f_\alpha - f)$ ensures that $\lim_\alpha \|(f_\alpha - f)(y h^{1/2})\|_2 = 0$ and hence that $\lim_\alpha \|i^{(1)}(y^* f_\alpha y) - i^{(1)}(y^* f y)\|_1 = \lim_\alpha \mathrm{tr}(|(h^{1/2} y^* (f_\alpha - f) y h^{1/2})|) = 0$. Since for each α we have $\|y^* f_\alpha y\|_\infty \leq \|y\|_\infty^2$ and $\|h^{1/2} y^* f_\alpha y h^{1/2}\|_1 \leq \|h^{1/2} y^*\|_2^2$, it is clear that $\sup_\alpha \|y^* f_\alpha y\|_{(\cap)} < \infty$. Lemma 11.7 now ensures that we also have $y^* f_\alpha y \to y^* f y$ in the σ-weak topology. Thus, as required $y^* f_\alpha y \to y^* f y$ in the $(\sigma - \|\cdot\|)$-topology. $\qquad\square$

Lemma 11.13 *Let ρ be a linear functional on $\mathscr{L}^1 \cap \mathscr{L}^\infty$ which is $(\sigma - \|\cdot\|)$-continuous on $\|\cdot\|_{(\cap)}$-bounded subsets of $\mathscr{L}^1 \cap \mathscr{L}^\infty$, and suppose that for some $\vartheta \in \mathcal{M}^*$ we have that $\rho(\iota^{(\infty)}(y)) = \vartheta(y)$ for all $y \in \mathfrak{m}_\varphi$. Then, there exists $\widetilde{\vartheta} \in \mathcal{M}_*$ with $\|\widetilde{\vartheta}\| \leq \|\vartheta\|$ such that $\rho(\iota^{(\infty)}(a)) = \widetilde{\vartheta}(a)$ for all $a \in \mathcal{M}$ satisfying $\iota^{(\infty)}(a) \in \mathscr{L}^1 \cap \mathscr{L}^\infty$. (So, loosely speaking, ρ is the 'restriction' of $\widetilde{\vartheta}$ to $\mathscr{L}^1 \cap \mathscr{L}^\infty$.)*

Proof. Let (e_α) be a right approximate identity for $\mathfrak{n}_\varphi$. For each α, define ϑ_α on $\mathcal{M}$ by setting $\vartheta_\alpha(a) = \vartheta(e_\alpha^* a e_\alpha)$ for all $a \in \mathcal{M}$. We clearly have $\|\vartheta_\alpha\| \leq \|\vartheta\|$ for each α.

Now, observe that since each $e_\alpha^* a e_\alpha$ belongs to $\mathfrak{m}_\varphi$, we have by hypothesis that $\vartheta_\alpha(a) = \rho(\iota^{(\infty)}(e_\alpha^* a e_\alpha))$ for all $a \in \mathcal{M}$. But then by Lemma 11.12, each ϑ_α belongs to $\mathcal{M}_*$.

Now, let the pair (π_u, H_u) be the universal representation of the C^*-algebra $\overline{\mathfrak{m}_\varphi}^{\|\cdot\|}$. We may then find vectors $\xi, \zeta \in H_u$ such that $\vartheta(a) = \langle \pi_u(y)\xi, \zeta \rangle_{H_u}$ for all $y \in \mathfrak{m}_\varphi$. The fact that $(\pi_u(e_\alpha))$ is now an approximate identity for $\pi_u(\overline{\mathfrak{m}_\varphi}^{\|\cdot\|})$ ensures $\pi_u(e_\alpha) \to \mathbb{1}$ strongly in $B(H_u)$. This ensures that as functionals on $\overline{\mathfrak{m}_\varphi}^{\|\cdot\|}$, $(\vartheta_\alpha \lceil \overline{\mathfrak{m}_\varphi})$ converges in norm to $\vartheta \lceil \overline{\mathfrak{m}_\varphi}$.

The C^*-algebra $\overline{\mathfrak{m}_\varphi}^{\|\cdot\|}$ is of course σ-weakly dense in $\mathcal{M}$. Using the fact that each ϑ_α belongs to $\mathcal{M}_*$, we will therefore for any pair of indices α_0, α_1 have that $\|\vartheta_{\alpha_0}\lceil\overline{\mathfrak{m}_\varphi} - \vartheta_{\alpha_1}\lceil\overline{\mathfrak{m}_\varphi}\|_{\overline{\mathfrak{m}_\varphi}} = \|\vartheta_{\alpha_0} - \vartheta_{\alpha_1}\|_{\mathcal{M}}$. Thus, (ϑ_α) is a Cauchy net in $\mathcal{M}_*$ and must therefore converge in norm to some $\widetilde{\vartheta} \in \mathcal{M}_*$. This norm convergence ensures that $\|\widetilde{\vartheta}\| = \lim_\alpha \|\vartheta_\alpha\| \leq \|\vartheta\|$ as required. From what we have already verified, we will for any $y \in \mathfrak{m}_\varphi$ then also have $\widetilde{\vartheta}(y) = \lim_\alpha \vartheta_\alpha(y) = \vartheta(y)$ and hence that $\widetilde{\vartheta}(y) = \rho(\iota^{(\infty)}(y))$.

Finally, let $g \in \mathscr{L}^1 \cap \mathscr{L}^\infty$ be given and select $(g_\beta) \subseteq \mathfrak{m}_\varphi$ such that it satisfies the hypothesis of Theorem 11.8. Then, (g_β) converges to g in the σ-weak topology and $(\iota^{(\infty)}(g_\beta))$ to $\iota^{(\infty)}(g)$ in the $(\sigma - \|\cdot\|)$-topology on a $\|\cdot\|_{(\cap)}$-bounded subset of $\mathscr{L}^1 \cap \mathscr{L}^\infty$. Therefore, $\rho(\iota^{(\infty)}(g)) = \lim_\beta \rho(\iota^{(\infty)}(g_\beta)) = \lim_\beta \widetilde{\vartheta}(g_\beta) = \widetilde{\vartheta}(g)$ as required. $\qquad\square$

Proof. Proof of Theorem 11.11   (i)$\Rightarrow$(ii): Suppose that the functional ρ is of the form $\rho = \langle\cdot, a\rangle_{\cap,+}$ for some $a \in \mathscr{L}^1 + \mathscr{L}^\infty$. Given $b_a \in \mathcal{M}$ and $c_a \in L^1(\mathcal{M})$ such that $a = \iota^{(\infty)}(b_a) + \iota^{(1)}(c_a)$, we will by definition have that $\rho(g) = \langle g, a\rangle_{\cap,+} = \mathrm{tr}(g_1 b_a) + \mathrm{tr}(g_\infty c_a)$ for all $g \in \mathscr{L}^1 \cap \mathscr{L}^\infty$, where $g_\infty \in \mathcal{M}$ and $g_1 \in L^1(\mathcal{M})$ have been chosen so that $\iota^{(\infty)}(g_\infty) = \iota^{(1)}(g_1) = g$. The functional $x \mapsto \mathrm{tr}(x b_a)$ is norm continuous on $L^1(\mathcal{M})$ and $y \mapsto \mathrm{tr}(y c_a)$ σ-weakly continuous on $\mathcal{M}$. Hence, the pair $(\mathrm{tr}(\cdot c_a), \mathrm{tr}(\cdot b_a))$ corresponds to a functional on $\mathcal{M} \times L^1(\mathcal{M})$, which is continuous with respect to the product topology generated by the σ-weak topology of $\mathcal{M}$ and norm topology of $L^1(\mathcal{M})$. The $(\sigma - \|\cdot\|)$-continuity of ρ now follows from Definition 11.10, and the discussion preceding it.

(ii)$\Rightarrow$(iii): This implication is obvious.

(iii)$\Rightarrow$(i): Let ρ be $(\sigma - \|\cdot\|)$-continuous on $\|\cdot\|_{(\cap)}$-bounded subsets of $\mathscr{L}^1 \cap \mathscr{L}^\infty$. Since the $\|\cdot\|_{(\cap)}$-topology is clearly stronger than the $(\sigma - \|\cdot\|)$-topology on $\mathscr{L}^1 \cap \mathscr{L}^\infty$, the functional is then a bounded linear functional on $(\mathscr{L}^1 \cap \mathscr{L}^\infty, \|\cdot\|_{(\cap)})$. As noted in the discussion preceding Definition 11.10, the space $(\mathscr{L}^1 \cap \mathscr{L}^\infty, \|\cdot\|_{(\cap)})$ may be viewed as the 'diagonal' of $(\mathcal{M} \times L^1(\mathcal{M}), \|\cdot\|_{max})$. So, under that identification, we may use the Hahn–Banach theorem to extend ρ to a bounded linear functional ρ_{ext} on $(\mathcal{M} \times L^1(\mathcal{M}), \|\cdot\|_{max})$ satisfying $\|\rho\| = \|\rho_{ext}\|$. The dual of $(\mathcal{M} \times L^1(\mathcal{M}), \|\cdot\|_{max})$ may, of course, be realised as $(\mathcal{M}^* \times \mathcal{M}, \|\cdot\|_{sum})$, where $\|(\vartheta, f)\|_{sum} = \|\vartheta\| + \|f\|$. So, there must exist $(\vartheta_\rho, b_\rho) \in (\mathcal{M}^* \times \mathcal{M}, \|\cdot\|_{sum})$ such that $\rho_{ext}(a, b) = (\vartheta_\rho, b_\rho)(x, y) = \vartheta_\rho(x) + \mathrm{tr}(b_\rho y)$

for all $(a, b) \in (\mathcal{M} \times L^1(\mathcal{M}))$, with $\|\rho\| = \|\rho_{ext}\| = \|\vartheta_\rho\| + \|b_\rho\|_\infty$. For any $g \in \mathscr{L}^1 \cap \mathscr{L}^\infty$, we specifically have that

$$\rho(g) = \rho_{ext}(g_\infty, g_1) = \vartheta_\rho(g_\infty) + \mathrm{tr}(b_\rho g_1) = \vartheta_\rho(g_\infty) + \langle g, \iota^{(\infty)}(b_\rho)\rangle_{\cap,+}$$

where $g_\infty \in \mathcal{M}$ and $g_1 \in L^1(\mathcal{M})$ are those unique elements for which $\iota^{(\infty)}(g_\infty) = g = \iota^{(1)}(g_1)$. The functional $\rho - \langle \cdot, \iota^{(\infty)}(b_\rho)\rangle_{\cap,+}$ therefore agrees with $\vartheta_\rho \circ (\iota^{(\infty)})^{-1}$ on $\mathscr{L}^1 \cap \mathscr{L}^\infty$. Therefore, we may apply Lemma 11.13 to obtain a functional $\widetilde{\vartheta}_\rho \in \mathcal{M}_*$ for which $\widetilde{\vartheta}_\rho(a) = \rho(\iota^{(\infty)}(a)) - \langle \iota^{(\infty)}(a), \iota^{(\infty)}(b_\rho)\rangle_{\cap,+}$ for all $a \in \mathcal{M}$ satisfying $\iota^{(\infty)}(a) \in \mathscr{L}^1 \cap \mathscr{L}^\infty$, and for which we further have $\|\widetilde{\vartheta}_\rho\| \leq \|\vartheta_\rho\|$. By the (L^1, L^∞)-duality, this functional is, of course, of the form $\widetilde{\vartheta}_\rho = \mathrm{tr}(c_\rho \cdot)$ for some $c_\rho \in L^1(\mathcal{M})$, where $\|c_\rho\|_1 = \|\widetilde{\vartheta}_\rho\|$. Therefore, it follows that ρ is precisely the functional $\langle \cdot, \iota^{(\infty)}(b_\rho) + \iota^{(1)}(c_\rho)\rangle_{\cap,+}$. Thus, (i) must hold.

It remains to prove the final claim. To see this, observe that it follows from the computations in the preceding paragraph that $\|b_\rho\|_\infty + \|c_\rho\|_1 = \|b_\rho\|_\infty + \|\widetilde{\vartheta}_\rho\| \leq \|b_\rho\|_\infty + \|\vartheta_\rho\| = \|\rho\|$. However, since $\rho = \langle \cdot, a\rangle_{\cap,+}$ where $a = \iota^{(\infty)}(b_\rho) + \iota^{(1)}(c_\rho)$, it follows from Theorem 11.1 and the definition of the norm on $\mathscr{L}^1 + \mathscr{L}^\infty$ that $\|\rho\| \leq \|g\|_{(+)} \leq \|b_\rho\|_\infty + \|c_\rho\|_1$. Hence, we must have $\rho = \|g\|_{(+)}$ as required. $\qquad\square$

11.2 A description of the context

Our primary objective in the next section is to show that the diagram $\mathscr{L}^1 \cap \mathscr{L}^\infty \hookrightarrow \mathscr{L}^p \hookrightarrow \mathscr{L}^1 + \mathscr{L}^\infty$ makes sense in the sense of continuous embeddings. Extensive preparation is necessary to lay the groundwork for proving this fact. We start by gathering some technical information, which will be used in the development of the theory.

Theorem 11.14 (Hadamard's three line theorem) *Let $a < b$ be given and let f be complex function which is continuous and bounded on the strip $\{z \in \mathbb{C} : a \leq \mathrm{Re}(z) \leq b\}$, and analytic on the interior of the strip. Then, the function $\log \circ M$ is convex on $[a, b]$, where M is the function $M : [a, b] \to [0, \infty)$ defined by $M(x) = \sup_{y \in \mathbb{R}} |f(x + iy)|$ for each $x \in [a, b]$. In other words, if for some $0 \leq t \leq 1$ we have $x = ta + (1-t)b$, then $M(x) \leq M(a)^t M(b)^{(1-t)}$.*

The above theorem is by now part of folklore and needs no proof. Armed with the above result and a deeper understanding of $\mathscr{L}^1 \cap \mathscr{L}^\infty$

and $\mathscr{L}^1 + \mathscr{L}^\infty$, we are now ready to describe the spaces within which we shall perform the interpolation. We will use $\mathcal{F}(\mathcal{M}, L^1(\mathcal{M}))$ to denote the set of functions f from the strip $S = \{z \in \mathbb{C} : 0 \le \mathrm{Re}(z) \le 1\}$ into $\mathscr{L}^1 + \mathscr{L}^\infty$, for which:

(i) f is bounded and continuous on S;

(ii) f is analytic on the interior S^o;

(iii) for every $t \in \mathbb{R}$, $f(it) \in \mathscr{L}^\infty$ with $\mathbb{R} \to \mathscr{L}^\infty : t \mapsto f(it)$ continuous and bounded;

(iv) for every $t \in \mathbb{R}$, $f(1+it) \in \mathscr{L}^1$ with $\mathbb{R} \to \mathscr{L}^1 : t \mapsto f(1+it)$ continuous and bounded.

For any $f \in \mathcal{F}(\mathcal{M}, L^1(\mathcal{M}))$, and any $z \in S^o$, we have that $S \to \mathbb{C} : z \mapsto \langle g, f(z) \rangle_{\cap,+}$ is analytic, with $|\langle g, f(it) \rangle_{\cap,+}| = |\mathrm{tr}(g_1 (\iota^{(\infty)})^{-1} (f(it)))| \le \|g_1\|_1 \cdot \|f(it)\|_{\mathscr{L}^\infty} \le \|f(it)\|_{\mathscr{L}^\infty}$ for each norm one element $g \in \mathscr{L}^1 \cap \mathscr{L}^\infty$. A similar argument shows $|\langle g, f(it) \rangle_{\cap,+}| \le \|f(1+it)\|_{\mathscr{L}^1}$. We may now use Hadamard's three line theorem to conclude that

$$|\langle g, f(z) \rangle_{\cap,+}| \le \max \left(\sup_{t \in \mathbb{R}} \|f(it)\|_{\mathscr{L}^\infty}, \sup_{t \in \mathbb{R}} \|f(1+it)\|_{\mathscr{L}^1} \right)$$

for each $f \in \mathcal{F}(\mathcal{M}, L^1(\mathcal{M}))$, each $z \in S$, and each norm one element $g \in \mathscr{L}^1 \cap \mathscr{L}^\infty$. Applying Theorem 11.11 yields the fact that

$$\|f(z)\|_{(+)} \le \max(\sup_{t \in \mathbb{R}} \|f(it)\|_{\mathscr{L}^\infty}, \sup_{t \in \mathbb{R}} \|f(1+it)\|_{\mathscr{L}^1}) \qquad (11.1)$$

for every $f \in \mathcal{F}(\mathcal{M}, L^1(\mathcal{M}))$ and every $z \in S$. In view of the above, the quantity $\|f\|_{\mathcal{F}} = \max(\sup_{t \in \mathbb{R}} \|f(it)\|_{\mathscr{L}^\infty}, \sup_{t \in \mathbb{R}} \|f(1+it)\|_{\mathscr{L}^1})$ is a natural norm to put on $\mathcal{F}(\mathcal{M}, L^1(\mathcal{M}))$. By classical results (see [**BL76**] for the proof strategy), the space $\mathcal{F}(\mathcal{M}, L^1(\mathcal{M}))$ is a Banach space under this norm.

We write $\mathcal{F}_0(\mathcal{M}, L^1(\mathcal{M}))$ for the closed subspace of $\mathcal{F}(\mathcal{M}, L^1(\mathcal{M}))$ whose elements additionally satisfy

(iii)$'$ $\|f(it)\|_{\mathscr{L}^\infty} \to 0$ as $|t| \to \infty$;

(iv)$'$ $\|f(1+it)\|_{\mathscr{L}^1} \to 0$ as $|t| \to \infty$.

With these identifications, we for each $p \in (1, \infty)$ then define $\mathscr{V}^p$ to be the space $\mathscr{V}^p = \{f(1/p) : f \in \mathcal{F}(\mathcal{M}, L^1(\mathcal{M})), f(1/p) = x\}$, equipped with the norm

$$\|x\|_{\mathcal{V}_p} = \inf\{\|f\|_{\mathcal{F}} \colon f \in \mathcal{F}_0(\mathcal{M}, L^1(\mathcal{M}))\}.$$

There is, of course, nothing special about the pair $(\mathscr{L}^1, \mathscr{L}^\infty)$. For any compatible couple (X_0, X_1) of complex Banach spaces, one may equally write $\mathcal{F}(X_0, X_1)$ for the set of functions f from the strip $S = \{z \in \mathbb{C} \colon 0 \le \mathrm{Re}(z) \le 1\}$ into $X_0 + X_1$, for which:

- (i) f is bounded and continuous on S;
- (ii) f is analytic on the interior S^o;
- (iii) for every $t \in \mathbb{R}$, $f(it) \in X_0$ with $\mathbb{R} \to X_0 : t \mapsto f(it)$ continuous and bounded;
- (iv) for every $t \in \mathbb{R}$, $f(1 + it) \in X_1$ with $\mathbb{R} \to X_1 : t \mapsto f(1 + it)$ continuous and bounded.

Then, repeat the above analysis. For any $0 \le \theta \le 1$, one may then define $(X_0, X_1)_\theta$ to be the space $\{f(\theta) \colon f \in \mathcal{F}(X_0, X_1), f(\theta) = x\}$. For future reference, we take note of two very deep and important theorems, which are foundational to this framework.

Theorem 11.15 (BL76, Theorem 4.1.2) *Given two compatible couples of Banach spaces (X_0, X_1) and (Y_0, Y_1), the pair $((X_0, X_1)_\theta, (Y_0, Y_1)_\theta)$ is an exact interpolation pair of exponent θ, that is if $T \colon X_0 + X_1 \to Y_0 + Y_1$ is a linear operator which is bounded from X_j to Y_j for $j = 0, 1$, then T is bounded from $(X_0, X_1)_\theta$ to $(Y_0, Y_1)_\theta$ with the norm satisfying $\|T\|_\theta \le \|T\|_0^{1-\theta} \|T\|_1^\theta$.*

Theorem 11.16 Complex reiteration theorem (see [BL76, Theorem 4.6.1]) *Let (X_0, X_1) be a compatible couple of complex Banach spaces and assume that $X_0 \cap X_1$ is dense in both X_0 and X_1. Let $A_0 = (X_0, X_1)_{\theta_0}$ and $A_1 = (X_0, X_1)_{\theta_1}$ where $0 \le \theta_0 \le \theta_1 \le 1$. If further $X_0 \cap X_1$ is dense in $A_0 \cap A_1$, then $(A_0, A_1)_\theta = (X_0, X_1)_\eta$ where $\eta = (1 - \theta)\theta_0 + \theta\theta_1$.*

The density condition is always satisfied if either $X_0 \subset X_1$ or $X_1 \subset X_0$.

Returning to our main objective of analysing the pair $(\mathcal{M}, L^1(\mathcal{M}))$, our immediate goal is to show that $\mathcal{V}^p$ is an isometric copy of $L^p(\mathcal{M})$. (Note that our notational convention deviates somewhat from that of [**BL76**]. In their analysis of the complex method, Bergh and Löfström write $\mathcal{F}(\mathcal{M}, L^1(\mathcal{M}))$ for what we call $\mathcal{F}_0(\mathcal{M}, L^1(\mathcal{M}))$ and do not really use our space $\mathcal{F}(\mathcal{M}, L^1(\mathcal{M}))$.

The next theorem is an implicit part of the foundational theory of the complex interpolation method. Here, we shall offer an outline of a proof.

Theorem 11.17 *Let X be a dual Banach space, and f a function from the strip $S = \{z \in \mathbb{C}\colon 0 \leq \mathrm{Re}(z) \leq 1\}$ into X for which the following conditions hold.*

> (i) *f is norm-bounded and point to σ-weak continuous on S.*
> (ii) *f is analytic on the interior S^o of S.*
> (iii) *Both $\mathbb{R} \to \mathbb{C} : t \mapsto f(it)$ and $\mathbb{R} \to \mathbb{C} : t \mapsto f(1 + it)$ are point to norm continuous.*

Then, f is point to norm continuous on S.

Outline of proof.   Let (P, Q) be the complementary pair of Poisson kernels for the strip S. (The interested reader may find a precise description of these kernels in [**Wid61**]. For some general theoretical background on these objects, see [**Kra06**].) The boundedness of f coupled with the point to norm continuity of each of $\mathbb{R} \to \mathbb{C} : t \mapsto f(it)$ and $\mathbb{R} \to \mathbb{C} : t \mapsto f(1 + it)$, ensures that the prescription

$$g(x + iy) = \int_{-\infty}^{\infty} f(it)P(x + iy, t)\, dt + \int_{-\infty}^{\infty} f(1 + it)Q(x + iy, t)\, dt$$

yields a well-defined point to norm continuous X-valued function on S. Now, let $v \in X_*$ be a σ-weakly continuous linear functional on X. For any $(x + iy) \in S$, we have that

$$(v \circ g)(x + iy) = \int_{-\infty}^{\infty} (v \circ f)(it)P(x + iy, t)\, dt$$
$$+ \int_{-\infty}^{\infty} (v \circ f)(1 + it)Q(x + iy, t)\, dt.$$

But by assumption (i) of the hypothesis, the above integral formula must yield $(v \circ f)(x + iy)$. In other words, for every $v \in X_*$ and every $(x + iy) \in S$, we have $(v \circ g)(x + iy) = (v \circ f)(x + iy)$. It is then clear that $g = f$, which proves the theorem. $\square$

11.3 L^p spaces as intermediate spaces of $\mathscr{L}^1 \cap \mathscr{L}^\infty$ and $\mathscr{L}^1 + \mathscr{L}^\infty$

We proceed with the microanalysis aimed at achieving the objective of showing that $\mathscr{V}^p$ as described above is an isometric copy of $L^p(\mathcal{M})$.

Lemma 11.18 *Let h be the density $h = \frac{d\widetilde{\varphi}}{d\tau}$. For any $a \in \mathfrak{n}_\varphi$ and any $2 \le r < \infty$, we have that $\|ah^{1/r}\|_r \le \mathbf{m}_{[af^{1\cap\infty}(h)^{1/2}]}(1) < \infty$.*

We remind the reader that the τ-measurability of $af^{1\cap\infty}(h)^{1/2}$ was verified in Remark 11.2.

Proof.   Let $2 \le r < \infty$ be given. Recall that $\mathbf{f}^{1\cap\infty}(t) = \max(1,t)$ for all $t \ge 0$. For any $1 \le p < \infty$, we clearly have $t^{1/p} \le \mathbf{f}^{1\cap\infty}(t)$. Setting $p = r/2$, the Borel functional calculus then yields the conclusion that $h^{2/r} \le \mathbf{f}^{1\cap\infty}(h)$. For any $a \in \mathfrak{n}_\varphi$, we therefore have that

$$\|ah^{1/r}\|_r = \|h^{1/r}a^*\|_r$$
$$= \|ah^{2/r}a^*\|_{r/2}^{1/2}$$
$$= \mathbf{m}_{ah^{2/r}a^*}(1)^{1/2}$$
$$\le \mathbf{m}_{af^{1\cap\infty}(h)a^*}(1)^{1/2}$$
$$= \mathbf{m}_{f^{1\cap\infty}(h)^{1/2}a^*}(1)$$
$$= \mathbf{m}_{[af^{1\cap\infty}(h)^{1/2}]}(1).$$

To see the final inequality, note that we saw in Remark 11.2 that $\mathbf{f}^{1\cap\infty}(h)^{1/2} \le \mathbb{1} + h^{1/2}$. Therefore, $\mathbf{m}_{[af^{1\cap\infty}(h)^{1/2}]}(1) = \mathbf{m}_{af^{1\cap\infty}(h)a^*}(1)^{1/2} \le \mathbf{m}_{a(\mathbb{1}+h^{1/2})^2a^*}(1)^{1/2} = \mathbf{m}_{[a(\mathbb{1}+h^{1/2})]}(1) \le \mathbf{m}_a(1/2) + \mathbf{m}_{[ah^{1/2}]}(1/2) = \|a\|_\infty + 2\|[ah^{1/2}]\|_2$. $\square$

Lemma 11.19   *Let $h = \frac{d\widetilde{\varphi}}{d\tau}$, and let $x \in \mathfrak{n}_\varphi$ be given. Then, the mappings $z \mapsto [xh^{z/2}]$ and $z \mapsto h^{z/2}x^*$ are differentiable on the open strip $S^o = \{z \in \mathbb{C}: 0 < \mathrm{Re}(z) < 1\}$, where for $z = s + it$ we understand $[xh^{z/2}]$ and $h^{z/2}x^*$ to be $[xh^{z/2}] = [xh^{s/2}]h^{it/2}$ and $h^{z/2}x^* = h^{it/2}(h^{s/2}x^*)$.*

Proof.   The proofs being similar, we only show that $z \mapsto h^{z/2}x^*$ is differentiable on the open strip S^o. To prove this claim, we first note that by Remark 11.2, we may write $h^{z/2}x^*$ as $[h^{z/2}\mathbf{f}^{1\cap\infty}(h)^{-1/2}](\mathbf{f}^{1\cap\infty}(h)^{1/2}x^*)$. Now, observe that $[h^{1/2}\mathbf{f}^{1\cap\infty}(h)^{-1/2}]$ is in fact a bounded operator in $\mathfrak{M}$. So, by Lemma 10.31, we find that $z \mapsto [h^{1/2}\mathbf{f}^{1\cap\infty}(h)^{-1/2}]^z$ is differentiable on the open right half-plane. But $\mathbf{f}^{1\cap\infty}(h)^{-1/2}$ is also an element of $\mathfrak{M}$. Moreover, if $z \in S^o$, then $1 - z$ will still belong to the open right half-plane. Hence, another application of Lemma 10.31 therefore shows that $z \mapsto [\mathbf{f}^{1\cap\infty}(h)^{-1/2}]^{1-z}$ is differentiable on the strip S^o. Thus, by the product rule, $z \mapsto [h^{1/2}\mathbf{f}^{1\cap\infty}(h)^{-1/2}]^z[\mathbf{f}^{1\cap\infty}(h)^{-1/2}]^{1-z} = [h^{z/2}\mathbf{f}^{1\cap\infty}(h)^{-1/2}]$ is differentiable on the open strip S^o. Thus, $z \mapsto h^{z/2}x^*$ is differentiable on the open strip S^o with derivative $\frac{1}{2}[\log(h)h^{z/2}\mathbf{f}^{1\cap\infty}(h)^{-1/2}](\mathbf{f}^{1\cap\infty}(h)^{1/2}x^*) = \frac{1}{2}[\log(h)h^{z/2}]x^*$. $\qquad\square$

Lemma 11.20   *Let h be the density $h = \frac{d\widetilde{\varphi}}{d\tau}$.*

- *For any $a \in \mathfrak{n}_\varphi$ and $t \in \mathbb{R}$, we have $\sigma_t^\varphi(a) \in \mathfrak{n}_\varphi$. The prescription $[ah^{1/2}] \mapsto [\sigma_t^\varphi(a)h^{1/2}]$ $(a \in \mathfrak{n}_\varphi)$ is, therefore, well-defined and for each $t \in \mathbb{R}$, extends to a unitary operator U_t on $L^2(\mathcal{M})$. The set $\{U_t\}$ is then a strongly continuous unitary group on $L^2(\mathcal{M})$, realised by the prescription $U_t(f) = h^{it}fh^{-it}$ $(f \in L^2(\mathcal{M}))$.*
- *For any $f \in L^1(\mathcal{M})$, we have $\mathrm{tr}(h^{it}fh^{-it}) = \mathrm{tr}(f)$ for all $t \in \mathbb{R}$. In addition, the formal prescription $L^1(\mathcal{M}) \to L^1(\mathcal{M}) : f \mapsto h^{it}fh^{-it}$ here also yields a strongly continuous group of isometric bijections on $L^1(\mathcal{M})$.*

Proof.   We will silently use Remark 10.41 throughout the proof. Let $a \in \mathfrak{n}_\varphi$ and $t \in \mathbb{R}$ be given. The first claim is a consequence of the fact that $\varphi \circ \sigma_t^\varphi = \varphi$ for all $t \in \mathbb{R}$ (see Theorem 8.10). We may similarly use this fact to conclude that

$$\begin{aligned}
\|[\sigma_t^\varphi(a)h^{1/2}]\|_2^2 &= \mathrm{tr}((h^{1/2}\sigma_t^\varphi(a^*))[\sigma_t^\varphi(a)h^{1/2}]) \\
&= \varphi(\sigma_t^\varphi(a^*)\sigma_t^\varphi(a)) \\
&= \varphi(a^*a) \\
&= \mathrm{tr}((h^{1/2}a^*)[ah^{1/2}]) \\
&= \|[ah^{1/2}]\|_2^2.
\end{aligned}$$

With a as before, we may next use Proposition 9.27 and Theorem 9.28 to see that $\sigma_t^\varphi(a)h^{1/2} = h^{it}ah^{-it}h^{1/2} = h^{it}ah^{1/2}h^{-it}$. We then clearly have $[\sigma_t^\varphi(a)h^{1/2}] = h^{it}[ah^{1/2}]h^{-it}$. Hence, for any fixed t, the prescription $[ah^{1/2}] \mapsto h^{it}[ah^{1/2}]h^{-it}$ yields an isometry on a dense subspace of $L^2(\mathcal{M})$. We denote the continuous extension of this isometry to all of $L^2(\mathcal{M})$ by U_t. It is obvious that the isometry U_{-t} is an inverse for U_t on the dense subspace $\{[ah^{1/2}] : a \in \mathfrak{n}_\varphi\}$, and hence on all of $L^2(\mathcal{M})$. This suffices to prove that U_t is in fact a unitary operator. By first considering their action on the dense subspace $\{[ah^{1/2}] : a \in \mathfrak{n}_\varphi\}$, and then extending by continuity, we similarly have that $U_t U_s = U_{t+s}$. We proceed to show that $t \mapsto U_t$ is strongly continuous. First, consider an element of the form $f = [ah^{1/2}]$ where $a \in \mathfrak{n}_\varphi$. In this case, KMS theory (Theorem 8.10) enables us to conclude that

$$\lim_{t \to 0} \|[[ah^{1/2}] - U_t([ah^{1/2}])]\|_2^2$$

$$= \lim_{t \to 0} \|[ah^{1/2}] - [\sigma_t^\varphi(a)h^{1/2}]\|^2$$

$$= \lim_{t \to 0} \operatorname{tr}(((h^{1/2}a^*) - (h^{1/2}\sigma_t^\varphi(a^*)))([ah^{1/2}] - [\sigma_t^\varphi(a)h^{1/2}]))$$

$$= \lim_{t \to 0}(\varphi(a^*a) - \varphi(a^*\sigma_t^\varphi(a)) - \varphi(\sigma_t^\varphi(a^*)a) + \varphi(\sigma_t^\varphi(a^*)\sigma_t^\varphi(a)))$$

$$= \lim_{t \to 0}[\varphi(a^*a) - \varphi(a^*\sigma_t^\varphi(a)) - \varphi(\sigma_t^\varphi(a^*)a) + \varphi(a^*a)]$$

$$= 0.$$

Next, let $f \in L^2(\mathcal{M})$ and $\epsilon > 0$ be given. We may then select $a \in \mathfrak{n}_\varphi$ so that $\|f - [ah^{1/2}]\|_2 < \frac{\epsilon}{2}$. The fact that each U_t is an isometry then enables us to conclude that $\|f - U_t(f)\|_2 \le \|f - [ah^{1/2}]\|_2 + \|[ah^{1/2}] - U_t([ah^{1/2}])\|_2 + \|U_t(f - [ah^{1/2}])\|_2 < \|[[ah^{1/2}] - U_t([ah^{1/2}])]\|_2 + \epsilon$. But then we must have $\limsup_{t \to 0} \|f - U_t(f)\|_2 \le \epsilon$. Since $\epsilon > 0$ was arbitrary, the only way this can be is if $\lim_{t \to 0} \|f - U_t(f)\|_2 = 0$.

We pass to the claims regarding $L^1(\mathcal{M})$. Notice that for $f \in L^1(\mathcal{M})$, $\|h^{it}fh^{-it}\|_1 = \mathbf{m}_{h^{it}fh^{-it}}(1) \le \mathbf{m}_f(1) = \|f\|_1$. Applying this inequality to $h^{-it}fh^{it}$ shows that in fact $\|h^{it}fh^{-it}\|_1 = \|f\|_1$ as claimed.

Next, observe that for any $a, b \in \mathfrak{n}_\varphi$ and $t \in \mathbb{R}$, we may again use KMS theory and what was shown in the first part of the proof, to see that

$$\operatorname{tr}(h^{it}(h^{1/2}b^*)[ah^{1/2}]h^{-it}) = \operatorname{tr}((h^{1/2}\sigma_t^\varphi(b^*))[\sigma_t^\varphi(a)h^{1/2}])$$

$$= \varphi(\sigma_t^\varphi(b^*a^*))$$

$$= \varphi(b^*a)$$

$$= \operatorname{tr}((h^{1/2}b^*)[ah^{1/2}]).$$

Thus, by Proposition 10.52, the first claim therefore holds on a dense subspace of $L^1(\mathcal{M})$. So, by continuity, it holds on all of $L^1(\mathcal{M})$.

Finally, let $g \in L^1(\mathcal{M})$ be given with polar decomposition $g = u|g|$. Then of course $u|g|^{1/2}, |g|^{1/2} \in L^2(\mathcal{M})$. Since then

$$g - h^{it}gh^{-it} = g - u|g|^{1/2}h^{it}|g|^{1/2}h^{-it} + u|g|^{1/2}h^{it}|g|^{1/2}h^{-it}$$
$$- h^{it}u|g|^{1/2}h^{-it}h^{it}|g|^{1/2}h^{-it} = u|g|^{1/2}(|g|^{1/2}$$
$$- U_t(|g|^{1/2})) + (u|g|^{1/2} - U_t(u|g|^{1/2}))U_t(|g|^{1/2}),$$

we may use Hölder's inequality to see that

$$\|g - h^{it}gh^{-it}\|_1 \leq \|u|g|^{1/2}\|_2 \| |g|^{1/2} - U_t(|g|^{1/2})\|_2$$
$$+ \|u|g|^{1/2} - U_t(u|g|^{1/2})\|_2 \| |g|^{1/2}\|_2.$$

It then clearly follows from what we have already proven that $\lim_{t \to 0} \|g - h^{it}gh^{-it}\|_1 = 0$. $\square$

Theorem 11.21 *Let $h = \frac{d\tilde{\varphi}}{d\tau}$, $x, y \in \mathfrak{n}_\varphi$ and $a \in L^p(\mathcal{M})$ ($1 < p < \infty$) be given with $\|a\|_p = 1$. With u denoting the partial isometry of the polar form $a = u|a|$ of a, the function $F : z \mapsto \mathrm{tr}(u|a|^{pz}(h^{(1-z)/2}y^*)[xh^{(1-z)/2}])$ is a well-defined bounded and continuous map on $S = \{z \in \mathbb{C} : 0 \leq \mathrm{Re}(z) \leq 1\}$, which is analytic on the interior S^o of S.*

Proof. We start the proof by clarifying exactly what needs to be done to prove the theorem. It is easy enough to check that $u|a|^{pz}(h^{(1-z)/2}y^*)$ $[xh^{(1-z)/2}]$ satisfies the membership criteria for L^1. Thus, F is well-defined. To prove the claims regarding F, we will show that each of the maps $z \mapsto |a|^{pz/2}(h^{(1-z)/2}y^*)$ and $z \mapsto [xh^{(1-z)/2}])u|a|^{pz/2}$ is then a well-defined bounded and continuous $L^2(\mathcal{M})$-valued map on $S = \{z \in \mathbb{C} : 0 \leq \mathrm{Re}(z) \leq 1\}$, which is analytic on the interior S^o of S. Then, $z \mapsto |a|^{pz/2}(h^{(1-z)/2}y^*)[xh^{(1-z)/2}])u|a|^{pz/2}$ is analytic on S^o and bounded on S, with the continuity on S following from the fact that for some constant $C > 0$, we will then have that

$$\| |a|^{pz/2}(h^{(1-z)/2}y^*)[xh^{(1-z)/2}])u|a|^{pz/2}$$
$$- |a|^{pz_0/2}(h^{(1-z_0)/2}y^*)[xh^{(1-z_0)/2}])u|a|^{pz_0/2}\|_1$$
$$\leq \| |a|^{pz/2}(h^{(1-z)/2}y^*)[xh^{(1-z)/2}])u|a|^{pz/2}$$

$$- |a|^{pz/2}(h^{(1-z)/2}y^*)[xh^{(1-z_0)/2}])u|a|^{pz_0/2}\|_1$$

$$+ \| |a|^{pz/2}(h^{(1-z)/2}y^*)[xh^{(1-z_0)/2}])u|a|^{pz_0/2}$$

$$- |a|^{pz_0/2}(h^{(1-z_0)/2}y^*)[xh^{(1-z_0)/2}])u|a|^{pz_0/2}\|_1$$

$$\leq \| |a|^{pz/2}(h^{(1-z)/2}y^*)\|_2 . \|[xh^{(1-z)/2}])u|a|^{pz/2}$$

$$- [xh^{(1-z_0)/2}])u|a|^{pz_0/2}\|_2 \| |a|^{pz/2}(h^{(1-z)/2}y^*)$$

$$- |a|^{pz_0/2}(h^{(1-z_0)/2}y^*)\|_2 . \|[xh^{(1-z_0)/2}])u|a|^{pz_0/2}\|_2$$

$$\leq C\|[xh^{(1-z)/2}])u|a|^{pz/2} - [xh^{(1-z_0)/2}])u|a|^{pz_0/2}\|_2$$

$$+ C\| |a|^{pz/2}(h^{(1-z)/2}y^*) - |a|^{pz_0/2}(h^{(1-z_0)/2}y^*)\|_2 .$$

To see that this suffices to prove the claim, observe that for $z = s + it$ with $0 \leq s \leq 1$, it follows that $u|a|^{ps}|a|^{ipt} = u|a|^{ipt/2}|a|^{ps}|a|^{ipt/2} = |a^*|^{ipt/2}u|a|^{ps}|a|^{ipt/2}$. We may now apply the second part of Lemma 11.20, to see that

$$\mathrm{tr}(u|a|^{pz}(h^{(1-z)/2}y^*)[xh^{(1-z)/2}])$$

$$= \mathrm{tr}(|a^*|^{ipt/2}u|a|^{ps}|a|^{ipt/2}h^{-it/2}(h^{(1-s)/2}y^*)[xh^{(1-s)/2}]h^{-it/2})$$

$$= \mathrm{tr}(h^{-it/2}|a^*|^{ipt/2}u|a|^{ps}|a|^{ipt/2}h^{-it/2}(h^{(1-s)/2}y^*)[xh^{(1-s)/2}]).$$

Since for any $s \in \mathbb{R}$ we have $\theta_s(h^{-it/2}|a^*|^{ipt/2}) = \theta_s(h)^{-it/2}|\theta_s(a^*)|^{ipt/2} = (e^{-s}h)^{-it/2}|e^{-s/p}a^*|^{ipt/2} = h^{-it/2}|a^*|^{it/2}$, it is clear that in fact $h^{-it/2}|a^*|^{it/2} \in L^\infty(\mathcal{M})$. Similarly, $|a|^{it/2}h^{-it/2} \in L^\infty(\mathcal{M})$. Thus, we may use Theorem 10.34 to conclude from the above that

$$\mathrm{tr}(u|a|^{pz}(h^{(1-z)/2}y^*)[xh^{(1-z)/2}])$$

$$= \mathrm{tr}(h^{-it/2}|a^*|^{ipt/2}u|a|^{ps}|a|^{ipt/2}h^{-it/2}(h^{(1-s)/2}y^*)[xh^{(1-s)/2}])$$

$$= \mathrm{tr}(|a|^{ps/2}|a|^{ipt/2}h^{-it/2}(h^{(1-s)/2}y^*)[xh^{(1-s)/2}]h^{-it/2}h^{-it/2}$$

$$|a^*|^{ipt/2}u|a|^{ps/2})$$

$$= \mathrm{tr}(|a|^{ps/2}|a|^{ipt/2}h^{-it/2}(h^{(1-s)/2}y^*)[xh^{(1-s)/2}]$$

$$h^{-it/2}h^{-it/2}u|a|^{ipt/2}|a|^{ps/2})$$

$$= \mathrm{tr}(|a|^{pz/2}(h^{(1-z)/2}y^*)[xh^{(1-z)/2}]u|a|^{pz/2}). \tag{11.2}$$

We now prove that each of the maps $z \mapsto |a|^{pz/2}(h^{(1-z)/2}y^*)$ and $z \mapsto [xh^{(1-z)/2}])u|a|^{pz/2}$ is a well-defined bounded and continuous $L^2(\mathcal{M})$-valued map on $S = \{z \in \mathbb{C}: 0 \leq \mathrm{Re}(z) \leq 1\}$, which is analytic on the interior S^o of S. The proofs being similar, we will only prove the claims made regarding the map $z \mapsto |a|^{pz/2}(h^{(1-z)/2}y^*)$. First, we investigate the claim regarding analyticity. We know from Lemma 10.31 that $z \mapsto |a|^{pz/2}$ is differentiable on the open right half-plane, and from Lemma 11.19, that $z \mapsto h^{(1-z)/2}y^*$ is similarly differentiable on the open strip S^o. A similar argument to that used in Lemma 10.32 then shows that $z \mapsto |a|^{pz/2}(h^{(1-z)/2}y^*)$ is an L^2-valued analytic function on the open strip S^o.

Our next task is to show that $z \mapsto |a|^{pz/2}(h^{(1-z)/2}y^*)$ is bounded on S. To see this, recall that for $z \in S$ with $z = s + it$, the expression $|a|^{pz/2}(h^{(1-z)/2}y^*)$ can be interpreted as $|a|^{ps/2}|a|^{ipt/2}h^{-it/2}(h^{(1-s)/2}y^*)$. Consequently, by Hölder's inequality and Lemma 11.18, we have that

$$\| |a|^{pz}(h^{(1-z)/2}y^*)\|_2$$

$$= \| |a|^{ps/2}|a|^{ipt/2}h^{-it/2}(h^{(1-s)/2}y^*)\|_2$$

$$\leq \| |a|^{ps/2}\|_{2/s}.\| |a|^{ipt/2}h^{-it/2}\|_{\infty}.\|h^{(1-s)/2}y^*\|_{(s-1)/2}$$

$$\leq \mathbf{m}_{|a|^{ps/2}}(1)\|h^{(1-s)/2}y^*\|_{(s-1)/2}$$

$$\leq \mathbf{m}_{|a|}(1)^{ps/2}\mathbf{m}_{\mathbf{f}^{1\cap\infty}(h)^{1/2}y^*}(1)$$

$$= \|a\|_p^{ps/2}\mathbf{m}_{\mathbf{f}^{1\cap\infty}(h)^{1/2}y^*}(1)$$

$$= \mathbf{m}_{\mathbf{f}^{1\cap\infty}(h)^{1/2}y^*}(1)$$

as required.

If we can now show that $z \mapsto |a|^{pz/2}(h^{(1-z)/2}y^*)$ is point–weakly continuous on S and norm continuous on the edges of S, the point to norm continuity of this map on all of S will follow from Theorem 11.17.

We first indicate how the norm continuity on the edges may be verified. First, consider the map $z \mapsto |a|^{pz/2}(h^{(1-z)/2}y^*)$ restricted to $z = 1 + it$ ($t \in \mathbb{R}$). By Stone's theorem, each of $t \mapsto |a|^{it/2}$ and $t \mapsto h^{-it/2}$ is strongly continuous on the underlying Hilbert spaces and hence so is $t \mapsto w_t = |a|^{it/2}h^{-it/2}$. But $\{w_t\} \subseteq L^\infty(\mathcal{M})$. So, by the Haagerup standard form for $L^2(\mathcal{M})$ (see Theorem 10.59), this strong continuity means that for any $f \in L^2(\mathcal{M})$, the map $t \mapsto w_t f$ is point to L^2-norm continuous. If we combine this fact with the first part of Lemma 11.20, it is clear that the map

$$t \mapsto |a|^{it/2}|a|^{p/2}h^{-it/2}y^* = |a|^{it/2}h^{-it/2}(h^{it/2}|a|^{p/2}h^{-it/2})y^*$$
$$= w_t(U_{t/2}(|a|^{p/2}))y^*$$

is a continuous map from $\mathbb{R}$ to $L^2(\mathcal{M})$ Now, consider the restriction of the map $z \mapsto |a|^{pz/2}(h^{(1-z)/2}y^*)$ to $z = it$ ($t \in \mathbb{R}$). The continuity of this map follows by similar reasoning once we have noted that

$$it \mapsto |a|^{ipt/2}h^{-it/2}(h^{1/2}y^*) = U_{t/2}(v_t(U_{-t/2}(h^{1/2}y^*)))$$

where in this case $\{v_t\} \subseteq L^\infty(\mathcal{M})$ is of the form $v_t = h^{-it/2}|a|^{ipt/2}$ for each t.

It remains to prove that $z \mapsto |a|^{pz/2}(h^{(1-z)/2}y^*)$ is weakly continuous on S. Since this map is known to be norm continuous on S^o, we only need to show that as $z \to z_0$ where $\mathrm{Re}(z_0) \in \{0,1\}$, we also have weak continuity. That is, for any $f \in L^2(\mathcal{M})$ we have that $\langle |a|^{pz/2}(h^{(1-z)/2}y^*), f\rangle$ tends to $\langle |a|^{pz_0/2}(h^{(1-z_0)/2}y^*), f\rangle$ as $z \to z_0$ (equivalently $\mathrm{tr}(f^*|a|^{pz/2}(h^{(1-z)/2}y^*)$ tends to $\mathrm{tr}(f^*|a|^{pz_0/2}(h^{(1-z_0)/2}y^*)$ as $z \to z_0$. The proofs of the cases where $s + it \to it_0$ and $s + it \to 1 + it_0$ are very similar, and hence, we will only prove the first case. We will show that $\mathrm{tr}(f^*|a|^{pz/2}(h^{(1-z)/2}y^*)) = \mathrm{tr}(f^*|a|^{ps/2}|a|^{ipt/2}h^{-it/2}(h^{(1-s)/2}y^*))$ converges uniformly to $\mathrm{tr}(f^*|a|^{ipt/2}h^{-it/2}(h^{1/2}y^*))$ with respect to t, as s decreases to 0. The uniformity of this convergence combined with the norm convergence of the boundary mappings will then enable us to conclude that

$$\lim_{(s+it)\to it_0} \mathrm{tr}(f^*|a|^{ps/2}|a|^{ipt/2}h^{-it/2}(h^{(1-s)/2}y^*)) \text{ exists}$$

with

$$\lim_{(s+it)\to it_0} \mathrm{tr}(f^*|a|^{ps/2}|a|^{ipt/2}h^{-it/2}(h^{(1-s)/2}y^*))$$
$$= \lim_{t\to t_0} \lim_{s\to 0} \mathrm{tr}(f^*|a|^{ps/2}|a|^{ipt/2}h^{-it/2}(h^{(1-s)/2}y^*))$$
$$= \mathrm{tr}(f^*|a|^{ipt_0/2}h^{-it_0/2}(h^{1/2}y^*)).$$

Observe that we may write

$$|a|^{ps/2}|a|^{ipt/2}h^{-it/2}(h^{(1-s)/2}y^*) - |a|^{ipt/2}h^{-it/2}(h^{1/2}y^*)$$
$$= (|a|^{ps/2} - \chi_{(0,\infty)}(|a|^{p/2}))|a|^{ipt/2}h^{-it/2}(h^{(1-s)/2}y^*))$$
$$+ |a|^{ipt/2}h^{-it/2}((h^{(1-s)/2}y^*) - (h^{1/2}y^*)).$$

We proceed with proving the uniform convergence of $\mathrm{tr}(f^*|a|^{pz/2}$ $(h^{(1-z)/2}y^*)) = \mathrm{tr}(f^*|a|^{ps/2}|a|^{ipt/2}h^{-it/2}(h^{(1-s)/2}y^*))$ to $\mathrm{tr}(f^*|a|^{ipt/2}$ $h^{-it/2}(h^{1/2}y^*))$ with respect to t, as s decreases to 0. As $s \in (0,1)$ decreases to 0, the function $t \mapsto t^s$ will increase to the constant function 1 on $(0,1)$ and, for any $\delta > 1$, will converge uniformly to 1 on $[1,\delta]$. Now, let $\epsilon > 0$ be given, and select $\delta > 0$ so that $\tau(\chi_{(\delta,\infty)}(|a|^{p/2})) < \epsilon$. (This is possible since $|a|^{p/2} \in \widetilde{\mathfrak{M}}$.) By the Borel functional calculus, $|a|^{sp/2}\chi_{[1,\delta]}(|a|^{p/2})$ will then converge uniformly to $\chi_{[1,\delta]}(|a|^{p/2})$ as $s \to 0$. This, in particular, shows that $|a|^{sp/2}\chi_{[1,\infty)}(|a|^{p/2})$ converges in measure to $\chi_{[1,\infty)}(|a|^{p/2})$ as $s \searrow 0$. The Borel functional calculus also shows that $|a|^{sp/2}\chi_{(0,1]}(|a|^{p/2})$ increases to $\chi_{(0,1]}(|a|^{p/2})$. Now, let $f \in L^2_+(\mathcal{M})$ be given. We know from Lemma 10.23 that $\mathbf{m}_f(t) = t^{-1/2}\mathbf{m}_f(1)$ for any $t > 0$, and hence that $\mathbf{m}_{f^{1/2}}(t) \to 0$ as $t \to \infty$. Applying Lemma 5.27, it now follows that $f^{1/2}(\chi_{(0,1]}(|a|^{p/2}) - |a|^{sp/2}\chi_{(0,1]}(|a|^{p/2}))$ converges to 0 in measure. *Observation 1:* If we combine that with what we have shown regarding $|a|^{sp/2}\chi_{[1,\infty)}(|a|^{p/2})$, it follows that

$$f^{1/2}(\chi_{(0,\infty)}(|a|^{p/2}) - |a|^{sp/2}\chi_{(0,1]}(|a|^{p/2})) \text{ converges to 0 in measure.}$$

With f as before, we now show that $((h^{(1-s)/2}y^*) - (h^{1/2}y^*))f^{1/2}$ also converges to 0 in measure as s decreases to 0. Using the same trick as in Lemma 11.19, we may write $((h^{(1-s)/2}y^*) - (h^{1/2}y^*))f^{1/2}$ as

$$([h^{(1-s)/2}\mathbf{f}^{1\cap\infty}(h)^{-1/2}] - [h^{1/2}\mathbf{f}^{1\cap\infty}(h)^{-1/2}])(\mathbf{f}^{1\cap\infty}(h)^{1/2}y^*).$$

The claim can then be verified by analysing the behaviour of the term $([h^{(1-s)/2}\mathbf{f}^{1\cap\infty}(h)^{-1/2}] - [h^{1/2}\mathbf{f}^{1\cap\infty}(h)^{-1/2}])$. To this end, consider the function $t \mapsto t^{(1-s)/2}\mathbf{f}^{1\cap\infty}(t)^{1/2} = t^{(1-s)/2}\max(1, t^{1/2})$. As s decreases to 0, $t^{(1-s)/2}\mathbf{f}^{1\cap\infty}(t)^{1/2}$ increases to $t^{1/2}\mathbf{f}^{1\cap\infty}(t)^{1/2}$ on $[1,\infty)$ and converges uniformly to $t^{1/2}\mathbf{f}^{1\cap\infty}(t)^{1/2}$ on $[0,1]$. The Borel functional calculus now enables us to conclude that as s decreases to 0, $[h^{(1-s)/2}\mathbf{f}^{1\cap\infty}(h)^{-1/2}]\chi_{[0,1]}(h)$ converges uniformly to $[h^{1/2}\mathbf{f}^{1\cap\infty}(h)^{-1/2}]\chi_{[0,1]}(h)$, with $[h^{(1-s)/2}\mathbf{f}^{1\cap\infty}(h)^{-1/2}]\chi_{(1,\infty)}(h)$ increasing to $[h^{1/2}\mathbf{f}^{1\cap\infty}(h)^{-1/2}]\chi_{(1,\infty)}(h)$. It is an easy exercise to see that each of $\mathbf{f}^{1\cap\infty}(h)^{-s/2}$ and $[h^{(1-s)/2}\mathbf{f}^{1\cap\infty}(h)^{(s-1)/2}]$ is, in fact, a contractive map (and so elements of $\mathfrak{M}$). Hence, the same is true of each $[h^{(1-s)/2}\mathbf{f}^{1\cap\infty}(h)^{-1/2}] = [h^{(1-s)/2}\mathbf{f}^{1\cap\infty}(h)^{(s-1)/2}]\mathbf{f}^{1\cap\infty}(h)^{-s/2}$. The inequality

$$\mathbf{m}_{(\mathbf{f}^{1\cap\infty}(h)^{1/2}y^*)f^{1/2}}(t) \leq \mathbf{m}_{(\mathbf{f}^{1\cap\infty}(h)^{1/2}y^*)}(t/2)\mathbf{m}_{f^{1/2}}(t/2)$$

then ensures that $\mathbf{m}_{(\mathbf{f}^{1\cap\infty}(h)^{1/2}y^*)f^{1/2}}(t) \to 0$ as $t \to \infty$.

Observation 2: Arguing as before (with an argument based on Lemma 5.27 and the fact noted earlier that $\mathbf{m}_{f^{1/2}}(t) \to 0$ as $t \to \infty$), it now follows that $([h^{(1-s)/2}\mathbf{f}^{1\cap\infty}(h)^{-1/2}] - [h^{1/2}\mathbf{f}^{1\cap\infty}(h)^{-1/2}])(\mathbf{f}^{1\cap\infty}(h)^{1/2}y^*)f^{1/2} = ((h^{(1-s)/2}y^*) - (h^{1/2}y^*))f^{1/2}$ converges to zero in measure as s decreases to 0.

Notice that

$$f^{1/2}(|a|^{ps/2}|a|^{ipt/2}h^{-it/2}(h^{(1-s)/2}y^*) - |a|^{ipt/2}h^{-it/2}(h^{1/2}y^*))f^{1/2}$$
$$= f^{1/2}(|a|^{ps/2} - \chi_{(0,\infty)}(|a|^{p/2}))|a|^{ipt/2}h^{-it/2}(h^{(1-s)/2}y^*))f^{1/2}$$
$$+ f^{1/2}(|a|^{ipt/2}h^{-it/2})((h^{(1-s)/2}y^*) - (h^{1/2}y^*))f^{1/2}. \qquad (11.3)$$

We will use this formula to show that $f^{1/2}(|a|^{ps/2}|a|^{ipt/2}h^{-it/2}(h^{(1-s)/2}y^*) - |a|^{ipt/2}h^{-it/2}(h^{1/2}y^*))f^{1/2}$ converges to 0 in measure as s decreases to 0, and that this convergence is uniform with respect to t.

In our proof of this fact, we will for the sake of clarity momentarily depart from our usual convention and write $t \mapsto \mathbf{m}(t; g)$ for the decreasing rearrangement of an element $g \in \widetilde{\mathfrak{M}}$, rather than $t \mapsto \mathbf{m}_g(t)$. Observe that for any $r > 0$, we will have

$$\mathbf{m}(r; f^{1/2}(|a|^{ps/2} - \chi_{(0,\infty)}(|a|^{p/2}))|a|^{ipt/2}h^{-it/2}(h^{(1-s)/2}y^*))f^{1/2})$$
$$\leq \mathbf{m}(r/2; f^{1/2}(|a|^{ps/2} - \chi_{(0,\infty)}(|a|^{p/2}))).$$
$$\mathbf{m}(r/2; |a|^{ipt/2}h^{-it/2}(h^{(1-s)/2}y^*))f^{1/2})$$
$$\leq \mathbf{m}(r/2; f^{1/2}(|a|^{ps/2} - \chi_{(0,\infty)}(|a|^{p/2}))).\| |a|^{ipt/2}h^{-it/2}\|.$$
$$\mathbf{m}(r/2; (h^{(1-s)/2}y^*))f^{1/2})$$
$$\leq \mathbf{m}(r/2; f^{1/2}(|a|^{ps/2} - \chi_{(0,\infty)}(|a|^{p/2}))).\mathbf{m}(r/2; (h^{(1-s)/2}y^*))f^{1/2})$$
$$= \mathbf{m}(r/2; f^{1/2}(|a|^{ps/2} - \chi_{(0,\infty)}(|a|^{p/2}))).$$
$$\mathbf{m}(r/2; [h^{(1-s)/2}\mathbf{f}^{1\cap\infty}(h)^{-1/2}](\mathbf{f}^{1\cap\infty}(h)^{1/2}y^*)f^{1/2})$$
$$\leq \mathbf{m}(r/2; f^{1/2}(|a|^{ps/2} - \chi_{(0,\infty)}(|a|^{p/2}))).\|[h^{(1-s)/2}\mathbf{f}^{1\cap\infty}(h)^{-1/2}]\|.$$
$$\mathbf{m}(r/2; (\mathbf{f}^{1\cap\infty}(h)^{1/2}y^*)f^{1/2})$$
$$\leq \mathbf{m}(r/2; f^{1/2}(|a|^{ps/2} - \chi_{(0,\infty)}(|a|^{p/2}))).\mathbf{m}(r/2; \mathbf{f}^{1\cap\infty}(h)^{1/2}y^*)f^{1/2})$$

and that

$$\mathbf{m}(r; f^{1/2}(|a|^{ipt/2}h^{-it/2})((h^{(1-s)/2}y^*) - (h^{1/2}y^*))f^{1/2})$$
$$= \mathbf{m}(r/2; f^{1/2}(|a|^{ipt/2}h^{-it/2})).$$
$$\mathbf{m}(r/2; ((h^{(1-s)/2}y^*) - (h^{1/2}y^*))f^{1/2})$$
$$\leq \mathbf{m}(r/2; f^{1/2}).\| \, |a|^{ipt/2}h^{-it/2}\|.$$
$$\mathbf{m}(r/2; ((h^{(1-s)/2}y^*) - (h^{1/2}y^*))f^{1/2})$$
$$\leq \mathbf{m}(r/2; f^{1/2}).\mathbf{m}(r/2; ((h^{(1-s)/2}y^*) - (h^{1/2}y^*))f^{1/2}).$$

If we apply these inequalities to the decomposition in equation (11.3), it follows that

$$\mathbf{m}(r; f^{1/2}(|a|^{ps/2}|a|^{ipt/2}h^{-it/2}(h^{(1-s)/2}y^*) - |a|^{ipt/2}h^{-it/2}$$
$$(h^{1/2}y^*))f^{1/2})$$
$$\leq \mathbf{m}(r/2; f^{1/2}(|a|^{ps/2} - \chi_{(0,\infty)}(|a|^{p/2}))|a|^{ipt/2}h^{-it/2}$$
$$(h^{(1-s)/2}y^*))f^{1/2}) + \mathbf{m}(r/2; f^{1/2}(|a|^{ipt/2}h^{-it/2})((h^{(1-s)/2}y^*)$$
$$- (h^{1/2}y^*))f^{1/2})$$
$$\leq \mathbf{m}(r/4; f^{1/2}(|a|^{ps/2} - \chi_{(0,\infty)}(|a|^{p/2}))).\mathbf{m}(r/4; \mathbf{f}^{1\cap\infty}(h)^{1/2}y^*)f^{1/2})$$
$$+ \mathbf{m}(r/4; f^{1/2}).\mathbf{m}(r/4; ((h^{(1-s)/2}y^*) - (h^{1/2}y^*))f^{1/2}).$$

It now clearly follows from Observations 1 and 2 and Proposition 5.23 that

$$f^{1/2}(|a|^{ps/2}|a|^{ipt/2}h^{-it/2}(h^{(1-s)/2}y^*) - |a|^{ipt/2}h^{-it/2}(h^{1/2}y^*))f^{1/2}$$

converges to 0 in measure as s decreases to 0, and that the rate of convergence is controlled by the terms $\mathbf{m}(r/4; f^{1/2}(|a|^{ps/2} - \mathbb{1}))$ and $\mathbf{m}(r/4; ((h^{(1-s)/2}y^*) - (h^{1/2}y^*))f^{1/2})$, which are independent of t. Since the norm topology on $L^1(\mathcal{M})$ is just the relative topology of convergence in measure inherited from $\widetilde{\mathfrak{M}}$, we have that

$$f^{1/2}(|a|^{ps/2}|a|^{ipt/2}h^{-it/2}(h^{(1-s)/2}y^*) - |a|^{ipt/2}h^{-it/2}(h^{1/2}y^*))f^{1/2}$$

converges to zero in the $L^1(\mathcal{M})$ norm uniformly with respect to t as s decreases to 0. Thus

$$\mathrm{tr}(f(|a|^{ps/2}|a|^{ipt/2}h^{-it/2}(h^{(1-s)/2}y^*) - |a|^{ipt/2}h^{-it/2}(h^{1/2}y^*)))$$

$$= \mathrm{tr}(f^{1/2}(|a|^{ps/2}|a|^{ipt/2}h^{-it/2}(h^{(1-s)/2}y^*)$$

$$- |a|^{ipt/2}h^{-it/2}(h^{1/2}y^*))f^{1/2})$$

converges to 0 as s decreases to 0, and does so uniformly with respect to t. We have now proved our claim about weak convergence for all $f \in L^2_+(\mathcal{M})$. The fact that the claim also holds for general $f \in L^2(\mathcal{M})$ now follows from the fact that each such element may be written as a linear combination of four elements of $L^2_+(\mathcal{M})$. $\qquad\square$

Theorem 11.22 *For any $p \in (1,\infty)$, the map $\mathfrak{u}_p : \mathscr{L}^1 \cap \mathscr{L}^\infty \supseteq \iota^{(\infty)}(\mathfrak{m}_\varphi) \to L^p(\mathcal{M}) : \iota^{(\infty)}(f) \mapsto h^{1/2p}fh^{1/2p}$ is a contractive injection with dense range. This map continuously extends to a contractive injection of all of $\mathscr{L}^1 \cap \mathscr{L}^\infty$ into $L^p(\mathcal{M})$ and is $(\sigma - \|\cdot\|)$ to weakly-L^p continuous on $\|\cdot\|_{(\cap)}$-bounded subsets of $\mathscr{L}^1 \cap \mathscr{L}^\infty$.*

Proof. The claim regarding the density of the range is a consequence of Proposition 10.52. Select $q \in (1,\infty)$ so that $\frac{1}{p} + \frac{1}{q} = 1$. We first show that for any given $f \in \mathfrak{m}_\varphi$, we have that

$$\|h^{1/2p}fh^{1/2p}\|_p \leq \|f\|_\infty^{1/q}\cdot\|h^{1/2}fh^{1/2}\|_1^{1/p}. \tag{11.4}$$

So, let $f \in \mathfrak{m}_\varphi$ and $a \in L^q(\mathcal{M})$ be given with $\|a\|_q = 1$, and consider the map $F : z \mapsto \mathrm{tr}(u|a|^{qz}(h^{(1-z)/2}fh^{(1-z)/2}))$ on $S = \{z \in \mathbb{C}: 0 \leq \mathrm{Re}(z) \leq 1\}$ analysed in Theorem 11.21. It clearly follows from Lemma 11.20 and Hölder's inequality that

$$|F(it)| = |\mathrm{tr}(u|a|^{iqt}h^{-it/2}(h^{1/2}fh^{1/2})h^{-it/2})|$$

$$= |\mathrm{tr}(h^{-it/2}u|a|^{iqt}h^{-it/2}(h^{1/2}fh^{1/2}))|$$

$$= |\mathrm{tr}((h^{-it/2}|a^*|^{iqt/2})u(|a|^{iqt/2}h^{-it/2})(h^{1/2}fh^{1/2}))|$$

$$\leq \|(h^{-it/2}|a^*|^{iqt/2})u(|a|^{iqt/2}h^{-it/2})\|_\infty\cdot\|h^{1/2}fh^{1/2}\|_1$$

$$\leq \|h^{1/2}fh^{1/2}\|_1.$$

Here, we silently used the facts (verified in the proof of the preceding proposition) that $(h^{-it/2}|a^*|^{iqt/2})$ and $(|a|^{iqt/2}h^{-it/2})$ both belong to $L^\infty(\mathcal{M})$. Similarly

$$\begin{aligned}
|F(1+it)| &= |\mathrm{tr}(u|a|^{iqt}|a|^q h^{-it/2} f h^{-it/2})| \\
&= |\mathrm{tr}(|a^*|^{iqt/2} u |a|^q |a|^{iqt/2} h^{-it/2} f h^{-it/2})| \\
&= |\mathrm{tr}(h^{-it/2}|a^*|^{iqt/2} u |a|^q |a|^{iqt/2} h^{-it/2} f)| \\
&\leq \| h^{-it/2}|a^*|^{iqt/2} u \|_\infty \cdot \| |a|^q \|_1 \cdot \| (|a|^{iqt/2} h^{-it/2}) f \|_\infty \\
&\leq \| |a|^q \|_1 \cdot \| f \|_\infty \\
&= \| a \|_q^q \cdot \| f \|_\infty \\
&= \| f \|_\infty.
\end{aligned}$$

It therefore follows from the Hadamard three line theorem that

$$|\mathrm{tr}(a(h^{1/2q} f h^{1/2q}))| = |F(1/q)| \leq \| f \|_\infty^{1/q} \| h^{1/2} f h^{1/2} \|_1^{1/p}.$$

Since this inequality holds for an arbitrary norm 1 element of $L^q(\mathcal{M})$, the validity of inequality (11.4) follows. This inequality then ensures the contractivity of $\mathbf{u}_p$ on $\iota^{(\infty)}(\mathbf{m}_\varphi)$. Since $\iota^{(\infty)}(\mathbf{m}_\varphi)$ is dense in $\mathscr{L}^1 \cap \mathscr{L}^\infty$ in the manner described in Theorem 11.8, $\mathbf{u}_p$ clearly extends to a contraction on all of $\mathscr{L}^1 \cap \mathscr{L}^\infty$.

We proceed to proving the injectivity of $\mathbf{u}_p$. Let $a_1, a_2, b_1, b_2 \in \mathbf{n}_\varphi$ and $p, q \in (1, \infty)$ be given with $\frac{1}{p} + \frac{1}{q} = 1$. It then follows from Theorem 11.22 and Lemma 10.48 that

$$\begin{aligned}
\mathrm{tr}(&\mathbf{u}_p(\iota^{(\infty)}(a_2^* a_1)) \mathbf{u}_q(\iota^{(\infty)}(b_2^* b_1))) \\
&= \mathrm{tr}((h^{1/2p} a_2^*)[a_1 h^{1/2p}](h^{1/2q} b_2^*)[b_1 h^{1/2q}]) \\
&= \mathrm{tr}([b_1 h^{1/2q}](h^{1/2p} a_2^*)[a_1 h^{1/2p}](h^{1/2q} b_2^*)) \\
&= \mathrm{tr}([b_1 h^{1/2}] a_2^* a_1 (h^{1/2} b_2^*)) \\
&= \mathrm{tr}(a_2^* a_1 (h^{1/2} b_2^*)[b_1 h^{1/2}]).
\end{aligned}$$

That is $\mathrm{tr}(\mathbf{u}_p(\iota^{(\infty)}(x)) \mathbf{u}_q(\iota^{(\infty)}(y))) = \mathrm{tr}(x(h^{1/2} y h^{1/2}))$ for all $x, y \in \mathbf{m}_\varphi$. Now, let $b \in \mathscr{L}^1 \cap \mathscr{L}^\infty$ be given and let b_∞ be the unique element of $L^\infty(\mathcal{M})$ for which $b = \iota^{(\infty)}(b_\infty)$. By Theorem 11.8, the above equality then extends to the claim that

$$\mathrm{tr}(\mathbf{u}_p(b) \mathbf{u}_q(\iota^{(\infty)}(y))) = \mathrm{tr}(b_\infty(h^{1/2} y h^{1/2})) \text{ for all } y \in \mathbf{m}_\varphi. \tag{11.5}$$

So, if $\mathbf{u}_p(b) = 0$, we must then have that $\mathrm{tr}(b_\infty(h^{1/2} y h^{1/2})) = 0$ for all $y \in \mathbf{m}_\varphi$. Since $h^{1/2} \mathbf{m}_\varphi h^{1/2}$ is dense in $L^1(\mathcal{M})$, the only way this can be is if $b_\infty = 0$, which in turn guarantees that $b = \iota^{(\infty)}(b_\infty) = 0$ and hence that the extension of $\mathbf{u}_p$ is indeed injective.

Finally, suppose that we have a net (b_α) in the ball of radius K in $\mathscr{L}^1 \cap \mathscr{L}^\infty$ centred at the origin converging to some b in the $(\sigma - \|\cdot\|)$-topology. Writing $b_{\alpha,1}$ and $b_{\alpha,\infty}$ for the unique elements of L^1 and L^∞ satisfying $\iota^{(1)}(b_{\alpha,1}) = b = \iota^{(\infty)}(b_{\alpha,\infty})$, such convergence can, of course, only happen if $b_{\alpha,1} \to b_1$ in L^1-norm and $b_{\alpha,\infty} \to b_\infty$ σ-weakly. It now follows from equation (11.5) that $\mathrm{tr}(\mathfrak{u}_p(b_\alpha)\mathfrak{u}_q(\iota^{(\infty)}(y))) \to \mathrm{tr}(\mathfrak{u}_p(b)\mathfrak{u}_q(\iota^{(\infty)}(y)))$ for every $y \in \mathfrak{m}_\varphi$. The norm density of $\mathfrak{u}_q(\iota^{(\infty)}(\mathfrak{m}_\varphi))$ in L^q now ensures that for any $\epsilon > 0$ and any $a \in L^q$, there exists some $y_\epsilon \in \mathfrak{m}_\varphi$ such that $\|a - \mathfrak{u}_q(\iota^{(\infty)}(y_\epsilon))\| \leq \epsilon$. Since then $|\mathrm{tr}(\mathfrak{u}_p(b_\alpha - b)(a - \mathfrak{u}_q(\iota^{(\infty)}(y))))| \leq \|\mathfrak{u}_p(b_\alpha - b)\|_p \|(a - \mathfrak{u}_q(\iota^{(\infty)}(y))))\|_q \leq \|b_\alpha - b\|_{(\cap)} . \epsilon \leq 2K\epsilon$, it follows that $\limsup_\alpha |\mathrm{tr}(\mathfrak{u}_p(b_\alpha - b)(a))| \leq \limsup_\alpha |\mathrm{tr}(\mathfrak{u}_p(b_\alpha - b)(\mathfrak{u}_q(\iota^{(\infty)}(y))))| + 2K\epsilon = 2K\epsilon$. The fact that $\epsilon > 0$ is arbitrary ensures that $\limsup_\alpha |\mathrm{tr}(\mathfrak{u}_p(b_\alpha - b)(a)| = 0$ and hence that $\limsup_\alpha \mathrm{tr}(\mathfrak{u}_p(b_\alpha)(a)) = \mathrm{tr}(\mathfrak{u}_p(b)(a))$ as was required. $\qquad\square$

Theorem 11.23   *Let $p, q \in (1, \infty)$ be given with $\frac{1}{p} + \frac{1}{q} = 1$. For any $a \in L^p(\mathcal{M})$, there exists a unique $\mathfrak{v}_p(a) \in \mathscr{L}^1 + \mathscr{L}^\infty$ satisfying*

$$\langle b, \mathfrak{v}_p(a)\rangle_{\cap,+} = \mathrm{tr}(a\mathfrak{u}_q(b)) \text{ for any } b \in \mathscr{L}^1 \cap \mathscr{L}^\infty.$$

Moreover, the map $\mathfrak{v}_p : L^p(\mathcal{M}) \to \mathscr{L}^1 + \mathscr{L}^\infty : a \mapsto \mathfrak{v}_p(a)$ is a contractive linear injection.

Proof.   Consider the adjoint $(\mathfrak{u}_q)^*$ of the map $b \mapsto \mathfrak{u}_q(b)$ in the preceding theorem. The contractivity and density of the range of $\mathfrak{u}_q$ ensure that $(\mathfrak{u}_q)^*$ is a contractive injection into $(\mathscr{L}^1 \cap \mathscr{L}^\infty)^*$. For any $a \in L^p(\mathcal{M})$, we moreover have that

$$((\mathfrak{u}_q)^*(a))(b) = \mathrm{tr}(a\mathfrak{u}_q(b)) \text{ for all } b \in \mathscr{L}^1 \cap \mathscr{L}^\infty.$$

We may now apply the above formula to the final fact recorded in Theorem 11.22 to see that the prescription $b \mapsto ((\mathfrak{u}_q)^*(a))(b)$ is $(\sigma - \|\cdot\|)$-continuous on norm bounded subsets of $\mathscr{L}^1 \cap \mathscr{L}^\infty$. Since for any $a \in L^p(\mathcal{M})$ we will (again by Theorem 11.22) also have that

$$|((\mathfrak{u}_p)^*(a))(b)| = |\mathrm{tr}(a\mathfrak{u}_q(b))| \leq \|a\|_p . \|\mathfrak{u}_q(b)\|_q \leq \|a\|_p . \|b\|_{(\cap)}$$

for all $b \in \mathscr{L}^1 \cap \mathscr{L}^\infty$, it now follows from Theorem 11.11 that there is a unique element of $\mathscr{L}^1 + \mathscr{L}^\infty$ corresponding to $(\mathfrak{u}_q)^*(a)$ which we may define to be $\mathfrak{v}_p(a)$. That then proves the theorem. $\qquad\square$

Remark 11.24 Recall that in the last part of the proof of Theorem 11.22, we showed that $\mathrm{tr}(\mathfrak{u}_p(\iota^{(\infty)}(x))\mathfrak{u}_q(\iota^{(\infty)}(y))) = \mathrm{tr}(x(h^{1/2}yh^{1/2}))$ for all $x, y \in \mathfrak{m}_\varphi$. According to Theorem 11.1, this fact may be reformulated as the claim that $\mathrm{tr}(\mathfrak{u}_p(\iota^{(\infty)}(x))\mathfrak{u}_q(\iota^{(\infty)}(y))) = \langle \iota^{(\infty)}(x), \iota^{(\infty)}(x)\rangle_{\cap,+}$ for all $x, y \in \mathfrak{m}_\varphi$. An application of Theorem 11.8 now reveals that

$$\mathrm{tr}(\mathfrak{u}_p(f)\mathfrak{u}_q(g)) = \langle f, g\rangle_{\cap,+} \text{ for all } f, g \in \mathscr{L}^1 \cap \mathscr{L}^\infty.$$

This fact in turn ensures that the diagram below commutes:

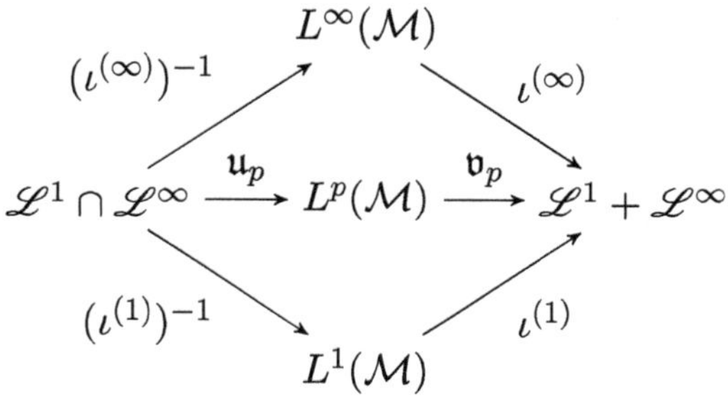

$$(11.6)$$

11.4 L^p spaces as interpolation spaces of L^1 and L^∞

Lemma 11.25 *Let $a \in L^p(\mathcal{M})$ $(1 \le p < \infty)$ be given with $\|a\|_p \le 1$ and with polar decomposition $a = u|a|$. For any $z \in S = \{w \in \mathbb{C}: 0 \le \mathrm{Re}(w) \le 1\}$ with $z = s + it$, there then exists a unique element $f(z) \in \mathscr{L}^1 + \mathscr{L}^\infty$ that satisfies*

$$\langle x, f(z)\rangle_{\cap,+} = \mathrm{tr}(u|a|^{pz}(h^{-it/2}\mathfrak{u}_{(1-s)^{-1}}(x)h^{-it/2})) \text{ for all } x \in \mathscr{L}^1 \cap \mathscr{L}^\infty.$$
$$(11.7)$$

The mapping $S \to \mathscr{L}^1 + \mathscr{L}^\infty : z \mapsto f(z)$ satisfies the following properties.

    (i) $\|f(z)\|_{(+)} \le 1$ *for all* $z \in S$.

    (ii) *For any* $a \in \mathfrak{m}_\varphi$, $z \mapsto \langle \iota^{(\infty)}(a), f(z)\rangle_{\cap,+}$ *is continuous on S and analytic on S°.*

    (iii) *For any* $t \in \mathbb{R}$, *we have that:*

        (a) $f(it) \in \mathscr{L}^\infty$, $\|f(it)\|_{\mathscr{L}^\infty} \le 1$, *and that* $t \mapsto (\iota^{(\infty)})^{-1}(f(it))$ *is point to σ-weak continuous;*

(b) $f(1+it) \in \mathscr{L}^1$, $\|f(1+it)\|_{\mathscr{L}^1} \leq 1$, and that $t \mapsto f(1+it)$ is point to $\|\cdot\|_{\mathscr{L}^1}$ continuous.

Proof. Suppose that $1 \leq p < \infty$, and let $a \in L^p(\mathcal{M})$ ($1 \leq p < \infty$) be given with $\|a\|_p \leq 1$. Notice that equation (11.4) may be reformulated as the claim that $\|u_p(\iota^{(\infty)}(f))\|_p \leq \|f\|_\infty^{1/q} \cdot \|h^{1/2}fh^{1/2}\|_1^{1/p}$. For a general element $x \in \mathscr{L}^1 \cap \mathscr{L}^\infty$, we may use Theorem 11.8 to conclude from the above that

$$\|u_p(x)\|_p \leq \|x_\infty\|_\infty^{1/q} \cdot \|x_1\|_1^{1/p}$$

where $x_\infty \in L^\infty(\mathcal{M})$ and $x_1 \in L^1(\mathcal{M})$ are the unique elements for which $x = \iota^{(\infty)}(x_\infty) = \iota^{(1)}(x_1)$. In the case where $p = 1$, $u_1(x)$ is nothing but x_1, with the inequality reducing to the trivial claim that $\|u_1(x)\|_1 = \|x_1\|_1$. Similarly, in the case where $p = \infty$, $u_\infty(x)$ is just x_∞, with the inequality reducing to the trivial observation that $\|u_\infty(x)\|_\infty = \|x_\infty\|_\infty$.

Given $z = s + it$ with $0 \leq \mathrm{Re}(z) \leq 1$, we define the functional $F_z : \mathscr{L}^1 \cap \mathscr{L}^\infty \to \mathbb{C}$ by $F_z(x) = \mathrm{tr}(u|a|^{pz}(h^{-it/2}u_{(1-s)^{-1}}(x)h^{-it/2}))$. By Lemma 11.20, we may rewrite this formula as

$$\begin{aligned}
F_z(x) &= \mathrm{tr}(u|a|^{pz}(h^{-it/2}u_{(1-s)^{-1}}(x)h^{-it/2})) \\
&= \mathrm{tr}(|a^*|^{ipt/2}u|a|^{ps}|a|^{ipt/2}(h^{-it/2}u_{(1-s)^{-1}}(x)h^{-it/2})) \\
&= \mathrm{tr}((h^{-it/2}|a^*|^{ipt/2}u)|a|^{ps}(|a|^{ipt/2}h^{-it/2})u_{(1-s)^{-1}}(x))).
\end{aligned}$$
$$(11.8)$$

Recall that we noted in the proof of Theorem 11.21 that $(h^{-it/2}|a^*|^{ipt/2}u)$ and $(|a|^{ipt/2}h^{-it/2})$ both belong to $L^\infty(\mathcal{M})$. We may therefore use Hölder's inequality and what we noted at the start of the proof, to see that

$$\begin{aligned}
|F_z(x)| &\leq \|(h^{-it/2}|a^*|^{ipt/2}u)\|_\infty \cdot \| \, |a|^{ps}\|_{s^{-1}} \cdot \|(|a|^{ipt/2}h^{-it/2}) \\
&\quad \|_\infty \cdot \|u_{(1-s)^{-1}}(x))\|_{(1-s)^{-1}} \\
&\leq \| \, |a|^{ps}\|_{s^{-1}} \cdot \|u_{(1-s)^{-1}}(x))\|_{(1-s)^{-1}} \\
&\leq \|a\|_p^{ps} \cdot \|u_{(1-s)^{-1}}(x))\|_{(1-s)^{-1}} \\
&\leq \|x_\infty\|_\infty^s \cdot \|x_1\|_1^{1-s} \\
&\leq \|x\|_{(\cap)}.
\end{aligned}$$

Thus, $x \mapsto F_z(x)$ realises a contractive functional on $\mathscr{L}^1 \cap \mathscr{L}^\infty$. We may next apply equation (11.8) to the final claim of Theorem 11.22 to see that F_z is moreover $(\sigma - \|\cdot\|)$-continuous on norm bounded subsets of $\mathscr{L}^1 \cap \mathscr{L}^\infty$. Therefore, it follows from Theorem 11.11 that there exists $f(z) \in \mathscr{L}^1 + \mathscr{L}^\infty$ so that $\langle \cdot, f(z) \rangle_{\cap,+} = F_z$, and $\|f(z)\|_{(+)} = \|F_z\| \le 1$.

Property (ii) follows from Theorem 11.21. To conclude the proof, we therefore need to show that property (iii) holds. First, consider the case where $z = 1 + it$. Given $x \in \mathscr{L}^1 \cap \mathscr{L}^\infty$, the inequality $|\langle x, f(z) \rangle_{\cap,+}| = |F_z(x)| \le \|x_\infty\|_\infty^s \cdot \|x_1\|_1^{1-s}$ in this case reduces to $|\langle x, f(1 + it) \rangle_{\cap,+}| = |F_{1+it}z(x)| \le \|x_\infty\|_\infty$. Thus, $x_\infty \mapsto \langle \iota^{(\infty)}(x_\infty), f(1 + it) \rangle_{\cap,+}$ is a contractive functional in its action on $(\iota^{(\infty)})^{-1}(\mathscr{L}^1 \cap \mathscr{L}^\infty) \subseteq L^\infty(\mathcal{M})$. The subspace $(\iota^{(\infty)})^{-1}(\mathscr{L}^1 \cap \mathscr{L}^\infty)$ of course contains $\mathfrak{m}_\varphi$, and hence is $(\sigma$-weakly) weak* dense in $L^\infty(\mathcal{M})$. It is also clear from equation (11.8) that in this case $\langle \iota^{(\infty)}(x_\infty), f(1 + it) \rangle_{\cap,+} = \mathrm{tr}((h^{-it/2}|a^*|^{ipt/2}u)|a|^p(|a|^{ipt/2}h^{-it/2})x_\infty)$. Thus, $x_\infty \mapsto \langle \iota^{(\infty)}(x_\infty), f(1 + it) \rangle_{\cap,+}$ is in this case just a restriction of the functional $b \mapsto \mathrm{tr}((h^{-it/2}|a^*|^{ipt/2}u)|a|^p(|a|^{ipt/2}h^{-it/2})b)$ to $(\iota^{(\infty)})^{-1}(\mathscr{L}^1 \cap \mathscr{L}^\infty)$. Since (as noted before) $(h^{-it/2}|a^*|^{ipt/2}u)$ and $(|a|^{ipt/2}h^{-it/2})$ both belong to $L^\infty(\mathcal{M})$, it is clear that $(h^{-it/2}|a^*|^{ipt/2}u)|a|^p(|a|^{ipt/2}h^{-it/2}) \in L^1(\mathcal{M})$. Therefore, we may set $f(1 + it) = \iota^{(1)}((h^{-it/2}|a^*|^{ipt/2}u)|a|^p(|a|^{ipt/2}h^{-it/2}))$. The assumption that $\|a\|_p \le 1$ then ensures that $\|f(1 + it)\|_{\mathscr{L}^1} \le 1$. Finally observe that

$$(h^{-it/2}|a^*|^{ipt/2}u)|a|^p(|a|^{ipt/2}h^{-it/2})$$

$$= [h^{it/2}(h^{-it}|a^*|^{ipt}u|a|^{p/2})h^{-it/2}] \cdot [h^{it/2}(|a|^{p/2})h^{-it/2}].$$

As a map into $L^2(\mathcal{M})$, the map $t \mapsto [h^{it/2}|a|^{p/2}h^{-it/2}]$ is point-norm continuous by Lemma 11.20. We show that $t \mapsto [h^{it/2}(h^{-it}|a^*|^{ipt}|a|^{p/2})h^{-it/2}]$ is similarly point to norm continuous as a map into $L^2(\mathcal{M})$. To see this, recall that we showed in Theorem 11.21 that $(h^{-it}|a^*|^{ipt})_{t\in\mathbb{R}} \subset L^\infty(\mathcal{M})$. By Stone's theorem, $t \mapsto h^{-it}|a^*|^{ipt}$ is σ-strongly continuous. On applying this fact to the Haagerup standard form for $L^2(\mathcal{M})$, it follows that $t \mapsto (h^{-it}|a^*|^{ipt}|a|^{p/2})$ is $L^2(\mathcal{M})$-norm continuous. If at this point we call on the strong continuity of the map $t \mapsto U_{t/2}$ (see Lemma 11.20) the point to norm continuity of $t \mapsto [h^{it/2}(h^{-it}|a^*|^{ipt}|a|^{p/2})h^{-it/2}]$ is then clear.

On arguing as in the final part of the proof of Lemma 11.20, it now follows that the 'product map',

$$t \mapsto (h^{-it/2}|a^*|^{ipt/2}u)|a|^p(|a|^{ipt/2}h^{-it/2})$$

$$= h^{it/2}((h^{-it}|a^*|^{ipt}u|a|^p)h^{-it/2})$$

is point to norm continuous as a map into $L^1(\mathcal{M})$. This translates to the claim that $t \mapsto f(1+it)$ is point-$\|\cdot\|_{\mathscr{L}^1}$ continuous.

Next, consider the map $t \mapsto f(it)$. Given $x \in \mathscr{L}^1 \cap \mathscr{L}^\infty$, the inequality $|\langle x, f(z)\rangle_{\cap,+}| = |F_z(x)| \leq \|x_\infty\|_\infty^s \cdot \|x_1\|_1^{1-s}$ in this case reduces to $|\langle x, f(it)\rangle_{\cap,+}| = |F_{it}(x)| \leq \|x_1\|_1$. Thus, in its action on $(\iota^{(1)})^{-1}(\mathscr{L}^1 \cap \mathscr{L}^\infty) \subseteq L^1(\mathcal{M})$, $x_1 \mapsto \langle \iota^{(1)}(x_1), f(it)\rangle_{\cap,+}$ is a contractive functional. The subspace $(\iota^{(1)})^{-1}(\mathscr{L}^1 \cap \mathscr{L}^\infty)$ contains the dense subspace $h^{1/2}\mathfrak{m}_\varphi h^{1/2}$ of $L^1(\mathcal{M})$ and therefore is itself dense in $L^1(\mathcal{M})$. In this case, we may use equation (11.8) to see that

$$\langle \iota^{(1)}(x_1), f(it)\rangle_{\cap,+} = \mathrm{tr}(|a^*|^{ipt/2}(u|a|^{ipt/2}h^{-it/2})(h^{1/2}bx_1)$$

$$= \mathrm{tr}((h^{-it/2}|a^*|^{ipt/2})u(|a|^{ipt/2}h^{-it/2})x_1).$$

We saw in the proof of Theorem 11.21 that each of $(h^{-it/2}|a^*|^{ipt/2})$ and $(|a|^{ipt/2}h^{-it/2})$ belongs to $L^\infty(\mathcal{M})$. Then, $(h^{-it/2}|a^*|^{ipt/2})u(|a|^{ipt/2}h^{-it/2})$ also does. By (L^1, L^∞) duality, this ensures that

$$f(it) = \iota^{(\infty)}((h^{-it/2}|a^*|^{ipt/2})u(|a|^{ipt/2}h^{-it/2}))$$

and hence that $f(it) \in \mathscr{L}^\infty$ with $\|f(it)\|_{\mathscr{L}^\infty} \leq 1$. By Stone's theorem, both $t \mapsto (|a|^{ipt/2}h^{-it/2})$ and $t \mapsto (h^{-it/2}|a^*|^{ipt/2})$ are point to strongly continuous. Therefore, so is $t \mapsto (h^{-it/2}|a^*|^{ipt/2})u(|a|^{ipt/2}h^{-it/2}) = (\iota^{(\infty)})^{-1}(f(it))$. The continuity claim in (iii)(a) now follows from the fact that the strong topology is stronger than the σ-weak topology on $L^\infty(\mathcal{M})$. $\qquad\square$

Lemma 11.26 *Let $f : S \to \mathscr{L}^1 + \mathscr{L}^\infty$ be a function satisfying conditions (i)–(iii) in the preceding lemma. For each $n \in \mathbb{N}$ and each $z \in S$, define $f_n(z)$ to be that element of $\mathscr{L}^1 + \mathscr{L}^\infty$ for which*

$$\langle \iota^{(\infty)}(b), f_n(z)\rangle = \sqrt{\frac{n}{\pi}} \int_{\mathbb{R}} e^{-nt^2} \langle \iota^{(\infty)}(b), f(z-it)\rangle \, dt \text{ for all } b \in \mathfrak{m}_\varphi.$$

$$(11.9)$$

Then, each f_n belongs to $\mathcal{F}(\mathcal{M}, L^1(\mathcal{M}))$. Moreover, $\|f_n\|_{\mathcal{F}} \leq 1$ for each $n \in \mathbb{N}$ with $\lim_{n\to\infty} \|f(z) - f_n(z)\|_{(+)} = 0$ for every $z \in S^0$.

Proof.   Let f be a function satisfying conditions (i)–(iii) in the preceding lemma. Condition (ii) represents what one in Banach space terms may call weak analyticity. However, when weak analyticity is combined with for example criteria such as those provided by condition (i), then weak analyticity equals strong analyticity (that is norm analyticity). This may be seen as follows. Let z_0 be an arbitrary point in S^o, and let M be a bound for $\|f(z)\|$ on a disc of radius r around z_0, where $r > 0$ is small enough to ensure that the disc lies entirely in S^o. Let $h, k \in \mathbb{C}$ be given with $|h|, |k| < r/2$. For any $g \in \mathscr{L}^1 \cap \mathscr{L}^\infty$, the function $z \mapsto \langle g, f(z) \rangle_{\cap,+}$ is analytic by condition (ii). So, we may apply the Cauchy integration formulae to conclude that

$$|\langle g, (f(z_0 + t) - f(z_0)))\rangle_{\cap,+} t^{-1} - \langle g, (f(z_0 + k) - f(z_0)))\rangle_{\cap,+} k^{-1}|$$
$$= \frac{1}{2\pi} \left| \int_C \frac{\langle g, f(z)\rangle_{\cap,+} (t - k)}{(z - s - t)(z - s - k)(z - s)} \, dz \right|$$

where C is the positively oriented circle of radius r centred at z_0. On taking the supremum over all elements of $\mathscr{L}^1 \cap \mathscr{L}^\infty$ of norm 1 and estimating the integral, it follows from Theorem 11.11 that $\|(f(z_0+t) - f(z_0))t^{-1} - (f(z_0+k) - f(z_0))k^{-1}\| < 4M|t-k|r^{-2}$, where M is a bound for $\|f(z)\|$ on a disc of radius r around z_0. Therefore, the quotients $((f(z_0 + t) - f(z_0))t^{-1})$ are a Cauchy net in $\mathscr{L}^1 \cap \mathscr{L}^\infty$ as t decreases to 0, and hence must have a limit in $\mathscr{L}^1 \cap \mathscr{L}^\infty$, i.e., $f'(z_0)$ exists as an element of $\mathscr{L}^1 \cap \mathscr{L}^\infty$. The norm analyticity of the function $z \mapsto f(z)$ on S^o of course ensures that this function is norm continuous on S^o. This in turn implies that for each fixed $z \in S^o$, $\mathbb{R} \to \mathscr{L}^1 + \mathscr{L}^\infty : t \mapsto f(z - it)$ is $\| \cdot \|_{(+)}$-continuous. Since it is also $\| \cdot \|_{(+)}$-bounded for each fixed $z \in S^0$ and each $n \in \mathbb{N}$, the integrals $f_n(z) = \sqrt{\frac{n}{\pi}} \int_\mathbb{R} e^{-nt^2} f(z - it) \, dt$ are well-defined, with the resultant element $f_n(z)$ satisfying equation (11.9).

Now, consider the case where $\mathrm{Re}(z) = 1$. In this case $\mathbb{R} \to \mathscr{L}^1 + \mathscr{L}^\infty : t \mapsto f(z - it)$ is $\| \cdot \|_{\mathscr{L}^1}$-continuous and bounded by condition (iii)(b) in Lemma 11.25. Hence, here, $f_n(z) = \sqrt{\frac{n}{\pi}} \int_\mathbb{R} e^{-nt^2} f(z - it) \, dt$ exists as an element of $\mathscr{L}^1$ which once again satisfies equation (11.9).

Finally, consider the case where $\mathrm{Re}(z) = 0$. Recall that in this case $t \mapsto (\iota^{(\infty)})^{-1}(f(z - it))$ is an $\mathcal{M}$-valued $\| \cdot \|_\infty$-bounded and σ-weakly continuous function. We claim that these conditions are enough to ensure that there must exist some $F(n, z) \in M$ such that $\mathrm{tr}(gF(n, z)) = \sqrt{\frac{n}{\pi}} \int_\mathbb{R} e^{-nt^2} \mathrm{tr}(g(\iota^{(\infty)})^{-1}(f(z - it))) \, dt$ for all $g \in L^1(\mathcal{M})$. We provide a brief outline of the proof of this fact. Readers who want full details should consider [**Arv74**, Proposition 1.2] alongside the duality theory for the pair $(L^1(\mathcal{M}), \mathcal{M})$. First, note that the assumptions on $t \mapsto (\iota^{(\infty)})^{-1}(f(z - it))$ are sufficient to ensure that for each fixed $g \in L^1(\mathcal{M})$, $t \mapsto \operatorname{tr}(g(\iota^{(\infty)})^{-1}(f(z - it)))$ is a continuous and bounded function. This ensures the well-definiteness of $\sqrt{\frac{n}{\pi}} \int_{\mathbb{R}} e^{-nt^2} \operatorname{tr}(g(\iota^{(\infty)})^{-1}(f(z - it))) \, dt$ for all $g \in L^1(\mathcal{M})$. The next step is to check that the prescription $L^1(\mathcal{M}) \to \mathbb{C} : g \mapsto \sqrt{\frac{n}{\pi}} \int_{\mathbb{R}} e^{-nt^2} \operatorname{tr}(g(\iota^{(\infty)})^{-1}(f(z - it))) \, dt$ is indeed a bounded linear functional. Once that is done, the duality theory for the pair $(L^1(\mathcal{M}), \mathcal{M})$, then guarantees the existence of the $F(n, z)$ we seek. With the existence of such an $F(n, z)$ assured, we now set $f_n(z) = \iota^{(\infty)}(F(n, z))$ and restrict the action of the formula

$$\operatorname{tr}(gF(n, z)) = \sqrt{\frac{n}{\pi}} \int_{\mathbb{R}} e^{-nt^2} \operatorname{tr}(g(\iota^{(\infty)})^{-1}(f(z - it))) \, dt$$

$$\text{for all } g \in L^1(\mathcal{M})$$

to the subspace $h^{1/2} \mathfrak{m}_\varphi h^{1/2}$ of $L^1(\mathcal{M})$. It is then an instructive exercise to show that the above formula may in this case be written as the claim that $\langle \iota^{(\infty)}(b), f_n(z) \rangle = \sqrt{\frac{n}{\pi}} \int_{\mathbb{R}} e^{-nt^2} \langle \iota^{(\infty)}(b), f(z - it) \rangle_{\cap,+} \, dt$ for all $b \in \mathfrak{m}_\varphi$. But this is precisely the formula we want.

So, in summary, we have now, for each $z \in S$ and each $n \in \mathbb{R}$, shown the existence of a well-defined element $f_n(z) \in \mathscr{L}^1 + \mathscr{L}^\infty$, satisfying equation (11.9).

Having ensured the existence of each f_n, we show that the sequence (f_n) belongs to $\mathcal{F}(\mathcal{M}, L^1(\mathcal{M}))$. We first show that each f_n is bounded. Given $z \in S^o$, the above integral representation ensures that

$$\|f_n(z)\|_{(+)} \leq \sqrt{\frac{n}{\pi}} \int_{\mathbb{R}} e^{-nt^2} \|f(z - it)\|_{(+)} \, dt \leq \sqrt{\frac{n}{\pi}} \int_{\mathbb{R}} e^{-nt^2} \, dt \leq 1.$$

In the case where $\operatorname{Re}(z) = 1$, we similarly have that

$$\|f_n(z)\|_{\mathscr{L}^1} \leq \sqrt{\frac{n}{\pi}} \int_{\mathbb{R}} e^{-nt^2} \|f(z - it)\|_{\mathscr{L}^1} \, dt \leq \sqrt{\frac{n}{\pi}} \int_{\mathbb{R}} e^{-nt^2} \, dt \leq 1.$$

Finally, in the case $\operatorname{Re}(z) = 0$, we will for any $g \in L^1(\mathcal{M})$ have that

$$|\langle \iota^{(1)}(g), f_n(z) \rangle_{\cap,+}| = |\operatorname{tr}(gF(n, z))|$$

$$\leq \sqrt{\frac{n}{\pi}} \int_{\mathbb{R}} e^{-nt^2} |\operatorname{tr}(g(\iota^{(\infty)})^{-1}(f(z - it)))| \, dt$$

$$\leq \|g\|_1 \cdot \int_{\mathbb{R}} e^{-nt^2}\, dt$$

$$\leq \|g\|_1.$$

This ensures that $\|f_n(z)\|_{\mathscr{L}^\infty} \leq 1$. Therefore, we have shown that $\|f_n(z)\|_{(+)} \leq 1$ for each n. (Once we know that $f_n \in \mathcal{F}(\mathcal{M}, L^1(\mathcal{M}))$, this will also ensure that $\|f_n\|_{\mathcal{F}} \leq 1$.)

Next, we verify the continuity on S. First, suppose that $z_0 \in S^o$. For any $b \in \mathfrak{m}_\varphi$, $n \in \mathbb{N}$, and any $z \in S^o$, one may then show that

$$\langle \iota^{(\infty)}(b), f_n(z)\rangle_{\cap,+} = \sqrt{\frac{n}{\pi}} \int_{\mathbb{R}} e^{-nt^2} \langle \iota^{(\infty)}(b), f(z - it)\rangle_{\cap,+}\, dt$$

$$= \sqrt{\frac{n}{\pi}} \int_{\mathbb{R}} e^{-n(t - i(z - z_0))^2} \langle \iota^{(\infty)}(b), f(z_0 - it)\rangle_{\cap,+}\, dt.$$

$$(11.10)$$

The second equality in the above claim deserves some comment. Given any $R > \max(|z|, |z_0|)$, the function $v \mapsto e^{nv^2} \langle \iota^{(\infty)}(b), f(z - v)\rangle_{\cap,+}$ is continuous on the parallelogram with vertices $\pm iR$ and $(z - z_0) \pm iR$, and analytic on the interior. So, by the Cauchy–Goursat theorem, the integral of this function around the perimeter is 0. The equality claimed above will therefore follow if we are able to show that the integrals along the line segments $v_+(s) = iR + s(z - z_0)$ ($s \in [0,1]$) and $v_-(s) = -iR + s(z - z_0)$ ($s \in [0,1]$) tend to 0 as $R \to \infty$. To see this for the contour $v_+(s)$ ($s \in [0,1]$), observe that $v_+(s)^2 = -nR^2 + 2iRs(z - z_0) + s^2(z - z_0)^2$. Hence

$$|\exp(nv(s)^2)| = |\exp(-nR^2)\exp(2iRs(z - z_0) + s^2(z - z_0)^2))|$$

$$\leq \exp(-nR^2)\exp(2R|z - z_0| + |z - z_0|^2).$$

This ensures that

$$\left| \int_0^1 e^{nv(s)^2} \langle \iota^{(\infty)}(b), f(z - v(s))\rangle_{\cap,+}(z - z_0)\, ds \right|$$

$$\leq e^{-nR^2} e^{2R|z - z_0| + |z - z_0|^2} \|\iota^{(\infty)}(b)\|_{(\cap)}|z - z_0|$$

with the final term converging to 0 as $R \to \infty$. Thus, as required,

$$\int_0^1 e^{nv(s)^2} \langle \iota^{(\infty)}(b), f(z - v(s))\rangle_{\cap,+}(z - z_0)\, ds \to 0 \text{ as } R \to \infty.$$

The fact noted in equation (11.10) now enables us to conclude that

$$
\left|\langle \iota^{(\infty)}(b), f_n(z_0) - f_n(z)\rangle_{\cap,+}\right|
$$

$$
= \left|\sqrt{\frac{n}{\pi}} \int_{\mathbb{R}} (e^{-nt^2} - e^{-n(t-i(z-z_0))^2})\langle \iota^{(\infty)}(b), f(z_0 - it)\rangle_{\cap,+}\, dt\right|
$$

$$
\leq \sqrt{\frac{n}{\pi}} \int_{\mathbb{R}} \left|(e^{-nt^2} - e^{-n(t-i(z-z_0))^2})\langle \iota^{(\infty)}(b), f(z_0 - it)\rangle_{\cap,+}\right|\, dt
$$

$$
\leq \|\iota^{(\infty)}(b)\|_{(\cap)} \cdot \|f\|_{\mathcal{F}} \cdot \sqrt{\frac{n}{\pi}} \int_{\mathbb{R}} \left|(e^{-nt^2} - e^{-n(t-i(z-z_0))^2})\right|\, dt
$$

$$
\leq \|\iota^{(\infty)}(b)\|_{(\cap)} \cdot \sqrt{\frac{n}{\pi}} \int_{\mathbb{R}} \left|(e^{-nt^2} - e^{-n(t-i(z-z_0))^2})\right|\, dt.
$$

By Theorems 11.8 and 11.11, we then have that

$$
\|f_n(z_0) - f_n(z)\|_{(+)} \leq \sqrt{\frac{n}{\pi}} \int_{\mathbb{R}} \left|(e^{-nt^2} - e^{-n(t-i(z-z_0))^2})\right|\, dt.
$$

This inequality then ensures that $\|f_n(z_0) - f_n(z)\|_{(+)} \to 0$ as $z \to z_0$. Similar arguments show that

$$
\|f_n(it_0) - f_n(it)\|_{\mathscr{L}^\infty} \leq \sqrt{\frac{n}{\pi}} \int_{\mathbb{R}} \left|(e^{-nt^2} - e^{-nt_0^2})\right|\, dt
$$

and

$$
\|f_n((1+it_0) - f_n(1+it)\|_{\mathscr{L}^1} \leq \sqrt{\frac{n}{\pi}} \int_{\mathbb{R}} \left|(e^{-nt^2} - e^{-nt_0^2})\right|\, dt,
$$

which ensure that each of the maps $\mathbb{R} \to \mathscr{L}^\infty : t \mapsto f_n(it)$ and $\mathbb{R} \to \mathscr{L}^1 : t \mapsto f_n(1+it)$ are continuous. The boundedness of these maps has already been shown.

We next show that each f_n is analytic on S^o. If we can show that for each $b \in \mathfrak{m}_\varphi$ and each $n \in \mathbb{N}$, the map $z \mapsto \langle \iota^{(\infty)}(b), f_n(z)\rangle_{\cap,+}$ is analytic on S^o, then the same argument used at the beginning of this proof shows that f_n is in fact norm-analytic on S^o. Hence, let $b \in \mathfrak{m}_\varphi$ and $n \in \mathbb{N}$ be given, and let Γ be a closed piecewise-C^1 curve in S^o. Since $\sqrt{\frac{n}{\pi}}e^{-nt^2}\, dt$ is a finite measure on $\mathbb{R}$, we may use Fubini's theorem to see that

$$\int_\Gamma \langle \iota^{(\infty)}(b), f_n(z)\rangle_{\cap,+}\, dz$$

$$= \sqrt{\frac{n}{\pi}} \int_\Gamma \left(\int_{\mathbb{R}} e^{-nt^2} \langle \iota^{(\infty)}(b), f(z-it)\rangle_{\cap,+}\, dt \right) dz$$

$$= \sqrt{\frac{n}{\pi}} \int_{\mathbb{R}} e^{-nt^2} \left(\int_\Gamma \langle \iota^{(\infty)}(b), f(z-it)\rangle_{\cap,+}\, dz \right) dt$$

Applying the Cauchy–Goursat theorem to the analyticity of $z \mapsto \langle \iota^{(\infty)}(b), f(z)\rangle_{\cap,+}$, now yields the conclusion that

$$\int_\Gamma \langle \iota^{(\infty)}(b), f_n(z)\rangle_{\cap,+}\, dz = \sqrt{\frac{n}{\pi}} \int_{\mathbb{R}} e^{-nt^2}.0\, dt = 0.$$

Thus, by Morera's theorem, $z \mapsto \langle \iota^{(\infty)}(b), f_n(z)\rangle_{\cap,+}$ is analytic on S^o, as claimed.

Having checked all the conditions for membership, we now have $(f_n) \subseteq \mathcal{F}(\mathcal{M}, L^1(\mathcal{M}))$ as claimed.

It remains to show that for each $z \in S^o$ and each $b \in \mathfrak{m}_\varphi$, we have $\lim_{n\to 0}\langle \iota^{(\infty)}(b), f_n(z)\rangle_{\cap,+} = \langle \iota^{(\infty)}(b), f(z)\rangle_{\cap,+}$. This is a simple consequence of the manner in which each f_n was defined, and the known fact that in its action on continuous bounded functions p on $\mathbb{R}$, the limit $\lim_{n\to\infty} \sqrt{\frac{n}{\pi}} \int_{\mathbb{R}} e^{-nt^2} p(t)\, dt$ yields $\lim_{n\to\infty} \sqrt{\frac{n}{\pi}} \int_{\mathbb{R}} e^{-nt^2} p(t)\, dt = p(0)$. This in turn is a consequence of the fact that the increasingly rapid exponential decay of e^{-nt^2} away from 0 tends to 'concentrate' the integral around the point 0. $\square$

To prove the next proposition, we need the following fact from the general theory of the complex method. The interested reader may find a proof thereof for the case of general Banach couples in [**BL76**, Lemma 4.2.3].

Lemma 11.27   *The linear subspace* $\mathcal{F}_{00}(\mathcal{M}, L^1(\mathcal{M}))$ *of* $\mathcal{F}_0(\mathcal{M}, L^1(\mathcal{M}))$ *consisting of all functions* $f : S \to \mathscr{L}^1 \cap \mathscr{L}^\infty$ *for which we have that:*

- *f is bounded and continuous on S;*
- *f is analytic on S^o;*
- *for any $z \in S$, $\|f(z)\|_{(\cap)} \to 0$ uniformly in $\mathrm{Re}(z)$ as $\mathrm{Im}(z) \to 0$;*

is dense in $\mathcal{F}_0(\mathcal{M}, L^1(\mathcal{M}))$.

Proposition 11.28 *Let $p \in (1, \infty)$ be given. Then $\mathscr{V}^p \subseteq \mathscr{L}^p$, with $\|\zeta\|_{\mathscr{L}^p} \leq \|\zeta\|_{\mathscr{V}^p}$ for all $\zeta \in \mathscr{V}^p$.*

Proof. We need to prove that for all $g \in \mathcal{F}_0(\mathcal{M}, L^1(\mathcal{M}))$ we have $g(1/p) \in \mathscr{L}^p$ with $\|g(1/p)\|_{\mathscr{L}^p} \leq \|g\|_{\mathcal{F}}$. We shall first prove this claim for $g \in \mathcal{F}_{00}(\mathcal{M}, L^1(\mathcal{M}))$, before using the lemma to show that the general claim then follows. The first step in proving this restricted claim is to prove that for any $g \in \mathcal{F}_{00}(\mathcal{M}, L^1(\mathcal{M}))$ and any $a \in L^q(\mathcal{M})$ (where $1 = p^{-1} + q^{-1}$), we have that

$$|\langle g(1/p), \mathfrak{v}_q(a)\rangle_{\cap,+}| \leq \|g\|_{\mathcal{F}} \cdot \|a\|_q. \tag{11.11}$$

We may without loss of generality, assume that $\|a\|_q = 1$, and take f to be the function defined as in Lemma 11.25 (with q instead of p). For each $n \in \mathbb{N}$, let F_n be the function defined by

$$F_n : S \to \mathbb{C} : z \mapsto \langle g(1 - z), f_n(z)\rangle_{\cap,+}$$

where f_n is defined as in Lemma 11.26. The content of Lemma 11.26 then ensures that F_n is bounded and continuous on S and analytic on S^o. For all $t \in \mathbb{R}$, we have

$$\begin{aligned}
|F_n(it)| &= |\langle g(1 - it), f_n(it)\rangle_{\cap,+}| \\
&= |\mathrm{tr}((\iota^{(\infty)})^{-1}(f_n(it))(\iota^{(1)})^{-1}(g(1 - it)))| \\
&\leq \|(\iota^{(\infty)})^{-1}(f_n(it))\|_\infty \cdot \|(\iota^{(1)})^{-1}(g(1 - it))\|_1 \\
&= \|f_n(it)\|_{\mathscr{L}^\infty} \cdot \|g(1 - it)\|_{\mathscr{L}^1} \\
&\leq \|f_n\|_{\mathcal{F}} \cdot \|g\|_{\mathcal{F}} \\
&\leq \|g\|_{\mathcal{F}}
\end{aligned}$$

and

$$\begin{aligned}
|F_n(1 + it)| &= |\langle g(-it), f_n(1 + it)\rangle_{\cap,+}| \\
&= |\mathrm{tr}((\iota^{(1)})^{-1}(f_n(1 + it))(\iota^{(\infty)})^{-1}(g(-it)))| \\
&\leq \|(\iota^{(1)})^{-1}(f_n(1 + it))\|_\infty \cdot \|(\iota^{(\infty)})^{-1}(g(-it))\|_1 \\
&= \|f_n(1 + it)\|_{\mathscr{L}^1} \cdot \|g(-it)\|_{\mathscr{L}^\infty} \\
&\leq \|f_n\|_{\mathcal{F}} \cdot \|g\|_{\mathcal{F}} \\
&\leq \|g\|_{\mathcal{F}}.
\end{aligned}$$

So, by Hadamard's three-line theorem, we must then have that

$$|\langle g(1/p), f_n(1/q)\rangle_{\cap,+}| = |F_n(1/q)| \le \|g\|_{\mathcal{F}}.$$

It is an exercise (using the definition of f and Theorem 11.23) to see that $f(1/q) = \mathfrak{v}_q(a)$. So, since $f_n(1/q) \to f(1/q)$ as $n \to \infty$, we obtain

$$|\langle g(1/p), \mathfrak{v}_q(a)\rangle_{\cap,+}| = |\langle g(1/p), f(1/q)\rangle_{\cap,+}| \le \|g\|_{\mathcal{F}}$$

as required.

Equation (11.11) ensures that $a \mapsto \langle g(1/p), \mathfrak{v}_q(a)\rangle_{\cap,+}$ is a bounded linear functional on $L^q(\mathcal{M})$, and hence by L^p-duality, there must exist some $b \in L^p(\mathcal{M})$ such that $\langle g(1/p), \mathfrak{v}_q(a)\rangle_{\cap,+} = \text{tr}(ba)$ for all $a \in L^q(\mathcal{M})$. By Remark 11.24, this ensures that

$$\text{tr}(b\mathfrak{u}_q(x)) = \langle g(1/p), x\rangle_{\cap,+} = \text{tr}(\mathfrak{u}_p(g(1/p))\mathfrak{u}_q(x)) \text{ for all } x \in \mathscr{L}^1 \cap \mathscr{L}^\infty.$$

This suffices to ensure that $b = \mathfrak{u}_p(g(1/p))$ (equivalently that $g(1/p) = \mathfrak{v}_p(b) \in \mathscr{L}^p$) with $\|g(1/p)\|_{\mathscr{L}^p} \le \|g\|_{\mathcal{F}}$, thereby proving the claim for all $g \in \mathcal{F}_{00}(\mathcal{M}, L^1(\mathcal{M}))$.

In conclusion, we prove that the validity of the claim for $\mathcal{F}_{00}(\mathcal{M}, L^1(\mathcal{M}))$, ensures that it holds for $\mathcal{F}_0(\mathcal{M}, L^1(\mathcal{M}))$. Here, we shall make extensive use of the diagram in Remark 11.24. First, note that for any $g \in \mathcal{F}_{00}(\mathcal{M}, L^1(\mathcal{M}))$, we have by the above that $\|\mathfrak{u}_p(g(1/p))\|_p \le \|g\|_{\mathcal{F}}$ (or equivalently $\|g(1/p)\|_{\mathscr{L}^p} \le \|g\|_{\mathcal{F}}$). Hence, the map $T_p : \mathcal{F}_{00}(\mathcal{M}, L^1(\mathcal{M})) \to L^p(\mathcal{M}) : g \mapsto \mathfrak{u}_p(g(1/p))$ is continuous. By continuity, this map extends to all of $\mathcal{F}_0(\mathcal{M}, L^1(\mathcal{M}))$. We shall retain the same notation for the extension of T_p. So, for any $g \in \mathcal{F}_0(\mathcal{M}, L^1(\mathcal{M}))$, we have $T_p(g) \in L^p(\mathcal{M})$. Furthermore, if we select $(g_n) \subseteq \mathcal{F}_{00}(\mathcal{M}, L^1(\mathcal{M}))$ so that $\|g - g_n\|_{\mathcal{F}} \to 0$ as $n \to 0$, then by continuity, $\mathfrak{u}_p(g_n(1/p)) \to T_p(g)$ as $n \to \infty$. When composing the map T_p with the continuous embedding $\mathfrak{v}_p : L^p(\mathcal{M}) \to \mathscr{L}^1 + \mathscr{L}^\infty$, we obtain that $g_n(1/p) = \mathfrak{v}_p(\mathfrak{u}_p(g_n(1/p))) \to \mathfrak{v}_p(T_p(g))$ as $n \to \infty$. (We remind the reader that $\mathfrak{v}_p(\mathfrak{u}_p(g_n(1/p)))$ is just $g_n(1/p)$ considered as an element of $\mathscr{L}^1 + \mathscr{L}^\infty$ rather than $\mathscr{L}^1 \cap \mathscr{L}^\infty$.) But we know from equation (11.1) that $\|g(1/p) - g_n(1/p)\|_{(+)} \le \|g - g_n\|_{\mathcal{F}}$. Thus, we also have $g_n(1/p) \to g(1/p)$ in $\mathscr{L}^1 + \mathscr{L}^\infty$, thereby ensuring that $g(1/p) = \mathfrak{v}_p(T_p(g))$, or equivalently that $\mathfrak{u}_p(g(1/p)) = T_p(g) \in L^p(\mathcal{M})$ as required. This fact then also ensures that $\|g(1/p)\|_{\mathscr{L}^p} = \lim_{n\to\infty} \|g_n(1/p)\|_{\mathscr{L}^p} \le \lim_{n\to\infty} \|g_n(1/p)\|_{\mathcal{F}} = \|g(1/p)\|_{\mathcal{F}}$, which proves the proposition. $\square$

Lemma 11.29   *Let $p \in (1, \infty)$ be given. Given $\zeta \in \mathscr{L}^p$ with $\|\zeta\|_{\mathscr{V}^p} \le 1$, there exists a sequence $(\zeta_n) \subseteq \mathscr{V}^p$ with $\|\zeta_n\|_{\mathscr{V}^p} \le 1$ for each $n \in \mathbb{N}$, such that $\lim_{n\to\infty} \|\zeta - \zeta_n\|_{\mathscr{V}^p} = 0$.*

Proof.   We trivially have that $\zeta = \mathfrak{v}_p(a)$ for some $a \in L^p(\mathcal{M})$ with $\|a\|_p \le 1$. Now define $f : S \to \mathbb{C}$ corresponding to a as in Lemma 11.25. Next, construct the sequence $(f_n) \subseteq \mathcal{F}(\mathcal{M}, L^1(\mathcal{M}))$ of approximating functions as in Lemma 11.26. We now modify this sequence by setting $\widetilde{f}_n(z) = e^{z^2/n} f_n(z)$ for each $n \in \mathbb{N}$ and each $z \in S$. It is an exercise to see that then $(\widetilde{f}_n) \subseteq \mathcal{F}_0(\mathcal{M}, L^1(\mathcal{M}))$, with $\|\widetilde{f}_n\|_{\mathcal{F}} \le e^{(1/n)}$. (All the properties of the functions f_n—bar the norm estimates—are inherited by the $\widetilde{f}_n$, with the factor $e^{z^2/n}$ ensuring that each $\widetilde{f}_n$ also satisfies the two additional membership criteria for $\mathcal{F}_0(\mathcal{M}, L^1(\mathcal{M}))$.) Therefore, we have $(\widetilde{f}_n(1/p)) \subseteq \mathscr{V}^p$ with $\|\widetilde{f}_n\|_{\mathscr{V}^p} \le e^{(1/n)}$. Now set $\zeta_n = e^{(-1/n)} \widetilde{f}_n(1/p)$ for each $n \in \mathbb{N}$. Then, by construction, $(\zeta_n) \subseteq \mathscr{V}^p$ with $\|\zeta_n\|_{\mathscr{V}^p} \le 1$ for all $n \in \mathbb{N}$. Since by Lemmata 11.25 and 11.26 $f_n(1/p) \to f(1/p) = \mathfrak{v}_p(a) = \zeta$ in $\mathscr{L}^1 \cap \mathscr{L}^\infty$ as $n \to \infty$, it follows that

$$\zeta_n = e^{(-1/n)} \widetilde{f}_n(1/p) = e^{(-1/n)} e^{1/(np^2)} f_n(1/p) \to \zeta \text{ in } \mathscr{L}^1 \cap \mathscr{L}^\infty$$

as $n \to \infty$. $\qquad\square$

Proposition 11.30   *For any $p \in (1, \infty)$, the closed unit ball of $\mathscr{V}^p$ is dense in the closed unit ball of $\mathscr{L}^p$.*

Proof.   Let $\zeta \in \mathscr{L}^p$ be given with $\|\zeta\|_{\mathscr{L}^p} \le 1$, and let (ζ_n) be as in the preceding lemma. The sequence (ζ_n) is then in the unit ball of $\mathscr{V}^p$, and hence also in the unit ball of $\mathscr{L}^p$ by Proposition 11.28. If we combine equation (11.1) with the definition of the norm $\| \cdot \|_{\mathscr{V}^p}$, it is clear that $\|\zeta - \zeta_n\|_{(+)} \le \|\zeta - \zeta_n\|_{\mathscr{V}^p}$ for all n, and hence that $\zeta_n \to \zeta$ in the $\| \cdot \|_{(+)}$ norm. This fact ensures that (ζ_n) also converges to ζ in the $\sigma((\mathscr{L}^1 + \mathscr{L}^\infty), (\mathscr{L}^1 + \mathscr{L}^\infty)^*)$ topology. The continuity of the injection $\mathfrak{v}_p : L^p(\mathcal{M}) \to \mathscr{L}^1 + \mathscr{L}^\infty$, ensures that for each $F \in (\mathscr{L}^1 + \mathscr{L}^\infty)^*$, $F \circ \mathfrak{v}_p$ will by (L^p, L^q)-duality correspond to an element of $L^q(\mathcal{M})$ where $1 = p^{-1} + q^{-1}$. This ensures that the topology on $L^p(\mathcal{M})$ inherited from $\sigma((\mathscr{L}^1 + \mathscr{L}^\infty), (\mathscr{L}^1 + \mathscr{L}^\infty)^*)$ by means of $\mathfrak{v}_p$ is coarser than the $\sigma(L^p(\mathcal{M}), L^q(\mathcal{M}))$-topology. We shall denote the former topology by $\mathscr{T}_p$. Let $\mathscr{B}(L^p(\mathcal{M}))$ be the closed unit ball of $L^p(\mathcal{M})$. The fact that $\mathscr{T}_p$ is coarser than $\sigma(L^p(\mathcal{M}), L^q(\mathcal{M}))$ ensures that the identity map

$$(\mathscr{B}(L^p(\mathcal{M})), \sigma(L^p(\mathcal{M}), L^q(\mathcal{M}))) \to (\mathscr{B}(L^p(\mathcal{M})), \mathscr{T}_p)$$

is continuous. Now note that the reflexivity of $L^p(\mathcal{M})$ ensures that $\mathscr{B}(L^p(\mathcal{M}))$ is $\sigma(L^p(\mathcal{M}), L^q(\mathcal{M}))$-compact. Since the topology $\mathscr{T}_p$ is a Hausdorff topology, a basic compactness result then ensures that the identity map

$$(\mathscr{B}(L^p(\mathcal{M})), \sigma(L^p(\mathcal{M}), L^q(\mathcal{M}))) \to (\mathscr{B}(L^p(\mathcal{M})), \mathscr{T}_p)$$

is in fact a homeomorphism (see [**Wil04**, Theorem 17.14]). But then $(\mathfrak{v}_p^{-1}(\zeta_n))$ converges to $\mathfrak{v}_p^{-1}(\zeta)$ in the $\sigma(L^p(\mathcal{M}), L^q(\mathcal{M}))$ topology. With $\mathscr{B}(\mathscr{V}^p)$ denoting the unit ball of $\mathscr{V}^p$, this ensures that $\mathfrak{v}_p^{-1}(\mathscr{B}(\mathscr{V}^p))$ is $\sigma(L^p(\mathcal{M}), L^q(\mathcal{M}))$-dense in $\mathscr{B}(L^p(\mathcal{M}))$. Since both of these sets are convex, the $\sigma(L^p(\mathcal{M}), L^q(\mathcal{M}))$-density is equivalent to $\|\cdot\|_p$-density. So, as required, $\mathfrak{v}_p^{-1}(\mathscr{B}(\mathscr{V}^p))$ is $\|\cdot\|_p$-dense in $\mathscr{B}(L^p(\mathcal{M}))$, or equivalently $\mathscr{B}(\mathscr{V}^p)$ is $\|\cdot\|_{\mathscr{V}^p}$-dense in $\mathscr{B}(\mathscr{L}^p)$. $\square$

Theorem 11.31 *For any $p \in (1, \infty)$, we have $\mathscr{V}^p = \mathscr{L}^p$ with equality of norms.*

Proof. Let j_p be the continuous embedding of $\mathscr{V}^p$ into $\mathscr{L}^p$ with $\mathscr{B}(\mathscr{V}^p)$ and $\mathscr{B}(\mathscr{L}^p)$, respectively, denoting the closed unit balls of $\mathscr{V}^p$ and $\mathscr{L}^p$. The preceding proposition informs us that $\mathscr{B}(\mathscr{L}^p) \subseteq \overline{j_p(\mathscr{B}(\mathscr{V}^p))}$. But by a well-known lemma to the Open Mapping Theorem [**Gol66**, Lemma II.1.7], we then have $\mathscr{B}(\mathscr{L}^p)^0 \subseteq j_p(\mathscr{B}(\mathscr{V}^p))$, where $\mathscr{B}(\mathscr{L}^p)^0$ is the open unit ball. So $\mathscr{L}^p$ is clearly contained in $\mathscr{V}^p$. In addition, given $\zeta \in \mathscr{L}^p$ with $\|\zeta\|_{\mathscr{L}^p} = 1$, we will for any $\epsilon > 0$ have that $\|\zeta\|_{\mathscr{V}^p} \leq (1 + \epsilon)$, and hence that $\|\zeta\|_{\mathscr{V}^p} \leq 1$. So, in general, $\|\zeta\|_{\mathscr{V}^p} \leq \|\zeta\|_{\mathscr{L}^p}$ for any $\zeta \in \mathscr{L}^p$. The claim now follows by considering these facts alongside the conclusion of Proposition 11.28. $\square$

We have finally reached the conclusion that the spaces $L^p(\mathcal{M})$ $(1 \leq p \leq \infty)$ (more precisely the spaces $\mathscr{L}^p$ where $1 \leq p \leq \infty$) form a complex interpolation scale. That fact then enables us to use the full power of the abstract complex method of interpolation to prove noncommutative versions of classical interpolation theorems for L^p-spaces. By way of example, we refer the reader to noncommutative Riesz convexity theorem (see Theorem 12.10) proved in the next chapter.

Chapter 12
Extensions of maps to $L^p(\mathcal{M})$ spaces and applications

The material in this chapter is based on material taken from [**GL99**, **HJX10**, **SV19**]. Two main extension theorems are presented in this chapter. Given a von Neumann algebra $\mathcal{M}$ admitting a group action, the first extension theorem presents criteria under which a completely bounded operator on $\mathcal{M}$ canonically extends to the crossed product with that group action. The second extension theorem is built around the notion of integrability of a (completely) positive map on a given von Neumann algebra—a notion which ensures that such a map extends canonically to the associated L^p-spaces. The content of this theorem is that any positive map T for which we have that $\varphi \circ T \leq \varphi$ where φ is the reference weight on $\mathcal{M}$ is automatically integrable. In closing, a complete theory of Markov operators is developed.

12.1 Complete boundedness and $L^p(\mathcal{M})$ spaces

We shall have occasion to use the theory of completely bounded maps on $L^p(\mathcal{M})$-spaces in this chapter and hence take the time to briefly review the elements of this theory. We shall restrict ourselves to the setting of operators on von Neumann algebras because of the relative simplicity of the theory in that setting and also because we shall not need a more general theory. The theory of operator spaces and completely bounded maps is, however, a very broad and burgeoning theory extending far beyond the reaches of von Neumann algebras. Readers interested in a detailed account of this theory are referred to [**BLM04**, **ER00**, **Pis03**].

Given a von Neumann algebra $\mathcal{M}$ acting on a Hilbert space H and equipped with a faithful normal semifinite weight φ, the matrix algebra $M_n(\mathcal{M}) = \{[a_{ij}]\colon a_{ij} \in \mathcal{M}, 1 \leq i,j \leq n\}$ will for any $n \in \mathbb{N}$ also be a von Neumann algebra with a faithful normal semifinite weight given by $\varphi^{(n)}([a_{ij}]) = \sum_{i=1}^{n} \varphi(a_{ii})$. Any bounded operator $T : \mathcal{M}_1 \to \mathcal{M}_2$

Noncommutative measures and L^p and Orlicz Spaces, with Applications to Quantum Physics. Stanisław Goldstein and Louis Labuschagne, Oxford University Press. © Stanisław Goldstein and Louis Labuschagne (2025).
DOI: 10.1093/oso/9780198950202.003.0014

induces a bounded map T_n from $M_n(\mathcal{M}_1)$ to $M_n(\mathcal{M}_2)$ by means of the prescription $T_n([a_{ij}]) = [Ta_{ij}]$. For an operator to be *completely bounded*, we additionally demand that $\sup_{n \in \mathbb{N}} \|T_n\| < \infty$ with this supremum then being called the completely bounded norm. For such an operator, we shall simply write $\|T\|_{cb}$ for the completely bounded norm. Similarly, a positive map $T : \mathcal{M}_1 \to \mathcal{M}_2$ will be called completely positive (or just CP for short) if not only T, but each T_n is positive. If all we have is that for some fixed $k \in \mathbb{N}$ the maps T_n are positive for all $1 \le n \le k$, we shall call T k-positive. The famous Russo-Dye theorem for positive linear maps [**RD66**, Corollary 1] assures us that the norm of a positive map $T : \mathcal{M}_1 \to \mathcal{M}_2$ is attained at $T(\mathbb{1})$. Thus, if T is CP, then the norm of each T_n will be determined by $T_n \mathbb{1}_n$, which is just the matrix with $T(\mathbb{1})$ on the main diagonal and 0s everywhere else. Thus, CP maps are automatically completely bounded.

Given $a \in \mathcal{M}_1$ is clear that $\begin{bmatrix} \mathbb{1} & a \\ 0 & 0 \end{bmatrix}^* \cdot \begin{bmatrix} \mathbb{1} & a \\ 0 & 0 \end{bmatrix} = \begin{bmatrix} \mathbb{1} & a \\ a^* & a^*a \end{bmatrix}$ is a then a positive element of $M_2(\mathcal{M}_1)$. For any 2-positive map $T : \mathcal{M}_1 \to \mathcal{M}_2$ with $T(\mathbb{1}) \le \mathbb{1}$, it then follows that $\begin{bmatrix} \mathbb{1} & T(a) \\ T(a^*) & T(a^*a) \end{bmatrix} \ge \begin{bmatrix} T(\mathbb{1}) & T(a) \\ T(x^*) & T(a^*a) \end{bmatrix} \ge 0$. But now basic matrix analysis reveals that $T(a^*a) - T(a^*)T(a) = \begin{bmatrix} -T(a) \\ \mathbb{1} \end{bmatrix}^* \begin{bmatrix} \mathbb{1} & T(a) \\ T(a^*) & T(a^*a) \end{bmatrix} \cdot \begin{bmatrix} -T(a) \\ \mathbb{1} \end{bmatrix} \ge 0$. This is the so-called *operator Schwarz inequality* for 2-positive maps. As long ago as 1952 Kadison proved that for general positive maps with $T(\mathbb{1}) \le \mathbb{1}$, this inequality still holds for self-adjoint elements of $\mathcal{M}_1$ [**Kad52**, Theorem 1]. If for a general element $a \in \mathcal{M}_1$ we now apply Kadison's result to the two self-adjoint operators $(a^* + a)$ and $i(a^* - a)$ and sum the results, this yields the appropriate formulation of the operator Schwarz inequality for general contractive positive maps on von Neumann algebras, namely that

$$T(a^*a + aa^*) \ge T(a^*)T(a) + T(a)T(a^*) \text{ for all } a \in \mathcal{M}_1.$$

The algebra $M_n(\mathcal{M})$ may, of course, equivalently be realised as the algebra $\mathcal{M} \overline{\otimes} M_n(\mathbb{C})$ acting on $H \otimes \mathbb{C}$. In this setting, T_n corresponds to $T \otimes \mathrm{Id}_n$, with the derived weight on $\mathcal{M} \overline{\otimes} M_n(\mathbb{C})$ given by $\varphi \otimes \mathrm{Tr}_n$, where Tr_n is the non-normalised trace on $M_n(\mathbb{C})$. One may also show that a linear map $T : \mathcal{M}_1 \to \mathcal{M}_2$ is completely bounded if and only if for any infinite-dimensional Hilbert space K, the map $T \otimes \mathbb{1} : \mathcal{M}_1 \overline{\otimes} B(K) \to \mathcal{M}_2 \overline{\otimes} B(K)$ (where we used the von Neumann algebra tensor product) is a bounded map, with $\|T \otimes \mathbb{1}\| = \|T\|_{cb}$ [**Tom83**]. When this fact is compared to Proposition 9.8, the potential utility of this concept for crossed products becomes clear.

These ideas had their origin in the work of Stinespring [**Sti55**] who used what is essentially a vector-valued version of the GNS construction, to show that for any CP map $T : \mathcal{A} \to B(H)$ (where $\mathcal{A}$ is a C^*-algebra), there exists a Hilbert space H_T, a representation $\pi_T : \mathcal{A} \to B(H_T)$ of $\mathcal{A}$, and a bounded operator $V_T : H \to H_T$ with $\|V_T\|^2 = \|T\|$ so that $T = V_T^* \pi_T(\cdot) V_T$. There moreover exists a minimal version of this representation in the sense that if $\pi' : \mathcal{A} \to B(H')$ is another representation of $\mathcal{A}$ and $W : H \to H'$ a bounded operator with $T = W^* \pi'(\cdot) W$, then there exists an isometry $U : H_T \to H'$ onto a subspace invariant under π' and intertwining π' and π_T, so that $W = UV_T$.

We pass to analysing the behaviour of $L^p(\mathcal{M})$ spaces with respect to these structures. Throughout this discussion, we shall assume that $\mathcal{M}$ is equipped with the faithful normal semifinite weight φ, with h denoting the density $\frac{d\widetilde{\varphi}}{d\tau}$ of the dual weight $\widetilde{\varphi}$ with respect to the canonical faithful normal semifinite trace τ on $\mathfrak{M} = \mathcal{M} \rtimes_\varphi \mathbb{R}$. As usual, the dual action of $\mathbb{R}$ on $\mathfrak{M}$ will be denoted by θ_t ($t \in \mathbb{R}$). Using the weight $\varphi \otimes \mathrm{Tr}_n$ on $\mathcal{M} \overline{\otimes} M_n(\mathbb{C})$, one obtains that

$$\sigma_t^{\varphi \otimes \mathrm{Tr}_n} = \sigma_t^\varphi \otimes \sigma_t^{\mathrm{Tr}_n} = \sigma_t^\varphi \otimes \mathrm{Id}_n \text{ for every } n \in \mathbb{N}.$$

With $M_n(\mathcal{M}) \equiv \mathcal{M} \overline{\otimes} M_n(\mathbb{C})$ acting on $H \otimes \mathbb{C}^n$, the crossed product $\mathfrak{M}_n = (\mathcal{M} \overline{\otimes} M_n(\mathbb{C})) \rtimes_{\varphi \otimes \mathrm{Tr}_n} \mathbb{R}$ will then act on $(H \otimes \mathbb{C}^n) \otimes L^2(\mathbb{R}) \equiv (H \otimes L^2(\mathbb{R})) \otimes \mathbb{C}^n$ and may equivalently be identified with the algebra $M_n(\mathfrak{M}) = M_n(\mathcal{M} \rtimes_\varphi \mathbb{R}) \equiv (\mathcal{M} \rtimes_\varphi \mathbb{R}) \overline{\otimes} M_n(\mathbb{C})$. With this identification, the canonical trace on $M_n(\mathcal{M} \rtimes_\varphi \mathbb{R})$ is just $\tau \otimes \mathrm{Tr}_n$ with the dual weight of $\varphi \otimes \mathrm{Tr}_n$ corresponding to $\widetilde{\varphi} \otimes \mathrm{Tr}_n$. The existence of a well-developed theory of tensor products of affiliated operators [**KR83**, Section 11.2] allows us to now identify the density $h_n = \frac{d\widetilde{\varphi} \otimes \mathrm{Tr}_n}{d\tau \otimes \mathrm{Tr}_n}$, with the operator $h \otimes \mathbb{1}_n$. One may further use this theory of tensor products of unbounded operators to $*$-isomorphically identify $\widetilde{\mathfrak{M}_n}$ with $M_n(\widetilde{\mathfrak{M}})$

Fixing a system $(e_{ij})_{1 \leq i,j \leq n}$ of matrix units for M_n, this may be done by associating each element $a \in \widetilde{\mathfrak{M}_n}$ with the matrix $[a_{ij}] \in M_n(\widetilde{\mathfrak{M}})$ where for each $1 \leq i,j \leq n$ the term $a_{ij} \otimes e_{ij}$ corresponds to $[(e_{jj} \otimes \mathbb{1}) \cdot a \cdot (e_{ii} \otimes \mathbb{1})]$. Conversely, a matrix $[a_{ij}] \in M_n(\widetilde{\mathfrak{M}})$ may be associated with the term $\left[\sum_{i,j=1}^n e_{ij} \otimes a_{ij} \right]$, where the square brackets indicate the closure being taken after summation.

We pause to introduce some new concepts before presenting some useful identifications of these concepts.

Definition 12.1 With $\mathcal{M}$ and φ as above, we will for any $q \geq 2$ and $p \geq 1$ set $\mathfrak{n}_\varphi^{(q)} = \{a \in \mathcal{M}: ah^{1/q} \text{ is closable and } [ah^{1/q}] \in L^q(\mathcal{M})\}$ and $\mathfrak{m}_\varphi^{(p)} = \mathrm{span}\{b^*a: a, b \in \mathfrak{n}^{(2p)}\}$.

Remark 12.2

- It is clear from Lemma 10.39 and Proposition 10.40 that $\mathfrak{n}^{(2)} = \mathfrak{n}_\varphi$ and hence that $\mathfrak{m}_\varphi^{(1)} = \mathfrak{m}_\varphi$. Therefore, Proposition 10.40 also ensures that $\mathfrak{n}_\varphi^{(q)} \subset \mathfrak{n}_\varphi$ and that $\mathfrak{m}_\varphi^{(p)} \subset \mathfrak{m}_\varphi$ for each $2 \leq q \leq \infty$ and $1 \leq p < \infty$.

- Note that if $a \in \mathfrak{n}_\varphi^{(q)}$, then $h^{1/q}a^*$ is closed with the adjoint being a closed extension of $ah^{1/q}$. Since $ah^{1/q}$ is premeasurable and densely defined by assumption, the uniqueness of the closed extension of such maps (see Proposition 3.71) ensures that $h^{1/q}a^* = [ah^{1/q}]^*$. Thus, $h^{1/q}a^* \in L^q(\mathcal{M})$. Conversely, if $h^{1/q}a^* \in L^q(\mathcal{M})$, then it is in particular densely defined. The operator $ah^{1/q}$ is always densely defined with a domain that includes at least the domain of $h^{1/q}$. Its adjoint is $h^{1/q}a^*$. Basic operator theory now informs us that if this adjoint is densely defined, then $ah^{1/q}$ itself must be closable, with the closure $[ah^{1/q}]$ corresponding to $h^{1/q}a^*$. Thus, if $h^{1/q}a^* \in L^q(\mathcal{M})$, then $a \in \mathfrak{n}_\varphi^{(q)}$. It follows that $\mathfrak{n}_\varphi^{(q)} = \{a \in \mathcal{M}: h^{1/q}a^* \in L^q(\mathcal{M})\}$.

We may now mimic Definition 10.47 and for each $\mathfrak{n}_\varphi^{(q)}$ define the map $\mathfrak{n}_\varphi^{(q)} \ni a \mapsto [ah^{1/q}] \in L^q(\mathcal{M})$. It is clear from the remark that this map is a natural extension of the map $\mathrm{j}^{(q)}$ introduced in Definition 10.47. We may similarly define an embedding of $\mathfrak{m}_\varphi^{(p)}$ into $L^p(\mathcal{M})$, which turns out to be a natural extension of the map $\mathrm{i}^{(p)}$ introduced in Theorem 10.50. Therefore, we will use the same notation for the extension. Technical adjustments to the proofs of Propositions 10.49 and Theorem 10.50 that amount to little more than notational changes now yield the following results. First, we note that $\mathfrak{n}_\varphi^{(q)}$ allows for the following analogue of Lemma 10.48.

Lemma 12.3 *Let $a \in \mathfrak{n}_\varphi^{(q)}$ and $b \in \mathfrak{n}_\varphi$ with q, q^* chosen so that $2 \leq q, q^* < \infty$ and $\frac{1}{q} + \frac{1}{q^*} = \frac{1}{2}$. Then*

$$\mathrm{tr}([ah^{1/q}](h^{1/q^*}b^*)) = \mathrm{tr}(a(h^{1/2}b^*)).$$

Proposition 12.4

(i) *For $q \in [2, \infty)$, each of the maps $\mathfrak{j}^{(q)}$ (see Proposition 10.49) extends to a linear injective map from $\mathfrak{n}_\varphi^{(q)}$ into $L^q(\mathcal{M})$ realised by the prescription $a \mapsto [ah^{1/q}]$. We shall use the same notation for the extension.*

(ii) *For each $p \in [1, \infty)$, the map*

$$\mathfrak{i}^{(p)} : (\mathfrak{m}_\varphi)_+ \to L^p(\mathcal{M})$$

introduced in Theorem 10.50 canonically extends to a map $\mathfrak{i}^{(p)} : (\mathfrak{m}_\varphi^{(p)})_+ \to L^p(\mathcal{M})$ realised by the prescription of setting

$$\mathfrak{i}^{(p)}(a^*a) = (h^{1/2p}a^*)[ah^{1/2p}] = \mathfrak{j}^{(q)}(a)^*\bar{\mathfrak{j}}^{(q)}(a)$$

for any $a \in \mathfrak{n}_\varphi^{(2p)}$. Using the same notation for the extended map, the map thus defined is additive and injective and has a unique extension to a linear map (also called $\mathfrak{i}^{(p)}$) from $\mathfrak{m}_\varphi^{(p)}$ into $L^p(\mathcal{M})$. The extension is injective and positivity preserving and for any $a, b \in \mathfrak{n}_\varphi^{(p)}$ satisfies the formula

$$\mathfrak{j}^{(2p)}(a)^*\bar{\mathfrak{j}}^{(2p)}(b) = \mathfrak{i}^{(p)}(a^*b).$$

As a positive map, it is normal in the sense that if a net $(a_\lambda) \subset (\mathfrak{m}_\varphi^{(p)})_+$ increases to $a \in (\mathfrak{m}_\varphi^{(p)})_+$, then $\mathfrak{i}^{(p)}(a_\lambda)$ increases to $\mathfrak{i}^{(p)}(a)$.

For these spaces and embeddings, we have the following useful identifications (see also [**SV19**, Lemma A.2]).

Proposition 12.5

(1) *For every $0 < p < \infty$, the space $L^p(M_n(\mathcal{M}))$ may be identified with the subspace $L^p(\mathcal{M}) \otimes M_n(\mathbb{C})$ of $\widetilde{\mathfrak{M}} \otimes M_n(\mathbb{C})$.*

(2) *For every $q \in [2, \infty)$, we have $\mathfrak{n}_{(\varphi \otimes \mathrm{Tr}_n)}^{(q)} \equiv M_n(\mathfrak{n}_\varphi^{(q)})$. Furthermore, for every $a = [a_{ij}]_{1 \leq i,j \leq n}$ in $\mathfrak{n}_{(\varphi \otimes \mathrm{Tr}_n)}^{(q)}$, the element $\mathfrak{j}_{(\varphi \otimes \mathrm{Tr}_n)}^{(q)}(a) \in L^q(M_n(\mathcal{M}))$ can be identified with $\left[\mathfrak{j}_\varphi^{(q)}(a_{ij})\right]$.*

(3) *For every $1 \leq p < \infty$, we have $\mathfrak{m}_{(\varphi \otimes \mathrm{Tr}_n)}^{(p)} \equiv M_n(\mathfrak{m}_\varphi^{(p)})$. Moreover, for every $a = [a_{ij}]_{1 \leq i,j \leq n}$ in $\mathfrak{m}_{(\varphi \otimes \mathrm{Tr}_n)}^{(p)}$, the element $\mathfrak{i}_{(\varphi \otimes \mathrm{Tr}_n)}^{(p)}(a) \in L^p(M_n(\mathcal{M}))$ may be identified with $\left[\mathfrak{i}_\varphi^{(p)}(a_{ij})\right]$.*

(4) *Under the identifications made at the start of this section, the tracial functional* tr_n *on* $L^1(M_n(\mathcal{M})) = M_n(L^1(\mathcal{M}))$ *is of the form* $\mathrm{tr}_n = \mathrm{tr} \otimes \mathrm{Tr}_n$.

Proof.   **(1):** This follows on noticing that the action of $\mathbb{R}$ on $M_n(\widetilde{\mathfrak{M}}) = \widetilde{M_n(\mathfrak{M})}$, which is dual to $\sigma^{(\varphi \otimes \mathrm{Tr}_n)} = \sigma^\varphi \otimes \mathrm{Id}_n$, is given by is $\theta_t^\varphi \otimes \mathrm{Id}_n$.

(2): Let $a = [a_{ij}] \in M_n(\mathcal{M})$ be given. If $a \in \mathfrak{n}_\varphi^{(q)} \otimes M_n(\mathbb{C})$, then by the remark $h^{1/q}a_{ji}^* \in L^q(\mathcal{M})$ for every $1 \le i, j \le n$, which in turn ensures that

$$h_n^{1/q}a^* = (h^{1/q} \otimes \mathbb{1}_n) \sum_{1 \le i,j \le n} a_{ij}^* \otimes e_{ji}$$

$$\supseteq \sum_{1 \le i,j \le n} (h^{1/q} \otimes \mathbb{1}_n)(a_{ij}^* \otimes e_{ji})$$

$$= \sum_{1 \le i,j \le n} h^{1/q}a_{ij}^* \otimes e_{ji}.$$

By part (1), each $h^{1/q}a_{ij}^* \otimes e_{ji}$ belongs to $L^q(M_n(\mathcal{M}))$, and hence, $h_n^{1/q}a^*$ is therefore a closed extension of a formal sum of finitely many elements of $L^q(M_n(\mathcal{M}))$. By the theory of τ-measurable operators, the operator $h_n^{1/q}a^*$ must therefore agree with the closure of this formal sum and must therefore itself belong to $L^q(M_n(\mathcal{M}))$. Hence, by the remark, $a \in \mathfrak{n}_{(\varphi \otimes \mathrm{Tr}_n)}^{(q)}$.

Finally, note that the fact that $h_n^{1/q}a^* = \left[\sum_{1 \le i,j \le n} h^{1/q}a_{ij}^* \otimes e_{ji} \right]$ corresponds to the claim that $\mathrm{j}_{(\varphi \otimes \mathrm{Tr}_n)}^{(q)}(a)^*$ may be identified with $\left[\mathrm{j}_\varphi^{(q)}(a_{ij}) \right]^*$.

Conversely, if $a \in \mathfrak{n}_{(\varphi \otimes \mathrm{Tr}_n)}^{(q)}$, then for fixed $1 \le i, j \le n$, the fact that $\mathfrak{n}_{(\varphi \otimes \mathrm{Tr}_n)}^{(q)}$ is a left ideal ensures that $a_i := (\mathbb{1} \otimes e_{ii})a$ also belongs to $\mathfrak{n}_{(\varphi \otimes \mathrm{Tr}_n)}^{(q)}$. But for any $1 \le j \le n$, the operator $(\mathbb{1} \otimes e_{jj})(h_n^{1/q}a_i^*)$ will then be the product of an element of $M_n(\mathcal{M})$ and of $L^q(M_n(\mathcal{M}))$ and must therefore be τ_n-densely defined. Since $(\mathbb{1} \otimes e_{jj})h_n^{1/q} = (\mathbb{1} \otimes e_{jj})(h^{1/q} \otimes \mathbb{1}_n) \subseteq h^{1/q} \otimes e_{jj}$, we surely have $(\mathbb{1} \otimes e_{jj})h_n^{1/q}a_i^* \subseteq (h^{1/q} \otimes e_{jj})a_i^* = h^{1/q}a_{ij}^* \otimes e_{ji}$, thereby showing that these operators also have τ_n-dense domains. Since $\tau_n = \tau \otimes \mathrm{Tr}_n$, standard trace calculations show that each $h^{1/q}a_{ij}^*$ is itself τ-densely defined and hence an element of $L^q(\mathcal{M})$ by part (1). The remark now ensures that $a_{ij} \in \mathfrak{n}_\varphi^{(q)}$ as required.

(3): This follows readily from part (2) and the definition of the embeddings $\mathrm{i}_{(\varphi \otimes \mathrm{Tr}_n)}^{(p)}$ and $\mathrm{j}_\varphi^{(p)}$.

(4): To start with, let $[a_{ij}] \in M_n(\mathfrak{m}_\varphi^{(p)})_+$. Since for each $1 \leq i \leq n$ we have that $a_{ii} \geq 0$, we may use Remark 10.41 to see that

$$\mathrm{tr}_n(\mathrm{i}^{(1)}_{(\varphi \otimes \mathrm{Tr}_n)}([a_{ij}])) = (\varphi \otimes \mathrm{Tr}_n)([a_{ij}]) = \sum_{i=1}^{n} \varphi(a_{ii})$$

$$= \sum_{i=1}^{n} \mathrm{tr}(\mathrm{i}^{(1)}_\varphi(a_{ii})) = (\mathrm{tr} \otimes \mathrm{Tr}_n)([a_{ij}]).$$

By linearity, this equality then holds on all of $M_n(\mathfrak{m}_\varphi^{(p)})$. The norm continuity of $\mathrm{tr} : L^1(\mathcal{M}) \to \mathbb{C}$ ensures that $(\varphi \otimes \mathrm{Tr}_n)$ is norm continuous on $M_n(L^1(\mathcal{M})) = L^1(M_n(\mathcal{M}))$. The claim therefore follows from the density of $M_n(\mathfrak{m}_\varphi^{(p)})$ in $M_n(L^1(\mathcal{M}))$ (Theorem 10.46). $\qquad\square$

We end this section by reproducing a very important theorem of Haagerup, Junge and Xu, which clearly demonstrates the utility of complete boundedness for crossed products previously only hinted at. The final observation is new. All other claims are as in [**HJX10**, Theorem 4.1].

Theorem 12.6 *Let $\mathcal{M}$ and $\mathcal{N}$ be two von Neumann algebras acting on the same Hilbert space H, G a locally compact abelian group and α and β *-automorphic actions of G on $\mathcal{M}$ and $\mathcal{N}$, respectively. Suppose that $T : \mathcal{M} \to \mathcal{N}$ is a completely bounded normal map such that*

$$T \circ \alpha_g = \beta_g \circ T, \qquad g \in G. \tag{12.1}$$

Then, the prescription

$$\widetilde{T}(\lambda_g \pi_\alpha(x)) = \lambda_g \pi_\beta(Tx), \qquad x \in \mathcal{M}, \; g \in G \tag{12.2}$$

allows for the unique completely bounded normal extension of T to an operator $\widetilde{T}$ from $\mathcal{M} \rtimes_\alpha G$ into $\mathcal{N} \rtimes_\beta G$ such that $\|\widetilde{T}\|_{cb} = \|T\|_{cb}$. The extension $\widetilde{T}$ satisfies the following properties.

   (i) *With $\mathcal{A}$ denoting the von Neumann subalgebra on $L^2(G, H)$ generated the operators λ_g $(g \in G)$, we will have that*

$$\widetilde{T}(a\pi_\alpha(x)b) = a\pi_\beta(Tx)b, \qquad x \in \mathcal{M}, \; a, \; b \in \mathcal{A}. \tag{12.3}$$

(ii) $\widetilde{T} \circ \widehat{\alpha}_\gamma = \widehat{\beta}_\gamma \circ \widetilde{T}$, $\gamma \in \widehat{G}$.

(iii) *If T is a homomorphism, $*$-homomorphism or CP map, so is $\widetilde{T}$.*

(iv) *If $\mathcal{N}$ is a subalgebra of $\mathcal{M}$, $\beta = \alpha\!\restriction\!\mathcal{N}$ and T is a (faithful) normal conditional expectation from $\mathcal{M}$ onto $\mathcal{N}$, then $\widetilde{T}$ is a (faithful) normal conditional expectation from $\mathcal{M} \rtimes_\alpha G$ onto $\mathcal{N} \rtimes_\beta G$.*

(v) *Let φ (respectively, ψ) be a faithful normal semifinite weight on $\mathcal{M}$ (respectively, $\mathcal{N}$) such that*

$$T \circ \sigma_t^\varphi = \sigma_t^\psi \circ T, \qquad t \in \mathbb{R}. \tag{12.4}$$

Then

$$\widetilde{T} \circ \sigma_t^{\widetilde{\varphi}} = \sigma_t^{\widetilde{\psi}} \circ \widetilde{T}, \qquad t \in \mathbb{R}. \tag{12.5}$$

(vi) *Additionally, assume that $T \geq 0$. Let φ (respectively, ψ) be a faithful normal semifinite weight on $\mathcal{M}$ (respectively, $\mathcal{N}$) such that $\psi \circ T \leq C\,\varphi$ for some $C > 0$. Then $\widetilde{\psi} \circ \widetilde{T} \leq C\,\widetilde{\varphi}$.*

(vii) *Suppose that $\sigma^\varphi\!\restriction\!\mathcal{N} = \sigma^\psi$, and let T be a CP map on $\mathcal{M}$ satisfying*

$$T \circ \sigma_t^\varphi = \sigma_t^\psi \circ T, \qquad t \in \mathbb{R}$$

and $\varphi \circ T \leq C\,\varphi$ for some $C > 0$. Then, the extension $\widetilde{T} : \mathcal{M} \rtimes_\varphi \mathbb{R} \to \mathcal{N} \rtimes_\psi \mathbb{R}$ described above additionally satisfies $\tau_\psi \circ \widetilde{T} \leq C\,\tau_\varphi$, where τ_ψ and τ_φ are, respectively, the canonical traces on $\mathcal{N} \rtimes_\psi \mathbb{R}$ and $\mathcal{M} \rtimes_\varphi \mathbb{R}$, as guaranteed by Theorem 9.28.

Proof.   Since T is completely bounded and normal, so is

$$T \otimes \mathrm{Id}_{B(L^2(G))} : \mathcal{M} \overline{\otimes} B(L^2(G)) \to \mathcal{N}\,\overline{\otimes} B(L^2(G)).$$

Moreover, $\|T \otimes \mathrm{Id}_{B(L^2(G))}\|_{cb} = \|T\|_{cb}$. We know from Proposition 9.8 that $\mathcal{M} \rtimes_\alpha G$ and $\mathcal{N} \rtimes_\beta G$ are, respectively, subspaces of $\mathcal{M} \overline{\otimes} B(L^2(G))$ and $\mathcal{N}\,\overline{\otimes} B(L^2(G))$. The proof of the theorem consists in showing that the restriction of $T \otimes \mathrm{Id}_{B(L^2(G))}$ to $\mathcal{M} \rtimes_\alpha G$ is of the form claimed in the hypothesis. We proceed with that verification.

Fix an orthonormal basis $(f_i)_{i \in I}$ in $L^2(G)$. Recall that in the proof of Theorem 9.9, we saw that for any $x \in \mathcal{M}$, $\pi_\alpha(x)$ is then of the form $\pi_\alpha(x) = \sum_{i,j} \pi_\alpha(a)_{ij} \otimes u_{i,j}$ where $\pi_\alpha(x)_{ij} = \int_G \alpha_{s^{-1}}(x)\overline{f_i}(s)f_j(s)\,ds$ and $u_{i,j} = \langle \cdot, f_j \rangle\,f_i$, with a similar claim holding for elements of $\mathcal{N}$. By the normality of $T \otimes \mathrm{Id}_{B(L^2(G))}$ we see that $T \otimes \mathrm{Id}_{B(L^2(G))}(\pi_\alpha(x)) = T \otimes \mathrm{Id}_{B(L^2(G))}(\sum_{i,j} \pi_\alpha(x)_{ij} \otimes u_{i,j}) = \sum_{i,j} T(\pi_\alpha(x)_{ij}) \otimes u_{i,j}$. But the normality of T and equation (12.1) ensure that

$$T\big((\pi_\alpha(x))_{ij}\big) = \int_G T(\alpha_{s^{-1}}(x))\overline{f_i}(s) f_j(s)\, ds$$

$$= \int_G \beta_{s^{-1}}(Tx)\overline{f_i}(s) f_j(s)\, ds$$

$$= \big(\pi_\beta(Tx)\big)_{ij}.$$

Thus, $T \otimes \mathrm{Id}_{B(L^2(G))}$ maps $\pi_\alpha(x)$ onto $\pi_\beta(T(x))$.

On the other hand, since $\lambda_g = \mathbb{1} \otimes \ell_g$, left multiplication by λ_g clearly commutes with $T \otimes \mathrm{Id}_{B(L^2(G))}$. It therefore follows that we will for all $x \in \mathcal{M}$ and $g \in G$ have that

$$T \otimes \mathrm{Id}_{B(L^2(G))}\big(\lambda_g \pi_\alpha(x)\big) = \lambda_g T \otimes \mathrm{Id}_{B(L^2(G))}\big(\pi_\alpha(x)\big) = \lambda_g \pi_\beta(Tx).$$
$$(12.6)$$

Recall that the family of all finite linear combinations of $\lambda_g \pi_\alpha(x)$, where $g \in G$ and $x \in \mathcal{M}$, is a w*-dense involutive subalgebra of $\mathcal{M} \rtimes_\alpha G$ (see Remark 9.7) (and similarly for $\mathcal{N} \rtimes_\alpha G$). The fact verified above therefore shows that $T \otimes \mathrm{Id}_{B(L^2(G))}$ maps a σ-weakly dense subspace of $\mathcal{M} \rtimes_\alpha G$, into a σ-weakly dense subspace of $\mathcal{N} \rtimes_\alpha G$. The normality of $T \otimes \mathrm{Id}_{B(L^2(G))}$ therefore ensures that $T \otimes \mathrm{Id}_{B(L^2(G))}$ maps $\mathcal{M} \rtimes_\alpha G$ into $\mathcal{N} \rtimes_\alpha G$, with the restriction $\widetilde{T}$ of $T \otimes \mathrm{Id}_{B(L^2(G))}$ to $\mathcal{M} \rtimes_\alpha G$, amounting to an extension of the prescription given in equation (12.3). Since $T \otimes \mathrm{Id}_{B(L^2(G))}$ is completely bounded and normal so is the restriction to $\mathcal{M} \rtimes_\alpha G$. Thus, $\widetilde{T}$ is the desired extension of T. The uniqueness of $\widetilde{T}$ follows from equation (12.2) and the normality noted earlier.

We proceed to check the other properties of $\widetilde{T}$. Given $g, h \in G$ and $x \in \mathcal{M}$, it follows from Proposition 9.5 that

$$\lambda_g \pi_\alpha(x) \lambda_h = \lambda_g \lambda_h \pi_\alpha(\alpha_{h^{-1}} x) = \lambda_{gh} \pi_\alpha(\alpha_{h^{-1}} x).$$

Thus, by equations (12.2) and (12.1),

$$\widetilde{T}\big(\lambda_g \pi_\alpha(x)\lambda_h\big) = \lambda_{gh} \pi_\beta\big(T(\alpha_{h^{-1}}(x))\big)$$

$$= \lambda_{gh} \pi_\beta\big(\alpha_{h^{-1}}(Tx)\big) = \lambda_g \pi_\beta(Tx)\lambda_h.$$

This yields equation (12.3) in the case where $a = \lambda_g$ and $b = \lambda_h$ for any given $g, h \in G$. The general case then follows from the Borel functional calculus and the normality of $\widetilde{T}$.

(ii) is a consequence of equations (9.2) and (12.1). Claim (iii) is an easy consequence of the fact that if T is a homomorphism, $*$-homomorphism or CP map, then so is $T \otimes \mathrm{Id}_{B(L^2(G))}$.

Now, suppose that the assumptions in (iv) hold. Careful checking of the definition of the crossed product reveals that if $\mathcal{N}$ is a subalgebra of $\mathcal{M}$ and $\beta = \alpha \!\restriction\! \mathcal{N}$, then $\pi_\beta \mathcal{N} \subset B(L^2(G, H))$ is a subalgebra of $\pi_\alpha(\mathcal{M})$, with $\mathcal{N} \rtimes_\beta G$ then being a subalgebra of $\mathcal{M} \rtimes_\alpha G$ generated by $\pi_\beta \mathcal{N}$, and the same set of translation operators used to generate $\mathcal{M} \rtimes_\alpha G$. Now, if T is a (faithful) normal conditional expectation from $\mathcal{M}$ to $\mathcal{N}$, then $T \otimes \mathrm{Id}_{B(L^2(G))}$ is also a (faithful) normal conditional from $\mathcal{M}\overline{\otimes}B(L^2(G))$ to $\mathcal{N}\,\overline{\otimes}B(L^2(G))$. The easiest way to see the claim about the faithfulness is to select an fns weight ψ on $\mathcal{N}$ and to note that $\psi \circ T$ must then be an fns weight on $\mathcal{M}$ which is preserved by T. By Theorem 8.39 and the comment preceding it, $(\psi \circ T) \otimes \mathrm{Tr}_{B(L^2(G))}$ is then an fns weight preserved by the expectation $T \otimes \mathrm{Id}_{B(L^2(G))}$. This can of course only be if $T \otimes \mathrm{Id}_{B(L^2(G))}$ is itself faithful. Hence, the restriction $\widetilde{T}$ is a (faithful) normal conditional expectation from $\mathcal{M} \rtimes_\alpha G$ to $\mathcal{N} \rtimes_\beta G$.

Claim (v) follows from Theorem 9.25 and equation (12.1).

To prove (vi), we first use normality to, respectively, extend T and $\widetilde{T}$ to the extended positive parts of $\mathcal{M}$ and $\mathcal{M} \rtimes_\alpha G$. To see that this is possible, recall by way of an example that any $a \in \widehat{\mathcal{M}}_+$ may be written as the supremum of an increasing net $(a_i) \subset \mathcal{M}_+$. We may then simply define $T(a)$ to be $\sup_i T(a_i)$ where the supremum is taken in $\widehat{\mathcal{M}}_+$. The normality of T then guarantees the uniqueness of $T(a)$ thus defined. For the extension of T, we will therefore clearly have that $\widehat{\psi} \circ T \leq C\,\widehat{\varphi}$, where $\widehat{\psi}$ and $\widehat{\varphi}$ respectively denote the extensions of ψ and φ to $\widehat{\mathcal{N}}$ and $\widehat{\mathcal{N}}$. Let $\mathscr{W}_\alpha$ and $\mathscr{W}_\beta$, respectively, be the operator-valued weights from $\mathcal{M} \rtimes_\alpha G$ to $\mathcal{M}$ and from $\mathcal{N} \rtimes_\beta G$ to $\mathcal{N}$. Using the normality of these extensions, we may then conclude from (i) that the equality $\pi_\beta^{-1} \circ \widetilde{T} = T \circ \pi_\alpha^{-1}$ holds on $\widehat{\pi_\alpha(\mathcal{M})}_+$ and from Definition 9.20 that

$$\mathscr{W}_\beta \circ \widetilde{T}(a) = \int_G \widehat{\beta}_\gamma(\widetilde{T}(a))\, d\gamma$$

$$= \int_G \widetilde{T}\widehat{\alpha}_\gamma((a))\, d\gamma$$

$$= \widetilde{T} \circ \int_G \widehat{\alpha}_\gamma((a))\, d\gamma$$

$$= \widetilde{T} \circ \mathscr{W}_\alpha.$$

We therefore obtain

$$\widetilde{\psi} \circ \widetilde{T} = \widehat{\psi} \circ \pi_\beta^{-1} \circ \mathscr{W}_\beta \circ \widetilde{T}$$

$$= \widehat{\psi} \circ \pi_\beta^{-1} \circ \widetilde{T} \circ \mathscr{W}_\alpha$$

$$= \widehat{\psi} \circ T \circ \pi_\alpha^{-1} \circ \mathscr{W}_\alpha$$

$$\leq C\,\widehat{\varphi} \circ \pi_\alpha^{-1} \circ \mathscr{W}_\alpha$$

$$= \widetilde{\varphi}.$$

Hence, (vi) is proved. Therefore, it remains to prove (vii). Recall that in the proof of (iv), we noted that $\mathcal{N} \rtimes_\psi \mathbb{R}$ will under the given assumptions in all respects be a subalgebra of $\mathcal{M} \rtimes_\varphi \mathbb{R}$, with the two algebras sharing the same translation operators λ_g. By Theorem 9.28, we therefore have that $\frac{d\widetilde{\psi}}{d\tau_\beta} = \frac{d\widetilde{\varphi}}{d\tau_\alpha}$. We will write h for this derivative. Theorem 9.28 also assures us that for any $\epsilon > 0$, the operator $(\epsilon\mathbb{1} + h)^{-1/2}$ belongs to the algebra $\mathcal{A}$ described in (i). By equation (12.3) and part (vi), we will therefore have $\widetilde{\psi}((\epsilon\mathbb{1} + h)^{-1/2}\widetilde{T}(\cdot)(\epsilon\mathbb{1} + h)^{-1/2}) = \widetilde{\psi}(\widetilde{T}((\epsilon\mathbb{1} + h)^{-1/2}(\cdot)(\epsilon\mathbb{1} + h)^{-1/2})) \leq C\widetilde{\varphi}((\epsilon\mathbb{1} + h)^{-1/2}\widetilde{T}(\cdot)(\epsilon\mathbb{1} + h)^{-1/2})$. On again appealing to Theorem 9.28, we see that letting ϵ decrease to zero yields the inequality $\tau_\beta \circ \widetilde{T} \leq C\tau_\alpha$, as required. $\qquad\square$

Remark 12.7 Let $\mathcal{L}$ be a third von Neumann algebra, γ an automorphic representation of G on $\mathcal{L}$ and $S : \mathcal{N} \to \mathcal{L}$ a completely bounded normal map such that $S \circ \beta_g = \gamma_g \circ S$ for all $g \in G$. It is easy to check that the extension in the preceding theorem satisfies the functorial property $\widetilde{S \circ T} = \widetilde{S} \circ \widetilde{T}$.

12.2 Integrability of operators on $\mathcal{M}$

Definition 12.8 Let T be an operator with domain $\mathrm{dom}(T) \subset \mathcal{M}$. Then, T is said to be *p-integrable* (with respect to φ) for some $1 \leq p \leq \infty$ if:

(i) $\mathrm{dom}(T) \supseteq \mathfrak{m}_\varphi$;

(ii) $T(\mathfrak{m}_\varphi) \subset \mathfrak{m}^{(p)}$ (see Definition 12.1);

(iii) the operator $T_0^{(p)}$ defined by $T_0^{(p)} : \mathfrak{i}^{(p)}(a) \mapsto \mathfrak{i}^{(p)}(Ta)$ is L^p-continuous.

If T is p-integrable, we write $T^{(p)}$ for the unique continuous extension of $T_0^{(p)}$ to $L^p(\mathcal{M})$. A 1-integrable operator will simply be said to be *integrable*. If T is integrable, then we shall write $T^\dagger$ for $T^{(1)*}$.

Observe that for any $b_0, c_0, b_1, c_1 \in \mathfrak{n}_\varphi$ and pairs of conjugate indices $1 \le p_0, q_0 \le \infty$ and $1 \le p_1, q_1 \le \infty$, we may use Lemma 10.48 to see that

$$\mathrm{tr}(((h^{1/2p_0}c_0^*))[b_0 h^{1/2p_0}])((h^{1/2q_0}c_1^*)[b_1 h^{1/2q_0}]))$$

$$= \mathrm{tr}(([b_0 h^{1/2q_0}])((h^{1/2p_0}c_1^*)[b_1 h^{1/2p_0}])(h^{1/2q_0}c_0^*)))$$

$$= \mathrm{tr}(([b_0 h^{1/2q_1}])((h^{1/2p_1}c_1^*)[b_1 h^{1/2p_1}])(h^{1/2q_0}c_1^*)))$$

$$= \mathrm{tr}(((h^{1/2p_1}c_0^*))[b_0 h^{1/2p_1}])((h^{1/2q_1}c_1^*)[b_1 h^{1/2q_1}])).$$

For any $a_0, a_1 \in \mathfrak{m}_\varphi = \mathrm{span}\{b^*c \colon b, c \in \mathfrak{n}_\varphi\}$ and (p_0, q_0) and (p_1, q_1) as before, we therefore have that

$$\mathrm{tr}(\mathfrak{i}^{(p_0)}(a_0)\mathfrak{i}^{(q_0)}(a_1)) = \mathrm{tr}(\mathfrak{i}^{(p_1)}(a_0)\mathfrak{i}^{(q_1)}(a_1)). \tag{12.7}$$

The following fact is key to the ensuing development.

Proposition 12.9 *The left-ideals $\left\{\mathfrak{n}_\varphi^{(q)} \colon q \in [2, \infty)\right\}$ are increasing:*

$$q_1 \le q_2 \Rightarrow \mathfrak{n}_\varphi^{(q_1)} \subset \mathfrak{n}_\varphi^{(q_2)}.$$

The subspaces $\left\{\mathfrak{m}_\varphi^{(p)} \colon p \in [1, \infty)\right\}$ are therefore similarly increasing.

Proof. Let $a \in \mathfrak{n}^{(q_1)}$ and $q_2 \ge q_1$. Then, $h^{1/q_2}a^*$ is closed and affiliated to $\mathfrak{M} = \mathcal{M} \rtimes_\varphi \mathbb{R}$, and its domain contains that of $h^{1/q_1}a^*$ and so is τ-dense. Therefore, $h^{1/q_2}a^* \in \widetilde{\mathfrak{M}}$, and so it remains to show that it belongs to $L^{q_2}(\mathcal{M})$.

Put $s = \left(q_1^{-1} - q_2^{-1}\right)^{-1}$, and let $y = \theta_t\left(h^{1/q_2}a^*\right) - e^{-t/q_2}h^{1/q_2}a^*$. For $b \in \mathfrak{n}$, using Lemma 10.48 twice, we have

$$\left[bh^{1/s}\right]^{\bar{\cdot\top}}\theta_t\left(h^{1/q_2}a^*\right) = e^{t/s}{}^\top\theta_t\left(\left[bh^{1/s}\right]^{\bar{\cdot}}h^{1/q_2}a^*\right)$$

$$= e^{t/s}{}^\top\theta_t\left(b^{\bar{\cdot}}h^{1/q_1}a^*\right)$$

$$= e^{-t/q_2}b^{\bar{\cdot}}h^{1/q_1}a^*$$

$$= e^{-t/q_2}\left[bh^{1/s}\right]^{\bar{\cdot\top}}h^{1/q_2}a^*$$

so $\left[bh^{1/s}\right]^{\cdot}y = 0$. But the strong product $L^s \times L^{q_2} \ni (x,y) \mapsto x^{\cdot}y \in L^{q_1}$ is continuous and, by Theorem 10.46, $\left\{\left[bh^{1/s}\right] : b \in \mathfrak{n}\right\}$ is dense in L^s. Therefore, $y = 0$; in other words, $h^{1/q_2}a^* \in L^{q_2}$, or equivalently $a \in \mathfrak{n}^{(q_2)}$. $\qquad\qquad\qquad\qquad\qquad\qquad\qquad\qquad\qquad\qquad\qquad\quad\square$

One of the key theorems in this section is the following noncommutative analogue of the famous Riesz convexity theorem.

Theorem 12.10 (Noncommutative Riesz–Thorin convexity theorem) *Let $\mathcal{M}_i$ $(i = 1,2)$ be von Neumann algebras equipped with faithful normal weights φ_i, and let $1 \leq p_0 \leq p_1 \leq \infty$ be given With h_k denoting the densities $h_k = \frac{d\widetilde{\varphi}_k}{d\tau_k}$ $(k = 1,2)$, let $T : \mathcal{M}_1 \to \mathcal{M}_2$ be a linear operator that is both p_0 and p_1-integrable, where $1 \leq p_0 < p_1 \leq \infty$, with $\|T^{(p_0)}\| = C_0$ and $\|T^{(p_1)}\| = C_1$. Then, T is p-integrable for any $p_0 < p < p_1$, with the norm of $T^{(p)}$ not exceeding $C_0^{1-\theta}.C_1^{\theta}$, where $0 \leq \theta \leq 1$ is chosen so that $\frac{1}{p} = \frac{1-\theta}{p_0} + \frac{\theta}{p_1}$.*

Notice that by Proposition 12.9, the p_0-integrability of T already ensures that $T(\mathfrak{m}_{\varphi_1}) \subset \mathfrak{m}_{\varphi_2}^{(p)}$. Thus, all that needs to be checked is if the prescription $\mathfrak{i}^{(p)}(\mathfrak{m}_{\varphi_1}) \to L^p(\mathcal{M}_2) : \mathfrak{i}^{(p)}(a) \mapsto \mathfrak{i}^{(p)}(T(a))$ yields a continuous operator.

Proof. For the sake of simplicity, we will prove the theorem for the case $p_0 = 1$ and $p_1 = \infty$. With all notation and concepts as in Chapter 11, the general case will then follow from the fact that by the complex reiteration theorem (Theorem 11.16), the spaces $\mathscr{L}^p(\mathcal{M}_k)$ $(k \in \{0,1\})$ where $p_0 \leq p \leq p_1$ are again a complex interpolation scale, the structure of which perfectly aligns with the structure of the scale $\mathscr{L}^p(\mathcal{M}_k)$ $(k \in \{0,1\})$ where $1 \leq p \leq \infty$.

The map T of course induces a corresponding map $\widetilde{T} = \iota^{(\infty)} \circ T \circ (\iota^{(\infty)})^{-1}$ from $\mathscr{L}^\infty(\mathcal{M}_1)$ to $\mathscr{L}^\infty(\mathcal{M}_2)$ with the same norm C_1. The criteria in the hypothesis ensure that the restriction of $\widetilde{T}$ to $(\mathscr{L}^1 \cap \mathscr{L}^\infty)(\mathcal{M}_1)$ extends to a bounded linear map from $\mathscr{L}^1(\mathcal{M}_1)$ to $\mathscr{L}^1(\mathcal{M}_2)$ with norm C_0. By classical arguments, the map $\widetilde{T}$ may then be extended to a bounded map from $(\mathscr{L}^1 + \mathscr{L}^\infty)(\mathcal{M}_1)$ to $(\mathscr{L}^1 + \mathscr{L}^\infty)(\mathcal{M}_2)$. On applying Theorem 11.15, it follows that this extension restricts to a map from $\mathscr{L}^p(\mathcal{M}_1)$ to $\mathscr{L}^p(\mathcal{M}_2)$, with norm not exceeding $C_1^{1-1/p}.C_0^{1/p}$. This fact in turn translates to the conclusion of the theorem. $\qquad\square$

Proposition 12.11   *Let $T : \mathcal{M}_1 \supseteq \mathrm{dom}(T) \to \mathcal{M}_2$ be an integrable operator such that $T^\dagger : \mathcal{M}_2 \to \mathcal{M}_1$ is also integrable. Then, T is σ-weakly continuous on $\mathfrak{m}_{\varphi_1} = \mathfrak{m}_1$ and $T^{\dagger\dagger}$ a normal extension of $T{\upharpoonright}\mathfrak{m}_1$ to all of $\mathcal{M}$.*

Proof.   Since both T and $T^\dagger$ are integrable, respectively, they will map $\mathfrak{m}_{\varphi_1}$ into $\mathfrak{m}_{\varphi_2}$ and $\mathfrak{m}_{\varphi_2}$ into $\mathfrak{m}_{\varphi_1}$. So, for any $a_0 \in \mathfrak{m}_{\varphi_1}$ and $a_1 \in \mathfrak{m}_{\varphi_2}$, we have that

$$
\begin{aligned}
\mathrm{tr}(Ta_0\overline{\mathrm{i}}^{(1)}(a_1)) &= \mathrm{tr}(T^{(}1)\mathrm{i}^{(1)}(a_0)^{-}a_1) \\
&= \mathrm{tr}(\mathrm{i}^{(1)}(a_0)^{-}T^\dagger a_1) \\
&= \mathrm{tr}(a_0^{-}T^{\dagger(1)}\mathrm{i}^{(1)}(a_1)) \\
&= \mathrm{tr}(T^{\dagger\dagger}a_0\overline{\mathrm{i}}^{(1)}(a_1)).
\end{aligned}
$$

The density of $\mathrm{i}^{(1)}(\mathfrak{m}_2)$ in $L^1(\mathcal{M}_2)$ then ensures that on $\mathfrak{m}_{\varphi_1}$, the operator $T^{\dagger\dagger}$ agrees with T as claimed. $\square$

The above proposition suggests that normality is somehow endemic to operators exhibiting good integrability properties. *We shall in the rest of this section therefore restrict ourselves to everywhere defined normal operators.* Under that assumption, we obtain the following refinement of the above proposition.

Proposition 12.12   *Let $T : \mathcal{M}_1 \to \mathcal{M}_2$ be a normal (σ-weakly continuous) integrable operator. If $T^\dagger(\mathfrak{m}_{\varphi_2}) \subset \mathfrak{m}_{\varphi_1}$, $T^\dagger$ will be integrable.*

Proof.   Let T_* be the pre-adjoint of T. For any $a_0 \in \mathfrak{m}_{\varphi_2}$ and $a_1 \in \mathfrak{m}_{\varphi_1}$, a similar argument to the one used before shows that

$$
\begin{aligned}
\mathrm{tr}(\mathrm{i}^{(1)}(T^\dagger(a_0))^{-}a_1) &= \mathrm{tr}(T^\dagger(a_0)\overline{\mathrm{i}}^{(1)}(a_1)) \\
&= \mathrm{tr}(a_0^{-}T^{(1)}(\mathrm{i}^{(1)}(a_1))) \\
&= \mathrm{tr}(a_0\overline{\mathrm{i}}^{(1)}(T(a_1))) \\
&= \mathrm{tr}(\mathrm{i}^{(1)}(a_0)^{-}T(a_1)) \\
&= \mathrm{tr}(T_*(\mathrm{i}^{(1)}(a_0))^{-}a_1).
\end{aligned}
$$

The density of $\mathfrak{i}^{(1)}(\mathfrak{m}_1)$ in $L^1(\mathcal{M}_1)$ then ensures that $T_* : L^1(\mathcal{M}_2) \to L^1(\mathcal{M}_1)$ is indeed an extension of the action $\mathfrak{i}^{(1)}(a) \mapsto \mathfrak{i}^{(1)}(T^\dagger(a))$ on $\mathfrak{m}_{\varphi_1}$, thereby proving the claim. $\qquad\square$

The above proposition suggests that for a class of sufficiently 'regular' operators, conditions that ensure that T maps $\mathfrak{m}_{\varphi_1}$ into $\mathfrak{m}_{\varphi_2}$ may also ensure that such an operator is integrable. The next theorem gives conditions under which this is true. This theorem is a sharpening of Yeadon's ergodic theorem for positive maps [**Yea80**]. It was originally stated in [**GL95**]. The result was later independently verified by Haagerup, Junge and Xu, using a different approach which we present below (see [**HJX10**, Theorem 5.1]).

Lemma 12.13 ([Haa75a, Lemma 1.2]) *Let x be a self-adjoint element of $\mathfrak{m}_\varphi$. Then*

$$\|\mathfrak{i}^{(1)}(x)\|_1 = \inf\left\{\varphi(a) + \varphi(b) : x = a - b,\ a,\ b \in \mathfrak{p}_\varphi\right\}.$$

Proof. In the proof, we shall simply write $\mathfrak{m}$ for $\mathfrak{m}_\varphi$ and denote the self-adjoint portion of this algebra by $\mathfrak{m}_h$. Denote the infimum on the right-hand side by $\rho(x)$. Then, $x \mapsto \rho(x)$ defines a seminorm on $\mathfrak{m}_h$. Recall that by Remark 10.41

$$\rho(x) = \varphi(x) = \mathrm{tr}(\mathfrak{i}^{(1)}(x)) = \|\mathfrak{i}^{(1)}(x)\|_1 \text{ for any } x \in \mathcal{M}_+.$$

Now, let $x \in \mathfrak{m}_h$ and let $a, b \in \mathfrak{p}$ be given with $x = a - b$. We then have that

$$\|\mathfrak{i}^{(1)}(x)\|_1 \le \|\mathfrak{i}^{(1)}(a)\|_1 + \|\mathfrak{i}^{(1)}(b)\|_1 = \varphi(a) + \varphi(b).$$

Taking the infimum over all such decompositions then yields the fact that

$$\|\mathfrak{i}^{(1)}(x)\|_1 \le \rho(x), \quad x \in \mathfrak{m}_h.$$

To prove the converse inequality, we fix $x_0 \in \mathfrak{m}_h$. By the Hahn–Banach theorem, there exists a linear functional $f : \mathfrak{m}_h \to \mathbb{R}$ such that

$$f(x_0) = \rho(x_0) \text{ and } |f(x)| \le \rho(x), \text{ for all } x \in \mathfrak{m}_h.$$

The functional f may now be extended to all of $\mathfrak{m}$ by the process of complexification. The extended functional will still be denoted by f and appears as a Hermitian functional on $\mathfrak{m}$ satisfying $-\varphi \le f \le \varphi$ on

$\mathfrak{p}$. Thus, by Theorem 10.50 and the Cauchy–Schwarz inequality, we will for any $x,\, y \in \mathfrak{m}$ have that

$$
\begin{aligned}
|f(y^*x)| &\leq \frac{1}{2}\big[|(\varphi + f)(y^*x)| + |(\varphi - f)(y^*x)|\big] \\
&\leq \frac{1}{2}\Big[\big((\varphi + f)(y^*y)\big)^{1/2}\big((\varphi + f)(x^*x)\big)^{1/2} \\
&\qquad + \big((\varphi - f)(y^*y)\big)^{1/2}\big((\varphi - f)(x^*x)\big)^{1/2}\Big] \\
&\leq \big(\varphi(y^*y)\big)^{1/2}\big(\varphi(x^*x)\big)^{1/2} \\
&= \mathrm{tr}(\mathfrak{i}^{(1)}(y^*y)))^{1/2}.\mathrm{tr}(\mathfrak{i}^{(1)}(x^*x)))^{1/2} \\
&= \mathrm{tr}(\mathfrak{j}^{(2)}(y)^*\mathfrak{j}^{(2)}(y))^{1/2}.\mathrm{tr}(\mathfrak{j}^{(2)}(x)^*\mathfrak{j}^{(2)}(x))^{1/2} \\
&= \|\mathfrak{j}^{(2)}(y)\|_2\,\|\mathfrak{j}^{(2)}(x)\|_2.
\end{aligned}
$$

Therefore, we may regard f as a bounded sesquilinear form on $\mathfrak{j}^{(2)}(\mathfrak{m}) \subset L_2(\mathcal{M})$ defined by $(\mathfrak{j}^{(2)}(x), \mathfrak{j}^{(2)}(y)) \mapsto |f(y^*x)|$. The density of $\mathfrak{j}^{(2)}(\mathfrak{m})$ in $L_2(\mathcal{M})$ now allows us to extend this sesquilinear form to all of $L_2(\mathcal{M})$. The basic theory of sesquilinear forms on Hilbert spaces therefore guarantees the existence of a contraction B on $L_2(\mathcal{M})$ such that

$$
\langle B(\mathfrak{j}^{(2)}(x)),\, \mathfrak{j}^{(2)}(y)\rangle = f(y^*x), \quad x,y \in \mathfrak{m}.
$$

Now, we assume that $\mathcal{M}$ is in the standard form described in Theorem 10.59. We claim that B belongs to the commutant of $\mathcal{M}$. To see this, note that for $a \in \mathcal{M}$ and $x,y \in \mathfrak{m}$, we have that

$$
\begin{aligned}
\langle Ba(\mathfrak{j}^{(2)}(x)),\, \mathfrak{j}^{(2)}(y)\rangle &= \langle B(a\mathfrak{j}^{(2)}(x)),\, \mathfrak{j}^{(2)}(y)\rangle \\
&= \langle B(\mathfrak{j}^{(2)}(ax)),\, \mathfrak{j}^{(2)}(y)\rangle \\
&= f(y^*ax) \\
&= \langle B(\mathfrak{j}^{(2)}(x)),\, \mathfrak{j}^{(2)}(a^*y)\rangle \\
&= \langle B(\mathfrak{j}^{(2)}(x)),\, a^*\mathfrak{j}^{(2)}(y)\rangle \\
&= \langle aB(\mathfrak{j}^{(2)}(x)),\, \mathfrak{j}^{(2)}(y)\rangle.
\end{aligned}
$$

By the density of $\mathfrak{j}^{(2)}(\mathfrak{m})$ in $L_2(\mathcal{M})$, this is sufficient to guarantee that $Ba = aB$, as claimed. Thus, by Theorem 10.59, B coincides with right multiplication on $L_2(\mathcal{M})$ by an element $b \in \mathcal{M}$. Hence, we deduce that

$$
f(y^*x) = \langle \mathfrak{j}^{(2)}(x)b,\, \mathfrak{j}^{(2)}(y)\rangle = \mathrm{tr}(\mathfrak{j}^{(2)}(y)^*\mathfrak{j}^{(2)}(x)b), \text{ for all } x,y \in \mathfrak{m}.
$$

Consequently,

$$\rho(x_0) = f(x_0) = \mathrm{tr}(\mathrm{j}^{(2)}(x_0)^{1/2}\mathrm{j}^{(2)}(x_0)^{1/2}b) = \mathrm{tr}(\mathrm{i}^{(1)}(x_0)b)$$

$$\leq \|\mathrm{i}^{(1)}(x_0)\|_1\|b\|_\infty \leq \|\mathrm{i}^{(1)}(x_0)\|_1$$

thereby proving the lemma. $\qquad\square$

Theorem 12.14 *Let $\mathcal{M}$ and $\mathcal{N}$ be von Neumann algebras equipped, respectively, with faithful normal semifinite weights φ and ψ, and let $T : \mathcal{M} \to \mathcal{N}$ be a positive normal linear map satisfying $\psi \circ T \leq C\,\varphi$. Then, T is integrable with the norm of the induced map $T^{(1)}$ from $L^1(\mathcal{M}_1)$ to $L^1(\mathcal{M}_2)$ majorised by C. The operator $T^\dagger$ is again a positive operator for which we similarly have $\varphi \circ T^\dagger \leq \|T\|_\infty\,\psi$. If T is CP, then so is $T^\dagger$. For each n we then have $(T_n)^\dagger = (T^\dagger)_n$. Finally, if additionally $\sigma_t^\psi \circ T = T \circ \sigma_t^\varphi$ for all $t \in \mathbb{R}$, then also $T^\dagger \circ \sigma_t^\psi = \sigma_t^\varphi \circ T^\dagger$.*

Proof. As in the proof of the lemma, we shall write $\mathfrak{m}$ for $\mathfrak{m}_\varphi$ and denote the self-adjoint portion of this algebra by $\mathfrak{m}_h$. Let $x \in \mathfrak{p}$ be given. Assume first that $x \geq 0$. Since both T and $\mathrm{i}^{(1)}$ are positivity preserving, we then clearly have $\mathrm{i}^{(1)}(x) \geq 0$, and so $T^{(1)}(\mathrm{i}^{(1)}(x)) = \mathrm{i}^{(1)}(T(x)) \geq 0$. Therefore, it follows from Remark 10.41 that

$$\|T^1(\mathrm{i}^{(1)}(x))\|_1 = \mathrm{tr}(T^1(\mathrm{i}^{(1)}(x))) = \mathrm{tr}(\mathrm{i}^{(1)}(T(x))) = \varphi_2(T(x))$$

$$\leq C\,\varphi_1(x) = C\,\|\mathrm{i}^{(1)}(x)\|_1.$$

Next, consider the case where $x \in \mathfrak{m}_h$. Given $\epsilon > 0$, it follows from Lemma 12.13 that there exist $a, b \in \mathfrak{p}$ such that $x = a - b$ and

$$\|\mathrm{i}^{(1)}(a)\|_1 + \|\mathrm{i}^{(1)}(b)\|_1 \leq \|\mathrm{i}^{(1)}(x)\|_1 + \epsilon.$$

It is clear that

$$T^{(1)}(\mathrm{i}^{(1)}(x)) = \mathrm{i}^{(1)}(T(x)) = \mathrm{i}^{(1)}(T(a)) - \mathrm{i}^{(1)}(T(b))$$

and hence that

$$\|T^{(1)}(\mathrm{i}^{(1)}(x))\|_1 \leq \|\mathrm{i}^{(1)}(T(a))\|_1 + \|\mathrm{i}^{(1)}(T(b))\|_1$$

$$\leq C\left(\|\mathrm{i}^{(1)}(a)\|_1 + \|\mathrm{i}^{(1)}(b)\|_1\right)$$

$$\leq C\left(\|\mathrm{i}^{(1)}(x)\|_1 + \epsilon\right).$$

Since $\epsilon > 0$ was arbitrary, we have

$$\|T^{(1)}(\mathfrak{i}^{(1)}(x))\|_1 \leq C\|\mathfrak{i}^{(1)}(x)\|_1$$

as required.

For the general case $x \in \mathfrak{m}$, we may decompose x into its real and imaginary parts to get

$$\|T^{(1)}(\mathfrak{i}^{(1)}(x))\|_1 \leq 2\,C\,\|\mathfrak{i}^{(1)}(x)\|_1.$$

Thus, T induces an L^1-bounded operator from $\mathfrak{i}^{(1)}(\mathfrak{m})$ to $L^1(\mathcal{M}_2)$. Since $\mathfrak{i}^{(1)}(\mathfrak{m})$ is dense in $L_1(\mathcal{M}_1)$, $T^{(1)}$ extends to a bounded map from $L_1(\mathcal{M}_1)$ into $L_1(\mathcal{M}_2)$ with $\|T^{(1)}\| \leq 2C_1$. Since $T^{(1)}$ is positive on $\mathfrak{i}^{(1)}(\mathfrak{m})$, so is the extension to $L_1(\mathcal{M}_1)$ (see Theorem 10.46).

It remains to sharpen the norm estimate for $T^{(1)}$. To this end, we consider the adjoint $T^\dagger : \mathcal{N} \to \mathcal{M}$. The operator $T^\dagger$ is positive since $T^{(1)}$ is positivity preserving. Thus, by the Russo–Dye theorem $T^\dagger$ will attain its norm at the identity of $\mathcal{M}_2$. Therefore

$$\|T^{(1)}\| = \|T^\dagger\| = \|T^\dagger(\mathbb{1})\|_\infty.$$

The theorem will therefore follow if we are able to show that $\|T^\dagger(\mathbb{1})\|_\infty \leq C$. Given $x \in \mathfrak{p}$, we can use Remark 10.41 to see that

$$\mathrm{tr}(T^\dagger(\mathbb{1})\,\mathfrak{i}^{(1)}(x)) = \mathrm{tr}(T^{(1)}(\mathfrak{i}^{(1)}(x))) = \mathrm{tr}(\mathfrak{i}^{(1)}(Tx)) = \psi(T(x))$$

$$\leq C\,\varphi(x) = C\,\|\mathfrak{i}^{(1)}(x)\|_1.$$

Since $\mathfrak{i}^{(1)}(\mathfrak{p})$ is dense in the positive cone of $L^1(\mathcal{M}_1)$, the desired estimate of $\|T^\dagger(\mathbb{1})\|_\infty$ follows from L^p-duality.

We pass to proving the second claim. If we now refer to part (6) of Theorem 4.4, it is clear that the partial sums in that representation yield a net $(\varphi_\alpha) \subset \mathcal{M}_*^+$ increasing to φ. By Proposition 9.33, the densities $h_\alpha = \frac{d\widetilde{\varphi}_\alpha}{d\tau}$ then increase to $h_\varphi = \frac{d\widetilde{\varphi}}{d\tau}$. We may next invoke Proposition 10.42 to establish the existence of a net $(b_\alpha) \subset \mathfrak{n}_\varphi$ of contractive elements such that $h_\varphi^{1/2} b_\alpha = h_\alpha^{1/2}$. Given $a \in \mathfrak{p}_\psi$, it follows from L^p-duality and Remark 10.41 that

$$\begin{aligned}
\varphi_\alpha(T^\dagger(a)) &= \mathrm{tr}(h_\alpha^{1/2}T^\dagger(a)h_\alpha^{1/2})\\
&= \mathrm{tr}((h_\varphi^{1/2}b_\alpha)[b_\alpha h_\varphi^{1/2}]T^\dagger(a))\\
&= \mathrm{tr}(i^{(1)}(|b_\alpha|^2)T^\dagger(a))\\
&= \mathrm{tr}(T^{(1)}(i^{(1)}(|b_\alpha|^2))a)\\
&= \mathrm{tr}(i^{(1)}(T(|b_\alpha|^2)))a)\\
&= \mathrm{tr}(T(|b_\alpha|^2)i^{(1)}(a)))\\
&\le \|T(|b_\alpha|^2)\|\,tr(i^{(1)}(a)))\\
&\le \|T\|_\infty\,\psi(a).
\end{aligned}$$

We therefore have that $\varphi(T^\dagger(a)) = \sup_\alpha \varphi_\alpha(T^\dagger(a)) \le \|T\|_\infty\,\psi(a)$ as was required.

Next, suppose that T is CP. It is then an easy exercise to see that $\psi \circ T \le C\,\varphi$ ensures that $(\psi \otimes \mathrm{Tr}_n) \circ T_n \le C\,(\varphi \otimes \mathrm{Tr}_n)$, where T_n is the induced map on $M_n(\mathcal{M})$. Therefore, it follows from the first part of the proof that $(T_n)^\dagger$ is a positive normal operator satisfying $(\varphi \otimes \mathrm{Tr}_n) \circ (T_n)^\dagger \le \|T\|\,(\varphi \otimes \mathrm{Tr}_n)$. Showing that $(T_n)^\dagger = (T^\dagger)_n$ for all n will show that $T^\dagger$ is CP, and hence establish the claim. To see that this is the case, let $[a_{ij}] \in M_n(\mathfrak{m}_\psi)$ and $[b_{ij}] \in M_n(\mathfrak{m}_\varphi)$ be given. We may then use Proposition 12.5 to see that

$$\begin{aligned}
\mathrm{tr}_n((T_n)^\dagger([a_{ij}]).i^{(1)}_{(\varphi\otimes\mathrm{Tr}_n)}([b_{ij}])) &= \mathrm{tr}_n([a_{ij}].(T_n)^{(1)}(i^{(1)}_{(\varphi\otimes\mathrm{Tr}_n)}([b_{ij}])))\\
&= \mathrm{tr}_n([a_{ij}].(i^{(1)}_{(\psi\otimes\mathrm{Tr}_n)}([T(b_{ij})])))\\
&= \mathrm{Tr}_n([\mathrm{tr}(\sum_{k=1}^{n} a_{ij}i^{(1)}_\psi(T(b_{kj})))])\\
&= \mathrm{Tr}_n([\mathrm{tr}(\sum_{k=1}^{n} a_{ij}T^{(1)}(i^{(1)}_\varphi(b_{kj})))])\\
&= \mathrm{Tr}_n([\mathrm{tr}(\sum_{k=1}^{n} T^\dagger(a_{ij})(i^{(1)}_\varphi(b_{kj})))])\\
&= \mathrm{tr}_n([T^\dagger(a_{ij})].i^{(1)}_{(\varphi\otimes\mathrm{Tr}_n)}([b_{ij}]))\\
&= \mathrm{tr}_n((T^\dagger)_n([a_{ij}]).i^{(1)}_{(\varphi\otimes\mathrm{Tr}_n)}([b_{ij}])).
\end{aligned}$$

By the density of $i^{(1)}_{(\varphi \otimes \mathrm{Tr}_n)}(M_n(\mathfrak{m}_\varphi))$ in $L^1(M_n(\mathcal{M}))$, we have by L^p-duality that $(T_n)^\dagger$ and $(T^\dagger)_n$ agree on $M_n(\mathfrak{m}_\psi)$. Since this space is σ-weakly dense in $M_n(\mathcal{N})$ and both operators normal, we are done.

To see the final claim, notice that on writing h_ψ for $\frac{d\widetilde{\psi}}{d\tau}$, it follows from Propositions 9.5 and 9.27 and Theorem 9.28 that $\sigma_t^\psi = h_\psi{}^{it}(\cdot)h_\psi{}^{-it}$ and $\sigma_t^\varphi = h_\varphi^{it}(\cdot)h_\varphi^{-it}$ for all $t \in \mathbb{R}$. Let $a \in \mathcal{N}$ and $b \in \mathfrak{p}_\varphi$ be given. Notice that since $\varphi \circ \sigma_t^\varphi = \varphi$ for all $t \in \mathbb{R}$, $\mathfrak{p}_\varphi$ is invariant under the action of σ_t^φ. We therefore have that

$$
\begin{aligned}
\mathrm{tr}_\mathcal{M}(\sigma_t^\varphi(T^\dagger(a))i^{(1)}(b)) &= \mathrm{tr}_\mathcal{M}(h_\varphi^{it}(T^\dagger(a))h_\varphi^{-it}(h_\varphi^{1/2}b^{1/2})[b^{1/2}h_\varphi^{1/2}]) \\
&= \mathrm{tr}_\mathcal{M}(T^\dagger(a)h_\varphi^{-it}(h_\varphi^{1/2}b^{1/2})[b^{1/2}h_\varphi^{1/2}]h_\varphi^{it}) \\
&= \mathrm{tr}_\mathcal{M}(T^\dagger(a)(h_\varphi^{1/2}h_\varphi^{-it}b^{1/2}h_\varphi^{it})[h_\varphi^{-it}b^{1/2}h_\varphi^{it}h_\varphi^{1/2}]) \\
&= \mathrm{tr}_\mathcal{M}(T^\dagger(a)i^{(1)}(h_\varphi^{it}bh_\varphi^{-it})) \\
&= \mathrm{tr}_\mathcal{M}(aT^{(1)}(i^{(1)}(\sigma_{-t}^\varphi(b)))) \\
&= \mathrm{tr}_\mathcal{N}(a(i^{(1)}(T\sigma_{-t}^\varphi(b)))) \\
&= \mathrm{tr}_\mathcal{N}(aT^{(1)}(i^{(1)}(\sigma_{-t}^\psi(T(b))))) \\
&= \mathrm{tr}_\mathcal{N}(ai^{(1)}(h_\psi{}^{it}(Tb)h_\psi{}^{-it})) \\
&= \mathrm{tr}_\mathcal{N}(a(h_\psi{}^{1/2}(h_\psi{}^{-it}(Tb)^{1/2}h_\psi{}^{it})) \\
&\qquad [(h_\psi{}^{-it}(Tb)^{1/2}h_\psi{}^{it})h_\psi{}^{1/2}]) \\
&= \mathrm{tr}_\mathcal{N}((h_\psi{}^{it}ah_\psi{}^{-it})(h_\psi{}^{1/2}((Tb)^{1/2}h_\psi{}^{it})) \\
&\qquad [(h_\psi{}^{-it}(Tb)^{1/2})h_\psi{}^{1/2}]) \\
&= \mathrm{tr}_\mathcal{N}(\sigma_t^\psi(a)i^{(1)}(Tb)) \\
&= \mathrm{tr}_\mathcal{N}(\sigma_t^\psi(a)T^{(1)}(i^{(1)}(b))) \\
&= \mathrm{tr}_\mathcal{M}(T^\dagger(\sigma_t^\psi(a))i^{(1)}(b)).
\end{aligned}
$$

The density of $\mathfrak{p}_\varphi$ in $L^1_+(\mathcal{M})$ (Theorem 10.46) when combined with L^p-duality now ensures that $\sigma_t^\varphi(T^\dagger(a)) = T^\dagger(\sigma_t^\psi(a))$ as was claimed. $\square$

In the case of normal CP maps, we may use the results of the previous section to obtain a more powerful conclusion. Specifically, if we are given von Neumann algebras $\mathcal{M}_k$ ($k = 1, 2$) equipped with faithful normal semifinite weights φ_k, and a normal CP linear map $T : \mathcal{M}_1 \to \mathcal{M}_2$ such that $T \circ \sigma_t^{\varphi_1} = \sigma_t^{\varphi_2} \circ T$ and $\varphi_2 \circ T \le C\,\varphi_2$ for some $C > 0$, then the extension $\widetilde{T}$ to $\mathfrak{M}_1 \to \mathfrak{M}_2$ (where $\mathfrak{M}_k = \mathcal{M}_k \rtimes_{\varphi_k} \mathbb{R}$) satisfies $\tau_2 \circ \widetilde{T} \leq C\tau_1$ where τ_k is the canonical trace on $\mathfrak{M}_k$. Theorem 12.14 therefore proves that the extended operator is integrable with respect to $(\mathfrak{M}_1, \tau_1)$ and $(\mathfrak{M}_2, \tau_2)$. Therefore, it canonically induces a bounded map from $L^1(\mathfrak{M}_1, \tau_1)$ to $L^1(\mathfrak{M}_2, \tau_2)$. Since here we are working with semifinite von Neumann algebras, this gives us access to real interpolation, which is the tool that enables us to draw sharper conclusions in this setting.

Proposition 12.15 *Let $\mathcal{M}$ and $\mathcal{N}$ be von Neumann algebras acting on the same Hilbert space, respectively, equipped with faithful normal semifinite weights φ and ψ. Let $T : \mathcal{M} \to \mathcal{N}$ be a CP map satisfying $\psi \circ T \leq \varphi$ and $T \circ \sigma_t^{\varphi} = \sigma_t^{\psi} \circ T$ for each $t \in \mathbb{R}$, and let $\widetilde{T}$ be the completely positive extension to $\mathfrak{M}$ guaranteed by Theorem 12.6. Then, $\widetilde{T}$ canonically induces a bounded map from $(L^{\infty} + L^1)(\mathfrak{M}, \tau_{\mathfrak{M}})$ to $(L^{\infty} + L^1)(\mathfrak{N}, \tau_{\mathfrak{N}})$ with norm majorised by $\max\{\|T\|, C\}$ where $\mathfrak{M} = \mathcal{M} \rtimes_{\varphi} \mathbb{R}$ and $\mathfrak{N} = \mathcal{N} \rtimes_{\psi} \mathbb{R}$.*

Proof. Recall that under the stated conditions, Theorem 12.6 guarantees that $\tau_{\mathfrak{N}} \circ \widetilde{T} \leq C\tau_{\mathfrak{M}}$. If we apply Theorem 12.14 to this fact, it is clear that $\widetilde{T}$ canonically induces a map from $L^1(\mathfrak{M}, \tau_{\mathfrak{M}})$ to $L^1(\mathfrak{N}, \tau_{\mathfrak{N}})$. The claim will then follow from the interpolation theory presented in Chapter 7. By that theory, the norm of the induced operator on $(L^{\infty} + L^1)(\mathfrak{M}, \tau_{\mathfrak{M}})$ will be majorised by $\max\{\|\widetilde{T}\|_{\infty}, \|\widetilde{T}\|_1\}$ where $\|\widetilde{T}\|_{\infty}$ and $\|\widetilde{T}\|$ are the norms on $L^{\infty}(\mathfrak{M}, \tau_{\mathfrak{M}})$ and $L^1(\mathfrak{M}, \tau_{\mathfrak{M}})$, respectively. But by Theorems 12.6 and 12.14, we respectively have that $\|\widetilde{T}\|_{\infty} \leq \|T\|_{cb} = \|T\|$ and $\|\widetilde{T}\|_1 \leq C$. $\square$

The final phase in our construction is to extract the information we need from what we have already shown for the action of $\widetilde{T}$ on $(L^{\infty} + L^1)(\mathfrak{M}, \tau_{\mathfrak{M}})$. We now investigate this point.

Proposition 12.16 *Let T and $\widetilde{T}$ be as before. For the sake of simplicity, we will also write $\widetilde{T}$ for the extension of $\widetilde{T}$ to $(L^{\infty} + L^1)(\mathfrak{M}, \tau_{\mathfrak{M}})$. In its action on $(L^{\infty} + L^1)(\mathfrak{M}, \tau)$, the operator satisfies the condition that $\widetilde{T}(ab) = a\widetilde{T}(b)$ for all $a \in \mathcal{A}$ and all $b \in (L^{\infty} + L^1)(\mathfrak{M}, \tau_{\mathfrak{M}})$, where $\mathcal{A}$ is von Neumann subalgebra generated by the $\lambda_t s$ ($t \in \mathbb{R}$) used in contructing $\mathfrak{M}$.*

Proof. The stated property is known to hold for the action of $\widetilde{T}$ on $\mathcal{M}$ (see Theorem 12.6). Since $\mathfrak{M} \cap L^1(\mathfrak{M}, \tau_{\mathfrak{M}})$ is norm-dense in $L^1(\mathfrak{M}, \tau_{\mathfrak{M}})$, the continuity of $\widetilde{T}$ on $L^1(\mathfrak{M}, \tau_{\mathfrak{M}})$ then ensures that it also holds for $L^1(\mathfrak{M}, \tau_{\mathfrak{M}})$ and therefore for $(L^\infty + L^1)(\mathfrak{M}, \tau_{\mathfrak{M}})$. $\square$

The reason for proving the above is to have access to the Orlicz spaces $L^\Psi(\mathcal{M})$ with upper fundamental index less than 1, which by Proposition 10.20 we know to be subspaces of $(L^\infty + L^1)(\mathcal{M}, \tau)$. This fact is fully exploited in the following result.

Theorem 12.17   *Let $\mathcal{M}$ and $\mathcal{N}$ be as in proposition 12.15 and let Ψ be a Young's function. Let $T : \mathcal{M} \to \mathcal{N}$ be a CP normal map such that $\psi \circ T \leq C\,\varphi$, and*

$$T \circ \sigma_t^\varphi = \sigma_t^\psi \circ T, \quad t \in \mathbb{R}$$

and let $\widetilde{T}$ denote the map induced by T on $(L^\infty + L^1)(\mathfrak{M}, \tau_{\mathfrak{M}})$. If the fundamental index $\overline{\beta}_{L^\Psi}$ satisfies $\overline{\beta}_{L^\Psi} < 1$, then

- *$\widetilde{T}$ restricts to a bounded map $T^{(\Psi)}$ from $L^\Psi(\mathcal{M})$ to $L^\Psi(\mathcal{N})$;*
- *if $\Psi(t) = t^p$ where $1 < p < \infty$, then $T^{(\Psi)}$ is a exactly $T^{(p)}$ as guaranteed by Theorems 12.10 and 12.14.*

Proof.   If $\overline{\beta}_{L^\Psi} < 1$, then $L^\Psi(\mathcal{M})$ lives inside $(L^\infty + L^1)(\mathfrak{M}, \tau_{\mathfrak{M}})$ and also gets its topology from $(L^\infty + L^1)(\mathfrak{M}, \tau_{\mathfrak{M}})$ (see Proposition 10.20). So, all one needs to do to prove the first claim is to note

- that if $a \in (L^\infty + L^1)(\mathfrak{M}, \tau_{\mathfrak{M}})$, then also $v_s^{1/2} a v_s^{1/2} \in (L^\infty + L^1)(\mathfrak{M}, \tau_{\mathfrak{M}})$ (where the operators v_s are as in Lemma 10.7),
- and that then $\widetilde{T}(v_s^{1/2} a v_s^{1/2}) = v_s^{1/2}\widetilde{T}(a)v_s^{1/2}$ by Proposition 12.16,

in which case the fact that $\widetilde{T}(L^\Psi(\mathcal{M})) \subseteq L^\Psi(\mathcal{N})$ will follow by definition.

If $\Psi(t) = t^p$ for some $1 < p < \infty$, then of course $L^\Psi = L^p$. Note that the inequality $\psi \circ T \leq C\varphi$ ensures that $T(\mathfrak{p}_\varphi) \subseteq \mathfrak{p}_\psi$. We shall silently use this fact below. Since by assumption $\mathcal{M}$ and $\mathcal{N}$ act on the same Hilbert space, it is clear from the definition of the operators λ_t that both crossed products share the same shift operators. Therefore, it follows from Theorem 9.28 that $\frac{d\widetilde{\varphi}}{d\tau_{\mathfrak{M}}} = \frac{d\widetilde{\psi}}{d\tau_{\mathfrak{M}}}$. We shall denote this operator by h. Given any $a \in \mathfrak{p}_\varphi$, we have $(h^{1/2p}a^{1/2})[a^{1/2}h^{1/2p}] = \mathfrak{i}^{(p)}(a) \in L^p(\mathcal{M}) \subset (L^1 + L^\infty)(\mathfrak{M})$. Note that by Theorem 9.28, each

$h^{1/2p}\chi_{[0,n]}(h)$ belongs to $\mathcal{A}$. So, on applying Proposition 12.16 to $\widetilde{T}$, it follows that

$$\chi_{[0,n]}(h)\widetilde{T}((h^{1/2p}a^{1/2})[a^{1/2}h^{1/2p}])\chi_{[0,n]}(h)$$
$$= \widetilde{T}((h^{1/2p}\chi_{[0,n]}(h))a(h^{1/2p}\chi_{[0,n]}(h)))$$
$$= (h^{1/2p}\chi_{[0,n]}(h))\widetilde{T}(a)(h^{1/2p}\chi_{[0,n]}(h))$$
$$= (h^{1/2p}\chi_{[0,n]}(h))T(a)(h^{1/2p}\chi_{[0,n]}(h))$$
$$= \chi_{[0,n]}(h)((h^{1/2p}T(a)^{1/2})[T(a)^{1/2}h^{1/2p}])\chi_{[0,n]}(h).$$

On letting $n \to \infty$, it follows that $\widetilde{T}(\mathfrak{i}^{(p)}(a)) = \mathfrak{i}^{(p)}(T(a))$ for all $a \in \mathfrak{p}_\varphi$, which is enough to prove the claim. $\qquad\square$

We pause to collate the behaviour of conditional expectations with respect to the L^p-structures implicit in the foregoing analysis.

Remark 12.18 With Theorem 12.14 at our disposal, we are now able to make more refined statements about the action of a normal conditional expectation E from $\mathcal{M}$ onto $\mathcal{N}$ for which $\varphi \circ \mathrm{E} = \varphi$. Specifically, for each $1 \le p < \infty$, the prescription given there yields a contractive map $\mathrm{E}^{(p)}$ on $L^p(\mathcal{M})$, for which we have $\mathrm{E}^{(p)} \circ \mathrm{E}^{(p)} = \mathrm{E}^{(p)}$ and $\mathrm{E}^{(p)}(f^*) = \mathrm{E}^{(p)}(f)^*$. Since $\varphi \circ \mathrm{E} = \varphi$, it is clear that $\mathrm{E}(\mathfrak{p}_\varphi) = \mathfrak{p}(\mathcal{N})_\varphi$. So, for any $b \in \mathfrak{p}_\varphi$, we will also have $\mathrm{tr}(\mathrm{E}^{(1)}(\mathfrak{i}^{(1)}(b)) = \mathrm{tr}(\mathfrak{i}^{(1)}(\mathrm{E}(b))) = \varphi(\mathrm{E}(b)) = \varphi(a) = \mathrm{tr}(\mathfrak{i}^{(1)}(b))$. The density of $\mathrm{span}(\mathfrak{i}^{(1)}(\mathfrak{p}_\varphi))$ in $L^1(\mathcal{M})$, then ensures that $\mathrm{tr}(\mathrm{E}^{(1)}(b)) = tr(b)$ for each $b \in L^1(\mathcal{M})$.

It is now clear from the above and the discussion in the first part of the proof of Theorem 10.54 that in the setting discussed there, each $L^\Psi(\mathcal{N})$ will by definition (see Definition 10.8) be a subspace of $L^\Psi(\mathcal{M})$, and also that for any $0 < p < \infty$, each $L^p(\mathcal{N})$ is a subspace of $L^p(\mathcal{M})$ with $L^p(\mathcal{N})$ being an expected subspace (of $\mathrm{E}^{(p)}$) of $L^p(\mathcal{M})$. Moreover, when the form of the quasi-norm on L^Ψ (see Theorem 10.11) is considered alongside Remark 5.8, it is clear that the quasi-norm on $L^\Psi(\mathcal{N})$ is just a restriction of the quasi-norm on $L^\Psi(\mathcal{M})$. Theorems 12.10 and 12.14 now ensure that E will for any $1 \le p < \infty$ canonically induce a contractive map from $L^p(\mathcal{M})$ to $L^p(\mathcal{N})$. In fact, for any Orlicz space L^Ψ with upper fundamental index strictly less than 1, Propositions 10.20 and 12.15 considered alongside Theorem 12.17 show that these spaces can be equivalently renormed in such a way that $\mathbb{E}$ will here also induce a contractive map from $L^\Psi(\mathcal{M})$ to $L^\Psi(\mathcal{N})$.

There is one more fact we may note regarding the action of conditional expectations with respect to L^p structures, but to see this we need to work a bit harder. We shall need the following consequence of Lemma 10.44.

Lemma 12.19 *Let $\mathfrak{n}_\varphi^\infty$ be the collection of all entire analytic elements a of $\mathfrak{n}_\varphi \cap \mathfrak{n}_\varphi^*$ for which $\sigma_w^\varphi(a) \in \mathfrak{n}_\varphi \cap \mathfrak{n}_\varphi^*$ for all $w \in \mathbb{C}$. Then, $\mathfrak{n}_\varphi^\infty$ is σ-strongly dense in $\mathcal{M}$ and for each $q \geq 2$, $\mathrm{j}^{(q)}(\mathfrak{n}_\varphi^\infty)$ is dense in $L^q(\mathcal{M})$. Furthermore, given $z \in \mathbb{C}$ with $0 \leq \mathrm{Re}(z) \leq 1/2$, we have that*

$$[ah^z] = h^z \sigma_{iz}(a) \quad \text{for all } a \in \mathfrak{n}_\infty$$

where $h = \frac{d\widetilde{\varphi}}{d\tau}$.

Proof. To see the first claim, select $(f_\gamma) \subset \mathfrak{n}_\varphi(\mathcal{M}) \cap \mathfrak{n}_\varphi(\mathcal{M})^*$ as in Proposition 10.43. Note that (f_γ) is norm bounded. For any entire analytic $b \in \mathcal{M}$ we will therefore have that each $f_\gamma b f_\gamma$ is an entire analytic element of $\mathfrak{m}_\varphi(\mathcal{M}) \subset (\mathfrak{n}_\varphi \cap \mathfrak{n}_\varphi^*)$ (see Theorem 8.13). The net $(f_\gamma b f_\gamma)$ will moreover converge strongly to b, since for any ξ in the underlying Hilbert space, we will have that

$$\|b\xi - f_\gamma b f_\gamma \xi\| \;\leq\; \|(\mathbb{1} - f_\gamma)b(\mathbb{1} - f_\gamma)\xi + (\mathbb{1} - f_\gamma)b\xi + b(\mathbb{1} - f_\gamma)\xi\|$$
$$\leq\; 2\|b\|_\infty \|\xi - f_\gamma \xi\| + \|b\xi - f_\gamma b\xi)\|.$$

Since for each γ we have $\|f_\gamma b f_\gamma\| \leq \|b\|$, the convergence is actually σ-strongly. We therefore have that $\mathfrak{n}^\infty$ is σ-strongly dense in the collection of entire analytic elements of $\mathcal{M}$, which in turn is known to be σ-weakly dense in $\mathcal{M}$ by Theorem 8.13.

The claim regarding the density of $\mathrm{j}^{(q)}(\mathfrak{n}_\varphi^\infty)$ in $L^q(\mathcal{M})$ can be verified using essentially the same argument as was used to prove Theorem 10.46. The only change that needs to be made in the proof is that instead of selecting b in $\mathcal{M}$, one selects b in $\mathfrak{n}_\varphi^\infty$. The σ-strong density of $\mathfrak{n}_\varphi^\infty$ in $\mathcal{M}$ ensures that the argument still goes through.

The final claim corresponds to Corollary 10.45. $\qquad\square$

Proposition 12.20 *Let $\mathcal{N}$ be a von Neumann subalgebra of $\mathcal{M}$ for which there exists a faithful normal conditional expectation $\mathbb{E}$ from*

$\mathcal{M}$ onto $\mathcal{N}$ satisfying $\varphi \circ \mathrm{E} = \varphi$. Given $1 \leq p, q, r \leq \infty$ such that $\frac{1}{p} + \frac{1}{q} = \frac{1}{r}$ we will for $a \in L^p(\mathcal{N})$ and $b \in L^q(\mathcal{M})$ have that $\mathrm{E}^{(r)}(ab) = a\mathrm{E}^{(q)}(b)$ and $\mathrm{E}^{(r)}(ba) = \mathrm{E}^{(q)}(b)a$.

Proof. Let $f \in \mathfrak{n}_\varphi^\infty$ and $a, b \in \mathfrak{n}(\mathcal{N})_\varphi^\infty$ be given. We shall prove the claim for the case where $1 \leq p, q < \infty$. The first step in the proof is to show that E preserves analyticity. We shall prove this by using the description of analyticity given in Definition 8.14. Let δ be the infinitesimal generator of the modular group σ_t^φ and let $a \in \mathrm{dom}(\delta)$. By definition, this means that the expressions $\frac{1}{t}(\sigma_t^\varphi(a) - a)$ converge σ-weakly to $\delta(a)$. But since E is normal and also commutes with σ_t^φ, we see that $\frac{1}{t}\mathrm{E}(\sigma_t^\varphi(a) - a) = \frac{1}{t}(\sigma_t^\varphi(\mathrm{E}(a)) - \mathrm{E}(a))$ converges σ-weakly to $\delta(a)$. So, by definition, $\mathrm{E}(a) \in \mathrm{dom}(\delta)$ with $\delta(\mathrm{E}(a)) = \mathrm{E}(\delta(a))$. The claim is now obviously a fairly direct consequence of Definition 8.14.

With h denoting $\frac{d\widetilde{\varphi}}{d\tau}$, repeated applications of Lemmata 10.48 and 12.19 now show that

$$
\begin{aligned}
&(h^{1/2p}a)[bh^{1/2p}]\mathrm{E}^{(q)}(\mathrm{i}^{(q)}(f^*f)) \\
=\ &(h^{1/2p}a)[bh^{1/2p}]\mathrm{i}^{(q)}(\mathrm{E}(f^*f)) \\
=\ &(h^{1/2p}a)[bh^{1/2p}](h^{1/2q}\sqrt{\mathrm{E}(f^*f)})[\sqrt{\mathrm{E}(f^*f)}h^{1/2q}] \\
=\ &(h^{1/2p}a)[bh^{1/2r}]\sqrt{\mathrm{E}(f^*f)}[\sqrt{\mathrm{E}(f^*f)}h^{1/2q}] \\
=\ &(h^{1/2p}a)[bh^{1/2r}][\mathrm{E}(f^*f)h^{1/2q}] \\
=\ &(h^{1/2p}a)[bh^{1/2r}](h^{1/2q}\sigma_{i/2q}(\mathrm{E}(f^*f))) \\
=\ &(h^{1/2p}a)[bh^{1/2q}](h^{1/2r}\sigma_{i/2q}(\mathrm{E}(f^*f))) \\
=\ &(h^{1/2p}a)[bh^{1/2q}][\sigma_{-i/2r}(\sigma_{i/2q}(\mathrm{E}(f^*f)))h^{1/2r}] \\
=\ &[\sigma_{-i/2p}(a)h^{1/2p}](h^{1/2q}\sigma_{i/2q}(b))[\sigma_{-i/2p}(\mathrm{E}(f^*f))h^{1/2r}] \\
=\ &[\sigma_{-i/2p}(a)h^{1/2r}]\sigma_{i/2q}(b)[\sigma_{-i/2p}(\mathrm{E}(f^*f))h^{1/2r}] \\
=\ &(h^{1/2r}\sigma_{i/2r}(\sigma_{-i/2p}(a)))\sigma_{i/2q}(b)[\sigma_{-i/2p}(\mathrm{E}(f^*f))h^{1/2r}] \\
=\ &\mathrm{i}^{(r)}(\sigma_{i/2q}(a)\sigma_{i/2q}(b)\sigma_{-i/2p}(\mathrm{E}(f^*f))) \\
=\ &\mathrm{i}^{(r)}(\mathrm{E}(\sigma_{i/2q}(a)\sigma_{i/2q}(b)\sigma_{-i/2p}(f^*f))) \\
=\ &\mathrm{E}^{(r)}(\mathrm{i}^{(r)}(\sigma_{i/2q}(a)\sigma_{i/2q}(b)\sigma_{-i/2p}(f^*f))).
\end{aligned}
$$

A similar argument to the one used above shows that

$$
(h^{1/2p}a)[bh^{1/2p}]\mathrm{i}^{(q)}(f^*f) = \mathrm{i}^{(r)}(\sigma_{i/2q}(a)\sigma_{i/2q}(b)\sigma_{-i/2p}(f^*f))
$$

and hence that

$$(h^{1/2p}a)[bh^{1/2p}]\mathrm{E}^{(q)}(\mathrm{i}^{(q)}(f^*f)) = \mathrm{E}^{(r)}((h^{1/2p}a)[bh^{1/2p}]\mathrm{i}^{(q)}(f^*f)).$$

The density assertion in Lemma 12.19 now leads to the conclusion that $\mathrm{E}^{(r)}(ab) = a\mathrm{E}^{(q)}(b)$ for all $a \in L^p(\mathcal{N})$ and $b \in L^q(\mathcal{M})^+$. By linearity, the claim holds for all $b \in L^q(\mathcal{M})$. $\qquad\square$

12.3 Integrable, Markov and KMS pairs of operators

We are now ready to introduce the properties of Markov and KMS-symmetric operators. Throughout, $\mathcal{M}$ will be a von Neumann algebra equipped with a faithful normal semifinite weight φ, with the density of the dual weight $\widetilde{\varphi}$ denoted by h.

Definition 12.21 Let $p \in [1, \infty]$ and let S be an operator on $L^p(\mathcal{M})$ with domain $\mathrm{dom}\,(S)$. Then:

- S is *reality preserving* if for every $x \in \mathrm{dom}(S)$, we have that $x^* \in \mathrm{dom}(S)$ and $S(x^*) = (Sx)^*$;
- S is *positivity preserving* if (i) $\mathrm{dom}(S) = \mathrm{span}(\mathrm{dom}(S)_+)$ and (ii) $Sx \geq 0$ for all $x \in \mathrm{dom}(S)_+$ (where $\mathrm{dom}(S_+ = \mathrm{dom}(S) \cap L^p_+(\mathcal{M}))$;
- S is *Markov* with respect to $(L^p(\mathcal{M}), h^{1/p})$, if:

  (i) any $x \in L^p$ for which $0 \leq x \leq h^{1/p}$ holds, belongs to $x \in \mathrm{dom}(S)$;
  (ii) $0 \leq S(x) \leq h^{1/p}$;

- S is *Markov* with respect to $(\mathcal{M}, \varphi)$ if:
  - (i) any $x \in \mathfrak{m}_\varphi$ for which $0 \leq x \leq \mathbb{1}$ holds, belongs to $\mathrm{dom}(S)$;
  - (ii) for any $x \in \mathrm{dom}(S)$ satisfying $0 \leq x \leq \mathbb{1}$, we have that $0 \leq S(x) \leq \mathbb{1}$.

We adopt the convention of only saying that an operator T is Markov on $(L^p(\mathcal{M}), h^{1/p})$ if the domain includes all of $L^p(\mathcal{M})$.

As a consequence of the Russo–Dye theorem and the Operator Schwarz inequality for positive operators on normal elements, the everywhere defined L^∞-Markov operators are the positivity preserving contractions.

Definition 12.22 A pair of partially defined operators (T, S) on $\mathcal{M}$ is a *KMS pair* (with respect to φ) if:

(i) $\mathrm{dom}(T) \cap \mathrm{dom}(S) \supseteq \mathfrak{m}_\varphi$;

(i) $\mathrm{tr}(Ta\bar{\cdot}\mathrm{i}(b)) = \mathrm{tr}(\mathrm{i}(a)\bar{\cdot}Sb)$ for all $a, b \in \mathfrak{m}_\varphi$.

We say that T is *KMS-symmetric* if (T, T) is a *KMS pair*.

We pause to present a refinement of equation (12.7). For $b_0, c_0 \in \mathfrak{n}_\varphi^{(2)}$ and $b_1, c_1 \in \mathfrak{n}_\varphi$, we may use Lemma 12.3 to see that

$$\mathrm{tr}(((h^{1/4}c_0^*)[b_0 h^{1/4}])((h^{1/4}c_1^*)[b_1 h^{1/4}])) = \mathrm{tr}((c_0^* b_0)((h^{1/2}c_1^*)[b_1 h^{1/2}]))$$

from which it then follows that

$$\mathrm{tr}(\mathrm{i}^{(2)}(a_0)\mathrm{i}^{(2)}(a_1)) = \mathrm{tr}(a_0 \mathrm{i}^{(1)}(a_1)) \tag{12.8}$$

for each $a_0 \in \mathfrak{m}_\varphi^{(2)}$ and $a_1 \in \mathfrak{m}_\varphi$. In the ensuing discussion, we shall repeatedly use equalities (12.7) and (12.8) without mention.

Among other consequences, these equalities now easily yield the following important estimate:

$$\|\mathrm{i}^{(2)}(a)\|^2 \leq \|a\| . \|\mathrm{i}^{(1)}(a)\| \text{ for all } a \in \mathfrak{m}. \tag{12.9}$$

Proposition 12.23 *Let T be a reality preserving integrable operator with domain $\mathrm{dom}(T) \subset \mathcal{M}$ such that $T^\dagger$ is also integrable. Then, $(T, T^\dagger)$ is a KMS pair, and T is both σ-weakly continuous on $\mathfrak{m}$ and 2-integrable. Moreover*

$$\left\| T^{(2)} \right\| \leq \left\| T^\dagger T^{\dagger\dagger} \right\|^{1/2}.$$

Proof. The claim regarding the σ-weak continuity of T on $\mathfrak{m}$ has already been noted in Proposition 12.11. From the first part of that proof, it is clear on arguing as before we will for any $a_0, a_1 \in \mathfrak{m}_\varphi$ have that

$$\mathrm{tr}(Ta_0\bar{\cdot}\mathrm{i}^{(1)}(a_1)) \;\; = \;\; \mathrm{tr}(T^{(}1)\mathrm{i}^{(1)}(a_0)\bar{\cdot}a_1) = \mathrm{tr}(\mathrm{i}^{(1)}(a_0)\bar{\cdot}T^\dagger a_1)$$

which in turn suffices to show that $(T, T^\dagger)$ is a KMS pair. However, by the density of $\mathrm{i}^{(1)}(\mathfrak{m})$ in $L^1(\mathcal{M})$, we may also use these equalities to see that

$$\mathrm{tr}(Ta^-x) = \mathrm{tr}(a^- T^{\dagger(1)}x) = \mathrm{tr}(T^{\dagger\dagger}a^-x)$$

for all $a \in \mathfrak{m}$ and $x \in L^1(\mathcal{A})$, so that $T{\restriction}\mathfrak{m} \subseteq T^{\dagger\dagger}$ which is σ-weakly continuous. Let $S = T^{\dagger}T^{\dagger\dagger}$. By reality preservation and the equality noted earlier, we see that if $a \in \mathfrak{m}$, then

$$
\begin{aligned}
\left\| i^{(2)}(Ta) \right\|^2 &= \mathrm{tr}\left(i^{(2)}(Ta)^* i^{(2)}(Ta) \right) \\
&= \mathrm{tr}\left(i^{(2)}(Ta^*) i^{(2)}(Ta) \right) \\
&= \mathrm{tr}\left(Ta^* i^{(1)}(Ta) \right) \\
&= \mathrm{tr}\left(T^{(1)*}Ta^* i^{(1)}(a) \right) \\
&= \mathrm{tr}\left(i^{(2)}(Sa^*) i^{(2)}(a) \right) \leq \left\| i^{(2)}(Sa^*) \right\| \left\| i^{(2)}(a) \right\|.
\end{aligned}
$$

Now, S is bounded, reality preserving and integrable, and $S^{\dagger} = S$. Iterating the above inequality gives

$$
\left\| i^{(2)}(Ta) \right\|_2 \leq \left\| i^{(2)}(a) \right\|_2^{1/2} \left\| i^{(2)}(Sa) \right\|_2^{1/2}
$$

$$
\vdots
$$

$$
\leq \left\| i^{(2)}(a) \right\|_2^{1-2^{-k}} \left\| i^{(2)}(S^{2^{k-1}}a) \right\|_2^{2^{-k}}.
$$

But, by equation (12.9),

$$
\begin{aligned}
\left\| i^{(2)}(S^{2^{k-1}}a) \right\|_2 &\leq \left\| S^{2^{k-1}}a \right\|_\infty^{1/2} \left\| i^{(1)}(S^{2^{k-1}}a) \right\|_1^{1/2} \\
&\leq \|S\|^{2^{k-2}} \left\| S^{(1)} \right\|^{2^{k-2}} \left(\|a\|_\infty \left\| i^{(1)}(a) \right\|_1 \right)^{1/2} \\
&\leq \|S\|^{2^{k-1}} \left(\|a\|_\infty \left\| i^{(1)}(a) \right\|_1 \right)^{1/2}
\end{aligned}
$$

and so $\left\| i^{(2)}(Ta) \right\|_2 \leq \left\| i^{(2)}(a) \right\|_2^{1-2^{-k}} \|S\|^{1/2} \left(\|a\|_\infty \left\| i^{(1)}(a) \right\|_1 \right)^{2^{-k-1}}.$ Letting $k \to \infty$ leads to the desired inequality. $\qquad\square$

Corollary 12.24 *Let T and S be integrable reality preserving operators with domains contained in $\mathcal{M}$. Then, the following are equivalent.*

(a) T and S are both 2-integrable with $(T^{(2)})^* = S^{(2)}$.

(b) (T, S) is a KMS pair.

(c) S and $T^\dagger$ agree on $\mathfrak{m}$.

Proof. Since T and S are integrable, $\mathfrak{m}$ is contained in both $\mathrm{dom}(T)$ and $\mathrm{dom}(S)$, with in addition $T(\mathfrak{m}) \subset \mathfrak{m}$ and $S(\mathfrak{m}) \subset \mathfrak{m}$.

(a$\Rightarrow$b): For any $a, b \in \mathfrak{m}$, we have that

$$
\begin{aligned}
\mathrm{tr}(Ta\,\bar{\imath}(b)) &= \mathrm{tr}(\mathfrak{i}^{(2)}(Ta)\bar{\imath}^{(2)}(b)) \\
&= \langle T^{(2)}\mathfrak{i}^{(2)}(a^*), \mathfrak{i}^{(2)}(b) \rangle \\
&= \langle \mathfrak{i}^{(2)}(a^*), S^{(2)}\mathfrak{i}^{(2)}(b) \rangle \\
&= \mathrm{tr}(\mathfrak{i}(a)Sb).
\end{aligned}
$$

So, (T, S) is a KMS pair.

(b$\Rightarrow$c): Let $a \in \mathfrak{m}$. Then

$$
\begin{aligned}
\mathrm{tr}(T^\dagger a\,\bar{\imath}^{(1)}(b)) &= \mathrm{tr}(a\,T^{(1)}\mathfrak{i}^{(1)}(b)) \\
&= \mathrm{tr}(a\,\bar{\imath}^{(1)}(Tb)) \\
&= \mathrm{tr}(\mathfrak{i}^{(1)}(a)Tb) \\
&= \mathrm{tr}(Sa\,\bar{\imath}^{(1)}(b))
\end{aligned}
$$

for all $b \in \mathfrak{m}$. So, by the density $\mathfrak{i}^{(1)}(\mathfrak{m})$ in $L^1(\mathcal{M})$, we have $T^\dagger a = Sa$ as claimed.

(c$\Rightarrow$a): The 2-integrability of T and S follows from Proposition 12.23. With that fact established, it is then an exercise to see that

$$
\begin{aligned}
\langle T^{(2)}\mathfrak{i}^{(2)}(a), \mathfrak{i}^{(2)}(b) \rangle &= \mathrm{tr}(\mathfrak{i}^{(2)}(b^*)\,T^{(2)}\mathfrak{i}^{(2)}(a)) \\
&= \mathrm{tr}(\mathfrak{i}^{(2)}(b^*)\bar{\imath}^{(2)}(Ta)) \\
&= \mathrm{tr}(\mathfrak{i}^{(1)}(b^*)\,Ta) \\
&= \mathrm{tr}(S(b^*)\bar{\imath}^{(1)}(a)) \\
&= \mathrm{tr}(\mathfrak{i}^{(2)}(Sb)^*\bar{\imath}^{(2)}(a)) \\
&= \langle \mathfrak{i}^{(2)}(a), S^{(2)}\mathfrak{i}^{(2)}(b) \rangle.
\end{aligned}
$$

The density of $\mathfrak{i}^{(2)}(\mathfrak{m})$ in $L^2(\mathcal{M})$ then ensures that $(T^{(2)})^* = S^{(2)}$ as claimed. $\qquad\square$

Proposition 12.25 *Let (T, S) be a KMS pair of operators with respect to $(\mathcal{M}, \varphi)$.*

(a) *If S is integrable, then $T{\upharpoonright}\mathfrak{m}$ is σ-weakly continuous.*
(b) *If T is everywhere defined and σ-weakly continuous, then T is bounded.*
(c) *If T is bounded and $T(\mathfrak{m}) \subset \mathfrak{m}$ then T is integrable.*

Proof. (a) Given $a, b \in \mathfrak{m}$, we then have that

$$\operatorname{tr}\left(T a \bar{\cdot} \mathfrak{i}^{(1)}(b)\right) = \operatorname{tr}\left(\mathfrak{i}^{(1)}(a)\bar{\cdot}Sb\right) = \operatorname{tr}\left(a\bar{\cdot}S^{(1)}\mathfrak{i}^{(1)}(b)\right) = \operatorname{tr}\left(S^{(1)*}a\bar{\cdot}\mathfrak{i}^{(1)}(b)\right).$$

Since $\mathfrak{i}^{(1)}(\mathfrak{m})$ is dense in $L^1(\mathcal{A})$, we have the inclusion $S^{(1)*} \supseteq T{\upharpoonright}\mathfrak{m}$, and so T is σ-weakly continuous on $\mathfrak{m}$.
(b) Apply the Closed Graph Theorem.
(c) Let $a, b \in \mathfrak{m}$, then

$$\left|\operatorname{tr}\left(\mathfrak{i}^{(1)}(Ta)\bar{\cdot}b\right)\right| = \left|\operatorname{tr}\left(Ta\bar{\cdot}\mathfrak{i}^{(1)}(b)\right)\right| = \left|\operatorname{tr}\left(\mathfrak{i}^{(1)}(a)\bar{\cdot}Sb\right)\right| \le \|\mathfrak{i}(a)\|_1 \, \|S\| \, \|b\|.$$

Hence, by Kaplansky's density theorem,

$$\|T_0^{(1)}\mathfrak{i}^{(1)}(a)\| = \sup\left\{|\operatorname{tr}(\mathfrak{i}(Ta)\bar{\cdot}b)| : b \in \mathfrak{m},\ \|b\| \le 1\right\} \le \|\mathfrak{i}(a)\|_1 \, \|S\|.$$

So, $T_0^{(1)}$ is bounded and T therefore integrable. $\qquad\square$

Proposition 12.26 *Let (T, S) be a bounded positivity preserving KMS-pair of operators with respect to $(\mathcal{M}, \varphi)$. Then, each of T and S is integrable.*

Proof. Let (f_γ) be the net in $\mathfrak{n}_\varphi$ guaranteed by Proposition 10.43. For each $z \in \mathbb{C}$, let $(\kappa_z) = \sup_{\gamma \in \Lambda} \|\sigma_z^\varphi(f_\gamma)\|$ the corresponding constants. If we set $c_\gamma = |\sigma_{i/2}(f_\gamma)^*|^2$, we will by Lemma 10.39, Proposition 10.40, and the preceding lemma then have that

$$f_\gamma\bar{\cdot}h\bar{\cdot}f_\gamma^* = [f_\gamma h^{1/2}]\, h^{1/2}f_\gamma^* = h^{1/2}\sigma_{i/2}(f_\gamma)[\sigma_{i/2}(f_\gamma)^*h^{1/2}] = \mathfrak{i}(c_\gamma).$$

Therefore, for any $a \in \mathfrak{p}$, we will by Remark 10.41 have that

$$
\begin{aligned}
\varphi(f_\gamma^*(Ta)f_\gamma) &= \operatorname{tr}([h^{1/2}f_\gamma^*]((Ta)f_\gamma h^{1/2})) \\
&= \operatorname{tr}([h^{1/2}f_\gamma^*]Ta(f_\gamma h^{1/2})) \\
&= \operatorname{tr}((\sigma_{i/2}(f_\gamma^*)h^{1/2})Ta[h^{1/2}\sigma_{i/2}(f_\gamma^*)]) \\
&= \operatorname{tr}([h^{1/2}\sigma_{i/2}(f_\gamma^*)](\sigma_{i/2}(f_\gamma^*)h^{1/2})Ta) \\
&= \operatorname{tr}\left(\mathfrak{i}^{(1)}(c_\gamma)^{\cdot}Ta\right) \\
&= \operatorname{tr}\left(Sc_\gamma^{-}\mathfrak{i}^{(1)}(a)\right) \\
&\leq \|S\|(\kappa_{i/2})^2 \left\|\mathfrak{i}^{(1)}(a)\right\|_1.
\end{aligned}
$$

Since $f_\gamma^*(Ta)f_\gamma \to Ta$ in the weak topology and $\|f_\gamma^*(Ta)f_\gamma\| \leq \|Ta\|$ for each γ, we have that $f_\gamma^*(Ta)f_\gamma \to Ta$ weak* (σ-weakly). Hence, by the σ-weak lower semicontinuity of φ,

$$
\varphi(Ta) \leq \liminf \varphi(e_\lambda^*(Sa)e_\lambda) \leq \|S\|(\kappa_{i/2})^2\varphi(a) < \infty,
$$

Thus, $T(\mathfrak{p}) \subset \mathfrak{p}$, with the result now following from Proposition 12.25.
$\square$

We now give an intrinsic definition of KMS-symmetry for the class of operators that are of most interest here.

Proposition 12.27 *Let T and S be σ-weakly continuous and positivity preserving linear maps on $\mathcal{M}$. Then, (T, S) is a KMS pair on $(\mathcal{M}, \varphi)$ if and only if T and S satisfy:*
(i) $T\mathfrak{m} \subset \mathfrak{m}$ and $S\mathfrak{m} \subset \mathfrak{m}$;
(ii) $\varphi(Ta\sigma_{-i/2}(b)) = \varphi(\sigma_{i/2}(a)Sb)$ for all $a, b \in \operatorname{span}(\mathfrak{n}_\infty^* . \mathfrak{n}_\infty)$.

Proof.   Let (T, S) be a pair of mutually σ-weakly continuous and positivity preserving operators on $\mathcal{M}$. If (T, S) is a KMS pair, then each of T and S is bounded and leaves $\mathfrak{m}$ invariant by Propositions 12.25 and 12.26. It remains to prove that (T, S) satisfies (ii). Let c and b be given with $c \in \mathfrak{n}$, and $b = b_1^*b_2$ where $b_1, b_2 \in \mathfrak{n}^\infty$.

We may now combine Remark 10.41, Lemma 12.19, and equation (12.7) to see that

$$
\begin{aligned}
\varphi(\sigma_{i/2}(b)c) &= \operatorname{tr}((h^{1/2}\sigma_{i/2}(b))[ch^{1/2}]) \\
&= \operatorname{tr}([bh^{1/2}][ch^{1/2}]) \\
&= \operatorname{tr}([ch^{1/2}][b_1^* b_2 h^{1/2}]) \\
&= \operatorname{tr}([ch^{1/2}]b_1^*[b_2 h^{1/2}]) \\
&= \operatorname{tr}(c^-[h^{1/2}b_1^*][b_2 h^{1/2}]) \\
&= \operatorname{tr}(ci^{(1)}(b)) \\
&= \operatorname{tr}(i^{(1)}(b)c).
\end{aligned}
$$

So, by linearity and taking adjoints, we have that

$$
\varphi(\sigma_{i/2}(b)c) = \operatorname{tr}(i^{(1)}(b)c) \quad \text{and} \quad \varphi(c^*\sigma_{-i/2}(b)) = \operatorname{tr}(a^-i^{(1)}(b)) \tag{12.10}
$$

for all $c \in \mathfrak{n}$ and $b \in \operatorname{span}(\mathfrak{n}_\infty^* \mathfrak{n}_\infty)$. Given $a, b \in \operatorname{span}(\mathfrak{n}_\infty^* \mathfrak{n}_\infty)$, we may use the above equalities along with equation (12.7) to see that

$$
\varphi(T(a)\sigma_{-i/2}(b)) = \operatorname{tr}(Ta^-i^{(1)}(b)) = \operatorname{tr}(i^{(1)}(a)^-Sb) = \varphi(\sigma_{i/2}(a)Sb)
$$

for all $a, b \in \operatorname{span}(\mathfrak{n}_\infty^* \mathfrak{n}_\infty)$.

Conversely, if (T, S) satisfies (i) and (ii), we may similarly use equation (12.10) to see that $\operatorname{tr}(Ta^-i^{(1)}(b)) = \operatorname{tr}(i^{(1)}(a)^-Sb)$ for all $a, b \in \operatorname{span}(\mathfrak{n}_\infty^* \mathfrak{n}_\infty)$. The σ-weak density of $\mathfrak{n}_\infty$ in $\mathcal{M}$ ensures that $\operatorname{span}(\mathfrak{n}_\infty^* . \mathfrak{n}_\infty)$ is also σ-weakly dense and in particular σ-weakly dense in $\mathfrak{m}$. The σ-weak continuity of T and S therefore ensures that $\operatorname{tr}(Ta^-i^{(1)}(b)) = \operatorname{tr}(i^{(1)}(a)^-Sb)$ for all $a, b \in \mathfrak{m}$ and hence that (T, S) is a KMS pair. $\qquad\square$

We need one more preparatory result and one new concept before we are able to put all these results together to obtain the basic correspondence. We now introduce the notion of integrability for positivity preserving operators on $L^p(\mathcal{M})$. The first stage in introducing this concept is the following definition.

Definition 12.28 Given $1 < p, p^* < \infty$ with $\frac{1}{p} + \frac{1}{p^*} = 1$, we define $\mathfrak{n}_\varphi(L^p)$ to be the subspace consisting of all elements $f \in L^{2p}(\mathcal{M})$ such that $fh^{1/2p^*}$ and $h^{1/2p^*}f^*$ are both closable with closures $[fh^{1/2p^*}]$ and $[h^{1/2p^*}f^*]$ belonging to $L^2(\mathcal{M})$ and satisfying $[fh^{1/2p^*}]^* = [h^{1/2p^*}f^*]$. We then further define $\mathfrak{m}_\varphi(L^p)$ to be $\mathrm{span}\{f^*g\colon f,g \in \mathfrak{n}_\varphi(L^p)\}$.

As before, minor technical adjustments to the proofs of Proposition 10.49 and Theorem 10.50 yield the following result. First, we note that $\mathfrak{n}_\varphi^{(q)}$ allows for the following analogue of Lemma 12.3.

Proposition 12.29   *Let $1 < p, p^* < \infty$ be given with $\frac{1}{p} + \frac{1}{p^*} = 1$.*

(i) *The prescription $f \mapsto [fh^{1/2p^*}]$, where $f \in \mathfrak{n}_\varphi(L^p)$, produces a linear injective map from $\mathfrak{n}_\varphi(L^p)$ into $L^2(\mathcal{M})$.*

(ii) *The prescription $f^*f \mapsto [fh^{1/2p^*}]^*[fh^{1/2p^*}]$, where $f \in \mathfrak{n}_\varphi(L^p)$, yields an additive and injective map*

$$\mathfrak{k}^{(p)} : \mathfrak{m}_\varphi(L^p)^+ \to L^1_+(\mathcal{M})$$

which canonically extends to a linear map (also called $\mathfrak{k}^{(p)}$) from $\mathfrak{m}_\varphi(L^p)$ into $L^1(\mathcal{M})$. The extension is injective and positivity preserving and, for any $a, b \in \mathfrak{n}_\varphi(L^p)$, satisfies the formula

$$[ah^{1/2p^*}]^*[bh^{1/2p^*}] = \mathfrak{k}^{(p)}(a^*b).$$

Armed with this technology, we now introduce the following concept.

Definition 12.30   Given $1 < p < \infty$, let $T : L^p(\mathcal{M}) \mapsto L^p(\mathcal{M})$ be a positivity preserving operator. Then, T is said to be integrable:

- if T maps for any $\mathfrak{i}^{(p)}(\mathfrak{m}_\varphi)$ into $\mathfrak{m}_\varphi(L^p)$;
- and if the formal prescription

$$\mathfrak{i}^{(1)}(a) \mapsto \mathfrak{k}^{(p)}(T(\mathfrak{i}^{(p)}(a))) \qquad a \in \mathfrak{m}_\varphi$$

extends to a bounded map on $L^1(\mathcal{M})$.

We pause to justify the sensibility of the definition.

Remark 12.31   Let $T : L^2(\mathcal{M}) \to L^2(\mathcal{M})$ be a positivity preserving operator such that for any $a \in \mathfrak{p}_\varphi$, $\sqrt{T(\mathfrak{i}^{(2)}(a))}h^{1/4}$ and $h^{1/4}\sqrt{T(\mathfrak{i}^{(2)}(a))}$ are densely defined premeasurable elements with

$[\sqrt{T(\mathrm{i}^{(2)}(a))}h^{1/4}]^* = [h^{1/4}\sqrt{T(\mathrm{i}^{(2)}(a))}]$. With $e_n = \chi_{[0,n]}(h^{1/2p})$ and $n(t) = e^{t/2p}n$, we have, as in the proof of Proposition 10.42, that $\theta_t(e_n) = e_{n(t)}$. For any $t \in \mathbb{R}$, we have $\theta_t([\sqrt{T(\mathrm{i}^{(2)}(a))}h^{1/4}])e_{n(t)} = \theta_t([\sqrt{T(\mathrm{i}^{(2)}(a))}h^{1/4}].e_n) = \theta(\sqrt{T(\mathrm{i}^{(2)}(a))}.h^{1/4}e_n) = e^{-t/2}\sqrt{T(\mathrm{i}^{(2)}(a))}.h^{1/4}e_{n(t)} = e^{-t/2}[\sqrt{T(\mathrm{i}^{(2)}(a))}h^{1/4}]e_{n(t)}$. If we now let $n \to \infty$, it will follow that $\theta_t([\sqrt{T(\mathrm{i}^{(2)}(a))}h^{1/4}]) = e^{-t/2}[\sqrt{T(\mathrm{i}^{(2)}(a))}h^{1/4}]$ for all t and hence that $[\sqrt{T(\mathrm{i}^{(2)}(a))}h^{1/4}] \in L^2(\mathcal{M})$. Thus, the prescription

$$\mathrm{i}^{(1)}(a) \mapsto [h^{1/4}\sqrt{T(\mathrm{i}^{(2)}(a))}][\sqrt{T(\mathrm{i}^{(2)}(a))}h^{1/4}]$$

yields an affine map on a dense subset of $L^1_+(\mathcal{M})$.

We are now finally able to formulate the basic correspondence.

Theorem 12.32 *Let (T, S) be a bounded KMS pair of Markov operators with respect to $(\mathcal{M}, \varphi)$ with domain $\mathfrak{m}$. Then, both T and S are σ-weakly continuous and p-integrable, for $p = 1, 2$. The operators T and S extend to $S^\dagger$ and $T^\dagger$, respectively, and both $T^{(2)}$ and $S^{(2)}$ are integrable Markov operators on $(L^2(\mathcal{M}), h^{1/2})$ with $(T^{(2)})^* = S^{(2)}$. Conversely, let S be a bounded operator on $L^2(\mathcal{M})$ for which (S, S^*) is a pair of integrable Markov operators on $(L^2(\mathcal{M}), h^{1/2})$. Then, there is a unique KMS pair of σ-weakly continuous Markov operators (T_0, T_1) on $(\mathcal{M}, \varphi)$ that are defined everywhere, that map $\mathfrak{m}_\varphi$ into $\mathfrak{m}_\varphi$, and which satisfy $\mathrm{i}^{(2)} \circ T_0 = S \circ \mathrm{i}^{(2)}$ and $\mathrm{i}^{(2)} \circ T_1 = S^* \circ \mathrm{i}^{(2)}$ on $\mathfrak{m}_\varphi$. Moreover, $T_0^\dagger = T_1$ and $T_1^\dagger = T_0$.*

Proof. Let (T, S) be as in the theorem. By Proposition 12.26, T and S are both integrable. So, by Proposition 12.25, both T and S are σ-weakly continuous with $T^\dagger \supseteq S$ and $S^\dagger \supseteq T$ by Corollary 12.24. In particular, as unique σ-weakly continuous extensions of T and S, $T^\dagger$ and $S^\dagger$ are also integrable, with both T and S 2-integrable and $(T^{(2)})^* = S^{(2)}$. We prove the Markov property of $T^{(2)}$ and $S^{(2)}$. Let $x \in L^2(\mathcal{M})$ be given with $0 \leq x \leq h^{1/2}$. By Proposition 10.42, there exists $b \in \mathfrak{n}^{(2)}$ so that $\|b\| \leq 1$ with $x = [h^{1/4}b^*](bh^{1/4})$. By Theorem 10.59, we may further assume the underlying Hilbert space to be $L^2(\mathcal{M})$ itself. In this setting, we then let $(f_\gamma) \subset \mathfrak{n}_\varphi$ be as in Proposition 10.43.

Given $a \in L^2(\mathcal{M})$, we may then use Lemma 12.19 to see that

$$
\begin{aligned}
\operatorname{tr}(T^{(2)}x.a) &= \operatorname{tr}(T^{(2)}([h^{1/4}b^*](bh^{1/4}))a) \\
&= \operatorname{tr}([h^{1/4}b^*](bh^{1/4})(T^{(2)})^*a) \\
&= \lim_\gamma \operatorname{tr}([h^{1/4}b^*](bh^{1/4})f_\gamma(T^{(2)})^*a)f_\gamma) \\
&= \lim_\gamma \operatorname{tr}([h^{1/4}\sigma^\varphi_{i/4}(f_\gamma)b^*](b\sigma^\varphi_{-i/4}(f_\gamma)h^{1/4})(T^{(2)})^*a) \\
&= \lim_\gamma \operatorname{tr}(T^{(2)}([h^{1/4}\sigma^\varphi_{i/4}(f_\gamma)b^*](b\sigma^\varphi_{-i/4}(f_\gamma)h^{1/4}))a) \\
&= \lim_\gamma \operatorname{tr}(\mathfrak{i}^{(2)}(T(|b\sigma^\varphi_{-i/4}(f_\gamma)|^2))a).
\end{aligned}
$$

This ensures that $\mathfrak{i}^{(2)}(T(|b\sigma^\varphi_{-i/4}(f_\gamma)|^2)) \to T^{(2)}x$ weakly in $L^2(\mathcal{M})$. Since $0 \le \mathfrak{i}^{(2)}(T(|b\sigma^\varphi_{-i/4}(f_\gamma)|^2)) \le \|T(|b\sigma^\varphi_{-i/4}(f_\gamma)|^2)\|h^{1/2} \le \|b\|h^{1/2} \le h^{1/2}$, we will therefore also have $0 \le T^{(2)}x \le h^{1/2}$ as required. The proof that $S^{(2)}$ is Markov follows similarly.

It remains to show that each of $T^{(2)}$ and $S^{(2)}$ is integrable. This is, however, a fairly straightforward consequence of the fact that T and S are integrable. We shall show this for $T^{(2)}$. Given any $a \in \mathfrak{p}_\varphi$, we clearly have $T(a) \in \mathfrak{p}_\varphi$, with $T^{(2)}(\mathfrak{i}^{(2)}(a)) = \mathfrak{i}^{(2)}(T(a)) = (h^{1/4}\sqrt{T(a)})[\sqrt{T(a)}h^{1/4}]|^2$ where each of $[\sqrt{T(a)}h^{1/4}]$ and $(h^{1/4}\sqrt{T(a)})$ belongs to $L^4(\mathcal{M})$. We further have $[\sqrt{T(a)}h^{1/4}]h^{1/4} \supseteq \sqrt{T(a)}h^{1/2}$ with closure in $L^2(\mathcal{M})$. To see this, observe that $h^{1/4}(h^{1/4}\sqrt{T(a)}) = h^{1/2}\sqrt{T(a)}$ is an element of $L^1(\mathcal{M})$ by Proposition 10.40. The adjoint $[\sqrt{T(a)}h^{1/2}]$ of $h^{1/4}.(h^{1/4}\sqrt{T(a)})$ is an extension of $[\sqrt{T(a)}h^{1/2}]h^{1/4} = (h^{1/4}\sqrt{T(a)})^*.h^{1/4}$ for which we also have $[\sqrt{T(a)}h^{1/2}]h^{1/4} \supseteq \sqrt{T(a)}h^{1/2}$. Hence, $[\sqrt{T(a)}h^{1/4}]h^{1/4}$ is closable and τ-densely defined (since $\sqrt{T(a)}h^{1/2}$ is closable and τ-densely defined) with the unique τ-measurable closure being $[\sqrt{T(a)}h^{1/2}]$. Thus $[h^{1/4}\sqrt{T^{(2)}(\mathfrak{i}^{(2)}(a))}][\sqrt{T^{(2)}(\mathfrak{i}^{(2)}(a))}h^{1/4}] = T^{(1)}(\mathfrak{i}^{(1)}(a))$, which suffices to prove the claim.

Conversely, let S be as in the hypothesis of the theorem, and let $a \in \mathfrak{p}_\varphi$ be given. We then have $0 \le \mathfrak{i}^{(2)}(a) \le \|a\|h^{1/2}$. Since S is Markov, we must have $0 \le S(\mathfrak{i}^{(2)}(a)) \le \|a\|h^{1/2}$. The integrability assumption ensures that we will then find that $h^{1/4}\sqrt{S(\mathfrak{i}^{(2)}(a))}$ and $\sqrt{S(\mathfrak{i}^{(2)}(a))}h^{1/4}$ are closable with both closures belonging to $L^2(\mathcal{M})$, and $[h^{1/4}\sqrt{S(\mathfrak{i}^{(2)}(a))}]^* = [\sqrt{S(\mathfrak{i}^{(2)}(a))}h^{1/4}]$. We will then also have $0 \le [h^{1/4}\sqrt{S(\mathfrak{i}^{(2)}(a))}][\sqrt{S(\mathfrak{i}^{(2)}(a))}h^{1/4}] \le \|a\|h$. But by Proposition 10.42 there must then exist some $y \in \mathfrak{p}_\varphi$ such that

$[h^{1/4}\sqrt{S(\mathfrak{i}^{(2)}(a))}][\sqrt{S(\mathfrak{i}^{(2)}(a))}h^{1/4}] = \mathfrak{i}^{(1)}(y)$. To see this, notice that for $w = [[h^{1/4}\sqrt{S(\mathfrak{i}^{(2)}(a))}][\sqrt{S(\mathfrak{i}^{(2)}(a))}h^{1/4}]]$, we have $0 \le w^{1/2} \le \|a\|^{1/2}h^{1/2}$. The proposition then guarantees the existence of some $x \in \mathfrak{n}_\varphi$ such that $w^{1/2} = [xh^{1/2}] = h^{1/2}x^*$. The element we seek is $y = x^*x$.] We will for any $z \in \mathfrak{p}_\varphi$ then have that

$$
\begin{aligned}
\mathrm{tr}(S(\mathfrak{i}^{(2)}(a))\overline{\mathfrak{i}}^{(2)}(z)) &= \mathrm{tr}((z^{1/2}h^{1/2})S(\mathfrak{i}^{(2)}(a))[h^{1/4}z^{1/2})) \\
&= \mathrm{tr}(z^{1/2}[h^{1/4}\sqrt{S(\mathfrak{i}^{(2)}(a))}][\sqrt{S(\mathfrak{i}^{(2)}(a))}h^{1/4}]z^{1/2}) \\
&= \mathrm{tr}(z^{1/2}\mathfrak{i}^{(1)}(y)z^{1/2}) \\
&= \mathrm{tr}(\mathfrak{i}^{(1)}(y)z) \\
&= \mathrm{tr}(\mathfrak{i}^{(2)}(y)\overline{\mathfrak{i}}^{(2)}(z)).
\end{aligned}
$$

The density of $\mathrm{span}(\mathfrak{i}^{(2)}(\mathfrak{p}_\varphi))$ in $L^2(\mathcal{M})$ ensures that $S(\mathfrak{i}^{(2)}(a)) = \mathfrak{i}^{(2)}(y)$, and hence that S maps $\mathfrak{i}^{(2)}(\mathfrak{m}_\varphi)$ back into itself. Thus, S induces a map $\widetilde{T}_0$ on $\mathfrak{m}_\varphi$. Given $a \in \mathfrak{m}_\varphi$ with $0 \le a \le 1$, we, of course, have $0 \le \mathfrak{i}^{(2)}(a) \le \|a\|.h^{1/2}$. The fact that S is Markov ensures that then $0 \le S(\mathfrak{i}^{(2)}(a)) = \mathfrak{i}^{(2)}(\widetilde{T}_0 a) \le \|a\|.h^{1/2}$, which translates to the fact that $0 \le \widetilde{T}_0 a \le 1$. Thus, $\widetilde{T}_0$ is Markov. Of course, a similar claim holds for S^* with the operator induced by S^* denoted by $\widetilde{T}_1$.

The integrability assumption on S and S^* ensures that, respectively, they canonically induce bounded maps $\mathscr{S}_0$ and $\mathscr{S}_1$ on $L^1(\mathcal{M})$. We now set $T_0 = (\mathscr{S}_1)^*$ and $T_1 = (\mathscr{S}_0)^*$. The fact that each of S and S^* is Markov ensures that these maps are reality preserving on $\mathfrak{m}_\varphi$. For any $a, b \in \mathfrak{m}_\varphi$, we will therefore have that

$$
\begin{aligned}
\mathrm{tr}(\mathfrak{i}^{(1)}(a))^{-}T_1 b) &= \mathrm{tr}(\mathscr{S}_0(\mathfrak{i}^{(1)}(a))^{-}b) \\
&= \mathrm{tr}(\mathfrak{i}^{(2)}(S(\mathfrak{i}^{(2)}(a)))^{-}b) \\
&= \mathrm{tr}(S(\mathfrak{i}^{(2)}(a))^{-}\mathfrak{i}^{(2)}(b)) \\
&= \langle S(\mathfrak{i}^{(2)}(a)), \mathfrak{i}^{(2)}(b)^* \rangle \\
&= \langle \mathfrak{i}^{(2)}(a), S^*(\mathfrak{i}^{(2)}(b)^*) \rangle \\
&= \mathrm{tr}(\mathfrak{i}^{(2)}(a)^{-}S^*(\mathfrak{i}^{(2)}(b))).
\end{aligned}
$$

These equalities allow us to make two conclusions. We first have that

$$
\begin{aligned}
\mathrm{tr}(\mathfrak{i}^{(1)}(a))^{-}T_1 b) &= \mathrm{tr}(a\overline{\mathfrak{i}}^{(2)}(S^*(\mathfrak{i}^{(2)}(b)))) \\
&= \mathrm{tr}(a\overline{\mathfrak{i}}^{(1)}(\widetilde{T}_1 b)) \\
&= \mathrm{tr}(\mathfrak{i}^{(1)}(a)^{-}\widetilde{T}_1 b).
\end{aligned}
$$

Second, the density of $i^{(1)}(\mathfrak{m}_\varphi)$ in $L^1(\mathcal{M})$ now suffices to ensure that $T_1 \supseteq \widetilde{T}_1$. Similarly, $T_0 \supseteq \widetilde{T}_0$. In particular, both T_0 and T_1 must then preserve $\mathfrak{m}_\varphi$. The same set of equalities as before also yields the conclusion that

$$
\begin{aligned}
\operatorname{tr}(i^{(1)}(a)^-T_1 b) &= \operatorname{tr}(a^-i^{(2)}(S^*(i^{(2)}(b)))) \\
&= \operatorname{tr}(a^-\mathscr{S}_1(i^{(1)}(b))) \\
&= \operatorname{tr}(T_0 a^-i^{(1)}(b))
\end{aligned}
$$

for all $a, b \in \mathfrak{m}_\varphi$. Thus, (T_0, T_1) is a KMS pair. It also follows from the above that $i^{(2)}(T_k a) = S_k(i^{(2)}(a))$ and $i^{(1)}(T_k a) = \mathscr{S}_k(i^{(1)}(a))$ for any $a \in \mathfrak{m}_\varphi$ and any $k \in \{0, 1\}$ and hence that T_k is both 1- and 2-integrable.

It remains to show that T_0 and T_1 are Markov. We do this for T_0. Fix $a \in \mathcal{M}_+$, and let (a_γ) be a net in $\{b \in \mathfrak{m}_\varphi : 0 \leq b \leq \|a\|\mathbb{1}\}$ converging to a in the σ-weak topology. The fact that $\widetilde{T}_0$ is Markov ensures that $(T_0 a_\gamma) = (\widetilde{T}_0 a_\gamma) \subset \{b \in \mathfrak{m}_\varphi : 0 \leq b \leq \|a\|\mathbb{1}\}$. The σ-weak continuity of T_0 now ensures $T_0 a_\gamma \to T_0 a$ and also that $0 \leq T_0 a \leq \|a\|.\mathbb{1}$, as required. $\qquad\square$

A byproduct of this theorem is that adjoint pairs of integrable Markov operators on $(L^2(\mathcal{M}), h^{1/2})$ are necessarily contractions. To see this, note that on considering Corollary 12.24 alongside the preceding theorem, all such pairs are of the form $(T^{(2)}, (T^{(2)})^*)$ where T is an integrable Markov map on $\mathcal{M}$ for which $T^\dagger$ is also an integrable Markov map on $\mathcal{M}$. The contractivity of $T^{(2)}$ then follows from the norm estimate in Proposition 12.23. We may therefore restate the theorem in the following simpler form.

Theorem 12.33 *There is a bijective correspondence between φ-KMS-pairs (T, S) of positive normal contractions on $\mathcal{M}$ and adjoint pairs of positive self-adjoint integrable contractions on $L^2(\mathcal{M})$ which are Markov with respect to $h^{1/2}$.*

We shall call an operator T 2-Markov with respect to $(L^p(\mathcal{M}), h^{1/p})$ if the induced operator $T_2 = \left[\begin{smallmatrix} T & 0 \\ 0 & T \end{smallmatrix}\right]$ is Markov with respect to $(M_2(L^p(\mathcal{M})), h^{1/p} \otimes \mathbb{1}_2)$. It is an easy exercise to see that any 2-Markov operator is Markov. In the case of 2-Markov operators, the integrability requirement for the induced operators on $(L^2(\mathcal{M}), h^{1/2})$ may be lifted from the above theorem. We close this section by proving this claim. The proof uses the following identification.

The embedding $\mathrm{i}^{(2)}$ determines a bijection between the class of operators S on $L^2(\mathcal{A})$ which have domain $\mathrm{i}^{(2)}\left(\mathfrak{m}_\varphi\right)$ and range in $\mathrm{i}^{(2)}\left(\mathfrak{m}^{(2)}\right)$, and the class of operators T with domain $\mathfrak{m}$ and range in $\mathfrak{m}^{(2)}$, by means of the equality:

$$\mathrm{i}^{(2)} \circ T = S \circ \mathrm{i}^{(2)}. \tag{12.11}$$

Lemma 12.34 *Let S and T be in correspondence as in equation (12.11). If S (equivalently T) is reality preserving, then:*

(a) *T is positivity preserving if and only if S is positivity preserving;*

(b) *T is Markov if and only if S is L^2-Markov.*

Proof. Both (a) and (b) follow quite easily once we know that for any $a \in \mathfrak{m}^{(2)}$, we have

$$0 \le a \le \mathbb{1} \text{ if and only if } 0 \le \mathrm{i}^{(2)}(a) \le h^{1/2}.$$

The 'only if' part of this statement is clear. To see the 'if' part, suppose that $a \in \mathfrak{p}_\varphi$ is given with $0 \le \mathrm{i}^{(2)}(a) \le h^{1/2}$. For every $b \in \mathfrak{p}_\varphi$, we will then have that

$$\begin{aligned}
\operatorname{tr}(a\mathrm{i}^{(1)}(b)) &= \operatorname{tr}((b^{1/2}h^{1/2})a[h^{1/2}b^{1/2}]) \\
&= \operatorname{tr}((b^{1/2}h^{1/4})\mathrm{i}^{(2)}(a)[h^{1/4}b^{1/2}]) \\
&\le \operatorname{tr}((b^{1/2}h^{1/4}){\cdot}h^{1/2}{\cdot}[h^{1/4}b^{1/2}]) \\
&= \operatorname{tr}((b^{1/2}{\cdot}h{\cdot}b^{1/2}) \\
&= \operatorname{tr}((b^{1/2}h^{1/4})[h^{1/4}b^{1/2}]) \\
&= \operatorname{tr}(\mathrm{i}^{(1)}(b)).
\end{aligned}$$

Since $\mathrm{i}^{(1)}(\mathfrak{p})$ is dense in $L^1_+(\mathcal{M})$, we have $a \le \mathbb{1}$ as required. $\square$

Theorem 12.35 *Let (T, S) be a bounded KMS pair of 2-Markov operators with respect to $(\mathcal{M}, \varphi)$ with domain $\mathfrak{m}$. Then, both T and S are σ-weakly continuous and p-integrable, for $p = 1, 2$. The operators T and S extend to $S^\dagger$ and $T^\dagger$, respectively, and both $T^{(2)}$ and $S^{(2)}$ are*

2-Markov operators on $(L^2(\mathcal{M}), h^{1/2})$ with $(T^{(2)})^* = S^{(2)}$. Conversely, let S be a bounded operator on $L^2(\mathcal{M})$ for which (S, S^*) is a pair of 2-Markov operators on $(L^2(\mathcal{M}), h^{1/2})$. Then, there is a unique KMS pair of bounded σ-weakly continuous, 2-Markov operators (T_0, T_1) on $(\mathcal{M}, \varphi)$ which are everywhere defined, which map $\mathfrak{m}_\varphi$ into $\mathfrak{m}_\varphi^{(2)}$, and which satisfy $\mathfrak{i}^{(2)} \circ T_0 = S \circ \mathfrak{i}^{(2)}$ and $\mathfrak{i}^{(2)} \circ T_1 = S^* \circ \mathfrak{i}^{(2)}$ on $\mathfrak{m}_\varphi$.

Proof. The first part of the proof runs along similar lines as before. We therefore pass to the proof of the converse implication. Since S is a Markov operator on $(L^2(\mathcal{M}), h^{1/2})$, we will for any $a \in \mathfrak{p}_\varphi$ have $0 \leq S(\mathfrak{i}^{(2)}(a)) \leq \|a\| h^{1/2}$. By Proposition 10.42, there then exists some $b_a \in \mathfrak{m}_\varphi^{(2)}$ such that $S(\mathfrak{i}^{(2)}(a)) = \mathfrak{i}^{(2)}(b_a)$. Thus, S induces a linear map $\widetilde{T}_0$ from $\mathrm{span}(\mathfrak{p}_\varphi) = \mathfrak{m}_\varphi$ into $\mathrm{span}(\mathfrak{m}^{(2)})$ such that $\mathfrak{i}^{(2)} \circ \widetilde{T}_0 = S \circ \mathfrak{i}^{(2)}$ on $\mathfrak{m}_\varphi$. Similarly, S^* induces a linear map $\widetilde{T}_1$ from $\mathfrak{m}_\varphi$ into $\mathrm{span}(\mathfrak{m}^{(2)})$ such that $\mathfrak{i}^{(2)} \circ \widetilde{T}_1 = S^* \circ \mathfrak{i}^{(2)}$ on $\mathfrak{m}_\varphi$. Using the fact that both S and S^* are 2-Markov, a similar argument to the one used above shows that $S \otimes \mathbb{1}_2$ and $S^* \otimes \mathbb{1}_2$ both induce positivity preserving maps from $M_2(\mathfrak{i}^{(2)}(\mathfrak{m}_\varphi))$ to $M_2(\mathfrak{i}^{(2)}(\mathfrak{m}^{(2)}))$ for which we have that $(\mathfrak{i}^{(2)} \circ \widetilde{T}_0) \otimes \mathbb{1}_2 = (S \circ \mathfrak{i}^{(2)}) \otimes \mathbb{1}_2$ and $(\mathfrak{i}^{(2)} \circ \widetilde{T}_1) \otimes \mathbb{1}_2 = (S^* \circ \mathfrak{i}^{(2)}) \otimes \mathbb{1}_2$ on $M_2(\mathfrak{i}^{(2)}(\mathfrak{m}_\varphi))$. Lemma 12.34 now informs us that the maps $\widetilde{T}_0$ and $\widetilde{T}_1$ are both positivity preserving and 2-Markov. Our primary task is to show that the maps $\widetilde{T}_0$ and $\widetilde{T}_1$ are bounded. Once that is done, the rest of the proof follows directly.

Let $a \in \mathfrak{m}_\varphi$ and let (f_γ) be the net in $\mathfrak{n}_\varphi^+$ guaranteed by Proposition 10.43. Recall that we then have that $f_\gamma \leq \mathbb{1}$ for all γ, that $(f_\gamma a)$ converges to a strongly, and that $(\sigma_{-\frac{i}{4}}(f_\gamma))$ is bounded.

We claim that $\mathfrak{i}^{(2)}(f_\gamma a) \xrightarrow[\gamma \in \mathcal{I}]{} \mathfrak{i}^{(2)}(a)$ weakly. To see this, notice that, by Lemma 12.19, we have $h^{1/4} f_\gamma = [\sigma_{-i/4}(f_\gamma) h^{1/4}]$. So, for any $a \in \mathfrak{p}_\varphi$, we have $\mathfrak{i}^{(2)}(f_\gamma a) = (h^{1/4} f_\gamma a^{1/2})[a^{1/2} h^{1/4}] = ([\sigma_{-i/4}(f_\gamma) h^{1/4}] a^{1/2})[a^{1/2} h^{1/4}] = \sigma^\varphi_{-i/4}(f_\gamma)(h^{1/4} a^{1/2})[a^{1/2} h^{1/4}] = \sigma^\varphi_{-i/4}(f_\gamma) \mathfrak{i}^{(2)}(a)$. Now, recall that by Proposition 10.43, the net $(\sigma^\varphi_{-i/4}(f_\gamma))$ is σ-weakly convergent to $\mathbb{1}$. For any $g \in L^2(\mathcal{M})$, we therefore have that

$$\langle \mathfrak{i}^{(2)}(f_\gamma a), g \rangle = \mathrm{tr}(g^* \mathfrak{i}^{(2)}(f_\gamma a))$$

$$= \mathrm{tr}(\sigma^\varphi_{-i/4}(f_\gamma) \mathfrak{i}^{(2)}(a) g^*)$$

$$\xrightarrow[\gamma \in \mathcal{I}]{} \mathrm{tr}(\mathfrak{i}^{(2)}(a) g^*)$$

$$= \langle \mathfrak{i}^{(2)}(a), g \rangle$$

as claimed.

Since $\widetilde{T}_0$ is Markov and positivity preserving, we will have that $0 \leq \widetilde{T}_0(a^*a) \leq \|a\|^2 \mathbb{1}$, that $0 \leq \widetilde{T}_0(f\gamma^2) \leq \mathbb{1}$, and that $\widetilde{T}_0(f_\gamma a)^* = \widetilde{T}_0(a^*f_\gamma)$ for every γ. Furthermore, $\left[\begin{smallmatrix} g_\gamma & a \\ 0 & 0 \end{smallmatrix}\right] \in \mathfrak{n}_{(\varphi \otimes \mathrm{Tr}_2)}$, whence $0 \leq \left[\begin{smallmatrix} f_\gamma^2 & f_\gamma a \\ a^* f_\gamma & a^*a \end{smallmatrix}\right] = \left[\begin{smallmatrix} g_\gamma & a \\ 0 & 0 \end{smallmatrix}\right]^* \left[\begin{smallmatrix} g_\gamma & a \\ 0 & 0 \end{smallmatrix}\right] \in \mathfrak{m}_{(\varphi \otimes \mathrm{Tr}_2)}$. Since $\widetilde{T}_0 \otimes \mathrm{Id}_2$ is also Markov and positivity preserving, we must have that $\left[\begin{smallmatrix} \widetilde{T}_0(f_\gamma^2) & \widetilde{T}_0(f_\gamma a) \\ \widetilde{T}_0(a^*f_\gamma) & \widetilde{T}_0(a^*a) \end{smallmatrix}\right] \geq 0$. The positivity of this matrix ensures that we will for any $\zeta, \eta \in H$ have that

$$0 \leq \left\langle \left[\begin{smallmatrix} \widetilde{T}_0(f_\gamma^2) & \widetilde{T}_0(f_\gamma a) \\ \widetilde{T}_0(a^*f_\gamma) & \widetilde{T}_0(a^*a) \end{smallmatrix}\right] \cdot \left[\begin{smallmatrix} \zeta \\ \eta \end{smallmatrix}\right], \left[\begin{smallmatrix} \zeta \\ \eta \end{smallmatrix}\right] \right\rangle$$
$$\leq \langle \widetilde{T}_0(f_\gamma^2)\zeta, \zeta \rangle + 2\mathrm{Re}(\langle \widetilde{T}_0(f_\gamma a)\eta, \zeta \rangle) + \langle \widetilde{T}_0(a^*a)\zeta, \zeta \rangle.$$

By now replacing η with $e^{i\alpha}\eta$ where $0 \leq \alpha < 2\pi$ is chosen so that $e^{i\alpha}\langle \widetilde{T}_0(f_\gamma a)\eta, \zeta \rangle = -|\langle \widetilde{T}_0(f_\gamma a)\eta, \zeta \rangle|$, we obtain $0 \leq \langle \widetilde{T}_0(f_\gamma^2)\zeta, \zeta \rangle - 2|\langle \widetilde{T}_0(f_\gamma a)\eta, \zeta \rangle| + \langle \widetilde{T}_0(a^*a)\zeta, \zeta \rangle$ for all $\zeta, \eta \in H$ and hence that

$$|\langle \widetilde{T}_0(f_\gamma a)\eta, \zeta \rangle| \leq \frac{1}{2}\left(\langle \widetilde{T}_0(f_\gamma^2)\zeta, \zeta \rangle + \langle \widetilde{T}_0(a^*a)\zeta, \zeta \rangle\right) \leq \frac{\|a\|}{2}(\|\zeta\|^2 + \|\eta\|^2).$$

Consequently, $\|\widetilde{T}_0(f_\gamma a)\| \leq \|a\|$ for any γ. Thus, by passing to a subnet if necessary, we may assume that this net is σ-weakly convergent to some $c \in \mathcal{M}$ with $\|c\| \leq \|a\|$. Given any $b \in \mathfrak{m}_\varphi$, we then have that

$$\mathrm{tr}\left(\mathfrak{i}^{(2)}(\widetilde{T}_0(f_\gamma a))\overline{\mathfrak{i}}^{(2)}(b)\right) = \mathrm{tr}\left(\widetilde{T}_0(f_\gamma a)\overline{\mathfrak{i}}^{(1)}(b)\right) \xrightarrow[\gamma \in I]{} \mathrm{tr}\left(c\overline{\mathfrak{i}}^{(1)}(b)\right).$$

Since S^* is positivity and hence adjoint preserving with $\mathfrak{i}^{(2)} \circ \widetilde{T}_0 = S \circ \mathfrak{i}^{(2)}$ on $\mathfrak{m}_\varphi$, the expression on the left-hand side is also equal to

$$\mathrm{tr}\left(S(\mathfrak{i}^{(2)}(f_\gamma a))\mathfrak{i}^{(2)}(b)\right) = \mathrm{tr}\left(\mathfrak{i}^{(2)}(f_\gamma a)S^*(\mathfrak{i}^{(2)}(b))\right)$$
$$\xrightarrow[\gamma \in I]{} \mathrm{tr}\left(\mathfrak{i}^{(2)}(a)\overline{\cdot}S^*(\mathfrak{i}^{(2)}(b))\right) = \mathrm{tr}\left(S(\mathfrak{i}^{(2)}(a))\overline{\mathfrak{i}}^{(2)}(b)\right)$$
$$= \mathrm{tr}\left(\mathfrak{i}^{(2)}(T_0 a)\overline{\mathfrak{i}}^{(2)}(b)\right) = \mathrm{tr}\left(T_0 a \overline{\mathfrak{i}}^{(1)}(b)\right).$$

(see equation (12.8). By the density of $\mathfrak{i}^{(2)}(\mathfrak{m}_\varphi)$ in $L^2(\mathcal{M})$, we therefore have that $c = \widetilde{T}_0 a$ and hence that $\left\|\widetilde{T}_0 a\right\| \leq \|a\|$, as desired. A similar proof shows that $\widetilde{T}_1$ is also contractive.

Given $a, b \in \mathfrak{m}_\varphi$, it is now clear that

$$
\begin{aligned}
\operatorname{tr}(\mathfrak{i}^{(1)}(a)^*\widetilde{T}_1 b) &= \operatorname{tr}(\mathfrak{i}^{(2)}(a)^*\bar{\mathfrak{i}}^{(2)}(\widetilde{T}_1 b)) \\
&= \operatorname{tr}(\mathfrak{i}^{(2)}(a)^* S^*(\mathfrak{i}^{(2)}(b))) \\
&= \operatorname{tr}(S(\mathfrak{i}^{(2)}(a))^*\bar{\mathfrak{i}}^{(2)}(b)) \\
&= \operatorname{tr}(\mathfrak{i}^{(2)}(\widetilde{T}_0 a)^*\bar{\mathfrak{i}}^{(2)}(b)) \\
&= \operatorname{tr}(\widetilde{T}_0 a\,\bar{\mathfrak{i}}^{(1)}(b)).
\end{aligned}
$$

Thus, $(\widetilde{T}_0, \widetilde{T}_1)$ is a KMS pair of 2-Markov operators on $\mathfrak{m}_\varphi$. Proposition 12.26 then ensures that both of these operators are integrable. On setting $T_0 = (\widetilde{T}_1)^\dagger$ and $T_1 = (\widetilde{T}_0)^\dagger$, we will then for any $a, b \in \mathfrak{m}_\varphi$ have that

$$
\begin{aligned}
\operatorname{tr}(\mathfrak{i}^{(1)}(a))^* T_1(b)) &= \operatorname{tr}(\mathfrak{i}^{(1)}(a))^*(\widetilde{T}_0)^\dagger(b)) \\
&= \operatorname{tr}(\mathfrak{i}^{(1)}(\widetilde{T}_0 a))^* b) \\
&= \operatorname{tr}(\mathfrak{i}^{(2)}(\widetilde{T}_0 a)^*\bar{\mathfrak{i}}^{(2)}(b)) \\
&= \operatorname{tr}(S(\mathfrak{i}^{(2)}(a))^*\bar{\mathfrak{i}}^{(2)}(b)) \\
&= \operatorname{tr}(\mathfrak{i}^{(2)}(a)^* S^*(\mathfrak{i}^{(2)}(b))) \\
&= \operatorname{tr}(\mathfrak{i}^{(2)}(a)^*\bar{\mathfrak{i}}^{(2)}(\widetilde{T}_1 b)) \\
&= \operatorname{tr}(\mathfrak{i}^{(1)}(a)^*\widetilde{T}_1(b)).
\end{aligned}
$$

Thus, $T_k \supseteq \widetilde{T}_k$ for $k = 0, 1$, showing that both $\widetilde{T}_0$ and $\widetilde{T}_1$ extend to σ-weakly continuous operators on all of $\mathcal{M}$. The positivity of each $\widetilde{T}_k$ combined with the fact that $\mathfrak{p}_\varphi$ is σ-weakly dense in $\mathcal{M}_+$ then ensures that each T_k is positive. For each T_k, we may moreover use the Markovness of $\widetilde{T}_k$ to see that $0 \le T(\mathbb{1}) = \lim_\gamma T_k(f_\gamma) = \lim_\gamma \widetilde{T}_k(f_\gamma) \le \mathbb{1}$, where (f_γ) is as before. Thus, each T_k is Markov as claimed.

A similar argument applied to $\widetilde{T}_k \otimes \mathbb{1}_2$ (where $k = 0, 1$) at the level of $M_2(\mathcal{M})$ shows that these operators too extend to σ-weakly continuous Markov operators $\mathcal{T}_k$ on $M_2(\mathcal{M})$. But each $T_k \otimes \mathbb{1}_2$ is already a σ-weakly continuous extension of the corresponding $\widetilde{T}_k \otimes \mathbb{1}_2$ to all of $M_2(\mathcal{M})$. So, by the uniqueness of the extension, we must have $T_k \otimes \mathbb{1}_2 = \mathcal{T}_k$ for each $k = 0, 1$. It follows that each T_k is 2-Markov, thereby establishing the theorem. $\qquad\square$

By Theorem 12.33, finding integrable Markov operators on L^2 corresponds to finding KMS pairs of Markov operators on $\mathcal{M}$. The results regarding positive normal maps in the previous section now yield the following.

Proposition 12.36 *Let T be a contractive normal positive map on $\mathcal{M}$ satisfying $\varphi \circ T \leq \varphi$. With $T^\dagger$ denoting the operator guaranteed by Theorem 12.14, $(T, T^\dagger)$ is a KMS pair of Markov maps. If T is CP, then $(T, T^\dagger)$ is a KMS pair of complete Markov maps.*

Proof. The positivity and contractivity of $T^\dagger$ follow from Theorem 12.33. The fact that $(T, T^\dagger)$ is a KMS pair was verified in Proposition 12.23. If additionally T is CP and $T \circ \sigma_t^\varphi = \sigma_t^\varphi \circ T$ for all $t \in \mathbb{R}$, then by the same argument $(T, T^\dagger)$ is a KMS pair of complete Markov maps. $\square$

Chapter 13
Haagerup's reduction theorem

Although only formally published in 2010 [**HJX10**], Haagerup's reduction theorem first saw the light of day as a handwritten copy date marked 15 May 1978, which for many decades has been circulated among a select few individuals. Haagerup's intention with this theorem was to establish a framework within which a σ-finite von Neumann algebra $\mathcal{M}$ could be indirectly approximated with a sequence of finite von Neumann algebras equipped with faithful normal tracial states. This theorem provides a framework within which one can first enlarge a given σ-finite von Neumann algebra $\mathcal{M}$ to a larger von Neumann superalgebra $\mathcal{R}$ (still σ-finite) inside of which $\mathcal{M}$ appears as an expected subalgebra. It is then this enlarged algebra $\mathcal{R}$ rather than $\mathcal{M}$ itself that Haagerup showed can be approximated with a sequence of finite von Neumann subalgebras equipped with faithful normal tracial states.

Haagerup's hope was that this theorem could be used to extend results regarding noncommutative L^p-spaces from the easier to handle tracial setting to the more enigmatic setting of σ-finite von Neumann algebras.

When Haagerup's reduction theorem finally appeared in print [**HJX10**], only the σ-finite case was presented. However, with sufficient care, the theorem can be made to work for general von Neumann algebras. Although this fact was known to Haagerup, Junge and Xu at the time of printing of [**HJX10**], they elected to only publish the σ-finite version of the theorem. As part of an investigation into noncommutative Hardy spaces, the second named author of this book then subsequently independently re-proved the general version of this result. It is this proof that we will present here. In particular, we will show how an arbitrary von Neumann algebra $\mathcal{M}$ equipped with an arbitrary faithful normal semifinite weight φ may be enlarged to a superalgebra $\mathcal{R}$ inside of which $\mathcal{M}$ appears as an expected subalgebra, with the enlarged algebra $\mathcal{R}$ in this case being approximable by a sequence of semifinite von Neumann subalgebras, each equipped with a faithful normal semifinite trace.

Noncommutative measures and L^p and Orlicz Spaces, with Applications to Quantum Physics. Stanisław Goldstein and Louis Labuschagne, Oxford University Press. © Stanisław Goldstein and Louis Labuschagne (2025).
DOI: 10.1093/oso/9780198950202.003.0015

13.1 The construction of the superalgebra $\mathcal{R} = \mathcal{M} \rtimes \mathbb{Q}_D$

Write $\mathbb{Q}_D$ for the dyadic rationals. This group plays a crucial role in the construction, and for this reason, we pause to elucidate its structure. In so doing, we will take our cue from the helpful discussion on the site https://en.wikipedia.org/wiki/Dyadic_rational. When equipped with the discrete topology, the dual group of the dyadic rationals is the so-called dyadic solenoid (a compact group). To see this, note that the dyadic rationals are the direct limit of infinite cyclic subgroups of the rational numbers, namely $\varinjlim \left\{ 2^{-i}\mathbb{Z} \mid i = 0, 1, 2, \dots \right\}$ with the dual group then turning out to be the inverse limit of the dual groups of each of these subgroups, namely the unit circle group under the repeated squaring map $\zeta \mapsto \zeta^2$. It is this inverse limit that is commonly referred to as the dyadic solenoid. Elements of this group can be represented as an infinite sequence of complex numbers $q_0, q_1, q_2, \dots,$ with the properties that each q_i lies in the unit circle and that, for all $i > 0$, $q_i^2 = q_{i-1}$. The group operation on these elements multiplies any two sequences component-wise. Each element of the dyadic solenoid corresponds to a character of the dyadic rationals that maps $a/2^b$ to the complex number q_b^a. Conversely, every character γ of the dyadic rationals corresponds to an element of the dyadic solenoid given by $q_i = \gamma(1/2i)$.

Let $\mathcal{M}$ be a von Neumann algebra equipped with a faithful normal weight φ. The restriction of the mapping $t \mapsto \sigma_t^\varphi$ to $\mathbb{Q}_D$ determines an action $\mathbb{Q}_D \ni t \mapsto \alpha_t$ of the group $\mathbb{Q}_D$ on $\mathcal{M}$. We write $\mathcal{M} \rtimes \mathbb{Q}_D$ for the crossed product of $\mathcal{M}$ with respect to this action. By Corollary 9.22, there exists a $\widehat{\alpha}_t$-invariant faithful normal conditional expectation $\mathscr{W}_{\mathbb{Q}_D}$ from $\mathcal{M} \rtimes \mathbb{Q}_D$ onto $\mathcal{M}$.

In our presentation of the reduction theorem, we closely follow, but not clone, the presentation of this theorem in [**HJX10**]. The astute reader will notice some subtle but important differences. One such difference is that, in the case of a type III algebra which is not σ-finite, $\mathcal{M} \rtimes \mathbb{Q}_D$ will be approximated by a sequence of semifinite algebras rather than a net of finite algebras.

For the rest of this section, denote $\mathcal{M} \rtimes \mathbb{Q}_D$ by $\mathcal{R}$. We have previously observed that the subgroups $\left\{ 2^{-i}\mathbb{Z} \mid i = 0, 1, 2, \dots \right\}$ increase to $\mathbb{Q}_D$. We shall use this structure to construct a matching net of von Neumann algebras increasing to $\mathcal{R}$.

The rest of this section will be devoted to proving the version of the reduction theorem stated below. We shall prove this theorem by means of a series of not insignificant lemmata.

Theorem 13.1 *Let $\mathcal{M}$ be a von Neumann algebra equipped with a faithful normal semifinite weight φ. With $\mathcal{R}$ as above, there exists an increasing sequence $(\mathcal{R}_n)_{n\geq 1}$ of von Neumann subalgebras of $\mathcal{R}$ that satisfy the following properties.*

> *(1) Each $\mathcal{R}_n$ is semifinite. If φ is a state, each $\mathcal{R}_n$ is even finite admitting a faithful normal tracial state.*
> *(2) $\bigcup_{n\geq 1}\mathcal{R}_n$ is σ-strong* dense in $\mathcal{R}$.*
> *(3) For every $n \in \mathbf{N}$, there exists a faithful normal conditional expectation $\mathscr{W}_n$ from $\mathcal{R}$ onto $\mathcal{R}_n$ such that*
>
> $$\widetilde{\varphi} \circ \mathscr{W}_n = \widetilde{\varphi} \quad and \quad \sigma_t^{\widetilde{\varphi}} \circ \mathscr{W}_n = \mathscr{W}_n \circ \sigma_t^{\widetilde{\varphi}}, \quad t \in \mathbb{R}.$$
>
> *(Notice that the equality $\widetilde{\varphi} \circ \mathscr{W}_n = \widetilde{\varphi}$ ensures that $\mathscr{W}_n$ maps $\mathfrak{p}(\mathcal{R})_{\widetilde{\varphi}}$ onto $\mathfrak{p}(\mathcal{R}_n)_{\widetilde{\varphi}}$. Therefore, by the normality of $\mathscr{W}_n$, $\mathfrak{p}(\mathcal{R}_n)_{\widetilde{\varphi}}$ must be σ-weakly dense in $\mathcal{R}_n^+$ and hence $\widetilde{\varphi}{\restriction}\mathcal{R}_n$ semifinite.)*

Before proceeding with the proof, we have one technical issue to sort out. We computed the crossed product $\mathcal{R} = \mathcal{M} \rtimes \mathbb{Q}_D$ by using a *restriction* of the modular group $t \mapsto \sigma_t^\varphi$ to $\mathbb{Q}_D$. Let us denote this restriction by $t \mapsto \alpha_t$. The unitary group $t \mapsto \lambda_t$ $(t \in \mathbb{Q}_D)$ induces an automorphism group $t \mapsto \lambda_t(\cdot)\lambda_t$ $(\,t \in \mathbb{Q}_D)$ on $\mathcal{R}$. This then raises the question of how this automorphism group compares to the modular group $t \mapsto \sigma_t^{\widetilde{\varphi}}$ on $\mathcal{M} \rtimes \mathbb{Q}_D$. Since α_t is just the restriction of σ_t^φ, φ is clearly α_t invariant. This fact ensures that the above question has a very elegant answer.

Lemma 13.2 *For any $t \in \mathbb{Q}_D$ and any $x \in \mathcal{R}$, we have $\sigma_t^{\widetilde{\varphi}}(x) = \lambda_t x \lambda_t^*$.*

Proof. For the sake of clarity, we will here distinguish between $\mathcal{M}$, and the copy thereof inside $\mathcal{R}$, writing $\pi(\mathcal{M})$ for that copy. For any $a \in \mathcal{M}$ and any $t \in \mathbb{Q}_D$, Proposition 9.5 and Theorem 9.25 ensure that

$$\lambda_t \pi(a) \lambda_t^* = \pi(\alpha_t(a)) = \pi(\sigma_t^\varphi(a)) = \sigma_t^{\widetilde{\varphi}}(\pi(a)).$$

Theorem 9.25 also ensures that $\lambda_t \lambda_s \lambda_t^* = \lambda_s = \sigma_t^{\tilde{\varphi}}(\lambda_s)$ for any $s, t \in \mathbb{Q}_D$. By Remark 9.7, these observations are enough to prove the claim. $\square$

13.2 Approximating $\mathcal{R}$ with semifinite algebras $(\mathcal{R}_n)$

With $\mathbb{T}$ denoting the unit circle of the complex plane equipped with normalised Lebesgue measure $d\mathrm{m}$, we obtain the following very elegant formula.

Lemma 13.3 *For each $f \in L_\infty(\mathbb{T})$ and each $t \in \mathbb{Q}_D \setminus \{0\}$, we have that*

$$\mathscr{W}_{\mathbb{Q}_D}(f(\lambda_t))) = \int_{\mathbb{T}} f(z)\, d\mathrm{m}(z)\mathbb{1}. \tag{13.1}$$

Proof.   Let $t \in \mathbb{Q}_D \setminus \{0\}$ be given. We may then apply Corollary 9.22 to see that for each $t \in \mathbb{Q}_D$ and each $n \in \mathbb{Z}$, we have that

$$\mathscr{W}_{\mathbb{Q}_D}(\lambda_t^n) = \mathscr{W}_{\mathbb{Q}_D}(\lambda_{nt}) = \begin{cases} \mathbb{1}, & \text{if } n = 0, \\ 0, & \text{otherwise.} \end{cases}$$

We, of course, also have that

$$\int_{\mathbb{T}} z^n\, d\mathrm{m}(z) = \int_{|z|=1} z^{n-1}\, dz = \begin{cases} 1, & \text{if } n = 0, \\ 0, & \text{otherwise.} \end{cases}$$

Thus, (13.1) holds whenever f is a trigonometric polynomial. The set of trigonometric polynomials is, of course, σ-weakly dense in $L_\infty(\mathbb{T})$, and hence the normality of $\mathscr{W}_{\mathbb{Q}_D}$ ensures that (13.1) holds for all $f \in L_\infty(\mathbb{T})$. $\square$

Lemma 13.4

(1) $\lambda_t \in \mathcal{Z}(\mathcal{R}_{\tilde{\varphi}})$ *for any $t \in \mathbb{Q}_D$.*

(2) *For every $n \in \mathbb{N}$, there exists a unique $b_n \in \mathcal{Z}(\mathcal{R}_{\tilde{\varphi}})$ such that $0 \leq b_n \leq 2\pi$ and $e^{ib_n} = \lambda_{2^{-n}}$.*

Proof.   We first note that by Theorem 9.25, each λ_t $(t \in \mathbb{Q}_D)$ does indeed belong to $\mathcal{R}_{\tilde{\varphi}}$. For all elements of the centraliser, we therefore have by Lemma 13.2 that $x = \lambda_t x \lambda_t^*$ for all $t \in \mathbb{Q}_D$. This clearly ensures the validity of (1).

To prove (2), we use the branch of the complex logarithmic function $z \mapsto \log(z)$ given by $\log(z) = \ln|z| + i\arg(z)$ where $0 \leq \arg(z) < 2\pi$. For a unimodular complex number z, we have that $-i\log(z) = \arg(z)$ where $0 \leq \arg(z) \leq 2\pi$. Any unitary $u \in \mathcal{R}$ generates a commutative von Neumann subalgebra $\mathcal{R}_u$ of $\mathcal{R}$. From Theorem 1.54, we know that this subalgebra is $*$-isomorphic to some $L^\infty(X, \Sigma, \mu)$. Therefore, for convenience, we may identify $\mathcal{R}_u$ with $L^\infty(X, \Sigma, \mu)$. The unitaries in $L^\infty(X, \Sigma, \mu)$, of course, correspond to all the measurable functions on X that almost everywhere take values in $\mathbb{T}$. In particular, u corresponds to such a function. But then $\log(u)$ corresponds to a function for which $0 \leq -i\log(u) \leq 2\pi$. (This follows from the fact that the unimodularity of u ensures that $\log(u) = i\arg(u)$.) Thus, $-i\log(u)$ is then a bounded positive map (with $\|-i\log(u)\| \leq 2\pi$) belonging to the same commutative subalgebra $\mathcal{R}_u$. In the case of $u = \lambda_{2-n}$, the subalgebra $\mathcal{R}_u$ is, in fact, a subalgebra of $\mathcal{Z}(\mathcal{R}_{\widetilde{\varphi}})$. Therefore, setting $b_n = -i\log(\lambda_{2-n})$ will do the job once we note that (by construction) $e^{ib_n} = \lambda_{2-n}$.

The uniqueness of b_n follows from the fact that λ_{2-n} does not have a point spectrum by virtue of Lemma 13.3 and the faithfulness of $\widetilde{\varphi}$. $\square$

Now, let $a_n = 2^n b_n$ and define a sequence $(\varphi_n)_{n \geq 1}$ of weights on $\mathcal{R}$ by

$$\varphi_n(x) = \widetilde{\varphi}(e^{-a_n/2} x e^{-a_n/2}), \quad x \in \mathcal{R}_+, \ n \geq 1. \tag{13.2}$$

Lemma 13.5

(1) *Each φ_n is a faithful normal semifinite weight on $\mathcal{R}$.*

(2) *We have*

$$\sigma_t^{\varphi_n}(x) = e^{-ita_n} \sigma_t^{\widetilde{\varphi}}(x) e^{ita_n}, \quad x \in \mathcal{R}, \ t \in \mathbb{R}. \tag{13.3}$$

(3) *$\sigma_t^{\varphi_n}$ is 2^{-n}-periodic for all $n \geq 1$.*

(4) *On setting $\mathcal{R}_n = \mathcal{R}_{\varphi_n}$, $n \geq 1$, it follows that there exists a unique faithful normal conditional expectation $\mathcal{W}_n$ from $\mathcal{R}$ onto $\mathcal{R}_n$ such that*

$$\widetilde{\varphi} \circ \mathcal{W}_n = \widetilde{\varphi} \quad \text{and} \quad \sigma_t^{\widetilde{\varphi}} \circ \mathcal{W}_n = \mathcal{W}_n \circ \sigma_t^{\widetilde{\varphi}}, \quad t \in \mathbb{R}, \ n \geq 1.$$

(5) *$\mathcal{R}_n \subseteq \mathcal{R}_{n+1}$.*

(6) *For each $n \in \mathbb{N}$, φ_n is a faithful normal semifinite trace on $\mathcal{R}_n$. Moreover, if φ is a state, then each φ_n is a finite trace on $\mathcal{R}_n$.*

Proof.   (1): The normality of φ_n follows from the facts that $\widetilde{\varphi}$ is normal and that $\sup_\alpha e^{-a_n/2} x_\alpha e^{-a_n/2} = e^{-a_n/2} \sup_\alpha x_\alpha e^{-a_n/2}$ for increasing nets. Faithfulness similarly easily follows from the fact that for any $x \in \mathcal{R}_+$, $0 = \varphi_n(x) = \widetilde{\varphi}(e^{-a_n/2} x e^{-a_n/2}) \Rightarrow e^{-a_n/2} x e^{-a_n/2} = 0 \Rightarrow x = 0$. Semifiniteness follows from the fact that the σ-weak density of $\mathrm{span}\{x \in \mathcal{R}_+ : \widetilde{\varphi}(x) < \infty\}$ in $\mathcal{R}$, ensures the σ-weak density of $\mathrm{span}\{x \in \mathcal{R}_+ : \varphi_n(x) < \infty\} = \mathrm{span}\{e^{a_n/2} x e^{a_n/2} \in \mathcal{R}_+ : \widetilde{\varphi}(x) < \infty\}$.

(2): By part (1) of Lemma 13.4, $a_n \in \mathcal{Z}(\mathcal{R}_{\widetilde{\varphi}}) \subset \mathcal{R}_{\widetilde{\varphi}}$. Thus, this claim is a direct consequence of Theorem 8.23.

(3): We know from Lemma 13.2 that $\sigma_t^{\widetilde{\varphi}}(a) = \lambda_t a \lambda_t^*$ for every $t \in \mathbb{Q}_D$ and every $a \in \mathcal{R}$. Therefore, it follows from part (2) of Lemma 13.4 that

$$\sigma_{2-n}^{\varphi_n}(x) = e^{-i\,b_n} \sigma_{2-n}^{\widetilde{\varphi}}(x) e^{ib_n} = \lambda_{2-n}^* \lambda_{2-n} x \lambda_{2-n}^* \lambda_{2-n} = x$$

for all $x \in \mathcal{R}$, thereby ensuring the validity of (3).

(4): Define $\mathscr{W}_n$ by

$$\mathscr{W}_n(x) = 2^n \int_0^{2^{-n}} \sigma_t^{\varphi_n}(x)\, dt, \quad x \in \mathcal{R}.$$

By the 2^{-n} periodicity of σ^{φ_n}, we have that

$$\mathscr{W}_n(x) = \int_0^1 \sigma_t^{\varphi_n}(x)\, dt, \quad x \in \mathcal{R}.$$

It is then a routine matter to check that $\mathscr{W}_n$ is a faithful normal conditional expectation from $\mathcal{R}$ onto $\mathcal{R}_n$.

Next, let $x \in \mathcal{R}_+$ be given. First, suppose that $\widetilde{\varphi}(x) < \infty$. Since $a_n \in \mathcal{R}_{\widetilde{\varphi}}$, we may then apply Theorem 8.19 and use the $\sigma^{\widetilde{\varphi}}$-invariance of $\widetilde{\varphi}$ to conclude that

$$\widetilde{\varphi}\big(\sigma_t^{\varphi_n}(x)\big) = \widetilde{\varphi}\big(e^{-ita_n} \sigma_t^{\widetilde{\varphi}}(x) e^{ita_n}\big) = \widetilde{\varphi}\big(\sigma_t^{\widetilde{\varphi}}(x)\big) = \widetilde{\varphi}(x), \quad t \in \mathbb{R}.$$

If, on the other hand, $\widetilde{\varphi}\big(\sigma_t^{\varphi_n}(x)\big) < \infty$, we may apply what we have just proven to conclude that

$$\widetilde{\varphi}\big(\sigma_t^{\varphi_n}(x)\big) = \widetilde{\varphi}\big(\sigma_{-t}^{\varphi_n}(\sigma_t^{\varphi_n}(x))\big) = \widetilde{\varphi}(x), \quad t \in \mathbb{R}.$$

Thus, for any $x \in \mathcal{R}_+$ and any $t \in \mathbb{R}$, $\widetilde{\varphi}(x) < \infty$ if and only if $\widetilde{\varphi}\big(\sigma_t^{\varphi_n}(x)\big) < \infty$, in which case they are equal. We therefore clearly have that $\widetilde{\varphi} \circ \sigma_t^{\varphi_n} = \widetilde{\varphi}$ for all $t \in \mathbb{R}$ and all $n \geq 1$. But then

$$\widetilde{\varphi}(\mathscr{W}_n(x)) = \int_0^1 \widetilde{\varphi}(\sigma_t^{\varphi_n}(x))\,dt = \widetilde{\varphi}(x), \quad x \in \mathcal{R}_+;$$

that is $\widetilde{\varphi} \circ \mathscr{W}_n = \widetilde{\varphi}$. The uniqueness of $\mathscr{W}_n$ is now ensured by Theorem 10.54 with the claimed commutation relation following from equation (13.3) and the definition of $\mathscr{W}_n$.

(5): For every natural number n, a_n and a_{n+1} will commute, given that they both belong to $\mathcal{Z}(\mathcal{R}_{\widetilde{\varphi}})$. It is now an easy exercise to see that $\varphi_{n+1}(x) = \varphi_n(h_n x)$ for all $x \in \mathcal{R}$, where $h_n = e^{-a_{n+1}}e^{a_n} = e^{-(a_{n+1}-a_n)}$. If we are able to show that $h_n \in \mathcal{Z}(\mathcal{R}_n)$, it will then follow from Theorem 8.23 that $\sigma_t^{\varphi_{n+1}}(x) = e^{-ith_n}\sigma_t^{\varphi_n}(x)e^{ith_n} = \sigma_t^{\varphi_n}(x)$ for all $x \in \mathcal{R}_n$ and all $t \in \mathbb{R}$. This will clearly ensure that $\mathcal{R}_n \subset \mathcal{R}_{n+1}$ as claimed.

By part (2) and the fact that $a_k \in \mathcal{Z}(\mathcal{R}_{\widetilde{\varphi}})$, we have that $\mathcal{R}_{\widetilde{\varphi}} \subseteq \mathcal{R}_{\varphi_k}$ for each $k \in \mathbb{N}$. In particular, we will then have that $h_n \in \mathcal{R}_{\varphi_n} = \mathcal{R}_n$. As in Lemma 13.4, we now use the branch of the complex logarithmic function $z \mapsto \log(z)$ given by $0 \leq \arg(z) < 2\pi$. Then

$$a_n = -i2^n \log \lambda_{2^{-n}} = -i2^n \log\left(\lambda_{2^{-n-1}}^2\right)$$

whence

$$a_{n+1} - a_n = -i2^n\left[2\log \lambda_{2^{-n-1}} - \log\left(\lambda_{2^{-n-1}}^2\right)\right].$$

However, for any $z \in \mathbb{T}$,

$$2\log z - \log(z^2) = \begin{cases} 0, & \text{if } 0 \leq \arg z < \pi, \\ 2\pi i, & \text{if } \pi \leq \arg z < 2\pi. \end{cases}$$

Hence

$$a_{n+1} - a_n = 2^{n+1}\pi e_n,$$

where e_n is the spectral projection of $\lambda_{2^{-n-1}}$ corresponding to $\mathrm{Im}(z) < 0$. So, for all $x \in \mathcal{R}$ and all $t \in \mathbb{R}$,

$$\sigma_t^{\varphi_{n+1}}(x) = h_n^{it}\sigma_t^{\varphi_n}(x)h_n^{-it} = e^{-i2^{n+1}\pi t e_n}\sigma_t^{\varphi_n}(x)e^{i2^{n+1}\pi t e_n}.$$

Consequently, if $x \in \mathcal{R}_n$, the 2^{-n-1}-periodicity of $\sigma_t^{\varphi_{n+1}}$ will ensure that

$$x = e^{-i\pi e_n}\sigma_{2^{-n-1}}^{\varphi_n}(x)e^{i\pi e_n} = e^{-i\pi e_n}xe^{i\pi e_n}.$$

Now, observe that the Borel functional calculus ensures that $e^{-i\pi e_n}e_n = -e_n$ and that $e^{-i\pi e_n}(\mathbb{1} - e_n) = e^{-i\pi 0}(\mathbb{1} - e_n) = (\mathbb{1} - e_n)$ (using the fact that $\mathfrak{n}(e_n) = \mathbb{1} - e_n$). Hence, $e^{-i\pi e_n} = \mathbb{1} - 2e_n$. The above centred equality may therefore be reformulated as $(\mathbb{1} - 2e_n)x = x(\mathbb{1} - 2e_n)$, clearly showing that $\mathbb{1} - 2e_n \in \mathcal{Z}(\mathcal{R}_n)$, or equivalently that $e_n \in \mathcal{Z}(\mathcal{R}_n)$. Thus, $h_n \in \mathcal{Z}(\mathcal{R}_n)$, which then yields the desired inclusion $\mathcal{R}_n \subseteq \mathcal{R}_{n+1}$.

(6): The faithfulness of φ_n on $\mathcal{R}_n$ is a clear consequence of part (1). The fact that $\widetilde{\varphi} \circ \mathcal{W}_n = \widetilde{\varphi}$ ensures that the restriction of $\widetilde{\varphi}$ to $\mathcal{R}_n$ is normal and semifinite. We may now use a similar argument to that used in the proof of part (1) to conclude from this that φ_n is normal and semifinite on $\mathcal{R}_n$. Therefore, it remains to show that φ_n satisfies the trace property on $\mathcal{R}_n$. Thus, let $x \in \mathcal{R}_n$ be given.

First, suppose that $\varphi_n(x^*x) < \infty$. Thus, $x^*x = |x|^2 \in \mathfrak{m}_{\varphi_n}$. Let $x = v|x|$ be the polar decomposition of x. Then, we clearly have $v \in \mathcal{R}_n = \mathcal{R}_{\varphi_n}$. But in that case we may use Theorem 8.19 to conclude that $x^*xv^* = |x|^2v^* \in \mathfrak{m}_{\varphi_n}$ and hence that $\varphi_n(x^*x) = \varphi_n(|x|^2v^*v) = \varphi(v|x|^2v^*) = \varphi_n(xx^*)$. Therefore, it follows that $\varphi_n(x^*x) < \infty$ if and only if $\varphi_n(xx^*) < \infty$, in which case the two quantities are equal. But then $\varphi_n(x^*x) = \varphi_n(xx^*)$ for all $x \in \mathcal{R}_n$ as required.

Finally, suppose that φ is a state. Since in this particular case $\mathcal{W}_{\mathcal{Q}_D}$ is a faithful normal conditional expectation, we therefore have $\widetilde{\varphi}(\mathbb{1}) = \varphi \circ \mathcal{W}_{\mathcal{Q}_D}(\mathbb{1}) = \varphi(\mathbb{1}) = 1$. So, $\widetilde{\varphi}$ is also a state. But then $\varphi_n(\mathbb{1}) = \widetilde{\varphi}(e^{-a_n}) \leq \|e^{-a_n}\|_\infty < \infty$ as required. $\qquad\square$

It remains to show the σ-weak density of the union of the $\mathcal{R}_n$s in $\mathcal{R}$. The groundwork for this verification will be laid by the following pair of lemmas. The first is essentially just a special case of Lemma 8.16.

Lemma 13.6 *Let ψ be a faithful normal semifinite weight on the von Neumann algebra $\mathcal{N}$. In applying Lemma 8.16 to the pair $(\mathcal{N}, \psi)$, we may choose k to be $\|a\|$, if also $a \in \mathcal{N}_\psi$.*

Proof. Note that under the conditions of the hypothesis, the map $t \mapsto \sigma_t^\psi(a)$ $(t \in \mathbb{R})$ is the constant map $t \mapsto a$. The unique continuous extension of this map to the strip $\{z \in \mathbb{C}: 0 \geq \mathrm{Im}(z) \geq -1/2\}$, which is analytic on the interior of the strip, is of course again the constant map $z \mapsto a$. Thus, by the discussion preceding Lemma 8.16, we have $a = \sigma_{-i/2}^\varphi(a)$. Therefore, the claim follows directly from the Lemma 8.16. $\qquad\square$

In the following lemma, $[x,\,y]$ denotes the commutator of two operators x and y, i.e., $[x,\,y] = xy - yx$. If ψ is a normal weight on a von Neumann algebra $\mathcal{N}$, $\|x\|_\psi$ will for any $x \in \mathcal{N}$ denote the quantity $\psi(x^*x)^{1/2}$. In the case where ψ is also faithful and semifinite and $x \in \mathfrak{n}_\psi$, $\|x\|_\psi$ is of course nothing but $\|\eta_\varphi(x)\|$ where $\eta_\varphi(x)$ is as in Remark 8.2. For simplicity of notation, we will in the following theorem identify $\mathcal{M}$ with the canonical copy thereof inside $\mathcal{R}$.

Lemma 13.7 *Let b_n, $\widetilde{\varphi}$ and $\mathcal{R}$ be as in Lemma 13.4.*

(a) *The subspaces* $\operatorname{span}\{\lambda_t a\colon t \in \mathbb{Q}_D, a \in \mathfrak{n}_\varphi(\mathcal{M}) \cap \mathfrak{n}_\varphi(\mathcal{M})^*\}$ *and* $\operatorname{span}\{\lambda_t a\colon t \in \mathbb{Q}_D, a \in \mathfrak{m}_\varphi(\mathcal{M}), a \ \text{entire analytic}\}$ *are both σ-weakly dense subspaces of $\mathcal{R}$. Moreover,* $\operatorname{span}\{\lambda_t a\colon t \in \mathbb{Q}_D, a \in \mathfrak{n}_\varphi(\mathcal{M})\} \subseteq \mathfrak{n}_{\widetilde{\varphi}}$ *and* $\operatorname{span}\{\lambda_t a\colon t \in \mathbb{Q}_D, a \in \mathfrak{n}_\varphi^*(\mathcal{M})\} \subseteq \mathfrak{n}_{\widetilde{\varphi}}^*$. *Each element of* $\operatorname{span}\{\lambda_t a\colon t \in \mathbb{Q}_D, a \in \mathfrak{m}_\varphi(\mathcal{M}), a \ \text{entire analytic}\}$ *is moreover again entire analytic with respect to $\sigma_t^{\widetilde{\varphi}}$.*

(b) *For any w in either* $\operatorname{span}\{\lambda_t a\colon t \in \mathbb{Q}_D, a \in \mathfrak{n}_\varphi(\mathcal{M}) \cap \mathfrak{n}_\varphi(\mathcal{M})^*\}$ *or* $\operatorname{span}\{\lambda_t a\colon t \in \mathbb{Q}_D, a \in \mathfrak{m}_\varphi(\mathcal{M}), a \ \text{entire analytic}\}$, *we have*
(i) $\displaystyle\lim_{n\to\infty} \|[b_n,\,w]\|_{\widetilde{\varphi}} = 0$;
(ii) $\displaystyle\lim_{n\to\infty} \sup_{t\in[-1,\,1]} \left\|[e^{itb_n},\,w]\right\|_{\widetilde{\varphi}} = 0.$

(c) *For any $f \in \operatorname{span}\{\lambda_t a\colon t \in \mathbb{Q}_D, a \in \mathfrak{m}_\varphi(\mathcal{M}), a \ \text{entire analytic}\}$ and any $x \in \mathcal{R}$, we have that* $\displaystyle\lim_{n\to\infty} \sup_{t\in\mathbb{R}} \left\|(\sigma_t^{\varphi_n}(x) - x)f\right\|_{\widetilde{\varphi}} = 0.$

Proof. **(a):** Recall that in the proof of Lemma 12.19, we showed that the subspace $\{a \in \mathfrak{m}_\varphi(\mathcal{M})\colon a \ \text{entire analytic}\}$ is σ-weakly dense in $\mathcal{M}$. Thus, the σ weak closure of $\operatorname{span}\{\lambda_t a\colon t \in \mathbb{Q}_D, a \in \mathfrak{m}_\varphi(\mathcal{M})\}$ must include the σ-weak closure of $\operatorname{span}\{\lambda_t a\colon t \in \mathbb{Q}_D, a \in \mathcal{M}, a \ \text{entire analytic}\}$, which, by Theorem 8.13, must in turn include $\operatorname{span}\{\lambda_t a\colon t \in \mathbb{Q}_D, a \in \mathcal{M}\}$. But this latter space is known to be σ-weakly dense in the crossed product $\mathcal{R} = \mathcal{M} \rtimes \mathbb{Q}_D$. Hence, as required, the subspace

$$\operatorname{span}\{\lambda_t a\colon t \in \mathbb{Q}_D, a \in \mathfrak{m}_\varphi(\mathcal{M}), a \ \text{entire analytic}\}$$

is σ-weakly dense in $\mathcal{R}$. To see the second claim, select $(f_\gamma) \subset \mathfrak{n}_\varphi(\mathcal{M}) \cap \mathfrak{n}_\varphi(\mathcal{M})^*$ as in Proposition 10.43. Note that (f_γ) is norm-bounded. For any entire analytic $b \in \mathcal{M}$, we will therefore have that each $f_\gamma b f_\gamma$ is an entire analytic element of $\mathfrak{m}_\varphi(\mathcal{M})$ (see Theorem 8.13) with $(f_\gamma b f_\gamma)$ strongly converging to b (see Proposition 0.11).

Next, we show that $\mathrm{span}\{\lambda_t a\colon t \in \mathbb{Q}_D, a \in \mathfrak{n}_\varphi(\mathcal{M})\}$ is contained in $\mathfrak{n}_{\widetilde{\varphi}}$. For any $a \in \mathfrak{n}_\varphi(\mathcal{M})$ and any $\mathbb{Q}_D$, the fact that $\mathscr{W}_{\mathbb{Q}_D}$ is a conditional expectation, ensures that $\widetilde{\varphi}(|\lambda_t a|^2) = \widetilde{\varphi}(|a|^2) = \varphi(\mathscr{W}_{\mathbb{Q}_D}(|a|^2)) = \varphi(|a|^2) < \infty$ as required. If indeed $a \in \mathfrak{n}(\mathcal{M})_\varphi^*$, it follows similarly that $\widetilde{\varphi}(|(\lambda_t a)^*|^2) = \widetilde{\varphi}(|\lambda_t a^* \lambda_t^*|^2) = \widetilde{\varphi}(|\sigma_t^{\widetilde{\varphi}}(a^*)|^2) = \widetilde{\varphi}(\sigma_t^{\widetilde{\varphi}}(|a^*|^2)) = \widetilde{\varphi}(|a^*|^2) = \varphi(\mathscr{W}_{\mathbb{Q}_D}(|a^*|^2)) = \varphi(|a^*|^2) < \infty$. This proves the third claim. The fourth claim now follows by duality once we notice that Lemma 13.2 and Theorem 9.25 ensure that $\lambda_t a = [\lambda_{-t} \sigma_t^\varphi(a^*)]^*$ for every $t \in \mathbb{Q}_D$ and every $a \in \mathcal{M}$.

Since for any $a \in \mathcal{M}$ and $t \in \mathbb{Q}_D$ we have that $\sigma_t^\varphi(a) = \sigma_t^{\widetilde{\varphi}}(a)$ with $\lambda_t \in \mathcal{R}_{\widetilde{\varphi}}$ (see Lemma 13.4), it clearly follows that each term of the form $\lambda_t a$ will be entire analytic with respect to $\sigma_t^{\widetilde{\varphi}}$ if a is entire analytic with respect to σ_t^φ.

(b)(i): Let w be in either $\mathrm{span}\{\lambda_t a\colon t \in \mathbb{Q}_D, a \in \mathfrak{n}_\varphi(\mathcal{M}) \cap \mathfrak{n}_\varphi(\mathcal{M})^*\}$ or $\mathrm{span}\{\lambda_t a\colon t \in \mathbb{Q}_D, a \in \mathfrak{m}_\varphi(\mathcal{M}), a \text{ entire analytic}\}$, and let $k \in \mathbb{Z}$ be given. For any $t \in \mathbb{Q}_D$, we have by Lemma 13.4 that

$$\left\| [\lambda_{2^{-n}}^k,\, w] \right\|_{\widetilde{\varphi}}$$

$$= \left\| (\lambda_{k 2^{-n}} w - w \lambda_{k 2^{-n}}) \right\|_{\widetilde{\varphi}}$$

$$= \left\| (w - \lambda_{k 2^{-n}}^* w \lambda_{k 2^{-n}}) \right\|_{\widetilde{\varphi}}$$

$$= \left\| (w - \sigma_{-k 2^{-n}}^{\widetilde{\varphi}}(z)) \right\|_{\widetilde{\varphi}}$$

$$= \widetilde{\varphi}(w^* w) - \widetilde{\varphi}(\sigma_{-k 2^{-n}}^{\widetilde{\varphi}}(w^*) w) - \widetilde{\varphi}(w^* \sigma_{-k 2^{-n}}^{\widetilde{\varphi}}(w)) + \widetilde{\varphi}(|\sigma_{-k 2^{-n}}^{\widetilde{\varphi}}(w)|^2)$$

$$= 2\widetilde{\varphi}(w^* w) - \widetilde{\varphi}(\sigma_{-k 2^{-n}}^{\widetilde{\varphi}}(w^*) w) - \widetilde{\varphi}(w^* \sigma_{-k 2^{-n}}^{\widetilde{\varphi}}(w)).$$

(For the last equality, we used the fact that $\widetilde{\varphi} \circ \sigma_t^{\widetilde{\varphi}} = \widetilde{\varphi}$.) In the case where $w \in \mathrm{span}\{\lambda_t a\colon t \in \mathbb{Q}_D, a \in \mathfrak{n}_\varphi(\mathcal{M}) \cap \mathfrak{n}_\varphi(\mathcal{M})^*\}$, we may now apply Theorem 8.10 to see that

$$\lim_{n \to \infty} \left\| [P(\lambda_{2^{-n}}),\, w] \right\|_{\widetilde{\varphi}} = 0 \tag{13.4}$$

for any monomial, and hence also for any trigonometric polynomial P. In the case where $w = \lambda_t a$ where $t \in \mathbb{Q}_D$ and a is an entire-analytic element of $\mathfrak{m}_\varphi(\mathcal{M})$, the analyticity will, by definition, ensure the continuity of the function $t \mapsto \sigma_t^{\widetilde{\varphi}}(w)$ on $\mathbb{R}$ and hence also the validity of the above limit.

We show how to obtain the final conclusion for the case where $w \in \mathrm{span}\{\lambda_t a\colon t \in \mathbb{Q}_D, a \in \mathfrak{n}_\varphi(\mathcal{M}) \cap \mathfrak{n}_\varphi(\mathcal{M})^*\}$. The proof for the other case is entirely analogous. Let $\log$ be as in Lemma 13.4. On $\mathbb{T}$, $\log$ agrees with the bounded almost everywhere continuous function arg. Thus, restricted to $\mathbb{T}$, we have $\log \in L^\infty(\mathbb{T}) \subseteq L^2(\mathbb{T})$. It follows that there exists a trigonometric polynomial P such that $\|P + i\log\|_{L_2(\mathbb{T})} < \epsilon$.

Let $a \in \mathfrak{n}_\varphi(\mathcal{M}) \cap \mathfrak{n}_\varphi(\mathcal{M})^*$ and any $s \in \mathbb{Q}_D$ be given. (In the case left as an exercise, one would assume that a is an entire analytic element of $\mathfrak{m}_\varphi(\mathcal{M})$.) Taking into account that $\mathscr{W}_{\mathbb{Q}_D}$ is an expectation and that $\lambda_{2^{-n}}$, b_n and λ_s all belong to the same abelian von Neumann subalgebra, it follows from equation (13.1) that

$$
\begin{aligned}
\|(b_n - P(\lambda_{2^{-n}}))\lambda_s a\|_{\widetilde{\varphi}} &= \widetilde{\varphi}(a^* \lambda_s^* |b_n - P(\lambda_{2^{-n}})|^2 \lambda_s a)^{1/2} \\
&= \widetilde{\varphi}(a^* |b_n - P(\lambda_{2^{-n}})|^2 a)^{1/2} \\
&= \varphi(\mathscr{W}_{\mathbb{Q}_D}(a^* |b_n - P(\lambda_{2^{-n}})|^2 a))^{1/2} \\
&= \varphi(a^* \mathscr{W}_{\mathbb{Q}_D}(|b_n - P(\lambda_{2^{-n}})|^2)a)^{1/2} \\
&= \| - i\log - P\|_{L_2(\mathbb{T})}\varphi(|a|^2)^{1/2} \\
&< \epsilon\varphi(|a|^2)^{1/2}.
\end{aligned}
$$

Recall that in the proof of Lemma 13.5, we showed that $\mathcal{R}_{\widetilde{\varphi}} \subset \mathcal{R}_n$. Since $\lambda_{2^{-n}}$ is a unitary belonging to $\mathcal{Z}(\mathcal{R}_{\widetilde{\varphi}})$, each of $\lambda_{2^{-n}}$, a_n and b_n will moreover belong to the abelian von Neumann subalgebra of $\mathcal{R}_n$ generated by $\lambda_{2^{-n}}$. We may now use these facts alongside Lemmata 13.4 and 13.5 to similarly see that

$$
\begin{aligned}
\|\lambda_s a(b_n - P(\lambda_{2^{-n}}))\|_{\widetilde{\varphi}} &= \widetilde{\varphi}(|\lambda_s a(b_n - P(\lambda_{2^{-n}}))|^2)^{1/2} \\
&= \widetilde{\varphi}((b_n - P(\lambda_{2^{-n}}))^* |a|^2 (b_n - P(\lambda_{2^{-n}})))^{1/2} \\
&= \widetilde{\varphi}(\mathscr{W}_n((b_n - P(\lambda_{2^{-n}}))^* |a|^2 (b_n - P(\lambda_{2^{-n}}))))^{1/2} \\
&= \widetilde{\varphi}((b_n - P(\lambda_{2^{-n}}))^* \mathscr{W}_n(|a|^2)(b_n - P(\lambda_{2^{-n}})))^{1/2} \\
&= \varphi_n(e^{a_n/2}(b_n - P(\lambda_{2^{-n}}))^* \mathscr{W}_n(|a|^2) \\
&\qquad (b_n - P(\lambda_{2^{-n}}))e^{a_n/2})^{1/2} \\
&= \varphi_n(e^{a_n/2}|b_n - P(\lambda_{2^{-n}})|^2 \mathscr{W}_n(|a|^2)e^{a_n/2})^{1/2} \\
&= \widetilde{\varphi}(|b_n - P(\lambda_{2^{-n}}|^2 \mathscr{W}_n(|a|^2))^{1/2} \\
&= \widetilde{\varphi}(\mathscr{W}_n(|b_n - P(\lambda_{2^{-n}}|^2 |a|^2))^{1/2} \\
&= \widetilde{\varphi}(|b_n - P(\lambda_{2^{-n}}|^2 |a|^2)^{1/2} \\
&= \widetilde{\varphi}(\mathscr{W}_{\mathbb{Q}_D}(|b_n - P(\lambda_{2^{-n}}|^2 |a|^2))^{1/2}.
\end{aligned}
$$

(Here we silently use the fact that φ_n is a trace on $\mathcal{R}_n$.) On once again applying equation (13.1), it therefore follows that

$$\|a(b_n - P(\lambda_{2^{-n}}))\lambda_s\|_{\tilde{\varphi}} \leq \| - i \log -P\|_{L_2(\mathbb{T})} \varphi(|af|^2)^{1/2} < \epsilon\varphi(|a|^2)^{1/2}.$$

So, if w is of the form $z = \sum_{k=1}^m \lambda_{s_k} a_k$, we will have that

$$\|[b_n, w]\|_{\tilde{\varphi}} \leq \|[P(\lambda_{2^{-n}}), w]\|_{\tilde{\varphi}} + \|[b_n - P(\lambda_{2^{-n}}), w]\|_{\tilde{\varphi}}$$
$$\leq \|[P(\lambda_{2^{-n}}), w]\|_{\tilde{\varphi}} + \|(b_n - P(\lambda_{2^{-n}}))w\|_{\tilde{\varphi}}$$
$$+ \|w(b_n - P(\lambda_{2^{-n}}))\|_{\tilde{\varphi}}$$
$$\leq \|[P(\lambda_{2^{-n}}), w]\|_{\tilde{\varphi}} + \sum_{k=1}^m \|(b_n - P(\lambda_{2^{-n}}))\lambda_{s_k} a_k\|_{\tilde{\varphi}}$$
$$+ \sum_{k=1}^m \|\lambda_{s_k} a_k(b_n - P(\lambda_{2^{-n}}))\|_{\tilde{\varphi}}$$
$$\leq \|[P(\lambda_{2^{-n}}), w]\|_{\tilde{\varphi}} + \epsilon[2 \sum_{k=1}^m \varphi(|a_k|^2)^{1/2}].$$

Therefore, by the above calculations, we then have that

$$\limsup_{n \to \infty} \|[b_n, w]\|_{\tilde{\varphi}} \leq \epsilon[2 \sum_{k=1}^m \varphi(|a_k|^2)^{1/2}]$$

whence $\lim_{n \to \infty} \|[b_n, w]\|_{\tilde{\varphi}} = 0$ as required.

(b)(ii): Let w be an element of either $\text{span}\{\lambda_t a : t \in \mathbb{Q}_D, a \in \mathfrak{n}_\varphi(\mathcal{M}) \cap \mathfrak{n}_\varphi(\mathcal{M})^*\}$ or $\text{span}\{\lambda_t a : t \in \mathbb{Q}_D, a \in \mathfrak{n}_\varphi(\mathcal{M}), a$ entire analytic$\}$. By Lemma 13.6, the fact that $b_n \in \mathcal{R}_{\tilde{\varphi}}$ with $\|b_n\|_\infty \leq 2\pi$, similarly ensures that $b_n^* . \tilde{\varphi} . b_n \leq \|b_n\|_\infty^2 \tilde{\varphi} \leq (2\pi)^2 \tilde{\varphi}$. We claim that $\|[b_n^m, w]\|_{\tilde{\varphi}} \leq m(2\pi)^{m-1}\|[b_n, w]\|_{\tilde{\varphi}}$ holds for all $m \in \mathbb{N}$. The validity of the case $m = 1$ clearly follows from what we have noted above. Now, suppose that the claim holds for $m = k$. Since

$$[b_n^{k+1}, w] = b_n[b_n^k, w] + [b_n, w]b_n^k$$

we may again use the observations made above to see that

$$\|[b_n^{k+1},\, w]\|_{\widetilde{\varphi}} \le \|b_n[b_n^k,\, w]\|_{\widetilde{\varphi}} + \|[b_n,\, w]b_n^k\|_{\widetilde{\varphi}}$$

$$\le \|b_n\|.\|[b_n^k,\, w]\|_{\widetilde{\varphi}} + \|[b_n,\, w]b_n^k\|_{\widetilde{\varphi}}$$

$$\le k(2\pi)^k.\|[b_n,\, w]\|_{\widetilde{\varphi}} + \|[b_n,\, w]b_n^k\|_{\widetilde{\varphi}}$$

$$\le k(2\pi)^k.\|[b_n,\, w]\|_{\widetilde{\varphi}} + \|b_n\|_{\infty}^k\|[b_n,\, w]\|_{\widetilde{\varphi}}$$

$$= (k+1)(2\pi)^k\|[b_n,\, w]\|_{\widetilde{\varphi}}.$$

The claimed estimate therefore follows by induction. Hence, for any $z \in \mathbb{C}$,

$$\|[e^{zb_n},\, w]\|_{\widetilde{\varphi}} \le \sum_{k=1}^{\infty} \frac{|z|^k}{k!}\,\|[b_n^k,\, w]\|_{\widetilde{\varphi}}$$

$$\le \sum_{k=1}^{\infty} \frac{|z|^k}{(k-1)!}\,(2\pi)^{k-1}\|[b_n,\, w]\|_{\widetilde{\varphi}}$$

$$= |z|\,e^{2\pi|z|}\|[b_n,\, w]\|_{\widetilde{\varphi}}.$$

Therefore

$$\sup_{t\in[-1,\,1]} \|[e^{itb_n},\, w]\|_{\widetilde{\varphi}} \le e^{2\pi}\|[b_n,\, w]\|_{\widetilde{\varphi}}$$

which on the strength of (i) implies (ii).

(c): First let w be an element of either $\operatorname{span}\{\lambda_t a : t \in \mathbb{Q}_D, a \in \mathfrak{n}_\varphi(\mathcal{M}) \cap \mathfrak{n}_\varphi(\mathcal{M})^*\}$, or $\operatorname{span}\{\lambda_t a : t \in \mathbb{Q}_D, a \in \mathfrak{m}_\varphi(\mathcal{M}), a \text{ entire analytic}\}$, and let $\epsilon > 0$ be given. By (ii), there exists $n_0 \in \mathbb{N}$ such that

$$\|[e^{isb_n},\, w]\|_{\widetilde{\varphi}} \le \epsilon, \quad \text{for all } s \in [-1,\, 1], \text{ and all } n \ge n_0. \tag{13.5}$$

Next, observe that by the continuity inherent in the analyticity described in either Theorem 8.10(2) or the second bullet of Definition 8.11, we have that

$$\|(\sigma_s^{\widetilde{\varphi}}(w) - w)\|_{\widetilde{\varphi}}^2 = \widetilde{\varphi}(w^*w) - \widetilde{\varphi}(\sigma_s^{\widetilde{\varphi}}(w^*)w) - \widetilde{\varphi}(w^*\sigma_s^{\widetilde{\varphi}}(w)) + \widetilde{\varphi}(\sigma_s^{\widetilde{\varphi}}(w^*w))$$

$$= 2\widetilde{\varphi}(w^*w) - \widetilde{\varphi}(\sigma_s^{\widetilde{\varphi}}(w^*)w) - \widetilde{\varphi}(w^*\sigma_s^{\widetilde{\varphi}}(w))$$

$$\to 0.$$

Therefore, n_0 can be chosen so that in addition

$$\|(\sigma_s^{\widetilde{\varphi}}(w) - w)\|_{\widetilde{\varphi}} \le \epsilon, \quad |s| \le 2^{-n_0}. \tag{13.6}$$

Let $t \in \mathbb{R}$ and $n \in \mathbb{N}$ with $n \geq n_0$. Write $t = t_1 + t_2$, where $t_1 = k\,2^{-n}$ for some $k \in \mathbb{Z}$ and $0 \leq t_2 \leq 2^{-n}$. Then, for any $y \in \mathcal{R}$,

$$\sigma_{t_1}^{\widetilde{\varphi}}(y) = \lambda_{k2^{-n}} y \lambda_{k2^{-n}}^* = e^{ikb_n} y e^{-ikb_n}$$
$$= e^{ik2^{-n}a_n} y e^{-ik2^{-n}a_n} = e^{it_1 a_n} y e^{-it_1 a_n}.$$

Since $\|\cdot\|_{\widetilde{\varphi}}$ is invariant under $\sigma_{t_1}^{\widetilde{\varphi}}$ and $a_n \in \mathcal{Z}(\mathcal{R}_{\widetilde{\varphi}})$, we may deduce that

$$\|(\sigma_t^{\widetilde{\varphi}}(w) - e^{ia_n t} w e^{-ia_n t})\|_{\widetilde{\varphi}} = \|(\sigma_{t_2}^{\widetilde{\varphi}}(w) - e^{ia_n t_2} w e^{-ia_n t_2})\|_{\widetilde{\varphi}}$$
$$\leq \|(\sigma_{t_2}^{\widetilde{\varphi}}(w) - w)\|_{\widetilde{\varphi}} +$$
$$\|(w - e^{ia_n t_2} w e^{-ia_n t_2})\|_{\widetilde{\varphi}}$$
$$= \|(\sigma_{t_2}^{\widetilde{\varphi}}(w) - w)\|_{\widetilde{\varphi}} + \|[e^{-ia_n t_2}, \, w]\|_{\widetilde{\varphi}}.$$

Now, $a_n t_2 = (2^n t_2) b_n$ and $2^n t_2 \leq 1$. Hence, from equations (13.5) and (13.6), it follows that

$$\|e^{-ia_n t} \sigma_t^{\widetilde{\varphi}}(w) e^{ia_n t} - w\|_{\widetilde{\varphi}} = \|\sigma_t^{\widetilde{\varphi}}(w) - e^{ia_n t} w e^{-ia_n t})\|_{\widetilde{\varphi}} \leq 2\epsilon.$$

(Here, we used the fact that $e^{ia_n t}$ is in $\mathcal{R}_{\widetilde{\varphi}}$.) On the basis of Lemma 13.5, this then ensures that

$$\lim_{n \to \infty} \sup_{t \in \mathbb{R}} \|(\sigma_t^{\varphi_n}(w) - w_\alpha) f\|_{\widetilde{\varphi}} = 0. \tag{13.7}$$

Now, let $f \in \mathrm{span}\{\lambda_t a : t \in \mathbb{Q}_D, a \in \mathfrak{m}_\varphi(\mathcal{M}), a \text{ entire analytic}\}$, and at first assume that $x \in \mathrm{span}\{\lambda_t a : t \in \mathbb{Q}_D, a \in \mathfrak{n}_\varphi(\mathcal{M}) \cap \mathfrak{n}_\varphi(\mathcal{M})^*\}$. Since f is analytic, we know from Lemma 8.16 that there exists a positive constant c_f such that $f^* . \widetilde{\varphi} . f \leq c_f \widetilde{\varphi}$. This, in turn, ensures that

$$\|(\sigma_t^{\varphi_n}(x) - x) f\|_{\widetilde{\varphi}} \leq c_f^{1/2} \|\sigma_t^{\varphi_n}(x) - x\|_{\widetilde{\varphi}} \leq c_f^{1/2} 2\epsilon$$

which on the strength of equation (13.7) proves the claim for this case.

Next, let x be an arbitrary element of $\mathcal{R}$. The subspace $\mathrm{span}\{\lambda_t a : t \in \mathbb{Q}_D, a \in \mathfrak{n}_\varphi(\mathcal{M}) \cap \mathfrak{n}_\varphi(\mathcal{M})^*\}$ of $\mathcal{R}$ is both convex and σ-weakly dense. Hence, it is also strongly dense with respect to the GNS representation engendered by $\widetilde{\varphi}$. Thus, we may select a net $(x_\alpha) \subseteq \mathrm{span}\{\lambda_t a : t \in \mathbb{Q}_D, a \in \mathfrak{n}_\varphi(\mathcal{M}) \cap \mathfrak{n}_\varphi(\mathcal{M})^*\}$ that converges to x (GNS) strongly. We know from the first part of the proof that

$$\lim_{n\to\infty} \sup_{t\in\mathbb{R}} \left\|(\sigma_t^{\varphi_n}(x_\alpha) - x_\alpha)f\right\|_{\widetilde{\varphi}} = 0 \tag{13.8}$$

for each α. Since in the GNS representation of $\mathcal{R}$ the term $\|(x - x_\alpha)f\|_{\widetilde{\varphi}}$ is just $\|\pi_{\widetilde{\varphi}}(x - x_\alpha)\eta(f)\|$, the strong convergence noted earlier ensures that

$$\lim_\alpha \|(x - x_\alpha)f\|_{\widetilde{\varphi}} = 0. \tag{13.9}$$

We may next use the fact that each $e^{ia_n t}$ and each λ_t belong to $\mathcal{R}_{\widetilde{\varphi}}$, to see that

$$\begin{aligned}
\|(e^{-ia_n t}\sigma_t^{\widetilde{\varphi}}(x - x_\alpha)e^{ia_n t})f\|_{\widetilde{\varphi}}^2 &= \|(e^{-ia_n t}\lambda_t(x - x_\alpha)\lambda_t^* e^{ia_n t})f\|_{\widetilde{\varphi}}^2 \\
&= \|(x - x_\alpha)(\lambda_t^* e^{ia_n t}f)\|_{\widetilde{\varphi}}^2 \\
&= \|(x - x_\alpha)(\lambda_t^* e^{ia_n t}fe^{-ia_n t}\lambda_t)\|_{\widetilde{\varphi}}^2 \\
&= \|(x - x_\alpha)(e^{ia_n t}\sigma_{-t}^{\widetilde{\varphi}}(f)e^{-ia_n t})\|_{\widetilde{\varphi}}^2.
\end{aligned}$$

It therefore follows from Lemma 13.5 that

$$\|\sigma_t^{\varphi_n}(x - x_\alpha)f\|_{\widetilde{\varphi}}^2 = \|(x - x_\alpha)\sigma_{-t}^{\varphi_n}(f)\|_{\widetilde{\varphi}} \text{ for all } \alpha.$$

Since (x_α) is strongly and therefore also σ-weakly convergent, the net $(x - x_\alpha)$ must be norm bounded. So, there must exist some $K > 0$ such that $\|x - x_\alpha\| \le K$ for all α. We may now use these two facts to see that

$$\begin{aligned}
\|\sigma_t^{\varphi_n}(x - x_\alpha)f\|_{\widetilde{\varphi}} &= \|(x - x_\alpha)\sigma_{-t}^{\varphi_n}(f)\|_{\widetilde{\varphi}} \\
&= \|(x - x_\alpha)(\sigma_{-t}^{\varphi_n}(f) - f)\|_{\widetilde{\varphi}} + \|(x - x_\alpha)\sigma_{-t}^{\varphi_n}(f)\|_{\widetilde{\varphi}} \\
&= K\|\sigma_{-t}^{\varphi_n}(f) - f\|_{\widetilde{\varphi}} + \|(x - x_\alpha)f\|_{\widetilde{\varphi}}.
\end{aligned}$$

A combination of equations (13.7) and (13.9) now ensures that

$$\limsup_{n\to\infty} \sup_{t\in\mathbb{R}} \|\sigma_t^{\varphi_n}(x - x_\alpha)f\|_{\widetilde{\varphi}} \le \|(x - x_\alpha)f\|_{\widetilde{\varphi}}. \tag{13.10}$$

Given $\epsilon > 0$, we may select α so that $\|(x_\alpha - x)f\|_{\widetilde{\varphi}} \le \epsilon$. So, by equations (13.8) and (13.10), we will then have that

$$\limsup_{n\to\infty} \sup_{t\in\mathbb{R}} \|(\sigma_t^{\varphi_n}(x) - x)f\|_{\widetilde{\varphi}}$$

$$\leq \limsup_{n\to\infty} \sup_{t\in\mathbb{R}} [\|\sigma_t^{\varphi_n}(x - x_\alpha)f\|_{\widetilde{\varphi}} + \|(\sigma_t^{\varphi_n}(x_\alpha) - x_\alpha)f\|_{\widetilde{\varphi}}$$

$$+ \|(x_\alpha - x)f\|_{\widetilde{\varphi}}]$$

$$\leq \limsup_{n\to\infty} \sup_{t\in\mathbb{R}} [\|\sigma_t^{\varphi_n}(x - x_\alpha)f\|_{\widetilde{\varphi}} + \|(\sigma_t^{\varphi_n}(x_\alpha) - x_\alpha)f\|_{\widetilde{\varphi}} + \epsilon]$$

$$\leq \limsup_{n\to\infty} \sup_{t\in\mathbb{R}} \|\sigma_t^{\varphi_n}(x - x_\alpha)f\|_{\widetilde{\varphi}} + \limsup_{n\to\infty} \sup_{t\in\mathbb{R}} \|(\sigma_t^{\varphi_n}(x_\alpha) - x_\alpha)f\|_{\widetilde{\varphi}} + \epsilon$$

$$= \epsilon.$$

This clearly suffices to prove the claim. $\qquad\qquad\qquad\qquad\square$

We are now finally ready to prove the σ-weak density of $\cup_{n\in\mathbb{N}}\mathcal{R}_n$ in $\mathcal{R}$ and hence conclude the proof of the reduction theorem.

Lemma 13.8 *For any $x \in \mathcal{R}$, the sequence $(\mathscr{W}_n(x))$ is σ-weakly convergent to x. Consequently, $\bigcup_{n\in\mathbb{N}} \mathcal{R}_n$ is σ-strong* dense in $\mathcal{R}$.*

Proof. We claim that the subspace of $\mathcal{R}_*$ spanned by functionals of the form $\widetilde{\varphi}(y^* \cdot f)$ where $y \in \mathfrak{n}_{\widetilde{\varphi}}$ and $f \in \mathrm{span}\{\lambda_t a\colon t \in \mathbb{Q}_D, a \in \mathfrak{m}_{\varphi}(\mathcal{M}), a$ entire analytic$\}$ is norm dense in $\mathcal{R}_*$. To prove this, all we need to show is that if for some $g \in \mathcal{R}$, we have $\widetilde{\varphi}(y^*gf) = 0$ for all y and f as above, then $g = 0$. To see that this is the case, note that when given such a g, we may for each f as above select y to be $y = gf$. We then have $\widetilde{\varphi}(|gf|^2) = 0$ for all f, which, by the faithfulness of $\widetilde{\varphi}$, ensures that $gf = 0$ for all f. But we know from Lemma 13.7 that $\mathrm{span}\{\lambda_t a\colon t \in \mathbb{Q}_D, a \in \mathfrak{m}_{\varphi}(\mathcal{M}), a$ entire analytic$\}$ is σ-weakly dense in $\mathcal{R}$. Therefore, it follows that $gg^* = 0$ and hence that $g = 0$ as required.

We proceed to show that for all y and f as above, we will for any $x \in \mathcal{R}$ have that

$$\lim_{n\to\infty} \widetilde{\varphi}(y^*(\mathscr{W}_n(x) - x)f) = 0. \tag{13.11}$$

Since the subspace of $\mathcal{R}_*$ spanned by functionals of the form $\widetilde{\varphi}(y^* \cdot f)$ is norm dense in $\mathcal{R}_*$ and the sequence $(\mathscr{W}_n(x) - x)$ norm-bounded, it will then follow from this that $\lim_{n\to\infty} \rho(\mathscr{W}_n(x) - x) = 0$ for all $\rho \in \mathcal{R}_*$ as required. This will establish the σ-weak density of $\bigcup_{n\in\mathbb{N}} \mathcal{R}_n$ in $\mathcal{R}$. The fact that $\bigcup_{n\in\mathbb{N}} \mathcal{R}_n$ is convex will then ensure that it is also σ-strong* dense in $\mathcal{R}$.

We proceed to prove the validity of equation (13.11). With y and f as before, we may invoke the Cauchy–Schwarz inequality to see that

$$|\widetilde{\varphi}(y^*(\mathscr{W}_n(x) - x)f)| \leq \widetilde{\varphi}(y^*y)^{1/2}\widetilde{\varphi}(|(\mathscr{W}_n(x) - x)f|^2)^{1/2}$$
$$= \widetilde{\varphi}(y^*y)^{1/2}\|(\mathscr{W}_n(x) - x)f\|_{\widetilde{\varphi}}.$$

It now follows from the definition of the expectations $\mathscr{W}_n$, that $\|(\mathscr{W}_n(x) - x)f\|_{\widetilde{\varphi}} \leq \sup_{t\in\mathbb{R}}\|(\sigma_t^{\varphi_n}(x) - x)f\|_{\widetilde{\varphi}}$ for each n. If we combine the above facts, we have that

$$|\widetilde{\varphi}(y^*(\mathscr{W}_n(x) - x)f)| \leq \widetilde{\varphi}(y^*y)^{1/2}\cdot\Big[\sup_{t\in\mathbb{R}}\|(\sigma_t^{\varphi_n}(x) - x)f\|_{\widetilde{\varphi}}\Big].$$

The claimed convergence therefore follows from part (c) of Lemma 13.7. $\qquad\square$

13.3 Approximating $L^p(\mathcal{R})$ with the spaces $L^p(\mathcal{R}_n, \tau_n)$

We close this chapter with the following result.

Theorem 13.9 *Let $\mathcal{M}$ be a von Neumann algebra equipped with a faithful normal semifinite weight φ and let $1 \leq p < \infty$ be given. Then, for $\mathcal{R} = \mathcal{M} \rtimes_\varphi \mathbb{Q}_\mathrm{D}$, we have that $L^p(\mathcal{R})$ is a Banach superspace of $L^p(\mathcal{M})$ isometrically containing $L^p(\mathcal{M})$. Moreover, the sequence $(\mathcal{R}_n)_{n\geq 1}$ of semifinite von Neumann algebras, each equipped with a faithful normal semifinite trace τ_n, admits an accompanying sequence of isometric embeddings $J_n : L^p(\mathcal{R}_n, \tau_n) \to L^p(\mathcal{R})$ such that:*

(1) *the sequence $\big(J_n(L^p(\mathcal{R}_n, \tau_n))\big)_{n\geq 1}$ is increasing;*

(2) *$\bigcup_{n\geq 1} J_n(L^p(\mathcal{R}_n, \tau_n))$ is dense in $L^p(\mathcal{R})$;*

(3) *the extension $\mathscr{W}_n^{(p)}$ of $\mathscr{W}_n$ to $L^p(\mathcal{R})$ contractively maps $L^p(\mathcal{R})$ onto $J_n(L^p(\mathcal{R}_n, \tau_n))$, with $\mathscr{W}_n^{(p)}(f)$ converging weakly to f for each $f \in L^p(\mathcal{R})$;*

(4) *$L^p(\mathcal{M})$ and each $J_n(L^p(\mathcal{R}_n, \tau_n))$ are the images of contractive projections on $L^p(\mathcal{R})$.*

Here, $L^p(\mathcal{R}_n, \tau_n)$ is the tracial noncommutative L^p-space associated with $(\mathcal{R}_n, \tau_n)$.

Proof. The proof makes use of Theorem 13.1, and hence, we keep all the notation there. The L^p-spaces for $\mathcal{R}$ will be constructed using $\widetilde{\varphi}$

and for $\mathcal{M}$ using φ. For the subalgebras, we shall sometimes use their traces and sometimes the weight $\widetilde{\varphi}{\upharpoonright}\mathcal{R}_n$ to construct the associated L^p-spaces. These two variants will, respectively, be denoted by $L^p(\mathcal{R}_n, \tau_n)$ and $L_p(\mathcal{R}_n)$.

Since by construction $\mathcal{W}_{\mathbb{Q}_D}$ is a faithful normal conditional expectation from $\mathcal{R}$ onto $\mathcal{M}$ for which we have that $\varphi \circ \mathcal{W}_{\mathbb{Q}_D} = \widetilde{\varphi}$, it follows from the discussion in Remark 12.18 that $L_p(\mathcal{M})$ is a subspace of $L^p(\mathcal{R})$, which is the image of a contractive projection. On considering Remark 12.18 alongside part (3) of Theorem 13.1, it similarly follows that each $L^p(\mathcal{R}_n)$ is a subspace of $L^p(\mathcal{R})$, which is the image of a contractive projection. The fact that $\mathcal{R}_n \subset \mathcal{R}_{n+1} \subset \mathcal{R}$ clearly ensures that $\mathfrak{n}(\mathcal{R}_n)_{\widetilde{\varphi}} \subseteq \mathfrak{n}(\mathcal{R}_{n+1})_{\widetilde{\varphi}} \subseteq \mathfrak{n}(\mathcal{R})_{\widetilde{\varphi}}$ and hence that $\mathfrak{m}(\mathcal{R}_n)_{\widetilde{\varphi}} \subset \mathfrak{m}(\mathcal{R}_{n+1})_{\widetilde{\varphi}}$. So, the same discussion in Remark 12.18 shows that we will have that $\mathfrak{i}^{(p)}(\mathfrak{m}(\mathcal{R}_n)_{\widetilde{\varphi}}) \subset \mathfrak{i}^{(p)}(\mathfrak{m}(\mathcal{R}_{n+1})_{\widetilde{\varphi}})$ and hence that $L^p(R_n) \subset L^p(\mathcal{R}_{n+1})$. (This follows from Theorem 10.46 and the manner in which the embedding $\mathfrak{i}^{(p)}$ is defined—see Definition 10.47.)

For each $n \in \mathbb{N}$, the space $L_p(\mathcal{R}_n, \tau_n)$ is by Proposition 10.14 linearly isometric to $L^p(\mathcal{R}_n)$. By Remark 12.18 and Theorem 10.46, $\{\mathfrak{j}^{(p)}(a) : a \in \mathfrak{n}(\mathcal{R}_n)_{\widetilde{\varphi}}\}$ will for each n be dense in $L^{2p}(\mathcal{R}_n)$. The claim that $\cup_{n=1}^{\infty} L^p(\mathcal{R}_n)$ is dense in $L^p(\mathcal{R})$ will therefore follow if we are able to verify the claim about weak convergence and also show that $\{\mathfrak{j}^{(p)}(a) : a \in \cup_{n=1}^{\infty}\mathfrak{n}(\mathcal{R}_n)_{\widetilde{\varphi}}\}$ is dense in $L^{2p}(\mathcal{R})$. We first prove that $\{\mathfrak{j}^{(p)}(a) : a \in \cup_{n=1}^{\infty}\mathfrak{n}(\mathcal{R}_n)_{\widetilde{\varphi}}\}$ is dense in $L^{2p}(\mathcal{R})$, which we shall do by modifying the argument of part (a) of Theorem 10.46.

Let $r > 1$ be given so that $1 = \frac{1}{q} + \frac{1}{r}$ where $q = 2p$. Suppose that $z \in L^r(\mathcal{M})$. If we can show that we must have $z = 0$ whenever $tr(z[ah^{1/q}]) = 0$ for each $a \in \cup_{n=1}^{\infty}\mathfrak{n}(\mathcal{R}_n)_{\widetilde{\varphi}}$, then by L^p-duality, $\{\mathfrak{j}^{(p)}(a) : a \in \cup_{n=1}^{\infty}\mathfrak{n}(\mathcal{R}_n)_{\widetilde{\varphi}}\}$ will be weakly dense in $L^q(\mathcal{M})$ and hence norm dense by convexity. So, suppose that we do indeed have $tr(z[ah^{1/q}]) = 0$ for each $a \in \{\mathfrak{j}^{(p)}(a) : a \in \cup_{n=1}^{\infty}\mathfrak{n}(\mathcal{R}_n)_{\widetilde{\varphi}}\}$. Given $b \in \cup_{n=1}^{\infty}\mathcal{R}_n$, the fact that each $\mathfrak{n}(\mathcal{R}_n)_{\widetilde{\varphi}}$ is a left-ideal ensures that $ba \in \cup_{n=1}^{\infty}\mathfrak{n}(\mathcal{R}_n)_{\widetilde{\varphi}}$ for each $a \in \cup_{n=1}^{\infty}\mathfrak{n}(\mathcal{R}_n)_{\widetilde{\varphi}}$ and hence that $tr(z[bah^{1/q}]) = 0$ for each $a \in \cup_{n=1}^{\infty}\mathfrak{n}(\mathcal{R}_n)_{\widetilde{\varphi}}$. Now, notice that for any $a \in \cup_{n=1}^{\infty}\mathfrak{n}(\mathcal{R}_n)_{\widetilde{\varphi}}$ we have $b[ah^{1/q}] \supset bah^{1/q}$. By the uniqueness of the τ-measurable extension, we see that the τ-measurable operator corresponding to the product $b \cdot [ah^{1/q}]$ is just $[bah^{1/q}]$. That means that for any $a \in \cup_{n=1}^{\infty}\mathfrak{n}(\mathcal{R}_n)_{\widetilde{\varphi}}$ and any $b \in \cup_{n=1}^{\infty}\mathcal{R}_n$, we have $0 = tr(z[bah^{1/q}]) = tr([bah^{1/q}]z) = tr(b([ah^{1/q}]z))$. But $\cup_{n=1}^{\infty}\mathcal{R}_n$ is known to be σ-weakly dense in $\mathcal{R}$. Hence, the duality between $L^1(\mathcal{R})$

and $L^\infty(\mathcal{R})$ then ensures that this can only be the case if $[ah^{1/q}]z = 0$ for each $a \in \cup_{n=1}^\infty \mathfrak{n}(\mathcal{R}_n)_{\widetilde{\varphi}}$ or equivalently that

$$z^*(h^{1/q}a^*) = 0 \text{ for each } a \in \cup_{n=1}^\infty \mathfrak{n}(\mathcal{R}_n)_{\widetilde{\varphi}}.$$

We now momentarily fix $n \in \mathbb{N}$. If we apply Theorems 10.11 and 10.46 to the fact centred above, it follows that $z^* L^{2p}(\mathcal{R}_n) = 0$ and hence that $z^*[ah^{1/q}] = 0$ for each $a \in \mathfrak{n}(\mathcal{R}_n)_{\widetilde{\varphi}}$. Fixing $a \in \mathfrak{n}(\mathcal{R}_n)_{\widetilde{\varphi}}$, it is easy to see that $z^*[ah^{1/q}] = 0$ if and only if $|z^*|[ah^{1/q}] = 0$. Since trivially $z^* = 0$ if and only if $|z^*| = 0$, it follows that we may assume that $z^* \geq 0$. Having made this assumption, one may then further note that $z^* = 0$ if and only if $(z^*\chi_{[0,\gamma]}(z^*)) = 0$ for every $\gamma > 0$. Since in this setting the equality $z^*[ah^{1/q}] = 0$ ensures that $0 = \chi_{[0,\gamma]}(z^*)z^*[ah^{1/q}]$ and hence that $0 = (z^*\chi_{[0,\gamma]}(z^*))[ah^{1/q}]$, it follows that we may further assume z^* to be bounded. Hence, given $\xi \in \mathrm{dom}(h^{1/2p}) \subset \mathrm{dom}([ah^{1/q}])$, we then have $z^*a = 0$ in the range of $h^{1/2p}$, which must be dense by the fact that h is non-singular and positive. Therefore, $z^*a = 0$. Since both n and $a \in \mathfrak{n}(\mathcal{R}_n)_{\widetilde{\varphi}}$ were arbitrary, we have managed to show that $z^*a = 0$ for all $a \in \cup_{n=1}^\infty \mathfrak{n}(\mathcal{R}_n)_{\widetilde{\varphi}}$.

Now, recall that the semifiniteness of $\widetilde{\varphi}{\restriction}\mathcal{R}_n$ ensures that $\mathfrak{n}(\mathcal{R}_n)_{\widetilde{\varphi}}$ is σ-weakly dense in $\mathcal{R}_n$. So, $\cup_{n=1}^\infty \mathfrak{n}(\mathcal{R}_n)_{\widetilde{\varphi}}$ is σ-weakly dense in $\cup_{n=1}^\infty \mathcal{R}_n$, and hence in $\mathcal{R}$. This σ-weak density coupled with the boundedness assumption in z ensures that, in fact, $z^* = 0$ (equivalently, $z = 0$) as required.

It remains to show that for $f \in L^p(\mathcal{R})$, $(\mathscr{W}_n^{(p)}(f))$ is weakly convergent to f. The case $p = 1$ is fairly easy. Given $x \in \mathcal{R}$, we may in this case use the facts noted in Remark 12.18 to see that $tr(x\mathscr{W}_n^{(1)}(f)) = tr(\mathscr{W}_n^{(1)}(x\mathscr{W}_n^{(1)}(f))) = tr(\mathscr{W}_n(x)\mathscr{W}_n^{(1)}(f)) = tr(\mathscr{W}_n^{(1)}(\mathscr{W}_n(x)f)) = tr(\mathscr{W}_n(x)f)$. But we know that $(\mathscr{W}_n(x))$ is σ-weakly convergent to x. Hence, as required, $\lim_{n\to\infty} tr(x\mathscr{W}_n^{(1)}(f)) = \lim_{n\to\infty} tr(\mathscr{W}_n(x)f) = tr(xf)$.

Now, suppose that $1 < p < \infty$ and let $q > 1$ be given such that $\frac{1}{p} + \frac{1}{q} = 1$. Fix $n \in \mathbb{N}$ and let $k \in \mathbb{N}$ be given such that $k \geq n$. Since then $L^q(\mathcal{R}_n) \subset L^q(\mathcal{R}_k)$, we will for any $x \in L^q(\mathcal{R}_n)$ have that $tr(x\mathscr{W}_k^{(p)}(f)) = tr(\mathscr{W}_k^{(q)}(x)\mathscr{W}_k^{(p)}(f)) = tr(\mathscr{W}_n^{(1)}(\mathscr{W}_k^{(q)}(x)f)) = tr(\mathscr{W}_k^{(q)}(x)f) = tr(xf)$. This shows that $\lim_{k\to\infty} tr(x\mathscr{W}_k^{(p)}(f)) = tr(xf)$ for all x in the norm dense subspace $\cup_{k\geq 1} L^q(\mathcal{R}_k)$ of $L^q(\mathcal{R})$. Since the sequence $(\mathscr{W}_k^{(p)}(f))$ is norm bounded, this is enough to ensure that, in fact, $\lim_{k\to\infty} tr(x\mathscr{W}_k^{(p)}(f)) = tr(xf)$ for all $x \in L^q(\mathcal{R})$ $\qquad\square$

Remark 13.10 Despite the remarkable fact demonstrated by the preceding result, it should be noted that semifinite and type III algebras cannot produce the same L^p-spaces. This follows from the important work of David Sherman [**She05**] who showed if for some $1 \leq p < \infty$ ($p \neq 2$) the $L^p(\mathcal{M}_0)$ and $L^p(\mathcal{M}_1)$ are linearly isometric, then the underlying algebras themselves are Jordan *-isomorphic.

Chapter 14
Applications to quantum physics

The bulk of the material in this chapter was condensed from the papers [**LM11**, **LM22**, **ML14**, **ML20**].

14.1 Introductory comments

In order to ensure that a larger audience will be able to appreciate the selected applications presented later on in this chapter, we first pause to sketch the historical development of quantum theory and in particular of how the mathematical technology we consider here fits into the picture. Along the way, we will give some pointers to what we believe to be a healthy approach to the task at hand and argue that it is essential for the Heisenberg approach to be studied for its own sake, with von Neumann algebras forming an indispensable part of such a study. We shall ultimately see that for this task, all three types of von Neumann algebras need to be considered, with the Orlicz spaces associated with these algebras forming an integral part of such a study. In this introduction, we borrowed heavily from the insightful 'study guide' by Adam Majewski [**Maj17**]. Readers wishing to have a fuller introduction to the rudiments of mathematical modelling of quantum mechanics will find one of [**Dim11**] or [**Emc09**] to be helpful. Among these, [**Dim11**] is arguably more suitable for mathematicians and [**Emc09**] for physicists. More advanced readers who are looking for detailed introductions to the applications of von Neumann algebras to quantum statistical mechanics and quantum field theory (QFT) could, respectively, refer to [**BR87**, **BR97**] and [**Ara99**]. Here, [**BR87**] provides the standard algebraic 'kit' for a description of quantum systems, with [**BR97**] devoted to applications. In particular, in the first part of [**BR97**, Chapter VI], the algebraic machinery was used for a description of lattice systems, with the second part indicating that continuous systems need more sophisticated tools. This monograph has been designed to provide precisely the tools from noncommutative analysis, which appear to be very useful for

Noncommutative measures and L^p and Orlicz Spaces, with Applications to Quantum Physics. Stanisław Goldstein and Louis Labuschagne, Oxford University Press. © Stanisław Goldstein and Louis Labuschagne (2025).
DOI: 10.1093/oso/9780198950202.003.0016

a description of both statistical physics (in the thermodynamic limit) as well as QFT (Quantum Field Theory).

In order to accommodate readers who like to 'jump ahead' to see 'how the story ends', we have at various points included heuristic descriptions of some of the more advanced machinery used in this chapter. We hope that this will enable a wide audience to, at an intuitive level, appreciate the cut and thrust of the theory developed here without first having to wade through all of the preceding chapters. Of course, a full comprehension of the proofs and of the subtleties of the developed theory will require familiarity with various parts of the preceding chapters.

The first rigorous framework for the study of quantum mechanics to be formulated was the so-called Heisenberg approach, which uses noncommutativity as the guiding principle. Although it was Heisenberg himself who first noticed the importance of noncommutativity in quantum mechanics, it was Born, Jordan and Dirac who, independently of each other, almost simultaneously made the connection to matricial structures. Heisenberg had initially formulated his approach in terms of what he called 'tables'. Legend has it that while discussing this approach with Born on a train journey, Jordan—who was well-trained in mathematics—realised that these tables are nothing but matrices. This approach was then fleshed out in circa 1922–1925 with the combined efforts of Born, Dirac, Heisenberg and Jordan. The formulation of this approach was then very quickly followed by the appearance of Erwin Schrödinger's wave function approach in 1926. The nature of the two approaches can best be illustrated by some very simple examples. As far as the Heisenberg approach is concerned, for one particle in one dimension, the position and momentum operators ($\widehat{x}$ and $\widehat{p}$) on $L^2(\mathbb{R})$ are densely defined self-adjoint operators, defined respectively by $\widehat{x}(f)(x) = xf(x)$ and $\widehat{p}f = -\hbar\frac{df}{dx}$ where $\hbar$ is the reduced Planck constant. It is easy to see that the so-called *Heisenberg relation*, namely

$$\widehat{xp} - \widehat{px} = \hbar\mathbb{1},$$

holds on the domain of $\widehat{xp} - \widehat{px}$. Noncommutativity is therefore from the start a very prominent ingredient of this approach. To get an idea of the nature of the Schrödinger (wave function) approach, we mention the position-space Schrödinger equation for a non-relativistic particle in one dimension, namely

$$i\hbar\frac{\partial\psi}{\partial x}(x,t) = \left[-\frac{\hbar^2}{2m}\frac{\partial^2}{\partial x^2} + V(x,t)\right]\psi(x,t)$$

where t is the time variable, m the mass of the particle and V a potential. Here, partial differential equations are clearly much more prominent. When the Hamiltonian satisfies the requirements of the Stone–von Neumann theorem, these two approaches can be seen to be equivalent in the sense that for a given observable in a given state both will yield the same expected values. This fact together with the fact that in the early part of the 20th century the mathematical community had a clear grasp on partial differential equations while the theory of Operator Algebras was by contrast very much in its infancy lead to the Schrödinger approach very quickly achieving almost astronomical popularity in comparison with its noncommutative sibling. We are therefore left with the question of whether a direct study of the Heisenberg approach is still justified.

Although at the level of elementary quantum mechanics the average values predicted by the two approaches agree, a purely wave function approach lacks the rich algebraic structure inherent in the Heisenberg approach. In fact later in life, Dirac himself very strongly insisted that in the context of, for example, QFT, the two approaches should not be regarded as equivalent and that here the Heisenberg approach should certainly be preferred. Dirac's objection was based on the fact that here the proof that the two pictures yield the same average values requires a certain amount of regularity from the Hamiltonian. In more complex quantum systems, such as those that may be found in QFT, *such regularity need not hold*. He further pointed out that in many of these types of models, the Heisenberg approach still yields a workable theory, whereas the Schrödinger approach will by contrast leave one with an unsolvable differential equation. The interested reader will find a full account of a more mature Dirac's position on this matter in the introductory section of [**Dir66**]. Dirac did express the hope that some way would be found to bring the Schrödinger approach 'back into the game' so to speak as far as QFT was concerned. More recently, in his plenary address at the 2000 International Congress of Mathematical Physics, Alain Connes on a similar note pointed out the severe limitations of the Schrödinger approach as a formalism for dealing with his own personal novel approach to the topic. Hence, although the bulk of the quantum physical literature at present seems to favour the Schrödinger wave function approach, the Heisenberg approach should be studied for its own sake and dare not be ignored.

The appearance of Paul Dirac's influential monograph *The Principles of Quantum Mechanics* [**Dir47**] in 1930 very strongly pointed to operators on Hilbert space as a home for quantum theory. Inspired by the Canonical Commutation Relations of which the Heisenberg relation is a particular example, many mathematical physicists spent many years trying to find matricial objects a and b, satisfying equation $ab - ba = \mathbb{1}$. It was only in 1947 that Wintner [**Win47**] succeeded in showing that for $B(H)$, it is *impossible* to formulate such an equation in a 'bounded' context. Shortly thereafter, Wielandt [**Wie49**] gave an embarrassingly simple proof of this fact, valid in a much more general context. The reader will find an elegant analysis of developments around this discovery in [**Sak91**, §2.2]. For the sake of interest, we present Wielandt's proof below.

Proof.

1. Assume by way of contradiction that we can indeed find $n \times n$ matrices a and b for which $ab - ba = \mathbb{1}$. Then, a and b can clearly not be 0.

2. Then, $ab^2 - b^2 a = (ab - ba)b + b(ab - ba) = 2b$, from which it inductively follows that $ab^3 - b^3 a = (ab^2 - b^2 a)b + b^2(ab - ba) = 3b^2$, next $ab^4 - b^4 a = (ab^3 - b^3 a)b + b^3(ab - ba) = 4b^3$, etc. Therefore, we have $ab^n - b^n a = nb^{n-1}$ for all $n \in \mathbb{N}$.

3. Now, notice that for N with $N > 2\|a\|\,\|b\|$, the general inequality $n\|b^{n-1}\| = \|ab^n - b^n a\| \le 2\|a\|\,\|b\|\,\|b^{n-1}\|$ can only be true if $b^{N-1} = 0$. But if $b^{N-1} = 0$, then by (2), we have that $b^{N-2} = \frac{1}{N-1}(ab^{N-1} - b^{N-1}a) = 0$.

4. We may now continue inductively until we get $b = 0$, which contradicts the observation made in (1). $\qquad\square$

The clear message from the Wielandt–Wintner theorem is that bounded contexts cannot accommodate equations like the Heisenberg uncertainty relation and hence are unsuitable for the modelling of quantum theory. Put differently, if we are to be true to the Heisenberg approach, then everything that is quantum is noncommutative, but not everything that is noncommutative is quantum. We need a formalism which is infinite-dimensional and which can also accommodate unboundedness. In Dirac's conception of elementary quantum mechanics, observables are identified with the self-adjoint portion of $B(H)$ and the (pure) states with one-dimensional subspaces of H. For an observable $a = a^* \in B(H)$ and a state (one-dimensional subspace) spanned by a unit vector x, the expected value of observable a in state x is then given by $\omega_{x,x}(a) = \langle ax, x \rangle$. The dynamical evolution is determined by a self-adjoint operator $\mathbb{H}$ (the Hamiltonian) through either of the prescriptions: $a_t = e^{it\mathbb{H}} a e^{-it\mathbb{H}}$ for observables or $x_t = e^{-it\mathbb{H}} x$ for states. A mere two years after the appearance of Dirac's monograph von Neumann published his iconic *Mathematische Grundlagen der Quantenmechanik* [**vN96**] which by contrast put the emphasis on unbounded operators as far as observables are concerned, which is more in line with what Wielandt and Wintner later indicated. The identification of $L^{\cosh -1}$ as a home for regular observables in Section 14.4 and Subsection 14.5.3 addresses, in a sense, the same issue by similarly making room for unboundedness.

In the Dirac formalism, the (pure) states may, of course, be equally identified with the set of vector states of $B(H)$. To allow for convex combinations of states (mixed states) and limiting behaviour, one should properly pass to the σ-weakly closed convex hull of the vector states, which is the set of normal states of $B(H)$. Hence, we are led to $B(H)_*$ as a natural home for the full set of states. By means of Tr-duality (Theorem 2.19), $B(H)_*$ may be identified with $\mathscr{S}_1(H)$, with each normal state $\omega \in B(H)_*$ then realised by a corresponding 'density matrix' $h_\omega \in \mathscr{S}_1^+(H)$ in the sense that $\omega = \mathrm{Tr}(h_\omega \cdot)$. So, already at the elementary level, we have a von Neumann algebra $B(H)$, an a priori given f.n.s. weight (trace) Tr and a matching noncommutative L^1-space $\mathscr{S}_1(H)$ entering into the picture. As can be seen from chapter 2, the pair $(B(H), \mathscr{S}_1(H))$ plays the role of noncommutative $(\ell^\infty(S), \ell^1(S))$ with the spaces $\mathscr{S}_p(H)$ $(1 \le p < \infty)$ nothing more than $\{a \in B(H) : \mathrm{Tr}(|a|^p) < \infty\}$ equipped with the norm $\| \cdot \|_p = \mathrm{Tr}(|\cdot|^p)^{1/p}$.

But what about more complex systems? Dirac's $B(H)$-based formalism is very elegant. Can't we just stick to the $B(H)$ framework described above (suitably modified to allow for unbounded observables)? To answer this question, we turn to the quantum Gibbs Ansatz. The classical Gibbs Ansatz was designed to describe a (classical) canonical equilibrium state which, up to a normalised constant, is given by $e^{-\beta \mathcal{H}}$ where $\mathcal{H}$ stands for the Hamiltonian of the considered system and β the 'inverse' temperature. A quantisation of this formalism would then amount to simply replacing the classical Hamiltonian with its quantum counterpart. For the result to be a state in the sense described above, the operator $e^{-\beta \mathbb{H}}$ (where $\mathbb{H}$ is now a quantum Hamiltonian) must then be an element of $\mathscr{S}_1(H)$ - a compact operator. But this can only happen if $\mathbb{H}$ has a pure point spectrum with accumulation point at infinity. But not even the Hamiltonian of the hydrogen atom—arguably the simplest possible model—fulfils this requirement!

We therefore need something more general but still true to the underlying philosophy of Dirac's formalism. $B(H)$ is of course the iconic example of a noncommutative von Neumann algebra. So, at this point, it is very compelling to suggest that more general systems could possibly be described by a more general triple consisting of a von Neumann algebra $\mathcal{M}$, an f.n.s. weight φ and some sort of matching L^1 space constructed from the pair $(\mathcal{M}, \varphi)$. Von Neumann himself realised that further structure is needed if the challenges of quantum mechanics are to be successfully engaged, as can be seen from, for example, [**VN39**, **VN37**] where precisely such structures are introduced. In the latter paper, he introduced symmetric norms on matrix algebras in a manner which anticipates the modern theory of noncommutative L^p and related spaces. Von Neumann algebras are for the most part infinite-dimensional and through the device of affiliated operators certainly able to accommodate unbounded operators. So, when equipped with the kind of additional structure that von Neumann envisioned, these algebras are a compelling option for modelling quantum mechanics. But which von Neumann algebras can we reasonably expect to play a role in quantum theory?

When Heisenberg, Born, Jordan and Dirac realised that noncommutativity is the leitmotif of quantum mechanics, they introduced the so-called process of *canonical quantisation*. This is a formal process to quantify classical Hamiltonian mechanics. We pause to present a rudimentary overview of Hamiltonian mechanics. Given some open region $\mathcal{O}$ in $\mathbb{R}^n$, the associated *phase space* would be $\mathcal{O} \times \mathbb{R}^n$ where the points in the phase space have the form (x, p) with $x \in \mathcal{O}$ describing the location of various objects and $p \in \mathbb{R}^n$ the associated momenta. Time evolution is then described by *trajectories* in phase space which are functions of the form $\mathbb{R} \to \mathcal{O} \times \mathbb{R}^n : t \mapsto (x(t), p(t))$ that are required to obey Hamilton's equations, namely

$$\frac{dx_i}{dt} = \frac{\partial \mathcal{H}}{\partial p_i} \qquad \frac{dp_i}{dt} = -\frac{\partial \mathcal{H}}{\partial x_i}$$

where $\mathcal{H} = \mathcal{H}(x, p)$ is a function on the phase space called the *Hamiltonian*. In this framework, a classical observable F will just be a smooth function on the phase space. The validity of Hamilton's equations then ensure that

$$\frac{dF}{dt} = \sum_{i=1}^{n} \frac{\partial F}{\partial x_i}\frac{dx_i}{dt} + \frac{\partial F}{\partial p_i}\frac{dp_i}{dt} = \sum_{i=1}^{n} \frac{\partial F}{\partial x_i}\frac{\partial \mathcal{H}}{\partial p_i} - \frac{\partial F}{\partial p_i}\frac{\partial \mathcal{H}}{\partial x_i}.$$

In abbreviated form this formula is usually written as $\frac{dF}{dt} = \{F, \mathcal{H}\}$ (the so-called Poisson bracket) where the right-hand side is evaluated at $(x(t), p(t))$. The canonical quantisation of a Hamiltonian system with phase space $\mathbb{R}^n \times \mathbb{R}^n$ then consists of replacing the points (x, p) with the operators $(\widehat{x}, \widehat{p}) = (\widehat{x}_1, \ldots, \widehat{x}_n, \widehat{p}_1, \ldots, \widehat{p}_n)$, where the $\widehat{x}_i$s and $\widehat{p}_j$ are the position and momentum operators described earlier and then replacing the Poisson brackets with the commutators of these operators. So, in the process of canonical quantisation, the classical facts

$$\{x_i, x_j\} = 0 = \{p_i, p_j\}, \quad \{x_i, p_j\} = \delta_{ij}1, \qquad i, j = 1, 2, 3, \ldots \quad (14.1)$$

get replaced by

$$[\widehat{x}_i, \widehat{x}_j] = 0 = [\widehat{p}_i, \widehat{p}_j], \quad [\widehat{x}_i, \widehat{p}_j] = i\hbar\delta_{ij}, \qquad i, j = 1, 2, 3\ldots \quad (14.2)$$

with the classical Hamiltonian $\mathcal{H}(x, p)$ replaced by the quantum Hamiltonian $\mathbb{H} = \mathcal{H}(\widehat{x}, \widehat{p})$. For each such set of commutation relations, one can then construct an associated von Neumann algebra by first passing to the unitary groups $U_j(t) = e^{i\widehat{p}_j t}$ and $V_k(t) = e^{i\widehat{x}_k t}$ and then generating a von Neumann algebra from these operators. In terms of these unitaries, the commutation relations become the claim that

$$U_j(s)V_k(t) = e^{ist\delta_{jk}}V_k(t)U_j(s), \qquad [U_j(s), U_k(t)] = 0 = [V_j(s), V_k(t)].$$

The Borel functional calculus then ensures that the family of position and momentum operators we started with is affiliated to the resultant algebra.

A system will be called *small* if, as in the above example, the position and momentum 'variables' enjoy a finite number of degrees of freedom and *large* if they enjoy an infinite number of degrees of freedom. Alongside the process of canonical quantisation, one has the so-called uniqueness theorem, attributed to von Neumann, Weyl and Rellich, which states that for small systems the relations (14.2) will up to unitary equivalence have a unique representation. So, for small systems, the appropriate von Neumann algebra we should use for further analysis of quantum systems is provided by canonical quantisation. For large systems (encompassing both statistical mechanics and field theory), the situation is very different. There are *plenty* of inequivalent representations of the relations (14.2) when the number of degrees of freedom is infinite. This then raises the hope that, for large systems, the 'physically reasonable' representations will turn out to be more 'regular' algebras. We pause to clarify this rather enigmatic statement.

We know from Theorem 0.127 that von Neumann algebras come in three types with type III characterised by the fact that they do not admit an f.n.s. trace (see Theorem 3.38). At the level of dimension functions on projection lattices, this translates to very 'unnatural' behaviour of such functions on type III factors, as noted in Theorem 3.4. So, for a long time, type III algebras were, especially in mathematical physics, considered to be very exotic. The existence of f.n.s. traces on algebras of types I and II in contrast ensured that much of the $B(H)$ structure we noted earlier could by similar techniques be carried over to these algebras. More specifically, the existence of an f.n.s. trace τ for such an algebra $\mathcal{M}$ enables us to enlarge the algebra to the algebra $\widetilde{\mathcal{M}}$ of τ-measurable operators (see Sections 3.4 and 3.5), which is a concrete algebra representing a noncommutative version of the completion of L^∞ with respect to the topology of convergence in measure. Using the pair $(\widetilde{\mathcal{M}}, \tau)$ as the starting point, we may then in a similar fashion as before define $L^p(\mathcal{M}, \tau)$ $(1 \leq p < \infty)$ to be the space $\{a \in \widetilde{\mathcal{M}} : \tau(|a|^p) < \infty\}$ equipped with the norm $\|\cdot\|_p = \tau(|\cdot|^p)^{1/p}$. As before, we then have that $\mathcal{M}_*$ corresponds to $L^1(\mathcal{M}, \tau)$ in the sense described in Theorem 6.24. The structure we have noted for the pair $(B(H), \mathscr{S}_1(H))$ then becomes a special case of a theory valid for all algebras of type I and II. Hence, for this family, the intuition gained from Dirac's original $B(H)$-based model will serve us well. With large systems admitting many inequivalent representations, the hope therefore was that all 'physically reasonable' representations of such systems would turn out to be of type I or II.

However, the 1967 work of Powers [**Pow67**] on representations of uniformly hyperfinite algebras resulted in the assumption that type III algebras were exotic being completely abandoned. In 'physical' terms,

Powers' results can be expressed as an analysis of a one-dimensional spin chain. Such a model consists of an infinite number of sites, with the algebra $M_2(\mathbb{C})$ associated with each site. Thus, local observables associated with a site are given by elements from $M_2(\mathbb{C})$. The local equilibrium at each site is given by a 2×2 matrix of the form $Z^{-1}e^{-\beta \mathbb{H}_{loc}}$ where Z is the normalising constant and $\mathbb{H}_{loc} \in M_2(\mathbb{C})$ is the local Hamiltonian associated with a site. Powers in essence showed that the thermodynamic limit of such 'small' systems lead to type III von Neumann algebras. Subsequent results of Araki-Woods, Hugenholtz et al. then confirmed that type III von Neumann algebras are typical in the study of large systems; see [**Yng05**]. The paper by Yngvason [**Yng05**] refers to the local algebras of algebraic quantum field theory (AQFT). With regard to type III factors in AQFT, one actually does not even need all axioms of AQFT to obtain that local algebras are type III. The key reason behind this phenomenon is the independence conditions implementing Einstein's causal principle. For example, a pair $(\mathcal{M}, \mathcal{M}')$ of von Neumann factors is strictly local in a vector state sense if and only if $\mathcal{M}$ is of type III (see [**Ham03**, Corollary 11.4.4] and the discussion on this topic in the last chapter of the same reference).

The ubiquitousness of type III structures may also be seen from Glimm's theorem (see [**Ped79**, Theorem 6.8.7]) stating that any C^*-algebra which is not of type I as a C^*-algebra must have a factor representation of type III. Since representations have physical meaning (as sectors of physical systems), type III factors are therefore ubiquitous in physical models. The reader should take heed that the C^*-algebraic notion of 'type I' being referred to here is very different to the von Neumann algebraic notion bearing the same name. A type I von Neumann algebra is, in general, NOT type I as a C^* algebra. The definition of type I C^*-algebras being referred to here corresponds to what Blackadar calls 'internally type I' in [**Bla06**, Definition IV.1.1].

We further note the work of Summers and Werner [**SW87**] who showed that under mild restrictions Bell's inequalities are maximally violated by the local algebras of AQFT in every single vector state of a wedge-shaped region of Minkowski spacetime [**SW87**]. This discovery also motivated further development regarding these algebras formerly regarded as 'exotic'.

The necessity of type III algebras then confronts us with the challenge of showing that even for these algebras $\mathcal{M}_*$ may be realised as some sort of space of 'density matrices'. With no convenient trace to work with, this is akin to trying to integrate with no obvious well-behaved integral in sight. This challenge can therefore clearly only be met by the development of some very heavy-duty machinery which we shall now elucidate. In doing so, we shall mildly abuse notation and suppress some technicalities for the sake of clarity. The first step to achieve this objective is to enlarge (a copy of) $\mathcal{M}$ to the algebra $\mathfrak{M} = \mathcal{M} \rtimes_\varphi \mathbb{R}$. (Here, φ is a reference f.n.s. weight on $\mathcal{M}$ and $\mathfrak{M}$ the crossed product with the modular group of φ.) The extended algebra admits both a dual action $s \mapsto \theta_s$ of $\mathbb{R}$ and a canonical f.n.s. trace τ characterised by the equality $\tau \circ \theta_s = e^{-s}\tau$. This trace then enables us to further enlarge $\mathfrak{M}$ to the algebra $\widetilde{\mathfrak{M}}$ of τ-measurable operators. Although $\mathcal{M}$ is in general not an expected subalgebra of $\mathfrak{M}$, there does exist an 'expectation-like' operator $\mathscr{W}$ from the extended positive part of $\mathfrak{M}$ onto that of $\mathcal{M}$ which is faithful normal and semifinite. The prescription $\rho \mapsto \widetilde{\rho} = \rho \circ \mathscr{W}$ can then be shown to realise a bijection from semifinite normal weights on $\mathcal{M}$ to θ_s-invariant semifinite normal weights ψ on $\mathfrak{M}$ (Theorem 9.24). We may now use our earlier technology for traces on semifinite algebras to put in place a further bijection from such θ_s-invariant normal semifinite weights $\widetilde{\rho}$, to positive operators h_ρ affiliated to $\mathfrak{M}$ satisfying $\widetilde{\rho} = \tau(h_\rho^{1/2} \cdot h_\rho^{1/2})$ and $\theta_s(h_\rho) = e^{-s}h_\rho$ for all s (Proposition 9.33). For the reference weight φ on $\mathcal{M}$, we shall drop the subscript here and simply write h for the associated density. In fact, we see that the original weight ρ is a normal state of $\mathcal{M}$ if and only if $h_\rho \in \widetilde{\mathfrak{M}}$ (Corollary 9.37). Therefore, it makes sense to define $L^1(\mathcal{M})$ as $L^1(\mathcal{M}) = \{a \in \widetilde{\mathfrak{M}}: \theta_s(a) = e^{-s}a\}$ with positive normal functionals that then, by definition, correspond to the cone of positive elements in $L^1(\mathcal{M})$. The final two steps are to then show that $L^1(\mathcal{M})$ admits a tracial functional tr (different from τ) in terms of which $L^1(\mathcal{M})$ is a Banach space when equipped with the norm $\|\cdot\|_1 = tr(|\cdot|)$ and further that the prescription $\rho \mapsto h_\rho$ extends to a linear isometry from $\mathcal{M}_*$ to $L^1(\mathcal{M})$ for which we have that $\rho = tr(h_\rho \cdot)$ for each $\rho \in \mathcal{M}_*^+$ (Proposition 10.29).

At this point, we need to sound a warning. The construction method described above ensures that each $L^p(\mathcal{M})$ (where $0 < p < \infty$) is of the form $L^p(\mathcal{M}) = \{a \in \widetilde{\mathfrak{M}}: \theta_s(a) = e^{-s/p}a\}$. We may think of the action of θ_s on $L^p(\mathcal{M})$ as the 'phase' of this space as a subspace of $\widetilde{\mathfrak{M}}$. The fact that different $L^p(\mathcal{M})$-spaces have different 'phases', then ensures that from a purely set-theoretical point of view, we have $L^p(\mathcal{M}) \cap L^q(\mathcal{M}) = \{0\}$ whenever $p \neq q$. So, if in this context we want to identify an object that is a true analogue of say $L^1 \cap L^2$ for L^p-spaces of classical measure spaces, we first need to 'shift' the spaces $L^1(\mathcal{M})$ and $L^2(\mathcal{M})$ so that they have the same 'phase' before taking the intersection. Such a shift can be effected by multiplying on the left and right with a suitable operator constructed from the density h that corresponds to the reference weight φ. In Definition 14.3 and Proposition 14.14, we perform precisely such shifts in order to be able to properly compare the spaces under consideration there. (In Definition 14.3, we change L_+^{ent} so it appears as a subspace of L^1.)

As far as Orlicz spaces are concerned, we will in the remainder of this chapter consider various examples showing how Orlicz spaces feature in quantum theory. These examples will show that such spaces are not a 'fringe phenomenon', but that they rather prove to be crucial in dealing with deeply foundational issues such as identifying states that have 'good' entropy and also in finding a natural home for 'unbounded' observables. We shall further see that these kinds of applications of Orlicz spaces feature in both Quantum Statistical Mechanics and QFT and hence that they seem to be endemic to quantum theory as a whole.

14.2 Entropy for general quantum systems

The key mathematical result of this section is Proposition 14.1. Readers who are willing to take the facts in the introductory discussion at face value may, therefore, benefit from this section once they have mastered part II of the book, the semifinite setting. In fact, the comment following this proposition needs little more than a familiarity with the Schatten-von Neumann classes. The closing part of this section from Remark 14.2 onwards is concerned with extending the concept of entropy to also include the setting of type III von Neumann algebras. For this familiarity with part III—the general setting—is essential.

The first question we should answer before embarking on an analysis of entropy is exactly what entropy is. The answer to this question is not as clear as one might think. Claude Shannon—the father of Shannon entropy—relates the following anecdote: *'My greatest concern was what to call it. I thought of calling it "information", but the word was overly used, so I decided to call it "uncertainty". When I discussed it with John von Neumann, he had a better idea. Von Neumann told me "You should call it entropy, for two reasons. In the first place your uncertainty function has been used in statistical mechanics under that name, so it already has a name. In the second place, and more important, no one really knows what entropy really is, so in a debate you will always have the advantage'".*

Wikipedia offers the following very elegant and deeply satisfying definition of entropy: *In the modern microscopic interpretation of entropy in statistical mechanics, entropy is the amount of additional information needed to specify the exact physical state of a system, given its thermodynamic specification. Therefore, it is often said that entropy is an expression of the disorder, or randomness of a system, or of our lack of information about it.* More precisely, entropy is the measure of a system's thermal energy per unit temperature that is unavailable for doing useful work. Because work is obtained from ordered molecular motion, the amount of entropy is also a measure of the molecular disorder, or randomness, of a system.

We emphasise that this concept is a fundamental ingredient of the second law of thermodynamics and as such is indispensable for the description of the fundamental properties of nature.

The origins of a quantity representing something like entropy may be found in the work of Ludwig Boltzmann. In his study of the dynamics of rarefied gases, Boltzmann formulated the so-called *spatially homogeneous Boltzmann equation* as far back as 1872, namely

$$\frac{\partial f_1}{\partial t} = \int d\Omega \int d^3 v_2 I(g,\theta) |\mathbf{v}_2 - \mathbf{v}_1| (f_1' f_2' - f_1 f_2)$$

where $f_1 \equiv f(\mathbf{v}_1, t), f_2' \equiv f(\mathbf{v}_2', t), \ldots$, are velocity distribution functions, $I(g,\theta)$ denotes the differential scattering cross section, $d\Omega$ the solid angle element and $g = |\mathbf{v}|$.

The natural Lyapunov-type functional for this equation is the so-called Boltzmann H-function, which is

$$H_+(f) = \int f(x) \log f(x) dx$$

where f is a postulated solution of the Boltzmann equation. The connection to entropy may be seen in the fact that the classical description of continuous entropy S differs from the functional H only by sign. Hence, Boltzmann's H-functional may be viewed as the first formalisation of the concept of entropy.

The challenge we face is to find a mathematical formalism which will allow for the extension of these and other mathematical physical concepts to the most general quantum mechanical frameworks. Inspired by the work of Boltzmann, von Neumann responded to this challenge by, in the context of $B(H)$, expressing the entropy as $\mathrm{Tr}(\rho \log(\rho))$ where ρ is a norm 1 element of $L^1_+(B(H))_+ = \mathscr{S}^1_+(H)$ (a so-called density matrix) representing the state of the system, where $\mathscr{S}^1(H)$ denotes the trace class operators on H.

We remind the reader that in Dirac's elementary quantum-mechanical formalism, the observables are represented by self-adjoint elements of $B(H)$ and the states by norm 1 positive elements of $L^1(B(H)) = \mathscr{S}^1(H)$. Von Neumann's description of entropy for this context works reasonably well, in the sense that for any density matrix ρ, a fixed value may be assigned to $\mathrm{Tr}(\rho \log(\rho))$—albeit possibly infinite. However, even here further analysis is required: the ubiquitous presence of density matrices with infinite von Neumann entropy would at first sight seem to be an obstacle to a dynamics which allows a return to equilibrium, but at least in this setting each density matrix admits a value for entropy. Based on the relative success of the $B(H)$ model, it is tempting to for more general von Neumann algebras also use the pair $\langle L^1(\mathcal{M}), \mathcal{M} \rangle$ as the home for states and observables. However, there are some serious problems with this approach, namely that for more general tracial von Neumann algebras $\mathcal{M}$, the quantity $\tau(\rho \log(\rho))$ can be extremely badly behaved with respect to the L^1-topology. So, $B(H)$ is somewhat exceptional in this regard! The challenge we face is therefore to find a formalism, which is a natural extension of the Dirac-von Neumann picture, but which also allows for states with 'good' entropy. In some sense, the challenge is then to find the 'correct' home for states with 'good' entropy. As we shall see, a space that presents itself as a possible solution to this problem is the space $(L^1 \cap L\log(L+1)(M, \tau)$. (Here, $L\log(L+1)(M, \tau)$ is the Orlicz space corresponding to the Young function $\Psi(t) = t\log(t+1)$.)

Regarding the space $L\log(L+1)$, it is worth noting that DiPerna–Lions (see [**DL88**], [**DL89**] and [**AV02**]) showed that the estimates

$$f \in L^\infty_t([0, T]; L^1_{x,v}((1 + |v|^2 + |x|^2)\, dx\, dv) \cap L\log(L+1)) \qquad (14.3)$$

and

$$D(f) \in L^1([0, T] \times \mathbb{R}^N_x), \qquad (14.4)$$

where $D(f) = \frac{1}{4} \int d\Omega \int d^3v_1 d^3v_2 I(g,\theta)|\mathbf{v}_2 - \mathbf{v}_1|(f'_1 f'_2 - f_1 f_2) \log \frac{f'_1 f'_2}{f_1 f_2}$, are sufficient to build a mathematical theory of weak solutions to Boltzmann's equation. (Lions, who received the Fields medal in 1994 for his work on nonlinear partial differential equations, was the first person to give a complete solution to the Boltzmann equation with proof.) Furthermore, Villani announced, see [**Vil02**, Chapter 2, Theorem 9], that for particular cross sections (collision kernels in Villani's terminology), weak solutions of Boltzmann's equation are actually in $L\log(L+1)$.

Let $\mathcal{M}$ be a von Neumann algebra equipped with a faithful normal semifinite trace τ and let f be a positive operator affiliated with $\mathcal{M}$. By Remark 6.39, $f \in L\log(L+1)(\mathcal{M},\tau)$ if and only if we can find some $\epsilon > 0$ for which $\tau((f/\epsilon)\log((f/\epsilon)+\mathbb{1})) < \infty$. Similarly, $f \in L^1(\mathcal{M})$ if and only if $\tau(f) < \infty$. Therefore, we clearly have $|\epsilon\tau((f/\epsilon)\log((f/\epsilon)+\mathbb{1})) + \log(\epsilon)\tau(f)| < \infty$ if and only if $f \in (L^1 \cap L\log(L+1)(M,\tau))$. Notice that given such an $\epsilon > 0$, we may then use logarithmic rules and the Borel functional calculus to see that

$$\tau(f\log(f+\epsilon\mathbb{1})) = \epsilon\tau((f/\epsilon)\log((f/\epsilon)+\mathbb{1})) + \log(\epsilon)\tau(f).$$

Thus, it is precisely the non-negative elements of the Orlicz space $(L^1 \cap L\log(L+1)(M,\tau)$, for which $\tau(f\log(f+\epsilon\mathbb{1}))$ is finite for some ϵ. We pause to note that we may write the above equality as

$$\tau(f\log(f+\epsilon\mathbb{1}) = \epsilon I_{\log}(f/\epsilon) + \log(\epsilon)\|f\|_1 \tag{14.5}$$

where $I_{\log}(f) = \tau(|f|\log(|f|+\mathbb{1}))$ is the canonical modular for the space $L\log(L+1)$. This then identifies this space as the home for states with 'good' entropy; a suspicion which is borne out by the following proposition:

Proposition 14.1 ([ML14]) *Let $\mathcal{M}$ be a semifinite algebra equipped with a faithful normal semifinite trace τ and $f \in L^1 \cap L\log(L+1)$ $(\mathcal{M},\tau)$ with $f \geq 0$. Then $\{f\log(f+\epsilon\mathbb{1}): \epsilon > 0\} \subset L^1(\mathcal{M},\tau)$ with $f\log(f) \leq f\log(f+\epsilon_0\mathbb{1}) \leq f\log(f+\epsilon_1\mathbb{1})$ whenever $0 < \epsilon_0 \leq \epsilon_1$. The quantity $\tau(f\log(f)) = \tau(f\log(f+\epsilon\mathbb{1})) - \tau(f\log(f+\epsilon\mathbb{1}) - f\log(f))$ ($\epsilon > 0$) is therefore a well-defined element of $[-\infty,\infty)$ and independent of the choice of $\epsilon > 0$. Moreover*

$$\tau(f \log f) = \inf_{\epsilon > 0} \tau(f \log(f + \epsilon)).$$

Proof. Let $f \in L^1 \cap L\log(L+1)(\mathcal{M}, \tau)$ be given with $f \geq 0$. In the discussion preceding the statement of the proposition, we saw that there exists some $\delta > 0$ for which both $\tau((f/\delta)\log((f/\delta) + \mathbb{1}))$ and $\tau(f)$ are finite. One may then use logarithmic rules and the Borel functional calculus to see that

$$\begin{aligned}
\tau(|f\log(f + \delta\mathbb{1})|) &= \tau(|[\delta(f/\delta)\log((f/\delta) + \mathbb{1})] + [\log(\delta)f]|) \\
&\leq \delta\tau((f/\delta)\log((f/\delta) + \mathbb{1})) + |\log(\delta)|\tau(f) \\
&< \infty.
\end{aligned}$$

Given any $\epsilon > \delta$, we may use the fact that $t \mapsto t\log(t + 1)$ is non-negative and increasing, to see that $0 \leq \tau((f/\epsilon)\log((f/\epsilon) + \mathbb{1})) \leq \tau((f/\delta)\log((f/\delta) + \mathbb{1})) < \infty$. We show that we may assume $\delta > 0$ to be arbitrarily small, from which it then follows that $\tau((f/\epsilon)\log((f/\epsilon) + \mathbb{1})) < \infty$ for any $\epsilon > 0$. To see that this is the case, observe that for any $t \geq 0$, we will have that $2t\log(2t+1) \leq 2t\log(t^2 + 2t + 1) = 4t\log(t+1)$. But then $\tau((f/(\delta/2))\log((f/(\delta/2)) + \mathbb{1})) \leq 4\tau((f/\delta)\log((f/\delta) + \mathbb{1})) < \infty$. Thus, δ may be replaced by $\delta/2$. Iterating the process proves the claim.

Therefore, it follows that $(f/\epsilon)\log((f/\epsilon) + \mathbb{1}) \in L^1(\mathcal{M}, \tau)$ for any $\epsilon > 0$. Given that by assumption $f \in L^1(\mathcal{M}, \tau)$, this proves that for any $\epsilon > 0$, we have $\epsilon(f/\epsilon)\log((f/\epsilon)+\mathbb{1})+\log(\epsilon)f = f\log(f+\epsilon\mathbb{1}) \in L^1(\mathcal{M}, \tau)$ as required. The fact that $f\log(f) \leq f\log(f + \epsilon_0\mathbb{1}) \leq f\log(f + \epsilon_1\mathbb{1})$ whenever $0 < \epsilon_0 \leq \epsilon_1$, follows from the Borel functional calculus combined with the fact that for every $t > 0$, we then have $t\log(t) < t\log(t + \epsilon_0) \leq t\log(t + \epsilon_1)$.

To see the last claim, we first recall that it follows from the classical analysis that for any $t > 0$, we have $t\log(t) = \inf_{0 < \epsilon \leq 1} t\log(t + \epsilon)$. If we apply this fact to the Borel functional calculus, it will follow that $f\log(f + \mathbb{1}) - f\log(f) = \sup_{0 < \epsilon \leq 1}(f\log(f + \mathbb{1}) - f\log(f + \epsilon\mathbb{1}))$, with $(f\log(f + \mathbb{1}) - f\log(f + \epsilon\mathbb{1}))$ increasing to $f\log(f + \mathbb{1}) - f\log(f)$ as ϵ decreases. Since by the first part of the proof each of the operators $f\log(f + \mathbb{1}) - f\log(f)$ and $(f\log(f + \mathbb{1}) - f\log(f + \epsilon\mathbb{1}))$ ($\epsilon > 0$) is positive, the normality of the extension of the trace to $\widehat{\mathcal{M}}_+$ now ensures that $\tau(f\log(f+\mathbb{1})-f\log(f)) = \sup_{0 < \epsilon \leq 1} \tau(f\log(f+\mathbb{1})-f\log(f+\epsilon\mathbb{1}))$. This translates to the claim that

$$\begin{aligned}
\tau(f\log(f)) \;&=\; \tau(f\log(f+\mathbb{1})) - \tau(f\log(f+\mathbb{1}) - f\log(f)) \\
&=\; \tau(f\log(f+\mathbb{1})) - \sup_{0<\epsilon\le 1}\tau(f\log(f+\mathbb{1}) - f\log(f+\epsilon\mathbb{1})) \\
&=\; \inf_{0<\epsilon\le 1}[\tau(f\log(f+\mathbb{1})) - \tau(f\log(f+\mathbb{1}) - f\log(f+\epsilon\mathbb{1}))] \\
&=\; \inf_{0<\epsilon\le 1}\tau(f\log(f+\epsilon\mathbb{1})).
\end{aligned}$$

Since by the first part of the proof $\inf_{0<\epsilon\le 1}\tau(f\log(f+\epsilon\mathbb{1})) = \inf_{0<\epsilon}\tau(f\log(f+\epsilon\mathbb{1}))$, we are done. $\qquad\square$

We compare the above approach to the von Neumann entropy, namely $S(\varrho) = -\operatorname{Tr}(\varrho\log\varrho)$, where ϱ is a density matrix on a Hilbert space H (a positive trace class operator with trace norm 1) and Tr is the canonical trace on $B(H)$. Since in this particular case one can show that $\widetilde{B(H)} = B(H)$, it follows that the trace class operators are nothing but $L^1(B(H), \operatorname{Tr})$. So, here, the entropy is considered in the context of the von Neumann algebra $B(H)$, with entropy defined for positive norm 1 elements of $L^1(B(H), \operatorname{Tr})$.

Proposition 14.1 suggests that states with 'good' entropy should all belong to $(L^1 \cap L\log(L+1))(B(H))$. This then raises the question of how the two pictures compare.

Let $0 \le \varrho \in L^1$ be given. Then, for some orthonormal system $\{x_i\} \subset H$, we have $\varrho = \sum \lambda_i p_{x_i}$, $\lambda_i \ge 0$, $\sum \lambda_i < \infty$, where p_{x_i} is the orthogonal projection onto the subspace spanned by the unit vector $x_i \in H$. Additionally, we may assume that the λ_is are arranged in decreasing order. But then we must have that λ_i decreases to 0 (or else $\sum \lambda_i < \infty$ will fail). Since $\log$ increases on $[1, \infty)$, this in turn ensures that

$$0 \le \lambda_i \log(\lambda_i + 1) \le K\lambda_i \quad \text{for all}\quad i$$

where $K = \log(\lambda_1 + 1)$. But then

$$\sum \lambda_i \log(\lambda_i + 1) < \infty.$$

Since $\sum \lambda_i \log(\lambda_i + 1)$ is nothing but the trace norm of $\varrho\log(\varrho + \mathbb{1})$, this then ensures that $\varrho \in L\log(L+1)$. Thus, in this exceptional case, one has that $L^1 \subseteq L\log(L+1)$.

Thus, the approach presented in the proposition does indeed canonically extend the elementary quantum theory based on the pair

$\langle L^1(B(H), \mathrm{Tr}), B(H)\rangle$. A serendipitous consequence of this observation is that the functionals $S_\epsilon(\varrho) = -\mathrm{Tr}(\varrho \log(\varrho + \epsilon))$, $\epsilon > 0$, provide well-defined approximations of von Neumann entropy $S(\varrho)$.

The theory we have achieved thus far is very satisfying, but still glaringly incomplete in the sense that some very important quantum systems correspond to type III von Neumann algebras which do not have a faithful normal semifinite trace. Is there, therefore, a way to define a concept of entropy for general von Neumann algebras, which canonically extends the tracial case described above? The first step in setting up a theory of entropy for type III algebras is to identify a type III analogue of the space $L^1 \cap L\log(L+1)$. Classically, this space is an Orlicz space produced by the Young function

$$\Psi_{ent}(t) = \max(t, t\log(t+1)) = \begin{cases} t, & 0 \le t \le e - 1 \\ t\log(t+1), & e - 1 \le t. \end{cases}$$

The most logical place to search for states with 'good' entropy in the type III case is therefore in this space. With each of $L\log(L+1)(0,\infty)$ and $L^1 \cap L\log(L+1)(0,\infty)$ equipped with the Luxemburg norm, it is then an exercise to see that the fundamental function of $L^1 \cap L\log(L+1)(0,\infty)$ is of the form $\mathbf{f}^{ent}(t) = \max(t, \mathbf{f}^{\log}(t))$, where $\mathbf{f}^{\log}$ is the fundamental function of $L\log(L+1)(0,\infty)$. It is this fundamental function that we use to construct the type III analogue of $L^1 \cap L\log(L+1)$ according to the prescriptions given in Chapter 10. Given a von Neumann algebra equipped with faithful normal semifinite weight φ, let us for the sake of brevity denote this space by $L^{ent}(\mathcal{M})$. Some further analysis is necessary before we can formulate a definition of entropy valid for the type III case, based on this space.

Remark 14.2 We pause to show that $L^{ent}(\mathcal{M})$ does everything expected of an 'intersection' of the spaces L^1 and $L\log(L+1)$, in the sense of admitting canonical embeddings into both $L^1(\mathcal{M})$ and $L\log(L+1)(\mathcal{M})$. From the above computations, it is clear that the functions $\zeta_1(t) = \frac{t}{\mathbf{f}^{ent}(t)}$ and $\zeta_{\log}(t) = \frac{\mathbf{f}^{\log}(t)}{\mathbf{f}^{ent}(t)}$ are both continuous and bounded above (by 1) on $(0,\infty)$. Hence, for $h = \frac{d\tilde{\varphi}}{d\tau_{\mathfrak{M}}}$, the operators $\zeta_1(h)$ and $\zeta_{\log}(h)$ are both contractive elements of $\mathfrak{M}$. It is now an exercise to see that the prescriptions $\iota_1(x) = \zeta_1(h)^{1/2} x \zeta_1(h)^{1/2}$

and $\iota_{\log}(x) = \zeta_{\log}(h)^{1/2} x \zeta_{\log}(h)^{1/2}$ where $x \in L^{ent}(\mathcal{M})$ yield continuous embeddings of $L^{ent}(\mathcal{M})$ into $L^1(\mathcal{M})$ and $L\log(L+1)(\mathcal{M})$, respectively.

Now, recall that Theorem 10.2 basically shows that the type III analogue of the modular $I_{\Psi} = \tau(\Psi(|\cdot|))$ on $L^{\Psi}(\mathcal{M}, \tau)$ is precisely the quantity $I_{\Psi}(f) = d_f(1) = \tau_{\mathfrak{M}}(\chi_{(1,\infty)}(|f|))$ on $L^{\Psi}(\mathcal{M})$. If we apply these facts to equation (14.5), then the following definition can be seen to be a faithful translation of the content of Proposition 14.1 to the type III setting:

Definition 14.3 (Entropy for general algebras) A state $\vartheta \in \mathcal{M}_*$ on the von Neumann algebra $\mathcal{M}$ is called *regular* if for some element g of $L_+^{ent}(\mathcal{M})$, $\frac{d\vartheta}{d\tau}$ is of the form $\zeta_1(h)^{1/2} g \zeta_1(h)^{1/2}$. For such a regular state, we then define the entropy to be

$$
\begin{aligned}
\widetilde{S}(\vartheta) &= \inf_{\epsilon>0}[\epsilon I_{\log}(\iota_{\log}(g)) + \log(\epsilon)\|\iota_1(g)\|_1] \\
&= \inf_{\epsilon>0}[\epsilon d_{\epsilon^{-1}\iota_{\log}(g)}(1) + \log(\epsilon)\|\iota_1(g)\|_1] \\
&= \inf_{\epsilon>0}[\epsilon \tau(\chi_{(\epsilon,\infty)}(\iota_{\log}(g))) + \log(\epsilon)\|\iota_1(g)\|_1].
\end{aligned}
$$

(Here, h is the density $\frac{d\widetilde{\varphi}}{d\tau}$ of the dual weight $\widetilde{\varphi}$ and τ the canonical trace on $\mathfrak{M}$.)

The following is now a fairly direct consequence of the definition:

Theorem 14.4 *If ϑ is a regular state, then $\widetilde{S}(\vartheta)$ is well defined (although possibly infinite).*

In closing this section, we pass to the case where the canonical weight φ is a state. In that case, $h = \frac{d\widetilde{\varphi}}{d\tau}$ belongs to $L^1(\mathcal{M})$. We show that in this case, φ is a regular state and we compute its entropy. For $\widetilde{\varphi}$ to be regular, $h = \frac{d\widetilde{\varphi}}{d\tau}$ must be of the form $\zeta_1(h) g \zeta_1(h)$ for some $g \in L^{ent}(\mathcal{M})$. Hence, we wish to have

$$
h = \left(\frac{h}{\mathbf{f}^{ent}(h)}\right)^{\frac{1}{2}} g \left(\frac{h}{\mathbf{f}^{ent}(h)}\right)^{\frac{1}{2}} \tag{14.7}
$$

which can only be the case if $g = \mathbf{f}^{ent}(h)$. It is not difficult to check that indeed $\mathbf{f}^{ent}(h) \in L^{ent}(\mathcal{M})$. We therefore have that

$$
\begin{aligned}
\widetilde{S}(\varphi) &= \inf_{\epsilon > 0} \left[\epsilon\, I_{ent}(\iota_{\log}(\mathbf{f}^{ent}(h))) + \log \epsilon\, \|\iota_1(\mathbf{f}^{ent}(h))\|_1 \right] \\
&= \inf_{\epsilon > 0} \left[\epsilon\, \tau \left(\chi_{(\epsilon,\infty)}(\zeta_{\log}(h))^{\frac{1}{2}} \mathbf{f}^{ent}(h)(\zeta_{\log}(h))^{\frac{1}{2}} \right) \right. \\
&\qquad \left. + \log \epsilon\, \|\zeta_1(h)^{\frac{1}{2}} \mathbf{f}^{ent}(h)\zeta_1(h)^{\frac{1}{2}}\|_1 \right].
\end{aligned}
\tag{14.7}
$$

Recall that for any $f \in L^1(\mathcal{M})$, we have by Lemma 10.23 that

$$
\tau\left(\chi_{(\epsilon,\infty)}(|f|) \right) = d_{\epsilon^{-1}f}(1) = \epsilon^{-1}\|f\|_1.
\tag{14.8}
$$

Thus

$$
\begin{aligned}
\widetilde{S}(\varphi) &= \inf_{\epsilon > 0} \left[\epsilon\, \tau \left(\chi_{(\epsilon,\infty)}(\zeta_{\log}(h))^{\frac{1}{2}} \mathbf{f}^{ent}(h)(\zeta_{\log}(h))^{\frac{1}{2}} \right) \right. \\
&\qquad \left. + \epsilon \log \epsilon\, \tau \left(\chi_{(\epsilon,\infty)}(\zeta_1(h)^{\frac{1}{2}} \mathbf{f}^{ent}(h)\zeta_1(h)^{\frac{1}{2}}) \right) \right] \\
&= \inf_{\epsilon > 0} \left[\epsilon\, \tau \left(\chi_{(\epsilon,\infty)}(\mathbf{f}^{\log}(h)) \right) + \epsilon \log \epsilon\, \tau \left(\chi_{(\epsilon,\infty)}(h) \right) \right].
\end{aligned}
$$

Now, observe that $\mathbf{f}^{\log}$ is a continuous strictly increasing function which is 0 at 0. So $t \geq \epsilon > 0$ if and only if $\mathbf{f}^{\log}(t) \geq \mathbf{f}^{\log}(\epsilon) > 0$, with $\mathbf{f}^{\log}(t) \to \infty$ as $t \to \infty$. So, for any positive Borel function f on $[0,\infty)$, we have $f(t) > \epsilon$ if and only if $\mathbf{f}^{\log}(f(t)) > \mathbf{f}^{\log}(\epsilon)$. If we combine this fact with the Borel functional calculus, it is clear that $\chi_{(\epsilon,\infty)}(\mathbf{f}^{\log}(h)) = \chi_{((\mathbf{f}^{\log})^{-1}(\epsilon),\infty)}(h)$. The above equality therefore becomes

$$
\begin{aligned}
\widetilde{S}(\varphi) &= \inf_{\epsilon > 0} \left[\epsilon\, \tau \left(\chi_{((\mathbf{f}^{\log})^{-1}(\epsilon),\infty)}(h) \right) \right. \\
&\qquad \left. + \epsilon \log \epsilon\, \tau \left(\chi_{(\epsilon,\infty)}(h) \right) \right] \\
&= \inf_{\epsilon > 0} \left[\frac{\epsilon}{(\mathbf{f}^{\log})^{-1}(\epsilon)} + \log(\epsilon) \right] . \|h\|_1.
\end{aligned}
\tag{14.9}
$$

Since $\mathbf{f}^{\log}(t) = \frac{1}{\Psi_{\log}^{-1}(\frac{1}{t})}$ where $\Psi_{\log}(t) = t\log(t+1)$, we have

$$
(\mathbf{f}^{\log})^{-1}(t) = \frac{1}{\Psi_{\log}(\frac{1}{t})} = \frac{t}{\log(\frac{1}{t}+1)}.
\tag{14.10}
$$

So

$$
\widetilde{S}(\varphi) = \inf_{\epsilon > 0} \left[\log\left(\frac{1}{\epsilon} + 1 \right) + \log \epsilon \right] \|h\|_1 = \left[\inf_{\epsilon > 0} \log(1 + \epsilon) \right] \|h\|_1 = 0.
\tag{14.11}
$$

14.3 Relative entropy

We pass to describing relative entropy in the noncommutative context. From a philosophical point of view, the relative entropy of two states measures the likelihood of a change of state from one state to the other. To fully appreciate the subtleties of this section, some familiarity with Part III—the general setting—is essential. The key results are Theorem 14.6 (which gives a more dynamical definition of relative entropy), Theorem 14.8 (which posits a more 'classical' formula for computing relative entropy), and Theorem 14.9 (which shows how the entropy introduced in the previous section aligns with the relative entropy defined here).

Let μ and ν be probability measures on a set X, and assume that μ is absolutely continuous with respect to ν. In this context, the relative entropy (also known as Kullback–Leibler divergence) is defined as

$$S(\mu|\nu) = \int_X \log\left(\frac{d\mu}{d\nu}\right) d\mu = \int_X \frac{d\mu}{d\nu}\log\left(\frac{d\mu}{d\nu}\right) d\nu \equiv \left\langle \log\frac{d\mu}{d\nu}\right\rangle_\mu \tag{14.12}$$

provided that the integrals in the above formulae exist. Here, $\frac{d\mu}{d\nu}$ is the Radon–Nikodym derivative of μ with respect to ν. Assume additionally that ν (so also μ) is absolutely continuous with respect to the reference measure λ. Then,

$$\frac{d\mu}{d\nu} = \frac{d\mu}{d\lambda}\cdot\frac{d\lambda}{d\nu} \tag{14.13}$$

and so under some additional assumptions, one has the more familiar formula for the relative entropy, namely

$$S(\mu|\nu) = \int_X p\log\frac{p}{q}d\lambda \tag{14.14}$$

where $p = \frac{d\mu}{d\lambda}$ and $q = \frac{d\nu}{d\lambda}$.

Let $\mathcal{M}$ be a σ-finite von Neumann algebra in standard form and let ψ and ϕ be two faithful normal states with unit vector representatives $\Psi, \Phi \in \mathcal{H}$ in the natural cone of $\mathcal{H}$. The basic theory of the Tomita–Takesaki theory easily extends to show that the densely defined antilinear operator $S_{\phi,\psi}(a\Psi) = a^*\Phi$ is, in fact, closable. The operator $\Delta_{\phi,\psi}$ is defined as $[S_{\phi,\psi}]^*[S_{\phi,\psi}]$. In the case $\phi = \psi$, we, of course, simply write Δ_ϕ. Recall that the prescription $\sigma_t^\phi(a) = \Delta_\phi^{it}a\Delta_\phi^{-it}$ where $t \in \mathbb{R}$

and $a \in \mathcal{M}$ defines a one-parameter group of automorphisms on $\mathcal{M}$ which leaves ϕ invariant—the so-called modular automorphism group of ϕ. Therefore, the operator $\Delta_{\phi,\psi}$ in a very real sense encodes the manner in which the dynamics determined by the modular automorphism group of one state differs from the other. Using this fact, Araki then defined the relative entropy $S(\phi|\psi)$ of ψ and ϕ as $-\langle \log(\Delta_{\phi,\psi})\Psi, \Psi\rangle$. We refer the interested reader to [**Ara76**, **Ara78**] and the references therein for a survey of the basic properties of this entropy.

Another more dynamic way in which to define the relative entropy is to use the Connes cocycle derivative $(D\psi \colon D\phi)_t$. We remind the reader that the Connes cocycle derivative is a one-parameter family $u_t = (D\psi \colon D\phi)_t$ of unitaries which, among other properties, intertwines the actions of the modular groups of ϕ and ψ in the sense that $\sigma_t^\psi(a) = u_t \sigma_t^\phi(a) u_t^*$ for all t and $a \in \mathcal{M}$.

To see the clear connection between the two approaches, we pause to note that the Connes cocycle derivative $(D\psi \colon D\phi)_t$ may be written as $(D\psi \colon D\phi)_t = \Delta_{\psi,\phi}^{it} \Delta_\phi^{-it}$ (see, for example, [**AM82**, Appendix B] for details). The chain rule for cocycle derivatives informs us that $\mathbb{1} = (D\psi \colon D\psi)_t = (D\psi \colon D\phi)_t (D\phi \colon D\psi)_t$ and hence that we also have that

$$(D\psi \colon D\phi)_t = (D\phi \colon D\psi)_t^{-1} = (\Delta_{\phi,\psi}^{it} \Delta_\psi^{-it})^{-1} = \Delta_\psi^{\ it} \Delta_{\phi,\psi}^{-it}.$$

Note that, by construction, we have that Ψ is an eigenvector of Δ_ψ corresponding to the eigenvalue 1. Hence, Ψ must then be an eigenvector of Δ_ψ^{-it} corresponding to the eigenvalue $1^{-it} = 1$. So, for any t, we must then have that

$$\langle (D\psi \colon D\phi)_t \Psi, \Psi\rangle = \langle \Delta_\psi^{it} \Delta_{\phi,\psi}^{-it} \Psi, \Psi\rangle$$

$$= \langle \Delta_{\phi,\psi}^{-it} \Psi, \Delta_\psi^{-it} \Psi\rangle = \langle \Delta_{\phi,\psi}^{-it} \Psi, \Psi\rangle.$$

It is therefore trivially clear that

$$\lim_{t\to 0} \frac{1}{t} \psi[(D\psi \colon D\phi)_t - \mathbb{1}] = \lim_{t\to 0} \frac{1}{t} \langle (\Delta_{\phi,\psi}^{-it} - \mathbb{1})\Psi, \Psi\rangle.$$

Ohya and Petz [**OP93**, Theorem 5.7] proved the above inequality, but under the assumption that $-\langle \log(\Delta_{\phi,\psi})\Psi, \Psi\rangle$ is finite. Building on ideas of Majewski and Labuschagne [**ML20**] we show that we get full equivalence when the spectral behaviour of $\Delta_{\phi,\psi}$ is modulated near 0. We speficially have the following:

Theorem 14.5 *Let $\mathcal{M}$ be a σ-finite von Neumann algebra in the standard form described above, and let ψ and ϕ be two faithful normal states with unit vector representatives $\Psi, \Phi \in \mathcal{H}$. Then, allowing for ∞ as a possible value, the limit $\lim_{\epsilon \searrow 0} \lim_{t \to 0} \frac{-i}{t} \psi[((D\psi : D\phi)_t - \mathbb{1})\chi_{[\epsilon,\infty)}(\Delta_{\phi,\psi})]$ always exists and equals $-\langle \log(\Delta_{\phi,\psi})\Psi, \Psi \rangle$. In the case where this quantity is finite, we even have that $\lim_{t \to 0} \frac{-i}{t} \psi[(D\psi : D\phi)_t - \mathbb{1}] = -\langle \log(\Delta_{\phi,\psi})\Psi, \Psi \rangle$.*

On the basis of this theorem we therefore propose the following definition of relative entropy for faithful normal states:

Definition 14.6 Let ϕ and ψ be faithful normal states on $\mathcal{M}$. We define the relative entropy $S(\phi|\psi)$ to be $S(\phi|\psi) = \lim_{\epsilon \searrow 0} \lim_{t \to 0} \frac{-i}{t} \psi[((D\psi : D\phi)_t - \mathbb{1})\chi_{[\epsilon,\infty)}(\Delta_{\phi,\psi})]$.

Proof.   Recall that Araki's definition of entropy is

$$S(\phi|\psi) = -\langle \log(\Delta_{\phi,\psi})\Psi, \Psi \rangle$$

where the latter term is understood to be

$$-\int_0^\infty \log(\lambda) \, d\langle e_\lambda \Psi, \Psi \rangle$$

(here $\lambda \to e_\lambda$ is the spectral resolution of $\Delta_{\phi,\psi}$). As Araki points out, the value of this integral is either real (in the case where log is integrable) or ∞ otherwise. To see this, notice that if log is not integrable, the fact that $0 \leq \log(\lambda) \leq \lambda$ on $[1, \infty)$ ensures that

$$0 \leq \int_1^\infty \log(\lambda) \, d\langle e_\lambda \Psi, \Psi \rangle \leq \int_0^\infty \lambda \, d\langle e_\lambda \Psi, \Psi \rangle = \langle \Delta_{\phi,\psi}\Psi, \Psi \rangle < \infty$$

and hence that we nevertheless still have that log is integrable on $[1, \infty)$. Therefore, any nonintegrability of log must be derived from the fact that $-\int_0^1 \log(\lambda) \, d\langle e_\lambda \Psi, \Psi \rangle = \infty$.

First, suppose that log is integrable. For any $t > 0$ and any $\lambda > 0$, we have that

$$\left|\frac{1}{t}(\lambda^{-it} - 1)\right| \le \left|\frac{1}{t}(\lambda^{-it/2} - 1)(\lambda^{-it/2} + 1)\right| \le \left|\frac{2}{t}(\lambda^{-it/2} - 1)\right|.$$

Carrying on inductively leads to the conclusion that $|\frac{1}{t}(\lambda^{-it} - 1)| \le |\frac{2^k}{t}(\lambda^{-it/2^k} - 1)|$ for any $k \in \mathbb{N}$. If now we let $k \to \infty$, we obtain the inequality $|\frac{1}{t}(\lambda^{-it} - 1)| \le |\log(\lambda)|$, which holds for any t and any $\lambda > 0$. Hence, we may apply the dominated convergence theorem to see that for any sequence (t_n) converging to 0, we have that

$$\lim_{n \to \infty} \frac{-i}{t_n} \langle (\Delta_{\phi,\psi}^{-it_n} - \mathbb{1})\Psi, \Psi \rangle = \lim_{n \to \infty} \frac{-i}{t_n} \int_0^\infty (\lambda^{-it_n} - 1)\, d\langle e_\lambda \Psi, \Psi \rangle$$

$$= -\int_0^\infty \log(\lambda)\, d\langle e_\lambda \Psi, \Psi \rangle.$$

This fact is enough to enable us to conclude that

$$\lim_{t \to 0} \frac{-i}{t} \langle (\Delta_{\phi,\psi}^{-it} - \mathbb{1})\Psi, \Psi \rangle = -\int_0^\infty \log(\lambda)\, d\langle e_\lambda \Psi, \Psi \rangle.$$

We now consider the general case and leave it as an exercise to show that we will for any $\epsilon > 0$ have that

$$\frac{-i}{t} \psi[((D\psi : D\phi)_t - \mathbb{1})\chi_{[\epsilon,\infty)}(\Delta_{\phi,\psi})] = \frac{-i}{t} \int_\epsilon^\infty (\lambda^{-it} - 1)\, d\langle e_\lambda \Psi, \Psi \rangle.$$

Since $\lambda \mapsto \log(\lambda)$ is bounded and continuous on $[\epsilon, 1]$, an entirely analogous argument shows that we will for any sequence (t_n) converging to 0 have that

$$\lim_{n \to \infty} \frac{-i}{t_n} \psi[((D\psi : D\phi)_{t_n} - \mathbb{1})\chi_{[\epsilon,\infty)}(\Delta_{\phi,\psi})] = -\int_\epsilon^\infty \log(\lambda)\, d\langle e_\lambda \Psi, \Psi \rangle.$$

Since $\int_1^\infty \log(\lambda)\, d\langle e_\lambda \Psi, \Psi \rangle$ is finite and log decreasing without bound on $[0, 1]$, it is palpably clear that even for the case where the the full integral diverges, we have that $\lim_{\epsilon \searrow} \int_\epsilon^\infty \log(\lambda)\, d\langle e_\lambda \Psi, \Psi \rangle = -\int_0^\infty \log(\lambda)\, d\langle e_\lambda \Psi, \Psi \rangle$ which then proves the theorem. $\qquad\square$

We pass to showing that even in the noncommutative context, one is able to write down a formula for relative entropy closely mimicking the formula stated in equation (14.14).

Recall that any $\phi \in \mathcal{M}_*^+$ may be written in the form

$$\phi(x) = \operatorname{tr}(h_\phi^{\frac{1}{2}} x h_\phi^{\frac{1}{2}}) \equiv \langle x h_\phi^{\frac{1}{2}}, h_\phi^{\frac{1}{2}} \rangle_{L^2(\mathcal{M})} \qquad (14.15)$$

where $h_\phi = \frac{d\widetilde{\phi}}{d\tau}$ with $h_\phi \in L^1_+(\mathcal{M})$. In other words, $h_\phi^{\frac{1}{2}}$ is a vector in the natural cone $L^2_+(\mathcal{M})$, which represents the state ϕ. (This formula is just a reformulation of the formula $\langle x\Phi, \Phi \rangle$ in the crossed product setting.)

To achieve our objective of reproducing a noncommutative version of formula (14.14), we need to describe $(D\psi \colon D\phi)_t$ in terms of $L^p(\mathcal{M})$ technology. Specifically, we show that

$$(D\vartheta \colon D\psi)_t = h_\vartheta^{it} h_\psi^{-it}$$

where $h_\vartheta = \frac{d\widetilde{\vartheta}}{d\tau}$ and $h_\psi = \frac{d\widetilde{\psi}}{d\tau}$. To see that the above claim is true, we use the cocycle chain rule to see that

$$\mathbb{1} = (D\widetilde{\psi} \colon D\widetilde{\psi})_t = (D\widetilde{\psi} \colon D\tau)_t (D\tau \colon D\widetilde{\psi})_t.$$

Equivalently

$$(D\tau \colon D\widetilde{\psi})_t = (D\widetilde{\psi} \colon D\tau)_t^{-1}.$$

But from Proposition 9.30 and Theorem 9.28, we know that $(D\widetilde{\psi} \colon D\tau)_t = h_\psi^{it}$. Hence $(D\tau \colon D\widetilde{\psi})_t = h_\psi^{-it}$. Since also $(D\widetilde{\vartheta} \colon D\tau)_t = h_\vartheta^{it}$, it is now clear from Theorem 9.26 and the cocycle chain rule that

$$\begin{aligned}
(D\vartheta \colon D\psi)_t &= (D\widetilde{\vartheta} \colon D\widetilde{\psi})_t & (14.16) \\
&= (D\widetilde{\vartheta} \colon D\tau)_t (D\tau \colon D\widetilde{\psi})_t \\
&= h_\vartheta^{it} h_\psi^{-it}.
\end{aligned}$$

Another important factor to take into account is that tr is only defined on $L^1(\mathcal{M})$. Thus, to proceed with our objective of developing a noncommutative version of formula (14.14), we must show that, in some sense, $h_\vartheta \log h_\vartheta - h_\vartheta \log h_\phi$ is in L^1. As we shall see below, this can indeed be achieved in a limiting sense. For this, we shall need Lemma 10.32. Let S be the closed complex strip $\{z \in \mathbb{C} \colon 0 \le \operatorname{Re}(z) \le 1\}$ and

S^o the corresponding open strip. We recall from Lemma 10.32 that the function

$$S^o \ni z \mapsto h_\vartheta^z h_\phi^{1-z} \in L^1(\mathcal{M}) \tag{14.17}$$

(where obviously $h_\vartheta, h_\phi \in L^1(\mathcal{M})$) is a well-defined L^1-valued function which is analytic on the open strip S^o. On taking the derivative, we get that

$$z \mapsto h_\vartheta^z \cdot \log h_\vartheta \cdot h_\phi^{1-z} - h_\vartheta^z \cdot \log h_\phi \cdot h_\phi^{1-z} \in L^1(\mathcal{M}) \tag{14.18}$$

for all $z \in S^o$. More importantly, the analyticity ensures that this derivative varies continuously with respect to z in the L^1 norm. A fact which underlies the above very regular behaviour on this strip is that for any $0 < s$, $x^s \log(x)$ is a very well behaved continuous function which is 0 at 0, and for which $x^s \leq x^s \log(x) \leq x^{s+1}$ whenever $x \geq e$. This fact can be used to show that for any positive τ-measurable operator g, $g^s \log(g)$ will again be τ-measurable.

Using the above formula and letting $z \to 1$ leads us to the promised noncommutative analogue of formula (14.14). To understand how this is achieved, assume that $z = s + it$ where $0 < s < 1$. Then, the fact that

$$h_\vartheta^z h_\phi^{1-z} = h_\vartheta^s [h_\vartheta^{it} h_\phi^{-it}] h_\phi^{1-s}$$

leads to the conclusion that

$$
\begin{aligned}
-i\frac{d}{dt}(h_\vartheta^s (D\vartheta : D\phi)_t h_\phi^{1-s}) &= -i\frac{d}{dt}(h_\vartheta^s [h_\vartheta^{it} h_\phi^{-it}] h_\phi^{1-s}) \\
&= \frac{d}{dz} h_\vartheta^z h_\phi^{1-z} \\
&= h_\vartheta^z \cdot \log h_\vartheta \cdot h_\phi^{1-z} - h_\vartheta^z \cdot \log h_\phi \cdot h_\phi^{1-z}.
\end{aligned}
$$

To see the above, notice that when, for an analytic function, computing a limit of the form $\lim_{\Delta z \to 0} \frac{f(z+\Delta z) - f(z)}{\Delta z}$, we may as well assume that $\Delta z = i\Delta t$. In particular, when computing the derivative anywhere along the line segment $0 < s < 1$, $t = 0$, we always have that

$$-i\frac{d}{dt}(h_\vartheta^s (D\vartheta : D\phi)_t h_\phi^{1-s})|_{t=0} = h_\vartheta^{s^-} \log h_\vartheta \cdot h_\phi^{1-s} - h_\vartheta^{s^-} \log h_\phi \cdot h_\phi^{1-s}$$

$$\in L^1(\mathcal{M}).$$

With the groundwork having been done, we are now ready to present the promised result.

The concept that is crucial in guaranteeing the validity of the theorem is the following ordering defined by Takesaki and Connes. (See [**CT77**, Definition 4.1].)

Definition 14.7 For two normal weights ϑ and ϕ on a von Neumann algebra $\mathcal{M}$ and some positive δ, we say that $\vartheta \leq \phi(\delta)$ if the function $t \mapsto (D\vartheta : D\phi)_t = u_t$ extends to an $\mathcal{M}$-valued map $z \mapsto u_z$ which is point to σ-weak continuous and bounded on the closed strip $\{z \in \mathbb{C} \colon -\delta \leq \operatorname{Im}(z) \leq 0\}$ and analytic on the open strip $\{z \in \mathbb{C} \colon -\delta < \operatorname{Im}(z) < 0\}$.

We note that by Theorem 8.32 the case $\delta = \frac{1}{2}$ in the above ordering corresponds exactly to the 'standard' ordering. Thus, this ordering may be seen as a refinement of the standard ordering of weights.

Theorem 14.8 *Let $\mathcal{M}$ be a σ-finite von Neumann algebra in the standard form described in Theorem 10.59, and let ϑ and ϕ be two faithful normal states with unit vector representatives $h_\vartheta^{1/2}, h_\phi^{1/2} \in L^2(\mathcal{M})$. If $\phi \leq \vartheta(\delta)$, and $S(\phi|\vartheta)$ is finite, then the limit*

$$\lim_{s \nearrow 1} \operatorname{tr}(h_\vartheta^{s} \cdot \log h_\vartheta \cdot h_\phi^{1-s} - h_\vartheta^{s} \cdot \log h_\phi \cdot h_\phi^{1-s})$$

exists, and equals $S(\phi|\vartheta)$.

Proof. Suppose that $S(\phi|\vartheta)$ is finite and let $\epsilon > 0$ be given. The finiteness ensures that $S(\phi|\vartheta) = \lim_{t \to 0} \frac{-i}{t}\vartheta[(D\vartheta \colon D\phi)_t - \mathbb{1}] = \lim_{t \to 0} \frac{-i}{t}\operatorname{tr}(h_\vartheta^{1/2}[h_\vartheta^{it} h_\phi^{-it} - \mathbb{1}]h_\vartheta^{1/2})$. Next, let s be given with $\frac{1}{2} < s < 1$. So, for $t_\epsilon > 0$ small enough, we will have that:

- $\frac{-i}{t_\epsilon}\operatorname{tr}(h_\vartheta^{1/2}[h_\vartheta^{it_\epsilon} h_\phi^{-it_\epsilon} - \mathbb{1}]h_\vartheta^{1/2})$ is within ϵ of $S(\vartheta|\phi)$;

- $\frac{-i}{t_\epsilon}h_\vartheta^{s}[h_\vartheta^{it_\epsilon} h_\phi^{-it_\epsilon} - \mathbb{1}]h_\phi^{1-s}$ is with respect to L^1-norm within ϵ of $-i\frac{d}{dt}(h_\vartheta^{s}(D\vartheta : D\phi)_t h_\phi^{1-s})|_{t=0} = h_\vartheta^{s} \cdot \log h_\vartheta \cdot h_\phi^{1-s} - h_\vartheta^{s} \cdot \log h_\phi \cdot h_\phi^{1-s}$.

Notice that by the properties of the trace functional tr, we have $\operatorname{tr}(h_\vartheta^{s}[h_\vartheta^{it_\epsilon} h_\phi^{-it_\epsilon} - \mathbb{1}]h_\phi^{1-s}) = \operatorname{tr}(h_\vartheta^{1/2}[h_\vartheta^{it_\epsilon} h_\phi^{-it_\epsilon} - \mathbb{1}]h_\phi^{1-s}h_\vartheta^{s-1/2})$.

By assumption, $\phi \leq \vartheta(\delta)$. This means that $t \mapsto (D\phi : D\vartheta)_t$ extends to an $\mathcal{M}$-valued function $f(z)$, which is point to σ-weak continuous on the closed strip $\{z \in \mathbb{C}: -\delta \leq \mathrm{Im}(z) \leq 0\}$ and analytic on the open strip $\{z \in \mathbb{C}: -\delta < \mathrm{Im}(z) < 0\}$. For each z, the value $f(z)$ is essentially just an extension of $h_\phi^{iz} h_\vartheta^{-iz}$. (For details of this construction, see [**Kos92**]). In view of this, we will simply write $[h_\phi^{iz} h_\vartheta^{-iz}]$ for $f(z)$. So if we set $z = ir$ where $0 \leq r \leq \delta$, we obtain that as $r \searrow 0$ we will have $[h_\phi^{-r} h_\vartheta^{r}] \to \mathbb{1}$ in the σ-weak topology on $\mathcal{M}$.

Next, notice that for $0 \leq 1 - s \leq \delta$, we have $h_\phi^{1-s} h_\vartheta^{s-1/2} = [h_\phi^{1-s} h_\vartheta^{-(1-s)}] h_\vartheta^{1/2}$. Therefore, as $s \nearrow 1$ on the interval $[1 - \delta, 1]$, we must have that $\frac{-i}{t_\epsilon}\mathrm{tr}(h_\vartheta^{1/2}[h_\vartheta^{it_\epsilon} h_\phi^{-it_\epsilon} - \mathbb{1}]h_\phi^{1-s} h_\vartheta^{s-1/2}) = \frac{-i}{t_\epsilon}\mathrm{tr}(h_\vartheta[h_\vartheta^{it_\epsilon} h_\phi^{-it_\epsilon} - \mathbb{1}][h_\phi^{1-s} h_\vartheta^{-(1-s)}])$ converges to $\frac{-i}{t_\epsilon}\mathrm{tr}(h_\vartheta[h_\vartheta^{it_\epsilon} h_\phi^{-it_\epsilon} - \mathbb{1}]) = \frac{-i}{t_\epsilon}\mathrm{tr}(h_\vartheta^{1/2}[h_\vartheta^{it_\epsilon} h_\phi^{-it_\epsilon} - \mathbb{1}]h_\vartheta^{1/2})$.

Therefore, there must exist a $\widetilde{\delta} > 0$ such that for any s with $1 - \widetilde{\delta} < s < 1$, the term $\frac{-i}{t_\epsilon}\mathrm{tr}(h_\vartheta^{1/2}[h_\vartheta^{it_\epsilon} h_\phi^{-it_\epsilon} - \mathbb{1}]h_\phi^{1-s} h_\vartheta^{s-1/2})$ will be within ϵ of $\frac{-i}{t_\epsilon}\mathrm{tr}(h_\vartheta^{1/2}[h_\vartheta^{it_\epsilon} h_\phi^{-it_\epsilon} - \mathbb{1}]h_\vartheta^{1/2})$. If we combine all the above observations, it follows that, for any s with $1 - \widetilde{\delta} < s < 1$, $\mathrm{tr}(h_\vartheta^{s-}\log h_\vartheta \cdot h_\phi^{1-s} - h_\vartheta^{s-}\log h_\phi \cdot h_\phi^{1-s})$ will be within 3ϵ of $S(\phi|\vartheta)$. This proves the theorem. $\qquad\square$

We close our analysis of relative entropy by demonstrating a link with the entropy introduced in the previous section.

Theorem 14.9 *Let $\mathcal{M}$ be equipped with the faithful normal state φ and let ϑ be a faithful regular state with h_ϑ of the form $\zeta_1(h)^{1/2} g \zeta_1(h)^{1/2}$, where $g \in (L\log(L+1) \cap L^1)_+(\mathcal{M}) = L_+^{ent}(\mathcal{M})$ commutes strongly with $h = h_\varphi$. Then $\widetilde{S}(\vartheta) = S(\varphi|\vartheta)$.*

Proof. By assumption, g and h, and therefore h_ϑ and h, are strongly commuting affiliated operators. Hence so are $p = h_\vartheta h^{-1}$ and h. The proof makes extensive use of the kinds of techniques employed in Theorem 14.6. Given $a \in \mathcal{M}$ and $b \in L_+^1(\mathcal{M})$, we will for this reason once again employ the notational device of writing $\langle ab^{1/2}, b^{1/2}\rangle$ for $\mathrm{tr}(ba)$.

In this case, let $\lambda \mapsto e_\lambda$ be the spectral resolution of p. The fact that both h and h_ϑ belong to $L^1(\mathcal{M})$ ensures that $\theta_s(p) = \theta_s(h_\vartheta)\theta_s(h^{-1}) = h_\vartheta h^{-1} = p$ for every $s \in \mathbb{R}$. This in turn is enough to ensure that p is actually affiliated with the 'subalgebra' $\mathcal{M}$ of the crossed product $\mathfrak{M}$. Therefore, the spectral projections e_λ are all elements of $\mathcal{M}$. The first stage of the proof is to show that in general $S(\varphi|\vartheta)$ is finite if and only if the integral $\langle p\log(p)h^{1/2}, h^{1/2}\rangle = \int_0^\infty \lambda\log(\lambda)\, d\langle e_\lambda h^{1/2}, h^{1/2}\rangle$ converges, in which case they are equal. Additionally, that the only way the integral can diverge is by diverging to ∞.

One of the main challenges we need to overcome to achieve the stated objective, is to understand how to translate expressions like $\langle p\log(p)h^{1/2}, h^{1/2}\rangle$ to the GNS setting. Here we shall in passing make use of the equivalence of Haagerup L^p-spaces with Araki-Masuda and Connes-Hilsum L^p-spaces. We specifically need the technology used to demonstrate the equivalence of these two settings with the crossed product setting of Haagerup.

By Theorem 9.31 and Proposition 10.14 we may freely pass to the setting where ϑ rather than φ is being used as the reference weight. If having done this we then use the weight $\vartheta' = \vartheta(J \cdot J)$ on $\mathcal{M}'$ to construct Connes-Hilsum spaces, it follows from a combination of [**Hil81**, Theorem 2] and [**Hia21**, Proposition 11.6] that, in moving from the one context to another, $h_\varphi h_\vartheta^{-1}$ will be mapped onto $\Delta_{\varphi,\vartheta}\Delta_\vartheta^{-1}$ by means of a spatial $*$-isomorphism. The commutation of the pair h_φ and h_ϑ is of course carried over to the pair $\Delta_{\varphi,\vartheta}$ and Δ_ϑ. Therefore $p = h_\vartheta h_\varphi^{-1}$ gets carried over to $\Delta_\vartheta\Delta_{\varphi,\vartheta}^{-1}$. Now consider the quantity $\mathrm{tr}(h^{1/2}p\log(p)h^{1/2}) = \langle p\log(p)h^{1/2}, h^{1/2}\rangle$. We may make sense of this quantity as a spectral integral where $\langle p\log(p)h^{1/2}, h^{1/2}\rangle = \int_0^\infty \lambda\log(\lambda)\, d\langle e_\lambda h^{1/2}, h^{1/2}\rangle$. Since $x\log(x)$ induces a bounded continuous function on $[0,1]$, this integral diverges iff $\int_1^\infty \lambda\log(\lambda)\, d\langle e_\lambda h^{1/2}, h^{1/2}\rangle = \infty$. On now using the fact that $\Delta_{\varphi,\vartheta}^{1/2}\Theta = \Phi$ ([**Hia21**, Proposition 10.6]) and appealing to the Borel functional calculus for commuting densely defined operators, it now follows that

$$
\begin{aligned}
\langle p\log(p)h^{1/2}, h^{1/2}\rangle &= \langle \Delta_\vartheta\Delta_{\varphi,\vartheta}^{-1}\log(\Delta_\vartheta\Delta_{\varphi,\vartheta}^{-1})\Phi, \Phi\rangle \\
&= \langle \Delta_\vartheta\Delta_{\varphi,\vartheta}^{-1}\log(\Delta_\vartheta\Delta_{\varphi,\vartheta}^{-1})\Delta_{\varphi,\vartheta}^{1/2}\Theta, \Delta_{\varphi,\vartheta}^{1/2}\Theta\rangle \\
&= \langle \Delta_\vartheta(\log(\Delta_\vartheta) - \log(\Delta_{\varphi,\vartheta}))\Theta, \Theta\rangle \\
&= -\langle \log(\Delta_{\varphi,\vartheta})\Theta, \Theta\rangle \\
&= S(\varphi|\vartheta).
\end{aligned}
$$

Here we used the fact that Θ is an eigenvector of Δ_ϑ with eigenvalue 1 to obtain the second to last equality.

The next part of the proof is to show that (infinite values included)

$$\inf_{\epsilon>0}\int_0^\infty \lambda\log(\lambda+\epsilon)\,d\langle e_\lambda h^{1/2}, h^{1/2}\rangle = \int_0^\infty \lambda\log(\lambda)\,d\langle e_\lambda h^{1/2}, h^{1/2}\rangle$$

from which we will then be able to deduce the claim. To see this next fact, observe that for any $\lambda>0$ and $\epsilon>0$, we have that $\log(\lambda+\epsilon)>\log(\lambda)$ and that

$$0<\lambda(\log(\lambda+\epsilon)-\log(\lambda))=\lambda\log(1+(\epsilon/\lambda))\leq\epsilon.$$

This ensures that

$$\begin{aligned}
\int_0^\infty \lambda\log(\lambda)\,d\langle e_\lambda h^{1/2}, h^{1/2}\rangle &\leq \int_0^\infty \lambda\log(\lambda+\epsilon)\,d\langle e_\lambda h^{1/2}, h^{1/2}\rangle \\
&\leq \int_0^\infty (\lambda\log(\lambda)+\epsilon)\,d\langle e_\lambda h^{1/2}, h^{1/2}\rangle \\
&= \int_0^\infty \lambda\log(\lambda)\,d\langle e_\lambda h^{1/2}, h^{1/2}\rangle + \epsilon\varphi(\mathbb{1}) \\
&= \int_0^\infty \lambda\log(\lambda)\,d\langle e_\lambda h^{1/2}, h^{1/2}\rangle + \epsilon
\end{aligned}$$

which establishes the claim.

If we combine the two facts we have proved thus far, it yields the conclusion that (infinite values included) we always have that

$$S(\varphi|\vartheta)=\inf_{\epsilon>0}\int_0^\infty \lambda\log(\lambda+\epsilon)\,d\langle e_\lambda h^{1/2}, h^{1/2}\rangle.$$

We now use this formula to show that $S(\varphi|\vartheta)=\tilde{S}(\vartheta)$. This claim will follow if we can show that for any $\epsilon>0$, the equality $\epsilon\tau(\chi_{(\epsilon,\infty)}(\zeta_{\log}(h)^{1/2}g\zeta_{\log}(h)^{1/2})) + \log(\epsilon)\|\zeta_1(h)^{1/2}g\zeta_1(h)^{1/2}\|_1 = \int_0^\infty(\lambda\log(\lambda)+\epsilon)\,d\langle e_\lambda h^{1/2}, h^{1/2}\rangle$ holds true. Let $\epsilon>0$ be given. First, observe that by the definition of ζ_1 and $\zeta_{\log}$, we have $\zeta_{\log}(h)=\mathbf{f}^{\log}(h)\mathbf{f}^{ent}(h)^{-1}=\mathbf{f}^{\log}(h)h^{-1}\zeta_1(h)$. Thus, by the commutation assumption, we have that

$$\begin{aligned}
\zeta_{\log}(h)^{1/2}g\zeta_{\log}(h)^{1/2} &=\mathbf{f}^{\log}(h)h^{-1}\zeta_1(h)^{1/2}g\zeta_1(h)^{1/2}=\mathbf{f}^{\log}(h)h^{-1}h_\vartheta \\
&=\mathbf{f}^{\log}(h)p.
\end{aligned}$$

Hence, we may apply Lemma 10.1 to see that

$$
\begin{aligned}
&\epsilon\tau(\chi_{(\epsilon,\infty)}(\zeta_{\log}(h)^{1/2}g\zeta_{\log}(h)^{1/2})) + \log(\epsilon)\|\zeta_1(h)^{1/2}g\zeta_1(h)^{1/2}\|_1 \\
={}& \epsilon\tau(\chi_{(\epsilon,\infty)}(\mathbf{f}^{\log}(h)p)) + \log(\epsilon)\mathrm{tr}(hp) \\
={}& \epsilon\tau(\chi_{(1,\infty)}(\mathbf{f}^{\log}(h)(p/\epsilon))) + \log(\epsilon)\mathrm{tr}(hp) \\
={}& \epsilon\tau(\chi_{(1,\infty)}(h(p/\epsilon)\log((p/\epsilon)+\mathbb{1}))) + \log(\epsilon)\mathrm{tr}(hp).
\end{aligned}
$$

Since $h \in L^1(\mathcal{M})$ with p affiliated to $\mathcal{M}$, the operator $b = h(p/\epsilon)\log(p/\epsilon) + \mathbb{1})$ is a positive operator affiliated with the crossed product for which we have $\theta_s(b) = e^{-s}b$ for each $s \in \mathbb{R}$. By Proposition 9.33, b corresponds to a normal weight Φ_b on $\mathcal{M}$. If we now apply Proposition 9.36, it follows that $\Phi_b(\mathbb{1}) = \tau(\chi_{(1,\infty)}(h(p/\epsilon)\log((p/\epsilon)+\mathbb{1})))$. Writing e_N for $\chi_{[0,N]}(p)$, we next again appeal to Proposition 9.33 to see that for each $N > 0$, the weight $f \mapsto \Phi_b(e_N f e_N)$ corresponds to $e_N b e_N$. All of these observations may now be combined with the normality of Φ_b and the definition of the tracial functional tr, to see that

$$
\begin{aligned}
&\epsilon\tau(\chi_{(1,\infty)}(h(p/\epsilon)\log((p/\epsilon)+\mathbb{1}))) + \log(\epsilon)\mathrm{tr}(hp) \\
={}& \epsilon\Phi_b(\mathbb{1}) + \log(\epsilon)\mathrm{tr}(hp) \\
={}& \epsilon\lim_{N\to\infty}\Phi_b(e_N) + \log(\epsilon)\mathrm{tr}(hp) \\
={}& \epsilon\lim_{N\to\infty}\tau(\chi_{(1,\infty)}(h(e_N p/\epsilon)\log((e_N p/\epsilon)+\mathbb{1}))) + \log(\epsilon)\mathrm{tr}(hp) \\
={}& \epsilon\lim_{N\to\infty}\mathrm{tr}(h(e_N p/\epsilon)\log((e_N p/\epsilon)+\mathbb{1})) + \log(\epsilon)\mathrm{tr}(hp) \\
={}& \epsilon\lim_{N\to\infty}\langle(e_N p/\epsilon)\log((e_N p/\epsilon)+\mathbb{1})h^{1/2}, h^{1/2}\rangle + \log(\epsilon)\langle ph^{1/2}, h^{1/2}\rangle \\
={}& \lim_{N\to\infty}\int_0^N \lambda\log((\lambda/\epsilon)+1)\,d\langle e_\lambda h^{1/2}, h^{1/2}\rangle \\
& + \log(\epsilon)\int_0^\infty \lambda\,d\langle e_\lambda h^{1/2}, h^{1/2}\rangle \\
={}& \int_0^\infty \lambda\log((\lambda/\epsilon)+1)\,d\langle e_\lambda h^{1/2}, h^{1/2}\rangle + \log(\epsilon)\int_0^\infty \lambda\,d\langle e_\lambda h^{1/2}, h^{1/2}\rangle \\
={}& \int_0^\infty \lambda\log(\lambda+\epsilon)\,d\langle e_\lambda h^{1/2}, h^{1/2}\rangle.
\end{aligned}
$$

This proves the claim required to establish the theorem. In obtaining the final equality, we silently used the facts that $\int_0^\infty \lambda\log((\lambda/\epsilon)+1)$

$d\langle e_\lambda h^{1/2}, h^{1/2}\rangle$ either converges or diverges to ∞ and that we always have $\int_0^\infty \lambda \, d\langle e_\lambda h^{1/2}, h^{1/2}\rangle = \mathrm{tr}(hp) = \mathrm{tr}(h_\vartheta) = \vartheta(\mathbb{1}) = 1 < \infty.$ $\qquad\square$

14.4 Regular observables

For this section, the reader would be well advised to have at least some familiarity with Part III—the general setting. To fully appreciate the final cycle of results focussing on dynamics, some knowledge of at least the first half of Part IV is advised. The key results are Proposition 14.14 which shows that elements of the space of regular observables have all generalised moments finite, Proposition 14.15 which shows the link of the space of regular observables with the space of states having 'good' entropy identified in Section 14.2 and Theorem 14.17 which shows that complete Markov dynamics extends to the space of regular observables.

Having introduced the concept of regular states, we now introduce the concept of regular observables. In the classical setting, this concept was introduced in a very innovative paper by Pistone and Sempi [**PS95**]. We review key elements of the classical regular statistical model from that paper. Let $\{\Omega, \Sigma, \nu\}$ be a probability space with ν as the reference measure. The set of densities f of all probability measures equivalent to ν in the sense that $\mathbb{E}_1(f) = 1$ will be called the state space $\mathcal{S}_\nu$, i.e.

$$\mathcal{S}_\nu = \{f \in L^1(\nu) \colon f > 0 \quad \nu - a.s., \, \mathbb{E}_1(f) = 1\} \tag{14.19}$$

where, in general, $\mathbb{E}_g(f) \equiv \int f \cdot g \, d\nu.$

Definition 14.10 The classical statistical model consists of the measure space $\{\Omega, \Sigma, \nu\}$, the state space $\mathcal{S}_\nu$ and the set of measurable functions $L^0(\Omega, \Sigma, \nu)$.

Within this framework, we wish to identify the so-called regular random variables; that is, random variables which, among other properties, have all moments finite. Our task is then to identify such moment-generating functions. To do this, first fix $f \in \mathcal{S}_\nu$. Then, for a real random variable u on $(\Omega, \Sigma, f d\nu)$, we define

$$\widehat{u}_f(t) = \int \exp(tu) f \, d\nu, \qquad t \in \mathbb{R}. \tag{14.20}$$

In the sequel we will need the following properties of $\widehat{u}$ (for details see Widder, [**Wid46**]).

1. $\widehat{u}$ is analytic in the interior of its domain.
2. Its derivatives are obtained by differentiating under the integral sign.

With the above as background, we now make the following definition.

Definition 14.11 The set of all random variables on $(\Omega, \Sigma, f d\nu)$ such that:

1. $\widehat{u}_f$ is well defined in a neighbourhood of the origin 0,
2. the expectation of u is zero,

is called the set of regular random variables and will be denoted by L_f.

The key feature of random variables is that *all the moments of every* $u \in L_f$ *exist and may be realised as the values at 0 of the derivatives of* $\widehat{u}_f$. In other words, we have selected all random variables such that for each u, any moment $\mathbb{E}_f(u^n)$ is well defined. The set of regular random variables L_f as defined above can be characterised as follows:

Theorem 14.12 Pistone-Sempi, [PS95] L_f *is the closed subspace of the Orlicz space* $L^{\cosh-1}(f \cdot \nu)$ *of zero-expectation random variables.*

Inspired by the above, we now make the following definition.

Definition 14.13 For a general von Neumann algebra $\mathcal{M}$, we define the regular observables as the self-adjoint part of $L^{\cosh-1}(\mathcal{M})$

We show that in the σ-finite case (the case where the canonical faithful normal semifinite weight φ is a state), the regular observables as defined above do indeed fulfil the criterion of being 'moment-generating' functions.

Proposition 14.14   *Let the canonical weight* φ *on* $\mathcal{M}$ *be a state and let* $h = \frac{d\widetilde{\varphi}}{d\tau}$. *For any* $1 \le p < \infty$, *the prescription*

$a \to \gamma_p(h)^{1/2} a \gamma_p(h)^{1/2}$, *where* $\gamma_p(h) = h^{1/p} \mathbf{f}^{\cosh-1}(h)^{-1}$, *yields a well-defined continuous embedding of* $L^{\cosh-1}(\mathcal{M})$ *into* $L^p(\mathcal{M})$. *Hence, for any* $a \in L^{\cosh-1}(\mathcal{M})$ *and any* $1 \le p < \infty$, *the* generalised *moment* $\mathrm{tr}(|\gamma_p(h)^{1/2} a \gamma_p(h)^{1/2}|^p)$ *is finite.*

Proof. Given any $2 \le p < \infty$, select $m \in \mathbb{N}$ so that $2m \le p < 2(m+1)$. Then, of course, $t^p \le t^{2m} + t^{2(m+1)}$ for all $t \ge 0$. On considering the Maclaurin expansion of $\cosh(t) - 1$, it now trivially follows that $\frac{1}{(2(m+1))!} t^p \le \frac{1}{(2(m+1))!}(t^{2m} + t^{2(m+1)}) \le \cosh(t) - 1$ for all $t \ge 0$. It is a straightforward exercise to conclude from this fact that $\left(\frac{t}{(2(m+1))!}\right)^{1/p} \le [\cosh^{-1}(t^{-1} + 1)]^{-1} = \mathbf{f}^{\cosh-1}(t)$. In other words, the function $\gamma_p(t) = \frac{t^{1/p}}{\mathbf{f}^{\cosh-1}(t)}$ is a well-defined bounded continuous function on $(0, \infty)$. Hence, $\gamma_p(h)$ is a bounded operator. So, for any $a \in L^{\cosh-1}(\mathcal{M})$, $\gamma_p(h)^{1/2} a \gamma_p(h)^{1/2}$ will clearly be τ-measurable. It is now an exercise to verify that the dual action $\{\theta_s\}$ on such objects fulfil the membership criteria for $L^p(\mathcal{M})$.

It remains to consider the case $1 \le p < 2$. Since φ is a state, we must have $h \in L^1(\mathcal{M})$. So, given $r \ge 2$ with $\frac{1}{p} = \frac{1}{2} + \frac{1}{r}$, it is ensured that $h^{1/(2r)} \in L^{2r}(\mathcal{M})$ and (by Hölder's inequality) that the prescription $b \mapsto h^{1/(2r)} b h^{1/(2r)}$ yields a continuous embedding of $L^2(\mathcal{M})$ into $L^p(\mathcal{M})$. The next step is to compose this embedding with the embedding $a \mapsto \gamma_2(h)^{1/2} a \gamma_2(h)^{1/2}$ and to check that $h^{1/r} \gamma_2(h) = \gamma_p(h)$. That then yields the conclusion that here, too, $\gamma_p(h)^{1/2} a \gamma_p(h)^{1/2}$ will be a well-defined element of $L^p(\mathcal{M})$ for each $a \in L^{\cosh-1}(\mathcal{M})$, with $a \mapsto \gamma_p(h)^{1/2} a \gamma_p(h)^{1/2}$ continuous in the topology of convergence in measure. $\qquad\square$

There is actually a close connection between the spaces $L^{\cosh-1}$ and $L\log(L+1)$, which we now elucidate.

Proposition 14.15 *The space* $L\log(L+1)(\mathcal{M})$ *is isomorphic to* $L^{(\cosh-1)^*}(\mathcal{M})$ *where* $(\cosh-1)^*$ *is the conjugate Young function of* $\cosh-1$. *In the tracial case,* $L^{(\cosh-1)^*}(\mathcal{M})$ *is of course just the Köthe dual of* $L^{\cosh-1}(\mathcal{M}, \tau)$.

Proof. We note from Exercise 6.37, that the conjugate Young function of $\cosh-1$ is $\int_0^t \sinh^{-1}(s)\,ds = \int_0^t \log(s + \sqrt{s^2 + 1})\,ds$. But for any $s \ge 0$, we have $\log(s+1) \le \log(s + \sqrt{s^2 + 1}) \le \log(s^2 + s + \sqrt{s^2 + 2s + 1}) =$

$\log((s+1)^2) = 2\log(s+1)$, with $\int_0^t \log(s+1)\,ds = (t+1)\log(t+1) - t$. Thus, the conjugate Young function of $\cosh - 1$ is equivalent to the Young function $(t+1)\log(t+1) - t$. We show that $t\log(t+1)$ is in turn equivalent to $(t+1)\log(t+1) - t$. In this regard, note that

$$\frac{(t+1)\log(t+1) - t}{t\log(t+1)} = 1 + \frac{\log(t+1) - t}{t\log(t+1)} \text{ for all } t > 0.$$

Therefore

$$\lim_{t\to\infty} \frac{(t+1)\log(t+1) - t}{t\log(t+1)} = 1 + \lim_{t\to\infty}\left(\frac{1}{t} - \frac{1}{\log(t)}\right) = 1.$$

Using L'Hospital's rule twice, we also have that

$$\begin{aligned}
\lim_{t\searrow 0} \frac{(t+1)\log(t+1) - t}{t\log(t+1)} &= 1 + \lim_{t\searrow 0} \frac{\log(t+1) - t}{t\log(t+1)}\\[2mm]
&= 1 + \lim_{t\searrow 0} \frac{(1/(t+1)) - 1}{(t/(t+1)) + \log(t+1)}\\[2mm]
&= 1 + \lim_{t\searrow 0} \frac{-t}{t + (t+1)\log(t+1)}\\[2mm]
&= 1 + \lim_{t\searrow 0} \frac{-1}{2 + \log(t+1)}\\[2mm]
&= \frac{1}{2}.
\end{aligned}$$

Since $\frac{(t+1)\log(t+1) - t}{t\log(t+1)}$ is strictly positive on $(0,\infty)$, these two facts ensure that, as required, there exists a constant $1 \leq \alpha$, so that $\alpha^{-1}t\log(t+1) \leq (t+1)\log(t+1) - t \leq \alpha t\log(t+1)$. If we combine this with our earlier analysis, it is now clear that there must exist some $\beta > 1$ so that $\beta^{-1}t\log(t+1) \leq \int_0^t \sinh^{-1}(s)\,ds \leq \beta t\log(t+1)$. We shall, respectively, write Φ_1 and Φ_2 for $t\log(t+1)$ and $\int_0^t \sinh^{-1}(s)\,ds$. In the tracial case, the above inequalities are already enough to ensure that the corresponding Orlicz spaces are isomorphic. (See Remark 6.45.) We show that these spaces are also isomorphic in the type III case. Observe that the convexity of Φ_1 (combined with the fact that $\Phi_1(0) = 0$) ensures that $\Phi_1(\beta^{-1}t) \leq \beta^{-1}\Phi_1(t)$ for all $t > 0$ and hence that $\Phi_1(\beta^{-1}t) \leq \Phi_2(t) \leq \Phi_2(\beta t)$ Both these functions are strictly increasing, and hence, this inequality translates to the inequality $\beta^{-1}\Phi_2^{-1}(t) \leq \Phi_1^{-1}(t) \leq \beta\Phi_2^{-1}(t)$. This, in turn, yields $\beta^{-1}\mathbf{f}^{\Phi_1}(t) \leq \mathbf{f}^{\Phi_2}(t) \leq \beta\mathbf{f}^{\Phi_1}(t)$. But then $\beta^{-1}\mathbb{1} \leq \kappa(h) \leq \beta\mathbb{1}$

where $\kappa(h) = \mathbf{f}^{\Phi_2}(h)(\mathbf{f}^{\Phi_2}(h))^{-1}$ and $h = \frac{d\widetilde{\varphi}}{d\tau}$. Using these facts, it is now an exercise to see that the prescription $a \mapsto \kappa(h)^{1/2} a \kappa(h)^{1/2}$ defines an invertible map from $L^{\Phi_1}(\mathcal{M})$ onto $L^{\Phi_2}(\mathcal{M})$.

The final claim is a direct consequence of Theorem 6.54. $\qquad\square$

There is one more observation we can make regarding the space $L^{\cosh -1}(\mathcal{M})$. For this, we need the concept of a Markov operator for the pair $(\mathcal{M}, \varphi)$ as introduced in Definition 12.21. If a Markov operator T is in fact completely positive, we will refer to it as a completely Markov map.

Recalling that the upper fundamental index of the space $L^{\cosh -1}(0, \infty)$ is strictly less than 1 (see Example 10.21), the following result is now an immediate consequence of the results in Chapter 11. Therefore, we see that physically reasonable completely Markov dynamics on $\mathcal{M}$ extends to the space $L^{\cosh -1}(\mathcal{M})$.

Proposition 14.16 *Let $\{T_t\}$ be a completely Markov normal semi-group on $\mathcal{M}$ satisfying $\varphi \circ T_t \leq \varphi$ for all t with respect to the reference weight φ. Then, the semigroup $\{T_t\}$ allows for a canonical extension to the space $L^{\cosh -1}(\mathcal{M})$.*

The above results indicate that the concepts of regular observables and regular states are somehow dual to each other and hence that the pair of spaces $\langle L\log(L+1), L^{\cosh -1} \rangle$ offer a very attractive and elegant rigorous mathematical framework for the study of Quantum Statistical Mechanics.

14.5 Regularity for AQFT

Our final objective in this chapter is to show that the spaces $L\log(L+1)(\mathcal{M})$ and $L^{\cosh -1}(\mathcal{M})$ are not only relevant for Quantum Statistical Mechanics, but also for AQFT (Algebraic Quantum Field Theory). We specifically show that the space $L^{\cosh -1}(\mathcal{M})$ appears in a very natural way in AQFT. The key mathematical result in this section is Theorem 14.29. The focus of this result is to show that under very natural criteria, a large number of quantum fields which satisfy a so-called generalised $\mathbb{H}$-bound condition automatically allow one to realise the field operators as elements of the space of regular observables $L^{\cosh -1}$.

Readers familiar with QFT will know that the field operators are necessarily unbounded. So, the fact that this realisation is even possible is because the space $L^{\cosh-1}$ is in the case of nonatomic algebras large enough to admit unbounded elements. The rest of the section leading up to this result is devoted to describing the context of this result, with, for example, Theorem 14.21 describing how AQFT modelling and modular theory line up. Readers willing to take this contextualisation at face value may dive in once they are familiar with the definition of Orlicz spaces for general von Neumann algebras.

14.5.1 Two classical paradigms

We start with a brief review of the basic precepts of AQFT. Gårding and Wightman formulated the first axiomatic operator-theoretic framework for AQFT in the mid-1950s (see [**GW54a, GW54b, Wig57**]). Then, in 1964, Rudolf Haag and Daniel Kastler proposed what seemed like an alternative axiomatic approach to AQFT, based on local algebras (see [**Haa96**] for a more recent account of the theory). We will briefly survey each of these two approaches before describing how the two frameworks may be harmonised.

We emphasise that we are not interested in a detailed application of the resultant theory to a specific context, but rather in the overarching mathematical framework. Therefore, we will not discuss the finer nuances of the various postulates of the two approaches.

In some sense, both frameworks have the objective of formulating a rigorous framework for the description of the observables of Minkowski space-time. Hence, some rudimentary understanding of the basic features of Minkowski spacetime is essential.

Definition 14.17 d-dimensional Minkowski space is a combination of the $(d-1)$-dimensional Euclidean space $\mathbb{R}^{(d-1)}$ and time into a d-dimensional manifold. With the speed of light set to $c = 1$, it is structurally basically $\mathbb{R}^d$ equipped with the Minkowski metric defined by means of the bilinear form

$$(t_1, a_1, a_2, \ldots, a_{(d-1)}).(t_2, b_1, b_2, \ldots, b_{(d-1)}) = t_1 t_2 - a_1 b_1 - a_2 b_2 - \ldots$$
$$- a_{(d-1)} b_{(d-1)}.$$

The two points $(t_1, a_1, a_2, \ldots, a_{(d-1)})$ and $(t_2, b_1, b_2, \ldots, b_{(d-1)})$ are said to be time-like separated if $(t_1 - t_2)^2 > \sum_{k=1}^{d-1}(a_k - b_k)^2$ and space-like if $(t_1 - t_2)^2 < \sum_{k=1}^{d-1}(a_k - b_k)^2$. We shall where convenient denote Minkowski space by $\mathbb{M}$.

Given a point $x \in \mathbb{M}$, we say that another point $y \in \mathbb{M}$ causally precedes x if x and y are time-like seperated with the time coordinate of $y - x$ positive. The *forward light cone* of x is then defined as the set of all points y causally preceded by x, whereas the *rearward light cone* of x is defined as the set of all points y causally preceding x. Geometrically, these are indeed two cones radiating from the point x.

Though Minkowski space may formally be defined for any dimension larger than 1, we shall be concerned with four-dimensional Minkowski space. A crucial companion to the study of four-dimensional Minkowski space is the Poincaré group.

Definition 14.18 The Poincaré group $P_+^\uparrow$ is a ten-generator non-abelian Lie group made up of the group of isometries on four-dimensional Minkowski space. The ten degrees of freedom for the isometries of Minkowski space are

- translation through time or space (four degrees, one per dimension);
- reflection through a plane (three degrees, the freedom in orientation of this plane);
- or a 'boost' in any of the three spatial directions (three degrees).

The boosts are transformations that connect two uniformly moving bodies (the relative motion of two frames, with one frame F moving with velocity v relative to another frame F'). All the other generators involve only a single frame.

Remark 14.19 For the sake of completeness, we note that the Poincaré group $P_+^\uparrow$ may be written as the semidirect product $\mathbb{R}^4 \rtimes SL(2, \mathbb{C})$, where $\mathbb{R}^4$ represents the translation group of $\mathbb{M}$ and $SL(2, \mathbb{C})$ the complex 2×2 matrices with unit determinant.

We start by reviewing the Gårding-Wightman postulates. These are as follows.

(F1) *Quantum fields:* The operators $\phi_1(f), ..., \phi_n(f)$ are given for each C^∞-function with compact support in the Minkowski space $\mathbb{R}^4$. Each $\phi_i(f)$ and its Hermitian conjugate operator $\phi_j^*(f)$ are defined on at least a common dense linear subset $\mathcal{D}$ of the Hilbert space H and $\mathcal{D}$ satisfies 14.5.1

$$\phi_j(f)\mathcal{D} \subset \mathcal{D}, \quad \phi_j^*(f)\mathcal{D} \subseteq \mathcal{D}$$

for any f, $j = 1, ..., n$. For any $v, w \in \mathcal{D}$

$$f \mapsto \langle \phi_j(f)w, v \rangle$$

is a complex-valued distribution.

(F2) *Relativistic symmetry:* There is a well-defined strongly continuous unitary representation $U(a, A)$ of $P_+^\uparrow$ ($a \in \mathbb{R}^4$, $A \in SL(2, \mathbb{C})$) such that

$$U(a, A)\mathcal{D} = \mathcal{D}$$

and

$$U(a, A)\phi_j(f)U(a, A)^* = \sum S(A^{-1})_{jk}\phi_k(f_{(a,A)}),$$

where the matrix $(S(A)_{j,k})$ is an n-dimensional representation of $A \in SL(2, \mathbb{C})$, and $f_{(a,A)}(z) = f(\Lambda(A)^{-1}(z - a))$.

(F3) *Local commutativity:* if the supports of f and g are space-like separated, then for any vector $v \in \mathcal{D}$

$$[\phi_j(f)^\diamond, \phi_k(g)^\diamond]_\mp(v) = 0$$

where $\diamond$ denotes the following possibilities: no $*$, one $*$ and both operators ϕ have a $*$.

(F4) *Vacuum:* There exists a well-defined vacuum state, that is, a vector $\Omega \in H$, invariant with respect to the Poincaré group such that the following spectrum condition is satisfied: the spectrum of the translation group $U(a, \mathbb{1})$ on $\Omega^\perp$ is contained in $V_m = \{p: (p, p) \geq m^2, p^0 > 0\}$, $m > 0$.

It is worth pointing out that the spectral conditions mentioned in the point (L6) above are relevant only for transformations defined in terms of inertial frames with no relative motion. In particular, the spectral conditions mentioned there are not applicable to generators of boosts.

We pass to introducing the Haag–Kastler local algebra approach. The basic object of this approach is a net of von Neumann algebras $\mathcal{O} \mapsto \mathcal{M}(\mathcal{O})$ on a Hilbert space H, labelled by subsets $\mathcal{O}$ of d-dimensional Minkowski space $\mathbb{M}$. Each local algebra $\mathcal{M}(\mathcal{O})$ is home to the observables of that 'piece' of spacetime with the various local algebras $\mathcal{M}(\mathcal{O})$ interacting with each other in a way that respects the structure of Minkowski space. The local algebras satisfy the following postulates.

(L1) *Isotony:* $\mathcal{O}_1 \subseteq \mathcal{O}_2$ implies $\mathcal{M}(\mathcal{O}_1) \subseteq \mathcal{M}(\mathcal{O}_2)$.

(L2) *Covariance:* For $g = (a, \Lambda) \in P_+^\uparrow$, there is a representation α_g in $Aut(\mathcal{M})$ such that

$$\alpha_g(\mathcal{M}(\mathcal{O})) = \mathcal{M}(g\mathcal{O}), \quad g\mathcal{O} = \{\Lambda x + a : x \in \mathcal{O}\}$$

$P_+^\uparrow$ stands for the Poincaré group, where the Lorentz group is restricted, homogeneous, see [**Ara99**, Section 3.3]. This action is realised by a strong operator continuous unitary group $\{U(a, \Lambda)\} \subset \mathcal{M} = \overline{\bigcup_{\mathcal{O} \subset \mathbb{R}^4} \mathcal{M}(\mathcal{O})}^{w^*}$. (See [**Bor66**] for this last restriction.)

(L3) *Locality:* If $\mathcal{O}_1$ and $\mathcal{O}_2$ are space-like separated, then $\mathcal{M}(\mathcal{O}_1)$ and $\mathcal{M}(\mathcal{O}_2)$ commute.

(L4) *Weak additivity:*

$$\mathcal{M} = \left(\bigcup_{x \in \mathbb{R}^4} \mathcal{M}(\mathcal{O} + x) \right)''$$

for all open $\mathcal{O}$.

(L5) *Vacuum vector:* There is a normalised vector $\Omega \in H$, unique up to a phase factor that satisfies $(\Omega, \alpha_g(f)\Omega) = (\Omega, f\Omega)$ for each f in the global algebra $\mathcal{M} = \overline{\bigcup_{\mathcal{O} \subset \mathbb{R}^4} \mathcal{M}(\mathcal{O})}^{w^*}$ and each $g = (a, \Lambda) \in P_+^\uparrow$.

(L6) *Positivity:* The generator of translation has spectrum lying in the forward light cone.

It is worth recalling that such Haag–Kastler local algebras are in fact type III_1 factors! See [**Yng05**].

We pause to review the famous Reeh–Schlieder theorem, which provides some information on what would be a natural candidate for a 'reference weight' on the global algebra $\mathcal{M}$.

Theorem 14.20 (Reeh–Schlieder [Haa96, Theorems II.5.3.1 and II.5.3.2]) *For each open bounded region $\mathcal{O}$, the set of vectors $\mathcal{M}(\mathcal{O})\Omega$ is dense in H. If $\mathcal{O}$ has a non-empty causal complement, Ω is also separating.*

In general, the Reeh–Schlieder theorem provides criteria ensuring that the vacuum vector is both cyclic and separating for each local algebra $\mathcal{M}(\mathcal{O})$ corresponding to some bounded open region in $\mathbb{M}$. However, this fact does not ensure that it is also separating for the full algebra $\mathcal{M}$. The most we can say is that it is (trivially) also cyclic for this algebra. Therefore, some ingenuity will be needed when using the vector state $\omega = \langle \cdot\,\Omega, \Omega \rangle$ to study the full algebra. We shall revisit this point at a later stage.

Thus far, 'boosts' have not really featured in our modelling. These come into play via the modular group of $\mathcal{M}$ equipped with ω. We briefly explain how.

Let W denote the wedge

$$W = \{ r \in \mathbb{M} \colon r^1 > |r^0|,\ r^2, r^3 \ \ \text{arbitrary} \} \tag{14.21}$$

in Minkowski space $\mathbb{M}$ and $U(\Lambda(s))$ (respectively, $U(r)$) the unitary operators implementing the boosts $\Lambda(s)$ (respectively, the space-time translations). The corresponding generators will be denoted by K and P_μ, i.e.

$$U(\Lambda(s)) = e^{iKs}, \quad U(r) = e^{iP_\mu r^\mu}. \tag{14.22}$$

Finally, let Θ stand for the so-called CPT operator (cf. [**Haa96**, Chapter II]) and let $R_1(\pi)$ denote a special rotation through the angle of π around the 1-axis. Bisognano and Wichmann proved the following (see [**BW75**] and [**BW76**]).

Theorem 14.21 ([BW75], [BW76]) *Let J_W and Δ_W denote the modular conjugation and the modular operator for the pair $(\mathcal{M}(W), \Omega)$. Then*

$$J_W = \Theta U(R_1(\pi)), \quad \Delta_W = e^{-2\pi K}. \tag{14.23}$$

Moreover, the modular automorphisms σ_t, $t \in \mathbb{R}$, act geometrically as the boosts!

When passing to the canonical copy of $\mathcal{M}(W)$ inside the crossed product $\mathcal{M}(W) \rtimes_\omega \mathbb{R}$, the modular automorphism group is then of course of the form $\sigma_t(\cdot) = h^{it}(\cdot)h^{-it}$ where $h = \frac{d\tilde{\omega}}{d\tau}$. So here the role played by $\Delta_W = e^{-2\pi K}$ above is played by h.

We now consider the problem of finding natural conditions under which the two paradigms may be merged. Here, the Hamiltonian will play an important role.

Definition 14.22 We define the Hamiltonian $\mathbb{H}$ of the system to be that operator for which $i\mathbb{H}$ is the time derivative at 0 of the subgroup of the unitary group $\{U(a,\Lambda)\}$ corresponding to translations. The fact that $\{U(a,\Lambda)\} \subset \mathcal{M}$ ensures that $\mathbb{H}$ is affiliated to $\mathcal{M}$.

14.5.2 Merging the two paradigms

In general, the field operators $\phi_i(f)$ are unbounded. The key to merging the two paradigms lies in imposing some regularity restrictions on the extent of their unboundedness. These restrictions are, in turn, nothing more than requiring good behaviour from the field operators with respect to the Hamiltonian $\mathbb{H}$ (the time derivative of the translation group). The following example of such a condition may be found in [**BY90**]:

Definition 14.23 Let ϕ be a Wightman field and let $\mathbb{H}$ denote its Hamiltonian. The field satisfies a *generalised* $\mathbb{H}$ *bound* if there exists a non-negative number $\alpha < 1$, such that $\phi(f)^{**}e^{-\mathbb{H}^\alpha}$ is a bounded operator for all f.

The physical significance of such conditions is that they select models with slightly more regular high energy behaviour. The utility of such a condition is illustrated by Theorem 14.24. This theorem shows that under the above restrictions, one may, in fact, construct a local algebra from a Wightman field.

We pause to point out that when constructing a local algebra, it is enough from a physical point of view to construct a net of algebras using only 'causally complete' regions in Minkowski space. The prototype of such a causally complete region is a *double cone*, where a double cone is defined as the intersection of the forward light cone of some point $x_1 \in \mathbb{M}$, with the rearward light cone of another point x_2 causally preceded by x_1. The causal completion of an open bounded region $\mathcal{O} \subset \mathbb{M}$ turns out to be nothing more than the smallest double cone containing $\mathcal{O}$.

Theorem 14.24 shows that, under suitable restrictions, we may indeed construct a local algebra using causally complete regions from a given Wightman field. However, some preparation is necessary if we are to make sense of this theorem.

The objective of showing how a field operator can be associated with a net of von Neumann algebras was realised by deep contributions from, for example, [**DSW86**], [**Buc90**], [**Ara99**]. We briefly summarise these contributions.

Let $\mathcal{P}$ be a family of operators with a common dense domain of definition $\mathcal{D}$ in a Hilbert space H (cf. Wightman's rule presented above) such that if $\phi \in \mathcal{P}$, then also $\phi^*|_{\mathcal{D}} \equiv \phi^\dagger \in \mathcal{P}$. The weak commutant $\mathcal{P}^w$ of $\mathcal{P}$ is defined as the set of all bounded operators C on H such that $\langle C\phi w, v \rangle = \langle Cw, \phi^\dagger v \rangle$, for all $v, w \in \mathcal{D}$.

For simplicity of our arguments, we will restrict ourselves to one type of real scalar field ϕ, i.e., $\phi(f)^*$ coincides with $\phi(\overline{f})$ on $\mathcal{D}$. Furthermore, apart from the Wightman postulates, we assume that:

- (A1) $\mathcal{P}(\mathcal{O}_q^p)^w$ is an algebra for any double cone $\mathcal{O}_q^p$ $\equiv \{x \colon p - x \in V_+, x - q \in V_+\}$, where $V_+ = \{\text{positive timelike vectors}\}$;
- (A2) the vacuum vector Ω is cyclic for the union of $\mathcal{P}(D')^w$ over all double cones D, where D' is the causal complement of D.

The following theorem is taken from [**Ara99**], but stems from results given in [**DSW86**] and [**Buc90**].

Theorem 14.24 ([Ara99], cf. [Buc90] and [DSW86]) *Assume that both conditions (A1) and (A2) hold. For each double cone D, define*

$$\mathcal{M}(D) = (\mathcal{P}(D)^w)'. \tag{14.24}$$

Then, $\mathcal{M}(D)$ is a von Neumann algebra and the net $D \mapsto \mathcal{M}(D)$ satisfies conditions (L1)-(L3) for local algebras. When defining $\mathcal{M}$ to be the von Neumann algebra generated by $\cup_D \mathcal{M}(D)$, the state on $\mathcal{M}$ determined by Ω is then a pure vacuum state for which:

- *Ω is cyclic for each $\mathcal{M}(D)$;*
- *each operator $\phi \in \mathcal{P}(D)$ has a closed extension $\phi_e \subseteq \phi^{\dagger,*}$ which is affiliated with $\mathcal{M}(D)$. (Here, $\phi_e \subseteq A$ means that the domain of ϕ_e is contained in the domain of A and that $\phi_e = A$ on the domain of ϕ_e.)*

Theorem 14.24 yields the following corollary.

Corollary 14.25 *Field operators lead to operators affiliated to the von Neumann algebra $\mathcal{M}(D)$. We recall that this property is the starting point for the definition of τ-measurable operators.*

There are sufficient conditions, motivated by physical requirements, for conditions (A1) and (A2) (given prior to Theorem 14.24) to hold (see [**Ara99**], [**DSW86**], [**Buc90**], [**BY90**]). One such condition is to select models that satisfy the so-called *generalised* $\mathbb{H}$ *bound* condition. Then, [**BY90**, Lemma 1.5] will ensure that (A1) holds. If to this requirement one adds the notion of *central positivity*[1], that would then ensure the validity of Theorem 14.24 (see [**BY90**, Theorem 3.1]).

We shall henceforth assume that such a generalised $\mathbb{H}$-bound holds.

14.5.3 Finding a home for the field operators

Theorem 14.26 Observation *For any field operator $\phi(f)$ satisfying Wightman's postulates, one has*

$$\langle \phi^n(f)w, v \rangle \in \mathbb{C} \tag{14.25}$$

for any $v, w \in D \subset H$ and any $n \in \mathbb{N}$. In other words, the number $\langle \phi^n(f)w, v \rangle$ is finite for any $n \in \mathbb{N}$.

The above observation shows that, in some sense, the field operators have some of the character of 'moment-generating' functions. Taking Proposition 14.14 into account, this suggests that from a physical point of view, the space $L^{\cosh-1}(\mathcal{M}(D))$ ought to be a natural home for the

---

[1] A state ρ on a unital C^*-algebra $\mathcal{A}$ is said to be centrally positive if for every selfadjoint element a_0 of $\mathcal{A}$, ρ is non-negative valued on all terms of the form $\sum_{n=1}^{\infty} a_0^n a_n$ where $(a_n) \subset \mathcal{A}$ is a sequence such that $\sum_{n=1}^{\infty} \lambda^n a_n \in \mathcal{A}_+$ for all $\lambda \in \mathbb{R}$

field operators. Thus, a natural restriction to place on them would be to require them to embed into the space $L^{\cosh{-1}}(\mathcal{M}(D))$.

Definition 14.27 A field operator $\phi(f)$ affiliated to $\mathcal{M}(D)$ is said to satisfy an $L^{\cosh-1}$ *regularity restriction* if the strong product $\mathbf{f}^{\cosh-1}(h)^{1/2}\phi(f)_e\mathbf{f}^{\cosh-1}(h)^{1/2}$ (where $\phi(f)_e$ is the minimal closed extension of $\phi(f)$) is a closable operator for which the closure is τ-measurable, that is, the closure is an element of the space $\widetilde{\mathfrak{M}}$. For the sake of simplicity of notation, we will hereafter simply write $\mathbf{f}^{\cosh-1}(h)^{1/2}\phi(f)\mathbf{f}^{\cosh-1}(h)^{1/2}$ instead of $\mathbf{f}^{\cosh-1}(h)^{1/2}\phi(f)_e\mathbf{f}^{\cosh-1}(h)^{1/2}$.)

Remark 14.28 Given that each of the $\phi(f)$s are fixed points of the action of (θ_s), it is a simple matter to verify that an $L^{\cosh-1}$ regularity restriction on some $\phi(f)$ ensures that the closure of $\mathbf{f}^{\cosh-1}(h)^{1/2}\phi(f)\mathbf{f}^{\cosh-1}(h)^{1/2}$ satisfies the membership criteria for $L^{\cosh-1}(\mathcal{M}(D))$ described above. To see this observe that for $v_s = \mathbf{f}^{\cosh-1}(e^{-s}h)\mathbf{f}^{\cosh-1}(h)^{-1}$, we clearly have that

$$\theta_s(\mathbf{f}^{\cosh-1}(h)^{1/2}\phi(f)\mathbf{f}^{\cosh-1}(h)^{1/2})$$
$$= \mathbf{f}^{\cosh-1}(e^{-s}h)^{1/2}\phi(f)\mathbf{f}^{\cosh-1}(e^{-s}sh)^{1/2}$$
$$= v_s^{1/2}(\mathbf{f}^{\cosh-1}(h)^{1/2}\phi(f)\mathbf{f}^{\cosh-1}(h)^{1/2})v_s^{1/2}$$

for each $s \in \mathbb{R}$.

We now show that, under very natural criteria, a large number of fields satisfying a generalised $\mathbb{H}$-bound condition will automatically satisfy an $L^{\cosh-1}$ regularity restriction. The crucial assumption here is that the vacuum vector is an analytic vector for the Hamiltonian $\mathbb{H}$. Note that this assumption of analyticity is intimately related to the validity of the Reeh–Schlieder theorem, which is a key ingredient in Quantum Field Theoretic modelling. To see this, the reader may refer to [**Hor90**, 1.3.1-1.3.3].

Theorem 14.29 *Let a quantum field be given that satisfies a generalised $\mathbb{H}$(-power)-bound condition, in the sense that each $\phi(f)$ is a densely defined Hermitian operator with $\pm\phi(f) \leq K(f)(\mathbb{1}+\mathbb{H})^n$ for some positive integer n and $K(f) > 0$, where $K(f)$ may depend on f. Let e_ω be the support projection of the state $\omega = \langle\cdot\Omega_\omega,\Omega_\omega\rangle$.*

Then, the compression $e_\omega \phi(f) e_\omega$ of the field satisfies an $L^{\cosh{-1}}$ regularity restriction for the compressed algebra $e_\omega \mathcal{M} e_\omega$ in the sense that $\mathbf{f}^{\cosh{-1}}(g_\omega)^{1/2} e_\omega \phi(f) e_\omega \mathbf{f}^{\cosh{-1}}(g_\omega)^{1/2}$ is preclosed, with the closure being a τ-measurable operator affiliated to $e_\omega \mathcal{M} e_\omega \rtimes_\omega \mathbb{R}$. Here, g_ω is the density of the dual weight $\widetilde{\omega}$ on $e_\omega \mathcal{M} e_\omega \rtimes_\omega \mathbb{R}$.

We point out that in one of the first concrete examples of a field satisfying a generalised $\mathbb{H}$-bound, Driessler and Fröhlich [**DF77**] proved that for the so-called Osterwalder–Schrader reconstruction, a generalised $\mathbb{H}$-bound pertains in the sense that $\pm \phi(f) \le \|f\|(\mathbb{H} + \mathbb{1})$ for some continuous norm on the space of test functions. In general, such a linear bound is too restrictive with a much larger number of examples satisfying a condition of the form described above. (See the discussion after [**DSW86**, Definition 5.1].)

Remark 14.30 In AQFT, the net of algebras $\mathcal{O} \mapsto \mathcal{M}(\mathcal{O})$ constitutes the intrinsic mathematical description of the theory; cf Section 1. Hence, for the vacuum state ω, the essential information is provided by

$$\mathcal{M}(\mathcal{O}) \ni a \mapsto \omega(a). \tag{14.26}$$

We note that as $\omega(e_\omega a e_\omega) = \omega(a)$ for $a \in \cup_\mathcal{O} \mathcal{M}(\mathcal{O})$, the net (14.26) can be used to great effect in the compressed algebra setting. In other words, the passage to the compression determined by e_ω does not spoil the basic features of the theory. The advantage of using the compression $e_\omega \cdot e_\omega$ will be explained in Remark 14.31.

Proof of Theorem 14.29   Throughout the proof, we shall write $\mathcal{M}$ for $\mathcal{M}(\mathbb{M})$. We first note that we of course have $e_\omega \in \mathcal{M}$ (Proposition 0.97). The further fact that the unitary group of which $\mathbb{H}$ is the generator is contained in $\mathcal{M}$, ensures that $\mathbb{H}$ is affiliated with $\mathcal{M}$ and hence also with $\mathfrak{M}$. We denote this group by $U(a)$ $(a \in \mathbb{M})$ since only translation is in view.

Select a faithful normal semifinite weight δ on $(\mathbb{1} - e_\omega)\mathcal{M}(\mathbb{1} - e_\omega)$. The weight δ extends to a normal semifinite weight on $\mathcal{M}$ in an obvious way. We now define a weight φ on $\mathcal{M}$ by prescription

$$a \mapsto \omega(e_\omega a e_\omega) + \delta((\mathbb{1} - e_\omega)a(\mathbb{1} - e_\omega)), \qquad a \in \mathcal{M}.$$

It is an exercise to see that φ is a faithful normal semifinite weight on $\mathcal{M}$. (To see the faithfulness, notice that for $a \geq 0$, we have $\varphi(a) = 0 \Rightarrow e_\omega a e_\omega = (\mathbb{1} - e_\omega)a(\mathbb{1} - e_\omega) = 0 \Rightarrow a^{1/2}e_\omega = a^{1/2}(\mathbb{1} - e_\omega) = 0 \Rightarrow a^{1/2} = 0 \Rightarrow a = 0$.) Now, pass to the crossed product $\mathcal{M} \rtimes_\varphi \mathbb{R}$. Given a normal semifinite weight γ on $\mathcal{M}$, we will write h_γ for the density of the dual weight. From Remark 9.34, we know that $h_\varphi = h_\omega + h_\delta$ and that $\operatorname{supp}(h_\omega) = \operatorname{supp}(\omega) = e_\omega$ and $\operatorname{supp}(h_\delta) = \operatorname{supp}(\delta) = \mathbb{1} - e_\omega$. Thus, $h_\omega h_\delta = 0$, which ensures that $h_\varphi^{it} = h_\omega^{it} + h_\delta^{it}$ for all $t \in \mathbb{R}$. But then $\sigma_t^\varphi(e_\omega) = h_\varphi^{it} e_\omega h_\varphi^{-it} = (h_\omega^{it} + h_\delta^{it})e_\omega(h_\omega^{-it} + h_\delta^{-it}) = h_\omega^{it}e_\omega h_\omega^{-it} = e_\omega$ for all $t \in \mathbb{R}$.

Now, recall that the canonical normal state ω corresponds to the vacuum state. But the vacuum state is invariant under the action of the automorphisms α_a implemented by translation at the level of $\mathbb{M}$ [**Ara99**, Theorem 4.5]. This means that for any $f \in \mathcal{M}$ and any $a \in \mathbb{M}$, we must have $\omega(f) = \omega(\alpha_a(f)) = \omega(U(a)fU(a)^*)$. Next, observe that in the language of Haagerup L^p-spaces (see Remark 10.41), this fact corresponds to the claim that $\operatorname{tr}(h_\omega f) = \operatorname{tr}(h_\omega U(a)fU(a)^*) = \operatorname{tr}(U(a)^* h_\omega U(a)f)$ for all $f \in \mathcal{M}$, where h_ω is the density of the dual weight $\widetilde{\omega}$ on $\mathcal{M} \rtimes_\varphi \mathbb{R}$. The well-developed duality theory of Haagerup L^p-spaces, combined with the fact that both h_ω and $U(a)^* h_\omega U(a)$ belong to $L^1(\mathcal{M})$, now ensures that $h_\omega = U(a)^* h_\omega U(a)$, or equivalently that $U(a)h_\omega = h_\omega U(a)$ for each $a \in \mathbb{M}$. When expressed at the level of the generator $\mathbb{H}$ of the group $U(a)$ ($a \in \mathbb{M}$), this corresponds to the claim that $\mathbb{H}$ and h_ω are commuting affiliated operators of $\mathfrak{M}$.

By postulate (F4), the vacuum vector is in the kernel of the energy operator $\mathbb{H}$ and therefore is automatically analytic for all powers $\mathbb{H}^k$ of $\mathbb{H}$. We shall need this fact shortly. We also remind the reader that in the context of Haagerup L^p-spaces, the standard form of a von Neumann algebra $\mathcal{M}$, may be realised by representing $\mathcal{M}$ as left-multiplication operators on $L^2(\mathcal{M})$—see Theorem 10.59.

Using this theorem, the vector Ω that implements the canonical faithful normal state ω may be identified with $h_\omega^{1/2}$. Saying that $h_\omega^{1/2}$ is in the domain of, say, $\mathbb{H}^k$ in this setting means that $\mathbb{H}^k h_\omega^{1/2}$ is τ-pre-measurable and that the extension (for which we use the same notation) is in $L^2(\mathcal{M})$. The assumption of analyticity regarding the vacuum vector ensures that $\sum_{m=1}^\infty \frac{1}{m!} t^m \mathbb{H}^{mk}\Omega$ converges absolutely, which, in turn, can only be true if each $\mathbb{H}^{mk} h_\omega^{1/2}$ extends uniquely to an element of $L^2(\mathcal{M})$. The series $\sum_{m=0}^\infty \frac{1}{m!} t^m \mathbb{H}^{mk} h_\omega^{1/2}$ of such extensions then converges absolutely to an element in $L^2(\mathcal{M})$. We may then use the comparison test for the series to conclude that $\sum_{m=0}^{\infty} \frac{1}{(2m)!} \|t^{2m}\mathbb{H}^{2mk} h_\omega^{1/2}\|_2$ converges. The absolute convergence of the series $\sum_{m=0}^{\infty} \frac{1}{(2m)!} t^{2m}\mathbb{H}^{2mk} h_\omega^{1/2}$ in $L^2(\mathcal{M})$ ensures that it must correspond to a well-defined element of $L^2(\mathcal{M})$. But this series is just a Maclaurin series representation of $(\cosh -1)(t\mathbb{H}^k)h_\omega^{1/2}$. So $(\cosh -1)(t\mathbb{H}^k)h_\omega^{1/2} \in L^2(\mathcal{M})$. Now, recall that the fact that ω is a state ensures that $h_\omega \in L^1(\mathcal{M})$, or equivalently, that $h_\omega^{1/2} \in L^2(\mathcal{M})$. Therefore, we have $h_\omega^{1/2}(\cosh -1)(t\mathbb{H}^k)h_\omega^{1/2} \in L^1(\mathcal{M})$ for every $1 \le k \le n$.

The fact that $\mathbb{H}$ and h_ω are commuting affiliated operators now comes into play. It now follows from Lemma 10.1 that

$$\tau(\chi_{(1,\infty)}(h_\omega^{1/2}(\cosh -1)(t\mathbb{H}^k)h_\omega^{1/2}))$$
$$= \tau(\chi_{(1,\infty)}(\mathbf{f}^{\cosh -1}(h_\omega)^{1/2}(t\mathbb{H}^k)\mathbf{f}^{\cosh -1}(h_\omega)^{1/2})).$$

By Lemma 10.23, we have that

$$\tau(\chi_{(1,\infty)}(h_\omega^{1/2}(\cosh -1)(t\mathbb{H}^k)h_\omega^{1/2})) = \|h_\omega^{1/2}(\cosh -1)(t\mathbb{H}^k)h_\omega^{1/2}\|_1 < \infty$$

and hence that

$$\tau(\chi_{(1,\infty)}(\mathbf{f}^{\cosh -1}(h_\omega)^{1/2}(t\mathbb{H}^k)\mathbf{f}^{\cosh -1}(h_\omega)^{1/2})) < \infty.$$

Thus, by definition, $\mathbf{f}^{\cosh -1}(h_\omega)^{1/2}(t\mathbb{H}^k)\mathbf{f}^{\cosh -1}(h_\omega)^{1/2}$ is τ-measurable. Since this holds for every $1 \le k \le n$, it follows that $\mathbf{f}^{\cosh -1}(h_\omega)^{1/2}(1 + \mathbb{H})^n\mathbf{f}^{\cosh -1}(h_\omega)^{1/2}$ is τ-measurable.

Recall that for any test function f, we have $\pm\phi(f)_e \le K(f)(1+\mathbb{H})^n$. It is therefore clear that

$$0 \ \le \ \mathbf{f}^{\cosh -1}(h_\omega)^{1/2}(K(f)(1 + \mathbb{H})^n + \phi(f)_e)\mathbf{f}^{\cosh -1}(h_\omega)^{1/2}$$
$$\le \ 2K(f)\mathbf{f}^{\cosh -1}(h_\omega)^{1/2}(1 + \mathbb{H})^n\mathbf{f}^{\cosh -1}(h_\omega)^{1/2}.$$

We may now use this inequality and the fact that $\operatorname{supp}(h_\omega) = \operatorname{supp}(\omega) = e_\omega$, to conclude that each $\mathbf{f}^{\cosh -1}(h_\omega)^{1/2}(K(f)(1 + \mathbb{H})^n + \phi(f)_e)\mathbf{f}^{\cosh -1}(h_\omega)^{1/2}$ is τ-measurable and hence also each $\mathbf{f}^{\cosh -1}(h_\omega)^{1/2}\phi(f)_e\mathbf{f}^{\cosh -1}(h_\omega)^{1/2} = \mathbf{f}^{\cosh -1}(h_\omega)^{1/2}e_\omega\phi(f)_e e_\omega\mathbf{f}^{\cosh -1}(h_\omega)^{1/2}$.

But since $\sigma_t^\varphi(e_\omega) = e_\omega$ for all t, we know from the proof of Theorem 10.55 that the subalgebra of $\mathcal{M} \rtimes_\varphi \mathbb{R}$ generated by $e_\omega\mathcal{M}e_\omega$ and the compressed left-translation operators $e_\omega\lambda(t)e_\omega$ is an exact copy of $e_\omega\mathcal{M}e_\omega \rtimes_\omega \mathbb{R}$. Therefore, it follows from the above fact that

$\mathbf{f}^{\cosh^{-1}}(g_\omega)^{1/2}e_\omega\phi(f)_e e_\omega \mathbf{f}^{\cosh^{-1}}(g_\omega)^{1/2}$ is a τ-measurable operator affiliated to $e_\omega \mathcal{M} e_\omega \rtimes_\omega \mathbb{R}$. $\qquad\qquad\qquad\square$

Remark 14.31 Haagerup's approach to noncommutative integration is very well suited to type III factors and hence well suited to the present setting. Furthermore, in this approach, the Tomita–Takesaki modular theory plays a crucial role. Note that in the context of Theorem 14.29, Ω is a common cyclic and separating vector for each of the local algebras $\mathcal{M}(\mathcal{O})$ in the net (14.27). But this implies that for $\mathcal{O} \neq \mathcal{O}'$, one gets two pairs $(\mathcal{M}(\mathcal{O}), \Omega)$ and $(\mathcal{M}(\mathcal{O}'), \Omega)$ with, in general, $\mathcal{M}(\mathcal{O}) \neq \mathcal{M}(\mathcal{O}')$. Thus, the modular structures associated with $(\mathcal{M}(\mathcal{O}), \Omega)$ and $\mathcal{M}(\mathcal{O}'), \Omega)$ are different—see the warning given prior to the proof of [**BR87**, Theorem 6.3.28].

It is therefore an important point to observe that here the employment of the compression given in Theorem 14.29 radically changes the picture. Specifically, here, the subalgebra of $B(e_\omega H)$ generated by $e_\omega \mathcal{M}(\mathcal{O})e_\omega$ is strongly dense in $e_\omega \mathcal{M} e_\omega$. To see this, it is enough to carefully apply [**KR83**, Theorem 9.2.36].

Consequently, in the compressed structures, one gets a unique modular structure and hence a unique basis for noncommutative integration. This explains why the use of compression in Theorem 14.29 was both necessary and expected.

Bibliography

[AAP82]   Charles A. Akemann, Joel Anderson, and Gert K. Pedersen. Triangle inequalities in operator algebras. *Linear and Multilinear Algebra*, 11(2):167–178, 1982.

[AM82]   Huzihiro Araki and Tetsuya Masuda. Positive cones and L_P-spaces for von Neumann algebras. *Publ. Res. Inst. Math. Sci.*, 18(2):759–831 (339–411), 1982.

[Ara73]   Huzihiro Araki. Relative Hamiltonian for faithful normal states of a von Neumann Algebra. *Publ. Res. Inst. Math. Sci.*, 9:165–209, 1973.

[Ara99]   Huzihiro Araki. *Mathematical theory of quantum fields*, volume 101 of *International Series of Monographs on Physics*. Oxford University Press, New York, 1999. Translated from the 1993 Japanese original by Ursula Carow-Watamura.

[Ara76]   Huzihiro Araki. Relative entropy of states of von Neumann algebras. *Publ. Res. Inst. Math. Sci.*, 11(3):809–833, 1975/76.

[Ara78]   Huzihiro Araki. Relative entropy for states of von Neumann algebras. II. *Publ. Res. Inst. Math. Sci.*, 13(1):173–192, 1977/78.

[Arv74]   William Arveson. On groups of automorphisms of operator algebras. *J. Funct. Anal.*, 15:217–243, 1974.

[Arv76]   William Arveson. *An invitation to C^*-algebras*. Springer-Verlag, New York–Heidelberg, 1976. Graduate Texts in Mathematics, No. 39.

[Arv02]   William Arveson. *A short course on spectral theory*, volume 209 of *Graduate Texts in Mathematics*. Springer-Verlag, New York, 2002.

[ARZ07]   Maryam H. A. Al-Rashed and Bogusław Zegarliński. Noncommutative Orlicz spaces associated to a state. *Studia Math.*, 180(3):199–209, 2007.

[ARZ11]   Maryam H. A. Al-Rashed and Boguslaw Zegarlinski. Noncommutative orlicz spaces associated to a state II. *Linear Algebra Appl.*, 435(12):2999–3013, 2011.

[ARZ14]   Maryam H. A. Al-Rashed and Boguslaw Zegarlinski. Monotone norms and finsler structures in noncommutative spaces. *Infin. Dimens. Anal. Quantum Probab. Relat. Top.*, 17(4):1450029, 22 pp., 2014.

[AV02]   R. Alexandre and C. Villani. On the Boltzmann equation for long-range interactions. *Comm. Pure Appl. Math.*, 55(1):30–70, 2002.

[Bad54]   W. G. Bade. Weak and strong limits of spectral operators. *Pacific J. Math.*, 4:393–413, 1954.

[BdlH20] Bachir Bekka and Pierre de la Harpe. *Unitary Representations of Groups, Duals, and Characters*, volume 250 of *Mathematical Surveys and Monographs*. American Mathematical Society, Providence, RI, 2020.

[BGL22]   David Blecher, Stanisław Goldstein, and Louis Labuschagne. Abelian von Neumann algebras, measure algebras and L^∞-spaces. *Expo. Math.*, 40(3):758–818, 2022.

[Bik04]   A. M. Bikchentaev. The continuity of multiplication for two topologies associated with a semifinite trace on von Neumann algebra. *Lobachevskii J. Math.*, 14:17–24, 2004.

[Bik06]   A. M. Bikchentaev. Local convergence in measure on semifinite von Neumann algebras. *Proc. Steklov Inst. Math.*, 255:35–48, 2006.

[BK90]   Lawrence G. Brown and Hideki Kosaki. Jensen's inequality in semi-finite von Neumann algebras. *J. Operator Theory*, 23(1):3–19, 1990.

[BL76]   Jöran Bergh and Jörgen Löfström. *Interpolation spaces: An introduction.* Springer-Verlag, Berlin-New York, 1976. Grundlehren der Mathematischen Wissenschaften, No. 223.

[Bla06]   B. Blackadar. *Operator algebras*, volume 122 of *Encyclopaedia of Mathematical Sciences*. Springer-Verlag, Berlin, 2006. Theory of C^*-algebras and von Neumann algebras, Operator Algebras and Non-commutative Geometry, III.

[BLM04]   David P. Blecher and Christian Le Merdy. *Operator algebras and their modules—an operator space approach*, volume 30 of *London Mathematical Society Monographs. New Series.* The Clarendon Press, Oxford University Press, Oxford, 2004. Oxford Science Publications.

[Bor66]   H. J. Borchers. Energy and momentum as observables in quantum field theory. *Commun. Math. Phys.*, 2:49–54, 1966.

[BR87]   Ola Bratteli and Derek W. Robinson. *Operator algebras and quantum statistical mechanics. 1. Texts and Monographs in Physics*, 2nd edn. Springer-Verlag, New York, 1987. C^*- and W^*-algebras, symmetry groups, decomposition of states.

[BR97]   Ola Bratteli and Derek W. Robinson. *Operator algebras and quantum statistical mechanics. 2. Texts and Monographs in Physics*, 2nd edn, Springer-Verlag, Berlin, 1997. Equilibrium states. Models in quantum statistical mechanics.

[BS88]   Colin Bennett and Robert Sharpley. *Interpolation of operators*, volume 129 of *Pure and Applied Mathematics*. Academic Press, Inc., Boston, MA, 1988.

[BS21]   Airat M. Bikchentaev and Anatolij N. Sherstnev. Studies on noncommutative measure theory in Kazan University (1968–2018). *Internat. J. Theor. Phys.*, 60(2):585–596, 2021.

[Buc90]   Detlev Buchholz. On quantum fields that generate local algebras. *J. Math. Phys.*, 31(8):1839–1846, 1990.

[BW75]   Joseph J. Bisognano and Eyvind H. Wichmann. On the duality condition for a Hermitian scalar field. *J. Math. Phys.*, 16:985–1007, 1975.

[BW76]   Joseph J. Bisognano and Eyvind H. Wichmann. On the duality condition for quantum fields. *J. Math. Phys.*, 17(3):303–321, 1976.

[BY90]   H. J. Borchers and Y. Yngvason. Positivity of Wightman functionals and the existence of local nets. *Commun. Math. Phys.*, 127:607–615, 1990.

[Cas13]   M. Caspers. The L^p-fourier transform on locally compact quantum groups. *J. Oper. Theory*, 69(1):161–193, 2013.

[CdlS15]   Martijn Caspers and Mikael de la Salle. Schur and fourier multipliers of an amenable group acting on non-commutative L^p-spaces. *Trans. Amer. Math. Soc.*, 367(10):6997–7013, 2015.

[Cec78]   Carlo Cecchini. On two definitions of measurable and locally measurable operators. *Boll. Un. Mat. Ital. A (5)*, 15(3):526–534, 1978.

[Cec86]   Carlo Cecchini. Noncommutative integration for states on von Neumann algebras. *J. Oper. Theory*, 15(2):217–237, 1986.

[CFK14]   F. Cipriani, U. Franz, and A. Kula. Symmetries of Lévy processes on compact quantum groups, their Markov semigroups and potential theory. *J. Funct. Anal.*, 266(5):2789–2844, 2014.

[Coh13]   Donald L. Cohn. *Measure Theory: Second Edition.* Birkhäuser Advanced Texts. Springer, Heidelberg-Dordrecht–London, 2013.

[Com71]   François Combes. Poids et espérances conditionnelles dans les algèbres de von Neumann. *Bull. Soc. Math. France,* 99:73–112, 1971.

[Con73]   Alain Connes. Une classification des facteurs de type III. *Ann. Sci. École Norm. Sup. (4),* 6:133–252, 1973.

[Con80]   A. Connes. On the spatial theory of von Neumann algebras. *J. Funct. Anal.,* 35(2):153–164, 1980.

[CPPR15]   Martijn Caspers, Javier Parcet, Mathilde Perrin, and Éric Ricard. Noncommutative de Leeuw theorems. *Forum Math. Sigma,* 3:e21, 59, 2015.

[CT77]   Alain Connes and Masamichi Takesaki. The flow of weights on factors of type III. *Tohoku Math. J. (2),* 29(4):473–575, 1977.

[Cuc93]   Ioan Cuculescu. Some remarks on tensor products of standard forms of von Neumann algebras. *Boll. Un. Mat. Ital. B, (7),* 7(4):907–919, 1993.

[CZ76]   Ioana Ciorănescu and László Zsidó. Analytic generators for one-parameter groups. *Tohoku Math. J. (2),* 28(3):327–362, 1976.

[Dav96]   Kenneth R. Davidson. *C*-algebras by example,* volume 6 of *Fields Institute Monographs.* American Mathematical Society, Providence, RI, 1996.

[DDD+95]   P. G. Dodds, T. K. Dodds, P. N. Dowling, C. J. Lennard, and F. A. Sukochev. A uniform Kadec-Klee property for symmetric operator spaces. *Math. Proc. Cambridge Philos. Soc.,* 118(3):487–502, 1995.

[DDdP89]   Peter G. Dodds, Theresa K.-Y. Dodds, and Ben de Pagter. Noncommutative Banach function spaces. *Math. Z.,* 201(4):583–597, 1989.

[DDdP92]   Peter G. Dodds, Theresa K. Dodds, and Ben de Pagter. Fully symmetric operator spaces. *Integral Equ. Oper. Theory.,* 15(6):942–972, 1992.

[DDdP93]   Peter G. Dodds, Theresa K.-Y. Dodds, and Ben de Pagter. Non-commutative Köthe duality. *Trans. Amer. Math. Soc.,* 339(2):717–750, 1993.

[DDLS16]   H. G. Dales, F. K. Dashiell, Jr., A. T.-M. Lau, and D. Strauss. *Banach spaces of continuous functions as dual spaces.* CMS Books in Mathematics/Ouvrages de Mathématiques de la SMC. Springer, Cham, 2016.

[DdPS23]   P. G. Dodds, B. de Pagter, and F. Sukochev. *Noncommutative integration and Operator Theory,* volume 349 of *Progress in Mathematics.* Birkhäuser, Cham, Switzerland, 2023.

[DDS97]   P. G. Dodds, T. K. Dodds, and F. A. Sukochev. Lifting of Kadec-Klee properties to symmetric spaces of measurable operators. *Proc. Amer. Math. Soc.,* 125(5):1457–1467, 1997.

[DDST05]   P. G. Dodds, T. K. Dodds, F. A. Sukochev, and O. Ye. Tikhonov. A non-commutative Yosida-Hewitt theorem and convex sets of measurable operators closed locally in measure. *Positivity,* 9(3):457–484, 2005.

[DF77]   W. Driessler and J. Fröhlich. The reconstruction of local observable algebras from the Euclidean green's functions of relativistic quantum field theory. *Ann. Inst. H. Poincaré Sect. A (N.S.),* 27(3):221–236, 1977.

[DFSW16]   M. Daws, P. Fima, A. Skalski, and S. White. The Haagerup property for locally compact quantum groups. *J. Reine Angew. Math.,* 711:189–229, 2016.

[Dig75]   Trond Digernes. *Duality for weights on covariant systems and its applications.* ProQuest LLC, Ann Arbor, MI, 1975. Thesis (Ph.D.) – University of California, Los Angeles.

[Dim11]   Jonathan Dimock. *Quantum mechanics and quantum field theory: A mathematical primer*. Cambridge University Press, Cambridge, 2011.

[Dir47]   P. A. M. Dirac. *The Principles of Quantum Mechanics*, 3rd edn. Oxford, Clarendon Press, 1947.

[Dir66]   P. A. M. Dirac. *Lectures on Quantum Field Theory*, 1st edn. Belfer Graduate School of Science, Monographs Series No. 3, Yeshiva University, 1966.

[Dix51]   Jacques Dixmier. Sur certains espaces considérés par M.H. stone. *Summa Brasiliensis Math.*, 2:151–182, 1951.

[Dix53]   Jacques Dixmier. Formes linéaires sur un anneau d'opérateurs. *Bull. Soc. Math. France*, 81:9–39, 1953.

[Dix57]   Jacques Dixmier. *Les algèbres d'opérateurs dans l'espace hilbertien (Algèbres de von Neumann)*. Cahiers scientifiques, Fascicule XXV. Gauthier-Villars, Paris, 1957.

[Dix71]   P. G. Dixon. Unbounded operator algebras. *Proc. London Math. Soc. (3)*, 23:53–69, 1971.

[Dix81]   Jacques Dixmier. *von Neumann algebras*, volume 27 of *North-Holland Mathematical Library*. North-Holland Publishing Co., Amsterdam–New York, 1981. With a preface by E. C. Lance, Translated from the second French edition by F. Jellett.

[Dix96a]   Jacques Dixmier. *Les algèbres d'opérateurs dans l'espace Hilbertien (algèbres de von Neumann)*. Les Grands Classiques Gauthier-Villars. [Gauthier-Villars Great Classics]. Éditions Jacques Gabay, Paris, 1996. Reprint of the second (1969) edition.

[Dix96b]   Jacques Dixmier. *Les C*-algèbres et leurs représentations*. Les Grands Classiques Gauthier-Villars. [Gauthier-Villars Great Classics]. Éditions Jacques Gabay, Paris, 1996. Reprint of the second (1969) edition.

[DJT95]   Joseph Diestel, Hans Jarchow, and Andrew Tonge. *Absolutely summing operators*, volume 43 of *Cambridge studies in Advanced Mathematics*. Cambridge University Press, Cambridge, U.K., 1995.

[DKSS12]   M. Daws, P. Kasprzak, A. Skalski, and P. M. Soltan. Closed quantum subgroups of locally compact quantum groups. *Adv. Math.*, 231(6):3473–3501, 2012.

[DL88]   R. J. DiPerna and P.-L. Lions. On the fokker-Planck-Boltzmann equation. *Comm. Math. Phys.*, 120(1):1–23, 1988.

[DL89]   R. J. DiPerna and P.-L. Lions. On the Cauchy problem for Boltzmann equations: global existence and weak stability. *Ann. Math. (2)*, 130(2):321–366, 1989.

[dP07]   B. de Pagter. Non-commutative Banach function spaces. In *Positivity*, Trends in Mathematics, pages 197–227. Birkhäuser, 2007.

[DS88]   N. Dunford and J. T. Schwartz. *Linear operators. Part III. Spectral operators (Reprint of the 1971 original)*. Wiley Classics Library. Wiley-Interscience, John Wiley & Sons, Inc., New York, 1988.

[DSW86]   Wulf Driessler, Stephen J. Summers, and Eyvind H. Wichmann. On the connection between quantum fields and von Neumann algebras of local operators. *Comm. Math. Phys.*, 105(1):49–84, 1986.

[Emc09]   Gerard G. Emch. *Algebraic methods in Statistical mechanics and Quantum Field Theory*. Dover Books on Physics. Dover, 2009. Reprint of the 1972 Wiley-Interscience edition.

[ER00]   Edward G. Effros and Zhong-Jin Ruan. *Operator spaces*, volume 23 of *London Mathematical Society Monographs. New Series*. The Clarendon Press, Oxford University Press, New York, 2000.

[Fil96]   Peter A. Fillmore. *A user's guide to operator algebras.* Canadian Mathematical Society Series of Monographs and Advanced Texts. John Wiley & Sons, Inc., New York, 1996. A Wiley-Interscience Publication.

[FIW14]   M. Fragoulopoulou, A. Inoue, and M. Weigt. Tensor products of unbounded operator algebras. *Rocky Mountain J. Math.*, 44(3):895–912, 2014.

[FK52]   Bent Fuglede and Richard V. Kadison. Determinant theory in finite factors. *Ann. Math. (2)*, 55:520–530, 1952.

[FK86]   Thierry Fack and Hideki Kosaki. Generalized s-numbers of τ-measurable operators. *Pacific J. Math.*, 123(2):269–300, 1986.

[Fol16]   Gerald B. Folland. *A course in abstract harmonic analysis*, 2nd edn. *Textbooks in Mathematics.* CRC Press, Boca Raton, 2016.

[Fre89]   D. H. Fremlin. Measure algebras. In *The handbook of Boolean Algebras*, volume 3, pages 877–980. North-Holland, Amsterdam–New York, 1989.

[Fre04]   D. H. Fremlin. *Measure Theory.* Torres Fremlin, Colchester, 2003-2004.

[FS09]   U. Franz and A. Skalski. On idempotent states on quantum groups. *J. Algebra*, 322(5):1774–1802, 2009.

[Gar79]   L. Terrell Gardner. An inequality characterizes the trace. *Canadian J. Math.*, 31(6):1322–1328, 1979.

[GL95]   Stanisław Goldstein and J. Martin Lindsay. KMS-symmetric Markov semigroups. *Math. Z.*, 219(4):591–608, 1995.

[GL99]   Stanisław Goldstein and J. Martin Lindsay. Markov semigroups KMS-symmetric for a weight. *Math. Ann.*, 313(1):39–67, 1999.

[GL20]   S. J. Goldstein and Labuschagne L. E. *Notes on noncommutative L^p and Orlicz spaces.* Łódź University Press, Łódź, 2020.

[God51]   R. Godemont. Memoire sur la theorie des characteres dan les groupes localement compacts unimodulaires. *J. Math. Pures Appl.*, 30(1):1–110, 1951.

[Gol66]   Seymour Goldberg. *Unbounded linear operators: Theory and applications.* McGraw-Hill Book Co., New York-Toronto, Ontario-London, 1966.

[GP15]   Stanisław Goldstein and Adam Paszkiewicz. Infinite measures on von Neumann algebras. *Internat. J. Theor. Phys.*, 54(12):4341–4348, 2015.

[GW54a]   L. Gårding and A. Wightman. Representations of the anticommutation relations. *Proc. Natl. Acad. Sci. U.S.A.*, 40:617–621, 1954.

[GW54b]   L. Gårding and A. Wightman. Representations of the commutation relations. *Proc. Natl. Acad. Sci. U.S.A.*, 40:622–626, 1954.

[Haa75a]   Uffe Haagerup. Normal weights on W^*-algebras. *J. Funct. Anal.*, 19:302–317, 1975.

[Haa75c]   Uffe Haagerup. The standard form of von Neumann algebras. *Math. Scand.*, 37(2):271–283, 1975.

[Haa78a]   Uffe Haagerup. On the dual weights for crossed products of von Neumann algebras. I. Removing separability conditions. *Math. Scand.*, 43(1):99–118, 1978.

[Haa78b]   Uffe Haagerup. On the dual weights for crossed products of von Neumann algebras. II. Application of operator-valued weights. *Math. Scand.*, 43(1):119–140, 1978.

[Haa79a]   Uffe Haagerup. L^p-spaces associated with an arbitrary von Neumann algebra. In *Algèbres d'opérateurs et leurs applications en physique mathématique (Proc. Colloq., Marseille, 1977)*, volume 274 of *Colloques Internationaux du CNRS*, pages 175–184. CNRS, Paris, 1979.

[Haa79b]   Uffe Haagerup. Operator-valued weights in von Neumann algebras. I. *J. Funct. Anal.*, 32(2):175–206, 1979.

[Haa79c]   Uffe Haagerup. Operator-valued weights in von Neumann algebras. II. *J. Funct. Anal.*, 33(3):339–361, 1979.

[Haa96]   Rudolf Haag. *Local quantum physics*, 2nd edn, *Texts and Monographs in Physics*. Springer-Verlag, Berlin, 1996. Fields, particles, algebras.

[Ham03]   Jan Hamhalter. *Quantum measure theory*, volume 134 of *Fundamental Theories of Physics*. Kluwer Academic Publishers Group, Dordrecht, 2003.

[Hia21]   F. Hiai. *Lectures on Selected Topics in von Neumann Algebras*, volume 32 of *EMS Series of Lectures in Mathematics*. European Mathematical Society (EMS), 2021.

[Hil81]   Michel Hilsum. Les espaces L^p d'une algèbre de von Neumann définies par la derivée spatiale. *J. Funct. Anal.*, 40(2):151–169, 1981.

[HJX10]   Uffe Haagerup, Marius Junge, and Quanhua Xu. A reduction method for noncommutative L_p-spaces and applications. *Trans. Amer. Math. Soc.*, 362(4):2125–2165, 2010.

[HK86]   Herbert Halpern and Victor Kaftal. Compact operators in type III_λ and type III_0 factors. *Math. Ann.*, 273(2):251–270, 1986.

[HK87]   Herbert Halpern and Victor Kaftal. Compact operators in type III_λ and type III_0 factors. II. *Tohoku Math. J. (2)*, 39(2):153–173, 1987.

[HKZ91]   Herbert Halpern, Victor Kaftal, and László Zsidó. Finite weight projections in von Neumann algebras. *Pacific J. Math.*, 147(1):81–121, 1991.

[HLP29]   G. H. Hardy, J. E. Littlewood, and G. Pólya. Some simple inequalities satisfied by convex functions. *Messenger Math.*, 58:145–152, 1929.

[HLP88]   G. H. Hardy, J. E. Littlewood, and G. Pólya. *Inequalities*. Cambridge Mathematical Library. Cambridge University Press, Cambridge, 1988. Reprint of the 1952 edition.

[HM00]   Henryk Hudzik and Lech Maligranda. Amemiya norm equals Orlicz norm in general. *Indag. Math. (N.S.)*, 11(4):573–585, 2000.

[Hor90]   S. S. Horuzhy. *Introduction to algebraic quantum field theory*, volume 19 of *Mathematics and its Applications (Soviet Series)*. Kluwer Academic Publishers Group, Dordrecht, 1990. Translated from the Russian by K. M. Cook.

[Izu96]   Hideaki Izumi. Non-commutative L^p-spaces. Number 956, pages 24–37. 1996. Theory of paragroups and its applications (Japanese) (Kyoto, 1996).

[Izu97]   Hideaki Izumi. Constructions of non-commutative L^p-spaces with a complex parameter arising from modular actions. *Internat. J. Math.*, 8(8):1029–1066, 1997.

[Izu98a]   Hideaki Izumi. Natural bilinear forms, natural sesquilinear forms and the associated duality on non-commutative L^p-spaces. *Internat. J. Math.*, 9(8):975–1039, 1998.

[Izu98b]   Hideaki Izumi. *Non-commutative L^p-spaces constructed by the complex interpolation method*, volume 9 of *Tohoku Mathematical Publications*. Tohoku University, Mathematical Institute, Sendai, 1998. Dissertation, Tohoku University, Sendai, 1998.

[Jar81]   Hans Jarchow. *Locally convex spaces*. Math. Leitfäden [Mathematical Textbooks]. B.G.Teubner, Stuttgart, 1981.

[Jen06]   Anna Jenčová. A construction of a nonparametric quantum information manifold. *J. Funct. Anal.*, 239(1):1–20, 2006.

[JMP14]   Marius Junge, Tao Mei, and Javier Parcet. Smooth fourier multipliers on group von Neumann algebras. *Geom. Funct. Anal.*, 24(6):1913–1980, 2014.

[JNR09]   M. Junge, M. Neufang, and Z-J Ruan. A representation theorem for locally compact quantum groups. *Internat. J. Math.*, 20(3):377–400, 2009.

[Kad52]   Richard V. Kadison. A generalized Schwarz inequality and algebraic invariants for operator algebras. *Ann. Math.*, 56(3):494–503, 1952.

[Kaf77]   Victor Kaftal. On the theory of compact operators in von Neumann algebras. I. *Indiana Univ. Math. J.*, 26(3):447–457, 1977.

[Kaf78]   Victor Kaftal. On the theory of compact operators in von Neumann algebras. II. *Pacific J. Math.*, 79(1):129–137, 1978.

[Kap68]   Irving Kaplansky. *Rings of operators.* W. A. Benjamin, Inc., New York–Amsterdam, 1968.

[Kat04]   Yitzhak Katznelson. *An introduction to harmonic analysis*, 3rd edn. Cambridge Mathematical Library. Cambridge University Press, Cambridge, 2004.

[KN63]   J. L. Kelley and Isaac Namioka. *Linear topological spaces.* With the collaboration of W. F. Donoghue, Kenneth R. Lucas, B. J. Pettis, Ebbe Thue Poulsen, G. Baley Price, Wendy Robertson, W. R. Scott, Kennan T. Smith. The University Series in Higher Mathematics. D. Van Nostrand Co., Inc., Princeton, NJ, 1963.

[Kos84a]   Hideki Kosaki. Applications of the complex interpolation method to a von Neumann algebra: Noncommutative L^p-spaces. *J. Funct. Anal.*, 56(1):29–78, 1984.

[Kos84b]   Hideki Kosaki. On the continuity of the map $\varphi \to |\varphi|$ from the predual of a W^*-algebra. *J. Funct. Anal.*, 59(1):123–131, 1984.

[Kos92]   Hideki Kosaki. A remark on sakai's quadratic Radon-Nikodým theorem. *Proc. Amer. Math. Soc.*, 116(3):783–786, 1992.

[KPS82]   S. G. Kreĭn, Yu. Ī. Petunīn, and E. M. Semënov. *Interpolation of linear operators*, volume 54 of *Translations of Mathematical Monographs.* American Mathematical Society, Providence, R.I., 1982. Translated from the Russian by J. Szűcs.

[KR83]   Richard V. Kadison and John R. Ringrose. *Fundamentals of the theory of operator algebras.* Elementary Theory: Vol. I, *Elementary theory*, volume 100 of *Pure and Applied Mathematics.* Academic Press, Inc. [Harcourt Brace Jovanovich, Publishers]. New York, 1983.

[KR86]   Richard V. Kadison and John R. Ringrose. *Fundamentals of the theory of operator algebras.* Advanced Theory: Vol. II, *Advanced theory*, volume 100 of *Pure and Applied Mathematics.* Academic Press, Inc., Orlando, FL, 1986.

[Kra06]   Steven G. Krantz. *Geometric function theory: Explorations in complex analysis.* Cornerstones. Birkhäuser Boston, Inc., Boston, MA, 2006.

[Kun58]   R. A. Kunze. L_P fourier transforms on locally compact unimodular groups. *Trans. Amer. Math. Soc.*, 89:519–540, 1958.

[Kun90]   Wolfgang Kunze. Noncommutative Orlicz spaces and generalized Arens algebras. *Math. Nachr.*, 147:123–138, 1990.

[KV00]   Johan Kustermans and Stefaan Vaes. Locally compact quantum groups. *Ann. sci. École Norm. Sup. Série 4*, 33(6):837–934, 2000.

[KV03]   Johan Kustermans and Stefaan Vaes. Locally compact quantum groups in the von Neumann algebraic setting. *Math. Scand.*, 92(1):68–92, 2003.

[Lab13]   Louis E. Labuschagne. A crossed product approach to Orlicz spaces. *Proc. Lond. Math. Soc. (3)*, 107(5):965–1003, 2013.

[LM11]   Louis E. Labuschagne and Władysław A. Majewski. Maps on non-commutative Orlicz spaces. *Illinois J. Math.*, 55(3):1053–1081 (2013), 2011.

[LM20]   L. E. Labuschagne and W. A. Majewski. Dynamics on noncommutative Orlicz spaces. *Acta Math. Sci.*, 40B(5):1249–1270, 2020.

[LM22]   L. E. Labuschagne and W. A. Majewski. Integral and differential structures for quantum field theory. *Adv. Theor. Math. Phys.*, 26(6):1787–1836, 2022.

[LMM06]   Louis Labuschagne, W Adam Majewski, and Marcin Marciniak. On k-decomposability of positive maps. *Expo. Math.*, 24:103–125, 2006.

[Maj17]   W. A. Majewski. On quantum statistical mechanics: A study guide. *Adv. Math. Phys.*, 2017(2):Article ID 9343717, 2017.

[MC16]   M. A. Muratov and V. I. Chilin. Topological algebras of measurable and locally measurable operators. *Sovrem. Mat. Fundam. Napravl.*, 61:115–163, 2016.

[Med87]   A. M. Medzhitov. Symmetric spaces on semifinite von Neumann algebras. *Dokl. Akad. Nauk UzSSR*, (4):10–12, 1987.

[ML14]   W. Adam Majewski and Louis E. Labuschagne. On applications of Orlicz spaces to statistical physics. *Ann. Henri Poincaré*, 15(6):1197–1221, 2014.

[ML20]   W. A. Majewski and L. E. Labuschagne. On entropy for general quantum systems. *Adv. Theor. Math. Phys.*, 24(2):491–526, 2020.

[Mog]   Mogget. group von neumann algebra and its plancherel weight. Mathematics Stack Exchange. URL:https://math.stackexchange.com/q/4178999 (version: 2021-06-21).

[Mur78]   M. A. Muratov. Noncommutative Orlicz spaces. *Dokl. Akad. Nauk UzSSR*, (6):11–13, 1978.

[Mur79]   M. A. Muratov. The Luxemburg norm in an Orlicz space of measurable operators. *Dokl. Akad. Nauk UzSSR*, (1):5–6, 1979.

[Mur90]   Gerard J. Murphy. C^*-algebras and operator theory. Academic Press, Inc., Boston, MA, 1990.

[Mus83]   Julian Musielak. *Orlicz spaces and modular spaces*, volume 1034 of *Lecture Notes in Mathematics*. Springer-Verlag, Berlin, 1983.

[MvN36]   F. J. Murray and J. von Neumann. On rings of operators. *Ann. Math. (2)*, 37(1):116–229, 1936.

[MvN37]   F. J. Murray and J. von Neumann. On rings of operators. II. *Trans. Amer. Math. Soc.*, 41(2):208–248, 1937.

[MvN43]   F. J. Murray and J. von Neumann. On rings of operators. IV. *Ann. Math. (2)*, 44:716–808, 1943.

[Nai72]   M. A. Naĭmark. *Normed algebras*, 3rd edn. Wolters-Noordhoff Publishing, Groningen, 1972. Translated from the second Russian edition by Leo F. Boron, Wolters-Noordhoff Series of Monographs and Textbooks on Pure and Applied Mathematics.

[Nel74]   Edward Nelson. Notes on non-commutative integration. *J. Funct. Anal.*, 15:103–116, 1974.

[NP06]   Constantin P. Niculescu and Lars-Erik Persson. *Convex functions and their applications: A contemporary approach*, volume 23 of *CMS Books in Mathematics/Ouvrages de Mathématiques de la SMC*. Springer, New York, 2006.

[NR11]   S. Neuwirth and E. Ricard. Transfer of fourier multipliers into schur multipliers and sumsets in a discrete group. *Canad. J. Math.*, 63(5): 1161–1187, 2011.

[OP93]   Masanori Ohya and Dénes Petz. *Quantum entropy and its use. Texts and Monographs in Physics.* Springer-Verlag, Berlin, 1993.

[OZ99]   Robert Olkiewicz and Bogusław Zegarlinski. Hypercontractivity in non-commutative L_p spaces. *J. Funct. Anal.*, 161(1):246–285, 1999.

[Pav20]   D. Pavlov. Gelfand-type duality for commutative von Neumann algebras. arXiv:2005.05284v2, June 2020.

[Ped79]   Gert K. Pedersen. *C*-algebras and their automorphism groups*, volume 14 of *London Mathematical Society Monographs*. Academic Press, Inc. [Harcourt Brace Jovanovich, Publishers], London-New York, 1979.

[Ped89]   Gert K. Pedersen. *Analysis now*, volume 118 of *Graduate Texts in Mathematics*. Springer-Verlag, New York, 1989.

[Ped18]   Gert K. Pedersen. *C*-algebras and their automorphism groups*. Pure and Applied Mathematics (Amsterdam). Academic Press, London, 2018. Second edition of [MR0548006], Edited and with a preface by Søren Eilers and Dorte Olesen.

[Pis03]   Gilles Pisier. *Introduction to operator space theory*, volume 294 of *London Mathematical Society Lecture Note Series*. Cambridge University Press, Cambridge, 2003.

[Pow67]   Robert T. Powers. Representations of uniformly hyperfinite algebras and their associated von Neumann rings. *Ann. Math. (2)*, 86:138–171, 1967.

[PS75]   Jaak Peetre and Gunnar Sparr. Interpolation and non-commutative integration. *Ann. Mat. Pura Appl. (4)*, 104:187–207, 1975.

[PS95]   Giovanni Pistone and Carlo Sempi. An infinite-dimensional geometric structure on the space of all the probability measures equivalent to a given one. *Ann. Statist.*, 23(5):1543–1561, 1995.

[PT73]   Gert K. Pedersen and Masamichi Takesaki. The Radon-Nikodym theorem for von Neumann algebras. *Acta Math.*, 130:53–87, 1973.

[PX03]   Gilles Pisier and Quanhua Xu. Non-commutative L^p-spaces. In *Handbook of the geometry of Banach spaces*, volume 2, pages 1459–1517. North-Holland, Amsterdam, 2003.

[RD66]   B. Russo and H. A. Dye. A note on unitary operators in C^*-algebras. *Duke Math. J.*, 33(2):413–416, 1966.

[Rud74]   Walter Rudin. *Real and complex analysis*, 2nd edn, McGraw-Hill Book Co., New York–Düsseldorf–Johannesburg, 1974. McGraw-Hill Series in Higher Mathematics.

[RZ22]   C. Roberto and B. Zegarlinski. Hypercontractivity for markov semigroups. *J. Funct. Anal.*, 282(12):109439, 30 pp., 2022.

[Sad12]   Ghadir Sadeghi. Non-commutative Orlicz spaces associated to a modular on τ-measurable operators. *J. Math. Anal. Appl.*, 395(2):705–715, 2012.

[Sak56]   Shôichirô Sakai. A characterization of W^*-algebras. *Pacific J. Math.*, 6:763–773, 1956.

[Sak71]   Shôichirô Sakai. *C*-algebras and W*-algebras.* Springer-Verlag, New York–Heidelberg, 1971. Ergebnisse der Mathematik und ihrer Grenzgebiete, Band 60.

[Sak91]   Shôichirô Sakai. *Operator algebras in dynamical systems*, volume 41 of *Encyclopedia of Mathematics and its Applications*. Cambridge University Press, Cambridge, 1991. The theory of unbounded derivations in C^*-algebras.

[Sal20]   Dietmar Salamon. *Measure and integration*, 2nd edn. Unpublished preprint, 2020.

[San59]   S. Sankaran. The *-algebra of unbounded operators. *J. London Math. Soc.*, 34:337–344, 1959.

[Sch86]   Lothar M. Schmitt. The Radon-Nikodým theorem for L^p-spaces of W^*-algebras. *Publ. Res. Inst. Math. Sci.*, 22(6):1025–1034, 1986.

[Seg53]   I. E. Segal. A non-commutative extension of abstract integration. *Ann. Math. (2)*, 57:401–457, 1953.

[She74]   A. N. Sherstnev. On the general theory of states on von Neumann algebras. *Funkcional. Anal. Priložen.*, 8(3):89–90, 1974.

[She77]   A. N. Sherstnev. Every smooth weight is an l-weight. *Izv. Vysš. Učebn. Zaved. Matematika*, 8(183):88–91, 1977.

[She78]   A. N. Sherstnev. The noncommutative analogue of the space L_1. *Uspehi Mat. Nauk*, 33(1(199)):231–232, 1978.

[She05]   David Sherman. Noncommutative L^p structure encodes exactly Jordan structure. *J. Funct. Anal.*, 221(1):150–166, 2005.

[She07]   David Sherman. On the dimension theory of von Neumann algebras. *Math. Scand.*, 101(1):123–147, 2007.

[She08]   A. N. Sherstnev. *Methods of bilinear forms in non-commutative measure and integral theory*. Fizmatlit, Moskva, 2008.

[She12]   David Sherman. On cardinal invariants and generators for von Neumann algebras. *Canad. J. Math.*, 64(2):455–480, 2012.

[Sri98]   S. M. Srivasatava. *A course on Borel sets*, volume 180 of *Graduate Texts in Mathematics*. Springer-Verlag, New York, 1998.

[Sti55]   W. Forrest Stinespring. Positive functions on C^*-algebras. *Proc. Amer. Math. Soc.*, 6:211–216, 1955.

[Sto49]   M. H. Stone. Boundedness properties in function-lattices. *Can. J. Math.*, 1(2):176–186, 1949.

[Str81]   Şerban Strătilă. *Modular theory in operator algebras*. Editura Academiei Republicii Socialiste România, Bucharest; Abacus Press, Tunbridge Wells, 1981. Translated from the Romanian by the author.

[Str04]   R. F. Streater. Quantum Orlicz spaces in information geometry. *Open Syst. Inf. Dyn.*, 11(4):359–375, 2004.

[Str08]   R. F. Streater. The set of density operators modelled on an Orlicz space. In *Quantum stochastics and information*, pages 99–109. World Scientific Publishing, Hackensack, NJ, 2008.

[STS02]   A. I. Stolyarov, O. E. Tikhonov, and A. N. Sherstnev. Characterization of normal traces on von Neumann algebras by inequalities for the modulus. *Mat. Zametki*, 72(3):448–454, 2002.

[Sun87]   V. S. Sunder. *An invitation to von Neumann algebras*. Universitext. Springer-Verlag, New York, 1987.

[Sut78]   Colin E. Sutherland. Type analysis of the regular representation of a nonunimodular group. *Pacific J. Math.*, 79(1):225–250, 1978.

[SV19]   Adam Skalski and Ami Viselter. Convolution semigroups on locally compact quantum groups and noncommutative Dirichlet forms. *J. Math. Pure. Appl. (9)*, 124:59–105, 2019.

[SW87]   Stephen J. Summers and Reinhard Werner. Maximal violation of Bell's inequalities is generic in quantum field theory. *Comm. Math. Phys.*, 110(2):247–259, 1987.

[SW93]   Anton Ströh and Graeme P. West. τ-compact operators affiliated to a semifinite von Neumann algebra. *Proc. Roy. Irish Acad. Sect. A*, 93(1):73–86, 1993.

[SW99]   H. H. Schaefer and M. P. Wolff. *Topological vector spaces*, volume 3 of *Graduate Texts in Mathematics*, 2nd edn. Springer-Verlag, New York, 1999.

[SZ79]   Şerban Strătilă and László Zsidó. *Lectures on von Neumann algebras*. Editura Academiei, Bucharest; Abacus Press, Tunbridge Wells, 1979. Revision of the 1975 original, Translated from the Romanian by Silviu Teleman.

[Tak70]   M. Takesaki. *Tomita's theory of modular Hilbert algebras and its applications*, volume 128 of *Lecture Notes in Mathematics*, Springer-Verlag, Berlin-New York, 1970.

[Tak73]   M. Takesaki. Duality for crossed products and the structure of von Neumann algebras of type III. *Acta Math.*, 131:249–310, 1973.

[Tak02]   M. Takesaki. *Theory of operator algebras. I*, volume 124 of *Encyclopaedia of Mathematical Sciences*. Springer-Verlag, Berlin, 2002. Reprint of the first (1979) edition, Operator Algebras and Non-commutative Geometry, 5.

[Tak03a]   M. Takesaki. *Theory of operator algebras. II*, volume 125 of *Encyclopaedia of Mathematical Sciences*. Springer-Verlag, Berlin, 2003. Operator Algebras and Non-commutative Geometry, 6.

[Tak03b]   M. Takesaki. *Theory of operator algebras. III*, volume 127 of *Encyclopaedia of Mathematical Sciences*. Springer-Verlag, Berlin, 2003. Operator Algebras and Non-commutative Geometry, 8.

[Tar13]   Zs. Tarcsay. On form sums of positive operators. *Acta Math. Hungar.*, 140(1–2):187–201, 2013.

[Ter81]   M. Terp. L^p spaces associated with von Neumann algebras. Rapport No 3a, Københavns Universitet, Mathematisk Institut, 1981.

[Ter82]   Marianne Terp. Interpolation spaces between a von Neumann algebra and its predual. *J. Oper. Theory*, 8(2):327–360, 1982.

[Ter17]   Marianne Terp. L^p fourier transformation on non-unimodular locally compact groups. *Adv. Oper. Theory*, 2(4):547–583, 2017.

[Tik82]   O. E. Tikhonov. Spaces of L_P type with respect to a weight on a von Neumann algebra. *Izv. Vyssh. Uchebn. Zaved. Mat.*, (8):76–78, 1982.

[Tom58]   Jun Tomiyama. Generalized dimension function for W^*-algebras of infinite type. *Tohoku Math. J. (2)*, 10:121–129, 1958.

[Tom70]   J. Tomiyama. Tensor products and projections of norm one in von Neumann algebras. Lecture Notes, University of Copenhagen, 1970.

[Tom83]   Jun Tomiyama. Recent development of the theory of completely bounded maps between C^*-algebras. *Publ. Res. Inst. Math. Sci.*, 19(3):1283–1303, 1983.

[Tru78]   N. V. Trunov. Locally finite weights on von Neumann algebras (in Russian). Dep. VINITI No. 101-79, Kazan. Gos. Univ., Kazan', 1978. No. 101–79.

[Tru79]   N. V. Trunov. A noncommutative analogue of the space L_2. In *Constructive theory of functions and functional analysis, No. 2 (Russian)*, pages 93–114. Kazan. Gos. Univ., Kazan', 1979.

[Tru81]   N. V. Trunov. L_P spaces associated with a weight on a semifinite von Neumann algebra. In *Constructive theory of functions and functional analysis, No. 3*, pages 88–93. Kazan. Gos. Univ., Kazan', 1981.

[Tru82]   N. V. Trunov. On the theory of normal weights on von Neumann algebras (in Russian). *Izv. Vyssh. Uchebn. Zaved. Mat.*, (8):61–70, 1982.

[TS78a] N. V. Trunov and A. N. Sherstnev. On the general theory of integration with respect to a weight in algebras of operators. I. *Izv. Vyssh. Uchebn. Zaved. Mat.*, (7):79–88, 1978.

[TS78b] N. V. Trunov and A. N. Sherstnev. On the general theory of integration with respect to a weight in algebras of operators. II. *Izv. Vyssh. Uchebn. Zaved. Mat.*, (12):88–98, 1978.

[Vae01a] Stefaan Vaes. Examples of locally compact quantum groups through the bicrossed product construction. In *XIIIth International Congress on Mathematical Physics (London, 2000)*, pages 341–348. International Press, Boston, MA, 2001.

[Vae01b] Stefaan Vaes. A Radon-Nikodym theorem for von Neumann algebras. *J. Operator Theory*, 46(3, suppl.):477–489, 2001.

[vD74] Alfons van Daele. A new approach to the Tomita-Takesaki theory of generalized Hilbert algebras. *J. Funct. Anal.*, 15:378–393, 1974.

[vD78] A. van Daele. *Continuous crossed products and type III von Neumann algebras*, volume 31 of *London Mathematical Society Lecture Note Series*. Cambridge University Press, Cambridge-New York, 1978.

[vD07] Alfons van Daele. Locally compact quantum groups: the von Neumann algebra versus the C^*-algebra approach. *Bull. Kerala Math. Assoc. 2005.*, Special Issue:153–177, 2007.

[vD14] Alfons van Daele. Locally compact quantum groups: A von Neumann algebra approach. *SIGMA*, 10(82), 2014.

[Vil02] Cédric Villani. A review of mathematical topics in collisional kinetic theory. In *Handbook of mathematical fluid dynamics*, Vol. I, pages 71–305. North-Holland, Amsterdam, 2002.

[vN30] John von Neumann. Zur Algebra der funktionaloperationen und Theorie der normalen Operatoren. *Math. Ann.*, 102(1):370–427, 1930.

[VN37] John von Neumann. Some matrix-inequalities and metrization of matrix-space. *Tomsk Univ. Rev.*, 1:286–300, 1937.

[VN39] John von Neumann. On infinite direct products. *Compos. Math.*, 6(1): 1–77, 1939.

[vN40] John von Neumann. On rings of operators. III. *Ann. Math.* 41(2):94–161, 1940.

[vN49] John von Neumann. On rings of operators. Reduction theory. *Ann. Math.* 50(2):401–485, 1949.

[vN96] John von Neumann. *Mathematische Grundlagen der Quantenmechanik.* Springer Berlin, Heidelberg, 1996. Reprint of the original 1932 edition.

[Wei09] Martin Weigt. Jordan homomorphisms between algebras of measurable operators. *Quaest. Math.*, 32(2):203–214, 2009.

[Wid46] D. V. Widder. *The Laplace Transform.* Princeton University Press, Princeton, 1946.

[Wid61] D. V. Widder. Functions harmonic in a strip. *Proc AMS*, 12(1):62–72, 1961.

[Wie49] H. Wielandt. Über die unbeschränktheit der operatoren im Hilbertschen Raum. *Math. Ann.*, 121(1):21, 1949.

[Wig57] A. S. Wightman. Quelque problèmes mathématique de la théorie quantique relativiste. In *Lecture Notes*. Faculté de Sciences, Univ. de Paris, 1957.

[Wil04] S. Willard. *General Topology.* Dover Publications, 2004.

[Win47] A. Wintner. The unboundedness of quantum-mechanical matrices. *Phys. Rev.*, 71:738–739, 1947.

[WZ22]   Yifu Wang and Boguslaw Zegarlinski. Higher order coercive inequalities. *Potential Anal.*, 58(2):263–293, 2022.

[Xu07]   Quanhua Xu. Operator spaces and noncommutative L_p. Online, 2007. Summer School on Banach spaces and Operator spaces, Nankia University.

[Yea73]   F. J. Yeadon. Convergence of measurable operators. *Proc. Cambridge Philos. Soc.*, 74:257–268, 1973.

[Yea75]   F. J. Yeadon. Non-commutative L^p-spaces. *Math. Proc. Cambridge Philos. Soc.*, 77:91–102, 1975.

[Yea80]   F. J. Yeadon. Ergodic theorems for semifinite von Neumann algebras. II. *Math. Proc. Cambridge Philos. Soc.*, 88(1):135–147, 1980.

[Yea75]   F. J. Yeadon. On a result of P. g. Dixon. *J. London Math. Soc. (2)*, 9:610–612, 1974/75.

[Yng05]   Jakob Yngvason. The role of type III factors in quantum field theory. *Rep. Math. Phys.*, 55(1):135–147, 2005.

[Zeg90]   Bogusław Zegarliński. Log-sobolev inequalities for infinite one-dimensional lattice systems. *Comm. Math. Phys.*, 133(1):147–162, 1990.

[Zeg02]   Boguslaw Zegarlinski. Analysis of classical and quantum interacting particle systems. In *Quantum interacting particle systems (Trento, 2000)*, volume 14 of *QP–PQ: Quantum Probability and White Noise Analysis*, pages 241–336. World Scientific Publishing, River Edge, NJ, 2002.

[Zeg08]   Boguslaw Zegarlinski. Analysis in operator spaces. In *Quantum stochastics and information*, pages 121–140. World Scientific Publishing, Hackensack, NJ, 2008.

[Zeg15]   Boguslaw Zegarlinski. Framework, results and open problems for log sobolev inequalities in noncommutative spaces. https://www.birs.ca/events/2015/5-day-workshops/15w5098/videos, 2015. Part of: Hypercontractivity and Log Sobolev Inequalities in Quantum Information Theory (15w5098), Banff International Research Station Workshop, February 22–27, 2015.

[Zhu93]   Ke He Zhu. *An introduction to operator algebras*. Studies in Advanced Mathematics. CRC Press, Boca Raton, FL, 1993.

[Zol82]   A. A. Zolotarëv. L^P spaces with respect to a state on a von Neumann algebra, and interpolation. *Izv. Vyssh. Uchebn. Zaved. Mat.*, (8):36–43, 1982.

[Zol85]   A. A. Zolotarëv. On the interpolation theory of L_p spaces with respect to state on a von Neumann algebra. *Izv. Vyssh. Uchebn. Zaved. Mat.*, (1):59–61, 1985.

[Zol88]   A. A. Zolotarëv. Interpolation properties of a two-parameter scale of L_p-spaces with respect to a state on a semifinite von Neumann algebra. *Izv. Vyssh. Uchebn. Zaved. Mat.*, (2):71–73, 1988.

Notation index

The manufacturer's authorised representative in the EU for product safety is Oxford University Press España S.A. of el Parque Empresarial San Fernando de Henares, Avenida de Castilla, 2 – 28830 Madrid (www.oup.es/en or product. safety@oup.com). OUP España S.A. also acts as importer into Spain of products made by the manufacturer.

www.ingramcontent.com/pod-product-compliance
Ingram Content Group UK Ltd.
Pitfield, Milton Keynes, MK11 3LW, UK
UKHW022159180625
459817UK00012B/93